Third Edition

PHYSICAL CHEMISTRY

Principles and Applications in Biological Sciences

Ignacio Tinoco, Jr.
University of California, Berkeley

Kenneth Sauer
University of California, Berkeley

James C. Wang
Harvard University

PRENTICE HALL, UPPER SADDLE RIVER, NEW JERSEY 07458

Library of Congress Cataloging in Publication Data

Tinoco, Ignacio.
   Physical chemistry: principles and applications in biological
sciences / Ignacio Tinoco, Jr., Kenneth Sauer, James C. Wang. -- 3rd ed.
    p.   cm.
  Includes bibliographical references and index.
  1. Biochemistry.   2. Chemistry, Physical and theoretical.
I. Sauer, Kenneth. II. Wang, James C. III. Title.
QH345.T56 1995   541'.024574--dc20   94-37619
ISBN 0-13-186545-5

Production Editor:	Donna Young
Acquisitions Editor:	Deirdre Cavanaugh
Cover Designer:	Anthony Gemmellaro
Buyer:	Trudy Pisciotti
Editorial Assistant:	Veronica Wade

© 1995, 1985, 1978 by Prentice-Hall, Inc.
A Simon & Schuster Company
Upper Saddle River, New Jersey 07458

Printed in the United States of America

10 9 8 7 6 5 4

ISBN 0-13-186545-5

Prentice-Hall International (UK) Limited, *London*
Prentice-Hall of Australia Pty. Limited, *Sydney*
Prentice-Hall Canada Inc., *Toronto*
Prentice-Hall Hispanoamericana, S.A., *Mexico*
Prentice-Hall of India Private Limited, *New Delhi*
Prentice-Hall of Japan, Inc., *Tokyo*
Simon & Schuster Asia Pte. Ltd., *Singapore*
Editora Prentice-Hall do Brasil, Ltda., *Rio de Janeiro*

Contents

LIST OF TABLES

Preface to Third Edition

The third edition has been revised to make it obvious to students in the biological sciences why they need to learn physical chemistry. We start with a chapter which describes important problems in biology and medicine currently being studied by physical methods. Each succeeding chapter starts with sections on concepts and applications which show how the concepts the students learn in the chapter apply to biological and biochemical problems. Molecular interpretations of macroscopic properties—thermodynamics and transport properties—are emphasized in the next five chapters (Chapters 2–6). For example, in Chapter 3 we give an extensive discussion of entropy as a measure of disorder. The structures of proteins and nucleic acids are described in Chapter 3 so that they can be used in applications, examples, and problems freely afterwards.

Chapter 6 (MOLECULAR MOTION AND TRANSPORT PROPERTIES) has been changed extensively. Discussion of the Boltzmann energy distribution and the random walk has been introduced. The section on gel electrophoresis has been expanded to mirror the many new applications in separating proteins and very large DNAs, as well as in studying binding of proteins to nucleic acids.

Chapters 7 and 8 (KINETICS: RATES OF CHEMICAL REACTIONS, and ENZYME KINETICS) have been rearranged so that temperature-jump kinetics is now in the kinetics chapter where it belongs logically. The kinetics of competing substrates are discussed with application to the fidelity of DNA replication.

The chapter on MOLECULAR STRUCTURE AND INTERACTIONS: THEORY (Chapter 9, originally titled QUANTUM MECHANICS) now begins with intermolecular and intramolecular forces and the use of semi-empirical methods to calculate molecular conformations. This leads into the discussion of quantum mechanics.

The chapter on MOLECULAR STRUCTURE AND INTERACTIONS: SPECTROSCOPY (Chapter 10) has an increased emphasis on nuclear magnetic resonance. Two-dimensional NMR and magnetic resonance imaging are treated.

Chapter 11 (MOLECULAR DISTRIBUTIONS AND STATISTICAL THERMO-DYNAMICS) has been shortened because some of the material has been introduced in earlier chapters. Discussion of quantum mechanical distributions has been deleted.

Chapter 12 (MACROMOLECULAR STRUCTURES AND X-RAY DIFFRAC-TION) has a more detailed discussion of how a structure is obtained from a measured diffraction pattern. Scanning tunneling and atomic force microscopes are described.

There are now more problems at the end of each chapter; many old problems have been improved or deleted. Essentially all of the problems are biologically relevant. The references and suggested readings have been brought up to date and expanded.

Throughout the book we have tried to make the writing clearer and more explicit; the important equations are explained in words. The students, teaching assistants, and faculty in the biophysical chemistry courses at Berkeley pointed out sections that were unclear, and several reviewers of this new edition made very helpful comments. We are particularly grateful to our colleagues, Professor David Wemmer and Dr. Joseph Monforte, who made detailed comments and who contributed their homework and exam problems. Joan Tinoco prepared the Index. We would also like to acknowledge the following reviewers: James E. Davis, Harvard University; Eckard Münck, Carnegie Mellon University; Donald J. Nelson, Clark University; T. M. Schuster, University of Connecticut; Maurice Schwartz, University of Notre Dame; George Strauss, Rutgers University; and Jimmy W. Viers, Virginia Polytechnic Institute and State University.

1

Introduction

Physical chemistry is a group of principles and methods helpful in solving many different types of problems. In the following chapters we will present the principles of thermodynamics, transport properties, kinetics, quantum mechanics and molecular interactions, spectroscopy, and scattering and diffraction. We will also discuss various experimentally measurable properties, such as enthalpy, electrophoretic mobility, light absorption, and x-ray diffraction. All these experimental and theoretical methods can give useful information about the part of the universe you are interested in. We will emphasize the molecular interpretation of these methods, and stress biochemical and biological applications, but it is up to you to see how the methods presented here can be applied to the problems that interest you. By learning the principles behind the methods, you will be able to judge the conclusions obtained from them. This is the first step in inventing new methods, or discovering new concepts.

Chapters 2 through 5 of this book cover the fundamentals of thermodynamics, and their applications to chemical reactions and physical processes. Much of this should be a review of material covered in beginning chemistry courses, so the applications to biological macromolecules can be emphasized. Chapter 6 describes the effect of sizes and shapes of molecules on their motions in gases, liquids, and gels. The driving forces are either random thermal forces which cause diffusion, or the directed forces in sedimentation, flow, and electrophoresis. Chapter 6 thus covers transport properties. Chapter 7 describes general kinetics, and Chapter 8 concentrates on the kinetics of enzyme-catalyzed reactions. Chapter 9 deals with molecular structures and intermolecular interactions. Quantum mechanical principles are introduced so that bonding and spectroscopy can be understood. Calculations of protein and nucleic acid conformations by the use of classical force fields (Coulomb's law, van der Waals potential) are described. Chapter 10 includes the main spectroscopic methods used for studying molecules in solution: ultraviolet, visible, and infrared absorption; circular dichroism and optical rotatory dispersion; and nuclear magnetic resonance. Chapter 11 introduces the principles of statistical thermodynamics and describes their biochemical applications. Topics such as helix-coil transitions in polypeptides (proteins) and polynucleotides (nucleic acids), and binding of small molecules to macromolecules are

1

emphasized. Chapter 12 starts with the scattering of electromagnetic radiation from one electron, and proceeds through the diffraction of x rays by crystals. New scanning microscope methods are introduced. The Appendices contain numerical data used throughout the book, unit conversion tables, and the structures of many of the biological molecules mentioned in the text.

Other books will be useful as background sources and to provide more depth of coverage. For applications of physical chemistry to other areas, there are standard physical chemistry texts. Biochemistry and molecular biology texts can provide specific information about such areas as enzyme mechanisms, metabolic paths, and structures of membranes. Finally, a good physics textbook is useful for learning or reviewing the fundamentals of forces, charges, electromagnetic fields, and energy. A list of books is given at the end of this chapter.

In the next sections we present a few examples of problems that physical chemistry can solve. They should be read for pleasure, not for memorization. The aim is simply to illustrate some current research from the scientific literature, and to point out the principles and methods which are used. We hope to motivate you to learn the material discussed in the following chapters. It will help if you pick an article with a title that interests you from a journal such as *Nature* or *Science,* or even *Scientific American,* to learn how this book will make it easier for you to understand the article.

THE HUMAN GENOME

A goal of the Human Genome Project is to learn the complete sequence of all three billion (3×10^9) base pairs which make up the genetic information of humans—the human genome. Genes are sequences of base pairs in double-stranded DNA. In human sperm the DNA is packaged in 23 chromosomes: 22 autosomes plus a male Y chromosome or a female X chromosome. In human eggs the DNA is packaged in 22 autosomes plus a female X chromosome. Thus, each of us acquires 23 pairs of chromosomes; the XX pair makes us female, the XY male. Nobody knows how many human genes there are; estimates range from 50,000 to 100,000. It is known that less than 10% of the human DNA consists of genes. The rest may be structural regions that fold the DNA into a required three-dimensional shape, or sequences that have lost their function during evolution. There are long regions of highly repetitive sequences which are useful for distinguishing human DNA from other species, but do not have known biological functions. Identifying genes and learning their arrangement on the chromosomes is of great importance in medicine, but learning about the rest of the DNA may be a more fundamental advance.

The DNA sequence with its alphabet of four letters (A, C, G, T) is transcribed into the RNA alphabet (A, C, G, U). The RNA is translated into a protein sequence of 20 amino acids in a three-letter code.* As there are 4^3 (64) three-letter words with an alphabet of four letters, the code must be redundant. In the genetic dictionary most amino acids are coded

---

* The structures and names of the nucleic acid bases and the protein amino acids are given in Appendix A.9.

by two or four different words. Three amino acids have six words each (arginine, leucine, and serine) and two have only one word (tryptophan, methionine). Three of the words do not code for amino acids, but instead signal for protein synthesis to stop: UAA, UAG, UGA. One word, AUG, codes for the start of protein synthesis. (It also codes for methionine.) There are sequences before the starting AUG and after the terminating UAA, UAG, or UGA which control and regulate the synthesis of the protein.

The RNA that is translated into protein sequences is called messenger RNA. But not all RNA synthesized from the DNA code is translated into protein; some of it functions directly as RNA. Ribosomal RNA is a vital part of the ribosome where protein synthesis takes place, and transfer RNA is the molecule that reads the three-letter code and places the correct amino acid in the growing protein chain. In RNA viruses, including flu and HIV, the RNA acts as messenger RNA, but it also takes the place of DNA in storing the genetic information.

How is the transcription of DNA into RNA determined? In the DNA there are sequences called promoters, which signal the start of RNA synthesis, and sequences called terminators, which signal the stop of RNA synthesis. In addition there are other sequences (attenuators, enhancers), which further control and regulate the synthesis of the RNA molecules.

If we are to understand all this, we need to be able to determine the sequences of the proteins and nucleic acids. We need to know how they fold up into their biologically active forms. How do the proteins and nucleic acids interact with each other and with all the molecules they encounter? Which molecules bind, which react? How fast do they react? The answers to these questions can be obtained by the methods we will describe in this book. Of course, much of what is already known about

$$\text{DNA} \longrightarrow \text{RNA} \longrightarrow \text{Protein}$$

is based on these methods.

TRANSCRIPTION FACTORS

The control of gene expression in your body is essential for your life and good health. Although all your cells contain essentially the same DNA, insulin is made in the pancreas, hemoglobin in the bone marrow, and the growth hormone somatotropin in the pituitary gland. The amount of each of these proteins must be carefully controlled. For insulin the control must be immediately dependent on the blood sugar; for somatotropin the control can be slower.

Transcription factors are proteins that bind to DNA and control which genes are transcribed into RNA. Many of the transcription factors are hormone dependent; a particular steroid or thyroid hormone must bind to the protein before it binds to the DNA. A very common type of transcription factor contains zinc fingers—a folded structure of about 30 amino acids and 1 zinc ion. Zinc-finger proteins have been found to contain from 2 to 37 zinc fingers. It is estimated that about 1% of the DNA in human cells specifies zinc fingers, and that 8% of the proteins coded by chromosome 19 contains zinc fingers (see article by

Rhodes and Klug, 1993).* The three-dimensional solution structure of a zinc-finger DNA-binding domain was first determined by nuclear magnetic resonance (NMR). A peptide chain of 25 amino acids forms an α-helix next to a β-sheet with zinc coordinated in between as shown in Fig. 1.1. The zinc is held by two cysteines in the β-sheet and two histidines in the α-helix. Zinc fingers from steroid receptors often use four cysteines to bind the zinc. Three zinc fingers from a transcription factor from a mouse were cocrystalized with 11 base pairs of its DNA-binding site to learn how the zinc fingers interact with the DNA. X-ray diffraction studies provided the structure shown in Fig. 1.2a. The fingers position six amino acid side chains (five arginines and one histidine) to form hydrogen bonds with six guanines in one turn of the DNA double helix (see Fig. 1.2b). There is much more structural information available from the x-ray and NMR experiments than given in the figures. We learn about DNA-protein recognition in general, as well as gain insight into designing new DNA-binding proteins for controlling gene expression. You should be able to judge when these methods will be useful to your problems, and even more important, when

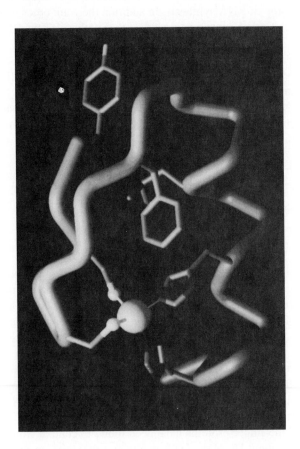

Fig. 1.1 The structure of a part of a protein that specifically binds DNA—a DNA-binding domain called a zinc finger. The peptide chain of 25 amino acids forms an α-helix (seen on the right) and a β-sheet (seen on the left). A zinc ion (the sphere in the center of the picture) is coordinated by two sulfur atoms of two cysteine residues of the β-sheet and two nitrogen atoms of two histidine residues of the α-helix. The six-membered rings shown are from tyrosine and phenylalanine. The structure was determined by nuclear magnetic resonance. [Photograph provided by Michael Pique and Peter E. Wright, Department of Molecular Biology, The Scripps Research Institute, La Jolla, California. (From *Science 245,* 635, 1989), using software from Ray-Tracing Corporation and Sun Microsystems.]

* References cited in the text will be found either in the References or Suggested Readings at the end of each chapter.

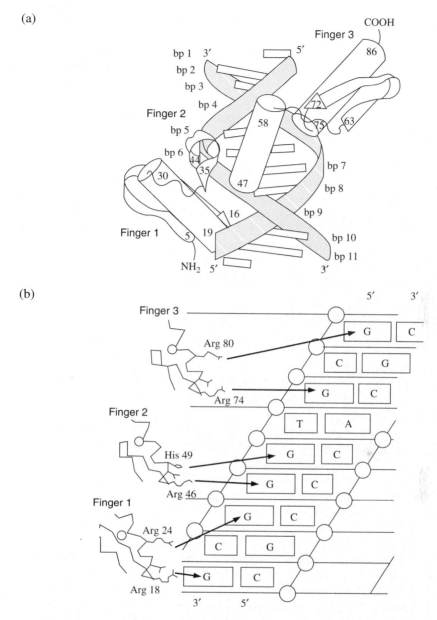

(a)

Finger 3
COOH

bp 1   3'   5'   86
bp 2
bp 3
bp 4   72
Finger 2   75   63
bp 5   58
bp 6   bp 7
        44   bp 8
    30 35
        47   bp 9
Finger 1   16
        19   bp 10
    5   bp 11
NH₂ 5'   3'

(b)

                                    5'   3'
Finger 3 _____ ○
                          ○   G   C
        Arg 80   ○
                          C   G
                       ○
                    G   C
        Arg 74
                       T   A
Finger 2
        His 49   G   C
                 G   C
        Arg 46
Finger 1
        Arg 24   G   C
                 C   G
        Arg 18   G   C
        3'   5'

Fig. 1.2 (a) A schematic diagram showing the interaction of three zinc fingers bound to one turn of DNA. The cylinders represent the α-helices; the folded ribbons are the β-sheets. (b) The specific contacts between the amino acids in the zinc fingers and the bases in the DNA which provide the specificity of binding. The basic amino acids arginine and histidine form hydrogen bonds with guanine bases in the DNA. The structure was determined by x-ray crystallography. [Figures from Pavletich, N. P., and C. O. Pabo, Zinc Finger-DNA Recognition: Crystal Structure of a Zif268-DNA Complex at 2.1 Å, *Science 252,* 809–817 (1991).]

you should believe the pretty pictures shown in the figures. How is the picture related to the experimental measurements? Is the structure of a fragment of a protein determined in a crystalline solid, or in an aqueous buffer, the same as that in a biological cell? To begin to answer these questions, we need to know how strong the forces are that hold the protein in a folded state. What is the energy for rotation about a single bond in a molecule? The torsion angles on either side of the amide group in a polypeptide characterize its conformation and distinguish an α-helix from a β-sheet. The strength of hydrogen bonds, electrostatic interactions, and hydrophobic interactions of nonpolar groups determine which sequences form which structures.

Gene regulation was first studied in bacteria. The regulation of the bacterial virus bacteriophage λ in *Escherichia coli* has been most extensively studied (see description by Ptashne et al., 1982). The bacteriophage λ DNA-binding protein, a repressor which turns off the bacteriophage genes, does not contain zinc fingers. Gene regulation in higher organisms was first studied extensively in the frog *Xenopus laevis*. Transcription factor IIIA (TFIIIA), a zinc-finger protein, is required to activate the gene that codes for a ribosomal RNA called 5S. This RNA is not a messenger RNA; it is part of the ribosome. The 5S RNA controls its own synthesis, because it competes with the DNA for binding the transcription factor. The reactions are

$$\text{DNA} + \text{TFIIIA} \rightleftharpoons \text{DNA} \cdot \text{TFIIIA complex} \qquad \text{(gene on)}$$

$$\text{DNA} \cdot \text{TFIIIA complex} + \text{5S RNA} \rightleftharpoons \text{5S RNA} \cdot \text{TFIIIA complex} + \text{DNA} \qquad \text{(gene off)}$$

When the concentration of free 5S RNA is high, the TFIIIA is bound to the RNA and the gene is off. When the concentration of 5S RNA drops, the concentration of free TFIIIA increases and the DNA complex forms, which activates the gene. Once the equilibrium constants are known, it is straightforward to calculate equilibrium concentrations. TFIIIA is only one of many proteins involved in the regulation; nevertheless you should be able to calculate equilibrium concentrations no matter how many simultaneous equilibria are involved. The kinetics of these interactions are important; maybe equilibrium is never reached because the reactions are slow compared to the rate of synthesis of the 5S RNA. How are the rates dependent on concentrations? Are enzyme-catalyzed reactions different from uncatalyzed ones? You should have a better understanding of kinetics and equilibrium (thermodynamics) when you finish this book.

To study either kinetics or equilibria you need to be able to measure the concentrations of the species involved. The easiest way is to use spectroscopy. Absorbance of visible light, or ultraviolet, or infrared is proportional to the concentration. Fluorescence is usually even more sensitive to concentration, if your species emits some of the absorbed light. For qualitative, or even quantitative, studies of protein-DNA binding, gel electrophoresis is the current method of choice. To discover new DNA-binding proteins you can do gel electrophoresis on a mixture of proteins and a radioactively labeled DNA. The different species separate on the gel according to size and charge, but only radioactive species are visualized by placing the gel in contact with a photographic sheet. Pure protein species will not be seen on the photograph, and free DNA will travel with the mobility of the pure reference DNA. Species seen on the photograph of the gel which travel slower than free DNA must be protein-DNA complexes. The complex can be removed from the gel and the bound protein characterized.

THE OZONE LAYER OF THE EARTH

Stratospheric ozone serves the important function of filtering out a large fraction of uv radiation of the sun. How ozone does this can be seen from the following reactions:

$$O_2 \xrightarrow{\text{hv (below 242 nm)}} O + O \tag{1}$$

$$O_2 + O + M \longrightarrow O_3 + M \tag{2}$$

$$O_3 \xrightarrow{\text{hv (190–300 nm)}} O_2 + O \tag{3}$$

Reaction (1) represents the splitting of oxygen molecules by the absorption of uv light shorter than 242 nm in wavelength; reaction (2) represents the formation of O_3 (ozone) by a three-body collisional process—M is an inert third molecule that dissipates some of the energy released when O_2 and O combine into O_3. Together, reactions (1) and (2) generate ozone. Reaction (3) represents the dissociation of ozone by the absorption of uv light in the wavelength range 190 to 300 nm. These reactions are important in maintaining a steady-state concentration of ozone; because both reactions (1) and (3) involve absorption of uv light, the stratospheric ozone helps the filtering of uv radiation. Ozone is particularly important in absorbing the biologically damaging radiation in the spectral region 240 to 300 nm.

If compounds such as NO and NO_2 (present in jet exhaust) or halogenated hydrocarbons (used as refrigerants among other things) get into the upper atmosphere, they reduce significantly the steady-state concentration of ozone. The molecule NO, for example, may reduce O_3 by the reactions

$$NO + O_3 \longrightarrow NO_2 + O_2 \tag{4}$$

$$O + NO_2 \longrightarrow NO + O_2 \tag{5}$$

Since NO is not used up in the pair of reactions, it will continue the catalytic removal of O_3 until it diffuses out of the ozone layer. A quantitative estimate of the magnitudes of the effects of releasing various chemicals into the atmosphere on the ozone concentration requires knowing the rates of the various reactions such as (1) through (5) under atmospheric conditions, and the rates with which various components diffuse from one atmospheric layer into another. Thus, studies of reaction kinetics and transport properties including diffusion provide a basis for making rational decisions on how the flying of supersonic jets in the upper atmosphere, or the use of halogenated hydrocarbons, should be regulated.

The molecule NO also has a very wide range of biological functions; it has been implicated in neurotransmission, immune regulation, and other physiological processes. NO is synthesized enzymatically from arginine and produces its effects by forming an iron-nitrosyl complex with a heme protein. Understanding the binding of iron to heme groups—and the binding of vital O_2, deadly CO, and helpful NO to the iron—depends on knowledge of atomic electron orbitals in Fe, and molecular electron orbitals in the ligand molecules.

INTRACELLULAR CONCENTRATIONS OF SMALL MOLECULES

What is the pH inside a cell? What are the concentrations of other small molecules, such as K^+, Na^+, ATP (adenosine triphosphate), and glucose? How do the concentrations change when the cell is under certain physiological stress, such as deprivation of oxygen? To answer these questions, one could take a concentrated suspension or paste of cells, break the cell walls by freezing and thawing of the cells repeatedly or by sonication, and then measure the various concentrations by chemical, physiochemical, or biochemical methods. This approach has been invaluable in giving us information about the cellular contents. It has several drawbacks, however. The disruption of cell walls, for example, will perturb the concentrations of certain species. Furthermore, in general, by this approach we can measure only the total concentrations of components such as ATP, and cannot decide what fraction of a measured total concentration is "free" or metabolically available, and what fraction is tightly bound to cellular components and is therefore metabolically unavailable.

Spectroscopy in general and nuclear magnetic resonance (NMR) spectroscopy in particular provide another powerful approach in studying intracellular metabolism.

A suspension of intact cells can be placed in an NMR spectrometer and the spectra of magnetic nuclei such as ^{31}P, ^{13}C, ^{1}H, ^{39}K, and ^{23}Na can be obtained. Concentrations of small molecules that are not tightly bound to macromolecules and are present at sufficiently high concentrations (for example, of the order of 10^{-3} M for ^{31}P resonances) can be deduced from the measured spectra. Although the concentration of H^+ is too low to be measured directly, it can be deduced from the effect of pH on the spectra of other measurable components, such as from the ratio of $H_2PO_4^-$ to HPO_4^{2-}. If a species, for example ^{23}Na, is present both inside and outside the cell, one can add a reagent that does not enter the cell to shift the resonances of the extracellular form, thus making the intracellular and extracellular forms distinguishable.

The development of special NMR instruments has also made it possible to measure the spatial distribution of particular molecules within a sample. Magnetic resonance imaging (MRI) is a very important diagnostic tool in medicine. The location and distribution of protons (mainly in water molecules) are determined. This provides very detailed images of soft tissues not available from x-ray imaging. Images based on other nuclei such as phosphorus or carbon are also possible.

REFERENCES

The following textbooks can be useful for the entire course.

Physical Chemistry

ALBERTY, R. A., and R. J. SILBEY, 1992. *Physical Chemistry,* John Wiley, New York.
ATKINS, P. W., 1990. *Physical Chemistry,* 4th ed., W. H. Freeman, New York.
BARROW, G. M., 1988. *Physical Chemistry,* 5th ed., McGraw-Hill, New York.
LEVINE, I. N., 1988. *Physical Chemistry,* 3rd ed., McGraw-Hill, New York.
MOORE, W. J., 1983. *Basic Physical Chemistry,* Prentice-Hall, Englewood Cliffs, New Jersey.

Biophysical Chemistry

CANTOR, C. R., and P. R. SCHIMMEL, 1980. *Biophysical Chemistry,* Parts I, II, III, W. H. Freeman, San Francisco.

Biochemistry

MATHEWS, C. K., and K. E. VAN HOLDE, 1990. *Biochemistry,* Benjamin/Cummings, Redwood City, California.

STRYER, L., 1994. *Biochemistry,* 4th ed., W. H. Freeman, San Francisco.

VOET, D., and J. G. VOET, 1990. *Biochemistry,* John Wiley, New York.

Molecular Biology

ALBERTS, B., D. BRAY, J. LEWIS, M. RAFF, K. ROBERTS, and J. D. WATSON, 1989. *Molecular Biology of the Cell,* 2nd ed., Garland Publishing, New York.

LEWIN, B., 1985. *Genes,* 2nd ed., John Wiley, New York.

WATSON, J. D., N. H. HOPKINS, J. W. ROBERTS, J. A. STEITZ, and A. M. WEINER, 1981. *Molecular Biology of the Gene,* Vols. I, II, 4th ed., Benjamin/Cummings, Redwood City, California.

For useful compilations of data, see *Handbook of Biochemistry and Molecular Biology,* 1976, 3rd ed., G. D. Fasman, ed., CRC Press, Cleveland, Ohio.

Physics

HALLIDAY, D., and R. RESNICK, 1988. *Fundamentals of Physics,* 3rd ed., John Wiley, New York.

SUGGESTED READINGS

BERG, P., and M. SINGER, 1992. *Dealing with Genes: The Language of Heredity,* University Science Books, Mill Valley, California.

ELLIOTT, S., and F. S. ROWLAND, 1987. Chlorofluorocarbons and Stratospheric Ozone, *J. Chem. Ed. 64,* 387–391.

GADIAN, D. G., and G. K. RADDA, 1981. NMR Studies of Tissue Metabolism, *Annu. Rev. Biochem. 50,* 69–83.

JOHNSON, A. D., A. R. POLEETE, G. LAUER, R. T. SAUER, G. K. ACKERS, and M. PTASHNE, 1981. λ Repressor and *cro*-Components of an Efficient Molecular Switch, *Nature 294,* 217–223.

JOHNSTON, H., 1971. Reduction of Stratospheric Ozone by Nitrogen Oxide Catalysts from Supersonic Transport Exhaust, *Science 173,* 517–522.

LEE, M. S., G. P. GIPPERT, K. V. SOMAIN, D. A. CASE, and P. E. WRIGHT, 1989. Three-Dimensional Solution Structure of a Single Zinc Finger DNA-Binding Domain, *Science 245,* 635–637.

MOLINA, M., and F. S. ROWLAND, 1974. Stratospheric Sink for Chlorofluoromethanes: Chlorine Atom-Catalyzed Destruction of Ozone, *Nature 249,* 810–812.

OGINO, T., J. A. DEN HOLLANDER, and R. G. SHULMAN, 1983. ^{39}K, ^{23}Na, and ^{31}P NMR Studies of Ion Transport in *Saccharomyces cerevisiae, Proc. Natl. Acad. Sci. USA 80,* 5185–5189.

PAVLETICH, N. P., and C. O. PABO, 1991. Zinc Finger-DNA Recognition: Crystal Structure of a Zif268-DNA Complex at 2.1 Å, *Science 252,* 809–817.

PTASHNE, M., A. D. JOHNSON, and C. O. PABO, 1982. A Genetic Switch in a Bacterial Virus, *Sci. Am. 247* (November), 128–140.

RHODES, D., and A. KLUG, 1993. Zinc Fingers, *Sci. Am. 268* (February), 56–65.

STAMLER, J. S., D. J. SINGEL, and J. LOSCALZO, 1992. Biochemistry of Nitric Oxide and its Redox-Activated Forms, *Science 258,* 1898–1902.

TOON, O. W., and R. P. TURCO, 1991. Polar Stratospheric Clouds and Ozone Depletion, *Sci. Am. 264* (June), 68–74.

PROBLEM

1. Read a paper in the scientific literature which sounds interesting to you.
   (a) Record the complete reference to it: authors, title, journal, volume, first and last pages, year.
   (b) Summarize the purpose of the paper and why it is worthwhile.
   (c) List the methods used and state how the measurements are related to the results.
   (d) What further experiments could be done to learn more about the problem being studied?

2

The First Law: Energy Is Conserved

CONCEPTS

A *scientific law* is an attempt to describe, in a few words, one aspect of nature. Therefore, in a sense all scientific laws will usually be "wrong": They are incomplete, approximate, or in error. In fact, the only useful scientific laws are those that in principle can be proved to be wrong. Many scientists spend their time testing theories in an attempt to disprove them or to discover their limitations. Other scientists spend their time trying to formulate more and more general laws—laws that always apply to all things. The cooperation and competition between scientists with these different approaches leads to progress in science.

Thermodynamics deals with interchanges among different forms of energy. The laws of thermodynamics are excellent examples of both the generality and the limitations of scientific laws. The *first law* states that energy is conserved; different forms of energy can interconvert, but the sum remains constant. The law was originally (about 1800) based mainly on experiments in which mechanical energy was converted into heat. A falling weight turned some paddles in a bucket of water; the water got hot. Other forms of energy, such as radiation, chemical bonding changes, and heats of phase transitions, were later recognized and included in the first law. For many years some biologists were unconvinced that living organisms are subject to the thermodynamic laws.

The form of energy most difficult to believe in was matter. In 1923, 18 years after Einstein postulated that $E = mc^2$, thermodynamicists were still not sure whether thermodynamics applied to radioactive materials. Now we think that the laws of conservation of mass and conservation of energy are each incomplete, but that the law of conservation of mass-energy is correct. Therefore, to make the first law of thermodynamics correct, we must in principle consider mass itself as a form of energy. In practice, this leads to significant effects only when nuclear reactions or radioactive decay processes are occurring.

One can see that the first law evolved from a simple description of a few experiments to a general statement about *all* forms of energy. Any new forms of energy that may be discovered can presumably be incorporated into the first law.

The *second law,* which states that the entropy of an isolated system always increases, has had a different history. It started (about 1820) with experiments on heat engines and led to a statement of the maximum efficiency for the conversion of heat into work. The law was later generalized as the first law was, but it was eventually found that the second law did not always apply to *very* small systems. Entropy was shown to be closely linked to probability; thus the second law could only be applied to a sample which contained enough molecules so that statistical predictions would work. Statistics can tell us that if we flip a coin 1000 times, we can be reasonably sure of obtaining around 500 heads, but it cannot predict whether one flip will be heads or tails. The second law is now known to be slightly limited in its application, but we think we know when and how to correct it.

The *third law* is the most recent addition to the principles of thermodynamics. It was clearly stated in the 1920s and essentially it has not been modified since. The third law states that the entropy of any pure, perfect crystal can be chosen as equal to zero at a temperature of 0 K (absolute zero). The law can be tested experimentally, and it is found to be correct for most chemicals. Whenever there has been a discrepancy, the problem has been traced to the lack of a perfectly ordered crystal at 0 K. The carbon monoxide molecule (CO), for example, forms what looks like a perfect crystal, but the molecule can fit into its crystal structure in two orientations, because the C and the O are similar in size and because the dipole moment of CO is nearly zero. This disorder remains at absolute zero and therefore the entropy is not zero. The relation of entropy and probability allows us to calculate the expected entropy, and we conclude that the entropy would be zero at 0 K if the CO crystal were perfectly ordered. Helium is an example of a substance that does not form a perfect crystal at absolute zero, but one whose entropy, nevertheless, becomes zero. Helium becomes a perfect superfluid with zero entropy at 0 K.

The additions, exceptions, and corrections to the third law, and to all other scientific laws, provide some of the reasons for our continuing study of science. We assume that new ideas will lead to new experiments and eventually to new laws. Presumably, the next two hundred years of science will be filled with as many new discoveries as were the past two hundred years.

APPLICATIONS

Thermodynamics applies to everything from black holes so massive that even light cannot escape from them, to massless neutrino particles. We will consider such questions as: How many grams of ice must you melt with your body heat to equal the calories you gain when you eat one gram of sucrose? How do you calculate the work done when a muscle contracts or expands? How can chemical reactions be used to do work, or to produce heat? A critical test of any explanation, proposal, or mechanism is to ask "Where does the energy come from and where does it end up?"

THE MECHANISM OF ENERGY CONSERVATION

Many experiments, done over a period of many years, have shown that energy is easily converted from one form to another but that the amount of energy remains constant. We shall eventually discuss this quantitatively, but a few examples will make the idea clearer.

Consider a brick on the window ledge of the fifth floor of an apartment building. Owing to its height above the sidewalk, it possesses gravitational potential energy. If the brick falls, most of the potential energy will become kinetic energy of motion (a small amount becomes heat due to air friction). What happens when the brick hits bottom? The kinetic energy is converted into many new forms of energy. Much heat will be produced. If the brick hits the sidewalk, there might be some light energy in the form of sparks. Some energy is used to make and break chemical bonds in the brick and sidewalk fragments. Some sound is produced. Although complicated changes in different forms of energy are involved, the first law tells us that the total energy will remain constant. We can also calculate how the potential energy of the brick will depend on its height above the sidewalk and the weight of the brick. This will tell us the maximum amount of damage to expect. It might also make us more careful of bricks (or flowerpots) on fifth-story ledges.

A more practical and important example is to consider the total amount of energy arriving from the sun, and to consider what forms of energy it is transformed into. Sunlight hitting a desert is mainly transformed into heat. However, sunlight striking a green leaf is partly transformed into useful chemical energy by photosynthesis. Sunlight absorbed by a solar energy cell generates electrical energy. Even when sunlight is converted directly into heat, this can be a useful outcome if the heat is generated in a solar collector that allows it to be transferred to warm the inside of a building. It is vitally important to us to know and understand the various kinds of energy that are available.

Systems and Surroundings

To be able to treat energy and its conversion quantitatively, we must define some new terms. Actually, we shall take common terms and give them specific meanings. We define *system* as the part of the universe that we are interested in. We must specify the boundaries of the object of interest before we can determine its energy gain or loss. The system we consider might be the sun, the earth, a person, the liver, a single cell, or a mole of liquid water at 15°C and 1 atm pressure. That is, we think of the part of the universe we are interested in, draw an enclosure around it, and label it the *system*. Everything else we call the *surroundings* (Fig. 2.1).

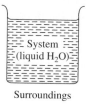

System
(liquid H₂O)

Surroundings

Fig. 2.1 In thermodynamics the *system* is what we focus our attention on. The *surroundings* is everything else in the universe, but we need to consider only the part that interacts with the system.

First Law of Thermodynamics

Energy can be transferred between the system and surroundings, but the total energy of the system plus surroundings is constant. This is a statement of the *first law of thermodynamics*. The first law can be stated more usefully as: The change in energy of a system equals the amount of energy that entered from the surroundings minus the energy that went out into the surroundings. All of this sounds obvious and even trivial, but it is not. For example, whenever we notice energy coming out of a system into the surroundings, we know that the system is losing energy, and we should think about the source of the energy. Nineteenth-century thermodynamicists who thought about the sun were worried. The sun shines about 80 kilojoules (kJ) of energy per square meter per minute on the upper atmosphere of the earth. This amounts to about 10^{15} kJ min^{-1} for the entire earth. The sun radiates its energy in all directions and we catch only a very small fraction. But the total amount of energy radiated by the sun per minute can be calculated. Not much energy comes into the sun, so the sun must be losing energy at a high rate. What is producing its energy?

The mass of the sun and its composition of roughly half hydrogen and half helium has long been known. Therefore, the early thermodynamicists could estimate how much energy was available from the sun for any chemical reaction that could occur. The answer was very discouraging. Either the sun should have burned out after a few million years, or thermodynamics did not work, or there was an unknown energy source. We are all happy to learn about the discovery of nuclear energy as the source of solar energy—which saved us, and thermodynamics, from an early death.

Energy Exchanges

We shall now concentrate on the system and how its energy can be changed. First, a few more definitions are useful. The simplest system to consider is one in which the system is defined to have no exchange of any kind with the surroundings: this is an *isolated system*. It is difficult, if not impossible, to construct such a system, but it is useful to think of one. Thermodynamics, like mathematics, defines ideal situations which can be obtained only approximately. The energy of an isolated system does not change, according to the first law. A *closed system* is defined to have no exchange of matter with the surroundings; it can exchange energy. A closed system can be made by actually putting a physical box around the system. Many chemical reactions are performed in a closed system, a stoppered flask being one example. The chemicals stay in the flask, but heat can come in or out. The most difficult type of system to consider (and of course also the most interesting and useful) is the open system. An *open system* can exchange both matter and energy with the surroundings. A fertilized egg being hatched by a hen is a good example of an open system. Oxygen comes in and carbon dioxide goes out of the egg. Heat is also exchanged between the egg and its surroundings.

We should emphasize that it is up to us to draw a box wherever we wish to separate our system from its surroundings. For example, if a solution containing the enzyme catalase is added to an open beaker containing a hydrogen peroxide solution, the enzyme will accelerate the reaction $H_2O_2 \rightarrow H_2O + \frac{1}{2}O_2$, and oxygen gas will come out of the beaker.

If we choose the liquid content in the beaker as our system, we have an open system. But if we choose the liquid content plus the oxygen evolved as our system, we have a closed system. We usually make our decision based on our specific interest and objectives, choosing boundaries that can be defined conveniently and unambiguously.

It is also important to realize that it is the *change* of energy which characterizes what has happened to the system. We are less concerned about the absolute value of energy for the system. We can add energy to a system in various ways; for example, to an open system, we can add matter. Adding matter to a system increases its chemical energy, because the matter can undergo various chemical reactions. But we do not have to think about the large amount of energy potentially available from nuclear reactions if we are only considering chemical reactions. That is, we do not have to include the $E = mc^2$ energy term, because it does not change significantly in an ordinary chemical reaction.

Work

It is convenient to divide energy exchange between system and surroundings into two types: heat and work. A system can do work on the surroundings; the energy of the system is decreased. Or the surroundings can do work on the system; the energy of the system is increased. *Work* is defined as the product of a force times a distance. To calculate the work done by, or on, the system, multiply the external force on the system by the distance moved—the displacement:

$$\text{work} \equiv \text{external force} \cdot \text{displacement} \qquad (2.1)$$

We must be careful about the sign of the work, because we will be combining heat and work in the application of the first law of thermodynamics. If the direction of the external force is the same as the direction of the displacement, the work is positive. Work is done by the surroundings on the system; the energy of the system increases. If the direction of the external force is opposite to the direction of the displacement, the work is negative. Work is done by the system on the surroundings; the energy of the system decreases. For example, if a spring (the system) is compressed or extended by an external force, the directions of the force and the displacement are the same. Therefore work is positive, which means work is done on the spring. If the compressed or extended spring returns towards its normal position against an external force, the directions of displacement and force are opposite to each other. Therefore work is negative, which means that the spring does work on the surroundings. To summarize:

Sign of work	Convention
+	Work done on system; energy of system increases
−	Work done by system; energy of system decreases

Work of compressing or extending a spring

The force that must be applied to extend or compress a spring is characterized by *Hooke's law,* which states that the force is directly proportional to the change in length of the spring.

To calculate the work of compressing or extending a spring we choose an x axis with one end of the spring fixed at $x = 0$. The other end is free to move along the x axis; its position when there are no forces acting on it is x_0; in general its position is at x.

According to Hooke's law,

$$\text{spring force} = -k \cdot (x - x_0) \tag{2.2}$$

where k is a constant for a given spring (if we do not extend or compress the spring too much). The magnitude of k will be different for different springs. The negative sign states that the direction of the spring force is opposite to the direction of displacement from the equilibrium position at x_0. The external force is the negative of the spring force, as the two are equal in magnitude but opposite in direction:

$$\text{external force} = k \cdot (x - x_0) \tag{2.3}$$

Because the force depends on the displacement, we must integrate the product of force times distance to calculate the work, w.

$$w = \int f \, dx$$

The work done on the spring when its length is changed from x_1 to x_2 is

$$
\begin{aligned}
w &= \int_{x_1}^{x_2} f \, d(x - x_0) \\[6pt]
&= \int_{x_1}^{x_2} k \cdot (x - x_0) \, d(x - x_0) \\[6pt]
&= \tfrac{1}{2} k [(x_2 - x_0)^2 - (x_1 - x_0)^2] \\[6pt]
&= k \cdot (x_2 - x_1)\left(\frac{x_2 + x_1}{2} - x_0\right)
\end{aligned}
\tag{2.4}
$$

We have thus quantitatively calculated the work done on the system by compressing or extending the spring. If the spring was originally compressed or extended, the system would do work on the surroundings when the spring returned to its equilibrium position. A muscle fiber works the same way. It can do work by stretching or contracting.

In *Standard International*, or *SI*, units work is expressed in joules (J) if k is in newtons meter^{-1} (N m^{-1}) and x is in meters (m): $1 \text{ N m} \equiv 1 \text{ J}$. One newton is the magnitude of the force that will cause an acceleration of 1 m s^{-2} when applied to a mass of 1 kg: 1 newton $\equiv 1 \text{ kg m s}^{-2}$. Other units for work are the erg: $1 \text{ erg} \equiv 10^{-7}$ J, and the calorie: $1 \text{ cal} = 4.184$ J. Additional energy conversion tables will be found in the Appendix. The trend now is to use SI units, which we will be using most often in this book. The basic units of length, mass, and time in the SI system are meter (m), kilogram (kg), and second (s).

Work of increasing or decreasing a volume

Consider a system such as a gas or liquid enclosed in a container with a movable wall or piston (Fig. 2.2). The system can expand and do work on the surroundings (if the external pressure, P_{ex}, is less than the pressure of the system, P), or the surroundings can do work

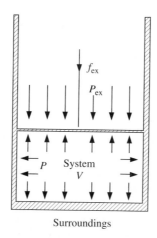

Fig. 2.2 Work done on a system by an external force, f, acting against an opposing force provided by the internal pressure, P, of the system. The work is positive when the volume, V, decreases.

Surroundings

on the system by compressing it (if the external pressure, P_{ex}, is greater than the pressure of the system, P). This example is very similar to the spring example that we discussed previously. The force per unit area, f/A, is the pressure, P; the volume change, dV, is $A\,dx$. Thus force times displacement is equivalent to pressure times change of volume.

$$f \cdot dx = (f/A) \cdot A\,dx = P \cdot dV$$

The work done is

$$w = -\int_{V_1}^{V_2} P_{op}\,dV \tag{2.5}$$

The change in volume, dV, can be positive or negative; it determines the sign of the work. The pressure, P_{op}, is always positive; it is the pressure opposing the change in volume. The opposing pressure, P_{op}, is equal to the pressure, P_{ex}, for an expansion; for a compression the opposing pressure is equal to the pressure, P, of the system. The negative sign is needed because compression—a decrease in volume, a negative dV—means that the surroundings have done work on the system; work is positive. Expansion—an increase in volume, a positive dV— means that the system does work on the surroundings; work is negative. To calculate work in joules, we must use pressure in N m^{-2} and volume in m^3. If the pressure is in atmospheres* and the volume in liters, the result in L atm must be converted using 1 L atm = 101.235 J.

In general the external pressure, P_{ex}, and the pressure, P, of the system will not be equal during an expansion or compression. However, if P_{ex} is always nearly equal to P (they differ only infinitesimally) the process is said to be *reversible*. A very small change in pressure will reverse the process from an expansion to a compression, and vice versa. The sign of the work will change from positive to negative depending on an infinitesimal change in the pressure. Similarly, in a reversible heating or cooling the temperature difference between system and surroundings is very small—infinitesimal. Heat will flow in or out depending on very small changes in temperature.

---

* Atmospheric pressure is the force per unit area exerted by the mass of air above the surface of the earth at sea level. The *standard atmosphere* is defined as 1.01325×10^5 Pa (pascals) = 1.01325×10^5 N m^{-2}.

Equation (2.5) tells us that we can calculate the work of expansion or compression of a system if we know the opposing pressure, P_{op}, and the volume change of the system. Often the pressure will change as the volume changes, so we write the equation as an integral. If the opposing pressure is kept constant, however, the integration can be done easily. For *constant pressure*,

$$w_P = -P_{op} \cdot (V_2 - V_1) \tag{2.6}$$

Expansion work done by a system containing only gases has been very important in thermodynamics as a model for heat engines. Engineers (and others) need to know how much work can be done by the expanding gas inside the cylinder of a car, for example. Biologists may be interested in the work done by the lungs on the air we breathe or the work done in systems containing liquids. The work done when liquid freezes is important because of the large forces that may be associated. The density of ice is less than that of liquid water, so work can be done by the system when water expands on freezing. The amount of work (and the freezing temperature) will depend on the external pressure. If the surroundings are a complete vacuum, no pressure-volume work is done in an expansion, because P_{ex} in Eq. (2.6) is exactly zero. Similarly, if a cylinder undergoing compression has previously been evacuated, no work can be done on the system.

Example 2.1 Calculate the pressure-volume work done when a system containing a gas expands from 1.0 L to 2.0 L against a constant external pressure of 10 atm.

Solution

$$w_P = -P_{ex} \cdot (V_2 - V_1) \qquad \text{(expansion)}$$
$$= -(10 \text{ atm})(2 \text{ L} - 1 \text{ L})$$
$$= -10 \text{ L atm}$$
$$w = -(10 \text{ L atm})\left(101.3 \ \frac{\text{J}}{\text{L atm}}\right)$$
$$= -1013 \text{ J}$$

The system does work; the sign of w is negative.

Example 2.2 Calculate the pressure-volume work done in joules when a sphere of water 1.00 micrometer ($1 \ \mu m = 10^{-6}$ m) in diameter freezes to ice at 0°C and 1 atm pressure.

Solution Find the volume of the sphere of water, then use the density of ice and liquid water at 0°C to calculate the volume of the frozen sphere.
The volume of the liquid water is

$$V_l = \frac{1}{6}\pi D^3 = \left(\frac{3.142}{6}\right)(10^{-6} \text{ m})^3$$
$$= 5.237 \times 10^{-19} \text{ m}^3$$

The density of liquid water at 0°C is 1.000 g cm^{-3}. The density of ice at 0°C is 0.915 g cm^{-3}. The volume of the solid water is

$$V_s = (5.237 \times 10^{-19} \text{ m}^3)\left(\frac{1.000}{0.915}\right)$$

$$= 5.723 \times 10^{-19} \text{ m}^3$$

The work done by the system is

$$w_P = P_{ex} \cdot (V_2 - V_1) \quad \text{(expansion)}$$

$$= -(1 \text{ atm})(5.723 \times 10^{-19} \text{ m}^3 - 5.237 \times 10^{-19} \text{ m}^3)$$

$$= -0.486 \times 10^{-19} \text{ m}^3 \text{ atm}$$

$$w = (-0.486 \times 10^{-19} \text{ m}^3 \text{ atm})(1.013 \times 10^5 \text{ J m}^{-3} \text{ atm}^{-1})$$

$$= -4.92 \times 10^{-15} \text{ J}$$

The system expands; therefore, work is done by the system and the sign of w is negative.

It should be clear that all the examples of work given so far have been very similar. In each case, work can be done on the system by changing it from its equilibrium state, or the system can do work on the surroundings by tending to return to its equilibrium state.

Friction

The force of friction causes an energy change whenever two surfaces in contact move relative to each other. The frictional force is opposite in direction to the force causing the motion. We will not discuss frictional effects in this chapter, but one should realize that they may be an important source of energy loss in real systems. Engineers, of course, try to maximize expansion work and minimize friction in engines. Frictional losses are also important in biological engines. As arteries get rougher and narrower with aging, the energy needed to circulate blood increases. The blood pressure must then also increase, and the heart must do more expansion and compression work. Friction is not all bad, however. Driving on icy roads is treacherous enough; think of what would happen if there were no friction at all!

Work in a gravitational field

All processes that occur on the earth are affected by the earth's gravitational field (and to a much lesser extent by the moon's and other astronomical bodies' gravitational fields). This means that all systems on earth can do gravitational work. If an object of mass m (the system) is lowered at a constant velocity from a height h_1 to a height h_2 above the earth's surface, the work done on the system is

$$w \text{ (gravitational)} = mg \cdot (h_2 - h_1) \qquad (2.7)$$

The work done on the system by the external force is negative ($h_2 - h_1$ is negative); in other words, the system is doing work on the surroundings.

For mass m, in kg, height h, in meters, and g, in m s^{-2} (the standard acceleration of gravity = 9.807 m s^{-2}), the work is obtained in joules.

Work in an electric field

If a system contains electrical charges, an electric field will produce a force on the charges which will cause them to move, and thus a current will flow. The work done on the system can be shown to be

$$w \text{ (electrical)} = -EIt \tag{2.8}$$

where E is the voltage ≡ potential difference ≡ electromotive force, I is the current, and t is the time. For E volts, I in amperes, and t in seconds, the units of work obtained are joules. The cost of electricity is for electrical work, and it is usually calculated in kilowatt-hours. Since a watt-second is a joule, 1 kWhr is 3.6×10^6 J.

Example 2.3   Calculate the electrical work done in joules and calories by a 12.0-V storage battery which discharges 0.1 A for 1.00 hr.

Solution   The electrical work is

$$w = -EIt \tag{2.8}$$
$$= -(12 \text{ V})(0.1 \text{ A})(1 \text{ hr})$$
$$= -1.2 \text{ V A hr} = -1.2 \text{ W hr}$$
$$= -(1.2 \text{ W hr})(3600 \text{ s hr}^{-1})$$
$$= -4.32 \times 10^3 \text{ W s}$$
$$= -4.32 \times 10^3 \text{ J}$$
$$= (-4.32 \times 10^3 \text{ J})(0.2389 \text{ cal J}^{-1})$$
$$= -1.03 \times 10^3 \text{ cal}$$

The minus sign tells us that the battery is doing work on the surroundings.

Heat

When two bodies are in contact with each other, their temperatures tend to become equal. Energy is being exchanged. The hot body will lose energy and cool; the cold body will gain energy and warm. The energy exchange is said to occur by *heat transfer*. Heat is defined in terms of temperature changes, but it is recognized in many processes. Chemical reactions, electrical currents, friction, and absorption of radiation all involve heat transfer.

The sign convention for heat is the same as the convention for work.

Sign of work	Convention
+	Heat added to system; energy of system increases
−	Heat lost by system; energy of system decreases

For a closed system, the amount of heat transferred, q, is proportional to the difference in temperature of the system before (T_1) and after (T_2) the heat exchange. The proportionality constant depends on the system and is called the *heat capacity* (C) of the system; in general, it will vary with temperature. We must therefore write the equation for the heat gained as an integral.

$$q = \int_{T_1}^{T_2} C \, dT \tag{2.9}$$

For a hot body in contact with a cold body, the heat capacity of each body must be known and Eq. (2.9) must be applied to each body separately. The heat capacity, C, is an extensive property that increases with amount of material in the body. For a pure chemical substance $C = n\overline{C}$, where n is the number of moles and $\overline{C}$ is the *molar heat capacity* or heat capacity per mole. Molar heat capacities at constant pressure, C_P, for several substances are tabulated in the first two columns of Table 2.1. In the third column values of *specific heat capacity, $C_P{}^*$*, or heat capacity per kilogram are listed.

Table 2.1 Heat capacities at constant pressure, 1 atm, of various substances near 25°C

Substance	$\overline{C}_P$	Substance	$\overline{C}_P$	Substance	$\overline{C}_P{}^*$
Gases	Molar heat capacities (J K^{-1} mol^{-1})	Liquids	Molar heat capacities (J K^{-1} mol^{-1})	Solids	Specific heat capacities (J K^{-1} kg^{-1})
He	20.8	Hg	28.0	Au	129
H_2	28.8	H_2O	75.2	Fe	452
O_2	29.4	Ethanol	111.4	C (diamond)	510
N_2	29.1	Benzene	136.1	Glass (Pyrex)	840
H_2O	33.6	n-Hexane	195.0	Brick	~800
CH_4	35.8			Al	902
CO_2	37.1			Glucose	1250
				Urea	2100
				H_2O (0°C)	2100
				Wood	~2000

Example 2.4 Calculate the heat in joules necessary to change the temperature of 100.0 g of liquid water by 50°C at constant pressure. The heat capacity of liquid water at constant pressure is 1.00 cal g^{-1} deg^{-1} and is nearly independent of temperature.

Solution   The heat absorbed by the system is

$$q = \int_{T_1}^{T_2} C \, dT = C \int_{T_1}^{T_2} dT$$

$$= C \cdot (T_2 - T_1)$$

$$= (100.0 \text{ g}) \left(1.00 \, \frac{\text{cal}}{\text{g deg}}\right) (50 \text{ deg}) \left(4.184 \, \frac{\text{J}}{\text{cal}}\right)$$

$$= 20.9 \text{ kJ}$$

Heat capacity for every material is an experimental quantity which characterizes how much heat is necessary to raise its temperature by 1 degree Celsius or 1 Kelvin. The units are often J K^{-1} mol^{-1} or cal K^{-1} mol^{-1}. It is easy to remember that the heat capacity of liquid water is about 1 cal K^{-1} g^{-1}. The heat capacity for a substance in a given phase (gas, liquid, or solid) will, in general, increase with increasing temperature. It will also depend on whether P or V is held constant during the heating. The symbol C_P means heat capacity at constant pressure, and C_V is heat capacity at constant volume. C_P is larger than C_V, although for solids and liquids they are nearly equal. The difference between them is significant for gases because of the extra energy required to expand the gas (the work done) when it is heated at constant pressure. For gases we will show later that the molar heat capacities differ approximately by R, the gas constant, and that $(\overline{C}_P \cong \overline{C}_V + R)$. Table 2.1 gives some representative values.

There are many practical applications of Eq. (2.9). We often want to know how much heat can be transferred from one system to another and what the final temperature will be. One of the main problems in solar heating is how to store the energy for use at night. It is simple to raise the temperature of a storage system such as water or rocks during the day and to transfer the heat at night to the cold air in your house. The amount of heat transferred depends on the temperature difference and the heat capacity of each system.

Example 2.5   Cold air at 0°C is passed through 100 kg of hot crushed rock that has been heated to 110°C. The air is heated to 20°C by the time it leaves the rock and is admitted into a house for heating. Calculate the total volume of 20°C air that can be obtained by this process. The heat capacity at constant pressure of the rock is 800 J K^{-1} kg^{-1} and of the air is 1000 J K^{-1} kg^{-1}. The density of air at 1 atm and 20°C is 1.20×10^{-3} g cm^{-3}; the density of the rock is 2.5 g cm^{-3}.

Solution   We need to calculate the heat transferred from the rock in cooling to 20°C. This is equal to the heat absorbed by the air. We can then calculate the weight of the air heated and therefore the volume of the air heated.

The heat lost by the rock is

$$q = C \cdot (T_2 - T_1)$$

$$= (100 \text{ kg}) (800 \text{ J K}^{-1} \text{kg}^{-1}) (90 \text{ K})$$

$$= 7.2 \times 10^6 \text{ J}$$

The weight of the air heated to 20°C by this amount of heat is

$$\text{air wt} = \frac{7.2 \times 10^6 \text{ J}}{(1000 \text{ J K}^{-1} \text{ kg}^{-1})(20 \text{ K})}$$

$$= 360 \text{ kg}$$

The volume of the air heated to 20°C is

$$\text{air vol} = \frac{3.6 \times 10^5 \text{ g}}{1.2 \times 10^{-3} \text{ g cm}^{-3}}$$

$$= 3 \times 10^8 \text{ cm}^3$$

$$= 3 \times 10^2 \text{ m}^3$$

This corresponds to the volume of a medium-sized room. The volume of crushed rock necessary is only 100×10^3 g/2.5 g cm^{-3}, or 4×10^4 cm^3 = 0.04 m^3.

Radiation

A very important method of energy exchange is that of radiation. Most of the earth's useful energy comes from the sun's radiation. It is important to be able to calculate the energy present in a given number (N) of photons of frequency ν, given in s^{-1}:

$$\text{radiation energy} = Nh\nu \tag{2.10}$$

where $h \equiv$ Planck's constant = 6.6262×10^{-34} J s.

Another useful equation, the Stefan-Boltzmann equation, describes how much energy is radiated by a body as a function of its temperature. The equation is for an ideal body that radiates and absorbs all wavelengths, a *blackbody:*

$$\text{radiation energy m}^{-2} \text{ s}^{-1} = \sigma T^4 \tag{2.11}$$

where $\sigma \equiv$ Stefan-Boltzmann constant = 5.67×10^{-8} J m^{-2} s^{-1} K^{-4}
T = absolute temperature

Example 2.6 The average surface temperature of the sun is about 6000 K; its diameter is about 1.4×10^9 m. Estimate the total energy radiated by the sun in J s^{-1}.

Solution The Stefan-Boltzmann equation provides the radiation rate per m^2. From the diameter of the sun we can obtain the surface area and thus the total energy radiated per second.

$$\text{radiation} = \sigma T^4 \tag{2.11}$$

$$= (5.67 \times 10^{-8} \text{ J m}^{-2} \text{ s}^{-1} \text{ K}^{-4})(6000 \text{ K})^4$$

$$= 7.35 \times 10^7 \text{ J m}^{-2} \text{ s}^{-1}$$

$$\text{area} = \pi D^2$$

$$= (\pi)(1.4 \times 10^9 \text{ m})^2$$

$$= 6.2 \times 10^{18} \text{ m}^2$$

energy radiation per second ≡ luminosity

$$= (7.35 \times 10^7 \text{ J m}^{-2} \text{ s}^{-1})(6.2 \times 10^{18} \text{ m}^2)$$

$$\text{luminosity} = 4.6 \times 10^{26} \text{ J s}^{-1}$$

This number agrees well with the measured luminosity of the sun.

By comparison, the intensity of bright summer sunlight at the surface of the earth is about 1000 J m^{-2} s^{-1}.

Example 2.7   Radiation can cause chemical reactions to occur. For some reactions each photon produces one molecule of product. How many photons are there per joule of red light? The wavelength is 700 nm; the frequency is 4.28×10^{14} s^{-1}.

Solution

$$N = \frac{\text{energy}}{h\nu} \tag{2.10}$$

$$= \frac{1 \text{ J}}{(6.63 \times 10^{-34} \text{ J s})(4.28 \times 10^{14} \text{ s}^{-1})}$$

$$= 3.5 \times 10^{18} \text{ photons}$$

VARIABLES OF STATE

We have been using the terms pressure, volume, and temperature without defining them further, because they are familiar to us. We must be careful about units, however. For pressure we will use atm, or Torr ≡ mm Hg, or pascals, Pa ≡ N m^{-2}.

$$1 \text{ atm} = 760 \text{ Torr} = 1.01325 \times 10^5 \text{ Pa}$$

For volume we use cm^3 = milliliter, or liter. For temperature we use degrees absolute ≡ Kelvin, K; deg means K. In the SI system the unit of pressure is the pascal. Whenever T occurs in thermodynamic equations, Kelvin is indicated. To convert from other temperature scales, use

$$K = °C + 273.1$$

$$K = \frac{°F - 32}{1.8} + 273.1$$

where °C = degrees Celsius, or centigrade
    °F = degrees Fahrenheit

We now want to describe P, V, and T as variables that help specify the state of a system. Consider a closed system in the absence of all external fields. This statement describes a system that we can only approximate on earth. We cannot turn off gravity, for example. However, for many practical purposes we can ignore the effects of gravity. An essential part of learning a science is in recognizing which approximations are useful and which are inappropriate.

If the system consists of a pure liquid, specifying the pressure (P), volume (V), and temperature (T) of the liquid is sufficient to specify many other properties of the liquid, such as the density, the energy, and so forth. These other properties of the system, and P, V, and T, are called *variables of state*. Such variables depend only on the state of the system, not on how the system arrived at its state. The useful characteristic of variables of state is that when a few (in general, more than P, V, and T) are used to specify a system, all other variables of state are determined implicitly. It is thus possible and practical to measure and tabulate certain properties of the system: those that are variables of state. The discovery of the first law of thermodynamics showed that *energy* ≡ *internal energy, E,* is a variable of state; it is a property of the system. Heat and work are *not* variables of state; they depend specifically on the method used to change from one state to another. In other words, they depend on the *path* between states. It is sometimes convenient for thermodynamicists to define new variables of state by combining previously defined ones. *Enthalpy, H,* is a variable of state defined as

$$H \equiv E + PV \tag{2.12}$$

H can be thought of as being no more than a shorthand notation for $E + PV$. The units must be the same as E, so the PV units must be converted to calories or joules before adding to E.

Enthalpy was defined as $E + PV$ because this definition makes the enthalpy change of a system at constant pressure equal to the heat absorbed or released (if only PV work is done). Thus the heat of a reaction at constant pressure is equal to the enthalpy change.

Variables of state are divided into two classes: extensive and intensive. An *extensive* variable of state is directly proportional to the mass of the system. If you double the mass of the system you double the magnitude of an extensive variable. An *intensive* variable of state is independent of the mass of the system. Of the variables that we have considered so far, P and T are intensive, whereas V, E, H, and the heat capacity C in Eq. (2.9) are extensive. However, we can always change an extensive variable to an intensive one by expressing it per-unit amount of material. Energy is extensive, but energy mol^{-1} or energy g^{-1} is intensive; mass is extensive, but density or mass volume^{-1} is intensive.

Equations of State

An *equation of state* is an equation that relates variables of state. A few variables of state are usually sufficient to specify all the others. This means that equations exist that can relate the variables of state. The simplest and most frequently used equations of state link P, V, and T.

Solids or liquids

The volume of a solid or a liquid does not change very much with either pressure or temperature. We are not considering changes from solid to liquid or any other change except P, V, and T. Therefore, a first approximation for the *equation of state of a solid or liquid is* $V \cong constant$. This means that to calculate the volume of the solid or liquid, one just finds the density or specific volume at one temperature and uses that value for any temperature and pressure. The volume is related to the density by

$$V = \frac{\text{mass}}{\text{density}}$$

Actually, the volume of a solid or liquid does change somewhat with T and P. Experimental data for 1 mol of liquid water are shown in Fig 2.3. The molar volume is plotted versus temperature at constant pressure in the upper half and the molar volume is plotted versus pressure at constant temperature in the lower half. Equations can be obtained for V as a function of P and T, $V(P, T)$, for liquid water from these data.

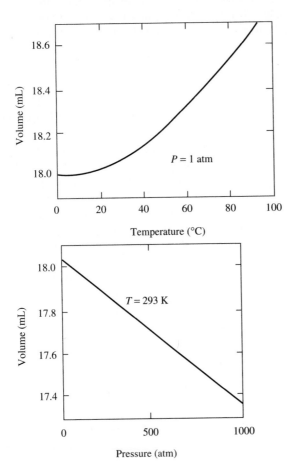

Fig 2.3 Volume of 1 mol of liquid water as a function of temperature and pressure. Note that the volume changes by less than 5% over the temperature and pressure ranges shown.

For example, at 293 K the linear empirical equation

$$\overline{V} = 18.07\,(1 - 45.9 \times 10^{-6}\,P)\ \text{mL mol}^{-1} \qquad (P \text{ in atm})$$

closely represents the plot shown in the lower part of Fig. 2.3. The number 45.9×10^{-6} atm^{-1} is the *isothermal compressibility* of liquid water at 293 K. Isothermal compressibilities, which represent the fractional decrease in volume per atm increase in pressure, are tabulated in handbooks of chemical and physical data. The dependence of the volume of liquid water on temperature (upper part of Fig. 2.3) is more complicated. Not only is the plot not a straight line, but the molar volume of water actually has a minimum value at 4°C. For most of our applications we shall use the first approximation, that V is independent of T and P for a solid or liquid.

Gases

For gases the volume varies greatly with T and P, but the variation is nearly independent of the type of gas. Thus there is a simple approximate equation of state for gases. The *ideal gas equation* is

$$PV = nRT \qquad (2.13)$$

where n = number of moles
$R \equiv$ universal gas constant = 0.08206 L atm deg^{-1} mol^{-1}
= 8.314 J K^{-1} mol^{-1}

A plot of how the volume of 1 mol of an ideal gas depends on pressure is shown in Fig. 2.4. Note that, although 1000 atm was necessary to change the volume of liquid water by 3%, a change from 1 atm to 1000 atm for an ideal gas will cause a change in volume by a factor of 1000. The ideal gas equation has the great advantage that it contains no constants applying to individual gases; it applies to all gases if the pressure is low enough. It is an exact limiting equation for all gases as P approaches zero. For higher pressures it is an approximation. The answers obtained using Eq. (2.13) are usually accurate within ± 10% for most gases near room temperature and atmospheric pressure. Of course, if the pressure causes the gas to liquify, the ideal gas equation cannot be used to calculate the volume of the liquid.

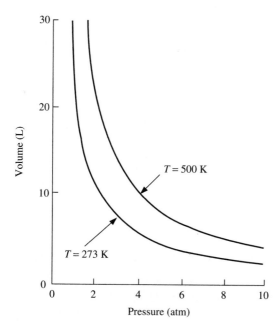

Fig. 2.4   Volume of 1 mol of an ideal gas as a function of temperature and pressure.

Many other more accurate equations of state for gases have been developed. They are corrections to the ideal gas equation which contain parameters relating to individual gases. One example is the *van der Waals gas equation:*

$$\left(P + \frac{n^2 a}{V^2}\right)(V - nb) = nRT \tag{2.14}$$

where a and b are constants that are different for each gaseous substance. The van der Waals a constant is a measure of the attractive forces between molecules, and the b constant is a measure of the intrinsic volume of the gas molecules themselves. When both of these are zero, Eq. (2.14) simplifies to the ideal gas equation.

The equations of state that we have been discussing have all been applied to systems containing only one component. For mixtures, the number of grams or moles of each component must be specified, and the equation of state will depend on the concentrations. For gases that can be approximated by the ideal gas equation, the results are particularly simple. The ideal gas equation can be applied to each gas in the mixture as if the others were not there. The partial pressure of each gas can be calculated as follows:

$$P_i = \frac{n_i RT}{V} \tag{2.15}$$

where P_i = partial pressure of component i in the ideal gas mixture
n_i = number of moles of component i
V = total volume of gases

The total pressure is just the sum of the partial pressures:

$$P_{\text{total}} = \sum_i P_i \tag{2.16}$$

$$P_{\text{total}} = \frac{n_{\text{total}} RT}{V} \tag{2.17}$$

The partial pressures can also be obtained from the total pressure and the mole fractions X_i of each component:

$$P_i = X_i P_{\text{total}}$$

$$X_i = \frac{n_i}{\sum_i n_i} \tag{2.18}$$

The equations for partial pressures are useful, because often we are interested only in one of the components of a gas mixture. For example, in the air we breathe, the partial pressure of oxygen or carbon dioxide (or even sulfur dioxide) is much more important than the total pressure.

Energy and Enthalpy Changes

It is important to be able to calculate the energy and enthalpy changes that occur in a system when the system changes from one state to another, or to be able to calculate the amount of energy necessary to cause a change in the state of the system.

We will discuss various methods of calculating energy (E) and enthalpy (H) changes. The most important fact to remember is that E and H are properties of the system; they are variables of state: they depend only on the state of the system. Therefore, the changes in E and H depend only on the initial and final states: they do not depend on the path that we find convenient to bring about the changes.

Heat and Work Changes

When heat is transferred to or from a system and work is done by or on a system, the energy of the system will change. For a closed system, if heat and work are the only forms of energy that the system exchanges with the surroundings, we can write

$$E_2 - E_1 = q + w \qquad (2.19)$$

where $E_2 - E_1 =$ change in energy from the initial state 1 to the final state 2
$\qquad\qquad q =$ net heat transferred *to* the system (the heat *in*)
$\qquad\qquad w =$ net work done *on* system (the work *in*)

This equation is one statement of the first law of thermodynamics. To use this equation we pick a convenient process, or path, to get from state 1 to state 2 and find the heat and work changes along the path. The heat and work will each depend on the path we choose, but their sum will not. We use the convention that both heat, q, and work, w, are positive when they increase the energy of the system. Some other books have work done *by* the system as positive; their equation corresponding to Eq. (2.19) would have $q - w$ in it, with w being the work done *by* the system.

Reversible path

There is an infinite number of paths to get from one state to another, but some are so convenient that they have received special attention. The most important path is the reversible path. In a *reversible path* the system always remains very near equilibrium.

As an example, let us consider the expansion of an ideal gas in a cylinder with a frictionless and weightless piston, as illustrated in Fig. 2.5a. The cylinder conducts heat well, so the temperature of the system (the gas in the cylinder) is always the same as the temperature of the surroundings, which we specify to be a constant. Suppose that the pressure outside the cylinder is always at 1 atm and the pressure inside the cylinder is initially at 2 atm. If we remove the stops that hold the piston in position, the gas will expand irreversibly until a final state is reached at which its pressure becomes 1 atm. During the course of expansion, the pressure of the system is always significantly greater than that of the surroundings (that is why the expansion is called *irreversible*). The two become the same only at the end of the expansion.

We can carry out the expansion in a different way. Instead of holding the piston in position with stops at the beginning, we put many small weights on top of the piston to make up for the pressure difference (Fig. 2.5b). The expansion is then carried out in a stepwise manner by removing one weight at a time. If the number of weights is very large (and the weight of each very small), the pressure of the system is always almost the same as that of

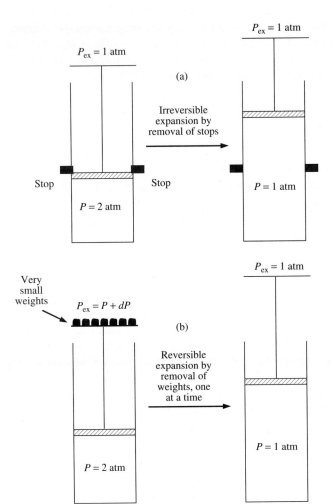

Fig. 2.5 Comparison of an irreversible expansion (a) and a reversible expansion (b). In a reversible expansion the internal pressure is always nearly equal to the external pressure.

the surroundings during the course of expansion. Furthermore, if we add rather than remove a weight, the process will be reversed.

A reversible path is one in which the process can be reversed at any instant by an infinitesimal change of the variable that controls the process. The path illustrated in Fig. 2.5a is not a reversible one; the path illustrated in Fig. 2.5b becomes a reversible one when the weights are very small and their number approaches infinity.

As a second example, let us consider the transition from liquid water to gaseous water (steam). The transition is reversible at 100°C and 1 atm. If we maintain the temperature at 100°C, liquid water will evaporate to steam if the pressure is lowered to slightly below 1 atm, and steam will condense to liquid water if the pressure is increased to slightly above 1 atm. Alternatively, if the pressure is maintained at 1 atm, the direction of the change can be reversed by causing a small energy flow—for example, by causing the temperature of the surroundings to differ slightly from that of the system.

There are many possible reversible paths for a process. Suppose, for example, that we want to use a reversible path to calculate the energy needed to evaporate 1 mol of liquid water at 25°C and 1 atm.* We can choose either of the reversible paths shown in Fig. 2.6. Each breaks the overall process into three successive steps. The first reversible path (1)

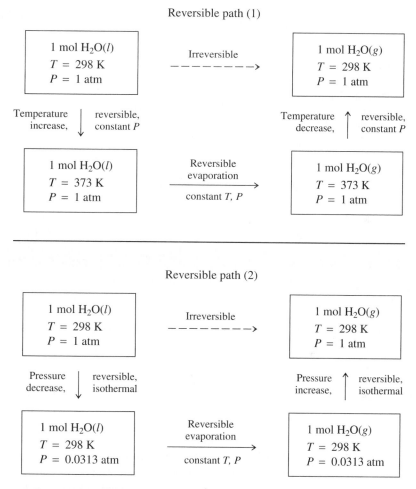

Reversible path (1)

| 1 mol H$_2$O(l) | Irreversible | 1 mol H$_2$O(g) |

Temperature increase, reversible, constant P | Temperature decrease, reversible, constant P

Reversible evaporation constant T, P

Reversible path (2)

Pressure decrease, reversible, isothermal | Pressure increase, reversible, isothermal

Reversible evaporation constant T, P

Fig. 2.6 Two possible reversible paths that can be taken between the same initial and final states. The letters in parentheses, (l) and (g), denote liquid and gas.

*   The reader may wonder about the practicability of calculating the energy of a process that is difficult or impossible to carry out. Supercooled gaseous water with a temperature of 25°C and a pressure of 1 atm has never been obtained experimentally. However, there are many valid reasons for wanting to know the difference in energy between liquid and gaseous water at 25°C. For example, the properties of gases at 1 atm pressure and 25°C are the standard values given in tables. One of the important points to remember is that thermodynamics does allow you to determine the energy changes for processes in which direct measurement is very difficult or impossible. The reader may prefer to think about real processes closer to equilibrium, such as the evaporation of superheated liquid water at 110°C and 1 atm, or the condensation of supercooled gaseous water at 25°C and 0.05 atm.

consists of raising the temperature of liquid water reversibly to 100°C, the normal boiling point, then evaporating the liquid reversibly at 1 atm and finally cooling the vapor back to 25°C. Each of these steps can be considered to be reversible. In the second reversible path (2) the pressure on the liquid water is decreased until it will evaporate reversibly at 25°C; the final step then is to bring the water vapor pressure back to 1 atm. The first reversible path (1) is a constant-pressure path; it is said to be *isobaric*. The second reversible path (2) is at constant temperature; it is said to be *isothermal*. Several other named paths are listed in the following table.

Restricted thermodynamic paths

Constant pressure	Isobaric	$\Delta P_{syst} = 0$
Constant temperature	Isothermal	$\Delta T_{syst} = 0$
Constant volume	Isochoric	$\Delta V = 0$
No heat transferred	Adiabatic	$q = 0$
Final state = initial state	Cyclic	No change in any property of state of system

Whichever path we choose, we would calculate the same value for $E_2 - E_1$. This statement also holds for the direct, irreversible path shown in Fig. 2.6. Reversible paths are stressed because it is often easier to calculate q and w along reversible paths; therefore, $E_2 - E_1$ can be easily calculated. Furthermore, it can be shown that the work that a system can do on the surroundings for a given change in energy is a maximum along a reversible path. Nevertheless, there is an important general property of variables of state that needs to be emphasized at this point. *The change in any variable of state of the system during a process depends only on the initial and final states of the system and not on the path over which the process occurs.* This statement applies to any variable of state, such as *P, V, T, E, H,* and so on; it does not apply to path-dependent properties such as q or w. A corollary of the statement above is: *For a cyclic path there is no change in any variable of state.*

Changes of Temperature and Pressure

It is important to be able to calculate the effects of temperature and pressure on chemical reactions, on melting and boiling points, on any process. In order to do this we must understand how each substance involved is affected by temperature and pressure. In the following sections we will consider the effects of *T* and *P* on the energy and enthalpy of pure solids, liquids, and gases. The results will apply directly to the effects of *T* and *P* on the enthalpies and energies of chemical reactions and of phase changes. The methods we use will be similar to those necessary to calculate (in the next chapter) the effect of *T* and *P* on entropies and free energies.

Temperature and Pressure Changes for a Liquid or Solid

Consider a system of one component which undergoes a change in *P, V,* and *T* only. External fields, friction, and surface effects are ignored. Only two of the three variables need be specified, because the equation of state allows calculation of the third. The number of moles, or grams, of the component does have to be specified. The goal is to calculate the energy change and enthalpy change for a variety of processes. One can get from the initial state to the final state by many paths. It is usually easiest to use a series of reversible steps, where either P, or V, or T is held constant in each one, and then add the energy or enthalpy contributions from each step. For illustration we will use one of the most important biological molecules there is, water. Some properties of water are given in Table 2.2.

First, let us consider heating or cooling some liquid water in an open flask at a constant pressure provided by the atmosphere. The reaction is

$$n \text{ mol } H_2O(l) \text{ at } T_1, P_1, V_1 \longrightarrow n \text{ mol } H_2O(l) \text{ at } T_2, P_1, V_2$$

Table 2.2 Physical properties of water, H_2O (mol wt = 18.016), at 1 atm*

Solid H_2O = ice
(at 0°C)

Density = 0.915 g cm^{-3}; specific volume = 1.093 cm^3 g^{-1}
Vapor pressure = 4.579 Torr
Heat of melting = 333.4 kJ kg^{-1} = 6.007 kJ mol^{-1}
Absolute molar entropy = 41.0 J K^{-1} mol^{-1}
Specific heat capacity = 2.113 kJ K^{-1} kg^{-1}
Molar heat capacity = 38.07 J K^{-1} mol^{-1}

Liquid H_2O

Temperature (°C)	Density (g cm^{-3})	Surface tension (mN m^{-1})	Vapor pressure (Torr)	Heat of vaporization (kJ kg^{-1})	Viscosity (mPa s)
0	0.9999	75.64	4.579	2493	1.7921
20	0.9982	72.75	17.535	2447	1.0050
40	0.9922	69.56	55.324	2402	0.6560
60	0.9832	66.18	149.38	2356	0.4688
80	0.9718	62.61	355.1	2307	0.3565
100	0.9584	58.85	760.00	2257	0.2838

Absolute molar entropy = 63.2 J K^{-1} mol^{-1} at 0°C
            = 87.0 J K^{-1} mol^{-1} at 100°C
Specific heat capacity = 4.18 kJ K^{-1} kg^{-1} between 0 and 100°C
Molar heat capacity = 75.4 J K^{-1} mol^{-1}
Heat of freezing = −333.4 J kg^{-1} at 0°C

Gaseous H_2O = steam
(at 100°C)

Density = 5.880 × 10^{-4} g cm^{-3}; specific volume = 1701 cm^3 g^{-1}
Absolute molar entropy = 196.2 J K^{-1} mol^{-1}
Specific heat capacity at constant pressure = 1.874 kJ K^{-1} kg^{-1}
Molar heat capacity at constant pressure = 33.76 J K^{-1} mol^{-1}
Heat of condensation = −2257 kJ kg^{-1} = −40.66 kJ mol^{-1}

* Some of the properties listed will be defined and discussed in later chapters.

The heat is calculated from Eq. (2.9); at constant P,

$$q_P = \int_{T_1}^{T_2} C_P \, dT \qquad (2.20)$$

The subscript P on the heat q and the heat capacity C remind us that the heat effects depend on the path. C_P is the heat capacity at constant pressure. In general, C_P will depend on the pressure and the temperature. However, we can usually neglect the effects of P and often neglect the effect of temperature. That is, C_P for 1 mol of liquid water is close to 75.4 J K^{-1} from 0 to 100°C and for pressures less than a few hundred atmospheres. Therefore, for a temperature change at constant P, if C_P is independent of T,

$$q_P = C_P \cdot (T_2 - T_1) \qquad (2.21)$$

Because values of heat capacity are often given per mole, we can write

$$q_P = n\overline{C}_P \cdot (T_2 - T_1) \qquad (2.22)$$

where $\overline{C}_P$ = molar heat capacity at constant P (the bar over the C means per mole)
$\quad\quad n$ = number of moles

If T_2 is greater than T_1, we know that heat is absorbed, which is consistent with the positive sign of q_P.

The work is calculated from Eq. (2.6); at constant P

$$w_P = -P_1 \cdot (V_2 - V_1) \qquad (2.6)$$

But if we assume that the volume change of the liquid water is negligible, then $V_2 - V_1 \cong 0$ and $w_P \cong 0$.

If the water is heated in a closed and very strong container which keeps the volume constant, the heat and work are, at constant V,

$$q_V = n\overline{C}_V \cdot (T_2 - T_1) \qquad (2.23)$$

$$w_V = 0 \qquad (2.24)$$

C_V is the heat capacity at constant volume; for a solid or a liquid, it is not very different in magnitude from C_P. The pressure-volume work for a constant volume process is obviously zero.

If the water is kept at constant temperature while the pressure is changed, there is only a negligible volume change, and there is no appreciable work done or heat transferred; for an isothermal process involving liquid or solid

$$q_T \cong 0 \qquad (2.25)$$

$$w_T \cong 0 \qquad (2.26)$$

Calculation of $E_2 - E_1$ and $H_2 - H_1$ for a solid or liquid

To obtain $E_2 - E_1$ for the changes discussed above, we just add q and w. To obtain $H_2 - H_1$ we use the definition of enthalpy:

$$H_2 - H_1 \equiv E_2 - E_1 + (P_2V_2 - P_1V_1) \qquad (2.12)$$

For any change of P_1, V_1, T_1 to P_2, V_2, T_2 we can obtain $E_2 - E_1$ and $H_2 - H_1$ by choosing a convenient path and then combining the q's and w's. For example, we could use an isothermal plus a constant-pressure path, or an isothermal plus a constant-volume path. Because the volume of a solid or a liquid does not change much with temperature or pressure, we find that, for a solid or a liquid,

$$E_2 - E_1 \cong H_2 - H_1 \qquad \text{(liquid, solid)}$$
$$C_P \cong C_V$$

The exact relations between C_P and C_V or $E_2 - E_1$ and $H_2 - H_1$ depend on the equation of state.

Example 2.8   Calculate $E_2 - E_1$ and $H_2 - H_1$ in joules for heating 1 mol of liquid water from 0°C and 1 atm to 100°C and 10 atm. The volume per gram of the water is essentially independent of pressure; it can be calculated from the average density of water given in Table 2.2, 0.98 g cm^{-3}.

Solution   Choose a path such as an isothermal path plus a constant-pressure path. First, the pressure is raised from 1 atm to 10 atm at 0°C. Then, the temperature is raised from 0°C to 100°C at 10 atm:

$$
\begin{aligned}
E_2 - E_1 &= \Delta E_T + \Delta E_P \\
&= q_T + w_T + q_P + w_P \\
&= 0 + 0 + n\overline{C}_P \cdot (T_2 - T_1) + 0 \\
&= (1 \text{ mol}) \left(75.4 \ \frac{\text{J}}{\text{mol deg}} \right) (100 \text{ deg}) \\
&= 7540 \text{ J}
\end{aligned}
$$

$$
\begin{aligned}
H_2 - H_1 &= E_2 - E_1 + P_2V_2 - P_1V_1 \\
&= E_2 - E_1 + (P_2 - P_1)V_1 \\
&= 7540 \text{ J} + (10 \text{ atm} - 1 \text{ atm}) \left(\frac{18.0 \text{ g}}{\text{mol}} \right) \left(\frac{\text{cm}^3}{0.98 \text{ g}} \right) (1 \text{ mol}) \\
&= 7540 \text{ J} + (165 \text{ cm}^3 \text{ atm}) \left(0.1013 \ \frac{\text{J}}{\text{cm}^3 \text{ atm}} \right) \\
&= 7540 \text{ J} + 16.7 \text{ J} \\
&= 7557 \text{ J}
\end{aligned}
$$

Temperature and Pressure Changes for a Gas

Let us now calculate $E_2 - E_1$ and $H_2 - H_1$ when H_2O as a gas changes from P_1, V_1, T_1 to P_2, V_2, T_2. The heat transferred at constant P or V has the same form as for a liquid:

$$q_P = n\overline{C}_P \cdot (T_2 - T_1) \tag{2.22}$$

$$q_V = n\overline{C}_V \cdot (T_2 - T_1) \tag{2.23}$$

Of course, the heat capacities for gaseous H_2O must be used here instead of those for the liquid.

The constant-pressure work deserves more discussion. For liquids, we can ignore the expansion work, but for gases it is important. The equation for calculating constant-pressure work is the usual one:

$$w_P = -P_{ex} \cdot (V_2 - V_1) \qquad \text{(expansion, constant } P_{ex})$$

Figure 2.7 illustrates the process. Note that for an expansion, the external pressure, P_{ex}, can be any pressure smaller than P. The volumes V_1 and V_2 are fixed by stops that hold the cylinder at these chosen values. The temperatures T_1 and T_2 are controlled by thermostats, and the pressure of the gas P depends on the number of moles of the gas and the equation of state. For the ideal gas $PV_1 = nRT_1$ and $PV_2 = nRT_2$. The maximum constant-pressure work would be done by the gas if the gas could be heated reversibly, keeping the external pressure just slightly less than the gas pressure P at all points during the expansion. For such a reversible expansion or compression,

$$P_{ex} \cong P \qquad \text{(reversible)}$$

Combining Eq. (2.6) and the ideal gas equation, we obtain

$$w_P = -P \cdot (V_2 - V_1) = -nR \cdot (T_2 - T_1) \tag{2.27}$$

Clearly, if the pressure is constant and the volume changes, the temperature must also change as required by the ideal gas law. The difference in volume $(V_2 - V_1)$ and the pressure determines the difference in temperature.

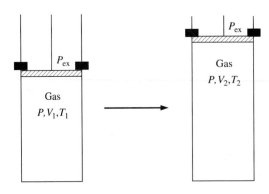

Fig. 2.7 Expansion of a gas at constant pressure, P_{ex}. For a reversible expansion, $P_{ex} = P$; for an irreversible expansion, $P_{ex} < P$.

Figure 2.8 illustrates that constant-pressure work can be thought of as the area of a rectangle of height P_{ex} and width $(V_2 - V_1)$ in a P versus V plot. If P_{ex} equals P for the gas, the constant-pressure work done by the gas is a maximum for a given change of volume, $V_2 - V_1$. If P_{ex} equals zero, the work is zero. For any P_{ex} between these extremes, the work done is intermediate. In practical engines the external pressure must be significantly less than the gas pressure inside the piston, however. Equation (2.27) thus represents the maximum work that would be available from an ideal engine of this type.

If a gas is expanded reversibly and isothermally from an initial volume V_1 to a final volume V_2, the pressure cannot be kept constant during the expansion. In this case we use the general expression for the work with P a function of volume.

$$w = -\int_{V_1}^{V_2} P_{ex}\, dV \qquad \text{(expansion)}$$

$$P_{ex} \cong P \qquad \text{(reversible)}$$

$$w = -\int_{V_1}^{V_2} P\, dV \qquad \text{(the same equation applies to a reversible compression)}$$

$$= -\int_{V_1}^{V_2} \frac{nRT}{V}\, dV \qquad \text{(ideal gas)}$$

$$= -nRT \int_{V_1}^{V_2} \frac{dV}{V} \qquad (T = \text{constant for an isothermal process, } n = \text{constant for a fixed number of moles of a gas)}$$

$$= -nRT \ln \frac{V_2}{V_1} \qquad \text{(reversible, isothermal expansion or compression of ideal gas)}$$

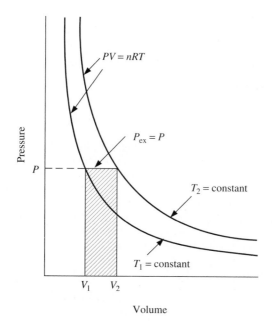

Fig. 2.8 Constant-pressure work is the area of rectangle of length P_{ex} and width $V_2 - V_1$ on a P versus V plot. The crosshatched area shown represents the work that is done by the system when $P_{ex} = P$. The curves represent paths of isothermal expansion.

This work is represented by the crosshatched area in Fig. 2.9. For an isothermal, reversible, ideal gas expansion or compression,

$$w_T = -nRT \ln \frac{V_2}{V_1} \tag{2.28a}$$

As volume is inversely proportional to pressure for an ideal gas at constant temperature,

$$w_T = -nRT \ln \frac{P_1}{P_2} \tag{2.28b}$$

For an ideal gas it seems reasonable that the energy is independent of the volume and depends only on temperature.* The ideal gas molecules do not interact with each other, so the energy of the gas will not depend on whether the molecules are close together (small volume) or far apart (large volume). Thus the heat absorbed during the isothermal expansion of an ideal gas is equal to minus the work done: for an isothermal, ideal gas,

$$q_T = -w_T$$

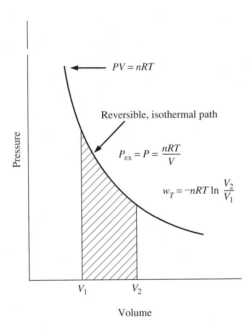

Fig. 2.9 Reversible isothermal expansion for an ideal gas. The crosshatched area represents the work done.

---

* Using a thermodynamic definition of temperature based on the second law of thermodynamics together with the equation of state for an ideal gas, we can show that the internal energy is dependent on only the temperature for an ideal gas. Thus, in the notation of partial derivatives (Chapter 3),

$$\left(\frac{\partial E}{\partial V} \right)_T \equiv 0 \quad \text{and} \quad \left(\frac{\partial E}{\partial P} \right)_T \equiv 0 \quad \text{(ideal gas)}$$

Real gases exhibit internal cohesive forces that result in the internal energy being dependent on volume and pressure as well as temperature.

For an isothermal, reversible expansion or compression of an ideal gas,

$$q_T = nRT \ln \frac{V_2}{V_1} = nRT \ln \frac{P_1}{P_2} \qquad (2.29)$$

Calculation of $E_2 - E_1$ and $H_2 - H_1$ for an ideal gas

For a change of P_1, V_1, T_1 to P_2, V_2, T_2 for an ideal gas, we can use any convenient path. Moreover, we need to specify only two of the three variables because $PV = nRT$. For example, we can choose the path $(T_1, P_1) \to (T_1, P_2) \to (T_2, P_2)$. In the first step the temperature is constant; in the second step the pressure is constant. We have chosen an isothermal path plus a constant-pressure path. Notice that we do not have to specify the volume at each step; we can calculate it from the values of P and T. To calculate the energy change and the enthalpy change, we use

$$E_2 - E_1 = q_T + w_T + q_P + w_P$$

$$= nRT_1 \ln \frac{P_1}{P_2} - nRT_1 \ln \frac{P_1}{P_2} + C_P \cdot (T_2 - T_1) - nR \cdot (T_2 - T_1)$$

$$= (C_P - nR)(T_2 - T_1)$$

$$H_2 - H_1 = E_2 - E_1 + P_2 V_2 - P_1 V_1$$

$$= E_2 - E_1 + nR \cdot (T_2 - T_1)$$

$$= C_P \cdot (T_2 - T_1)$$

For an isothermal path plus a constant-volume path, $(T_1, V_1) \to (T_1, V_2) \to (T_2, V_2)$.

$$E_2 - E_1 = q_T + w_T + q_V + w_V$$

$$= nRT_1 \ln \frac{V_2}{V_1} - nRT_1 \ln \frac{V_2}{V_1} + C_V \cdot (T_2 - T_1) + 0$$

$$= C_V \cdot (T_2 - T_1)$$

$$H_2 - H_1 = E_2 - E_1 + P_2 V_2 - P_1 V_1$$

$$= C_V \cdot (T_2 - T_1) + nR \cdot (T_2 - T_1)$$

$$= (C_V + nR) \cdot (T_2 - T_1)$$

Using the fact that $E_2 - E_1$ and $H_2 - H_1$ must be independent of path, we see that for an ideal gas,

$$
\begin{aligned}
C_P &= C_V + nR \\
\overline{C}_P &= \overline{C}_V + R
\end{aligned}
\qquad \text{(ideal gas)} \qquad (2.30)
$$

We also notice that *for an ideal gas, $E_2 - E_1$ and $H_2 - H_1$ depend only on T.* For *any* change of P, V, T for an ideal gas,

$$E_2 - E_1 = n\overline{C}_V \cdot (T_2 - T_1) \qquad \text{(ideal gas)} \qquad (2.31)$$

$$H_2 - H_1 = n\overline{C}_P \cdot (T_2 - T_1) \qquad \text{(ideal gas)} \qquad (2.32)$$

We have assumed for simplicity that $\overline{C}_P$ and $\overline{C}_V$ are independent of temperature in Eqs. (2.31) and (2.32). For real gases the heat capacity will be a function of temperature. This is often expressed in terms of a power series

$$\overline{C} = a + bT + cT^2$$

where values of the coefficients a, b, and c are determined empirically and tabulated for different gases. Using such data, which are often valid over a temperature range from 300 to 1500 K, energy and enthalpy changes can be calculated by integration.

$$E_2 - E_1 = n\int_{T_1}^{T_2} \overline{C}_V \, dT$$

$$H_2 - H_1 = n\int_{T_1}^{T_2} \overline{C}_P \, dT \qquad (2.33)$$

Properties of $E_2 - E_1$ and $H_2 - H_1$ Independent of Equation of State

Of course, the first law [Eq. (2.19)] and the definition of H [Eq. (2.12)] are independent of whether the system is a gas, a liquid, a solid, or a mixture of these. But there are also some special cases that are independent of the equation of state.

For a closed system at constant volume, the heat absorbed is equal to the energy change, because the work at constant volume is zero:

$$E_2 - E_1 = q_V \qquad (2.34)$$

For heating or cooling

$$E_2 - E_1 = n\int_{T_1}^{T_2} \overline{C}_V \, dT$$

$$= n\overline{C}_V \cdot (T_2 - T_1) \qquad (\overline{C}_V \text{ constant}) \qquad (2.35)$$

Note that this conclusion is obtained if the only type of work is the pressure-volume work. If the system can do other types of work (electrical, for example), the conclusion will not apply.

For a closed system at constant pressure, the heat absorbed is equal to the enthalpy change. The energy change is the heat absorbed plus the reversible expansion work done on the system:

$$E_2 - E_1 = q_P - P \cdot (V_2 - V_1) \qquad (2.36)$$

At constant pressure, using the definition of enthalpy [Eq. (2.12)] we obtain

$$H_2 - H_1 = E_2 - E_1 + P \cdot (V_2 - V_1) \qquad \text{(constant P)}$$

Thus, at constant pressure,

$$H_2 - H_1 = q_P$$

For heating or cooling

$$H_2 - H_1 = n \int_{T_1}^{T_2} \overline{C}_P \, dT \qquad (2.37)$$

$$= n\overline{C}_P \cdot (T_2 - T_1) \qquad (\overline{C}_P \text{ constant})$$

Note again that in obtaining Eq. (2.36), and therefore (2.37), it is assumed that the only type of work is the pressure-volume type. The enthalpy H is also known as the *heat content,* because $H_2 - H_1$ is equal to the heat absorbed at constant pressure. It should be clear that E and H are variables of state; only for certain special processes are $E_2 - E_1$ or $H_2 - H_1$ equal to the heat transferred to the system.

PHASE CHANGES

The preceding section has dealt with changes of P, V, and T only. Now we want to consider changes in phase; that is, the system changes from a solid to a liquid, for example. Names of phase changes are as follows:

Phase change	Name
Gas → liquid or solid	Condensation
Solid → liquid	Fusion, melting
Liquid → solid	Freezing
Liquid → gas	Vaporization
Solid → gas	Sublimation

There can also be phase changes between different solid phases and between different liquid phases. We are particularly interested in the thermodynamics of the reversible change that occurs at constant T and P. A phase change is usually a good way to store energy. It takes only 75.5 J to heat 1 mol of water from 99°C to 100°C, but it takes 40,660 J to change 1 mol of water from liquid to gas at 100°C.

Consider a reversible phase change from phase a to phase b at constant T and P. The work done on the system is

$$w_P = -P \cdot (\Delta V)$$

where ΔV = volume change $\equiv V$(phase b) $- V$(phase a). The heat absorbed by the system at constant P is q_P. It is equal to $\Delta H \equiv H$(phase b) $- H$(phase a), according to Eq. (2.37).

The enthalpy change for a reversible phase change at constant P is given as

$$\Delta H = q_P \qquad (2.38)$$

We have introduced the symbol Δ to represent a difference. To avoid confusion, we shall reserve this symbol for phase changes or chemical reactions. For changes caused solely by T or P, we will continue to use $V_2 - V_1$, $E_2 - E_1$, and so on. Thus we can use $\Delta H(T_2) - \Delta H(T_1)$ to represent the change in the enthalpy of a reaction with temperature.

Values of ΔH have been tabulated for various reversible phase changes. Because the change is at constant pressure and the only work involved is the pressure-volume type, the heat and enthalpy are equal. Thus one can speak about a heat of vaporization, or a heat of fusion, or an enthalpy of vaporization or fusion. The energy change of a phase change at constant P is

$$\Delta E = \Delta H - P \cdot (\Delta V) \tag{2.39}$$

Often a ΔH or a ΔE value is known at one T and P, but is needed at another. We know how to calculate the change of E and H with T and P so the calculation is easy. Suppose that we want to calculate the amount of heat removed when liquid water evaporates at human skin temperature (around 35°C). We want to know how effectively vaporization of sweat can cool us. If we know ΔH for vaporization of water only at 100°C, we can use an indirect path to calculate the ΔH for the desired reaction at 35°C:

$$H_2O(l),\ T_2 = 35°C,\ P = 1\ \text{atm} \quad \xrightarrow{\ \Delta H(35°C)\ } \quad H_2O(g),\ T_2 = 35°C,\ P = 1\ \text{atm}$$

constant P constant P

$$H_2O(l),\ T_1 = 100°C,\ P = 1\ \text{atm} \quad \xrightarrow{\ \Delta H(100°C)\ } \quad H_2O(g),\ T_1 = 100°C,\ P = 1\ \text{atm}$$

We can calculate ΔH for each step in the chosen path and add the ΔH's to obtain the overall ΔH:

$$\Delta H(35°C) = C_P(l) \cdot (373 - 308) + \Delta H(100°C) + C_P(g) \cdot (308-373)$$

$$= \Delta H(100°C) + n[\bar{C}_P(g) - \bar{C}_P(l)](-65)$$

The generalization for the temperature dependence of enthalpy of a phase change is

$$\Delta \bar{H}(T_2) = \Delta \bar{H}(T_1) + \Delta \bar{C}_P \cdot (T_2 - T_1) \tag{2.40}$$

where $\Delta \bar{H} = \bar{H}(\text{phase b}) - \bar{H}(\text{phase a})$
$\Delta \bar{C}_P = \bar{C}_P(\text{phase b}) - \bar{C}_P(\text{phase a})$

Example 2.9   Check the value for ΔH of vaporization of water at 20°C given in Table 2.2 by using Eq. (2.40) and ΔH (100°C) from the table.

Solution

$$\Delta \bar{H}(20°C) = \Delta \bar{H}(100°C) + \Delta \bar{C}_P \cdot (T_2 - T_1) \tag{2.40}$$

Equation (2.40) is written for 1 mol of substance, but it is easily adapted to 1 kg of substance.

$$\Delta H(20°C) = 2259\ \text{kJ kg}^{-1}$$

$$+ (1.874\ \text{kJ K}^{-1}\ \text{kg}^{-1} - 4.18\ \text{kJ K}^{-1}\ \text{kg}^{-1})(-80\ \text{K})$$

$$= 2259\ \text{kJ kg}^{-1} + 184.5\ \text{kJ kg}^{-1}$$

$$= 2443\ \text{kJ kg}^{-1}$$

The table gives 2447 kJ kg^{-1}. The temperature dependence of the heat capacities accounts for the small discrepancy.

The energy of a phase change at constant P can be obtained from Eq. (2.39). If one of the phases in the phase change is a gas (vaporization or sublimation), the volume of the solid or liquid is so much smaller than the volume of the gas that it can be ignored. Furthermore, the gas phase can be approximated as an ideal gas:

$$\Delta \bar{E} = \Delta \bar{H} - P\bar{V}(\text{gas}) - P\bar{V}(\text{liq})$$
$$\cong \Delta \bar{H} - P\bar{V}(\text{gas})$$
$$\Delta \bar{E} \cong \Delta \bar{H} - RT$$

Example 2.10   Calculate (a) the change of energy on freezing 1.00 kg of liquid water at 0°C and 1 atm, and (b) the change of energy on vaporizing 1.00 kg of liquid water at 0°C and 1 atm.

Solution

(a) $\Delta E = \Delta H - \text{P} \cdot (\Delta V)$ (2.39)

$\quad = -333.4 \text{ kJ kg}^{-1}$

$\quad\quad - (1 \text{ atm})(1.093 \text{ cm}^3 \text{ g}^{-1} - 1.000 \text{ cm}^3 \text{ g}^{-1})(1000 \text{ g kg}^{-1})$

$\quad = -333.4 \text{ kJ kg}^{-1} - 93 \text{ cm}^3 \text{ atm kg}^{-1}$

$\quad = -333.4 \text{ kJ kg}^{-1} - (93 \text{ cm}^3 \text{ atm kg}^{-1})\left(0.1013 \dfrac{\text{J}}{\text{cm}^3 \text{ atm}}\right)$

$\quad = -333.4 \text{ kJ kg}^{-1} - 0.009 \text{ kJ kg}^{-1}$

$\quad = -333.4 \text{ kJ kg}^{-1}$

(b) $\Delta E = \Delta H - nRT$

$\quad = 2493 \text{ kJ kg}^{-1}$

$\quad\quad - (8.314 \text{ J K}^{-1} \text{ mol}^{-1})(273.1 \text{ K})\left(\dfrac{1 \text{ mol}}{0.018016 \text{ kg}}\right)\left(\dfrac{1 \text{ kJ}}{1000 \text{ J}}\right)$

$\quad = 2493 \text{ kJ kg}^{-1} - 126.0 \text{ kJ kg}^{-1}$

$\quad = 2367 \text{ kJ kg}^{-1}$

Note that there is a significant difference between ΔH and ΔE when one of the phases is a gas [part (b)], but when neither phase is a gas [part (a)], the difference is insignificant.

CHEMICAL REACTIONS

We come next to the most important way the energy and enthalpy of a system can be changed: a chemical reaction can occur. Here the properties of variables of state are most useful. The fact that the enthalpy change when 1 teaspoon of sugar is burned to CO_2 and H_2O is the same whether the reaction occurs in a human being or in a calorimeter is very convenient. Of course, we do have to ensure that the initial and final states for the reaction are the same inside the human being and inside the calorimeter. Once this is done, however, it does not matter whether the path involves 10 enzyme-catalyzed steps or a direct reaction (combustion) with O_2.

The change in the system is represented by the general chemical reaction

$$n_A A + n_B B \longrightarrow n_C C + n_D D$$

That is, n_A moles of A react with n_B moles of B to give n_C moles of C and n_D moles of D. The conditions of P, V, and T must be specified for both the products and the reactants. The change in the variables of state, such as ΔE and ΔH, are desired. To repeat, it does not matter how the reaction actually takes place; only the initial and final states are important.

Heat Effects of Chemical Reactions

If the reaction takes place at constant P, the heat given off is equal to the decrease in enthalpy when there is no work other than the pressure-volume type:

$$q_P = \Delta H \tag{2.41}$$

A negative ΔH and a negative q_P means that heat is released (exothermic); a positive ΔH and a positive q_P means that heat is absorbed (endothermic). If the reaction takes place at constant V in a sealed container such as a bomb calorimeter, the heat given off is equal to the decrease in energy:

$$q_V = \Delta E \tag{2.42}$$

By measurement of the heat effects of chemical reactions at constant P or V, values of ΔH and ΔE for the reactions are obtained. The heat effect is usually measured by surrounding the reaction with a known amount of water and measuring the temperature rise of the water. If either ΔH or ΔE is measured, the other can be obtained from the definition of H:

$$\Delta H = \Delta E + \Delta(PV) \tag{2.43}$$

where $\Delta(PV) = PV(\text{products}) - PV(\text{reactants})$. If gases are involved in the reaction, we can ignore the volumes of the solids or liquids and use the ideal gas equation for the gases.

The amount of heat that can be obtained from a chemical reaction is obviously of great practical importance. We may be interested in obtaining the maximum amount of heat for a given weight of fuel. For example, if cost is unimportant, is hexane, methanol, benzene, or polyethylene the best fuel per unit weight? Or it may be that we want to minimize the heat. In a battery we want to convert chemical energy into electrical work and not waste the chemical energy in the form of heat released.

The biochemical reactions necessary to sustain life in a person produce about 6000 kJ day^{-1} of heat at constant pressure. This is the basal metabolic rate. Each person thus continually produces about 70 W (1 W $\equiv$ 1 J s^{-1}). If one does more than lie around in bed, further energy is produced. The 8000 to 12,000 kJ day^{-1} needed is normally replenished in the person in the form of food (remember that a nutritionist's Cal $\equiv$ a chemist's kcal = 4.184 kJ). Each gram of protein or carbohydrate provides about 15 kJ, and fat provides about 35 kJ g^{-1}.

The heat at constant pressure, and thus the enthalpy changes for many reactions, have been measured. These reactions and their enthalpies can be combined to calculate the enthalpies for many other reactions. For example, suppose that we know ΔH_1 for the oxidation of solid glycine at 25°C to form CO_2, ammonia, and liquid water:

(1)   3 O_2(g, 1 atm) + 2 NH_2CH_2COOH(s) $\longrightarrow$
              glycine

        4 CO_2(g, 1 atm) + 2 H_2O(l) + 2 NH_3(g, 1 atm) ΔH_1 = -1163.5 kJ mol^{-1}

The ΔH_2 for the hydrolysis of solid urea is also known.

(2) H_2O(l) + H_2NCONH_2(s) $\longrightarrow$ CO_2(g, 1 atm) + 2 NH_3(g, 1 atm)
              urea
                    ΔH_2 = 133.3 kJ mol^{-1}

If we subtract these two reactions, treating the chemicals and their ΔH's as algebraic quantities, we get

3 O_2(g, 1 atm) + 2 glycine(s) $-$ urea(s) $-$ H_2O(l) $\longrightarrow$

        4 CO_2(g, 1 atm) + 2 H_2O(l) + 2 NH_3(g, 1 atm)

                              $-$ CO_2(g, 1 atm) $-$ 2 NH_3(g, 1 atm)

Rearranging and canceling, we get

(3)   3 O_2(g, 1 atm) + 2 glycine(s) $\longrightarrow$ 1 urea(s) + 3 CO_2(g, 1 atm) + 3 H_2O(l)

                    ΔH_3 = ΔH_1 $-$ ΔH_2

                         = -1296.8 kJ mol^{-1}

This equation is of more biochemical interest, because urea rather than ammonia is the main oxidative metabolic product of amino acids.* However, the biological reaction does

---

* Students are often confused by the units of ΔH for an equation like reaction (3). The value of ΔH_3 given is the enthalpy change associated with the formation of 1 mol of urea, but at the same time it forms 3 mol of CO_2 and 3 mol of H_2O. To deal with this problem we adopt the convention of calculating thermodynamic quantities (ΔE, ΔH, etc.) per *mole of reaction* as it is written. In the present case a mole of reaction is defined for reaction (3) as the amount of reaction that produces 1 mol of urea, 3 mol of CO_2, and so on. Obviously, it is just as valid to describe the reaction as follows:

(3a) O_2(g, 1 atm) + $\frac{2}{3}$ glycine(s) $\longrightarrow$ $\frac{1}{3}$ urea(s) + CO_2(g, 1 atm) + H_2O(l)

For this reaction ΔH = $\frac{1}{3}(-1296.8)$ = -432.3 kJ mol^{-1}.

not involve solid glycine and solid urea, but rather aqueous solutions. Therefore, we use the reactions and heats for the dissolution of 1 mol of urea and of glycine.

(4) glycine(s) + ∞ H$_2$O(l) $\longrightarrow$ glycine(aq) ΔH_4 = 15.69 kJ (mol glycine)$^{-1}$

(5) urea(s) + ∞ H$_2$O(l) $\longrightarrow$ urea(aq) ΔH_5 = 13.93 kJ (mol urea)$^{-1}$

The enthalpies of solution will depend on concentration: here we will use enthalpies for very dilute solutions (designated aq) and assume that they do not depend on concentration. That means that we choose the number of moles of H$_2$O(l) to be a very large number, infinite ($\equiv \infty$), in the preceding equations. From reaction (3) we now subtract two times reaction (4) and add reaction (5):

(6) 3 O$_2$(g, 1 atm) + 2 glycine(s) − 2 glycine(s)

$$- \infty \text{ H}_2\text{O}(l) + \text{urea}(s) + \infty \text{ H}_2\text{O}(l) \longrightarrow$$

$$\text{urea}(s) + 3 \text{ CO}_2(g, 1 \text{ atm}) + 3 \text{ H}_2\text{O}(l) - 2 \text{ glycine}(aq) + \text{urea}(aq)$$

(6) 3 O$_2$(g, 1 atm) + 2 glycine(aq) $\longrightarrow$ urea(aq) + 3 CO$_2$(g, 1 atm) + 3 H$_2$O(l)

$$\Delta H_6 = \Delta H_3 - 2\,\Delta H_4 + \Delta H_5$$

$$= -1314.2 \text{ kJ mol}^{-1}$$

We now know the enthalpy for the reaction of a dilute aqueous solution of glycine with O$_2$ gas to form a dilute aqueous solution of urea plus CO$_2$ gas plus 3 mol of liquid H$_2$O. We cannot ignore the synthesized H$_2$O in the reaction, even though the product water is diluted into the "infinities" introduced from the dissolution reactions. The chemical bonds produced in forming the 3 mol of H$_2$O involve very significant enthalpy effects.

The enthalpy for reaction (6) will be quite close to that of the naturally occurring reaction in the human body. However, it should be clear that by adding or subtracting the enthalpies of other reactions we can calculate ΔH for *any* reaction we like. That is, if we think it is important, we can find the heat of solution of glycine in a defined buffer solution instead of pure water. We can specify that instead of CO$_2$(g, 1 atm) as a product, we have a carbonic acid solution of a certain pH. It may be very difficult to directly measure the ΔH of the reaction we want, but we can always find the ΔH by using a convenient alternative path. In other words, if we want the ΔH for reaction A → B, we may use a path that is a sum of many other reactions.

$$\Delta H = \Delta H_1 + \Delta H_2 + \Delta H_3 + \Delta H_4 + \Delta H_5 + \Delta H_6$$

It is convenient to remember that if

$$A \longrightarrow C \quad \text{has} \quad \Delta H_1$$

then

$$C \longrightarrow A \quad \text{has} \quad -\Delta H_1$$

and

$$nA \longrightarrow nC \quad \text{has} \quad n\Delta H_1$$

Temperature Dependence of ΔH

By the same reasoning used for phase changes, if ΔH is known at one temperature, it can be calculated at other temperatures.

$$A(T_2) \xrightarrow{\;\;\Delta H(T_2)\;\;} B(T_2)$$

$$-C_P^A \cdot (T_2 - T_1) \qquad\qquad C_P^B \cdot (T_2 - T_1)$$

$$A(T_1) \xrightarrow{\;\;\Delta H(T_1)\;\;} B(T_1)$$

$$\Delta H(T_2) = \Delta H(T_1) + \Delta C_P \cdot (T_2 - T_1) \tag{2.44}$$

where $\Delta H = H(\text{products}) - H(\text{reactants})$
$\quad\;\; \Delta C_P = C_P(\text{products}) - C_P(\text{reactants})$

and we have incorporated the assumption that C_P is not a function of temperature.

ΔE for a Reaction

For a reaction at constant pressure, ΔE can be calculated from the definition of H:

$$\Delta E = \Delta H - P \, \Delta V \tag{2.45}$$

If gases are involved, we ignore the volumes of solids and liquids, and approximate the gases as ideal gases.

$$\Delta E \cong \Delta H - \Delta n R T \tag{2.46}$$

where Δn is the number of moles of *gaseous* products minus the number of moles of *gaseous* reactants.

Standard Enthalpies (or Heats) of Formation

In combining chemical reactions and their enthalpies, we treated the products and reactants as algebraic quantities. For the reaction

$$n_A A + n_B B \longrightarrow n_C C + n_D D$$

the enthalpy change can be written

$$\Delta H = H(\text{products}) - H(\text{reactants})$$

$$\Delta H = n_C \overline{H}_C + n_D \overline{H}_D - n_B \overline{H}_B - n_A \overline{H}_A$$

(2.47)

where $\overline{H}$ = enthalpy mol^{-1}. However, enthalpies are all relative; only differences of enthalpy are defined. We can talk of the volume mol^{-1}, $\overline{V}$, but we can only specify an enthalpy difference. Therefore, we can arbitrarily choose a zero of enthalpy, just as we arbitrarily choose mean sea level as the reference point for measuring elevations on the surface of the earth. For example, in the reaction to form water vapor from its elements

$$H_2(g,\ 1\ \text{atm}) + \tfrac{1}{2} O_2(g,\ 1\ \text{atm}) \ \xrightarrow{\Delta H} \ H_2O(g,\ 1\ \text{atm})$$

$$\Delta H = 1\overline{H}(H_2O,\ g,\ 1\ \text{atm}) - \tfrac{1}{2}\overline{H}(O_2,\ g,\ 1\ \text{atm}) - 1\overline{H}(H_2,\ g,\ 1\ \text{atm})$$

If we choose $\overline{H}$ for the elements O_2 and H_2 equal to zero, the enthalpy of their compound H_2O is equal to the enthalpy of the formation reaction.

$$\Delta H = 1\overline{H}(H_2O,\ g,\ 1\ \text{atm}) - \tfrac{1}{2} \cdot 0 - 1 \cdot 0$$

$$\overline{H}(H_2O,\ g,\ 1\ \text{atm}) = \Delta H$$

Thermodynamicists have adopted the convention of assigning zero enthalpy to all elements in their most stable states at 1 atm pressure. These are called *standard states* and are designated by a superscript zero. *The standard enthalpy mol^{-1} of a compound is defined to be equal to the enthalpy of formation of 1 mol of the compound at 1 atm pressure from its elements in their standard states.*

$$\overline{H}^0(\text{compound}) \equiv \Delta \overline{H}_f^0$$

(2.48)

where $\Delta \overline{H}_f^0$ = enthalpy of formation of 1 mol from elements under standard conditions. The superscript zero means 1 atm pressure and the most stable form of the element. It should be clear that the standard enthalpy is a defined quantity that depends on the choice of the standard state. Assigning the elements *in their most stable state* at 1 atm pressure to have zero enthalpy is purely arbitrary.

The standard enthalpies of thousands of substances have been determined; the Appendix gives a few of them at 25°C. The enthalpies at 25°C and 1 atm pressure for very many reactions can be calculated from this table. One atmosphere pressure is part of the definition of the standard state, but the temperature is not specified in the definition. However, nearly all the tables available are for 25°C. The standard chemical reaction at 25°C is given as

$$\Delta H^0(298) = n_C \overline{H}_C^0 + n_D \overline{H}_D^0 - n_A \overline{H}_A^0 - n_B \overline{H}_B^0$$

where $\overline{H}^0$ = standard enthalpy mol^{-1} at 25°C = $\Delta \overline{H}_f^0$
$\Delta \overline{H}_f^0$ = 0 for all elements in their standard states

Example 2.11   Use the Appendix tables to calculate the value of ΔH for reacting 1 g of solid glycylglycine with oxygen to form solid urea, CO_2 gas, and liquid H_2O at 25°C, 1 atm.

Solution It is important to begin with a balanced stoichiometric equation for the reaction

$$3\,O_2(g) + C_4H_8N_2O_3(s) \longrightarrow CH_4N_2O(s) + 3\,CO_2(g) + 2\,H_2O(l)$$

glycylglycine   urea

$$\Delta H^0 = \overline{H}^0(\text{urea}) + 3\overline{H}^0(CO_2) + 2\overline{H}^0(H_2O, l) - \overline{H}^0(\text{glycylglycine}) - 3H^0(O_2)$$

$$= -333.17 + 3(-393.51) + 2(-285.83) - (-745.25) - 3(0)$$

$$= -1340.11 \text{ kJ mol}^{-1}$$

The enthalpy change calculated is for 1 mol of glycylglycine. To find ΔH per gram, we must divide by the molecular weight, 132.12 g mol^{-1}:

$$\Delta H = -10.14 \text{ kJ g}^{-1}$$

The negative sign means that heat is given off in the reaction.

What about other temperatures and pressures? The enthalpy at any temperature can be obtained from Eq. (2.44). The temperature dependence of the standard chemical reaction

$$\Delta H^0(T) = \Delta H^0(298) + \Delta C_P^0 \cdot (T - 298) \tag{2.49}$$

where $\Delta C_P^0 = n_C \overline{C}_P^0(C) + n_D \overline{C}_P^0(D) - n_A \overline{C}_P^0(A) - n_B \overline{C}_P^0(B)$. It is necessary to remember that ΔC_P^0 here must include all products and reactants. The value of $\overline{H}^0$ for elements is chosen as zero, but $\overline{C}_P^0$ is not zero. Furthermore, if we use a set of data where $\overline{H}^0$ for the elements is zero at 298 K, then their enthalpies at other temperatures will *not* be equal to zero.

$$\overline{H}_T^0(\text{elements}) = 0 + \overline{C}_P^0(\text{elements}) \cdot (T - 298)$$

The pressure dependence of ΔH is not large and we will ignore it. For ideal gases, H is independent of P; for solids or liquids, H is not very dependent on P. For geological processes where the pressures may become very large, these approximations will not be valid. We will also generally ignore the effect of concentration on ΔH.

It should be clear that a heat of reaction and a ΔH can be measured for any condition of concentration, solvent, pH, pressure, and so on. We make these approximations because they are accurate enough for most purposes and because they simplify the calculations. For a discussion of the pressure dependence of ΔH and of heats of dilution, the reader is referred to the standard thermodynamics textbooks listed in the References at the end of this chapter.

Bond Energies

Enthalpies of formation for many compounds have been determined very precisely. These are tabulated in extensive compilations of thermodynamic data. Some that are of particular interest to biologists and biochemists are listed with the References at the end of this chapter. There are many more compounds whose $\Delta \overline{H}_f^0$ is not known. However, we can approximate $\Delta \overline{H}_f^0$ and other heats of reaction by using bond dissociation energies, D. This essentially involves finding an alternative path for the reaction where the individual steps involve breaking and making chemical bonds. The *bond dissociation energy* is the enthalpy at 25°C and 1 atm for the reaction

$$A-B(g) \longrightarrow A(g) + B(g)$$

The bond dissociation energy should be called a bond dissociation enthalpy, but the tradition for energy is strong. In many cases the amount of energy necessary to break a particular type of bond in a molecule is not too dependent on the molecule. For example, the average energy necessary to break a C—H bond is 415 kJ mol^{-1} ± 10% in a wide range of organic compounds.

Some average bond dissociation energies are given in Table 2.3. We expect the value for ΔH estimated from the table to be reasonable. One notable exception is for molecules which are stabilized greatly by electron delocalization. For example, the data that were used to obtain values for C—C and C=C bonds in Table 2.3 came from molecules with single bonds, and isolated double bonds. Molecules with conjugated double bonds (C=C bonds separated by only one C—C bond) are more stable than those containing the same number of isolated single and double bonds. This difference in energy is called the *resonance energy*.

Table 2.3   Average bond dissociation energies at 25°C

Bond	D (kJ mol^{-1})
C—C	344
C=C	615
C≡C	812
C—H	415
C—N	292
C—O	350
C=O	725
C—S	259
N—H	391
O—O	143
O—H	463
S—H	339
H_2	436.0
N_2	945.4
O_2	498.3
C (graphite)	716.7

Source: After L. Pauling and P. Pauling, *Chemistry*, W. H. Freeman, San Francisco, 1975.

Resonance energy is important for aromatic molecules such as benzene or phenylalanine, for the carboxyl groups in carboxylic acids, for amino acids, porphyrins, carotenoids, and so on. For such molecules the difference in heat of formation calculated from bond energies and that experimentally measured is an estimate of the resonance energy. In other cases, ring strain produces energy effects that need to be added to the average bond dissociation energies.

The easiest way to show how Table 2.3 can be used is by some examples.

Example 2.12 Calculate the heat of formation for gaseous cyclohexane using Table 2.3 and compare with the measured values in the Appendix.

Solution The reaction for the formation of cyclohexane is

$$6\ C(\text{graphite}) + 6\ H_2(g) \longrightarrow C_6H_{12}(g)$$

We can write it as a sum of bond-breaking and bond-forming reactions:

(1) $6\ C(\text{graphite}) \longrightarrow 6\ C(g)$

$$\Delta H_1 = 6D(\text{graphite}) = (6)(716.7) = 4300\ \text{kJ}$$

This is the enthalpy required to remove 6 mol of carbon atoms from a crystalline lattice of graphite, which is the standard state for elemental carbon.

(2) $6\ H_2(g) \longrightarrow 12\ H(g)$

$$\Delta H_2 = 6D(H_2) = (6)(436.0) = 2616.0\ \text{kJ}$$

(3) $6\ C(g) + 12\ H(g) \longrightarrow (6\ C\!-\!C + 12\ C\!-\!H) = C_6H_{12}(g)$

$$\Delta H_3 = -6D(C\!-\!C) - 12D(C\!-\!H) = -(6)(344) - (12)(415)$$

$$= -7044\ \text{kJ}$$

Note that bond formation energies are just the negative of bond dissociation energies.

$$\Delta H_f^0(C_6H_{12}) = \Delta H_1 + \Delta H_2 + \Delta H_3$$

$$= -128\ \text{kJ}$$

The value in the Appendix is -123.15 kJ for cyclohexane, which is in good agreement. This is because cyclohexane is a molecule that contains normal bonds; also, there is no significant bond angle strain in the six-membered ring.

Example 2.13 Calculate the heat of formation for gaseous benzene using Table 2.3 and compare with the measured value in the Appendix.

Solution Because of resonance effects, we expect the measured value for benzene to differ greatly from that calculated for the classical structure of benzene. An explanation of the large thermodynamic stability of benzene was an important goal for chemists interested in chemical bonding. The reaction for benzene is

$$6\,C(graphite) + 3\,H_2(g) \longrightarrow C_6H_6(g)$$

(1) $6\,C(graphite) \longrightarrow 6\,C(g)$

$$\Delta H_1 = (6)(716.7) = 4300\ kJ$$

(2) $3\,H_2(g) \longrightarrow 6\,H(g)$

$$\Delta H_2 = (3)(436.0) = 1308\ kJ$$

(3) $6\,C(g) + 6\,H(g) \longrightarrow (3\,C{-}C + 3\,C{=}C + 6\,C{-}H) = C_6H_6(g)$

$$\Delta H_3 = -(3)(344) - (3)(615) - (6)(415)$$

$$= -5367\ kJ$$

$$\Delta H_f^0(C_6H_6) = \Delta H_1 + \Delta H_2 + \Delta H_3$$

$$= 241\ kJ$$

The value in the Appendix is 82.92 kJ for $C_6H_6(g)$! Benzene is about 158 kJ lower in enthalpy than would be expected for a molecule made up of 3 C—C single bonds, 3 C=C double bonds, and 6 C—H single bonds. This energy is what we call the resonance energy.

MOLECULAR INTERPRETATIONS OF ENERGY AND ENTHALPY

We have considered various ways of changing the energy and enthalpy of a system. They include:
(1) Heat and work exchanges between the system and surroundings
(2) Chemical reactions or phase changes within the system
Let us consider what the molecules are doing when the energy and enthalpy change. For example, if the system consists of an ideal gas we find that its internal energy is a function only of temperature. We can increase the internal energy (and the temperature) only by a process of the first type—adding heat or doing work. The individual ideal gas molecules have higher translational, rotational, vibrational, and (at high enough temperatures) electronic energies at the higher temperature. The internal energy of an ideal gas is just the sum of the energies of all the individual gas molecules. There are no interactions between ideal gas molecules. The heat capacity is a measure of how much energy is required to raise the temperature by one degree. Its magnitude is a measure of how many ways the molecules

have of storing energy. Even from the limited set of data in Table 2.1, one can see that gases made up of larger and more complex molecules have larger heat capacities than those consisting of simpler molecules or monatomic gases. For normal temperatures monatomic gases have only increased translational kinetic energies—increased velocities—to store energy.

For real gases and all liquids and solids, interactions between molecules become significant. For these systems the energy and enthalpy can change at constant temperature. Energy can be stored in intermolecular interactions. Thus compressing the system at constant temperature can increase its energy by increasing the repulsion between molecules as the distances between them decrease. The energy of a system of liquids, solids, or real gases is the sum of the energies of all the molecules plus the energy of interactions among all the molecules. Raising the temperature of a liquid, for example, raises the energy of individual molecules and also changes the intermolecular interactions. The heat capacity of a substance in the liquid phase is thus greater than that in the gas phase.

Chemical reactions and phase changes cause abrupt changes in energy and enthalpy. For chemical reactions the changes are a consequence of making and breaking bonds. For phase changes the changes result from the different intermolecular interactions in solids, liquids, and gases. The energy and enthalpy changes in a chemical reaction can be hundreds of kilojoules per mole; in a phase change tens of kilojoules can be involved. These are both large compared to the effects of temperature alone; a temperature change of 100 degrees may result in no more than 1 kJ mol^{-1} energy change.

The energy stored in chemical bonds represents one of our greatest energy resources. We run automobiles on the energy derived from the chemical combustion (oxidation) of the hydrocarbons of gasoline. One would not attempt to run a vehicle on the energy released by the cooling of 20 gallons of water from 100°C, even if all of that thermal energy could be converted into mechanical energy. It is important to keep the relative magnitudes of these quantities in mind, especially when approximations are made in thermodynamic calculations.

SUMMARY

State Variables

Name	Symbol	Units	Definition
Volume	V	liters (L), mL, cm^3	(length)^3
Pressure	P	atm, Torr, mmHg, dyn cm^{-2}, pascal (Pa), N m^{-2}	force area^{-1}
Temperature	T	K, °C, °F	
Energy	E	J, erg, cal	
Enthalpy	H	J, erg, cal	$H \equiv E + PV$

Unit Conversions

Volume:

$$1 \text{ L} \equiv 1000 \text{ mL} = 1 \text{ cm}^3$$

Pressure (P = force area^{-1}):

$$1 \text{ atm} = 760 \text{ Torr} = 1.01325 \times 10^6 \text{ dyn cm}^{-2} = 1.01325 \times 10^5 \text{ Pa}$$

$$1 \text{ Torr} \equiv 1 \text{ mmHg}$$

$$1 \text{ dyn cm}^{-2} \equiv 1 \text{ g cm}^{-1} \text{ s}^{-2}$$

$$1 \text{ pascal} \equiv 1 \text{ N m}^{-2} \equiv 1 \text{ kg m}^{-1} \text{ s}^{-2}$$

Temperature:

$$\text{K} = {}^\circ\text{C} + 273.15$$

$$^\circ\text{C} = \frac{^\circ\text{F} - 32}{1.8}$$

Energy and enthalpy:

$$1 \text{ cal} = 4.184 \text{ J} = 4.184 \times 10^7 \text{ erg}$$

$$1 \text{ J} \equiv 1 \text{ kg m}^2 \text{ s}^{-2} = 1 \times 10^7 \text{ erg}$$

$$1 \text{ erg} = 1 \text{ g cm}^2 \text{ s}^{-2}$$

$$1 \text{ L atm} = 24.22 \text{ cal}$$

General Equations

Energy, E: closed system, heat and work are the only forms of energy the system exchanges with surroundings.

$$E_2 - E_1 = q + w \tag{2.19}$$

Enthalpy, $H \equiv E + PV$:

$$H_2 - H_1 = E_2 - E_1 + P_2V_2 - P_1V_1 \tag{2.12}$$

Heat, q (heat absorbed by system is positive):

$$q = \int_{T_1}^{T_2} C \, dT \tag{2.9}$$

$$C = \text{heat capacity} = \frac{dq}{dT}$$

Work, w (work done on system is positive)—stretching or compressing a spring:

$$w = k \cdot (x_2 - x_1)\left(\frac{x_2 + x_1}{2} - x_0\right) \qquad (2.4)$$

k = Hooke's law constant
x_0 = length of spring in the absence of a force
x_1, x_2 = initial and final lengths of the spring, respectively

Expansion or compression of a gas:

$$w = -\int_{V_1}^{V_2} P_{ex} \, dV \qquad (2.5)$$

P_{ex} = external (opposing) pressure

$$w_P = -P_1 \cdot (V_2 - V_1) \qquad \text{(constant pressure)} \qquad (2.6)$$

Electrical work done by a system:

$$w = -EIt \qquad (2.8)$$

E = voltage
I = current
t = time

Pressure-Volume Work Only

$$E_2 - E_1 = q_V = n \int_{T_1}^{T_2} \overline{C}_V \, dT \qquad \text{(constant volume)} \qquad (2.35)$$

$\overline{C}_V$ = molar heat capacity at constant volume

$$H_2 - H_1 = q_P = n \int_{T_1}^{T_2} \overline{C}_P \, dT \qquad \text{(constant pressure)} \qquad (2.37)$$

$\overline{C}_P$ = molar heat capacity at constant pressure

Solids and Liquids

We assume in these equations that the volume of a solid or liquid is independent of T and P, that $C_P = C_V = C$, and that they do not depend on T and P.

$$E_2 - E_1 = n\overline{C} \cdot (T_2 - T_1) \qquad \text{(any change of } P, T)$$

$$H_2 - H_1 = n\overline{C} \cdot (T_2 - T_1) + (P_2 - P_1) \cdot V \qquad \text{(any change of } P, T)$$

n = number of moles
$\overline{C}$ = heat capacity per mole

Gases

We assume that gas properties can be approximated by the ideal gas equation and that C_P and C_V are independent of T. $PV = nRT$. $\overline{C}_P = \overline{C}_V + R$.

$$E_2 - E_1 = n\overline{C}_V \cdot (T_2 - T_1) \qquad \text{(any change of } P, V, T\text{)} \qquad (2.31)$$

$$H_2 - H_1 = n\overline{C}_P \cdot (T_2 - T_1) \qquad \text{(any change of } P, V, T\text{)} \qquad (2.32)$$

n = number of moles
$\overline{C}_P$ = heat capacity per mole at constant P
$\overline{C}_V$ = heat capacity per mole at constant V

$$w_P = -nR \cdot (T_2 - T_1) \qquad \text{(reversible, constant } P\text{)} \qquad (2.27)$$

$$w_T = -nRT \ln \frac{V_2}{V_1} = -nRT \ln \frac{P_1}{P_2} \qquad \text{(reversible, constant } T\text{)} \qquad (2.28)$$

$$q_T = nRT \ln \frac{V_2}{V_1} = nRT \ln \frac{P_1}{P_2} \qquad \text{(reversible, constant } T\text{)} \qquad (2.29)$$

$\ln x = 2.303 \log x$
$R = 8.314 \text{ J K}^{-1} \text{ mol}^{-1}$

Phase Changes

For a phase change, phase a $\rightarrow$ phase b, which occurs at constant T and P,

$$\Delta H = q_P \qquad (2.38)$$

$\Delta H = H(\text{phase b}) - H(\text{phase a})$

$$\Delta E = \Delta H - P \cdot (\Delta V) \qquad (2.39)$$

$\Delta E = E(\text{phase b}) - E(\text{phase a})$
$\Delta V = V(\text{phase b}) - V(\text{phase a})$

$$\Delta H(T_2) = \Delta H(T_1) + n \Delta \overline{C}_P \cdot (T_2 - T_1) \qquad (2.40)$$

n = number of moles
$\Delta \overline{C}_P$ = heat capacity per mole at constant P of phase b
    − heat capacity per mole at constant P of phase a

$$\Delta H(P_2) \cong \Delta H(P_1)$$

$$w_P = -P \cdot \Delta V$$

Chemical Reactions

For a chemical reaction

$$n_A A + n_B B \longrightarrow n_C C + n_D D$$

which occurs at constant T and P,

$$\Delta H = n_C \bar{H}_C + n_D \bar{H}_D - n_A \bar{H}_A - n_B \bar{H}_B = q_P$$

$$\bar{H}^0_{298}(A) \equiv \Delta \bar{H}^0_{f, 298}(A) = \text{enthalpy (heat) of formation of A}$$
$$\text{per mole from the elements in their}$$
$$\text{most stable states at standard}$$
$$\text{conditions (1 atm) and 25°C}$$

$$\Delta E = \Delta H - \Delta(PV) \tag{2.43}$$

$\Delta(PV) = PV(\text{products}) - PV(\text{reactants})$
$1 \text{ L atm} = 101.325 \text{ J}$

$$\Delta E = \Delta H - \Delta nRT \tag{2.46}$$

$\Delta n = $ number of moles of *gaseous* products $-$ number of moles of *gaseous* reactants
$R = 8.314 \text{ J K}^{-1} \text{ mol}^{-1}$

$$\Delta H(T_2) = \Delta H(T_1) + \Delta C_P \cdot (T_2 - T_1) \tag{2.44}$$

$$\Delta C_P = n_C \bar{C}_P(C) + n_D \bar{C}_P(D) - n_A \bar{C}_P(A) - n_B \bar{C}_P(B)$$

$$\Delta H(P_2) \cong \Delta H(P_1)$$

MATHEMATICS NEEDED FOR CHAPTER 2

Students should be able to integrate simple powers of x.

Indefinite integral of ax^n:

$$\int ax^n \, dx = \frac{ax^{n+1}}{n+1} \qquad (n \neq -1) \tag{1}$$

$$\int ax^{-1} \, dx = a \int \frac{dx}{x} = a \ln x \tag{2}$$

Here a is a constant independent of x.

Definite integral of ax^n:

$$\int_{x_1}^{x_2} ax^n \, dx = \frac{a \cdot (x_2^{n+1} - x_1^{n+1})}{n + 1} \qquad (n \neq -1) \tag{3}$$

$$\int_{x_1}^{x_2} a \frac{dx}{x} = a \cdot (\ln x_2 - \ln x_1) = a \ln \frac{x_2}{x_1} \tag{4}$$

Remember that $\ln ab = \ln a + \ln b$; $\ln (a/b) = \ln a - \ln b$; $\ln a = 2.303 \log a$.

Example $P = aV + bV^2$, with a and b constant. Using Eqs. (1) and (3), we obtain

$$\int P \, dV = \int (aV + bV^2) \, dV = a \frac{V^2}{2} + b \frac{V^3}{3}$$

$$\int_{V_1}^{V_2} P \, dV = \frac{a}{2} \cdot (V_2^2 - V_1^2) + \frac{b}{3} \cdot (V_2^3 - V_1^3)$$

REFERENCES

Freshman chemistry texts are good for reviewing basic thermodynamics.

KOTZ, J. C., and K. F. PURCELL, 1991. *Chemistry and Chemical Reactivity,* 2nd ed., Saunders College, Philadelphia.

OXTOBY, D., and N. H. NACHTRIEB, 1990. *Principles of Modern Chemistry,* 2nd ed., Saunders, New York.

The following textbooks on thermodynamics can be useful as supplements to Chapters 2 through 5.

FENN, J. B., 1982. *Engines, Energy, and Entropy,* W. H. Freeman, New York.

KLOTZ, I. M., and R. M. ROSENBERG, 1986. *Chemical Thermodynamics,* 4th ed., Benjamin/ Cummings, Menlo Park, California.

MARTIN, M. C., 1986. *Elements of Thermodynamics,* Prentice-Hall, Englewood Cliffs, New Jersey.

ROCK, P. A., 1983. *Chemical Thermodynamics,* University Science Books, Mill Valley, California.

SMITH, E. B., 1990. *Basic Chemical Thermodynamics,* 4th ed., Oxford University Press, New York.

Thermodynamic data for inorganic and organic chemicals are given in

COX, J. D., D. D. WAGMAN, and V. A. MEDVEDEV, eds., 1989. *CODATA Key Values for Thermodynamics,* Hemisphere Pub. Corp., New York.

STULL, D. R., E. F. WESTRUM, JR., and G. C. SINKE, 1987. *The Chemical Thermodynamics of Organic Compounds,* Krieger, Malabar, Florida.

For articles on applications of thermodynamics to biological molecules, see

BROWN, H. D., ed., 1969. *Biochemical Microcalorimetry,* Academic Press, New York.

JONES, M. N., ed., 1988. *Biochemical Thermodynamics,* 2nd ed., Elsevier, Amsterdam.

SUGGESTED READINGS

Energy for Planet Earth, 1990. *Sci. Am. 263* (Special Issue, September).
DOSTROSVSKY, I., 1991. Chemical Fuels from the Sun, *Sci. Am. 265* (December), 102–107.

PROBLEMS

1. (a) A hiker caught in a rainstorm might absorb 1 liter of water in his clothing. If it is windy so that this amount of water is evaporated quickly at 20°C, how much heat would be required for this process?

   (b) If all this heat were removed from the hiker (no significant heat was generated by metabolism in this time), what drop in body temperature would the hiker experience? The clothed hiker weighs 60 kg and you can approximate the heat capacity of hiker and clothes as equal to that of water. (The conclusion from this calculation is to stay out of the wind if you get your clothes wet.)

   (c) How many grams of sucrose would the hiker have to metabolize (quickly) to replace the heat of evaporating 1 liter of water, so that his temperature did not change? You can use the heats of reaction at 25°C; the reaction is:

$$\text{sucrose}(s) \ + \ \text{oxygen}(g) \ \longrightarrow \ \text{carbon dioxide}(g) \ + \ \text{water}(l)$$

2. Photosynthesis by land plants leads to the fixation each year of about 1 kg of carbon on the average for each square meter of an actively growing forest. The atmosphere is approximately 20% O_2 and 80% N_2 but contains 0.046% CO_2 by weight.

   (a) What volume of air (25°C, 1 atm) is needed to provide this 1 kg of carbon?

   (b) How much carbon is present in the entire atmosphere lying above each square meter of the earth's surface? (*Hint:* Remember that atmospheric pressure is the consequence of the force exerted by all the air above the surface; 1 atm is equivalent to 1.033×10^4 kg m^{-2}.)

   (c) At the current rate of utilization, how long would it take to use all the CO_2 in the entire atmosphere directly above the forest?

   (This assumes that atmospheric circulation and replenishment from the oceans, rocks, combustion of fuels, respiration of animals, and decay of biological materials are cut off.)

3. Calculate the work (in joules) done on the system for each of the following examples. Specify the sign of the work. The *system* is given in italics.

   (a) A *box of groceries* weighing 10 kg is carried up three flights of stairs (10 m altogether).

   (b) A 6.0-V *storage battery* is charged by a power supply for 2 hr with a current of 5.5 A.

   (c) A *muscle* of 1 cm^2 cross section and 10.0 cm length is stretched to 11.0 cm by hanging a mass on it. The muscle behaves like a spring that follows Hooke's law. The Hooke's law constant for the muscle was determined by finding that the muscle exerts a force of 5.00 N when it is stretched from 10.0 cm to 10.5 cm.

   (d) The *volume* of an ideal gas changes from 1 L to 3 L at an initial temperature of 25°C and a constant pressure of 1 atm.

   (e) The *volume* of an ideal gas changes from 1 L to 3 L at an initial temperature of 25°C and a constant pressure of 10^{-6} atm.

   (f) The *volume* of an ideal gas changes from 1 L to 3 L at a constant temperature of 25°C and the expansion is done reversibly for an amount of gas corresponding to an initial pressure of 1 atm.

4. (a) As heat is added to a sample of pure ice (solid H_2O) at constant pressure, the temperature rises until the melting point of the ice is reached. The ice then melts at the (constant) melting temperature, T_m. After the ice is completely melted the temperature continues to rise. Complete the plots below for heat added, q, vs. T for the three stages of this process. The slope of q plotted vs. T at constant pressure is the heat capacity, C_P. Plot C_P vs. temperature for this process of melting ice.

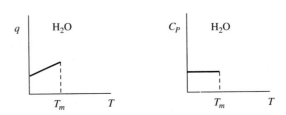

(b) As heat is added to a dilute buffer solution, the temperature rises. The change in temperature is approximately linear with added heat, as shown in the sketches.

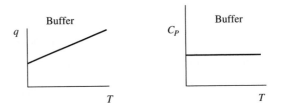

If a solution of DNA in dilute buffer is heated, the DNA double-stranded helix melts to single strands over a narrow temperature range. Heat is absorbed. The midpoint of the temperature range is called T_m. Sketch a q vs. T curve and a C_P vs. T curve for a DNA solution and describe how the heat necessary to melt (or denature) the DNA could be obtained from each curve.

5. Recently, biological organisms have been discovered living at great depths at the bottom of the ocean. The properties of common substances are greatly altered there because of enormous hydrostatic pressures caused by the weight of the ocean lying above. The organisms need to equilibrate their internal pressures with this external hydrostatic pressure to prevent being crushed.

(a) Calculate the hydrostatic pressure (in atm) in the ocean at depth 2500 m.

(b) Except near sources of heat, such as vents of hot gases coming from the earth's mantle, the temperature is close to 4°C. What is the percentage increase in density of liquid water under these conditions compared with that at the surface? The isothermal compressibility, β, of liquid water at 4°C is 49.5×10^{-6} atm^{-1}.

$$\beta = \frac{-\Delta V}{V_0 \Delta P} = \text{the fractional change in volume, } \Delta V/V_0, \text{ with change of pressure, } \Delta P.$$

(c) Consider a balloon filled with oxygen, O_2, at 1 atm pressure and occupying a 10-L volume at 293 K. Assuming ideal gas behavior, what would be its volume at a depth of 2500 m under the ocean surface at 277 K? What is the percentage increase in density of the O_2?

(d) Repeat the calculation of part (c) for O_2 gas taking into account its deviation from ideal behavior. The van der Waals constants for O_2 are: $a = 1.360 \text{ L}^2 \text{ atm mol}^{-2}$ and $b = 3.183 \times 10^{-2} \text{ L mol}^{-1}$.

6. Calculate the heat (in joules) absorbed by the system for each of the following examples. Specify the sign of the heat.
   (a) 100 g of liquid water is heated from 0°C to 100°C at 1 atm.
   (b) 100 g of liquid water is frozen to ice at 0°C at 0.01 atm.
   (c) 100 g of liquid water is evaporated to steam at 100°C at 1 atm.

7. One mole of an ideal gas initially at 27°C and 1 atm pressure is heated and allowed to expand reversibly at constant pressure until the final temperature is 327°C. For this gas, $\overline{C}_V = 20.8 \text{ J K}^{-1} \text{ mol}^{-1}$ and is constant over the temperature range.
   (a) Calculate the work, w, done on the gas in this expansion.
   (b) What are ΔE and ΔH for the process?
   (c) What is the amount of heat, q, absorbed by the gas?

8. One mole of an ideal monatomic gas initially at 300 K is expanded from an initial pressure of 10 atm to a final pressure of 1 atm. Calculate ΔE, q, w, ΔH, and the final temperature T_2 for this expansion carried out according to each of the following paths. The molar heat capacity at constant volume for a monatomic gas is $\overline{C}_V = \frac{3}{2}R$.
   (a) An isothermal, reversible expansion.
   (b) An expansion against a constant external pressure of 1 atm in a thermally isolated (adiabatic) system.
   (c) An expansion against zero external pressure (i.e., against a vacuum) in an adiabatic system.

9. For the following processes calculate the indicated quantities for a system consisting of 1 mol of N_2 gas. Assume ideal gas behavior.
   (a) The gas initially at 10 atm pressure is expanded tenfold in volume against a constant external pressure of 1 atm, all at 298 K. Calculate q, w, ΔE, and ΔH for the gas.
   (b) After the expansion in part (a), the volume is fixed and heat is added until the temperature reaches 373 K. Calculate q, w, ΔE, and ΔH.
   (c) What will be the pressure of the gas at the end of the process in part (b)?
   (d) If the gas at 373 K is next allowed to expand against 1 atm pressure under adiabatic conditions (no heat transferred), will the final temperature be higher, remain unchanged, or be lower than 373 K? Explain your answer.

10. For the following processes, state whether each of the thermodynamic quantities q, w, ΔE, and ΔH is greater than, equal to, or less than zero for the *system* described. Explain your answers briefly.
    (a) An *ideal gas* expands adiabatically against an external pressure of 1 atm.
    (b) An *ideal gas* expands isothermally against an external pressure of 1 atm.
    (c) An *ideal gas* expands adiabatically into a vacuum.
    (d) A *liquid* at its boiling point is converted reversibly into its vapor, at constant temperature and 1 atm pressure.
    (e) H_2 *gas* and O_2 *gas* are caused to react in a closed bomb at 25°C and the product water is brought back to 25°C.

11. One mole of liquid water at 100°C is heated until the liquid is converted entirely to vapor at 100°C and 1 atm pressure. Calculate q, w, ΔE, and ΔH for each of the following:
    (a) The vaporization is carried out in a cylinder where the external pressure on the piston is maintained at 1 atm throughout.
    (b) The cylinder is first expanded against a vacuum ($P_{ex} = 0$) to the same volume as in part (a), and then sufficient heat is added to vaporize the liquid completely to 1 atm pressure.

12. An ice cube weighing 18 g is removed from a freezer, where it has been at −20°C.
    (a) How much heat is required to warm it to 0°C without melting it?
    (b) How much additional heat is required to melt it to liquid water at 0°C?
    (c) Suppose that the ice cube was placed initially in a 180-g sample of liquid water at +20°C in an insulated (thermally isolated) container. Describe the final state when the system has reached equilibrium.

13. Describe what happens to the temperature, the volume, and the phase composition (number of moles of liquid or vapor present) following the removal of 400 J of heat from 1 mol of pure water initially in each of the following states. Assume that the pressure is maintained constant at 1 atm throughout and that the gas behaves ideally.
    (a) Water vapor initially at 125°C.
    (b) Liquid water initially at 100°C.
    (c) Water vapor initially at 100°C.
    (d) Which of the three processes above involves the largest amount of work done on or by the system? What is the sign of w? Explain your answer.

14. Consider the reaction $H_2O(l, \text{ 1 atm}) \rightarrow H_2O(g, \text{ 1 atm})$ at 298 K as illustrated in Fig. 2.6. Using the data in Table 2.2 and stating any important assumptions that you need to make:
    (a) Calculate ΔH^0_{298} for the process by reversible path (1).
    (b) Calculate ΔH^0_{298} for reversible path (2).
    (c) Compare your answers to parts (a) and (b) with the value obtained using the standard enthalpies of formation in Table A.5 of the Appendix.
    (d) Express the range of values that you obtained in the above calculations in terms of a percentage deviation. Which of the values is the most accurate? Why?

15. For the following processes, state whether each of the four thermodynamic quantities q, w, ΔE, and ΔH is greater than, equal to, or less than zero for the *system* described. Consider all gases to behave ideally. Each system is indicated by italic type. State explicitly any reasonable assumptions that you may need to make.
    (a) *Two copper bars*, one initially at 80°C and the other initially at 20°C, are brought into contact with one another in a thermally insulated compartment and then allowed to come to equilibrium.
    (b) A *sample of liquid* in a thermally insulated container (a calorimeter) is stirred for 1 hr by a mechanical linkage to a motor in the surroundings.
    (c) A *sample of H_2 gas* is mixed with *an equimolar amount of N_2 gas* at the same temperature and pressure under conditions where no chemical reaction occurs between them.

16. Some thermodynamic equations or relations discussed in this chapter are true in general; others are true only under restricted conditions, such as at constant volume, for an adiabatic process only, for an ideal gas, for PV work only, and so on. For each of the following equations, state what are the minimum conditions sufficient to make them true in the framework of chemical thermodynamics. (Some subscripts normally present have been omitted. You may assume that the system is uniform in properties: T, P, E, and so on, are everywhere the same.)

(a) $\Delta E = q + w$

(b) $q = \Delta H$

(c) $C_P = \dfrac{\Delta H}{\Delta T}$

(d) $\Delta H = \Delta E + \Delta(nRT)$

(e) $\left(P + \dfrac{n^2 a}{V^2}\right)(V - nb) = nRT$

(f) $w = -P_{ex}\,\Delta V$

17. If you set out to explore the surface of the moon, you would want to wear a space suit with thermal insulation. In such activity you might expect to generate roughly 4 kJ of heat per kilogram of mass per hour. If all this heat is retained by your body, by how much would your body temperature increase per hour owing to this rate of heat production? (Assume that your heat capacity is roughly that of water.) What time limit would you recommend for a moon walk under these conditions?

18. If a breath of air, with a volume of 0.5 L, is drawn into the lungs and comes to thermal equilibrium with the body at 37°C while the pressure remains constant, calculate the increase in enthalpy of the air if the initial air temperature is 20°C and the pressure is 1 atm. At a breathing rate of 12 per minute, how much heat is lost in this fashion in 1 day? Compare your answer (and that of Problem 17, assuming a body weight of 80 kg) with a typical daily intake of 12,000 kJ of food energy. What problems might you foresee in arctic climates, where the air temperature can reach −40°C and below? The heat capacity of air is about 30 J K^{-1} mol^{-1}.

19. Human beings expend energy during expansion and contraction of their lungs in breathing. Each exhalation from the lungs of an adult involves pushing out about 0.5 L of gas against 1 atm pressure. This occurs about 15,000 times in a 24-hr day.
    (a) Estimate the amount of work in breathing done by each person in the course of 24 hr.
    (b) To get a feeling for how much work this represents, imagine using it to raise a mass to the top of a 30-story building (about 100 m). First make a guess at how large a weight could be raised by this amount of work: 1 kg (2.2 lb), 10 kg, 100 kg, 1000 kg (about 1 ton). Now do the calculation and draw some conclusions about how much your body is working even when you are "resting."

20. One hundred grams of liquid H$_2$O at 55°C are mixed with 10 g of ice at −10°C. The pressure is kept constant and no heat is allowed to leave the system. The process is adiabatic. Calculate the final temperature for the system.

21. A reaction that is representative of those in the glycolytic pathway is the catabolism of glucose by complete oxidation to carbon dioxide and water:

$$C_6H_{12}O_6(s) + 6\,O_2(g) \longrightarrow 6\,CO_2(g) + 6\,H_2O(l)$$

Calculate the ΔH^0_{298} for the glucose oxidation.

22. Alcoholic fermentation by microorganisms involves the breakdown of glucose into ethanol and carbon dioxide by the reaction

$$glucose(s) \longrightarrow 2\,ethanol(l) + 2\,CO_2(g)$$

(a) Calculate the amount of heat liberated in a yeast brew upon fermentation of 1 mol of glucose at 25°C, 1 atm.

(b) What fraction is the heat calculated in part (a) of the amount of heat liberated by the complete combustion (reaction with O_2) of glucose to carbon dioxide and liquid water at 298 K calculated in Problem 21?

23. The enzyme catalase catalyzes the decomposition of hydrogen peroxide by the exothermic reaction

$$H_2O_2(aq) \xrightarrow{\text{catalase}} H_2O(l) + \frac{1}{2}O_2(g)$$

Estimate the minimum detectable concentration of H_2O_2 if a small amount of catalase (solid) is added to a hydrogen peroxide solution in a calorimeter. Assume that a temperature rise of 0.02°C can be distinguished. You can use a heat capacity of 4.18 kJ K^{-1} kg^{-1} for the hydrogen peroxide solution.

24. Consider the reaction

$$CH_3OH(l) \longrightarrow CH_4(g) + \frac{1}{2}O_2(g)$$

(a) Calculate ΔH^0_{298}

(b) Estimate ΔE^0_{298}

(c) Write an equation that would allow you to obtain ΔH at 500°C and 1 atm.

25. Use the standard enthalpies of formation tabulated in the Appendix to verify the value given in the text of this chapter for the oxidation of glycine to give urea and CO_2.

$$3\ O_2(g, 1\ atm) + 2\ NH_2CH_2COOH(s) \longrightarrow$$
$$H_2NCONH_2(s) + 3\ CO_2(g, 1\ atm) + 3\ H_2O(l)$$

26. Yeasts and other organisms can convert glucose ($C_6H_{12}O_6$) to ethanol or acetic acid. Calculate the change in enthalpy, ΔH, when 1 mol of glucose is oxidized to (a) ethanol, or (b) acetic acid by the following path at 298 K:

glucose $\longrightarrow$ glucose-6-phosphate $\longrightarrow$ fructose-6-phosphate ⌐

acetic acid $\longleftarrow$ acetaldehyde $\longleftarrow$ ethanol $\longleftarrow$ glyceraldehyde-3-phosphate $\longleftarrow$

You can ignore all heats of solution of products or reactants. The overall reactions are

$$C_6H_{12}O_6(s) \longrightarrow 2\ CH_3CH_2OH(l) + 2\ CO_2(g)$$

$$2\ O_2(g) + C_6H_{12}O_6(s) \longrightarrow 2\ CH_3COOH(l) + 2\ CO_2(g) + 2\ H_2O(l)$$

(c) Calculate the ΔH for the complete combustion of glucose to $CO_2(g)$ and $H_2O(l)$.

27. Estimate the change in ΔH if each reaction in Problem 26(a) and (b) is carried out by a thermophilic bacterium at 80°C. $\bar{C}_P$ for ethanol(l) = 111.5 J K^{-1} mol^{-1}, for acetic acid(l) = 123.5 J K^{-1} mol^{-1}, for glucose(s) $\cong$ 225 J K^{-1} mol^{-1}.

28. A good yield of photosynthesis for agricultural crops in bright sunlight is 20 kg of carbohydrate (e.g., sucrose) per hectare per hour. (1 hectare $= 10^4$ m^2.) The net reaction for sucrose formation in photosynthesis can be written

$$12 \text{ CO}_2(g) + 11 \text{ H}_2\text{O}(l) \xrightarrow[\text{light}]{} \text{C}_{12}\text{H}_{22}\text{O}_{11}(s) + 12 \text{ O}_2(g)$$

(a) Use standard enthalpies of formation to calculate ΔH^0 for the production of 1 mol of sucrose at 25°C by the reaction above.

(b) Calculate the energy equivalent of photosynthesis that yields 20 kg of sucrose (hectare hr)$^{-1}$.

(c) Bright sunlight corresponds to radiation incident on the surface of the earth at about 1 kW m^{-2}. What percentage of this energy can be "stored" as carbohydrate in photosynthesis?

29. (a) Calculate the enthalpy change on burning 1 g of $H_2(g)$ to $H_2O(l)$ at 25°C and 1 atm.

(b) Calculate the enthalpy change on burning 1 g of n-octane to $CO_2(g)$ and $H_2O(l)$ at 25°C and 1 atm.

(c) Compare $H_2(g)$ and n-octane(g) in terms of joules of heat available per gram.

30. Consider the reaction in which 1 mol of aspartic acid(s) is converted to alanine(s) and $CO_2(g)$ at 25°C and 1 atm pressure. The balanced reaction is

$$\text{H}_2\text{NCH(CH}_2\text{COOH)COOH} \rightleftharpoons \text{H}_2\text{NCH(CH}_3\text{)COOH} + \text{CO}_2$$

(a) How much heat is evolved or absorbed?

(b) Write a cycle you could use to calculate the heat effect for the reaction at 50°C. State what properties of molecules you would need to know and what equations you would use to calculate the answer.

31. What experiments would you have to do to measure the (a) energy, and (b) enthalpy change for the following reaction at 25°C and 1 atm:

$$\text{ATP}^{4-} + \text{H}_2\text{O} \rightleftharpoons \text{ADP}^{3-} + \text{HPO}_4^{2-} + \text{H}^+$$

The reaction takes place in aqueous solution using the sodium salts of each compound. Give as much detail as possible and show the equations you would need to use. ATP^{4-} is adenosine triphosphate; ADP^{3-} is adenosine diphosphate.

32. A household uses 22 kWh day^{-1} of electricity.

(a) How many kJ day^{-1} are used?

(2) About 1 kW m^{-2} of energy from the sun hits the surface of the earth. With a 10% efficient solar battery, what area (in m^2) of solar battery would be needed to provide sufficient solar energy to supply the household? Assume an average of 5 hr day^{-1} of sunshine. (A suitable energy-storage device would, of course, be needed to provide electricity at night.)

33. The enzyme catalase efficiently catalyzes the decomposition of hydrogen peroxide to give water and oxygen. At room temperature the reaction goes essentially to completion.

(a) Using heats of formation, calculate ΔH^0_{298} for the reaction

$$2 \text{ H}_2\text{O}_2(g) \longrightarrow 2 \text{ H}_2\text{O}(g) + \text{O}_2(g)$$

$\Delta H^0_f(298)$ for gaseous H_2O_2 is -133.18 kJ mol^{-1}.

(b) Calculate the bond dissociation energy for the O—O single bond.

(c) The enzyme normally acts on an aqueous solution of hydrogen peroxide, for which the equation is

$$2 \text{ H}_2\text{O}_2(aq) \longrightarrow 2 \text{ H}_2\text{O}(l) + \text{O}_2(g)$$

What is ΔH^0_{298} for this process?

(d) A solution initially 0.01 M in H_2O_2 and at 25.00°C is treated with a small amount of the enzyme. If all the heat liberated in the reaction is retained by the solution, what would be the final temperature? (Take the heat capacity of the solution to be 4.18 kJ K^{-1} kg^{-1}.)

34. Use bond energy data to calculate the enthalpy of formation for each of the following compounds at 25°C.
(a) *n*-Octane, $C_8H_{18}(g)$
(b) Naphthalene, $C_{10}H_8(g)$
(c) Formaldehyde, $H_2CO(g)$
(d) Formic acid, $HCOOH(g)$

Compare your answers with enthalpies of formation given in the tables in the Appendix. (The enthalpy of vaporization of formic acid is 46.15 kJ at 25°C.) Give the most likely reasons for the largest discrepancies between your calculated values and the ones in the tables.

35. The standard enthalpies of formation at 25°C of gaseous *trans*-2-butene and *cis*-2-butene are −11.1 and −7.0 kJ mol^{-1}, respectively. In the *cis* compound,

the relatively bulky methyl groups interact repulsively with one another, because they are both constrained to remain on the same side of the molecule. (There is no rotation about the C=C bond.) Calculate the thermodynamic enthalpy attributable to this steric repulsion in the *cis* compound relative to the *trans*

where the methyl groups are on the opposite sides of the molecule. This is an example of how detailed molecular information can be deduced from thermochemical data.

3

The Second Law:
The Entropy of the
Universe Increases

CONCEPTS

Entropy S is a measure of disorder; the greater the disorder, the greater the entropy. Disorder occurs spontaneously and thus entropy tends to increase. In a system kept at constant energy and volume, every change *increases* the entropy of the system. However, fluctuations may occur which decrease the entropy slightly for a short time. If the energy or the volume of the system are not constant, then the entropy of the system can decrease. The increase in entropy of the surroundings more than compensates for the decrease in entropy of the system; the entropy of the universe increases.

Gibbs free energy G is a combination of enthalpy H and entropy ($G = H - TS$) which always *decreases,* or does not change, in a process at constant temperature and pressure. Thus, by calculating a free energy change for a reaction, you can tell if it can occur at constant temperature and pressure.

APPLICATIONS

Everybody would like to be able to predict the future. Chemists, at least in their professional work, are interested mainly in predicting the future of chemical reactions. That is, they want to know (1) what reactions are impossible under given conditions, and (2) how the conditions can be changed so that impossible reactions become probable. A knowledge of thermodynamics will allow us to answer these questions. However, we can decide only *which* reactions are possible, not *when* they will actually occur. We can thus learn when it is useful to search for a catalyst (such as an enzyme) to make the reaction go, or when we should try a different reaction. Thermodynamics applies to other processes besides chemical reactions, and so we will consider very general changes and see whether they can occur.

An industrious inventor might ask the following questions during the course of a busy day. (1) Can diamonds be made out of pencil lead (graphite)? (2) Is it possible to make a heat shield that allows heat to go through one way but not the other? There would be many useful applications of this shield. (3) Can an engine be constructed that can convert heat completely into work?

We can answer the questions asked by the inventor once we understand entropy and the second law of thermodynamics. We do not even have to do the experiments suggested by the questions to determine the answers. Experiments alone might not convince the inventor. Just because a particular catalyst added to the pencil lead did not produce diamonds does not mean that some other catalyst will also not work. We shall jump ahead to our quantitative thermodynamic discussion and give the answers to question (1) as yes and to questions (2) and (3) as no.

After much effort and research, graphite at high pressures has been converted to diamonds. High temperatures are used to speed up the reaction, and certain transition metals act as catalysts. A one-way heat shield can never be found, and no engine operating in a cycle can be constructed that can convert heat into work without also discharging some heat to the surroundings.

HISTORICAL DEVELOPMENT OF THE SECOND LAW: THE CARNOT CYCLE

An important milestone in the development of the second law of thermodynamics was Carnot's analysis in the early nineteenth century of the efficiency of a heat engine. This led him to the discovery of a new thermodynamic variable of state—the entropy.

Consider the generalized cycle shown in Fig. 3.1. The system goes through a series of reversible steps and returns to its original state. At each step some heat or work is exchanged with the surroundings. This cycle is an idealized version of a real engine such as a steam engine or an automobile engine. In general, a hot gas at T_{hot} expands in a cylinder and does work ($w_1 + w_2$). As it does work it cools to T_{cold}. The gas is recompressed ($w_3 + w_4$) and returned to its original condition. For one complete cycle, the total work is $w = (w_1 + w_2 + w_3 + w_4)$ and the heat absorbed is $q = (q_1 + q_2 + q_3 + q_4)$. Because the initial and final states are the same, we know that there is no change in energy and that $q = -w$.

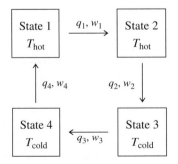

Fig. 3.1 Cycle of a heat engine. Each step is carried out reversibly.

For simplicity, Carnot considered an ideal gas in the engine, which underwent a particular cycle of four successive reversible steps to return to its original state. These four steps are shown as a PV diagram in Fig. 3.2a, and are described in detail in Fig. 3.2b. The total work done by the engine in the cyclic process is the area enclosed by the arrows in the PV diagram of Fig. 3.2a. It is easy to derive this result from the definition of work of compression or expansion

$$w = -\int P\,dV$$

From the figure it is obvious that the total amount of work done depends on the difference between T_{hot} and T_{cold}. In the first isothermal step ($1 \rightarrow 2$) heat is absorbed at T_{hot} and converted quantitatively to work done by the system; the energy of the system is unchanged. In an adiabatic step ($2 \rightarrow 3$) further work is done by the system; the energy of the system decreases. In the second isothermal step ($3 \rightarrow 4$) work is done on the system and an equal amount of heat is released; the energy of the system is unchanged. Finally, in an adiabatic step ($4 \rightarrow 1$) further work is done on the system to bring it back to its initial state. The energy of the system increases in this last step so that the net change in energy of the system is zero; as it must be for a cyclic process. In this cyclic process work has been done and heat has been absorbed at the high temperature. However, all the heat absorbed at the high temperature cannot be converted into expansion work because some heat must be released at the low temperature during the compression step necessary to return the engine to its initial state. Analysis of this process leads to the conclusion that heat cannot be converted into work by a cyclic process unless a temperature difference exists within the cycle. Furthermore, the maximum efficiency of conversion of heat into work is proportional to the difference in absolute temperatures between the high temperature at which the heat is absorbed, and the cold temperature at which some of the heat is released. Most importantly, a new thermodynamic variable of state—entropy—results from the analysis.

For the purpose of calculation we will consider an ideal gas to be the system. From the first law and the ideal gas law, $PV = nRT$, the q's and w's of each step can be calculated.

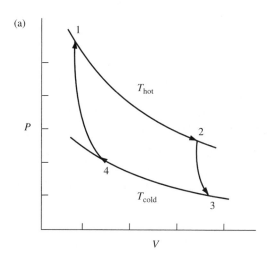

(a)

Fig. 3.2a A PV diagram for a Carnot cycle illustrating the four paths taken in moving between the states of Fig. 3.1. The area within the arrows is the total work done in the cyclic process. There are two reversible isothermal paths ($1 \rightarrow 2$ and $3 \rightarrow 4$), and two reversible adiabatic paths ($2 \rightarrow 3$ and $4 \rightarrow 1$).

(b)

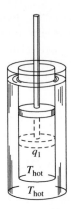

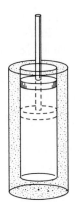

1. Isothermal reversible expansion
$P_1, V_1 \rightarrow P_2, V_2$
T_{hot} is constant
q_1 is positive
w_1 is negative

2. Adiabatic reversible expansion
$P_2, V_2 \rightarrow P_3, V_3$
$T_{hot} \rightarrow T_{cold}$
$q_2 = 0$
w_2 is negative

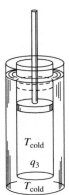

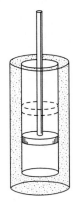

3. Isothermal reversible compression
$P_3, V_3 \rightarrow P_4, V_4$
T_{cold} is constant
q_3 is negative
w_3 is positive

4. Adiabatic reversible compression
$P_4, V_4 \rightarrow P_1, V_1$
$T_{cold} \rightarrow T_{hot}$
$q_4 = 0$
w_4 is positive

Fig. 3.2b The four paths, or steps, of the Carnot-cycle heat engine. The engine is in thermal contact with a hot heat reservoir, a cold heat reservoir, or is thermally isolated. The first two paths are expansions ($1 \rightarrow 2$ and $2 \rightarrow 3$) in which work is done by the engine. The next two paths are compressions ($3 \rightarrow 4$ and $4 \rightarrow 1$) in which work is done on the engine. The net result is that some of the heat absorbed from the hot reservoir at T_{hot} has been converted into work, and the engine has been returned to its initial state. The remaining heat has been released to the cold heat reservoir at T_{cold}.

For the first step (an isothermal reversible step),

$$w_1 = -\int_{V_1}^{V_2} P \, dV = -nRT_{\text{hot}} \ln \frac{V_2}{V_1}$$

$$E_2 - E_1 = 0 \qquad (E \text{ for an ideal gas is dependent only on temperature, as noted in Chapter 2})$$

$$q_1 + w_1 = E_2 - E_1 = 0 \qquad \text{(first law)}$$

$$q_1 = -w_1 = nRT_{\text{hot}} \ln \frac{V_2}{V_1}$$

For the second step (an adiabatic reversible step),

$$q_2 = 0 \qquad \text{(adiabatic)}$$

$$w_2 = E_2 - E_1$$

$$= C_V \cdot (T_{\text{cold}} - T_{\text{hot}}) \qquad \text{(See Eq. 2.31.)}$$

We can derive an additional important relation for an adiabatic reversible expansion of an ideal gas. For a small change dV in volume there will be a corresponding change in temperature dT. The energy change dE is $C_V \, dT$. Because the process is adiabatic, the energy change is equal to the work, $-P \, dV$:

$$C_V \, dT = -P \, dV$$

$$= -\frac{nRT}{V} \, dV$$

Dividing both sides by T and integrating, we have for step 2,

$$C_V \int_{T_{\text{hot}}}^{T_{\text{cold}}} \frac{dT}{T} = -nR \int_{V_2}^{V_3} \frac{dV}{V}$$

$$C_V \ln \frac{T_{\text{cold}}}{T_{\text{hot}}} = -nR \ln \frac{V_3}{V_2} = nR \ln \frac{V_2}{V_3} \qquad (3.1a)$$

Thus we have a relation between the temperature and volume changes for an adiabatic reversible expansion (compression) of an ideal gas. The third and fourth steps are similar to the first and second, respectively:

$$q_3 = -w_3 = nRT_{\text{cold}} \ln \frac{V_4}{V_3} \qquad \text{(isothermal)}$$

$$q_4 = 0 \qquad w_4 = C_V \cdot (T_{\text{hot}} - T_{\text{cold}}) \qquad \text{(adiabatic)}$$

$$-C_V \ln \frac{T_{\text{cold}}}{T_{\text{hot}}} = nR \ln \frac{V_4}{V_1} \qquad (3.1b)$$

The total heat absorbed is

$$q = q_1 + q_2 + q_3 + q_4 = nRT_{hot} \ln \frac{V_2}{V_1} + 0 + nRT_{cold} \ln \frac{V_4}{V_3} + 0$$

The total work done by the engine is

$$-w = -(w_1 + w_2 + w_3 + w_4) = nRT_{hot} \ln \frac{V_2}{V_1} + nRT_{cold} \ln \frac{V_4}{V_3}$$

because w_2 and w_4 cancel. Therefore, the total work done by the engine is just

$$-w = q$$

We could also have reached this conclusion by using the fact that $E_2 - E_1 = 0$ for any cyclic change. Therefore, from the first law, $q = -w$ for any cyclic change. Carnot noticed that the sum of the quantities (q_i/T_i) for all the steps of the cycle had an interesting property, that the sum is zero. We can derive this easily:

$$\frac{q_1}{T_{hot}} + 0 + \frac{q_3}{T_{cold}} + 0 = nR \ln \frac{V_2}{V_1} + nR \ln \frac{V_4}{V_3}$$

$$\frac{q_1}{T_{hot}} + \frac{q_3}{T_{cold}} = nR \ln \frac{V_2 V_4}{V_3 V_1}$$

We need a relation among the volumes of the cycle. For *reversible* adiabatic steps, such as steps 2 and 4 of the Carnot cycle, we have derived a relation between the initial and the final values of V and T. Adding Eqs. (3.1a) and (3.1b), we see that

$$nR \ln \frac{V_2 V_4}{V_3 V_1} = 0$$

which is what we wanted to prove. Therefore,

$$\frac{q_1}{T_{hot}} + \frac{q_3}{T_{cold}} = 0 \qquad (3.2)$$

Why is this conclusion important? Our system (gas in cylinder) has gone through a cycle and returned to its original state. The sum of the quantity q_{rev}/T for this cyclic path is zero (we use the subscript "rev" on q here to emphasize that all steps are carried out reversibly). This is what we learned about state functions such as E and H. For a cyclic path, state functions exhibit no change. Carnot recognized that by dividing q_{rev}, a path-dependent quantity, by T, he generated a path-independent function, a state function, now called *entropy*. The symbol S will be used for entropy:

$$\Delta S = \frac{q_{rev}}{T} \qquad (3.3)$$

Equation (3.3) can be considered to be a definition of entropy. We considered a heat engine containing an ideal gas, but the entropy is a property of state even when the system is not an ideal gas.

ENTROPY IS A STATE FUNCTION

The logic which led to the realization that entropy is a state function is as follows. The efficiency of a heat engine is expressed as the total work done by the engine divided by the heat absorbed at the high temperature; $q_1 = q_{hot}$

$$\text{efficiency} = \frac{-w}{q_{hot}} \tag{3.4}$$

The rest of the energy is discharged as heat $q_3 = q_{cold}$ at the low temperature and is wasted, as far as the efficiency for conversion into work is concerned.

With a little algebra we can solve for efficiency in terms of temperature. From Eq. (3.2),

$$\frac{q_{hot}}{T_{hot}} = -\frac{q_{cold}}{T_{cold}}$$

or

$$\frac{q_{hot}}{q_{cold}} = -\frac{T_{hot}}{T_{cold}} \tag{3.5}$$

From the first law, the net work is

$$-w = q_{hot} + q_{cold}$$

Then

$$\text{efficiency} = \frac{q_{hot} + q_{cold}}{q_{hot}} = 1 + \frac{q_{cold}}{q_{hot}}$$

From Eq. (3.5),

$$\text{efficiency} = 1 - \frac{T_{cold}}{T_{hot}}$$

or

$$\text{efficiency} = \frac{T_{hot} - T_{cold}}{T_{hot}} \tag{3.6}$$

For any engine operating under practical conditions the efficiency will always be less than 100%. Efficiency can be maximized, according to Eq. (3.6), by making T_{cold} as small as possible and T_{hot} as large as possible. *It is not possible in practice to bring the temperature of any system as low as absolute zero* (0 K). This is actually an alternative statement of the third law of thermodynamics, which we will consider presently. A corollary of this statement and Eq. (3.6) is the conclusion that no engine operating in a cycle can achieve an efficiency of 100% in converting heat into work.

Since all steps are reversible, the Carnot engine can also be operated in the reverse as a refrigerator or heat pump. In the forward direction as a heat engine, there is a net transfer of heat from the hot reservoir to the cold one, and the system does work on the surroundings. In the reverse direction, as a heat pump, the surroundings perform work on the system, and there is a net flow of heat from the cold reservoir to the hot one.

From our experience heat flows spontaneously only from a hot body to a cooler one. A consequence of this is that all heat engines operating with reversible cycles between T_{hot} and T_{cold} must have the same efficiency. Suppose this were not true. We can always operate the less efficient heat engine in reverse as a heat pump, and use the more efficient one as the heat engine. The work output of the engine is used to drive the heat pump. The combined effect of the two reversible engines operated in the fashion described is to cause a heat flow from the cold reservoir to the hot reservoir with no change in the surroundings (see Problem 2). Heat would be transferred spontaneously from a low temperature to a high temperature, which we do not believe can happen. Therefore, we must conclude that all reversible engines have the same efficiency.

Because all reversible engines have the same efficiency, it is straightforward to show that Eq. (3.5) and therefore Eq. (3.2), cannot be a consequence of using an ideal gas in our engine or of the particular paths chosen. In other words, we have defined a new state function, *entropy:*

$$\Delta S \equiv \frac{q_{rev}}{T} \qquad (3.3)$$

Entropy is an extensive variable of state; ΔS depends only on the initial and final states of the system. From Eq. (3.3), one sees that the dimensions of entropy are energy/temperature; we shall use units of J K^{-1}. The unit for entropy in the cgs system is cal deg^{-1}:

$$4.184 \text{ J K}^{-1} = 1 \text{ cal deg}^{-1}$$

A calorie per degree is also called an *entropy unit,* eu.

We have "proven" that entropy is a state function only because we accepted as true that heat flows spontaneously from a hot body to a cooler one. We can also consider Eq. (3.3) as *defining* a thermodynamic temperature scale. The temperature scale T is the one that makes q_{rev}/T a state function. The fact that Eq. (3.3) can also be derived for an ideal gas, as we have done, means that the thermodynamic temperature scale defined by Eq. (3.3) is identical to the temperature scale defined by the ideal gas law, $PV = nRT$.

THE SECOND LAW OF THERMODYNAMICS: ENTROPY IS NOT CONSERVED

When we calculate the entropy changes of both the system and the surroundings for simple processes that can occur spontaneously, such as the flow of heat from a hot body to a cooler one, the expansion of an ideal gas into a vacuum, or the flow of water down a hill, we find that the sum of the entropy changes is not zero. In other words, unlike energy, entropy is not conserved. The generalization of a great deal of experience, which is the second law of thermodynamics, is that the sum of the entropy changes of the system and the surroundings is always positive. Even zero values can be approached only as a limit, and negative values are never found. If there is a decrease in entropy in a system, there must be an equal or larger increase in entropy in the surroundings. The first two laws of thermodynamics can now be stated briefly: the total energy of the system plus the surroundings remains constant; the total entropy of the system plus surroundings never decreases. Equivalent statements of the second law are: (1) heat spontaneously flows from a hot body to a cold body, but work must be done to transfer heat from a cold body to a hot body; and (2) it is impossible by a *cyclic* process to remove heat from a hot body and convert it solely into work; some of the heat removed must be transferred to a cold body. The *second law* can be stated:

$$\Delta S(\text{system}) + \Delta S(\text{surroundings}) \geq 0 \qquad (3.7)$$

For an *isolated system,* since there is no energy or material exchange between such a system and the surroundings, there is no change in the surroundings. Therefore,

$$\Delta S(\text{isolated system}) \geq 0 \qquad (3.8)$$

The $\geq$ sign means greater than or equal to; the former applies for irreversible processes and the latter for reversible ones.

We can express the second law by Eqs. (3.7) and (3.8), because these definitions lead to general conclusions which are consistent with experiments. This is illustrated in the following example.

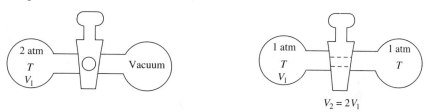

Example 3.1 One mole of an ideal gas initially at $P_1 = 2$ atm, T, and V_1 expands to $P_2 = 1$ atm, T, and $2\,V_1$. Consider two different paths: (a) the expansion occurs irreversibly into a vacuum as shown above, and (b) the expansion is reversible. Calculate $q_{\text{irreversible}}$, $\Delta S(\text{system})$, and $\Delta S(\text{surroundings})$ for (a) and q_{rev}, $\Delta S(\text{system})$, and $\Delta S(\text{surroundings})$ for (b). We know immediately the $\Delta S(\text{system})$ is independent of path and is the same for (a) and (b) because the initial and final stages of the system are the same for (a) and (b). However, the effect on the surroundings is quite different for (a) and (b); q depends on the path, and the change in the state of the surroundings is different for (a) and (b). Thus the values of $\Delta S(\text{surroundings})$ will be different.

Solution

(a)
$$w = 0 \ (\text{no work done against surroundings})$$

$$\Delta E = 0 \ (E \text{ for an ideal gas is independent of volume})$$

$$q_{\text{irreversible}} = \Delta E - w = 0$$

$$\Delta S(\text{surroundings}) = 0 \ (\text{no work or heat exchange between system and surroundings; surroundings not affected in any way by this process})$$

To calculate $\Delta S(\text{system})$, we take advantage of the fact that entropy is a property of state of the system, and that entropy changes between a given initial and final state are independent of the path connecting them. We choose a different path in which the gas expands isothermally and reversibly to the same final state. This is actually the path of (b); we choose it because we can readily calculate $\Delta S(\text{system})$ for this path.

$$w_{rev} = -\int_{V_1}^{V_2} P\, dV = -RT \int_{V_1}^{V_2} \frac{dV}{V} = -RT \ln \frac{V_2}{V_1}$$

$$= -RT \ln 2$$

$$\Delta E = 0$$

Therefore,

$$q_{rev} = \Delta E - w_{rev} = RT \ln 2$$

and

$$\Delta S(\text{system}) = \frac{q_{rev}}{T} = R \ln 2$$

Note that for this spontaneous, irreversible process,

$$\Delta S(\text{system}) + \Delta S(\text{surroundings}) = R \ln 2 + 0$$

$$> 0$$

consistent with Eqs. (3.7) and (3.8).

(b) We have already calculated that

$$q_{rev} = RT \ln 2$$

$$\Delta S(\text{system}) = R \ln 2$$

An amount q_{rev} of heat is transferred into the system from the surroundings. One way for this transfer to occur reversibly is to choose the surrounding temperature higher than T by only an infinitesimal amount. In other words, the surrounding temperature is also T. For the surroundings, the heat input is $-q_{rev}$ and

$$\Delta S(\text{surroundings}) = \frac{-q_{rev}}{T} = -R \ln 2$$

Note that for this reversible process,

$$\Delta S(\text{system}) + \Delta S(\text{surroundings}) = 0$$

consistent with Eq. (3.7).

MOLECULAR INTERPRETATION OF ENTROPY

The microscopic or molecular interpretation of entropy is that it is a measure of disorder. The more disorder, the higher the entropy. Therefore, the second law tells us that the universe is continually becoming more disordered. This does not mean that a small part of the universe (the system) cannot become more ordered. Every time we freeze an ice tray full of water in the refrigerator, we are increasing the order inside the ice tray. However, the disorder caused by the heat released outside the freezer compartment more than balances the order created in the ice. The total disorder in the universe has increased; the total entropy of the universe has increased. Many biological processes involve decreases of entropy for the organism itself but they are always coupled to other processes which increase the entropy, so that the sum is always positive. Photosynthesis in green plants results in the conversion of simple substances like CO_2 and H_2O into an enormously complex organism. There is a very large decrease in entropy associated with the growth of a plant, but it comes at the expense of an enormous increase in entropy in the sun to produce the light that is essential for photosynthesis to occur. It is a pessimistic, but accurate, view that anything we do will always add to the disorder of the universe.

Feynman (1963) defines disorder as the number of ways the insides can be arranged so that from the outside the system looks the same. For example, 1 mol of ice is more ordered than 1 mol of water at the same temperature, because the water molecules in the liquid may have many different arrangements but still have the properties of liquid water. The water molecules in the solid ice can have only the periodic arrangement corresponding to the crystal structure of ice to have the properties of ice.

The molar entropies of water at 0°C, 1 atm are:

	Entropy mol^{-1}
$H_2O(s)$	41.0 J K^{-1}
$H_2O(l)$	63.2 J K^{-1}
$H_2O(g)$	188.3 J K^{-1}

These values are consistent with the fact that the water molecules in ice are highly ordered, whereas in water vapor the molecules are free to move about in a relatively large space. The liquid state is intermediate, but it is more like the solid than the gas in terms of disorder. The entropy can be qualitatively related to the hardness of monatomic elements; hard solids are more ordered and have lower entropies than soft solids. This is illustrated for carbon at 25°C, 1 atm.

	Entropy mol^{-1}	Characteristic
Carbon (diamond)	2.4 J K^{-1}	Hard
Carbon (graphite)	5.7 J K^{-1}	Soft

All entropies increase as the temperature is raised, because increasing molecular motion increases disorder and thus increases the entropy.

Knowing that entropy is a measure of disorder allows us to understand and predict entropy changes in reactions. In the gas phase if the number of product molecules is less than the number of reactant molecules, the entropy will decrease. If the number increases, the entropy increases. Examples at 25°C, 1 atm are:

Reaction	ΔS^0 (J K^{-1} mol^{-1})
$2 H_2(g) + O_2(g) \rightarrow 2 H_2O(g)$	−88.9
$PCl_5(g) \rightarrow PCl_3(g) + Cl_2(g)$	181.9

If the number of molecules is unchanged by the reaction, a small change in entropy—either positive or negative—is expected. For reactions in solution it is more difficult to predict the entropy change because of the large effect of the solvent. Two solute molecules may react to form one product molecule, but unless we know what is happening to the solvent—whether it is being ordered or disordered—we cannot predict the change in entropy. However, we can understand and interpret the measured results. A few examples based on data in Table A.5 are as follows:

Reaction	ΔS^0 (J K^{-1} mol^{-1})
$CO_3^{2-}(aq) + H^+(aq) \rightarrow HCO_3^-(aq)$	+148.1
$NH_4^+(aq) + CO_3^{2-}(aq) \rightarrow NH_3(aq) + HCO_3^-(aq)$	+146.0
$OH^-(aq) + H^+(aq) \rightarrow H_2O(l)$	+80.7
$NH_4^+(aq) + HCO_3^-(aq) \rightarrow CO_2(aq) + H_2O(l) + NH_3(aq)$	+94.2
$NH_3(aq) + H^+(aq) \rightarrow NH_4^+(aq)$	+2.1
$HC_2O_4^-(aq) + OH^-(aq) \rightarrow C_2O_4^{2-}(aq) + H_2O(l)$	−23.1
$CH_4(aq) \rightarrow CH_4(CCl_4)$	+75

For the first four reactions, neutralization of charges occurs. Because a charged species tends to orient the water molecules around it, charge neutralization results in disorientation of some of the solvent molecules and an increase in entropy. The next two reactions involve transfer of charge but no neutralization. The entropy changes are small as a consequence. The last reaction involves the transfer of a nonpolar methane (CH_4) molecule from a polar solvent (H_2O) to a nonpolar solvent (CCl_4). The large positive ΔS^0 is primarily the result of the ordering of water molecules around a nonpolar molecule. Removal of the nonpolar molecule results in the randomization of the water molecules and hence a positive ΔS. This kind of effect is important for many reactions in aqueous media, such as the binding of a nonpolar substrate to the nonpolar region of an enzyme, or the folding of a protein.

A quantitative interpretation of entropy was obtained by Boltzmann. He derived the result that the entropy is proportional to the logarithm of the number of microscopic states of a system.

$$S = k \ln N \tag{3.9}$$

k = Boltzmann's constant (1.381×10^{-23} J K^{-1}); the gas constant R (8.314 J K^{-1} mol^{-1}) is Avogadro's number, N_0, times the Boltzmann constant

N = no. of microscopic states of the system

The number of microscopic states is just what Feynman said: the number of different ways the inside of the system can be changed without changing the outside. As an example consider an ordered deck of cards. There is only one microscopic state for the ordered deck: $N = 1$.

$$S(\text{ordered}) = k \ln 1 = 0$$

Shuffle thoroughly to get a disordered deck. All disordered decks are equivalent so now N is the number of possible disordered decks: $N = 52 \cdot 51 \cdot 50 \cdot 49 \ldots 2 \cdot 1 = 52!$

$$S\text{(disordered)} = k \ln 52! = 2.1 \times 10^{-21} \text{ J K}^{-1}$$

The entropy change on shuffling is positive

$$S\text{(disordered)} - S\text{(ordered)} = 2.1 \times 10^{-21} \text{ J K}^{-1}$$

This tells you that a deck of cards will spontaneously disorder (with a proper catalyst such as dropping the deck on the floor), but that it is very unlikely for a disordered deck to spontaneously order itself.

For simple systems (such as a deck of cards) we can count the number of microstates. All disordered decks are the same from the outside, and we know how many microstates are possible for a deck of cards. For simple molecular systems we can also calculate the number of microstates.

A pure, perfect crystal at 0 K has 1 microstate; its entropy is zero. However, in some crystals such as carbon monoxide the molecules can have two orientations at each site in the crystal lattice. Because of similar sizes and similar net charges on the carbon and oxygen in $C{=}O$, the molecule can orient in two equally probable ways ($C{=}O$ or $O{=}C$). The crystal is disordered with two possibilities at each site. The number of microstates per mol at 0 K is 2^{N_0}. Thus the molar entropy of the disordered crystal is $k \ln 2^{N_0} = R \ln 2$. The difference in entropy between the ordered crystal and the disordered crystal of 5.76 J K^{-1} mol^{-1} can be measured calorimetrically; these types of measurements provided evidence for Boltzmann's equation. For a pure ideal gas, because of the lack of interactions among the molecules, the entropy is the sum of translational, rotational, vibrational, and at high enough temperatures, electronic entropies. This is similar to the result for the energy of an ideal gas. However, the energy of an ideal gas depends only on temperature, whereas the entropy depends both on temperature and pressure (or volume). It is reasonable that the translational entropy of a gas is directly proportional to the logarithm of the volume. The larger the volume of the gas, the more ways it can be rearranged—the larger the value of N. Furthermore, in analogy with a disordered deck of cards, there is an increase of entropy when two ideal gases mix.

For real gases, for solutions, and for complex systems in general it becomes harder to calculate the entropy from the structures of the molecules and their interactions. However, we can use the measured entropy to calculate N, the number of microscopic states for any system. Solving for N in Eq. (3.9) we obtain

$$N = e^{S/k} \tag{3.10}$$

We can use Eq. (3.9) to calculate the entropy of mixing ideal gases; surely the entropy of the mixture will be greater than that of the unmixed gases. The result for the entropy of mixing of two ideal gases is

$$\Delta S\text{(mixing)} = -R \, [n_1 \ln X_1 + n_2 \ln X_2] \tag{3.11}$$

$n_1, n_2 = $ number of moles of each gas

$$X_1 = \frac{n_1}{n_1 + n_2}, X_2 = \frac{n_2}{n_1 + n_2} = \text{mole fractions}$$

Note that ΔS (mixing) will always be positive because X_1 and X_2 are always less than 1 and the logarithm of a number less than 1 is negative. Equation (3.11) can also be derived from the pressure dependence of entropy for an ideal gas. The formula for ΔS (mixing) applies approximately to mixing of liquids and of solutes in solution.

MEASUREMENT OF ENTROPY

The experimental measurement of entropy changes is based on the definition of entropy, Eq. (3.3). For a reaction at constant temperature the *reversible* heat transfer for the process is divided by the absolute temperature. If the temperature changes during the process, the reversible heat transferred at each different temperature must be divided by its temperature to calculate each entropy change. The total entropy change is just the sum of the individual entropy changes.

It is clear, then, that for any process we can write the entropy change as an integral:*

$$S_2 - S_1 = \int_1^2 dS = \int_1^2 \frac{dq_{rev}}{T} \tag{3.12}$$

We must be sure to understand entropy and its relation to the reversible heat given by Eq. (3.12). Remember that entropy depends only on the initial state (1) and the final state (2). It does not matter how we get from state 1 to state 2. However, the entropy difference can be measured as in Example 3.1 by finding a reversible path between states 1 and 2, measuring the heat change, and using Eq. (3.12). For an irreversible path the entropy change is *not* equal to the heat absorbed divided by T; furthermore, the entropy change is always greater than the irreversible heat divided by the temperature.

$$S_2 - S_1 > \int \frac{dq_{irreversible}}{T} \tag{3.13}$$

Exercise   Show that Example 3.1(a) is consistent with Eq. (3.13).

One-Way Heat Shield

We can now consider whether it is possible to place a one-way heat shield between two identical pieces of metal so that one piece gets hotter while the other gets cooler. In Fig. 3.3 the transparent side of the shield is facing left so that heat can move only from left to right. An equivalent but more general question is whether there is any method by which the change shown in the figure can occur without any change in the surroundings. Let us calculate the entropy change of the system and see. The entropy change of the surroundings must be zero, because there is no change in the surroundings. We must specify the initial

---

* Although we use the symbol dq_{rev} to express a differential, reversible heat change, we should keep in mind that there is an important difference between dq_{rev} and the differentials of state functions, such as dE and dH. For any cyclic path from state 1 back to state 1, the integrals $\int_1^1 dE$ and $\int_1^1 dH$ are always 0. But $\int_1^1 dq_{rev}$ is dependent on the particular path and is generally not zero. Mathematically, we say that dE and dH are exact differentials, but dq_{rev} is not. The factor $1/T$ that converts the inexact dq_{rev} into the exact differential dS is known as an integrating factor.

and final states of the system so that we can calculate the entropy change. We will choose the pressure to be constant, and from Fig. 3.3 we see that for half the system (left side) the temperature decreases from T_1 to T_A; for the other identical half (right side), the temperature increases from T_1 to T_B. From Eq. (3.12),

$$S_2 - S_1 = \Delta S \text{(left side)} + \Delta S \text{(right side)}$$
$$= \int_1^A \frac{dq_{\text{rev}}}{T} \text{(left)} + \int_1^B \frac{dq_{\text{rev}}}{T} \text{(right)}$$

Heat is being lost from the left side, so dq_{rev} (left) is negative and the integral from state 1 to state A is therefore negative. As heat is being gained by the right side, the integral from state 1 to B is positive. However, from the first law the magnitudes of the heats are identical; the heat lost by the left side must equal the heat gained by the right side:

$$| dq_{\text{rev}} \text{(left)}| = | dq_{\text{rev}} \text{(right)}|$$

The vertical lines, | |, indicate absolute magnitude; the magnitudes are equal, but the signs are different. As the heat flows from left to right, the temperature on the left decreases and the temperature on the right increases:

$$T_{\text{left}} < T_{\text{right}}$$

This means that as soon as the hypothetical process begins

$$\left|\frac{dq_{\text{rev}}}{T} \text{(left)}\right| > \left|\frac{dq_{\text{rev}}}{T} \text{(right)}\right|$$

Remembering that dq_{rev} (left) is negative, we see that

$$\int_1^A \frac{dq_{\text{rev}}}{T} \text{(left)}$$

is negative and greater in magnitude than

$$\int_1^B \frac{dq_{\text{rev}}}{T} \text{(right)}$$

We conclude that $S_2 - S_1$ would be negative for the process shown in Fig. 3.3! This is contrary to the second law, so we say that it cannot actually take place. Again, this is in agreement with our experience that says that *heat never flows spontaneously from a cold object to a hot object.*

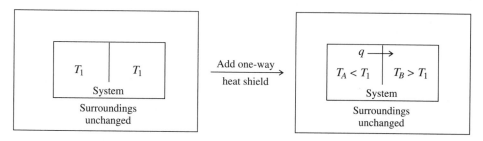

Fig. 3.3 One-way heat shield that will not work. The inventor, however, argues that, by analogy with one-way mirrors, only a few more months of experiments are necessary to develop the right material. The second law says "never."

What about the reverse of the process in Fig. 3.3? Can we start with a system in which the right side is hot (T_B), the left side is cool (T_A), and with no change in the surroundings reach a uniform temperature (T_1)? The answer from the second law and from experience is "yes." The only change in the previous discussion is that the left integral is now positive and greater in magnitude than the right integral, because heat is flowing to the cool side, so $S_2 - S_1$ is positive. The second law, therefore, tells us that a system will spontaneously tend to reach uniform temperature.

Fluctuations

As we have already shown (Example 3.1), if we assume that a system will spontaneously reach uniform pressure, we arrive at the conclusion that the sum of entropy changes of the system and surroundings is positive (the second law). Or if we start with the second law, we can predict that a system will spontaneously tend to reach uniform pressure if the temperature of the system is uniform (Fig. 3.4a). Similarly, starting with the second law, we predict that a system such as the one shown in Fig. 3.4b will tend to reach uniform composition spontaneously. That is, at constant pressure, gases originally separated in two halves of a system will tend to mix. Both of these processes can take place without changing the surroundings. The reverse of these reactions—unmixing gases, or going from a uniform pressure to unequal pressures—can only be done if the surroundings are also changed. These processes will not occur spontaneously. The quantitative statement of the second law was made to be consistent with experiment, so it does not surprise us to be able to draw experimentally correct conclusions from Eqs. (3.7) and (3.8).

But the second law was based on observations of relatively large systems and measurements averaged over time; it is not quite correct for very small systems and very short times. In a system of uniform pressure, there will be very slight increases of pressure on one side and decreases on the other side owing to the random motion of the molecules.

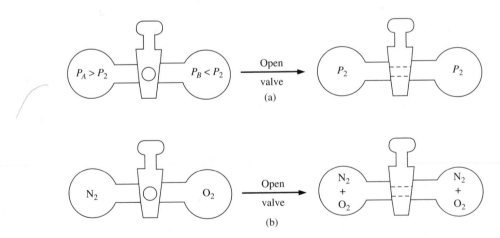

Fig. 3.4 The second law states that (a) pressures tend to become uniform, and (b) composition tends to become uniform.

These fluctuations in pressure are too small to be measured by a pressure gauge, but they can be measured indirectly by other methods, such as light scattering. Similarly, a system of uniform composition will have fluctuations in composition in any volume of the system. If, for a very short time, we watched a system containing 10 N_2 molecules and 10 O_2 molecules, we might notice that molecules segregated spontaneously so that all 10 molecules of N_2 were on one side of the system and all 10 molecules of O_2 were on the other side. We could then quickly say that this observation disproved the second law. We would have to say it very quickly because if we watched the system for a reasonable time, we would find that *on the average* the 20 molecules are randomly distributed.

The second law therefore does not apply to fluctuations. Fluctuations will be important only for very small systems and very short times. For large numbers of molecules and for small numbers of molecules observed for a long time, the second law always applies. This means that thermodynamics is a macroscopic theory; it applies to matter in bulk. Even if molecules were shown not to exist, the laws of thermodynamics would remain unchanged. However, we do believe in the existence of molecules and we can estimate the probability that a process will occur which is not consistent with the second law. From the relation between entropy and disorder [Eq. (3.9)], one can show that the probability of observing a violation of the second law depends exponentially on the number of molecules, N, in the system. The probability that you will observe a change contrary to the second law in a system containing N molecules is e^{-N}. We thus find that for five molecules, the probability of contradicting the second law is about $1/100$. For 10 molecules it is $1/10^4$; for 20 molecules, it is $1/10^8$; and for 100 molecules, it is $1/10^{43}$. Furthermore, observing five molecules for a long time is equivalent to observing a system of many molecules. Therefore, the second law is actually a very reliable and useful generalization, in spite of its slight limitations.

CHEMICAL REACTIONS

So far we have been discussing processes and changes that have not involved chemical reactions. One way to obtain the entropy for a chemical reaction is to know the entropies of the reactants and products. For a general reaction

$$n_A A + n_B B \longrightarrow n_C C + n_D D$$

at a chosen T and P, the entropy change is

$$\Delta S = n_C \bar{S}_C + n_D \bar{S}_D - n_A \bar{S}_A - n_B \bar{S}_B$$

where $\bar{S}_A$ is the entropy mol^{-1} of compound A at T and P. The standard molar entropies of compounds can be obtained from tables (see Appendix) which give the entropy mol^{-1} for some compounds at 25°C and 1 atm pressure. Note that, by contrast with the standard enthalpies, the entropies of the elements are not equal to zero at 25°C and 1 atm. We remember that by convention the enthalpies of the elements are taken as zero at 25°C and 1 atm. The values of the entropy given in the Appendix for both elements and compounds were obtained by using the third law of thermodynamics.

Example 3.2   Calculate the entropy change at 25°C and 1 atm for the decomposition of 1 mol of liquid water to form H_2 and O_2 gas.

Solution   The reaction for 1 mol of H_2O is

$$H_2O(l) \longrightarrow H_2(g) + \tfrac{1}{2} O_2(g)$$

The entropy change at 25°C and 1 atm is

$$\Delta S^0(25°C) = \tfrac{1}{2} \overline{S}^0_{O_2(g)} + \overline{S}^0_{H_2(g)} - \overline{S}^0_{H_2O(l)}$$

$$= \tfrac{1}{2}(205.14) + 130.68 - 69.91$$

$$= 163.34 \text{ J K}^{-1}$$

The entropy change is positive, which is consistent with our expectations for a reaction that involves the formation of two gases from one liquid. There is an increase
in disorder.

THIRD LAW OF THERMODYNAMICS

The *third law* states that the entropy of all pure, perfect crystals is zero at a temperature of
0 K (absolute zero):

$$S_A(0 \text{ K}) \equiv 0$$

where A is any pure, perfect crystal. The third law is an experimental one, as the first two
laws are, but we can understand it in terms of the relation of entropy and disorder. Near
0 K the disorder of a substance can approach zero; the number of microstates approaches
1. For this to happen, the substance must be crystalline; a liquid or a gas is still disordered
near 0 K.* Also, the substance must be pure; in a mixture the entropy could be reduced by
separating the mixture. So, for a perfect crystal of any pure compound, we know the entropy at absolute zero. We can obtain the entropy at any other temperature if we know how
entropy changes with temperature.

Temperature Dependence of Entropy

The entropy change for heating or cooling a system is easy to calculate. The heating or
cooling can be done essentially reversibly so that the entropy change is

$$S_2 - S_1 = \int \frac{dq_{\text{rev}}}{T} = \int_{T_1}^{T_2} \frac{C \, dT}{T}$$

At constant P,

$$S_2 - S_1 = \int_{T_1}^{T_2} \frac{C_P \, dT}{T} \tag{3.14}$$

* One notable exception is liquid helium (F. London, *Superfluids*, Vol. II, Dover, New York, 1964).

$$S_2 - S_1 = C_P \ln \frac{T_2}{T_1} \qquad \text{(if } C_P \text{ is independent of } T) \qquad (3.15)$$

where C_P is the heat capacity at constant P. At constant V,

$$S_2 - S_1 = \int_{T_1}^{T_2} \frac{C_V \, dT}{T} \qquad (3.16)$$

$$S_2 - S_1 = C_V \ln \frac{T_2}{T_1} \qquad \text{(if } C_V \text{ is independent of } T) \qquad (3.17)$$

where C_V = heat capacity at constant V. Because C_P and C_V are always positive, Eqs. (3.15) and (3.17) show that raising the temperature will always increase the entropy; this is expected from the increase in disorder.

Example 3.3 Calculate the change in entropy at constant P when 1 mol of liquid water at 100°C is brought in contact with 1 mol of liquid water at 0°C. Assume that the heat capacity of liquid water is independent of temperature and is equal to 75 J mol^{-1} deg^{-1}. No heat is lost to the surroundings.

Solution   As we bring equal amounts of the water in contact, C_P is constant, and no heat is lost; the first law tells us that the final temperature of the mixture will be the average of the two temperatures, 50°C. We use Eq. (3.15) to calculate the change in entropy of the hot water and of the cold water and add the results:

Hot water:

$$S(50°C) - S(100°C) = \bar{C}_P \ln \frac{323}{373}$$

$$= -10.79 \text{ J K}^{-1}$$

Cold water:

$$S(50°C) - S(0°C) = \bar{C}_P \ln \frac{323}{273}$$

$$= 12.61 \text{ J K}^{-1}$$

Sum for $H_2O(100°C) + H_2O(0°C) \longrightarrow 2 H_2O(50°C)$:

$$S_2 - S_1 = 1.82 \text{ J K}^{-1}$$

The entropy change is positive, as it must be for a spontaneous process in an isolated system.

Temperature Dependence of the Entropy Change for a Chemical Reaction

To calculate the entropy change for a chemical reaction at 1atm and some temperature other than 25°C, we can use Eq. (3.15). We consider a cycle in which products and reactants are heated or cooled to the new temperature and then add the entropy change for the heating or cooling to $\Delta S^0 (25°C)$:

$$A\,(T_2) \quad \xrightarrow{\quad \Delta S^0(T_2) \quad} \quad B\,(T_2)$$

$$A\,(25°C) \quad \xrightarrow{\quad \Delta S^0(25°C) \quad} \quad B\,(25°C)$$

$$\Delta S^0(T_2) = \Delta S^0(25°C) + \int_{T_2}^{298} C_P(A)\,\frac{dT}{T} + \int_{298}^{T_2} C_P(B)\,\frac{dT}{T}$$

We can generalize the result and rewrite the equation in a more compact form by using the identity

$$\int_a^b = -\int_b^a$$

$$\Delta S^0(T_2) = \Delta S^0(T_1) + \int_{T_1}^{T_2} \Delta C_P\,\frac{dT}{T} \tag{3.18}$$

where $\Delta C_P = C_P\,(\text{products}) - C_P\,(\text{reactants})$.

Example 3.4 If a spark is applied to a mixture of $H_2(g)$ and $O_2(g)$, an explosion occurs and water is formed. Calculate the entropy change when 2 mol of gaseous H_2O is formed at 100°C and 1 atm from $H_2(g)$ and $O_2(g)$ at the same temperature and each at a partial pressure at 1 atm.

Solution The reaction for 2 mol of H_2O is

$$2\,H_2(g) + O_2(g) \longrightarrow 2\,H_2O(g)$$

The entropy change at 25°C, 1 atm, is

$$\Delta S^0(25°C) = 2\bar{S}^0_{H_2O(g)} - \bar{S}^0_{O_2(g)} - 2\bar{S}^0_{H_2(g)}$$

$$= 2(188.72) - 205.04 - 2(130.57)$$

$$= -88.74\ \text{J K}^{-1}$$

To find $\Delta S^0 (100°C)$, we need to use Eq. (3.18); therefore, we need to know the heat capacities of $H_2(g)$, $O_2(g)$, and $H_2O(g)$. They can be taken as constants over the temperature range 25°C to 100°C with the values given in Table 2.1.

$$\Delta C_P^0 = 2\bar{C}^0_{PH_2O(g)} - \bar{C}^0_{PO_2(g)} - 2\bar{C}^0_{PH_2(g)}$$

$$= 2(33.6) - 29.4 - 2(28.8)$$

$$= -19.8\ \text{J K}^{-1}$$

From Eq. (3.18),

$$\Delta S^0(100°C) = \Delta S^0(25°C) + \int_{298}^{373} \Delta C_P \frac{dT}{T}$$

$$= \Delta S(25°C) + \Delta C_P \ln \frac{373}{298}$$

$$= -88.74 - (19.8)(0.224)$$

$$= -93.18 \text{ J K}^{-1}$$

The entropy of the system decreased; the second law thus requires that the entropy of the surroundings must have increased. The reaction is exothermic and heat is lost to the surroundings at constant temperature.

Entropy Change for a Phase Transition

On heating many compounds from 0 K to room temperature, various phase transitions, such as melting and boiling, may occur. The reversible heat absorbed divided by the *equilibrium* temperature of the transition gives the entropy change for the phase transition. It is important to stress equilibrium transition temperature. Otherwise, the transition is not reversible and the heat absorbed is not the reversible heat. For a phase transition, at constant P and T,

$$q_{\text{rev}} = \Delta H_{\text{tr}} \tag{2.37}$$

$$\Delta S_{\text{tr}} = \frac{\Delta H_{\text{tr}}}{T_{\text{tr}}} \tag{3.19}$$

where ΔS_{tr} = entropy of phase transition
ΔH_{tr} = entropy of phase transition at equilibrium temperature T_{tr}

Third law entropies or absolute entropies, such as those tabulated in the Appendix, are obtained by adding together the (1) entropy of the substance at 0 K (which will be zero if it is a perfect crystal), (2) the entropy increase associated with increasing the temperature, and (3) the entropy changes associated with any phase changes. For example, to obtain the third law entropy of a liquid compound at 25°C and 1 atm, we would use the equation

$$\overline{S}^0(25°C, 1 \text{ atm}) = S^0(0 \text{ K}) + \int_0^{T_m} \overline{C}_P(s) \frac{dT}{T} + \frac{\Delta H_m^0}{T_m} + \int_{T_m}^{298} \overline{C}_P(l) \frac{dT}{T}$$

where T_m is the melting temperature at 1 atm and we have assumed that there are no solid-solid transitions. If there were, we would add $(\Delta H_{\text{tr}}^0/T_{\text{tr}})$ for each transition and use an appropriate $\overline{C}_P$ for each solid phase.

Pressure Dependence of Entropy

The pressure dependence of entropy will not be so important for us as the temperature dependence. Raising the pressure usually lowers the entropy. For solids and liquids we will ignore the direct effect of pressure on entropy:

$$S(P_2) - S(P_1) \cong 0 \tag{3.20}$$

For gases we will approximate the effect by the effect of pressure at constant temperature on an ideal gas:

$$S(P_2) - S(P_1) = -nR \ln \frac{P_2}{P_1} \tag{3.21}$$

(The origin of this equation is given at the end of the chapter.) We notice in Eq. (3.21) that if either P_2 or P_1 is zero, the logarithm and the entropy become infinite. This should not surprise us because zero pressure means infinite volume, and infinite volume implies infinite disorder of the molecules in the volume. The point to remember is that as the pressure of a gas decreases at constant temperature, the entropy will increase.

Equations (3.20) and (3.21) could be used to calculate the entropy change for a chemical reaction at some pressure other than 1 atm.

Spontaneous Chemical Reactions

We began this chapter by wondering if pencil lead (graphite) could be converted into diamonds. One way to answer this question is to learn if the entropy of the universe increases when the reaction occurs. We can easily learn from the Appendix whether the entropy of the system increases, but it is not so easy to learn about the change in entropy of the surroundings. *Entropy is a useful criterion for spontaneity in isolated systems.* Then the surroundings do not change, and we can limit our attention to the system. If neither the energy nor the volume of the system changes during a reaction, the sign of the entropy change tells whether the reaction can occur. Very few reactions occur at constant energy and volume. Racemization of an optically active molecule to a racemic mixture is an example. The D form and the L form have identical energies, volumes, and entropies, but the mixture has a larger entropy and nearly the same energy and volume. The reaction is spontaneous. Other examples are the mixing of two components to form an ideal solution and the flow of heat from a hot object to a cold one.

GIBBS FREE ENERGY

For most reactions there are large changes in energy and enthalpy, and we are interested in whether the reaction will occur at some constant T and P. These are the conditions under which most experiments are done in the laboratory: the system is free to exchange heat with the surroundings to remain at room temperature, and it can expand or contract in volume to remain at atmospheric pressure. We need a criterion of spontaneity that applies to the system for these conditions. A new thermodynamic variable of state, the Gibbs free energy, is useful in this situation. The *Gibbs free energy, G,* is an extensive variable of state defined as a combination of enthalpy, temperature, and entropy:

$$G \equiv H - TS \qquad (3.22)$$

The Gibbs free energy has the same units as enthalpy or energy; it also depends only on initial and final states. Its definition in Eq. (3.22) was chosen because it can thus characterize whether a process will occur spontaneously at constant temperature and pressure. If ΔG for a reaction at constant T and P is negative, the reaction can occur spontaneously; if ΔG at constant T and P is positive, the reaction will not occur spontaneously; if ΔG at constant T and P is zero, the reaction is at equilibrium.

The *molar Gibbs free energy* (Gibbs free energy per mole) is an intensive property that has special properties and is known as the *chemical potential,* μ. For pure substances and for mixtures of ideal gases

$$\mu = \overline{G}$$

but for real gas mixtures and liquid solutions there is a more precise definition (see Chapter 4) that takes into account the concentration dependence of the free energies of the components. The chemical potential is a measure of the tendency of a chemical system to change by undergoing a reaction, forming a new phase, going into solution or similar transformations. When the chemical potential difference, $\Delta\mu$, between an initial and final state is negative, then the reaction or process can occur spontaneously at constant temperature and overall pressure. If $\Delta\mu$ is positive, the process will not occur spontaneously in the forward direction. For a system at equilibrium, $\Delta\mu = 0$, and no net change will occur. In this respect μ for a chemical system is analogous to gravitational potential or to electrical potential. Chemical systems have a tendency to move from states of high chemical potential to states of low chemical potential.

Spontaneous Reactions at Constant *T* and *P*

Let us derive the Gibbs free energy criterion for a spontaneous reaction at constant *T* and *P.* Combining Eqs. (3.12) and (3.13), we see that for any reaction at constant *T,*

$$S_2 - S_1 \geq \frac{q}{T} \qquad (3.23)$$

The equal sign applies to a reversible reaction; that is, one at equilibrium. The "greater than" sign applies to an irreversible reaction; that is, a spontaneous reaction. We can replace q by using the first law:

$$S_2 - S_1 \geq \frac{E_2 - E_1 - w}{T} \qquad \text{(constant } T\text{)}$$

If we consider only expansion and compression work, at constant P we have

$$S_2 - S_1 \geq \frac{E_2 - E_1 + P \cdot (V_2 - V_1)}{T} \qquad \text{(constant } T, P\text{)}$$

But the numerator of the right-hand side of the equation is $H_2 - H_1$ at constant P:

$$S_2 - S_1 \geq \frac{H_2 - H_1}{T} \qquad \text{(constant } T, P) \qquad (3.24)$$

Rearranging, we obtain

$$T \cdot (S_2 - S_1) - (H_2 - H_1) \geq 0$$

$$(H_2 - H_1) - T \cdot (S_2 - S_1) \leq 0$$

The left-hand side of the equations is just the difference in Gibbs free energy between final and initial states at the same temperature, as seen from Eq. (3.22). We thus obtain the desired criterion for a spontaneous reaction at *constant temperature and pressure*:

$$\Delta G < 0 \qquad \text{(a spontaneous reaction)} \qquad (3.25)$$

$$\Delta G = 0 \qquad \text{(an equilibrium reaction)} \qquad (3.26)$$

$$\Delta G > 0 \qquad \text{(no spontaneous reaction)} \qquad (3.27)$$

Calculation of Gibbs Free Energy

The Gibbs free energy change for a reaction at constant temperature can be obtained from the enthalpy and entropy changes. At constant temperature,

$$\Delta G = \Delta H - T\,\Delta S \qquad (3.28)$$

The Appendix gives values of $\bar{H}^0 \equiv \Delta H_f^0$ and $\bar{S}^0$ for various substances at 25°C and 1 atm. These values can therefore be used to calculate ΔH^0 and ΔS^0, and hence ΔG^0, at 25°C and 1 atm.

Values of ΔG_f^0, the standard free energy of formation, of various substances are also tabulated in the Appendix. The molar standard free energy of formation is defined as the free energy of formation of 1 mol of any compound at 1 atm pressure from its elements in their standard states at 1 atm. In a manner completely analogous to our discussion on standard enthalpy of formation, we arbitrarily assign the elements in their most stable state at 1 atm to have zero free energy.

Example 3.5 Calculate the Gibbs free energy for the following reaction at 25°C and 1 atm. Will the reaction occur spontaneously?

$$H_2O(l) \longrightarrow H_2(g) + \tfrac{1}{2}O_2(g)$$

Solution

$$\Delta G^0(25°C) = \bar{G}_{H_2(g)}^0 + \tfrac{1}{2}\bar{G}_{O_2(g)}^0 - \bar{G}_{H_2O(l)}^0$$

$$= 0 + 0 - (-237.19)$$

$$= 237.19 \text{ kJ mol}^{-1}$$

The Gibbs free energy change is just the negative of the Gibbs free energy of formation of liquid water.

Alternatively, we can calculate ΔH^0 and ΔS^0 from the Appendix:

$$\Delta H^0 = \bar{H}^0_{H_2(g)} + \tfrac{1}{2}\bar{H}^0_{O_2(g)} - \bar{H}^0_{H_2O(l)}$$

$$= 0 + 0 - (-285.83)$$

$$= 285.83 \text{ kJ mol}^{-1}$$

$$\Delta S^0 = \bar{S}^0_{H_2(g)} + \tfrac{1}{2}\bar{S}^0_{O_2(g)} - \bar{S}^0_{H_2O(l)}$$

$$= 130.57 + \tfrac{1}{2}(205.04) - 69.95$$

$$= 163.14 \text{ J K}^{-1} \text{ mol}^{-1}$$

ΔG^0 can then be obtained:

$$\Delta G^0 = \Delta H^0 - T\,\Delta S^0$$

$$= 285.83 \text{ kJ mol}^{-1} - (298)(163.14 \times 10^{-3}) \text{ kJ mol}^{-1}$$

$$= 237.19 \text{ kJ mol}^{-1}$$

The same answer is obtained. The reaction will *not* occur spontaneously, because ΔG is positive.

A proposed method of storing solar energy is to use sunlight to decompose water. The sunlight provides the driving force to overcome the large positive free energy. The hydrogen and oxygen gas produced make an excellent fuel. The light reactions of green plant photosynthesis involve a positive free energy change that is almost exactly the same as that for the reaction to make an equal amount of oxygen gas by the direct decomposition of water. The difference is that green plants do not make hydrogen gas; instead, they reduce carbon dioxide to carbohydrates and other products. The driving force that provides for this large increase in chemical potential is the sunlight absorbed by chlorophyll and the other plant pigments.

Example 3.6   We wish to know whether proteins in aqueous solution are unstable with respect to their constituent amino acids. As an example, let us calculate the standard free energy of hydrolysis for the dipeptide glycylglycine at 25°C and 1 atm in dilute aqueous solution.

Solution   The reaction is

$$^+H_3NCH_2CONHCH_2COO^-(aq) + H_2O(l) \longrightarrow 2\ ^+H_3NCH_2COO^-(aq)$$

glycylglycine   glycine

The standard free energy change when solid glycine dissolves has been measured and is small. We shall assume that this is also true for solid glycylglycine. Therefore, we use the free energy values in the Appendix for solid glycine and glycylglycine in the following calculation.

$$\Delta G^0(25°C) = 2\bar{G}^0(\text{glycine, } s) - \bar{G}^0(\text{glycylglycine, } s) - \bar{G}^0(H_2O, l)$$

$$= 2(-377.69) - (-490.57) - (-237.19)$$

$$= -27.62 \text{ kJ mol}^{-1}$$

The Second Law: The Entropy of the Universe Increases   **91**

The reaction is spontaneous, but luckily it normally occurs slowly. However, appropriate catalysts such as proteolytic enzymes can cause the reaction to occur rapidly. If the catalyst found its way into your bloodstream, the effects would be very unpleasant.

It is clear from Eq. (3.28) that ΔG depends explicitly on T. Both ΔH and ΔS may be independent of T, but ΔG will depend on T with slope equal to $-\Delta S$. Often, as an approximation for temperatures not too different from 25°C, Eq. (3.28) can be used to calculate ΔG at other temperatures:

$$\Delta G(T) \cong \Delta H(25°C) - T \cdot \Delta S(25°C) \qquad (3.29)$$

For example, values of ΔH and ΔS at 25°C from the Appendix can be used to calculate the approximate free energy of the reaction at physiological temperatures. A useful equation that is easily derived from Eq. (3.29) is

$$\Delta G(T) - \Delta G(25°C) \cong -(T - 298) \cdot \Delta S(25°C) \qquad (3.30)$$

This equation shows that the sign of ΔS for a reaction indicates how ΔG will change with temperature. If ΔS is negative, ΔG increases with increasing temperature. Another useful equation is

$$\frac{\Delta G(T)}{T} - \frac{\Delta G(25°C)}{298} \cong \left(\frac{1}{T} - \frac{1}{298} \right) \cdot \Delta H(25°C) \qquad (3.31)$$

which can be derived from Eqs. (3.28) and (3.30). These equations can be used to calculate the temperature dependence of ΔG, or they can obviously be used to calculate ΔH and ΔS if the temperature dependence of ΔG is known. It is clear that Eqs. (3.30) and (3.31) apply to any two temperatures; the approximation involves the assumption that ΔH and ΔS are independent of temperature. If ΔH or ΔS depend greatly on temperature, Eqs. (3.30) and (3.31) must be replaced by integrals. The most useful one to use is

$$\frac{\Delta G(T_2)}{T_2} - \frac{\Delta G(T_1)}{T_1} = - \int_{T_1}^{T_2} \frac{\Delta H(T)}{T^2} \, dT \qquad (3.32)$$

called the *Gibbs-Helmholtz equation*.

Example 3.7 What is the free energy of hydrolysis of glycylglycine at 37°C and 1 atm?

Solution We can use Eq. (3.30) to find $\Delta G^0(37°C)$ if we assume that ΔS is independent of temperature. Or we can use Eq. (3.31) if we assume that ΔH is independent of temperature. We shall use both equations and thus show that the two equations are indeed equivalent. Because the temperature range is small, the constancy of ΔH and ΔS is a good approximation. To calculate ΔS and ΔH, we use the data for the solid compounds.

$$\Delta S^0(25°C) = 2\bar{S}^0(\text{glycine}) - \bar{S}^0(\text{glycylglycine}) - \bar{S}^0(H_2O, l)$$

$$= 2(103.51) - 190.0 - 69.95$$

$$= -52.9 \text{ J K}^{-1} \text{ mol}^{-1}$$

$$\Delta H^0(25°C) = 2(-537.2) - (-745.25) - (-285.83)$$

$$= -43.32 \text{ kJ mol}^{-1}$$

With Eq. (3.30),

$$\Delta G^0(37°C) = \Delta G^0(25°C) - 12 \cdot \Delta S^0(25°C)$$

$$= -27.62 \text{ kJ mol}^{-1} - (12 \text{ K})(-0.0529 \text{ kJ K}^{-1} \text{ mol}^{-1})$$

$$= -26.98 \text{ kJ mol}^{-1}$$

With Eq. (3.31),

$$\frac{\Delta G^0(37°C)}{310} - \frac{\Delta G^0(25°C)}{298} = \left(\frac{1}{310} - \frac{1}{298}\right) \cdot \Delta H^0(25°C)$$

$$\frac{\Delta G^0(37°C)}{310} = \frac{-27.62}{298} + (-1.299 \times 10^{-4})(-43.32)$$

$$= -0.0871$$

$$\Delta G^0(37°C) = -26.99 \text{ kJ mol}^{-1}$$

Both methods lead to a decrease in the magnitude of the free energy of hydrolysis and agree well.

Pressure Dependence of Gibbs Free Energy

From values in the Appendix we can calculate the Gibbs free energy change ΔG^0, exactly at 25°C and approximately at any temperature. However, the tabulated values of ΔG^0 are all defined for 1 atm pressure. To calculate free energy at some other pressure we must know its dependence on pressure. The change of Gibbs free energy with pressure at constant temperature is directly proportional to the volume:

$$G(P_2) - G(P_1) = \int_{P_1}^{P_2} V \, dP \tag{3.33}$$

(We will discover the origins of this equation later in this chapter.) If the volume is independent of pressure, it can be taken out of the integral. This is usually a good approximation for a solid or liquid:

$$G(P_2) - G(P_1) \cong V \cdot (P_2 - P_1) \tag{3.34}$$

For a gas we use the equation of state to write V as a function of P. For an ideal gas, and approximately for any gas, we can substitute the ideal gas equation in Eq. (3.33):

$$G(P_2) - G(P_1) = \int_{P_1}^{P_2} \frac{nRT}{P} \, dP = nRT \ln \frac{P_2}{P_1} \tag{3.35}$$

To calculate the effect of pressure on the free energy of a chemical reaction, we just apply Eqs. (3.34) and (3.35) to each product and reactant. If all products and reactants are solids or liquids, we use Eq. (3.34):

$$\Delta G(P_2) - \Delta G(P_1) = \Delta V \cdot (P_2 - P_1) \qquad (3.36)$$

where $\Delta V = V(\text{products}) - V(\text{reactants})$. Equation (3.36) shows that if the volume of products is greater than the volume of reactants (ΔV is positive), increasing the pressure will increase the free energy of the reaction. If at least one gaseous product or reactant is involved, we can ignore the volume of the solids or liquids compared to that of the gases and use Eq. (3.35).

$$\Delta G(P_2) - \Delta G(P_1) = \Delta n R T \ln \frac{P_2}{P_1} \qquad (3.37)$$

where Δn is the number of moles of *gaseous* products minus the number of moles of *gaseous* reactants.

Example 3.8   Can graphite (pencil lead) be converted spontaneously into diamond at 25°C and 1 atm?

Solution   The reaction at 25°C and 1 atm

$$C(s, \text{graphite}) \longrightarrow C(s, \text{diamond})$$

has a Gibbs free energy of $+2.84$ kJ mol^{-1} from the Appendix. Therefore, the answer is "no"; graphite will not spontaneously convert into diamond at 25°C and 1 atm. The result does mean that diamond will spontaneously convert into graphite. Experimentally, the reaction has been found to be very slow, but a catalyst may eventually be found which can speed it up. There has not been much economic incentive to pursue this particular research.

Example 3.9   Will increasing the pressure favor the conversion of graphite to diamond? If so, what is the minimum pressure necessary to make this reaction spontaneous at 25°C? Will increasing the temperature favor the reaction?

Solution   We can use Eq. (3.36) to answer the first two questions. We need to know the volume change for the reaction. We will use the densities of graphite and diamond at 25°C to calculate the volumes and assume that the volumes are independent of pressure. Just knowing that diamond is denser than graphite allows us to conclude that ΔV for the reaction is negative ($\overline{V}$ of diamond is less than $\overline{V}$ of graphite). This means *that increasing the pressure* will decrease the free energy and *favor* the reaction graphite to diamond.

To calculate the minimum pressure necessary to allow the spontaneous reaction, we calculate the pressure that produces $\Delta G = 0$. The densities at 25°C and 1 atm are as follows:

	Density (g cm^{-3})
C (graphite)	2.25
C (diamond)	3.51

The molar volumes at 25°C and 1 atm are obtained by dividing the atomic weight of carbon by the densities:

	$\overline{V}$ (cm^3 mol^{-1})
C (graphite)	5.33
C (diamond)	3.42

For the reaction

$$C\,(graphite) \longrightarrow C\,(diamond)$$

the Gibbs free energy at 25°C and P atm is, from Eq. (3.36),

$$\Delta G(P) = \Delta G(1\text{ atm}) + \Delta V \cdot (P - 1)$$
$$= 2.84\text{ kJ mol}^{-1} + (3.42 - 5.33)\text{ cm}^3\text{ mol}^{-1} \cdot (P - 1)\text{ atm}$$

We must use the same units throughout our equation, so we convert cm^3 atm to kJ by multiplying by a ratio of gas constants, $R/R = 8.314 \times 10^{-3}$ kJ/82.05 cm^3 atm $= 1.0133 \times 10^{-4}$ kJ cm^{-3} atm^{-1}:

$$\Delta G(P) = 2.84 - 1.935 \times 10^{-4} \cdot (P - 1)\text{ kJ mol}^{-1}$$

We want to find the pressure that makes $\Delta G(P) = 0$.

$$0 = 2.84 - 1.935 \times 10^{-4} \cdot (P - 1)$$

$$P - 1 = \frac{2.84}{1.935 \times 10^{-4}}$$

$$P = 15{,}000\text{ atm}$$

The assumption that $\Delta\overline{V}$ is constant is not valid over this large pressure range, and the actual pressure needed is higher than 15,000 atm. Nevertheless, the effect is real, and small diamonds have been made in this way for many years for industrial uses.

To decide whether increasing temperature will favor the reaction, we have to know the sign of ΔS [see Eq. (3.30)]. From the Appendix we find that $\Delta S = 2.43 - 5.74 = -3.31$ J K^{-1} mol^{-1}; therefore, *increasing the temperature* will *not favor* the formation of diamond. We do not use Eq. (3.30) to calculate at what low temperature ΔG becomes zero, because ΔS will change significantly with temperature over the wide temperature change. The industrial process does use high temperatures, but for kinetic reasons. The increase in pressure provides the necessary free energy change.

Phase Changes

For a phase change that takes place at its equilibrium temperature and pressure, the change in Gibbs free energy is zero:

$$\Delta G = 0 \qquad \text{(at equilibrium)}$$

To calculate the Gibbs free energy at other temperatures and pressures, we can use

$$\Delta G = \Delta H - T \Delta S \qquad (3.28)$$

if ΔH and ΔS are known for the phase change at arbitrary T and P. Otherwise, we can use Eq. (3.30), (3.31), (3.35), or (3.36) for the change of ΔG with T and P.

HELMHOLTZ FREE ENERGY

For a process that takes place at constant T and V, the sign of the Gibbs free energy is not a criterion for equilibrium. A new thermodynamic variable of state is needed; it is called the *Helmholtz free energy, A*. The Helmholtz free energy is defined as

$$A \equiv E - TS \qquad (3.38)$$

and the criterion for equilibrium is, *at constant T and V,*

$$\Delta A < 0 \qquad \text{(spontaneous reaction)} \qquad (3.39)$$

$$\Delta A = 0 \qquad \text{(reaction at equilibrium)} \qquad (3.40)$$

$$\Delta A > 0 \qquad \text{(no spontaneous reaction)} \qquad (3.41)$$

We will not use the Helmholtz free energy very much in this book, although for reactions at constant volume it is as useful as the Gibbs free energy is for reactions at constant pressure. The Helmholtz free energy is widely used for geochemical problems, where the pressure may vary widely and the constant-volume restriction is more appropriate.

NONCOVALENT REACTIONS

The examples that we have given so far of the changes in H, S, and G for chemical reactions involve the breaking and formation of covalent bonds. Many important biochemical processes involve reactions in which no covalent bonds are made or broken; only weaker bonds and interactions are involved. Examples include the reaction of an antigen with its antibody, the binding of many hormones and drugs to nucleic acids and proteins, the reading of the genetic message (codon-anticodon recognition), denaturation of proteins and nucleic acids, and so on. Thermodynamics can help us to understand these reactions and to decide which proposed mechanisms are reasonable and which are not. Table 3.1 gives some measured enthalpies for simple reactions which illustrate the forces involved.

Charged species have ionic interactions. In a NaCl crystal, for example, very strong ionic interactions exist between the positively charged Na^+ ions and the negatively charged Cl^- ions. This is reflected in the large positive enthalpy change when solid NaCl is separated into Na^+ and Cl^- ions:

$$NaCl(s) \longrightarrow Na^+(g) + Cl^-(g) \qquad \Delta H^0_{298} = 785 \text{ kJ mol}^{-1}$$

In spite of such strong ionic forces, solid NaCl nevertheless dissolves in water easily, with a small ΔH^0_{298} of only 4 kJ mol^{-1}:

$$NaCl(s) + \infty H_2O(l) \longrightarrow Na^+(aq) + Cl^-(aq) \qquad \Delta H^0_{298} = 4 \text{ kJ mol}^{-1}$$

The reason is that there are strong interactions between the charged ions and the water molecules, so the net ΔH change is small.

The electronic distribution in an uncharged water molecule

is such that the bonding electrons are localized more on the oxygen atom than on the hydrogen atoms. Therefore, the oxygen atom is slightly negative and the hydrogen atoms are slightly positive. Since the molecule is not linear, the geometric centers of the positive charges and the negative charges do not coincide. We say that H_2O has an electric dipole. The extent of charge separation is expressed in terms of the dipole moment. If two point charges $+e$ and $-e$ are separated by a distance r, the dipole moment is er. Experimentally, the dipole moment of H_2O is found to be equivalent to an electron and a unit positive charge separated by 0.38 angstrom (1 Å = 10^{-10} m). If water is modeled as three charges on the three atoms, the charge on the oxygen atom is -0.834 electronic charges and each hydrogen atom is $+0.417$. Interactions between a charged ion and a neutral molecule with a dipole moment are called *ion-dipole interactions*. Ions can also interact with neutral molecules with zero dipole moment. Take a molecule of CCl_4, for example. Though the bonding electrons are localized more on the chlorine atoms, the *permanent dipole moment* of the molecule is zero because the four Cl atoms are symmetrically located at the four corners of a tetrahedron, with the C atom occupying the center of the tetrahedron. However, if a charge is placed near a CCl_4 molecule, the charge will distort the electronic distribution. We say that the CCl_4 molecule becomes *polarized.* The centers of the positive and negative charges in the polarized CCl_4 molecule no longer coincide, and the molecule has now an *induced dipole moment*. Interactions between a charged ion and polarized molecules are called *charge-induced dipole interactions.*

Induced dipole-induced dipole interactions exist even between neutral molecules with no permanent dipole moments. This is because of fluctuations in the electronic distributions in the molecules. A molecule with no permanent dipole moment may acquire an instantaneous dipole moment because of a fluctuation. This instantaneous dipole can induce a dipole in a neighboring molecule. Interactions between such fluctuation dipole-induced dipoles are called *London interactions.* London interaction is always present and is always an attractive force between molecules. It is the only attractive force acting between identical rare gas atoms. It is responsible for the 8 kJ mol^{-1} enthalpy change necessary to vaporize solid argon to gaseous argon, for example. The force depends on the *polarizability* of the interacting molecules, which is a measure of how easy it is to distort the electron clouds of the molecule. In addition to the London interaction, uncharged molecules have van der Waals interactions. *Van der Waals interactions* include permanent dipole-permanent dipole interactions, permanent dipole-induced dipole attractions, and steric repulsions. The London-van der Waals interactions are usually nonspecific forces which contribute to the energies of all reactions. For example, the heat of vaporization of liquid *n*-butane is 21.5 kJ

Table 3.1 Some enthalpies of noncovalent bonds and interactions*

Reaction	Characteristic interaction	ΔH^0 (kJ mol^{-1})
$Na^+(g) + Cl^-(g) \rightarrow NaCl(s)$	Ionic	-785
$NaCl(s) + \infty H_2O(l) \rightarrow Na^+(aq) + Cl^-(aq)$	Ionic and ion-dipole	4
$Argon(g) \rightarrow argon(s)$	London	-8
n-Butane$(g) \rightarrow n$-butane(l)	London-van der Waals	-20
$Acetone(g) \rightarrow acetone(l)$	London-van der Waals	-30

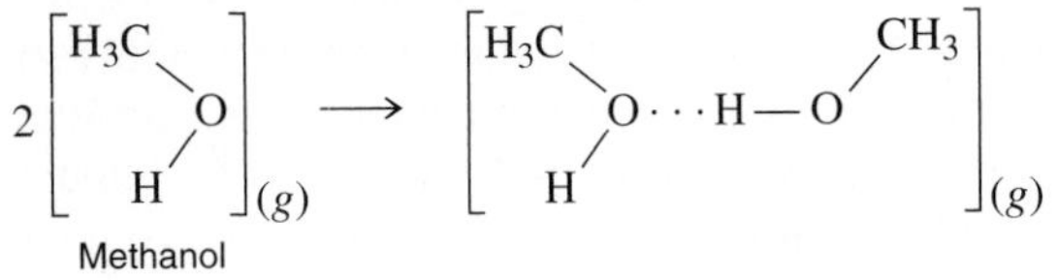

Methanol

Hydrogen bond — -20

Ammonia

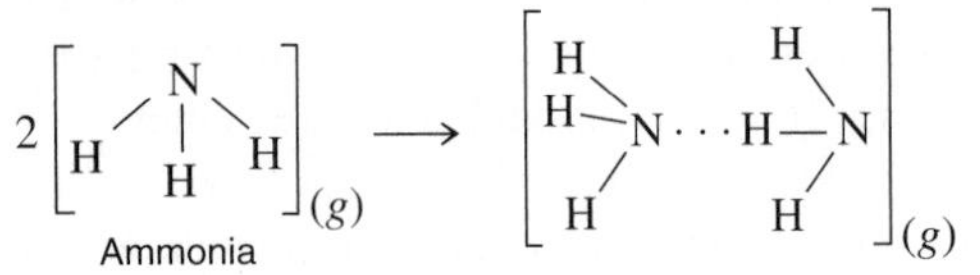

Hydrogen bond — -15

N-Methyl formamide

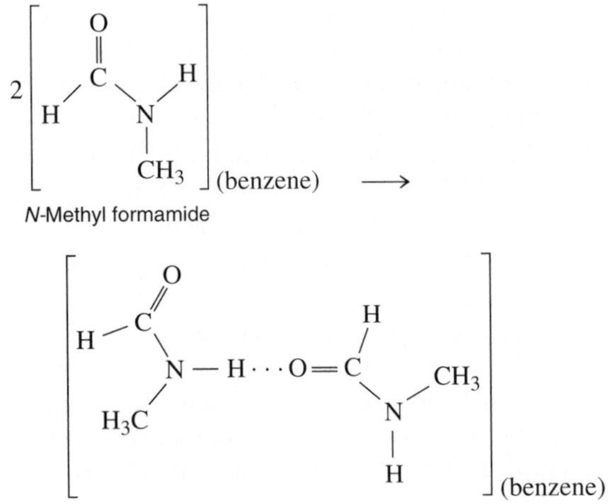

Hydrogen bond — -15

mol^{-1}. The London-van der Waals forces can become specific when careful fitting of molecules is required. Binding of a particular substrate to an enzyme, antigen-antibody binding, and the function of specific membrane lipids may be dominated by London-van der Waals interactions.

The *hydrogen bond* is one of the important bonds that determines the three-dimensional structures of proteins and nucleic acids. A hydrogen atom covalently attached to one oxygen or nitrogen can form a weak bond to another oxygen or nitrogen. For O—H $\cdots$ O, N—H $\cdots$ N, and N—H $\cdots$ O hydrogen bonds, the bond enthalpy of 15 to 20 kJ mol^{-1} can

Table 3.1 Some enthalpies of noncovalent bonds and interactions *(cont.)*

Reaction	Characteristic interaction	ΔH^0 (kJ mol^{-1})
[urea structure with N—H⋯O hydrogen bond] (aq) + [urea structure with O—H⋯O] (aq) ⟶		
[urea⋯O=C structure N—H⋯O=C] (aq) + [H₂O⋯O—H⋯O] (aq)	Hydrogen bond (aqueous)	−5
$C_3H_6(l) + \infty H_2O(l) \rightarrow C_3H_6(aq)$	Hydrophobic	−10
Benzene$(l) + \infty H_2O(l) \rightarrow$ benzene(aq)	Hydrophobic	0

*The enthalpies were obtained near room temperature except for the vaporization of solid argon; the standard state is the dilute solution extrapolated to 1 *M*. Data are from various sources and are rounded to the nearest integer. For the precise values, see: G. C. Pimentel and A. L. McClellan, *The Hydrogen Bond*, W. H. Freeman, San Francisco, 1960; W. Kauzmann, *Adv. Protein Chem. 14*, 1 (1959); and J. A. A. Ketalaar, *Chemical Constitution*, Elsevier, Amsterdam, 1958.

be compared with values of about 400 kJ mol^{-1} for covalent bonds. This weak bond becomes even weaker in aqueous solution, *because of competition between solute-solute hydrogen bonds and solute-water and water-water hydrogen bonds.* Between urea molecules in water the hydrogen-bond strength is just 5 kJ mol^{-1}. Urea can be considered as a model for a peptide bond, so this is the magnitude of the heat that is expected for breaking the hydrogen bond involving the peptide links in a protein. Therefore, the free energy for breaking a peptide hydrogen bond in aqueous solution at 25°C is expected to be close to zero.

Hydrophobic Interactions

One type of interaction that is important in aqueous solutions is the *hydrophobic* (fear-of-water) interaction. Water molecules have a strong attraction for each other, primarily as a consequence of hydrogen-bond formation. The oxygen atom of most molecules of liquid water is hydrogen-bonded to two hydrogen atoms of two other H_2O molecules, and the hydrogen atoms of most molecules are hydrogen-bonded to the oxygen atoms of two other water molecules. Therefore, the molecules of liquid water form a mobile network: most water molecules are primarily interacting, through hydrogen bonds, with four tetrahedrally oriented neighbors. The network is not a rigid one, and change of neighbors occurs rapidly because of thermal motions.

Let us consider what happens if a molecule such as propane (C_3H_8) is introduced into this network. A hole is created; some hydrogen bonds in the original network are broken. The C_3H_8 does not interact with water strongly; it does not form hydrogen bonds. The water molecules around the C_3H_8 molecule must orient themselves in a way which reforms the hydrogen bonds that were disrupted by the hydrocarbon molecule. The net result is that water molecules around the C_3H_8 actually become more ordered. Since there is little change in the number of hydrogen bonds, the enthalpy change is small. The ordering of water molecules around the hydrocarbon molecule, however, is associated with a negative entropy change. These interpretations are consistent with experimental data:

$$C_3H_8(l) \longrightarrow C_3H_8(aq)$$

$$\Delta H_{298}^0 \approx -8 \text{ kJ mol}^{-1}$$

$$\Delta S_{298}^0 \approx -80 \text{ J K}^{-1} \text{ mol}^{-1}$$

$$\Delta G_{298}^0 \approx +16 \text{ kJ mol}^{-1}$$

Now let us consider two such hydrocarbon groups, R. Each separate group when exposed to liquid water will, from the data above, cause an unfavorable free energy change. If the two groups cluster together, the disruptive effect on the solvent network will be less than the combined effects of two separate groups. Therefore, the association of the groups will be thermodynamically favored:

$$\underset{\text{separate}}{R(aq) + R(aq)} \longrightarrow \underset{\text{clustered}}{R-R(aq)}$$

The clustering of the groups is not because they like each other, but because they are both disliked by water. The clustered arrangement results in a decrease in the overall free energy of the system in comparison with the separate dispersion of unlike molecules. Such hydrophobic interactions are important in many biological systems. For example, hydrocarbon groups in a water-soluble protein are usually found to cluster in the interior of the protein. Similarly, polar lipid molecules form bilayer sheets or membranes in water, in which the hydrocarbon portions are buried inside and the polar or charged portions are on the surface, exposed to the water. Such molecules are called *amphiphilic*. Hydrophobic interactions are characterized by low enthalpy changes and are entropy-driven. We should note that hydrophobic interaction is a term that we use to describe the combined effects of London, van der Waals, and hydrogen-bonding interactions in certain processes in aqueous solutions; it is not a "force" different from the others we have discussed. In particular, there is no such thing as a hydrophobic bond.

Proteins and Nucleic Acids

Proteins are polypeptides of 20 naturally occurring amino acids; the structures of the amino acids are given in Appendix A.9. The amino acids are linked by amide bonds from the carboxyl group of one amino acid to the amino group of the next. The sequence of the polypeptide is called the primary structure. The polypeptides fold into different secondary structures held together by hydrogen bonds and other noncovalent interactions. The amide group

$$
\begin{array}{c}
O \qquad\qquad C \\
\backslash\backslash \qquad\quad / \\
C-N \\
/ \qquad\quad \backslash \\
C \qquad\qquad H
\end{array}
$$

is planar (because of partial double bond character of the central C—N bond) and has a *trans* conformation as shown. The different folded conformations of a polypeptide depend on rotation about the two single bonds attached to the amide group, as shown in Fig. 3.5. The α-helix is a common secondary structural element in proteins; its structure is shown in Fig. 3.6. It is a right-handed helix with hydrogen bonds between the C=O of each amide to the N—H of the amide four units ahead.

Nucleic acids are polynucleotides of four naturally occurring nucleotides, with structures as given in Appendix A.9. The nucleotides are composed of a sugar group connected to one of four bases and to a phosphate group. The nucleotides are linked by phosphodiester bonds (see Fig. 3.7b). The sequence of the bases in the polynucleotide is the primary structure. In DNA two strands form a double helix held together by Watson-Crick base pairs as shown in Fig. 3.7a. The stacking of the base pairs, one above the other, plus the hydrogen bonds between bases provide the stabilizing enthalpy and free energy of the helix. Both cross-strand and same-strand London-van der Waals interactions among the bases are important. The charged phosphate groups repel each other by Coulomb's law repulsion between like charges. The balance of these forces as well as interactions with the solvent determine the conformation of the double helix. Two different double helices—secondary structures—are shown in Fig. 3.8. The B-form helix is the one found for DNA

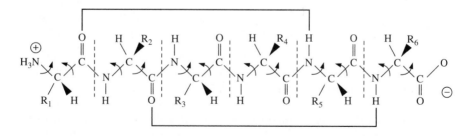

Fig. 3.5 A polypeptide chain showing the hydrogen bonding between the C=O and the NH groups in an α-helix. Residues are numbered starting from the amino terminal; the residues are separated by dotted lines. Arrows show the two bonds (on either side of each α-carbon) around which rotation can occur.

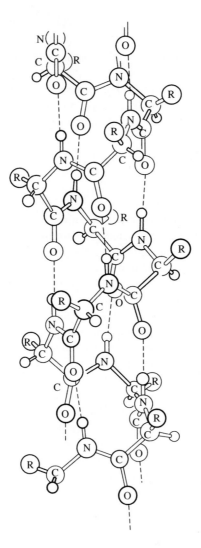

Fig. 3.6 The α-helix. (From Linus Pauling, *The Nature of the Chemical Bond,* Cornell University Press, Ithaca, New York, 1960, p. 500. Copyright © 1939 and 1940, 3rd ed., © 1960 by Cornell University. Used by permission of Cornell University Press.)

Right-handed α-helix

(deoxyribonucleic acid) in biological systems. The A-form helix is the secondary structure found in double-stranded regions of RNA (ribonucleic acid).

Table 3.2 lists some measured enthalpies for biochemical reactions involving changes in shape (conformation) of molecules. By changes in conformation we mean changes in the secondary structure of the molecule. (The primary structure involves the covalent bonds.) Examples include the change in a polypeptide from a rigid helix to a flexible coil. Denaturation of proteins involves this type of change. The corresponding change in nucleic acids and polynucleotides is the change from a two-strand helix to two single strands. Hydrogen bonds between the amides, the nucleic acid bases, and the solvent are important, but so are London-van der Waals types of interactions. The magnitudes of ΔH^0 and ΔS^0 can help us to understand the various interactions involved.

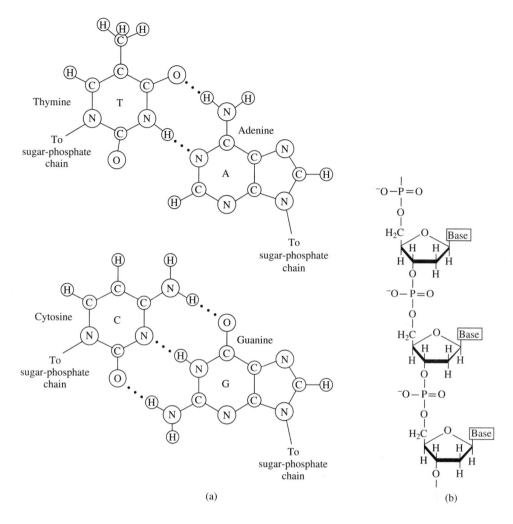

Fig. 3.7 (a) The Watson-Crick base pairs found in DNA. Adenine is complementary to thymine (A·T) and guanine is complementary to cytosine (G·C). In RNA uracil substitutes for thymine (A·U). (b) The sugar-phosphate chain in RNA differs only in that the sugar is ribose instead of deoxyribose as shown above.

The ΔH^0 for breaking a hydrogen bond between two urea molecules in water and forming urea-water hydrogen bonds is about 5 kJ mol^{-1}. From Table 3.2 we see that this is consistent with ΔH^0 for the helix-coil transition of poly-L-glutamate in aqueous solution. In the α-helix the peptide groups are hydrogen-bonded as shown in Fig. 3.6; in the coil they are hydrogen-bonded to water. The enthalpy increase for denaturation of a protein such as lysozyme depends greatly on the method used for denaturation (pH, urea, or temperature), but the large value of ΔH^0 clearly indicates many peptide-peptide hydrogen bonds are being broken.

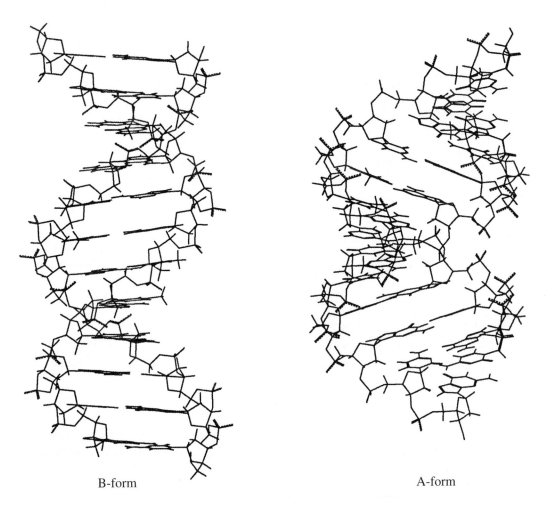

B-form   A-form

Fig. 3.8 Double-stranded, right-handed nucleic acid helices. The two polynucleotide strands in each double helix are antiparallel. One strand is as shown in Fig. 3.7b; the other runs in the opposite direction. The base pairs are tilted relative to the helix axis in A-form; they are perpendicular to the helix axis in B-form. DNA is B-form in biological cells; double-stranded regions of RNA are in A-form.

For DNA and RNA double-strand to single-strand transitions, we need to break two hydrogen bonds per A·T base pair (in DNA) or A·U base pair (in RNA) and three hydrogen bonds per G·C base pair (see Fig. 3.7). The ΔH^0 for dissociating a G·C base pair is larger than for A·T or A·U as expected, but the magnitude of ΔH^0 depends on the sequence of bases. The main interactions that stabilize the double strands are the stacking of the base pairs on each other; therefore the nearest neighbor sequences affect the thermodynamics. In DNA the range of ΔH^0 values for "melting" a base pair ranges from 30 kJ mol^{-1} for an A·T in a sequence of alternating –A–T–A–T–A–T–A– to 48 kJ mol^{-1} for a

Table 3.2 Thermodynamics of biochemical conformational transitions and noncovalent reactions studied near room temperature and neutral pH

Transition or noncovalent reaction	ΔH^0 (kJ mol^{-1})	ΔS^0 (J K^{-1} mol^{-1})	Reference*
Poly-L-glutamate helix-coil transition in 1.0 M KCl	4.5/amide	—	a
Lysozyme denaturation	≈450	—	a
RNA double-strand to single-strand transition in 1 M NaCl	≈40/base pair	≈104/base pair	b
DNA double-strand to single-strand transition in 1 M NaCl	≈35/base pair	≈88/base pair	b
Unstacking of bases in polyadenylic acid in 0.1 M KCl	36/nucleotide	113/nucleotide	a
Binding of Mg-ATP by tetrahydrofolate synthetase	31	182	c
Binding of Netropsin by poly dA · poly dT	−9.2	141	b

* (a) *Handbook of Biochemistry and Molecular Biology,* Vol. 1, 3rd ed., G. D. Fasman, ed. CRC Press, Cleveland, (1976); (b) Landolt-Börnstein Numerical Data and Functional Relationships in Science and Technology, Group VII, *Biophysics,* Vol. 1d, Nucleic Acids (1990); (c) N. P. Curthoys and J. C. Rabinowitz, *J. Biol. Chem. 246,* 6942 (1971).

G · C in a sequence of alternating −G−C−G−C−G−C−. A measure of the thermodynamics of unstacking is illustrated by the data for polyadenylic acid in Table 3.2. Polyadenylic acid unstacking refers to the transition between an ordered helical molecule with the adenine bases stacked on top of each other and a much more disordered molecule with the adenine bases not oriented relative to each other. This change in stacking does not involve hydrogen bonds directly, but it does have a ΔH^0 of 36 kJ per mole of nucleotide. This ΔH^0 is mainly London-van der Waals interactions among the bases and between the bases and the solvent.

The entropy changes for the double-strand to single-strand transition in nucleic acids are also consistent with the conclusions above. About 100 J K^{-1} mol^{-1} of entropy is gained for each base pair melted in DNA or RNA; this mainly corresponds to the increased possibility of rotation around single bonds in the nucleotides.

Two examples of binding of small molecules by a macromolecule are given in Table 3.2. The ΔH^0 values can be easily explained; heat must be added to favor the binding of the Mg-ATP complex to the enzyme tetrahydrofolate synthetase, but heat is released when the antibiotic Netropsin binds to double-stranded polydeoxyadenylic acid · polythymidylic acid, poly dA · poly dT. The entropy increases for both, however, might be surprising because of the loss of translational entropy when two molecules form a complex. The explanation is that water must be released when the molecules are bound; there is thus a large net gain in translational entropy.

USE OF PARTIAL DERIVATIVES

So far we have considered eight thermodynamic variables of state: E, H, S, G, A, P, V, and T. We can choose a few parameters as independent variables and express the others as functions of these independent variables. For example, if our system is a fixed amount of a pure substance, and if we choose P and T as independent variables, we can express V as a function of P and T. This is an example of an equation of state:

$$V = V(P, T)$$

The specific form of the function $V(P, T)$ depends on the nature of the substance. If it is an ideal gas,

$$V = V(P, T) = \frac{nRT}{P}$$

When we are interested in how V changes with T at constant P, we can take the derivative of V with respect to T, keeping P constant:

$$\frac{dV}{dT} = \frac{nR}{P}$$

Instead of stating explicitly that $P = $ constant, we can use the notation of partial derivatives:

$$\left(\frac{\partial V}{\partial T}\right)_P = \frac{nR}{P} \tag{3.42}$$

$(\partial V/\partial T)_P$ means simply the derivative of V with respect to T at constant P. Similarly, instead of writing

$$\frac{dV}{dP} = nRT \frac{d}{dP}\left(\frac{1}{P}\right) \qquad \text{at } T = \text{constant}$$

$$= -\frac{nRT}{P^2}$$

we can write

$$\left(\frac{\partial V}{\partial P}\right)_T = -\frac{nRT}{P^2} \tag{3.43}$$

It troubles some students in relations like Eqs. (3.42) and (3.43) that the partial derivative is actually a function of the variable that is held constant. What this means is that the slope of the curve of V as a function of T in Eq. (3.42), for example, has a different value at different pressures. However, at each pressure the slope dV/dT is determined with the pressure held constant.

Although in this book the usage of partial derivatives will be limited, we should point out that many thermodynamic relationships can be thus derived. For the state functions used in thermodynamics, it is generally true that if

$$V = V(P, T)$$

the total differential of V is

$$dV = \left(\frac{\partial V}{\partial P}\right)_T dP + \left(\frac{\partial V}{\partial T}\right)_P dT \qquad (3.44)$$

This equation states that for small changes, dV can be treated as the sum of how V changes with P (at constant T) and how V changes with T (at constant P). Also, the order of differentiation is not important.

$$\left[\frac{\partial}{\partial T}\left(\frac{\partial V}{\partial P}\right)_T\right]_P = \left[\frac{\partial}{\partial P}\left(\frac{\partial V}{\partial T}\right)_P\right]_T \qquad (3.45)$$

An example of the use of these equations follows.

Example 3.10 Show that:

(a) $\left(\dfrac{\partial G}{\partial P}\right)_T = V$ (b) $\left(\dfrac{\partial V}{\partial T}\right)_P = -\left(\dfrac{\partial S}{\partial P}\right)_T$

Solution

(a) Since we want an expression for $(\partial G/\partial P)_T$, our first task is to express G as a function of T and P. Start with the definition of G.

$$G = H - TS$$
$$= E + PV - TS$$

We apply the standard rules of differentiation to obtain

$$dG = dE + d(PV) - d(TS)$$
$$= dE + P\,dV + V\,dP - T\,dS - S\,dT$$

Since G is a state function, if we can derive an expression $G = G(T, P)$ for a specific path, the expression will hold for any other path. For simplicity, we pick a reversible path with pressure-volume work only. Applying the first law to such a path,

$$dE = dq_{rev} + dw_{rev}$$
$$= dq_{rev} - P\,dV$$

but

$$dq_{rev} = T\,dS$$

Thus

$$dE = T\,dS - P\,dV \qquad (3.46)$$

and, by substitution,

$$dG = T\,dS - P\,dV + P\,dV + V\,dP - T\,dS - S\,dT$$
$$= V\,dP - S\,dT \qquad (3.47)$$

Comparing Eq. (3.47) with the relationship according to Eq. (3.44),

$$dG = \left(\frac{\partial G}{\partial P}\right)_T dP + \left(\frac{\partial G}{\partial T}\right)_P dT$$

we obtain

$$\left(\frac{\partial G}{\partial P}\right)_T = V \tag{3.48}$$

and

$$\left(\frac{\partial G}{\partial T}\right)_P = -S \tag{3.49}$$

Equation (3.48) has been given previously in this chapter [in the integral form Eq. (3.33)] without proof. Similarly, Eq. (3.49) can be integrated to provide an exact replacement for Eq. (3.30).

(b) Since, according to Eq. (3.45),

$$\left[\frac{\partial}{\partial T}\left(\frac{\partial G}{\partial P}\right)_T\right]_P = \left[\frac{\partial}{\partial P}\left(\frac{\partial G}{\partial T}\right)_P\right]_T \tag{3.50}$$

from Eqs. (3.48) and (3.49) we obtain

$$\left(\frac{\partial V}{\partial T}\right)_P = -\left(\frac{\partial S}{\partial P}\right)_T \tag{3.51}$$

Note that in the special case of an ideal gas, since

$$\left(\frac{\partial V}{\partial T}\right)_P = \frac{nR}{P}$$

we have

$$\left(\frac{\partial S}{\partial P}\right)_T = -\frac{nR}{P}$$

Integrating at constant temperature gives

$$\int_{S_1}^{S_2} dS = -\int_{P_1}^{P_2} \frac{nR}{P}\, dP = -nR \ln \frac{P_2}{P_1}$$

This is Eq. (3.21).

Many useful thermodynamic relations can be derived by the use of partial derivatives, including the Gibbs-Helmholtz equation (3.32), which we gave without proof.

Earlier in this chapter we introduced the chemical potential, μ, which is simply the molar free energy for a pure substance. For solutions, however, the chemical potential of a component A is the partial molal Gibbs free energy

$$\mu_A = \left(\frac{\partial G}{\partial n_A}\right)_{T, P, n_B, n_C \dots} = \bar{G}_A \qquad (3.52)$$

This partial derivative describes the change in the Gibbs free energy that occurs upon an infinitesimal change dn_A in the number of moles of A in a solution, while keeping the temperature, pressure, and the number of moles of all other constituents constant. Substance A may be either a solute or the solvent in the solution. We will see in the next chapter that chemical potentials describe very important properties of solutions.

Relations Among Partial Derivatives

We have already obtained some useful equations among partial derivatives; now we will generalize and summarize the results. The object is to obtain equations relating the state variables T, P, V, E, H, G, S. We consider a closed system with no external fields (such as gravitational or electrical fields), and no chemical reactions or phase changes. Therefore, the only reason that one of the above variables changes is because one or more of the other variables changes. We usually think of T, P, V as the independent variables and E, H, G, S as the dependent variables. Furthermore, because of the equation of state telling us how P, V, T are related, we need consider only how E, H, G, S depend on two of the three variables: P, V, T.

For example, we derived Eq. (3.47) for G as a function of P and T.

$$dG = V \, dP - S \, dT \qquad (3.47)$$

If you are asked to calculate the change of G resulting from a change of V, first find the change of P and T corresponding to this change of V, then use Eq. (3.47). To calculate $G_2 - G_1$ for a change from P_1, T_1 to P_2, T_2, you must integrate Eq. (3.47).

$$G_2 - G_1 = \int_{P_1}^{P_2} V \, dP - \int_{T_1}^{T_2} S \, dT$$

If you need to know how ΔG for a chemical reaction or phase change depends on T and P, you use

$$\Delta G_2 - \Delta G_1 = \int_{P_1}^{P_2} \Delta V \, dP - \int_{T_1}^{T_2} \Delta S \, dT$$

where ΔV and ΔS are the volume change and entropy change for the chemical reaction or phase change. To do the integration you must know ΔV as a function of P and ΔS as a function of T. For some systems they are approximately constant.

We can obtain the corresponding equations for H as a function of T, P exactly analogous to the derivation for $G(T, P)$. We use

$$dH = \left(\frac{\partial H}{\partial T}\right)_P dT + \left(\frac{\partial H}{\partial P}\right)_T dP$$

The change of enthalpy with temperature at constant pressure was found to be the heat capacity at constant pressure [Eq. (2.45)]; thus

$$\left(\frac{\partial H}{\partial T}\right)_P \equiv C_P$$

We obtain $(\partial H/\partial P)_T$ from the analog to Eq. (3.47).

$$H = G + TS$$

$$dH = dG + T\,dS + S\,dT$$

$$= V\,dP - S\,dT + T\,dS + S\,dT$$

$$dH = V\,dP + T\,dS \tag{3.53}$$

We obtain the correct expression for $(\partial H/\partial P)_T$ by writing the derivative form of Eq. (3.53) and specifying that T is constant. Thus

$$\left(\frac{\partial H}{\partial P}\right)_T = V + T\left(\frac{\partial S}{\partial P}\right)_T$$

and we already have an expression for $(\partial S/\partial P)_T$ in terms of P, V, T [Eq. (3.51)]. Therefore,

$$\left(\frac{\partial H}{\partial P}\right)_T = V - T\left(\frac{\partial V}{\partial T}\right)_P$$

and

$$dH = C_P\,dT + \left[V - T\left(\frac{\partial V}{\partial T}\right)_P\right]dP \tag{3.54}$$

Equation (3.54) can be integrated to give $H_2 - H_1$ for a finite change from P_1, T_1 to P_2, T_2. For S we have

$$dS = \left(\frac{\partial S}{\partial T}\right)_P dT + \left(\frac{\partial S}{\partial P}\right)_T dP$$

In Eq. (3.14) we used the integrated form of

$$\left(\frac{\partial S}{\partial T}\right)_P = \frac{C_P}{T}$$

Therefore, using Eq. (3.51)

$$dS = \frac{C_P}{T}\,dT - \left(\frac{\partial V}{\partial T}\right)_P dP \tag{3.55}$$

For E the most useful variables to use are T and V instead of T and P. We write

$$dE = \left(\frac{\partial E}{\partial T}\right)_V dT + \left(\frac{\partial E}{\partial V}\right)_T dV$$

From Eq. (2.43)

$$\left(\frac{\partial E}{\partial T}\right)_V \equiv C_V$$

From Eq. (3.46)

$$\left(\frac{\partial E}{\partial V}\right)_T = T\left(\frac{\partial S}{\partial V}\right)_T - P$$

We need to write $(\partial S/\partial V)_T$ in terms of P, V, T. We do this from an analog of Eq. (3.47).

$$dA = -P\, dV - S\, dT \tag{3.56}$$

The reader should be able to derive this equation from the definition of $A \equiv E - TS$. Applying the relation

$$\left[\frac{\partial}{\partial T}\left(\frac{\partial A}{\partial V}\right)_T\right]_V = \left[\frac{\partial}{\partial V}\left(\frac{\partial A}{\partial T}\right)_V\right]_T$$

we obtain

$$\left(\frac{\partial S}{\partial V}\right)_T = \left(\frac{\partial P}{\partial T}\right)_V$$

Therefore,

$$dE = C_V\, dT + \left[T\left(\frac{\partial P}{\partial T}\right)_V - P\right] dV \tag{3.57}$$

To summarize our results, we can now calculate the change in E, H, S, G for a system in which P_1, V_1, T_1 change to P_2, V_2, T_2. We need only be given two of these three variables, because we can calculate the third from the equation of state. We use the integrated forms of Eqs. (3.47), (3.54), (3.55), and (3.57).

The perceptive student will note that we were able to derive many useful results from the following four equations:

$$dE = -P\, dV + T\, dS \tag{3.46}$$

$$dH = V\, dP + T\, dS \tag{3.53}$$

$$dG = V\, dP - S\, dT \tag{3.47}$$

$$dA = -P\, dV - S\, dT \tag{3.56}$$

Many more relations among partial derivatives can be obtained. Standard thermodynamics texts discuss the most useful ones.

SUMMARY

New State Variables

Name	Symbol	Units	Definition
Entropy	S	J K^{-1}, cal deg^{-1}, eu	$dS \equiv dq_{rev}/T$
Gibbs free energy	G	J, erg, cal	$G \equiv H - TS$
Helmholtz free energy	A	J, erg, cal	$A \equiv E - TS$

Unit Conversions

Entropy:

1 cal deg^{-1} $\equiv$ 1 eu = 4.184 J K^{-1}

Free energy:

1 cal = 4.184 J = 4.184 $\times$ 10^7 erg

General Equations

Entropy, S:

$$S = k \ln N \tag{3.9}$$

$$k = \text{Boltzmann's constant } (1.381 \times 10^{-23} \text{ J K}^{-1})$$

$$N = \text{no. of microscopic states of system}$$

$$S_2 - S_1 = \int_1^2 \frac{dq_{rev}}{T} \quad \text{(reversible path)} \tag{3.12}$$

$$S_2 - S_1 = \int_{T_1}^{T_2} \frac{C_P \, dT}{T} \quad \text{(constant pressure)} \tag{3.14}$$

$$S_2 - S_1 = \int_{T_1}^{T_2} \frac{C_V \, dT}{T} \quad \text{(constant volume)} \tag{3.16}$$

Gibbs free energy, G:

$$G \equiv H - TS \tag{3.22}$$

$$G_2 - G_1 = H_2 - H_1 - T \cdot (S_2 - S_1) \quad \text{(constant temperature)}$$

$$G(P_2) - G(P_1) = \int_{P_1}^{P_2} V \, dP \tag{3.33}$$

1 L atm = 101.325 J

Carnot-cycle efficiencies, for a heat engine:

$$\text{efficiency} = \frac{-w}{q_{\text{hot}}} = \frac{T_{\text{hot}} - T_{\text{cold}}}{T_{\text{hot}}} \qquad (3.6)$$

Second law of thermodynamics and criterion for a spontaneous process:

In the following equations, the equal sign applies for reversible processes and the other sign applies for spontaneous (irreversible) processes:

$$\Delta S \text{ (system)} + \Delta S \text{ (surroundings)} \geq 0 \qquad (3.7)$$

$$\Delta S \text{ (isolated system)} \geq 0 \qquad (3.8)$$

$$\Delta S \text{ (system)} \geq \frac{q}{T} \qquad (3.3), (3.13)$$

$$\Delta G \leq 0 \qquad \text{(constant } T, P) \qquad (3.25), (3.26)$$

$$\Delta A \leq 0 \qquad \text{(constant } T, V) \qquad (3.39), (3.40)$$

Third law of thermodynamics:

$$S_A \text{ (0 K)} \equiv 0$$

A = any pure, perfect crystal

Solids and Liquids

We assume in these equations that the volume of a solid or liquid is independent of T and P; that $C_P = C_V$; and that they do not depend on T and P.

$$S\,(P_2, T_2) - S\,(P_1, T_1) = n\overline{C}_P \ln \frac{T_2}{T_1} = n\overline{C}_V \ln \frac{T_2}{T_1}$$

n = number of moles
$\overline{C}_P$ = heat capacity per mole at constant P
$\overline{C}_V$ = heat capacity per mole at constant V

Gases

We assume that gases can be approximated by the ideal gas equation and that C_P and C_V are independent of T. $PV = nRT$. $\overline{C}_P = \overline{C}_V + R$.

$$S_2 - S_1 = -nR \ln \frac{P_2}{P_1} = nR \ln \frac{V_2}{V_1} \qquad \text{(constant temperature)} \qquad (3.21)$$

$$G_2 - G_1 = nRT \ln \frac{P_2}{P_1} = -nRT \ln \frac{V_2}{V_1} \qquad \text{(constant temperature)} \qquad (3.35)$$

$$S(T_2) - S(T_1) = n\overline{C}_P \ln \frac{T_2}{T_1} \qquad \text{(constant pressure)} \qquad (3.15)$$

$$R = 8.314 \text{ J mol}^{-1} \text{ K}^{-1}$$

Phase Changes at Constant T and P

$$\Delta G = 0 \qquad \text{(at equilibrium)}$$

$$\Delta S = \frac{\Delta H_{\text{transition}}}{T_{\text{transition}}} \qquad \text{(at equilibrium)} \tag{3.19}$$

Chemical Reactions

For a chemical reaction

$$n_A A + n_B B \longrightarrow n_C C + n_D D$$

which occurs at constant T and P:

$$\Delta S^0 = n_C \bar{S}_C^0 + n_D \bar{S}_D^0 - n_A \bar{S}_A^0 - n_B \bar{S}_B^0$$

$$\Delta G^0 = n_C \bar{G}_C^0 + n_D \bar{G}_D^0 - n_A \bar{G}_A^0 - n_B \bar{G}_B^0$$

the superscript zero means standard conditions, which here is 1 atm.

$$\Delta S(T_2) = \Delta S(T_1) + \int_{T_1}^{T_2} \Delta C_P \frac{dT}{T} \tag{3.18}$$

$$\Delta C_P = n_C \bar{C}_{P,\,C} + n_D \bar{C}_{P,\,D} - n_A \bar{C}_{P,\,A} + n_B \bar{C}_{P,\,B}$$

$$\Delta G(T_2) = \Delta G(T_1) - (T_2 - T_1) \cdot \Delta S \qquad \text{(if } \Delta S \text{ is independent of } T)$$

$$\frac{\Delta G(T_2)}{T_2} = \frac{\Delta G(T_1)}{T_1} + \left(\frac{1}{T_2} - \frac{1}{T_1}\right) \cdot \Delta H \qquad \text{(if } \Delta H \text{ is independent of } T)$$

$$\frac{\Delta G(T_2)}{T_2} - \frac{\Delta G(T_1)}{T_1} = -\int_{T_1}^{T_2} \frac{\Delta H(T)}{T^2} \, dT \qquad \text{(if } \Delta H \text{ depends on } T) \tag{3.32}$$

$$\Delta G(P_2) - \Delta G(P_1) = \Delta V \cdot (P_2 - P_1) \qquad \begin{array}{l}\text{(for solids and liquids only,}\\ \text{if } \Delta V \text{ is independent of } T)\end{array}$$

$$\Delta V = n_C \bar{V}_C + n_D \bar{V}_D - n_A \bar{V}_A - n_B \bar{V}_B \tag{3.36}$$

$$\Delta G(P_2) - \Delta G(P_1) = \Delta n R T \ln \frac{P_2}{P_1} \qquad \text{(at least one gaseous product or reactant)} \tag{3.37}$$

Δn = number of moles of gaseous products − number of moles of gaseous reactants.

REFERENCES

In addition to the textbooks listed in Chapter 2, see
ATKINS, P. W., 1984. *The Second Law,* Scientific American Library, W. H. Freeman, New York.
KLOTZ, I. M., 1967. *Energy Changes in Biochemical Reactions,* Academic Press, New York.
MOROWITZ, H. J., 1970. *Entropy for Biologists,* Academic Press, New York.

SUGGESTED READINGS

FEYNMAN, R. P., R. B. LEIGHTON, and M. SANDS, 1963. *The Feynman Lectures on Physics,* Addison-Wesley, Reading, Massachusetts.

FRAUTSCHI, S., 1982. Entropy in an Expanding Universe, *Science 217,* 593–599.

MCIVER, R. T., JR., 1980. Chemical Reactions without Solvation, *Sci. Am. 243* (November), 186–196.

WILSON, S. S., 1981. Sadi Carnot, *Sci. Am. 245* (August), 134–145.

PROBLEMS

1. One mole of an ideal monatomic gas is expanded from an initial state at 2 atm and 400 K to a final state at 1 atm and 300 K.
   (a) Choose two different paths for this expansion, specify them carefully, and calculate w and q for each path.
   (b) Now calculate ΔE and ΔS for each path.
   (c) How do your answers to parts (a) and (b) illustrate the difference between exact and inexact differentials?

2. The temperature of the heat reservoirs for a Carnot-cycle (reversible) engine are $T_{hot} = 1200$ K and $T_{cold} = 300$ K. The efficiency of the engine is calculated to be 0.75, from Eq. (3.6).
   (a) If $w = -100$ kJ, calculate q_1 and q_3. Explain the meaning of the signs of w, q_1 and q_3.
   (b) The same engine can be operated in the reverse order. If $w = +100$ kJ, calculate q_1 and q_3. Explain the meaning of the signs.
   (c) Suppose that it were possible to have an engine with a higher efficiency, say 0.80. If $w' = -100$ kJ, calculate q_1' and q_3'. The superscript denotes quantities for this engine.
   (d) If we use all the work done by the engine in part (c) to drive the heat pump in part (b), calculate $(q_1 + q_1')$ and $(q_3 + q_3')$, the amount of heat transferred. Explain the signs. Note that the net effect of the combination of the two engines is to allow a "spontaneous" transfer of heat from a cooler reservoir to a hotter reservoir, which should not happen.
   (e) Show that if it were possible to have a reversible engine with a lower efficiency operating between the same two temperatures, heat could also be transferred spontaneously from the cooler reservoir to the hotter reservoir. (*Hint:* Operate the engine with lower efficiency as a heat pump.)

3. The second law of thermodynamics states that entropy increases for spontaneous processes and that an increase in entropy is associated with transitions from ordered to disordered states. Living organisms, even the simplest bacteria growing in cultures, appear to violate the second law, because they grow and proliferate spontaneously. They convert simple chemical substances into the highly organized structure of their descendants. Write a critical evaluation of the proposition that living organisms contradict the second law. Be sure to state your conclusion clearly and to present detailed, reasoned arguments to support it.

4. Consider the process where n_A moles of gas A initially at 1 atm pressure mix with n_B moles of gas B also at 1 atm to form 1 mol of a uniform mixture of A and B at a final total pressure of 1 atm, and all at constant temperature, T. Assume that all gases behave ideally.
   (a) Show that the entropy change, $\Delta\bar{S}_{mix}$, for this process is given by $\Delta\bar{S}_{mix} = -X_A R \ln X_A - X_B R \ln X_B$, where X_A and X_B are the mole fractions of A and B, respectively. (*Hint:* Entropy changes are additive for components A and B; that is, $\Delta\bar{S}_{mix} = \Delta S_A + \Delta S_B$.)

(b) What can you say about the value (especially the sign) of $\Delta \bar{S}_{mix}$, and how does this correlate with the second law?

(c) Derive an expression for $\Delta \bar{G}_{mix}$ for the conditions of this problem and comment on values (and signs) of its terms.

5. A system consisting of 0.20 mol of supercooled gaseous water (95°C, 1 atm) partially condenses to liquid water; the process is done adiabatically at constant pressure.

(a) At equilibrium what is the temperature of the system?

(b) How many mol of gaseous water have condensed?

(c) Calculate the entropy change for the process.

6. The technology to build microscopic-size motors is now available. Can microscopic motors take advantage of fluctuations to beat the second law, and do work at constant temperature? Consider the following fluctuation motor.

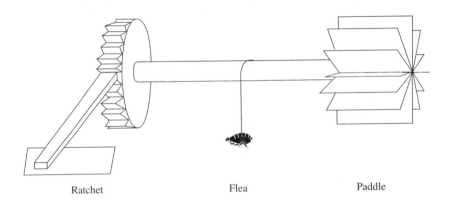

Ratchet   Flea   Paddle

Assume that the motor is so small that one gas molecule hitting a paddle can make it turn. The motor is placed in a box (at constant temperature) with a few gas molecules. A possible explanation of how the motor works is the following. Gas molecules randomly hitting the paddles will make the shaft turn clockwise sometimes and sometimes anti-clockwise. The flea will randomly be raised and lowered by the fluctuations, *if we don't consider the ratchet.* However, the ratchet is held down by a spring so it allows only clockwise rotations. Therefore the flea will be raised because the ratchet has converted the random fluctuations into a directed rotation. Work is done (the mass of the flea times the acceleration of gravity times the distance it is raised) at constant temperature. The energy comes from the thermal energy of the gas so the first law is not violated.

(a) Will the motor work? Discuss the action of the ratchet in converting random fluctuations into a directed clockwise motion.

(b) If the motor will not work as described, what simple change will allow it to work? Clearly, if the motor will work as described, you need not answer part (b).

This problem is discussed in *The Feynman Lectures on Physics,* Vol. 1.

7. For the following processes determine whether each of the thermodynamic quantities listed is greater than, equal to, or less than zero for the *system* described. Consider all gases to behave ideally. The system is shown in italic type in each case. Indicate your reasoning and state explicitly any reasonable assumptions or approximations that you need to make.

(a) A *sample of gas* is carried through a complete Carnot cycle (isothermal expansion, adiabatic expansion, isothermal compression, adiabatic compression—all reversible): ΔT, q, w, ΔE, ΔH, ΔS, ΔG.

(b) A sample of *hot water* is mixed with a sample of *cold water* in a thermally insulated, closed container of fixed volume: w, q, ΔE, ΔH, ΔS.

(c) An *ideal gas* expands adiabatically and reversibly: ΔV, ΔT, w, q, ΔE, ΔH, ΔS.

(d) A flask of *liquid nutrient solution inoculated with a small sample of bacteria* is maintained for several days in a thermostat until the bacteria have multiplied 1000-fold: ΔT, w, q, ΔE, ΔH, ΔG.

8. One suggestion for solving the fuel-shortage problem is to use electric power (obtained from solar energy) to electrolyze water to form $H_2(g)$ and $O_2(g)$. The hydrogen could be used as a pollution-free fuel.

(a) Calculate the enthalpy, entropy, and free energy change on burning 1 kg of $H_2(g)$ to $H_2O(l)$ at 25°C and 1 atm.

(b) Calculate the enthalpy, entropy, and free energy change on burning 1 kg of *n*-octane(g) to $H_2O(l)$ and $CO_2(g)$ at 25°C and 1 atm.

9. Use data from the Appendix to answer the following questions.

(a) A friend wants to sell you a catalyst that allows benzene to be formed by passing $H_2(g)$ over carbon (graphite) at 25°C and 1 atm. Should you buy? Why?

(b) Is the reaction $2 NO(g) + O_2(g) \rightarrow 2 NO_2(g)$ spontaneous at 25°C and 1 atm?

(c) Is the reaction to form solid alanine, CH_3CHNH_2COOH, and liquid H_2O from $CH_4(g)$, $NH_3(g)$, and $O_2(g)$ spontaneous at 25°C and 1 atm?

10. For the statements below, choose the word or words inside the parentheses that serve to make a correct statement. Each statement has at least one, and may have more than one, correct answer.

(a) According to the second law of thermodynamics, a spontaneous process, such as a balloon filled with hot gas cooling to the surroundings at constant pressure, will always occur (adiabatically, reversibly, irreversibly, without work done).

(b) Associated with such a process there is always an increase in entropy of (the system, the surroundings, the system plus the surroundings, none of these).

(c) For the example given, the heat gained by the surroundings is just equal to the negative of the (internal energy change, enthalpy change, entropy change, Gibbs free energy change) of the system.

(d) To return the system to its initial state requires from the surroundings an expenditure of entropy whose magnitude is (greater than, equal to, less than) that which is gained during the spontaneous process.

11. Consider the reaction

$$C_2H_5OH(l) \longrightarrow C_2H_6(g) + \tfrac{1}{2}O_2(g)$$

Calculate ΔH^0_{298}, ΔS^0_{298}, and ΔG^0_{298}. Estimate ΔE^0_{298}. State what further data would be needed to obtain ΔH^0 at 500°C and 1 atm.

12. The shells of marine organisms contain $CaCO_3$ largely in the crystalline form known as calcite. There is a second crystalline form of $CaCO_3$ known as aragonite.
    (a) Based on the thermodynamic and physical properties given below for these two crystalline forms, would you expect calcite in nature to convert spontaneously to aragonite given sufficient time? Justify your answer.
    (b) Will the conversion proposed in part (a) be favored or opposed by increasing the pressure? Explain.
    (c) What pressure should be just sufficient to make this conversion spontaneous at 25°C?
    (d) Will increasing the temperature favor the conversion? Explain.

Properties at 298 K	$CaCO_3$ (calcite)	$CaCO_3$ (aragonite)
$\Delta \overline{H}_f^0$ (kJ mol^{-1})	−1206.87	−1207.04
$\Delta \overline{G}_f^0$ (kJ mol^{-1})	−1128.76	−1127.71
$\overline{S}^0$ (J K^{-1} mol^{-1})	92.88	88.70
$\overline{C}_P$ (J K^{-1} mol^{-1})	81.88	81.25
Density (g cm^{-3})	2.710	2.930

13. Consider the reaction that converts pyruvic acid ($CH_3COCOOH$) into acetaldehyde (CH_3CHO) and gaseous CO_2, which is catalyzed in aqueous solution by the enzyme pyruvate decarboxylase. Assume ideal gas behavior for the CO_2.
    (a) Calculate ΔG_{298}^0 for this reaction.
    (b) What does Le Châtelier's principle predict to be the effect of increasing the pressure to 100 atm?
    (c) Calculate ΔG for this reaction at 298 K and 100 atm. State any important assumptions needed in addition to ideal gas behavior.

14. Consider the reversible, isothermal, constant-pressure freezing of 1 mol of water at 0°C and 1 atm.
    (a) Calculate ΔE in kJ.
    (b) Calculate ΔH in kJ.
    (c) Calculate ΔS in J K^{-1}.
    (d) Calculate ΔG in kJ.
    (e) Calculate q in kJ. Is heat absorbed or evolved?
    (f) Calculate w in kJ. Is work done by the system or on the system?

15. For the statements below, choose the word or words inside the parentheses that serve(s) to make a correct statement. More than one answer may be correct. At 100°C, the equilibrium vapor pressure of water is 1 atm. Consider the process where 1 mol of water vapor at 1 atm pressure is reversibly condensed to liquid water at 100°C by slowly removing heat into the surroundings.
    (a) In this process the entropy of the system will (increase, remain unchanged, decrease).
    (b) The entropy of the universe will (increase, remain unchanged, decrease).
    (c) Since the condensation process occurs at constant temperature and pressure, the accompanying free energy change for the system will be (positive, zero, negative).

(d) In practice, this process cannot be carried out reversibly. For the real process, compared with the ideal reversible one, different values will be observed for the (entropy change of the system, entropy change of the surroundings, entropy change of the universe, free energy change of the system).

At a lower temperature, of 90°C, the equilibrium vapor pressure of water is only 0.692 atm.
(e) If 1 mol of water is condensed reversibly at this temperature and pressure, the entropy of the system will (increase, remain unchanged, decrease).
(f) The entropy of the universe will (increase, remain unchanged, decrease).
(g) The free energy of the system will (increase, remain unchanged, decrease).
(h) Since the molar heat capacity of water vapor is only about one-half that of the liquid, the entropy change upon condensation at 90°C will be (more negative than, the same as, more positive than) that at 100°C.

16. Consider a system that undergoes a phase change from phase α to phase β at equilibrium temperature T_m, and equilibrium pressure P_m. At these equilibrium conditions the heat absorbed per mol of material undergoing this transition is q_m, and there is a molar volume change of ΔV_m. The molar heat capacities at constant pressure for the α and β phases are $C_{P,\alpha}$ and $C_{P,\beta}$; they are independent of temperature.
(a) Evaluate w, ΔE, ΔH, ΔS, and ΔG for converting one mol of the system from phase α to phase β at the equilibrium T_m and P_m. Your answer should be in terms of q_m, ΔV_m, etc.
(b) Evaluate ΔH and ΔS at temperature T^* different from T_m but at the same pressure P_m.
(c) The denaturation transition of a globular protein can be approximated as such a phase change α to β. Calculate ΔG, ΔH, and ΔS for a heat of transition of $q_m = 638$ kJ mol^{-1} at $T_m = 70$°C, $P_m = 1$ atm; $C_{P,\alpha} - C_{P,\beta} = 8.37$ kJ mol^{-1} K^{-1}.
(d) Calculate ΔG, ΔH, and ΔS for the same transition at 37°C and 1 atm. Assume that $C_{P,\alpha} - C_{P,\beta}$ is independent of temperature.
(e) If ΔV_m for the transition is +3 mL mol^{-1}, will the equilibrium transition temperature, T_m, be increased or decreased at a pressure of 1000 atm? Explain.
(f) Write a thermodynamic cycle which would allow you to calculate the equilibrium transition temperature corresponding to a pressure of 1000 atm. Use the cycle to write an equation which in principle could be solved to calculate T_m at 1000 atm.

17. Table 3.2 gives thermodynamic data for transitions from ordered, helical conformations of polypeptides and polynucleotides to disordered states. The enthalpy changes and entropy changes are positive as expected; heat is absorbed and entropy is gained. However, when the synthetic polypeptide polybenzyl-L-glutamate undergoes a transition from an ordered helix to a disordered coil in an ethylene dichloride-dichloroacetic acetic solvent at 39°C and 1 atm, $\Delta H^0 = -4.0$ kJ (per mol amide) and $\Delta S^0 = -12$ J K^{-1} (per mol amide).
(a) Give a possible explanation for the experimental result that heat is released and entropy is decreased for this transition. Does increasing the temperature favor the helix-coil transition?
(b) Is the reaction spontaneous at 39°C? What thermodynamic criterion did you use to reach this conclusion?
(c) At what temperature (°C) will the helix-coil reaction be reversible? This temperature is often called the "melting" temperature of the helix. Assume that ΔH^0 and ΔS^0 are independent of temperature.
(d) Can a reaction occur in an *isolated* system which leads to a decrease in the entropy of the system? If it can, give an example; if it cannot, state why not.

18. The following thermodynamic data at 25°C have been tabulated for the gas phase reaction shown below (dotted lines represent hydrogen bonds):

	ΔH_f^0 (kJ mol^{-1})	$\bar{S}^0$ (J K^{-1} mol^{-1})
HCOOH(g)	−362.63	251.0
(HCOOH)$_2$(g)	−785.34	347.7

(a) Write the equation for the standard heat of formation at 25°C, ΔH_f^0 of HCOOH(g) (specify the pressure and phase for each component).
(b) Calculate ΔH^0, ΔS^0, and ΔG^0 for the gas phase dimerization at 298 K. Is the formation of dimer from monomers spontaneous under these conditions?
(c) Calculate the enthalpy change per hydrogen bond formed in the gas phase. Why is not a similar calculation useful to estimate the entropy or free energy of hydrogen bond formation?

19. An electrochemical battery is used to provide 1 milliwatt of power for a (small) light. The chemical reaction in the battery is:

$$\tfrac{1}{2} N_2(g) + \tfrac{3}{2} H_2(g) \longrightarrow NH_3(g)$$

(a) What is the free energy change for the reaction at 25°C, 1 atm?
(b) Calculate the free energy change for the reaction at 50°C, 1 atm. State any assumptions made in the calculation.
(c) The limiting reactant in the battery is 100 grams of H_2. Calculate the maximum length of time (sec) the light can operate at 25°C.

20. Consider a fertilized hen egg in an incubator—a constant temperature and pressure environment. In a few weeks the egg will hatch into a chick.
(a) The egg is chosen as the system. Is the egg an open, an isolated, or a closed system? Define an isolated system.
(b) In the fertilized egg, hen proteins are formed into a highly ordered chick. Does the entropy of the system increase or decrease? Does this violate the second law of thermodynamics? Explain in two or three sentences why the development of the chick is or is not consistent with the second law.
(c) What happens to the energy of the system as the chick develops? What forms of energy contribute to the change in energy (if any) of the system?
(d) Does the free energy of the system increase, decrease, or remain the same? How do you know?

21. Calculate the entropy change when:
(a) Two moles of $H_2O(g)$ are cooled irreversibly at constant P from 120°C to 100°C.
(b) One mole of $H_2O(g)$ is expanded at constant pressure of 2 atm from an original volume of 20 L to a final volume of 25 L. You can consider the gas to be ideal.
(c) One hundred grams of $H_2O(s)$ at −10°C and 1 atm are heated to $H_2O(l)$ at +10°C and 1 atm.

22. You are asked to evaluate critically the following situations. Some of the proposals or interpretations are reasonable and others violate very basic principles of thermodynamics.

   (a) It is commonly known that one can supercool water and maintain it as a liquid at temperatures as low as $-10°C$. If a sample of supercooled liquid water is isolated in a closed, thermally insulated container, after a time it spontaneously changes to a mixture of ice and water at $0°C$. Thus it increases its temperature spontaneously with no addition of heat from outside, and furthermore, some low-entropy ice is produced.

   (b) An inventor proposed a new scheme for heating buildings in the winter in arctic climates. Since freezing temperatures reach only a few feet down into the soil, he proposes digging a well and immersing a coil of copper tubing in the water. He will then connect the ends of the coil to the radiators in the building and use a heat pump to transfer heat into the building. What would be your advice to an attorney who is assigned to evaluate this patent application? Base your evaluation on the relevant thermodynamics.

   (c) On a hot summer's day your laboratory partner proposes opening the door of the lab refrigerator to cool off the room.

   (d) A sample of air is separated from a large evacuated chamber, and the entire system is isolated from the surroundings. A small pinhole between the two chambers is opened and roughly half the gas is allowed to effuse into the second chamber before the pinhole is closed. Because nitrogen effuses faster than oxygen, the gas in the second chamber is richer in nitrogen and the gas in the first chamber is richer in oxygen than the original air. The gases have, thus, spontaneously unmixed (at least partially), and this is held to be a violation of the second law of thermodynamics.

   (e) Supercooled water at $-10°C$ has a higher entropy than does an equal amount of ice at $-10°C$. Therefore, supercooled water cannot go spontaneously to ice at the same temperature in an isolated system.

   (f) A volume of an aqueous solution of hydrogen peroxide is placed in a cylinder and covered with a tight-fitting piston. A small amount of the enzyme catalase is placed on a probe and inserted through an opening in the base of the cylinder. The catalase catalyzes the decomposition of the hydrogen peroxide, and the oxygen gas formed serves to raise the piston. The catalase is then withdrawn and the hydrogen peroxide is re-formed, causing the piston to return to its initial position. The piston is connected to an engine and net work is obtained indefinitely by cycles of simply inserting and withdrawing the catalase.

   (g) High temperatures can be achieved in practice by focusing the rays of the sun on a small object using a large parabolic reflector, such as is used in astronomical telescopes. Since the energy gathered increases as the square of the diameter of the reflector, it should be possible to produce temperatures higher than those of the sun by using a large-enough reflector.

   (h) Hydroelectric plants generate electric power (work) by using water falling in a gravitational potential. The water running down a mountain passes through a power plant; flows out to the ocean, where it subsequently evaporates; and gets re-precipitated in the mountains. This seems to be a perpetual-motion cycle and must, therefore, lie outside the realm of the second law of thermodynamics.

   (i) The maximum efficiency of a steam engine can be calculated using the second law of thermodynamics. If it operates between the boiling point of water and room temperature, the maximum efficiency is about

$$\frac{373 - 293}{373} = \frac{80}{373} = 0.215 \quad \text{or} \quad 21.5\%$$

Photosynthesis by green plants occurs almost entirely at ambient temperatures, yet recent publications report theoretical limits as high as 85% of the fraction of the light energy absorbed that can be converted into chemical energy (free energy or net work). Clearly, such estimates must be wrong, or else the second law cannot apply.

23. (a) Consider 1 mol of liquid water to be frozen reversibly to ice at 0°C by slowly removing heat to the surroundings. In this process the entropy of the system will (increase, remain unchanged, decrease).
    (b) If the freezing process occurs at 0°C and a pressure of 1 atm, the accompanying change in the Gibbs free energy for the system will be (positive, zero, negative).
    (c) For this process the enthalpy of the system will (increase, remain unchanged, decrease).

24. For the statements below, choose the word or words inside the parentheses that serve to make a correct statement. Each statement has at least one, and may have more than one, correct answer.
    (a) For a sample of an ideal gas, the product PV remains constant as long as the (temperature, pressure, volume, internal energy) is held constant.
    (b) The internal energy of an ideal gas is a function of only the (volume, pressure, temperature).
    (c) The second law of thermodynamics states that the entropy of an isolated system always (increases, remains constant, decreases) during a spontaneous process.
    (d) If the system is not isolated from the surroundings, then $T \Delta S$ for a system undergoing an isothermal, reversible process is (less than, equal to, greater than) the heat absorbed.
    (e) $T \Delta S$ for an isothermal, spontaneous irreversible process is (less than, equal to, greater than) the heat absorbed by the system.
    (f) When a sample of liquid is converted reversibly to its vapor at its normal boiling point, (q, w, ΔP, ΔV, ΔT, ΔE, ΔH, ΔS, ΔG, none of these) is equal to zero for the system.
    (g) If the liquid is permitted to vaporize isothermally and completely into a previously evacuated chamber that is just large enough to hold the vapor at 1 atm pressure, then (q, w, ΔE, ΔH, ΔS, ΔG) will be smaller in magnitude than for the reversible vaporization.

25. Starting with the definition of enthalpy, $H \equiv E + PV$, and additional relations based on the first and second laws of thermodynamics, derive the following:
    (a) dH as a function of T, S, V, and P.
    (b) Equations for $(\partial H / \partial P)_S$ and $(\partial H / \partial S)_P$.
    (c) The equation $(\partial T / \partial P)_S = (\partial V / \partial S)_P$.
    (d) The result of part (a) starting from Eq. (3.47) and the definition $G \equiv H - TS$.

26. The earth's atmosphere behaves as if it is approximately isentropic—the molar entropy of air is a constant independent of altitude up to about 10 km. It is well known that pressure and temperature vary with altitude. Using the isentropic model, calculate the temperature of the atmosphere 10 km above the earth, where the pressure is found to be 210 Torr. The temperature and pressure at the surface of the earth (sea level) are 25°C and 760 Torr, respectively. Assume that air behaves like an ideal gas with $C_P = \frac{7}{2} R$. You may ignore gravitation influences. (The temperature measured during rocket flights above New Mexico is −50°C at 10 km above the earth.)

27. The temperature of a typical laboratory deep-freeze unit is −20°C. If liquid water in a completely filled, closed container is placed in the deep-freeze, estimate the maximum pressure developed in the container at equilibrium. Assume liquid water and ice to be completely incompressible. The enthalpy of fusion of water may be taken as 333.4 kJ kg^{-1}, independent of temperature and pressure, and the densities of ice and liquid water at −20°C are 0.9172 and 1.00 g cm^{-3}, respectively.

4

Free Energy and Chemical Equilibria

CONCEPTS

Living organisms depend on spontaneous processes to maintain life. As spontaneous processes occur in a system, it tends towards a state of equilibrium where macroscopic change ceases. *Le Châtelier's principle* describes how a stress or perturbation on a system at equilibrium causes it to shift towards a new equilibrium state; the concentrations change so as to minimize the effect of the perturbation. Thermodynamics can help us to characterize and distinguish systems undergoing spontaneous processes from those at equilibrium; at constant temperature and pressure, free energy characterizes spontaneous processes. Chemical potential (partial molal Gibbs free energy) is analogous to gravitational potential energy. Imagine a countryside (potential energy surface) of hills and valleys. Rolling from the top of a hill into a valley is a spontaneous process that is accompanied by a decrease in potential energy. In the valley there is a local minimum in the potential energy. Other nearby valleys may have lower potential energies, but it will require a boost to get out of the potential minimum in the valley.

Rolling from the top of the hill into a valley may occur rapidly or slowly, depending on whether the surface is smooth rock or deep sand. The change in potential energy is the same in each case. Similarly, thermodynamics does not tell us how rapidly spontaneous processes occur, even though that is important for living organisms. It simply indicates the downhill direction.

In chemical systems, many factors such as concentration, temperature, pressure, and chemical bonding influence the chemical potential. However, increasing the concentration of any species always increases its chemical potential. Thus, by changing the concentrations of products or reactants we can favor the forward reaction or the backward reaction. At equilibrium the chemical potential of the products is equal to that of the reactants, and no further changes in concentrations will occur. If we change the temperature or pressure, this will tilt the potential surface so that the chemical potential of reactants and products is no longer equal. Now a change can occur spontaneously so as to restore equilibrium, but at different concentrations of reactants and products.

Although thermodynamics treats matter as macroscopic, it is also possible to consider equilibria from a microscopic or molecular point of view. A chemical system contains molecules characterized by their different kinetic energies (such as translational, rotational, vibrational energies) and potential energies (intermolecular and intramolecular interactions). The molecules can be highly ordered in crystals, for example, or they can be disordered in liquid and gaseous mixtures. These descriptions correlate with the enthalpy (energy) and entropy (order) contributions, respectively, to the free energy or chemical potential. The thermodynamic properties of macroscopic systems can be obtained by averaging over the distribution of energies available to the molecules. The energies of the molecules can be treated classically, or the quantized energy levels of the system can be treated using quantum mechanics (see Chapter 9). We must also consider the different arrangements of the molecules which correspond to each energy level. Each molecular arrangement is called a microscopic state. The distribution of many molecules among the available states is obtained using statistical analysis of large numbers (see Chapter 11). At constant temperature and pressure, spontaneous processes are associated with the tendency of large collections of molecules to reach a minimum in their enthalpy (total energy) while simultaneously increasing the entropy (disorder or randomness) of the collection to a maximum. When these two tendencies oppose one another, a balance between them is reached so that the chemical potential is minimized at the end of the spontaneous process. For example, if we spill a box of red marbles and a box of blue marbles down a hill, they will arrive at the bottom (minimum energy) all mixed up (maximum entropy). By examining the microscopic states at the beginning and at the end, we can evaluate the change in chemical potential that has occurred.

APPLICATIONS

It is challenging to describe a complex biological system in terms of thermodynamics, so we begin by separating the complex process into its components. Important components that can be analyzed thermodynamically include (1) metabolic reactions in which chemical bonds are broken and new ones formed, (2) dissociation of H^+ from acidic compounds and binding of H^+ to bases, (3) oxidation-reduction reactions in which electron transfer occurs, (4) interactions involving the aqueous medium in which metabolites and ionic species occur in the cytoplasm or other biological fluids, and (5) the assembly and disassembly of membranes and other multicomponent cellular structures.

By establishing the relation for how the free energy depends on the concentrations of species we can give quantitative expression to Le Châtelier's principle. The Gibbs free energy is a logarithmic function of the activities or concentrations of components of a reacting system. The free energy change for the reaction is the difference in free energies of products and reactants. Increasing the concentrations of reactant species causes the free energy change, ΔG, to become more negative (more spontaneous); increasing the product concentrations has the opposite effect. When the reaction proceeds to the equilibrium state, the ΔG goes to zero. Once the system is in the equilibrium state, it will not change further unless it is perturbed in some way. Associated with the equilibrium state is a fixed ratio of reactant and product concentrations, known as the *equilibrium constant*.

Metabolic Processes

Living organisms work hard to avoid the equilibrium state. Spontaneous processes in living organisms can be usefully described in terms of *stationary states,* where the concentrations, free energy changes, and so forth are not a function of time. Because the processes continue to occur spontaneously, however, the ΔG between reactants and products remains negative in such a stationary state, and the living system is not at equilibrium. The situation is somewhat like that for water in a lake in the mountains. So long as the streams feeding into it and the river leading out of it flow at constant rates, the lake level does not change with time. But it is not at equilibrium. This is easily demonstrated by blocking the inlets (or outlet) and watching the water level of the lake fall (or rise) accordingly.

Metabolic processes involve an enormous number of biochemical reactions operating together. The product of one reaction serves as the reactant for the next. The same compound may serve as a reactant for several parallel reactions which produce different products. To establish some degree of order in such a complex interacting metabolic network, we make use of a set of standard states for reference purposes. Just as it is usually more convenient to measure altitudes relative to sea level, rather than from the center of the earth, we chose *biochemical standard states* for convenience. Standard states specify the conditions for which the activities of all components are equal to 1 ($a = 1$); we establish standard states for solutes, as well as for the solvent, in solution. Biochemists prefer a standard state at pH = 7, which is more meaningful for cellular metabolism than the chemists' standard state at pH = 0 ($a_{H^+} = 1$). The temperature is not specified in the definition of a standard state, so it always needs to be stated; 25°C is used for most of the tables of standard thermodynamic values.

Once standard states are defined, it is straightforward to use equilibrium measurements to establish relations among the standard free energies, standard enthalpies, and standard entropies for each compound in its standard state. The mathematical relations between the thermodynamic properties of a metabolite and its activity in the living environment allow us to calculate its thermodynamic properties in the stationary metabolic state. We can then combine this information algebraically to describe the thermodynamics of metabolism. An advantage of working with state properties like G, H, and S is that we do not need to describe (or even know) all of the details of what is going on inside a cell. It is sufficient for thermodynamic purposes to know what is going into the system at the beginning and what is coming out at the end.

Ideal and Nonideal Behavior

Thermodynamic relations were developed first for ideal systems (gases, solutions, etc.), where many of the interactions between substances are ignored. The biochemical world is quite different. The behavior of DNA in an aqueous solution in distilled water is dramatically different from that in the presence of dissolved NaCl. Salts, acids, and bases that produce charged ions in aqueous solution have particularly large effects, owing to strong electrostatic interactions with other solutes. Biological macromolecules may tie up large amounts of solvent water in a hydration layer; small ions at large concentrations can produce a similar effect on solvent water. This tied-up water is less available as solvent for the other metabolites present, which increases their effective concentrations. Corrections need to be made using *activity coefficients* to describe how the real solution behavior differs from that of the ideal solution. Although these corrections are often not easy to make, they can be extremely important in describing biochemical processes realistically.

Oxidation-Reduction Reactions

Redox reactions involving electron transfer between donors and acceptors are widespread in biochemistry. O_2 from the air combines with carbohydrates, fats, and proteins in food to initiate a series of redox reactions that result in the production of metabolic energy. In green plants sunlight provides a source of chemical potential that drives the process of converting CO_2 and H_2O into carbohydrates and O_2 in a process that would otherwise be nonspontaneous. Because many of the essential components of such metabolic redox reactions can be studied electrochemically under carefully controlled conditions, they provide a convenient approach to determining the related thermodynamics in terms of electrochemical potentials. The *Nernst equation* allows us to make the connection to the corresponding Gibbs free energy changes.

Ligand Binding

Among essential dynamic processes in living cells is the class of noncovalent interactions that bind ligands like O_2 to hemoglobin, substrates to enzymes, protons to bases, complementary strands of DNA or RNA to one another, and repressor proteins to chromosomal DNA. Such interactions can be described at the molecular level using suitable statistical models. In many cases such models can be used to make useful predictions about macroscopic behavior, such as cooperativity in ligand binding. Interactions between neighboring sites in proteins or nucleic acids are important in understanding factors that govern the relative stability of different conformations (helix, random coil, and so forth) available to the biopolymers in solution.

IDEAL GASES

Free Energy Changes

Nitrogen is an essential element for the growth of all living organisms. Most organisms are unable to use N_2 directly from the atmosphere, but require it first to be reduced to ammonia

or oxidized to nitrite or nitrate. The nitrogen-fixing bacteria, some of which live symbioti-
cally with plants in root nodules, use a powerful reductant to convert N_2 metabolically (at
high energy cost) to reduced nitrogen compounds (NH_3), which can then be transferred to
the plant. To appreciate the thermodynamic relations involved in this complex process, we
will first look at the synthetic method developed by the chemist Fritz Haber to reduce atmospheric N_2 by H_2 to produce NH_3, which can then be used directly as a chemical fertil-
izer.

Because N_2, H_2, and NH_3 are all gases under the operating conditions for the Haber
process, we will express their concentrations as partial pressures. Consider a reaction
chamber where we can control the pressure of each gaseous reactant and product, such as
the flow reactor shown in Fig. 4.1. We flow into the reactor $N_2(g)$ and $H_2(g)$ at pressures
P_{N_2} and P_{H_2}. These pressures are completely under our control. The gases are both introduced at temperature T, which we choose. A reaction occurs spontaneously in the chamber
to produce some $NH_3(g)$. The $NH_3(g)$ is cooled, if necessary, to temperature T, and its partial pressure, P_{NH_3}, is determined. We can control P_{NH_3} by adjusting flow rates and experimental conditions in the reactor so as to maintain it in a stationary state.

We want to calculate the Gibbs free energy change for the reaction. We have speci-
fied the initial state (P_{N_2}, P_{H_2}, T) and the final state (P_{NH_3}, T), so the free energy change
is determined once we specify how many moles have reacted. Several points are worth
noting:

1. It does not matter what happens in the reactor. The temperature may be higher than T; it
   may be nonuniform. The partial pressures of reactants and products may change rapidly
   in an unknown way in the reactor. There may or may not be catalysts in the reactor.
   None of this matters as long as we know P_{N_2}, P_{H_2}, P_{NH_3}, and T.
2. We need not have stoichiometric amounts of N_2 and H_2. Any unreacted N_2 or H_2 is unchanged and therefore does not contribute directly to the change of free energy. Excess
   reactants may affect the partial pressure of NH_3, but we determine P_{NH_3}.
3. We emphasize once more that we can choose any partial pressures of products or reactants.

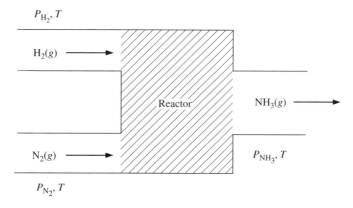

Fig. 4.1 Flow reactor in which $H_2(g)$ at P_{H_2} and T and $N_2(g)$ at P_{N_2} and T are
introduced, and the product $NH_3(g)$ is removed at P_{NH_3} and T.

If all of the pressures are 1 atm and the temperature is 25°C, then we can calculate the free energy change, ΔG_{298}^0, from the data in the Appendix. For any other temperature T we can use Eqs. (3.29) through (3.32) to find ΔG_T^0. What about other partial pressures? To obtain a ΔG_T that is general for any pressures, we must use an alternative path with individual steps that we can calculate. For the formation of 1 mol of $NH_3(g)$, the two paths are

$$\frac{1}{2} N_2(g, P_{N_2}) + \frac{3}{2} H_2(g, P_{H_2}) \xrightarrow{\Delta G_T} NH_3(g, P_{NH_3})$$

$$\downarrow \Delta G_1 \qquad\qquad \uparrow \Delta G_2$$

$$\frac{1}{2} N_2(g, 1\ atm) + \frac{3}{2} H_2(g, 1\ atm) \xrightarrow{\Delta G_T^0} NH_3(g, 1\ atm)$$

where $\Delta G_T = \Delta G_1 + \Delta G_T^0 + \Delta G_2$, and ΔG_1 and ΔG_2 are free energy changes due to changing the pressure of the (assumed) ideal gases. From Eq. (3.35),

$$G(P_2) - G(P_1) = nRT \ln \frac{P_2}{P_1} \qquad (3.35)$$

$$\Delta G_1 = \frac{1}{2} RT \ln \frac{1}{P_{N_2}} + \frac{3}{2} RT \ln \frac{1}{P_{H_2}}$$

$$= RT \ln \frac{1}{P_{N_2}^{1/2}} + RT \ln \frac{1}{P_{H_2}^{3/2}} \qquad (4.1)$$

We have used the algebraic identity $a \ln x = \ln x^a$. Similarly, we find

$$\Delta G_2 = RT \ln P_{NH_3} \qquad (4.2)$$

Equating the free energy changes along both paths linking initial and final states

$$\Delta G_T = \Delta G_T^0 + \Delta G_1 + \Delta G_2$$

$$\Delta G_T = RT \ln \frac{1}{P_{N_2}^{1/2}} + RT \ln \frac{1}{P_{H_2}^{3/2}} + \Delta G_T^0 + RT \ln P_{NH_3}$$

$$\Delta G_T = \Delta G_T^0 + RT \ln \frac{P_{NH_3}}{(P_{N_2})^{1/2}(P_{H_2})^{3/2}} \qquad (4.3)$$

Equation (4.3) is what we want. It relates the free energy change for any partial pressures, ΔG_T, to the standard free energy change at 1 atm partial pressures, ΔG_T^0. It should be clear now how important it is to note the presence or absence of the superscript "0". Its presence means standard conditions of 1 atm; its absence means different conditions, which must be specified. The superscript "0" is less important for ΔH, because enthalpy does not change very much with pressure, even for gases. However, for ΔG it is vital.

It is important to remember that the partial pressures in Eq. (4.3) must be in atmospheres. That is, the P's in the equation are actually the ratios of P atm to 1 atm, which is the standard state. It may help to remember that algebraically one cannot take the logarithm of a unit; therefore, one must be sure to use the correct units (here atm), which cancel those of the standard conditions.

Example 4.1 What will be the free energy, relative to that under standard conditions, of forming 1 mol of NH_3 at 298 K if (a) 10.0 atm of N_2 and 10.0 atm of H_2 are reacted to give 0.0100 atm of NH_3; (b) 0.0100 atm of N_2 and 0.0100 atm of H_2 are reacted to give 10.0 atm of NH_3? For part (b), NH_3 could be introduced with the reactants to give a high enough product partial pressure. The reaction is

$$\tfrac{1}{2} N_2(g, P_{N_2}) + \tfrac{3}{2} H_2(g, P_{H_2}) \longrightarrow NH_3(g, P_{NH_3})$$

Solution

(a)
$$\Delta G_{298} - \Delta G^0_{298} = RT \ln \frac{P_{NH_3}}{(P_{N_2})^{1/2}(P_{H_2})^{3/2}} \tag{4.3}$$

$$= (8.314 \text{ J K}^{-1} \text{ mol}^{-1})(298 \text{ K}) \ln \frac{0.010}{(3.16)(31.6)}$$

$$= -2.28 \times 10^4 \text{ J mol}^{-1}$$

The free energy is decreased (relative to the standard free energy) by having high pressures of reactants and low pressures of products; the reaction is favored relative to standard conditions.

(b)
$$\Delta G_{298} - \Delta G^0_{298} = (8.314 \text{ J K}^{-1} \text{ mol}^{-1})(298 \text{ K}) \ln \frac{10.0}{(0.1000)(0.0010)}$$

$$= 2.85 \times 10^4 \text{ J mol}^{-1}$$

The free energy is increased by having high pressures of products and low pressures of reactants; the reaction is not favored relative to the standard conditions. The calculated free energies (with units of J mol^{-1}) are for the numbers of mol in the chemical reaction shown in the equation. We can equally correctly give the ΔG in J. Equation 4.3 was derived from $\Delta G = nRT \ln P = RT \ln P^n$ so n, the numbers of mol, cancel the mol^{-1} in the units of R.

We can generalize Eq. (4.3) for any reaction involving only (ideal) gases. For a reaction of ideal gases at temperature T,

$$aA + bB \longrightarrow cC + dD$$

$$\Delta G_T = \Delta G^0_T + RT \ln Q \tag{4.4}$$

$$Q = \frac{(P_C)^c(P_D)^d}{(P_A)^a(P_B)^b} \tag{4.5}$$

where P_A, P_B, P_C, and P_D = partial pressure (in atm) of each reactant or product. The quotient, Q, is the ratio of the arbitrary partial pressures (in atm) of reactants and products, each raised to the power of its coefficient in the chemical equation. If we double the number of moles involved in the reaction, each of the coefficients is multiplied by 2, and Q is squared in Eq. (4.5). We see that Q will be large if product pressures are large or reactant pressures are small. A large Q means a positive (unfavorable) contribution to the free energy. Q is small if product pressures are small or reactant pressures are large. A small Q means a negative (favorable) contribution to the free energy. The following table gives quantitative values for the change of free energy with Q at 25°C. The reader can easily calculate values at any other temperature, T, by multiplying by $(T/298)$.

Q	$\Delta G_{298} - \Delta G^0_{298}$ (J)
100	11,410
10	5,705
1	0
0.1	−5,705
0.01	−11,410

We have emphasized here the distinction between standard free energy changes, ΔG^0_T, for defined concentrations, and actual free energy changes, ΔG_T, for any other concentrations specified by Q.

Equilibrium Constant

For every chemical reaction at any temperature there are partial pressures of products and reactants for which the system is at equilibrium. We can find these pressures by letting the reaction attain equilibrium. That is, we wait for equilibrium to be attained at a chosen T, then we measure the equilibrium partial pressures of products and reactants. When these equilibrium partial pressures are substituted into the expression for Q, we obtain the equilibrium constant, K. For the reaction

$$a\text{A} + b\text{B} \longrightarrow c\text{C} + d\text{D}$$

the equilibrium constant is

$$K = \frac{(P^{eq}_C)^c (P^{eq}_D)^d}{(P^{eq}_A)^a (P^{eq}_B)^b} \tag{4.6}$$

where P^{eq}_A, P^{eq}_B, and so on, are partial pressures (in atm) at equilibrium. Equation (4.4) of course still applies, but at equilibrium the free energy change for the reaction is zero. Therefore,

$$0 = \Delta G^0_T + RT \ln K \quad \text{(at equilibrium)} \tag{4.7}$$

$$\Delta G^0_T = -RT \ln K$$

Note that determination of the equilibrium constant experimentally permits calculation of the *standard* free energy change. Combining Eqs. (4.4) and (4.7), we have

$$\Delta G_T = -RT \ln K + RT \ln Q$$

$$= RT \ln \frac{Q}{K} \tag{4.8}$$

Equations (4.4) through (4.8) may be the most useful thermodynamic equations a bio-chemist learns. They relate the free energy change for a reaction to experimentally mea-surable quantities. Furthermore, Eq. (4.6), which defines the equilibrium constant, presents a ratio of equilibrium partial pressures that is constant for a chemical reaction at a given temperature. At this point the reader may not be convinced of the importance of these equa-tions, because they were all derived for ideal gases and the biochemist reader is presumably more interested in reactions in solution. The introduction of *activity* in the next section pro-vides the necessary connection between the real world and Eqs. (4.4) through (4.8).

Anticipating applications to reactions other than those involving ideal gases, we shall write Eq. (4.7) in different but equivalent forms. Solving for K, we obtain

$$K = e^{-\Delta G^0/RT}$$

or

$$K = 10^{-\Delta G^0/2.303RT} \tag{4.9}$$

Substituting for ΔG^0 gives

$$K = e^{\Delta S^0/R} e^{-\Delta H^0/RT}$$
$$= 10^{\Delta S^0/2.303R} 10^{-\Delta H^0/2.303RT} \tag{4.10}$$

(Remember that $e^a e^b = e^{a+b}$.) The superscript is vital on ΔG and ΔS, which depend markedly on concentration; but it is not so necessary for ΔH, which is approximately in-dependent of concentration.

Example 4.2   Calculate the equilibrium constant at 25°C for the decarboxylation of liquid pyruvic acid to form gaseous acetaldehyde and CO_2.

Solution   The reaction is

$$\underset{\substack{\text{O}\\ \|}}{CH_3CCOOH}(l) \longrightarrow \underset{\substack{\text{O}\\ \|}}{CH_3CH}(g) + CO_2(g)$$

From the Appendix,

$$\Delta G^0 = \Delta G_f^0(\text{acetaldehyde}) + \Delta G_f^0(CO_2) - \Delta G_f^0(\text{pyruvic acid})$$
$$= -133.30 + (-394.36) - (-463.38)$$
$$= -64.28 \text{ kJ}$$

From Eq. (4.9),

$$K = 10^{-\Delta G^0/2.303RT} \tag{4.9}$$

The factor $2.303RT = 5.708$ kJ mol^{-1} for 25°C; it will occur often and it is useful to remember this value for 25°C.

$$K = 10^{-\Delta G^0/5.708} \quad (\Delta G^0 \text{ in kJ})$$

$$= 10^{64.28/5.708}$$

$$= 10^{11.26}$$

$$K = 1.85 \times 10^{11}$$

Pyruvic acid is very unstable with respect to CO_2 and acetaldehyde, but a suitable catalyst is needed at room temperature to cause a rapid reaction leading to equilibrium.

SOLUTIONS

Nonideal Behavior

Many simple relations like the ideal gas law and the equilibrium relation between reactants and products do not hold exactly over a wide range of pressures, concentrations, or other conditions. For example, using the standard free energies of formation (Table A.5), the dissociation of water into $H^+(aq)$ and $OH^-(aq)$ at 298 K

$$H_2O(l) \;\rightleftharpoons\; H^+(aq) + OH^-(aq)$$

$$\Delta G^0_{298} = 0 - 157.244 - (-237.129) = 79.885 \text{ kJ}$$

corresponds to an equilibrium constant $K_w = 1.011 \times 10^{-14}$. When the concentrations $[H^+]$ and $[OH^-]$ are measured carefully in pure deionized water at 25°C, the product of the two concentrations, 1.008×10^{-14}, is very close to this value. Addition of a small amount of KCl, which is neutral salt and does not undergo hydrolysis, nevertheless changes the value of the product $[H^+][OH^-]$ to 1.24×10^{-14} in 0.01 M KCl and to 2.8×10^{-14} in 1 M KCl. This nonideal behavior of the self-ionization of water in solutions containing neutral salts is a consequence of coulombic interactions among the ions present. In addition, at the higher salt concentrations, a significant fraction of the bulk water (the solvent in this system) is tied up in the hydration spheres of the dissociated K^+ and Cl^-. For both of these reasons the *effective* concentrations of the H^+ and OH^- are decreased relative to their actual concentrations (in moles of solute per liter of solution). The effective (thermodynamic) concentration is called the *activity* of the species, a. Activities are defined in such a way that the thermodynamic equilibrium relation, such as $K_{298} = [a_{H^+}][a_{OH^-}] = 1.011 \times 10^{-14}$, is always maintained, even though the product of the corresponding concentrations in real solutions may differ significantly from this value.

The nonideal behavior of materials is of great practical importance. Consider, for example, the synthesis of NH_3 from N_2 and H_2 that we examined earlier in this chapter. Le Châtelier's principle clearly indicates the advantage of working at higher pressures to favor the formation of ammonia relative to the reactants. In fact, the commercial process is carried out at about 200 atm pressure. The gases exhibit significantly nonideal behavior under these conditions. This can be seen by comparing the gas concentrations (mol L^{-1}) calculated from the ideal gas equation with the experimentally observed concentrations at 200 atm. If we make this comparison at 298 K, we find that the "ideal concentrations" for each of the gases is the same.

$$[A] = \frac{n_A}{V} = \frac{P_A}{RT} = \frac{200 \text{ atm}}{(0.0821 \text{ L atm K}^{-1} \text{ mol}^{-1})(298 \text{ K})} = 8.18 \text{ mol L}^{-1}$$

For the real gases (considered separately, rather than in an equilibrium mixture) at 200 atm pressure and 25°C, the effective concentrations are quite different: $[H_2] = 7.2$ mol L^{-1}, $[N_2] = 7.6$ mol L^{-1}, and NH_3 is liquefied under these conditions.

Biological molecules also encounter significantly nonideal environments. Sea water, for example, contains dissolved Na^+, Mg^{2+}, Cl^-, and other ions such that it is equivalent to a 0.68 M NaCl solution. Many biological fluids have comparable ionic strengths. The *ionic strength*, μ, is defined as one-half the sum of the concentrations of each ion present multiplied by the square of its charge. Thus, divalent ions contribute four times as much to the ionic strength as univalent ions at the same concentrations. For solutions of uni-univalent electrolytes like NaCl, however, the ionic strength is equal to the concentration.

$$\mu = \tfrac{1}{2} \left([Na^+](1)^2 + [Cl^-](1)^2 \right)$$

Because of the high ionic strength of biological fluids (blood plasma, cytoplasm), nonideal behavior needs to be considered for comparison of many observed properties with those derived from thermodynamic relations.

Nonideal behavior is important for the solubility of proteins. All proteins have limited solubility in pure water. In dilute (10^{-2} M) salt solutions, the solubility is increased somewhat because of the effect of the added ions on the activity of the proteins. At higher concentrations (>1 M), however, this situation is reversed, and proteins are "salted out" of solution. This is the basis of the widely used method of adding high concentrations of ammonium sulfate, $(NH_4)_2SO_4$, to decrease the solubility of proteins and precipitate them from solution. Because the effect depends on the net charge on the protein, differential precipitation using added $(NH_4)_2SO_4$ can be used to separate different proteins from one another in complex biological fluids, and advantage can be taken of changing the pH of the precipitation medium to alter the net charges of different proteins relative to one another.

Other properties that are influenced by nonideal behavior include osmotic pressure, freezing point lowering, gas solubility, reduction potential, and pH of a buffer solution.

Activity and Chemical Potential

It was clear to thermodynamicists that Eqs. (4.4) through (4.8) were simple and easy to use; therefore, it would be convenient if they also applied to real gases, liquids, solids, and solutions. The thermodynamicists decided that the only way to ensure this was to define Eqs. (4.4), (4.7), and (4.8) to be generally true, and to change Eqs. (4.5) and (4.6) for Q and K so as to agree. This is done by introducing a new quantity, the activity, a. For our usual reaction of $aA + bB \rightarrow cC + dD$, the reaction quotient

$$Q \equiv \frac{(a_C)^c (a_D)^d}{(a_A)^a (a_B)^b} \tag{4.11}$$

the equilibrium constant

$$K \equiv \frac{(a_C^{eq})^c (a_D^{eq})^d}{(a_A^{eq})^a (a_B^{eq})^b} \tag{4.12}$$

where a_A, a_B, etc. = activities of each component

a_A^{eq}, a_B^{eq}, etc. = activities of each component at equilibrium

We often omit the superscript "eq" when writing the expression for K. However, *equilibrium* activities are understood. Equations (4.11) and Eq. (4.4) define activities. Activities are unitless numbers, so there is no problem with taking their logarithms. By substituting $c\bar{G}_C + d\bar{G}_D - a\bar{G}_A - b\bar{G}_B = \Delta G$ and a similar expression for ΔG^0 in Eq. (4.4) and using Eq. (4.11) for Q, we obtain

$$\bar{G}_A = \bar{G}_A^0 + RT \ln a_A \tag{4.13}$$

where $\bar{G}_A$ = free energy per mole of A in the mixture of molecules, A, B, C, D

$\bar{G}_A^0$ = standard free energy per mole of A

a_A = activity of A in the mixture (unitless)

Identical expressions hold for B, C, and D, so Eq. (4.13) defines the activity of any molecule. We repeat that activities were introduced so that Eqs. (4.4) through (4.8), which relate free energy to measurable quantities, would keep a simple form. Both $\bar{G}_A$ and $\bar{G}_A^0$ depend on temperature, and $\bar{G}_A$ depends on the concentration not only of A but, in general, to some extent on the concentrations of B, C, and D as well.

These molar free energies are also called *chemical potentials,* μ. The definition of chemical potential is

$$\mu_A \equiv \bar{G}_A \tag{4.14}$$

$\bar{G}_A$ is also called a *partial molal free energy,* because it can be defined in terms of a partial derivative. We shall return to partial molal quantities at the end of this chapter.

An important aspect of Eq. (4.13) is that activity is defined with respect to a standard state. The difference in free energy between two states, $\bar{G}_A - \bar{G}_A^0$, is a measurable quantity. But $\bar{G}_A$ can be obtained only after the standard conditions (standard state) are defined. Once $\bar{G}_A$ and $\bar{G}_A^0$ are determined, the activity, a_A, can be obtained. However, until the standard state is specified, the activity cannot be determined.

Standard States

Equation (4.13) tells us how to determine the activity for the substance A in a system. We determine the difference in free energy per mole between A in the system and A in a specified standard state. This difference allows us to calculate a_A.

$$\ln a_A = \frac{\bar{G}_A - \bar{G}_A^0}{RT} \tag{4.15}$$

It follows immediately that the *activity of a substance in its standard state is equal to unity.* In the following paragraphs we will give the standard states that are commonly used in specifying activities. However, the methods to measure free energies of molecules in solution will be discussed later.

Ideal gases

The standard state for an ideal gas is the gas with a partial pressure equal to 1 atm. The activity of an ideal gas is defined as its actual partial pressure divided by 1 atm, its partial pressure in its standard state:

$$a_A \equiv \frac{P_A \text{ (atm)}}{1 \text{ atm}} \tag{4.16}$$

where P_A = partial pressure of ideal gas (in atm). The standard conditions for a reaction involving only ideal gases are that each product and reactant has a partial pressure equal to 1 atm, as discussed in Chapter 3.

Real gases

The standard state for a real gas is defined as the extrapolated state where the pressure is 1 atm but the properties are those extrapolated from very low pressure. The activity of a real gas is a function of pressure, which we write

$$a_A = \gamma_A P_A \tag{4.17}$$

where γ_A = activity coefficient
P_A = partial pressure, atm

We know that all gases become ideal at low-enough pressures, so γ_A must become 1 as the total pressure in the system approaches zero. Near atmospheric pressure the activity coefficient of gases is close to 1, so we will treat real gases as if they behave ideally.

Pure solids or liquids

The standard state for a solid or liquid is the pure substance (solid or liquid) at 1 atm pressure. Therefore, the activity is equal to 1 for a pure solid or liquid at 1 atm. The free energy of a solid or liquid changes so slightly with pressure [see Eq. (3.33)] that we can usually neglect the change and use 1 for the activity of a solid or liquid at any pressure:

$$a_A = 1$$

This convention is familiar from elementary chemistry courses.

Solutions

By solution we mean a homogeneous mixture of two or more substances. We can have solid solutions, liquid solutions, or very complicated mixtures of components as might be found in a biological cell. The activity of each substance will depend on its concentration and the concentrations of everything else in the mixture. We write this dependence in a deceptively simple equation:

$$\text{activity} = (\text{activity coefficient}) \cdot (\text{concentration}) \tag{4.18}$$

The activity coefficient is not a constant; it incorporates all the complicated dependence of the activity of A on the concentrations of A, B, C, and so on. Another complication is that different concentration units are routinely used in Eq. (4.18). We will discuss each of the commonly used concentration units and discuss the standard state connected with each unit.

Mole fraction and solvent standard state

The standard state that uses mole fraction as a concentration unit is often called the solvent standard state. *The solvent standard state for a component of a solution defines the pure component as the standard state.* Let us choose the solvent standard state for component A in a solution. A useful concentration unit is the *mole fraction X,* the number of moles of species A divided by the total number of moles of all components present in the solution:

$$X_A = \frac{n_A}{n_T} \tag{4.19}$$

where n_A = number of moles of A
n_T = total number of moles of all components

The activity of A, then, from Eq. (4.18),

$$a_A = \gamma_A X_A \tag{4.20}$$

where a_A = activity of species A
γ_A = activity coefficient of A on the mole fraction scale
X_A = mole fraction of A

We choose the standard state so that the activity, a_A, becomes equal to the mole fraction, X_A, as the mole fraction of A approaches unity. Mathematically, we write this as

$$\lim_{X_A \to 1} a_A = X_A$$

which is read "in the limit as X_A approaches 1, $a_A = X_A$." Because $\gamma = a_A/X_A$, we can express the same idea as

$$\lim_{X_A \to 1} \gamma = 1$$

In the limit as X_A approaches 1, $\gamma = 1$. Choosing $a_A \to X_A \to 1$ defines the standard state, the state where the activity is 1. For a liquid solution, the standard state for the solvent is defined as the pure liquid. The best example is dilute aqueous solutions, for which $a_{H_2O} = 1$ is used in introductory chemistry courses. We ignore the water in writing equilibrium constants in aqueous solutions. The logic is that we are using an activity on the mole fraction scale for the H_2O and that the solution is dilute, so $X_{H_2O} \approx 1$ and $a_{H_2O} \approx 1$. The mole fraction scale for activities is traditionally used for the *solvent* in a solution.

We now summarize the useful equations for a solvent in a solution:

Very dilute solution: $\quad a_{solvent} = 1 \tag{4.21}$

Dilute solution or ideal solution: $\quad a_{solvent} = X_{solvent} \tag{4.22}$

Real solution: $\quad a_{solvent} = \gamma X_{solvent} \tag{4.23}$

Equation (4.22) defines an ideal solution: for any concentration the solvent has the properties of the solvent in a dilute solution. To determine the activity coefficient and therefore the activity in Eq. (4.23) we have to measure the free energy of the solvent in the solution. This can be done by methods involving measurements of the vapor pressure of the solvent in a solution, the freezing-point depression, the boiling-point elevation, or the osmotic pressure. These will be discussed in Chapter 5.

Solute standard states

For solutions in which certain components never become very concentrated, such as dilute aqueous salt solutions, we define a solute standard state whose free energy can be obtained from measurements on the dilute solution. *The solute standard state for a component is defined as the extrapolated state where the concentration is equal to 1 molar (M) or 1 molal (m), but the properties are those extrapolated from very dilute solution.* The solute standard state is a hypothetical state in the sense that it corresponds to a 1 M or 1 m *ideal* solution. For real solutions the ideal behavior is not approached except for concentrations that are much less than 1 M or 1 m. The definition of the solute standard state should become clearer after reading the following paragraphs.

Molarity

Concentrations are commonly measured in *molarity,* with units of mol L^{-1}. We shall use the symbol c for molarities. The activity of solute B on the molarity scale and with a solute standard state is, from Eq. (4.18)

$$a_B = \gamma_B c_B \qquad (4.24)$$

where a_B = activity of species B

γ_B = activity coefficient of B on the molarity scale

c_B = concentration of B, mol L^{-1}, M

One should immediately note that the activity of B, a_B, for a molecule B in a solution will be different if one uses Eq. (4.24) rather than Eq. (4.20). The free energy per mole of B in a specified solution has a definite value, but different choices of standard states produce different values of a_B. The standard state for the molarity scale is chosen so that the activity becomes equal to the concentration as the concentration approaches zero.

$$\lim_{c_B \to 0} a_B = c_B$$

and

$$\lim_{c_B \to 0} \gamma = 1$$

The molarity scale for activities is used mainly for *solutes.*

To determine the activity coefficient and therefore the activity in Eq. (4.24), we have to measure the free energy of component B in the solution. This can be done indirectly through its effect on the vapor pressure of the solvent in solution. Depending on the properties of B, various methods exist for measuring the free energy directly. If component B is volatile, its vapor pressure can be measured. Electrolytic cells provide easy methods for measuring certain activities.

The free energy in the standard state must also be measured to obtain a_B, so we must understand the definition of the standard state for the molarity scale. This is most easily seen in Fig. 4.2, in which a_B is plotted against c_B. This shows that the standard state for the molarity scale (solute standard state) is an extrapolated point. The dual requirement that $a_B = c_B$ in the limit as c_B approaches zero and that $a_B = 1$ in the standard state defines the extrapolated standard state. To obtain the free energy in the standard state, we measure $\bar{G}_B$ as a function of $\ln c_B$ in dilute solution and extrapolate linearly to $\ln c_B = 0$ ($c_B = 1$); we thus obtain $\bar{G}_B^0$, as shown in Fig. 4.3. The solute standard state is thus the state that has the properties of a very dilute solution extrapolated to a concentration of 1 M. It is a hypothetical state rather than an actual solution that can be prepared.

If a solute is an electrolyte, in aqueous solution it dissociates into its component ions. Therefore, when we speak of the partial molal free energy of NaCl in aqueous solution, we refer to the sum of the partial molal free energies of its component ions Na^+ and Cl^-:

$$
\begin{aligned}
\bar{G}_{NaCl} &\equiv \bar{G}_{Na^+} + \bar{G}_{Cl^-} \\
&= \bar{G}_{Na^+}^0 + \bar{G}_{Cl^-}^0 + RT \ln a_{Na^+} + RT \ln a_{Cl^-} \\
&= \bar{G}_{Na^+}^0 + \bar{G}_{Cl^-}^0 + RT \ln (a_{Na^+} \cdot a_{Cl^-})
\end{aligned} \qquad (4.25)
$$

We can also express

$$
\bar{G}_{NaCl} = \bar{G}_{NaCl}^0 + RT \ln a_{NaCl}
$$

in the usual way. A comparison of this expression with Eq. (4.30) shows that

$$
a_{NaCl} = a_{Na^+} \cdot a_{Cl^-} \qquad (4.26)
$$

Similarly, for Na_2SO_4, the activity in aqueous solution is

$$
a_{Na_2SO_4} = a_{Na^+}^2 \cdot a_{SO_4^{-2}} \qquad (4.27)
$$

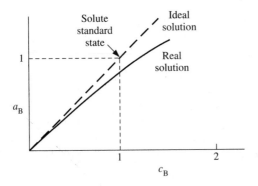

Fig. 4.2 Activity of a solute as a function of molarity for a real solution (solid curve) compared with that of an ideal solution (dashed line) extrapolated from very dilute conditions. The solute standard state lies on this extrapolated line, as shown.

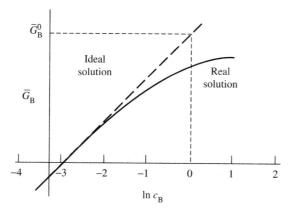

Fig. 4.3 Free energy per mole of a solute plotted against logarithm of molarity for a real solution (solid curve), compared with that of an ideal solution (dashed line) extrapolated from very dilute conditions.

Molality

Another concentration unit that is frequently used is the *molality, m,* with units of moles of solute per kilogram of solvent. The activity of B on the molality scale is

$$a_B = \gamma_B m_B \tag{4.28}$$

The discussion about molarity applies identically to molality. We have

$$\text{Dilute solution or ideal solution:} \quad a_B = m_B \tag{4.29}$$

$$\text{Real solution:} \quad a_B = \gamma_B m_B \tag{4.30}$$

The standard state is an extrapolated state; $\bar{G}_B^0$ is obtained by linearly extrapolating $\bar{G}_B$ measured in dilute solution to $\ln m_B = 0$ ($m_B = 1$). Molality is used instead of molarity for the most accurate thermodynamic measurements. Because molalities are defined by weight, not volume, they can be measured quite accurately, and they do not depend on temperature. For dilute aqueous solutions, 1 L of solution contains about 1 kg of water and therefore the numerical values of molarity and molality are very close.

Biochemist's Standard State

We have assumed in our discussion of activities and concentrations that we knew the concentration of each species involved in a system. For a molecule that dissociates in solution, this may be very difficult to determine. For example, a reaction may involve $H_2PO_4^-$; however, the species actually present in solution may include H_3PO_4, $H_2PO_4^-$, HPO_4^{2-}, and PO_4^{3-}. The distribution of these species will depend markedly on pH, so the concentration of $H_2PO_4^-$ will be difficult to specify. To simplify this situation, biochemists have chosen pH 7.0, which is near physiological pH, as their standard condition for a_{H^+}. This means that $a_{H^+} = 1$ for a concentration of $H^+ = 10^{-7} M$. The activity of each other molecule is set equal to the *total* concentration of all species of that molecule *at pH 7.0.*

$$\text{Dilute solution:} \quad a_A = \sum_i^{\text{species}} c_{i, A} \quad \text{(at pH 7.0)} \tag{4.31}$$

In our example, this sum over all species is the total concentration of phosphate added; that is, the concentration determined analytically. Knowledge of ionization constants is thus not needed, nor is it necessary to specify the concentration of each of the actual species involved in the reaction. When this biochemical standard state is used, the standard free energy is designated $\Delta G^{0'}$. $\Delta G^{0'}$ is the free energy change for a reaction at pH 7 when each product and reactant (except H^+ ion) has a total concentration of 1 M but the solution is ideal. The equilibrium is actually measured in dilute solution and the free energy is obtained by extrapolation to 1 M concentrations. The important difference between the biochemist's standard state and all the others is that the equilibrium constant applies only at pH 7. For a reaction involving, for example, the hydrolysis of adenosine triphosphate (ATP), pH can be considered a variable analogous to temperature. The biochemist's standard state is then a very practical and useful choice. The standard free energy and the equilibrium constant can be used directly from the tables at, or near, pH 7. For other pH's, one needs either to repeat the experiments to determine the new equilibrium concentrations, or to use known equilibrium constants to calculate how the concentrations of reactive species depend on pH.

To illustrate the use of the biochemist's standard state, consider the hydrolysis of ethyl acetate to produce acetic acid and ethanol.

$$CH_3\overset{\overset{\displaystyle O}{\|}}{C}OCH_2CH_3 + H_2O \rightleftharpoons CH_3\overset{\overset{\displaystyle O}{\|}}{C}OH + CH_3CH_2OH$$

$$EtOAc + H_2O \rightleftharpoons HOAc + EtOH$$

The equilibrium constant is

$$K = \frac{(a_{HOAc})(a_{EtOH})}{(a_{EtOAc})(a_{H_2O})}$$

We could use the dilute solution standard states for ethanol, acetic acid, and ethyl acetate. This means their activities would be replaced by molarities in dilute solution. For water we use the pure liquid standard state and replace a_{H_2O} by 1 in dilute solution. These standard states allow us to calculate a standard free energy from the equilibrium constant, $\Delta G^0 = -RT \ln K$. It provides information about the reaction at whatever pH occurs in the equilibrium mixture.

However, we may be interested in the reaction at other pH values; we want to consider the pH as an independent variable. We then write the reaction as

$$EtOAc + H_2O \rightleftharpoons OAc^- + H^+ + EtOH$$

$$K' = \frac{(a_{OAc^-})(a_{H^+})(a_{EtOH})}{(a_{EtOAc})(a_{H_2O})}$$

To use the biochemist's standard state we measure the equilibrium at pH 7 and set $a_{H^+} = 1$, a_{OAc^-} = sum of the concentrations of OAc^- and HOAc, and the other activities are the same as before ($a_{H_2O} = 1$, a_{EtOH}, a_{EtOAc} = concentrations). The main difference thus is that we use $a_{H^+} = 1$ (not 10^{-7} M) and a_{OAc^-} represents the sum of the acetate species (OAc^- + HOAc). The calculated $\Delta G^{0'} = -RT \ln K'$ applies at pH 7.

STANDARD FREE ENERGY AND THE EQUILIBRIUM CONSTANT

The easiest way to measure free energies of reactions, particularly complicated biochemical reactions, is to measure the equilibrium constant for the reaction. Once standard free energies are tabulated, it is easy to calculate equilibrium constants and it is straightforward to calculate equilibrium concentrations of reactants and products. We will repeat the appropriate equations (because they are so important) and give examples of how the various standard states are actually used.

$$\Delta G^0 = -RT \ln K \tag{4.7}$$

Experimental determination of the equilibrium constant allows us to calculate the *standard* free energy change for the reaction, with all products and reactants in their standard states. The free energy change, ΔG, for the reaction at equilibrium is of course zero. The free energy change at arbitrary concentrations specified by Q is

$$\Delta G = \Delta G^0 + RT \ln Q \tag{4.4}$$

We will consider several reactions and see how the equilibrium constant and the standard free energy depend on the choice of standard states. We begin with the dissociation of acetic acid, HOAc, into hydrogen ions, H^+, and acetate ions, OAc^-, in aqueous solutions:

$$HOAc(aq) \rightleftharpoons H^+(aq) + OAc^-(aq)$$

The equilibrium constant is

$$K = \frac{(a_{H^+})(a_{OAc^-})}{(a_{HOAc})} \tag{4.32}$$

If we choose the molarity scale for all species, $a = \gamma c$ and

$$K = \frac{(c_{H^+})(c_{OAc^-})}{(c_{HOAc})} \frac{\gamma_+ \gamma_-}{\gamma_{HOAc}} = K_c \frac{\gamma_+ \gamma_-}{\gamma_{HOAc}} \tag{4.33}$$

$$K_c = \frac{(c_{H^+})(c_{OAc^-})}{(c_{HOAc})}$$

The activity coefficients γ_{HOAc}, γ_+, and γ_- depend on concentration and approach 1 as the solution becomes very dilute. Therefore for very dilute solutions K and K_c are equal, and the experimental determination of one gives the other. For more concentrated solutions we need to know activity coefficients to calculate K_c from K, or vice versa.

In practice, it is not possible to determine γ_+ and γ_- individually. Consequently, a mean ionic activity coefficient is used instead. This is the geometric mean value, and for a 1–1 electrolyte such as HOAc, it is defined as

$$\gamma_\pm = (\gamma_+ \gamma_-)^{1/2}$$

The concentrations that require activity coefficients depend on the accuracy of the desired result and on the nature of the species in solution. Ions are very nonideal; for example, in 0.1 M NaCl, the mean ionic activity coefficient, $\gamma_\pm$, is 0.78, and in 0.01 M NaCl, it is 0.90. Therefore, 0.01 M solutions might be the upper limit for "dilute" ionic solutions.

From the thermodynamic equilibrium constant we can calculate the standard free energy, ΔG^0 [Eq. (4.7)]. This is the free energy change for 1 mol of HOAc at 1 M concentration in water (but with the properties of a very dilute solution) to dissociate to H^+ and OAc^- (each at 1 M concentration but with dilute solution properties):

$$HOAc(1\ M,\ aq) \xrightarrow{\ \Delta G^0\ } H^+(1\ M,\ aq)\ +\ OAc^-(1\ M,\ aq)$$

$$\Delta G^0 = \bar{G}^0(H^+, a = 1, aq) + \bar{G}^0(OAc^-, a = 1, aq) - \bar{G}^0(HOAc, a = 1, aq)$$

For the free energy change at any arbitrary concentration, we use Eq. (4.4). For example, for the dissociation of 1 mol of HOAc at 10^{-4} M concentration in water, the free energy change is

$$HOAc(10^{-4}\ M,\ aq) \xrightarrow{\ \Delta G\ } H^+(10^{-4}\ M,\ aq)\ +\ OAc^-(10^{-4}\ M,\ aq)$$

$$\Delta G = \Delta G^0 + RT \ln Q = \Delta G^0 + RT \ln \frac{(10^{-4})(10^{-4})}{10^{-4}}$$

At 25°C this is

$$\Delta G = \Delta G^0 + (8.314)(298) \ln (10^{-4})$$

$$= \Delta G^0 - 22{,}820\ \text{J}$$

From Le Châtelier's principle, we know that the lower the concentration, the more acetic acid dissociates. The equation tells us that dilution by a factor of 10^4 decreases the free energy by 22,820 J at 25°C.

A hydrolysis reaction illustrates the use of other standard states. Consider the hydrolysis of ethyl acetate:

$$H_2O + CH_3COOCH_2CH_3 \rightleftharpoons CH_3COOH + CH_3CH_2OH$$

$$H_2O + EtOAc \rightleftharpoons HOAc + EtOH$$

$$K = \frac{[a_{HOAc}][a_{EtOH}]}{[a_{H_2O}][a_{EtOAc}]}$$

We could choose the molarity standard state for all molecules, but the trouble with this choice is that in dilute solution, the concentration of water would be near 55.5 M (1000/18) and the activity coefficient of the water would not be 1. A better choice would be the mole fraction for the water:

$$K = \frac{(c_{HOAc})(c_{EtOH})}{(X_{H_2O})(c_{EtOAc})} \cdot \frac{\gamma_{HOAc}\gamma_{EtOH}}{\gamma_{H_2O}\gamma_{EtOAc}}$$

In dilute solution the mole fraction of H_2O approaches 1 and all the activity coefficients approach 1; hence, only concentrations appear in K. It should be clear that the numerical value of K will be very different, depending on our choice of standard state. Its value will depend on whether we use 55.5 or 1.0 for the concentration of water. The corresponding standard free energies will also be different; they refer to different standard reaction conditions. The free energy change, ΔG, for any particular concentration of reactants and products is a definite measurable quantity, however; it does not depend on choice of standard state.

Because one of the products, acetic acid, is ionizable, we might wish to choose the biochemist's standard state for HOAc. When we study the hydrolysis of ethyl acetate, we find that the equilibrium ratio of concentrations,

$$\frac{(c_{HOAc})(c_{EtOH})}{(c_{EtOAc})}$$

is very dependent on the concentration of HOAc. The reason for this is the ionization of HOAc to H^+ and OAc^-. However, if we are mainly interested in the reaction near physiological conditions, we can simplify our problem by choosing the biochemist's standard state. We study the hydrolysis in a pH 7 buffer. We now find that the equilibrium ratio of concentrations is nearly constant in dilute solution. The pH 7 buffer ensures that the acetic acid is essentially all in the form of acetate, and the HOAc concentration will be directly proportional to the acetate over a wide range of total concentration.

The decision among the possible standard states and equilibrium constants is simply to choose the most convenient for the system being studied.

Most of the time we will assume that the solutions are dilute enough so that we can set all activity coefficients equal to unity. However, we should realize that in concentrated solutions our calculations may be in error by factors of 2 or more, because of activity coefficients. Later we will discuss methods of measuring activity coefficients. Here it suffices to repeat that the activity coefficient of a molecule depends on the concentrations of *all* species in solution. If one is studying the ionization constant of dilute acetic acid in water, activity coefficients of the ions are nearly 1; however, the addition of NaCl to 1 *M* concentration will have a strong effect on the activity coefficients of the ions in the dilute acid. This is true even though NaCl does not play a direct role in the dissociation equilibrium.

Calculation of Equilibrium Concentrations: Ideal Solutions

One reason for studying equilibrium constants is to enable us to calculate concentrations at equilibrium. We want to know how to make buffer solutions of any pH, how much product is obtained from a reaction, how much metal ion is bound in a complex, and so on. Some of these problems are very simple, but many of them require a computer for complete analysis. We shall discuss here the general method, which in principle allows any equilibrium problem to be solved. We assume that we know the equilibrium constants for all of the equilibria involved and we assume that all solutions are ideal; that is, we ignore all activity coefficients.

The way most scientists solve equilibrium problems is to make intuitive approximations to simplify the arithmetic operations. The "intuition" comes either from having solved many similar problems or from a good memory of what we learned in introductory chemistry. The method we present here is a general one; it applies when intuition fails on simple problems or when the problems become more difficult.

The usual requirement is to learn the concentrations of all species present in a mixture. The problem is essentially an algebraic one of finding simultaneous solutions for a number of equations. So the first requirement is to have as many equations as unknowns. The equations are equilibrium expressions, plus two types of conservation equations. The conservation equations are the ones we tend to use intuitively, but they are always included explicitly in our thinking. They are *conservation of mass*—the total mass of each element is not altered by any chemical reaction—and *conservation of charge*—the number of positive charges must always equal the number of negative charges in the mixture. Once the number of equations equals the number of unknowns, we look for approximations that allow a simple solution. A solution, not necessarily simple, can always be obtained, however.

Let us consider a simple problem and see how the method works. The first equilibrium problem one usually encounters is the ionization of a weak acid; then one proceeds to buffers and to hydrolysis. Each problem is often treated as a separate case and one memorizes the appropriate simple method for each. Here let us do the general problem and explicitly see what approximations are made to simplify the arithmetic. Suppose that a solution is prepared by adding c_A moles of acetic acid and c_S moles of sodium acetate to water to form 1 L of aqueous solution. The solution will contain HOAc, OAc$^-$, Na$^+$, H$^+$, and OH$^-$. Even for this simple equilibrium we have five species in addition to the solvent water. We therefore need five equations. We use brackets around a species to denote its concentration in molarity; the five equations are:

Mass balance: $\quad$ $[Na^+] = c_S = $ constant

$\qquad\qquad\qquad$ $[HOAc] + [OAc^-] = c_A + c_S = $ constant

Charge balance: $\quad$ $[Na^+] + [H^+] = [OAc^-] + [OH^-]$

Equilibria: $\qquad$ $K_{HOAc} = \dfrac{[H^+][OAc^-]}{[HOAc]} = 1.8 \times 10^{-5}$

$\qquad\qquad\qquad$ $K_{H_2O} = [H^+][OH^-] = 1.0 \times 10^{-14}$

We are using the solute (1 M) standard state for H$^+$, OAc$^-$, HOAc, OH$^-$, and Na$^+$, and the mole fraction standard state for H$_2$O. These five equations allow us to solve any problem involving only these five species.

Example 4.3 What are the concentrations of all species in pure water?

Solution

Mass balance: $\quad$ $[Na^+] = 0$, $[HOAc] = 0$, $[OAc^-] = 0$

Charge balance: $\quad$ $[H^+] = [OH^-]$

Equilibrium: $\qquad$ $[H^+][OH^-] = 1.0 \times 10^{-14}$

Therefore,

$$[H^+]^2 = 1.0 \times 10^{-14}$$

$$[H^+] = 1.0 \times 10^{-7}$$

$$[OH^-] = 1.0 \times 10^{-7}$$

Example 4.4 What are the concentrations of all species in a 0.100 M HOAc solution?

Solution

Mass balance: $[Na^+] = 0$

$[HOAc] + [OAc^-] = 0.100$

Charge balance: $[H^+] = [OAc^-] + [OH^-]$

Equilibria: $K_{HOAc} = \dfrac{[H^+][OAc^-]}{[HOAc]} = 1.80 \times 10^{-5}$

$K_{H_2O} = [H^+][OH^-] = 1.00 \times 10^{-14}$

Except for pure water or solutions close to pH 7, either $[H^+]$ or $[OH^-]$ will be negligible in the charge-balance equation. The solutions will be acidic or basic, so either $[OH^-]$ or $[H^+]$ can be ignored. Even for a solution at pH 6.5, the $[H^+]$ concentration is 10 times the $[OH^-]$. We know that an acetic acid solution is acidic, so we ignore $[OH^-]$ in the charge-balance equation. Assuming that $[OAc^-] \gg [OH^-]$,

Charge balance: $[H^+] \cong [OAc^-]$

Remember that we can sometimes ignore species in sums, but we can never ignore them in products. We can never ignore either $[H^+]$ or $[OH^-]$ in K_{H_2O}.

If we set $[H^+] = x$, we also have $[OAc^-] = x$ and $[HOAc] = 0.100 - x$:

$$K_A = \frac{(x)(x)}{0.100 - x} = 1.80 \times 10^{-5}$$

We can solve for x in the quadratic equation

$$x^2 + 1.8 \times 10^{-5}x - 1.8 \times 10^{-6} = 0$$

$$x = \frac{-1.8 \times 10^{-5} + \sqrt{3.24 \times 10^{-10} + 7.2 \times 10^{-6}}}{2}$$

$$= \frac{-1.8 \times 10^{-5} + 2.68 \times 10^{-3}}{2}$$

$$= 1.33 \times 10^{-3}$$

$$[H^+] = [OAc^-] = 1.33 \times 10^{-3}$$

$$[HOAc] = 0.100 - 1.33 \times 10^{-3} = 9.87 \times 10^{-2}$$

$$[OH^-] = \frac{1.0 \times 10^{-14}}{1.33 \times 10^{-3}} = 7.52 \times 10^{-12}$$

Note that our assumption that $[OAc^-] \gg [OH^-]$ is well verified.

Alternatively, we can solve the problem using successive approximations. The secret in solving by approximations comes in setting x equal to concentrations of species that are *not* the largest involved in the equilibrium. In this case, we have set $x = [H^+]$ or $[OAc^-]$. We do not set $x = [HOAc]$, because it is the species that we expect would have the largest concentration in 0.1 M acetic acid. Even before knowing the exact answer, we can guess that $[H^+]$ and $[OAc^-]$ will have to be much smaller than $0.100\ M$ to satisfy the equilibrium constant $K_{HOAc} = 1.8 \times 10^{-5}$.

As a first approximation when $x \ll 0.1$,

$$K_A = \frac{x^2}{0.1 - x} \cong \frac{x^2}{0.1} = 1.80 \times 10^{-5}$$

$$x \cong 1.34 \times 10^{-3}$$

Indeed, we see that our initial approximation that $x \ll 0.1$ is justified. To obtain a more precise answer, we can use this first solution to generate a second approximation:

$$\frac{x^2}{0.1 - 0.0013} = 1.80 \times 10^{-5}$$

$$x = 1.33 \times 10^{-3}$$

Clearly, no further approximations are needed.

Example 4.5   What are the concentrations of all species in a $0.200\ M$ NaOAc solution?

Solution   This is a hydrolysis problem, because NaOAc is the salt of a weak acid and we might be tempted to write the hydrolysis reaction and derive a value for the hydrolysis equilibrium constant $K_h = K_{H_2O}/K_{HOAc}$. However, this is not necessary or particularly useful. Instead, our general method gives

Mass balance: $[Na^+] = 0.200$

$[HOAc] + [OAc^-] = 0.200$

Charge balance: $[H^+] + [Na^+] = [OH^-] + [OAc^-]$

Equilibria: $K_{HOAc} = \dfrac{[H^+][OAc^-]}{[HOAc]} = 1.80 \times 10^{-5}$

$K_{H_2O} = [H^+][OH^-] = 1.00 \times 10^{-14}$

When the salt of a weak acid hydrolyzes, the reaction

$$OAc^- + H_2O \rightleftharpoons HOAc + OH^-$$

occurs, and the resulting solution is basic; therefore, we an ignore $[H^+]$ in the charge balance. We already know that $[Na^+] = 0.200$; therefore, assuming that $[Na^+] \gg [H^+]$,

Charge balance: $0.200 = [OH^-] + [OAc^-]$

Comparing this with the mass-balance equation for $[HOAc] + [OAc^-]$, we see that

$$[HOAc] = [OH^-]$$

Now we set $x = [HOAc] = [OH^-]$ to facilitate making approximations. (The choice for x is between $[OAc^-]$ on the one hand and $[HOAc] = [OH^-]$ on the other. The equilibrium expression for K_{HOAc} tells us that $[OAc^-]/[HOAc] \gg 1$ because $[H^+] \ll 10^{-7}$ for a basic solution.) Thus

$$[OAc^-] = 0.200 - [HOAc] = 0.200 - x$$

$$[H^+] = \frac{10^{-14}}{x}$$

and

$$K_{HOAc} = \frac{(10^{-14}/x)(0.200 - x)}{x} = \frac{10^{-14}(0.200 - x)}{x^2} = 1.80 \times 10^{-5}$$

$$\frac{(0.200 - x)}{x^2} = 1.80 \times 10^9$$

We can solve the quadratic, but it is easier to use successive approximation. As our first approximation, when $x \ll 0.200$,

$$\frac{0.200}{x^2} = 1.80 \times 10^9$$

$$x = 1.05 \times 10^{-5} \quad \text{(consistent with assumption)}$$

No further approximations are needed.

$$[OH^-] = [HOAc] = 1.05 \times 10^{-5}$$

$$[H^+] = \frac{1.00 \times 10^{-14}}{1.05 \times 10^{-5}} = 9.52 \times 10^{-10}$$

$$[OAc^-] = 0.200$$

$$[Na^+] = 0.200$$

Note that $[Na^+] \gg [H^+]$, which justifies our early approximation in the charge-balance equation.

This general method works for any equilibrium. For more complicated examples we must be sure that the equilibrium expressions are independent. In the example above, instead of the $K_A = K_{HOAc}$ and K_{H_2O}, we could have used K_B:

$$K_B = \frac{[HOAc][OH^-]}{[OAc^-]}$$

and K_{H_2O}, or K_B and K_A. However, we cannot count K_A, K_B, and K_{H_2O} as three independent equations; only two are independent, because $K_A K_B$ always equals K_{H_2O}.

For buffer problems the computations are usually simpler, because both the acid and the salt (or the base and its corresponding salt) are present at an appreciable concentration. Furthermore, these concentrations are large compared with either $[H^+]$ or $[OH^-]$. In the case of a buffer involving acetic acid and sodium acetate, for example, we have:

Mass balance: $[Na^+] = c_S = \text{constant}$

$[HOAc] + [OAc^-] = c_A + c_S = \text{constant}$

Charge balance: $[Na^+] + [H^+] = [OAc^-] + [OH^-]$

which becomes

$$[Na^+] = [OAc^-] = c_S$$

because the $[H^+]$ and $[OH^-] \ll [Na^+]$ or $[OAc^-]$. Thus

$$[HOAc] = c_A$$

and the equilibrium constant is

$$K_{HOAc} = \frac{[H^+][OAc^-]}{[HOAc]} = \frac{[H^+]c_S}{c_A}$$

This equation is often written in the form

$$pH = pK_A + \log \frac{c_S}{c_A} \qquad (4.34)$$

where $pH = -\log[H^+]$
$pK_A = -\log K_A$

and is sometimes called the *Henderson-Hasselbalch equation*. (Note that for real solutions the proper definition is $pH = -\log a_{H^+}$.)

Example 4.6 What amount of solid sodium acetate is needed to prepare a buffer at pH 5.00 from 1 L of 0.10 M acetic acid?

Solution

$$pH = pK_{HOAc} + \log \frac{c_S}{c_A}$$

$$5.00 = 4.75 + \log \frac{c_S}{0.10} = 5.75 + \log c_S$$

$$\log c_S = -0.75$$

$$c_S = 0.18 \text{ mol L}^{-1}$$

$$\text{wt of NaOAc} = (0.18 \text{ mol})(82.0 \text{ g mol}^{-1}) = 14.8 \text{ g}$$

Of course, this answer assumes that the solutions are ideal and that the addition of sodium acetate has a negligible effect on the volume. In practice, one always makes the final adjustment of pH using a calibrated pH meter or other electrode system that measures a_{H^+} directly.

Temperature Dependence of the Equilibrium Constant

Once we know an equilibrium constant at one temperature, what can we say about the equilibrium constant at another temperature? Le Châtelier can help us if we think of heat as a component of the reaction. If heat is given off (an exothermic reaction, a negative ΔH), the products have a lower enthalpy than the reactants, and raising the temperature will favor reactants. The equilibrium constant will become smaller. If heat is absorbed (an endothermic reaction, a positive ΔH), raising the temperature will favor products. The equilibrium constant will become larger.

For example, we know from experience that a large amount of heat is generated upon mixing a 1 M acid solution (H^+) with 1 M base (OH^-) in the neutralization reaction to form water. The reverse reaction (the self-ionization of water) therefore absorbs heat.

$$\text{Heat} + H_2O(l) \longrightarrow H^+(aq) + OH^-(aq) \qquad \Delta H^0_{298} = 55.84 \text{ kJ}$$

Thus we expect that increasing the temperature of water will result in an increase in K_w, the equilibrium constant for the self-ionization of water. Experimentally, the value $K_{w,298} = 1.01 \times 10^{-14}$ at 25°C is increased to $K_{w,310} = 2.40 \times 10^{-14}$ at 37°C, which is the physiological temperature for humans.

The quantitative relation is easily obtained if ΔH is independent of temperature. We use

$$\Delta G^0 = \Delta H^0 - T\Delta S^0 = -RT \ln K$$

at two temperatures T_1 and T_2:

$$\ln K_2 = -\frac{\Delta H^0}{RT_2} + \frac{\Delta S^0}{R}$$

$$\ln K_1 = -\frac{\Delta H^0}{RT_1} + \frac{\Delta S^0}{R}$$

We have assumed that ΔH^0 and ΔS^0 are independent of T. Subtracting one equation from the other, we obtain the desired equation:

$$\ln \frac{K_2}{K_1} = -\frac{\Delta H^0}{R} \left(\frac{1}{T_2} - \frac{1}{T_1} \right) \tag{4.35}$$

This equation can be used to calculate an equilibrium constant K_2, at T_2 when K_1 and the standard enthalpy of the reaction, ΔH^0, are known. Alternatively, from the equilibrium constants measured at two different temperatures, one can calculate the enthalpy. The usual way of obtaining ΔH^0 from K as a function of T is to plot $\ln K$ or $\log K$ vs. $1/T$. A plot of $\log K$ vs. $1/T$ for the experimental values of K_w in the temperature range from 0° to 50°C is shown in Fig. 4.4. The slope is $-\Delta H^0/2.303R$:

$$\ln K(T) = -\frac{\Delta H^0}{RT} + \text{constant}$$

$$\log K(T) = -\frac{\Delta H^0}{2.303RT} + \text{constant} \tag{4.36}$$

The best fit line through the data shown in Fig. 4.4 gives a value of $\Delta H^0 = 55.76$ kJ, which agrees well with the value 55.84 kJ calculated from the standard enthalpies of formation at

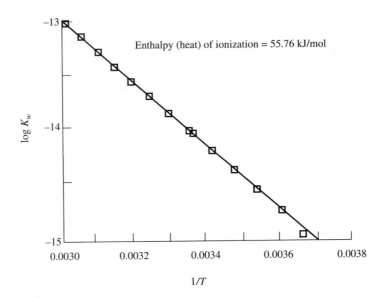

Enthalpy (heat) of ionization = 55.76 kJ/mol

Fig. 4.4 A plot of the logarithm of the ionization constant of water K_w versus $1/T$. This is a van't Hoff plot. The slope is equal to $-\Delta H^0/2.303RT$ [Eq. (4.36)]; the standard heat of ionization is thus found to be 55.76 kJ mol^{-1}. This is in excellent agreement with the value obtained from calorimetry.

298 K measured calorimetrically. Measurements of equilibrium constants at different temperatures are widely used to determine ΔH^0 values from plots of ln K vs. $1/T$; the plot is known as a van't Hoff plot. If we take the derivative of ln K in Eq. (4.36) with respect to $1/T$ we see that

$$\frac{\partial \ln K}{\partial (1/T)} = -\frac{\Delta H^0}{R}$$

This means that the slope of ln K vs. $1/T$ is equal to ΔH^0. A more rigorous derivation shows that Eq. (4.36) is correct even if ΔH^0 does change with temperature. For such a system the plot will show some curvature. The slope will change with temperature, but the slope at any temperature will equal $-\Delta H^0/R$ at that temperature.

Example 4.7 The equilibrium constant for ionization of 4-aminopyridine is 1.35×10^{-10} at 0°C and 3.33×10^{-9} at 50°C. Calculate ΔG^0 at 0°C and 50°C as well as ΔH^0 and ΔS^0.

Solution We assume that ΔH^0 is independent of temperature and use Eq. (4.35) with $K_2 = 1.35 \times 10^{-10}$, $K_1 = 3.33 \times 10^{-9}$, $T_2 = 273$, and $T_1 = 323$:

$$\ln \frac{K_2}{K_1} = -\frac{\Delta H^0}{R}\left(\frac{1}{T_2} - \frac{1}{T_1}\right) \tag{4.35}$$

$$\frac{1}{T_2} - \frac{1}{T_1} = \frac{1}{273} - \frac{1}{323} = (3.663 \times 10^{-3} - 3.096 \times 10^{-3})$$

$$= 5.67 \times 10^{-4} \text{ K}^{-1}$$

$$\ln \frac{K_2}{K_1} = \ln \frac{1.35}{33.3} = -3.205$$

$$\Delta H^0 = -\frac{(8.314 \text{ J K}^{-1} \text{ mol}^{-1})(-3.205)}{5.67 \times 10^{-4} \text{ K}^{-1}}$$

$$= 47,000 \text{ J mol}^{-1} = 47.0 \text{ kJ mol}^{-1}$$

This value of ΔH^0 is an average value over the temperature range $0°$ to $50°C$. From Eq. (4.7), we obtain ΔG^0 at each temperature:

$$\Delta G^0 = -RT \ln K \tag{4.7}$$

$$\Delta G^0(0°C) = -(8.314 \text{ J K}^{-1} \text{ mol}^{-1})(273 \text{ K}) \ln (1.35 \times 10^{-10})$$

$$= +51.58 \text{ kJ}$$

$$\Delta G^0(50°C) = -(8.314 \text{ J K}^{-1} \text{ mol}^{-1})(323 \text{ K}) \ln (3.33 \times 10^{-9})$$

$$= +52.42 \text{ kJ}$$

The value of ΔS^0, assumed to be independent of temperature between $0°$ and $50°C$, is

$$\Delta S^0 = \frac{\Delta H^0 - \Delta G^0}{T}$$

$$= \frac{47,000 - 51,580}{273}$$

$$= -16.78 \text{ J K}^{-1} \text{ mol}^{-1}$$

or, as a check,

$$\Delta S^0 = \frac{47,000 - 52,420}{323}$$

$$= -16.78 \text{ J K}^{-1} \text{ mol}^{-1}$$

For a small temperature range (less than $25°$), Eq. (4.35) should be adequate. However, in general we must consider the temperature dependence of ΔH.

An example where this is important is the reaction to form ammonia from N_2 and H_2, which has a favorable (negative) ΔG at room temperature, as we have seen. However, the reaction is impossibly slow, even in the presence of a good catalyst, unless the temperature is raised. The industrial process is usually carried out at about 800 K. Unfortunately this increase in temperature makes ΔG^0_{800} positive, because the reaction to form ammonia is exothermic.

$$\tfrac{1}{2} N_2(g) + \tfrac{3}{2} H_2(g) \longrightarrow NH_3(g)$$

$\Delta H^0_{298} = -46.11 \text{ kJ}$ (per mol of NH_3 formed)

To calculate K^{eq}_{800} for this reaction, we can use the Gibbs-Helmholtz equation, Eq. (3.32). This requires knowledge of the temperature dependence of $\Delta H^0(T)$, which in turn is obtained from Eq. (2.33) and the temperature-dependent heat capacity values, $C_P(T)$, for the reactants and products, which are tabulated for many simple gases. Using this approach we find that ΔH^0 for this reaction changes from -46.11 kJ at 298 K to -54.11 kJ at 800 K. If we ignore this change in ΔH^0 with temperature and use the method of Eq. (4.35) with a constant ΔH^0_{298}, we obtain $\Delta G^0_{800} = +33.5$ kJ mol^{-1} and log $K^{eq}_{800} = -2.2$. Using the heat capacities of the gases over the temperature range from 298 to 800 K, however, we obtain the more accurate values $\Delta G^0_{800} = +38.9$ kJ mol^{-1} and log $K^{eq}_{800} = -2.54$. Note that, even though the change in the value of ΔH^0 over this temperature range is significant, the error introduced into the calculation of the equilibrium constant amounts to a factor of only 2.2.

The accurate value $K^{eq}_{800} = 2.9 \times 10^{-3}$ indicates how unfavorable the reaction has become at 800 K, the temperature required to cause it to proceed at a reasonable rate. The yield of NH_3 under these conditions (800 K and 1 atm) would be prohibitively small. All is not lost, however. We saw at the beginning of this chapter that we can decrease the reaction quotient, Q, for this reaction by increasing the overall pressure. At 200 atm the value of Q is 200 times smaller than K for this reaction, which goes a long way to compensate for the positive standard free energy change at 800 K. At 200 atm and 800 K we need to reconsider the nonideal behavior of the reactant and product gases. Under these conditions the ideal gas equation predicts a concentration $[A] = 3.11$ mol L^{-1} for any gas, whereas the real gas values (again, for the pure gases at 200 atm and 800 K) are $[H_2] = 2.88$ mol L^{-1}, $[N_2] = 2.79$ mol L^{-1} and $[NH_3] = 3.00$ mol L^{-1}. (Note that even under these extreme conditions, the ideal gas equation is valid to within about 10% for each of these gases.) Experimental measurements made under these conditions show that an equilibrium mixture contains about 15 mol % NH_3. Not only is this an acceptable yield, but it is in good agreement with the best thermodynamic calculations. Following passage over the catalyst, the equilibrated mixture is cooled to liquefy the ammonia, and the unreacted N_2 and H_2 are recycled to the input reactant stream.

Despite the extreme temperature and pressure conditions required for the industrial process, liquid ammonia is still one of the most economical sources of biological nitrogen for agricultural purposes. Nevertheless, chemists continue to be intrigued by the prospect that, in the presence of a better catalyst (simply to make the reaction go faster, not to change the thermodynamics), this reaction could be induced to occur spontaneously at room temperature and 1 atm. Molecular biologists, in turn, are intrigued by the possibility of using genetic engineering to expand the applicability of biological nitrogen fixation to crops other than legumes.

This example, which we have now pursued in considerable detail, shows how the influence of temperature and pressure on thermodynamic properties can be used to influence the course of a reaction to produce a desirable result. There are many biological processes that are influenced in a similar fashion. Biologically relevant temperatures, even for organisms that have adapted to grow in hot springs, usually do not exceed 400 K, but pressures of 200 atm occur in the oceans at depths of 2000 meters, where unusual biological organisms have been observed to thrive in the vicinity of nutrient-rich thermal vents in the ocean floor. We may expect to find that the biological processes occurring under either of these extreme conditions are significantly different from those in the more common and familiar biological environments.

BIOCHEMICAL APPLICATIONS OF THERMODYNAMICS

Knowledge of equilibrium constants, free energies, and their dependence on concentration and temperature is very important in biochemistry and biology. Here we will discuss a few simple examples. Table 4.1 and Fig. 4.5 show equilibrium constants at 25°C and heats of ionization for various acids. A great deal of information about the acids is summarized in a small space. For example, some biochemical systems (such as human beings) work best at 37°C (98.5°F). Table 4.1 allows you to calculate the pK's at 37°C. We see that acetic acid is one of the few acids listed whose pK is the same at 37°C as at 25°C, because its $\Delta H \sim 0$. We usually consider that a neutral aqueous solution has a pH of 7, but this is true only at 25°C.

Example 4.8 What is the pH of pure water at 37°C?

Solution The necessary equation is

$$\ln \frac{K_2}{K_1} = -\frac{\Delta H^0}{R}\left(\frac{1}{T_2} - \frac{1}{T_1}\right) \tag{4.35}$$

The ionization constant of water is 1.00×10^{-14} at 25°C, its $\Delta H^0 = 55.84$ kJ (see Table 4.1), $K_1 = 10^{-14}$, $T_1 = 298$, and $T_2 = 310$. Then

$$\ln \frac{K_2}{10^{-14}} = \frac{-55,840}{8.314}\left(\frac{1}{310} - \frac{1}{298}\right)$$

$$= +0.87$$

$$\frac{K_2}{10^{-14}} = e^{+0.87} = 2.4$$

$$K_2 = 2.4 \times 10^{-14}$$

This is the ionization constant of water at 37°C. The H^+ and OH^- in pure water is surely low enough so that the activities of the ions equal concentrations: $[H^+] = [OH^-]$. At 37°C,

$$[H^+][OH^-] = 2.4 \times 10^{-14}$$

$$[H^+] = 1.55 \times 10^{-7} \, M$$

$$pH = -\log[H^+]$$

$$= 6.81$$

Example 4.9 A buffer is prepared by adding 26.8 mL of 0.200 M HCl to 50.0 mL of 0.200 M tris(hydroxymethyl)aminomethane, a weak base known as Tris and widely used as a buffer by biochemists. Tris has a $pK = 8.3$ at 20°C: its pK changes with temperature with a temperature coefficient of $\Delta pK / \Delta T = -0.029$ K^{-1}. The mixture is then diluted with pure water to a total volume of 200 mL. (a) What is the pH

Table 4.1 Ionization constants and enthalpies (heats) of ionization at 25°C

Compound	Ionizing species‡	pK*	ΔH^0 (kJ mol^{-1})
Acetic acid	$HOAc \rightarrow H^+ + OAc^-$	4.76	−0.25
Adenosine	$AH^+ \rightarrow A + H^+$	3.55	15.9
5′-Adenylic acid			
(adenosine phosphate)	$pAH^+ \rightarrow pA + H^+$	3.7	17.6
	$pA \rightarrow pA^- + H^+$	6.4	−7.5
	$pA^- \rightarrow pA^{2-} + H^+$	13.1	45.6
Adenosine triphosphate (ATP)	$pppAH^+ \rightarrow pppA + H^+$	4.0	15.5
	$pppA \rightarrow pppA^- + H^+$	7.0	−5.0
Alanine	$^+H_3NRCOOH \rightarrow {}^+H_3NRCOO^- + H^+$	2.35	2.9
	$^+H_3NRCOO^- \rightarrow H_2NRCOO^- + H^+$	9.83	45.2
Ammonia	$NH_4^+ \rightarrow NH_3 + H^+$	9.24	52.2
Aspartic acid	$^+H_3NRHCOOH \rightarrow {}^+H_3NRHCOO^- + H^+$	2.05	7.5
	$^+H_3NRHCOO^- \rightarrow {}^+H_3NR^-COO^- + H^+$	3.87	4.2
	$^+H_3NR^-COO^- \rightarrow H_2NR^-COO^- + H^+$	10.3	38.5
Carbonic acid	$H_2CO_3 \rightarrow HCO_3^- + H^+$	6.36	7.66
	$HCO_3^- \rightarrow CO_3^{2-} + H^+$	10.24	14.85
Fumaric acid	$R(COOH)_2 \rightarrow {}^-OOCRCOOH + H^+$	3.10	0.4
	$^-OOCRCOOH \rightarrow {}^-OOCRCOO^- + H^+$	4.6	−2.9
Glycine	$^+H_3NRCOOH \rightarrow {}^+H_3NRCOO^- + H^+$	2.35	3.92
	$^+H_3NRCOO^- \rightarrow H_2NRCOO^- + H^+$	9.78	44.2
Histidine	$^+H_3NRH^+COOH \rightarrow {}^+H_3NRH^+COO^- + H^+$	1.82	—
	$^+H_3NRH^+COO^- \rightarrow {}^+H_3NRCOO^- + H^+$	6.00	29.9
	$^+H_3NRCOO^- \rightarrow H_2NRCOO^- + H^+$	9.16	43.6
Hydrocyanic acid	$HCN \rightarrow CN^- + H^+$	9.21	43.5
Lysine	$^+H_3NRH^+COOH \rightarrow {}^+H_3NRH^+COO^- + H^+$	2.18	0.30
	$^+H_3NRH^+COO^- \rightarrow {}^+H_3NRCOO^- + H^+$	8.95	12.8
	$^+H_3NRCOO^- \rightarrow H_2NRCOO^- + H^+$	10.53	11.6
Phenol	$ROH \rightarrow RO^- + H^+$	9.98	23.6
Phosphoric acid	$H_3PO_4 \rightarrow H_2PO_4^- + H^+$	2.15	−7.95
	$H_2PO_4^- \rightarrow HPO_4^{2-} + H^+$	7.21	4.15
	$HPO_4^{2-} \rightarrow PO_4^{3-} + H^+$	12.35	14.7
Pyruvic acid	$RCOOH \rightarrow RCOO^- + H^+$	2.49	12.1
Tyrosine	$^+H_3NRHCOOH \rightarrow {}^+H_3NRHCOO^- + H^+$	2.20	—
	$^+H_3NRHCOO^- \rightarrow {}^+H_3NR^-COO^- + H^+$	9.11	—
	$^+H_3NR^-COO^- \rightarrow H_2NR^-COO^- + H^+$	10.05	25.1
Water	$H_2O \rightarrow OH^- + H^+$	14.00	55.84

* pK = −log K. The K values are thermodynamic equilibrium constants on the molarity scale. $\Delta G^0 = -RT \ln K$.

‡ The ionizable groups corresponding to the pK's in the table are shown in Fig. 4.5. Glycine is just like alanine with the methyl group repaced by a hydrogen. 5′-Adenylic acid is shown in the figure; adenosine is the same except that the phosphate is lacking. In ATP the single phosphate (OPO_3H_2) of adenylic acid is replaced by a triphosphate $(OPO_2)^-(OPO_2)^-(OPO_3H_2)$.

Source: Reprinted from *Handbook of Biochemistry and Molecular Biology*, 3rd ed., Physical and Chemical Data, Vol. I, CRC Press, Boca Raton, Florida, 1976; *The NBS Tables of Chemical Thermodynamic Properties,* D. D. Wagman et al., eds., *J. Phys. Chem. Ref. Data 11,* Suppl. 2 (1982).

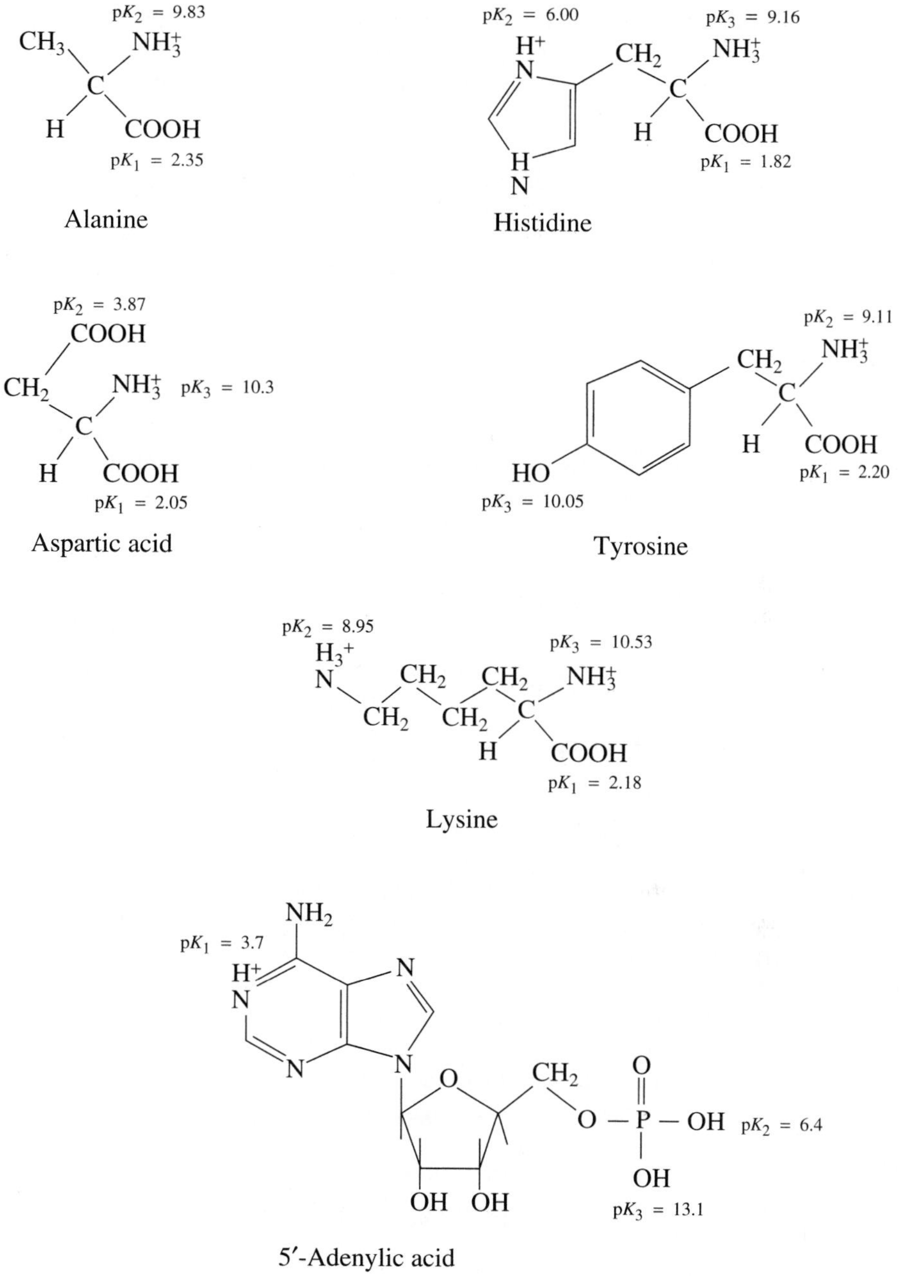

$pK_2 = 9.83$

CH_3 NH_3^+
C
H COOH

$pK_1 = 2.35$

Alanine

$pK_2 = 6.00$ $pK_3 = 9.16$

H^+
N CH$_2$ NH$_3^+$
C
H COOH
H
N

$pK_1 = 1.82$

Histidine

$pK_2 = 3.87$

COOH

CH_2 NH_3^+ $pK_3 = 10.3$
C
H COOH

$pK_1 = 2.05$

Aspartic acid

$pK_2 = 9.11$

CH_2 NH_3^+
C
H COOH

$pK_1 = 2.20$

HO

$pK_3 = 10.05$

Tyrosine

$pK_2 = 8.95$

H_3^+
N CH_2 CH_2 NH_3^+ $pK_3 = 10.53$
CH_2 CH_2 C
H COOH

$pK_1 = 2.18$

Lysine

NH_2

$pK_1 = 3.7$

H^+
N N

N

N N
O CH_2 O
O — P — OH $pK_2 = 6.4$
OH OH OH
$pK_3 = 13.1$

5′-Adenylic acid

Fig. 4.5 The pK values from Table 4.1 of several amino acids and of a nucleotide; p$K = -\log K$.

of the buffer solution at 20°C? (b) What is the "buffer capacity" of this solution, expressed as the change in pH that would occur upon the addition of 1 mmol of strong acid or base to 1 L of the buffer? (c) What would be the pH of the buffer solution of part (a) at 37°C?

Solution   The equilibrium involved for Tris buffer is

$$
\underset{\text{Tris·H}^+}{\begin{array}{c} \text{HOCH}_2 \\ | \\ \text{HOCH}_2-\overset{+}{\underset{|}{C}}-\text{NH}_3 \\ | \\ \text{HOCH}_2 \end{array}} \quad \overset{pK_{20}=8.3}{\rightleftharpoons} \quad \underset{\text{Tris}}{\begin{array}{c} \text{HOCH}_2 \\ | \\ \text{HOCH}_2-\underset{|}{C}-\text{NH}_2 \ + \ \text{H}^+ \\ | \\ \text{HOCH}_2 \end{array}}
$$

(a) Addition of the HCl solution (26.8 mL of 0.200 M HCl contains 5.36 mmol of HCl) to the Tris solution (50.0 mL of 0.200 M Tris contains 10.00 mmol of Tris) causes the conversion of 5.36 mmol of Tris to Tris·H$^+$. Thus the concentrations in the final 200 mL of buffer solution are

$$
[\text{Tris·H}^+] = \frac{5.36 \text{ mmol}}{200 \text{ mL}} = 26.8 \times 10^{-3} \, M
$$

$$
[\text{Tris}] = \frac{(10.00 - 5.36) \text{ mmol}}{200 \text{ mL}} = 23.2 \times 10^{-3} \, M
$$

Using Eq. (4.34)

$$
pH = 8.3 + \log \frac{23.2 \times 10^{-3}}{26.8 \times 10^{-3}} = 8.3 - 0.063 = 8.2
$$

If a more precise pH is desired, final adjustment of the buffer can be accomplished by the addition of a small volume of one or the other stock solution, monitoring the buffer solution with an accurate (calibrated) pH meter.

(b) The *buffer capacity* is a measure of the resistance to change in pH upon addition of H$^+$ or OH$^-$. Pure water has no buffer capacity. Addition of 1 mmol of HCl (or NaOH) to 1 L of water changes the pH from 7 to 3 (or 11). However, for 1 L of the buffer solution of part (a), addition of 1.0 mmol of HCl, for example, would result in the following changes

$$
[\text{Tris·H}^+] = (26.8 + 1.0) \times 10^{-3} = 27.8 \times 10^{-3} \, M
$$

$$
[\text{Tris}] = (23.2 - 1.0) \times 10^{-3} = 22.2 \times 10^{-3} \, M
$$

Thus

$$
pH = 8.3 + \log \frac{22.2 \times 10^{-3}}{27.8 \times 10^{-3}} = 8.3 - 0.097 = 8.2
$$

The change is only $\Delta pH = 0.097 - 0.063 = 0.034$ pH units. The buffer is still nominally at pH 8.2. This illustrates the ability of buffer solutions to resist changes in pH when, for example, H$^+$ is produced or consumed in metabolic reactions. Note that the buffer capacity is a direct function of the buffer concentration. This 50 mM Tris buffer solution has about 10 times the buffer capacity of a 5 mM buffer solution at pH 8.2.

(c) The pK for Tris decreases by 0.029 per degree (K) rise in temperature. Thus, at 37°C

$$pK_{37} = pK_{20} - (0.029)(37 - 20) = 8.3 - 0.49 = 7.8$$

and the pH of the solution prepared as described above would be

$$pH_{37} = pK_{37} - 0.063 = 7.7$$

This is a very significant change with temperature and should be kept in mind in all biochemical investigations. The temperature coefficient can be related to an enthalpy change upon dissociation, as tabulated for many weak acids in Table 4.1. Beginning with Eq. (4.35) we can readily derive the equation

$$\Delta H_T^0 = \frac{2.303RT_1T_2}{T_1 - T_2} \log \frac{K_2}{K_1} = \frac{2.303RT_1T_2}{T_2 - T_1} (pK_1 - pK_2)$$

For Tris

$$\Delta H_{298}^0 = \frac{(2.303)(8.314 \text{ J K}^{-1} \text{ mol}^{-1})(293 \text{ K})(294 \text{ K})}{1 \text{ K}} (+0.029)$$

$$= 48 \text{ kJ mol}^{-1}$$

From the pK's given in Table 4.1 or those that we calculate at other temperatures, we can calculate the activities and concentrations of various ionized species present in solution. For example, consider histidine in solution at 25°C. The equilibria are

We often need to know the concentration of each species at a given pH. Each ionic species has different chemical reactivity and physical properties; therefore, the properties of the solution will depend on the concentration of each species.

Example 4.10 Calculate the concentration of each species in a $0.10\ M$ solution of histidine at pH 7.

Solution The three equilibrium expressions are

$$K_1 = 1.51 \times 10^{-2} = \frac{[H^+][HisH_2^+]}{[HisH_3^{2+}]}$$

$$K_2 = 1.00 \times 10^{-6} = \frac{[H^+][HisH]}{[HisH_2^+]}$$

$$K_3 = 6.92 \times 10^{-10} = \frac{[H^+][His^-]}{[HisH]}$$

We assume that activities are equal to concentrations. If necessary, we could measure or estimate activity coefficients to obtain more accurate concentrations. The mass-balance equation is

$$[HisH_3^{2+}] + [HisH_2^+] + [HisH] + [His^-] = 0.100\ M$$

We now have four equations and four unknowns, so we can solve for the unknowns. As often happens, it will be convenient to solve by successive approximations. First, we find the ratios of species from the three equilibrium constants and the value of $[H^+] = 10^{-7}\ M$.

$$\frac{[HisH_2^+]}{[HisH_3^{2+}]} = 1.51 \times 10^5$$

$$\frac{[HisH]}{[HisH_2^+]} = 10.0$$

$$\frac{[His^-]}{[HisH]} = 6.92 \times 10^{-3}$$

From the ratios of species we see that a good first approximation will be to ignore $[HisH_3^{2+}]$ and $[His^-]$ in the mass-balance equation. That is, we need consider only the two species involved in the equilibrium whose pK's are closest to the pH of the solution.

The first-approximation mass balance is, assuming that $[HisH_3^{2+}] + [His^-] \ll [HisH] + [HisH_2^+]$,

$$[HisH] + [HisH_2^+] = 0.100\ M$$

but

$$[HisH] = [HisH_2^+] \cdot (10.0)$$

$$[HisH_2^+] \cdot (10.0) + [HisH_2^+] = 0.100$$

$$[HisH_2^+] = \frac{0.100}{11.0} = 9.1 \times 10^{-3} \, M$$

$$[HisH] = 10[HisH_2^+] = 9.1 \times 10^{-2} \, M$$

$$[His^-] = (6.92 \times 10^{-3})(9.1 \times 10^{-2}) = 6.3 \times 10^{-4} \, M$$

$$[HisH_3^{2+}] = \frac{9.1 \times 10^{-3}}{1.51 \times 10^5} = 6.0 \times 10^{-8} \, M$$

We have ignored activity coefficients, so the concentrations are at best accurate to ± 10%. If the second approximation does not change values by more than 10%, we can stop at the first approximation.

The second-approximation mass balance is

$$[HisH] + [HisH_2^+] = 0.100 - [His^-] - [HisH_3^{2+}]$$

$$= 0.0994 \, M$$

The first approximation is good enough; the concentrations are

$$[HisH_2^+] = 9.1 \times 10^{-3} \, M$$
$$[HisH] = 9.1 \times 10^{-2} \, M$$
$$[His^-] = 6.3 \times 10^{-4} \, M$$
$$[HisH_3^{2+}] = 6.0 \times 10^{-8} \, M$$

Note that the sum of the last two species is much less than the first two.

Using the methods illustrated in Example 4.9 we calculated the concentrations of all histidine species as a function of pH as shown in Fig. 4.6. The pK's correspond to the pH values where two species cross—where their concentrations are equal. Note that the concentration of each species reaches a maximum near 0.1 M (the total concentration of all histidine species).

These methods can, of course, be used to calculate the concentrations of species needed to product a buffer of a given pH and salt concentration. For common buffers the algebra has already been done, and buffer tables can be found in such handbooks as the *Handbook of Biochemistry and Molecular Biology* (CRC Press, Boca Raton, Florida). For new buffers or special requirements of salt concentrations you will have to be able to do the calculations yourself.

Thermodynamics of Metabolism

Metabolism refers to the process by which cells use energy from their environment to synthesize the building blocks of their macromolecules. Hundreds of reactions are involved, and the path to a single product can involve many intermediates. One of the advantages of thermodynamic analysis is that we can focus on the beginning and the end of any process and not worry about the steps in between. However, we can also learn about the driving force for each of the steps.

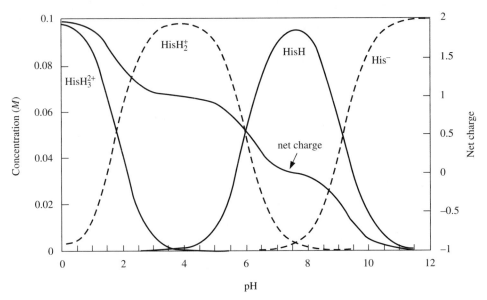

Fig. 4.6 The concentrations of histidine species as a function of pH; the values were calculated as shown in Example 4.9. Note the sharp change in concentration of species near their pK's (1.82, 6.00, 9.16), and the result that no more than two histidine species are present in significant concentrations at each pH. The net charge on an average histidine molecule is shown at each pH. The net charge decreases in steps from +2 to +1 (pH 4) to 0 (pH 8) to −1. This is analogous to what happens to a protein molecule as a function of pH. It will have a net positive charge at low pH, a zero charge at its isoelectric point, and a negative charge at high pH.

Figure 4.7 presents a scheme for glycolysis, which is the metabolic process that converts glucose into pyruvate, and produces adenosine triphosphate, ATP. In aerobic organisms, pyruvate then enters the mitochondria, where it is completely oxidized to CO_2 and H_2O in the citric acid cycle. These reactions produce additional ATP. Under less aerobic conditions, in muscle during active exercise, pyruvate is converted to lactate. Under anaerobic conditions in yeast, pyruvate is transformed into ethanol. Thus, Fig. 4.7 shows a major portion of metabolic activity in a variety of cellular environments.

The overall oxidative reaction (combustion) of glucose with $O_2(g)$ to form $CO_2(g)$ and $H_2O(l)$ is associated with a large negative Gibbs free energy change. Using ΔG_f^0 values at 298 K from the Appendix, we obtain for the reaction

$$C_6H_{12}O_6(s) + 6\,O_2(g) \longrightarrow 6\,CO_2(g) + 6\,H_2O(l) \qquad \Delta G_{298}^0 = -2878.41 \text{ kJ}$$

α-D-Glucose

Furthermore, most of this free energy change arises from the enthalpy contribution, $\Delta H_{298}^0 = -2801.6$ kJ, and relatively little from the entropy, $-T\,\Delta S_{298}^0 = -77.2$ kJ. Thus, when glucose is burned directly in oxygen, nearly all of the chemical potential is released as heat. By contrast, in living cells a significant fraction of the chemical potential is retained as chemical bond energy in the form of ATP and other synthesized molecules.

The free energies at 25°C and pH 7 for the steps in the metabolic degradation (catabolism) of glucose are given in Table 4.2 and illustrated in Fig. 4.7. The biochemist's

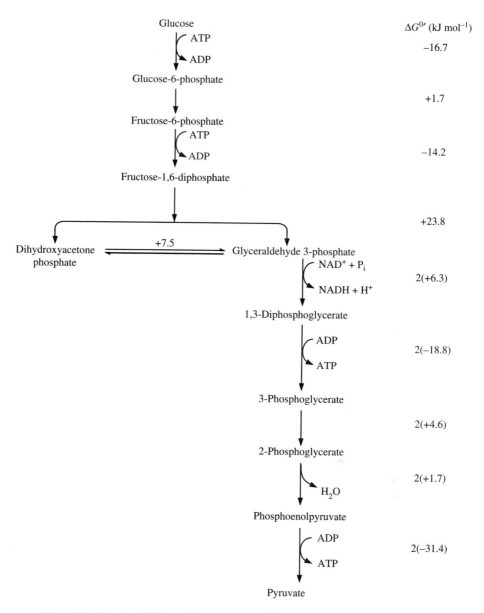

Fig. 4.7 The glycolytic path for the conversion of glucose to pyruvate in cells. The $\Delta G^{0\prime}$ for reactions from glyceraldehyde-3-phosphate to pyruvate are multiplied by 2 because 2 mol of these species are formed per mol of glucose.

standard state is used in this table. This means that the pH is 7, the activity of water is equal to 1, metabolites are solutes in aqueous solutions and the activities of all solutes are replaced by their total concentrations in mol L^{-1}. For example, ATP in the reaction refers to the sum of all ionized species of ATP present at pH 7. Standard states of all solutes have a total concentration equal to 1 M, but the solution is ideal.

Table 4.2   Standard free energies of reaction at 25°C, pH 7 for steps in the metabolism of glucose*

	$\Delta G^{0\prime}$ (kJ mol^{-1})
D-Glucose + ATP → D-glucose-6-phosphate + ADP	−16.7
D-Glucose-6-phosphate → D-fructose-6-phosphate	+1.7
D-Fructose-6-phosphate + ATP → D-fructose-1,6-diphosphate + ADP	−14.2
Fructose-1,6-diphosphate → dihydroxyacetone phosphate + glyceraldehyde-3-phosphate	+23.8
Dihydroxyacetone phosphate → glyceraldehyde-3-phosphate	+7.5
Glyceraldehyde-3-phosphate + phosphate + NAD$^+$ → 1,3-diphosphoglycerate + NADH + H$^+$	+6.3
1,3-Diphosphoglycerate + ADP → 3-phosphoglycerate + ATP	−18.8
3-Phosphoglycerate → 2-phosphoglycerate	+4.6
2-Phosphoglycerate → phosphoenolpyruvate + H$_2$O	+1.7
2-Phosphoenolpyruvate + ADP → pyruvate + ATP	−31.4
Pyruvate + NADH + H$^+$ → lactate + NAD$^+$	−25.1
Pyruvate → acetaldehyde + CO$_2$	−19.8
Acetaldehyde + NADH + H$^+$ → ethanol + NAD$^+$	−23.7

* An important reaction in many of these steps is the hydrolysis of ATP: ATP + H$_2$O → ADP + phosphate. $\Delta G^{0\prime}$ = −31.0 kJ mol^{-1}. $\Delta H^{0\prime}$ = −24.3 kJ mol^{-1}.

SOURCE: C. K. Mathews and K. E. van Holde, *Biochemistry,* Benjamin/Cummings (1990).

The first step in the metabolism of glucose (Table 4.2) is the formation of glucose-6-phosphate. The direct reaction of glucose with phosphate has a relatively large positive standard free energy change and will not occur significantly under physiological conditions.

(1)   glucose + phosphate ⟶ glucose-6-phosphate + H$_2$O

$\Delta G^{0\prime}$ = +14.3 kJ

In the first step of glycolysis, however, this reaction is coupled to the hydrolysis of ATP to form adenosine diphosphate, ADP, which is thermodynamically favorable.

(2)   ATP + H$_2$O ⟶ ADP + phosphate

$\Delta G^{0\prime}$ = −31.0 kJ

The sum of these two reactions has a negative standard free energy, and that process occurs spontaneously. Adding (1) and (2), we obtain

(3)   glucose + ATP ⟶ glucose-6-phosphate + ADP

$\Delta G^{0\prime}$ = −16.7 kJ

as shown in Table 4.2. In this manner the strongly spontaneous hydrolysis of ATP is coupled to the otherwise nonspontaneous glucose phosphorylation. This reaction is typical of the role played by ATP in metabolism.

All reactions in the glycolytic path are enzyme catalyzed; the first step is catalyzed by the enzyme hexokinase. The enzyme speeds up the reaction and serves to control the glycolytic path in response to the cell's demand for metabolic energy, but the enzyme does not affect the thermodynamics of the process. The free energy changes are associated with differences in chemical potential between reactants and products; they do not indicate how fast the processes occur.

The second step in the metabolism of glucose is the isomerization of glucose-6-phosphate to fructose-6-phosphate catalyzed by the enzyme phosphoglucose isomerase. This reaction is also nonspontaneous in the standard state, but the positive free energy change is relatively small, $\Delta G^{0\prime} = +1.7$ kJ mol^{-1}. The equilibrium constant associated with this free energy change at 25°C is 0.50, which means that significant amounts of product are formed under glycolytic conditions. Furthermore, the glycolytic process is drawn on by the third step in the process, which is the attachment of a second phosphate group to fructose-6-phosphate to form fructose-1,6-diphosphate. This step, like the first one, is driven by phosphate transfer from ATP and has a $\Delta \overline{G}^{0\prime} = -14.2$ kJ mol^{-1} that is strongly negative.

In the fourth step in the glycolytic path, the six-carbon sugar fructose-1,6-diphosphate is split into two three-carbon compounds, dihydroxyacetone phosphate and glyceraldehyde-3-phosphate, catalyzed by the enzyme aldolase. This reaction has a large positive free energy change,

(4) fructose-1,6-diphosphate $\longrightarrow$

dihydroxyacetone phosphate $+$ glyceraldehyde-3-phosphate

$\Delta G^{0\prime}_{298} = +23.8$ kJ

which would cause it to be very nonspontaneous under standard conditions. But the concentrations of these metabolites in the cell are actually much less than 1 M, and this results in a very small value for the reaction quotient, $Q \approx 10^{-4}$, for this reaction. The effect of such a small reaction quotient is to cause the actual free energy, ΔG, to be significantly less positive. At 37°C each factor of 0.1 in the reaction quotient decreases the free energy by about 6 kJ mol^{-1} ($RT \ln 0.1 = -5.9$ kJ mol^{-1}), so the effective value for the free energy change under actual metabolic conditions is $\Delta G \approx 0$. This is a direct consequence of the fact that in reaction (4), there are more product species than reactant species—all at low concentrations. Le Châtelier's principle tells us that a decrease in overall concentration for such a reaction will favor products. Even though the system is distinctly not at equilibrium under physiological conditions, that is the direction in which the spontaneous process is driven under low concentration conditions. An example of this behavior is presented in Example 4.1.

Dihydroxyacetone phosphate is in equilibrium with glyceraldehyde-3-phosphate. The free energy change for the reaction

dihydroxyacetone phosphate $\rightleftharpoons$ glyceraldehyde-3-phosphate

$\Delta G^{0\prime}_{298} = +7.5$ kJ

gives an equilibrium constant $K = 0.05$. Nevertheless, the next reactions in the glycolytic path drain the pool of glyceraldehyde-3-phosphate sufficiently that essentially all of the dihydroxyacetone phosphate becomes converted to glyceraldehyde-3-phosphate in the continuation of glycolysis.

The next step results in the simultaneous oxidation and phosphorylation of glyceraldehyde-3-phosphate to 1,3-diphosphoglycerate. The oxidation is carried out by the biological oxidant nicotinamide adenine dinucleotide, NAD, and the phosphate is taken up directly from the pool of inorganic phosphate, P_i.

glyceraldehyde-3-phosphate $+$ NAD^+ $+$ P_i $\longrightarrow$

$$1,3\text{-diphosphoglycerate} + NADH + H^+$$

$\Delta G^{0'}_{298} = +6.3$ kJ

Although this reaction has a positive standard free energy change, the process continues because of the next step in glycolysis, which is the transfer of one of the phosphates of 1,3-diphosphoglycerate to ADP to generate ATP and 3-phosphoglycerate. This reaction has a strongly negative standard free energy change.

$$1,3\text{-diphosphoglycerate} + ADP \longrightarrow 3\text{-phosphoglycerate} + ATP$$

$\Delta G^{0'}_{298} = -18.8$ kJ

The remaining three steps on the path to pyruvate involve two isomerization reactions and a final step in which the remaining phosphate is transferred to ADP to form ATP. No new principles are involved in these steps. The overall process by which one molecule of glucose is metabolized all the way to pyruvate can be summarized by the reaction

glucose $+$ $2\,NAD^+$ $+$ $2\,ADP$ $+$ $2\,P_i$ $\longrightarrow$

$$2\,\text{pyruvate} + 2\,NADH + 2\,ATP + 2\,H^+ + 2\,H_2O$$

$\Delta G^{0'}_{298} = -80.6$ kJ

In the overall glycolytic process to form pyruvate from glucose, there is a net formation of two moles of ATP and two moles of the physiological reductant NADH per mole of glucose consumed. In this way a significant fraction of the available chemical potential is retained as chemical energy for use in other biosynthetic or energy-requiring reactions. For example, under strenuous exercise in muscle cells, NADH serves to reduce pyruvate to lactate according to the reaction

pyruvate $+$ NADH $+$ H^+ $\longrightarrow$ lactate $+$ NAD^+ $\qquad$ $\Delta G^{0'}_{298} = -25.1$ kJ

Example 4.11 The enzyme aldolase catalyzes the conversion of fructose-1,6-diphosphate, FDP, to dihydroxyacetone phosphate, DHAP, and glyceraldehyde-3-phosphate, GAP. Under physiological conditions in red blood cells (erythrocytes), the concentrations of these species are [FDP] = 35 μM, [DHAP] = 130 μM and [GAP] = 15 μM. Will the conversion occur spontaneously under these conditions?

Solution The standard free energy change for this reaction is

$$FDP \longrightarrow DHAP + GAP \qquad \Delta G^{0'}_{298} = +23.8 \text{ kJ}$$

and

$$Q = \frac{[\text{DHAP}][\text{GAP}]}{[\text{FDP}]} = \frac{(130 \times 10^{-6})(15 \times 10^{-6})}{(35 \times 10^{-6})}$$

$$= 55.7 \times 10^{-6}$$

$$\Delta G = +23{,}800 + (8.314)(298) \ln (55.7 \times 10^{-6})$$

$$= +23{,}800 - 24{,}300$$

$$= -500 \text{ J} = -0.5 \text{ kJ}$$

Under the actual conditions in the cell the free energy of the reaction is negative, and the reaction will occur spontaneously.

Double Strand Formation in Nucleic Acids

The formation of a double-stranded helix from two single strands of DNA is crucial in identifying and locating specific sequences of nucleotides in a DNA. A probe oligo-nucleotide (a single strand of DNA containing 20 to 40 nucleotides) is added to the single-stranded target DNA to form a double helix with its complementary sequence. The double helix has Watson-Crick complementary base pairs in which adenine (A) pairs with thymine (T) and guanine (G) pairs with cytosine (C). This technique is used to locate genes in chromosomes, to diagnose infectious or genetic diseases, and to amplify DNA by the polymerase chain reaction (PCR). In PCR amplification the added oligonucleotide is called a primer. It binds to the DNA strand to be amplified and provides a starting site for the DNA polymerase enzyme to synthesize the complementary strand. A second primer is added to the newly synthesized strand so its complement can be made. Thus a double-stranded copy has been synthesized. As each cycle doubles the number of DNA molecules, rapid amplification can be obtained.

Knowledge of the equilibrium constant for formation of a DNA double strand of any sequence at any temperature is obviously of great interest. Systematic studies were done by mixing complementary single strands in 1 M NaCl at pH 7 (Breslauer et al., 1986). The amount of double helix—and thus the equilibrium constant—was determined from the decrease in ultraviolet light absorbance when a DNA double helix forms (see Chapter 10). The standard free energy changes were calculated from the equilibrium constants; calorimetric measurements provided the enthalpy changes.

When a series of similar reactions is studied, one looks for simple patterns in the data. The obvious question to ask for formation of DNA double strands is "Does the thermodynamics just depend on the number of A·T base pairs and the number of G·C base pairs formed?" The answer is no, this simple assumption does not fit the data. The next approximation is to assume that the sequence is important only through the base-pair nearest neighbors. This approximation is valid. We can rationalize it by noting that the main interactions involved in forming a double strand is hydrogen bonding between base pairs, plus London-van der Waals interaction between neighboring (stacked) base pairs. Thus ΔG^0, ΔH^0, ΔS^0—and also K as a function of temperature—can be calculated from 10 Watson-

Table 4.3 Thermodynamic parameters* for calculating double strand stability in DNA (pH 7, 1 M NaCl)

	ΔG^0 (kJ mol^{-1}) at 25°C	ΔH^0 (kJ mol^{-1})	ΔS^0 (J K^{-1} mol^{-1})
$5'-A \quad 3'-T \longrightarrow -A-A / -T-T$	−7.9	−38.1	−101.3
$5'-A \quad 3'-T \longrightarrow -A-T / -T-A$	−6.3	−36.0	−99.7
$5'-T \quad 3'-A \longrightarrow -T-A / -A-T$	−3.8	−25.1	−71.5
$5'-A \quad 3'-T \longrightarrow -A-C / -T-G$	−5.4	−27.2	−73.2
$5'-C \quad 3'-G \longrightarrow -C-A / -G-T$	−7.9	−24.3	−55.0
$5'-A \quad 3'-T \longrightarrow -A-G / -T-C$	−6.7	−32.6	−86.9
$5'-G \quad 3'-C \longrightarrow -G-A / -C-T$	−6.7	−23.4	−56.0
$5'-C \quad 3'-G \longrightarrow -C-G / -G-C$	−15.1	−49.8	−116.4
$5'-G \quad 3'-C \longrightarrow -G-C / -C-G$	−13.0	−46.4	−112.1
$5'-C \quad 3'-G \longrightarrow -C-C / -G-G$	−13.0	−46.0	−110.7

Initiation: If the strands contain no G·C base pairs use ΔG^0(A·T initiation) = +25.1 kJ mol^{-1}; ΔH^0(A·T initiation) = 0; ΔS^0(A·T initiation) = −84.1 J K^{-1} mol^{-1}. If the strands contain one or more G·C base pairs use ΔG^0(G·C initiation) = +20.9 kJ mol^{-1}; ΔH^0(G·C initiation) = 0; ΔS^0(G·C initiation) = −70.3 J K^{-1} mol^{-1}.

* Data are from Breslauer et al., *Proc. Natl. Acad. Sci. USA 83*, 3746–3750 (1986).

Crick nearest neighbor parameters, plus parameters for the formation of the first base pair of the double strand as shown in Table 4.3. The general expression is

$$\Delta G^0 = \Delta G^0(\text{initiation}) + \Sigma \, \Delta G^0(\text{nearest-neighbors}) \tag{4.36}$$

and similar expressions for ΔH^0 and ΔS^0. Consider the reaction

5′–ATAGCA–3′
+ $\rightleftharpoons$ 5′–ATAGCA–3′
                                      3′–TATCGT–5′
5′–TGCTAT–3′

The double strand contains antiparallel complementary strands; the structures of the molecules are shown in Fig. 3.7 and the Appendix. To calculate the thermodynamic values we use Eq. 4.36. The $\Delta G^0(\text{initiation})$ is positive—unfavorable. It represents the loss of entropy for bringing the two strands together; the enthalpy of forming the first base pair is negligible. For a DNA containing at least one $G \cdot C$ base pair, use $\Delta G^0(G \cdot C \text{ initiation})$; only if the strands contain no $G \cdot C$ is the $\Delta G^0(A \cdot T \text{ initiation})$ used. The $\Delta G^0(\text{nearest-neighbors})$ are all negative—favorable; they characterize the hydrogen bonding and stacking interactions of adding successive base pairs. It does not matter which base pair is assumed to form first. No matter where in the sequence you start adding base pairs, the nearest neighbors are the same as long as you do not confuse the 5′ to 3′ directions of the strands. However, it is convenient to start at the left-hand end for counting nearest neighbors. For the example given, the result is

$$\Delta G^0 = \Delta G^0(G \cdot C \text{ initiation}) + \Delta G^0\!\left(\begin{smallmatrix}A\,T\\ \cdot\ \cdot\\ T\,A\end{smallmatrix}\right) + \Delta G^0\!\left(\begin{smallmatrix}T\,A\\ \cdot\ \cdot\\ A\,T\end{smallmatrix}\right) + \Delta G^0\!\left(\begin{smallmatrix}A\,G\\ \cdot\ \cdot\\ T\,C\end{smallmatrix}\right) + \Delta G^0\!\left(\begin{smallmatrix}G\,C\\ \cdot\ \cdot\\ C\,G\end{smallmatrix}\right) + \Delta G^0\!\left(\begin{smallmatrix}C\,A\\ \cdot\ \cdot\\ G\,T\end{smallmatrix}\right)$$

Using Table 4.3

$$\Delta G^0(\text{kJ}) = 20.9 - 6.3 - 3.8 - 6.7 - 13.0 - 7.9 = -16.8 \text{ kJ}$$

$$K_{298} = e^{-\Delta G0/RT} = e^{-16,800/(8.314)(298)} = 881$$

For equilibrium constants at other temperatures, calculate ΔH^0 and ΔS^0 from Table 4.3, then ΔG^0 can be calculated at any temperature by assuming ΔH^0 and ΔS^0 are independent of temperature: $\Delta G_T^0 = \Delta H^0 - T \, \Delta S^0$.

GALVANIC CELLS

We have used free energies and presented tables of standard free energies without discussing the easiest and most accurate method of measuring free energies. If a chemical reaction can be made to occur in a galvanic cell, the voltage of the cell is a direct measure of the free energy of the reaction. Alternatively, from the known free energy change of a reaction, we can calculate the maximum voltage that can be obtained when the reaction takes place in an electrochemical cell. Clearly, this method can be applied to any reaction that involves oxidation and reduction, but it is even more general than that. As we shall see, it can be applied to equilibrium processes such as the solubility of a salt, the dissociation of a weak acid or base, the formation of complex ions, or concentration differences across osmotically active membranes. It is sufficient to carry out the thermodynamic studies in the presence of electrodes that are sensitive to the chemical or ionic species involved.

For a reversible process at constant T and P, one can show that the free energy change is equal to the maximum useful work; useful work is defined here as work other than pressure-volume work. From the definition of free energy in terms of energy and entropy [Eqs. (2.12) and (3.22)],

$$G = E + PV - TS$$

For a change in G at constant T and P

$$\Delta G = \Delta E + P\,\Delta V - T\,\Delta S$$

We can use the first law for a reversible process to obtain

$$\Delta G = q_{reversible} + w_{reversible} + P\,\Delta V - T\,\Delta S$$

But $q_{reversible}$ equals $T\,\Delta S$, so

$$\Delta G = w_{reversible} + P\,\Delta V$$

Because pressure-volume work equals $-P\,\Delta V$,

$$\Delta G = w_{reversible} - w_{pressure\text{-}volume}$$
$$\Delta G = w_{max.\ useful} \tag{4.37}$$

In a galvanic cell the maximum useful work is the reversible electric work:

$$-\Delta G = -w_{reversible\ electrical} \tag{4.38}$$

We use minus signs in the equation to emphasize that the *decrease* in free energy is equal to the reversible electrical work done *by* the system. Electrical work is equal to the voltage times the current multiplied by the time [Eq. (2.8)]. Because current multiplied by time is charge, the electrical work done by a chemical reaction in a galvanic cell is the voltage of the cell multiplied by the amount of charge transferred in the reaction. For n moles of electrons transferred, the free energy is

$$\Delta G\text{(electron volts)} = -n\,\mathscr{E} \tag{4.39}$$

where n = number of moles of electrons involved in the reaction
 $\mathscr{E}$ = reversible voltage of cell (the sign convention for a cell will be discussed shortly)

An electron volt is a unit of energy. One mole of electrons moving 1 cm in a field of 1 V cm^{-1} acquires an energy of 96.485 kJ mol^{-1}.

$$\Delta G\text{(kJ)} = -96.485\,n\,\mathscr{E} \tag{4.40}$$

The free energy is also often written

$$\Delta G\text{(J)} = -n\,\mathscr{E}F \tag{4.41}$$

where F = Faraday = 96,485 coulombs (C) mol^{-1} of electrons. Because the reversible voltage is proportional to free energy, it follows from $(\partial G/\partial T)_P = -S$ that the entropy change for the reaction in the cell depends on the temperature dependence of the voltage. See section on partial derivatives in Chapter 3.

$$\Delta S\,(\text{J K}^{-1}) = 96{,}485n \left(\frac{\partial \mathscr{E}}{\partial T} \right)_P \qquad (4.42)$$

Also, from the definition of $G \equiv H - TS$,

$$\Delta H\,(\text{kJ}) = 96.485n \left[-\mathscr{E} + T \left(\frac{\partial \mathscr{E}}{\partial T} \right)_P \right] \qquad (4.43)$$

Therefore, the thermodynamic characteristics of the cell reaction can be obtained from measurements of the reversible voltage of a galvanic cell at several temperatures.

When a reaction occurs spontaneously in a galvanic cell, the free energy change must be negative, and the voltage, by convention, is positive. By convention we write the oxidation reaction as occurring at the left electrode and reduction at the right electrode (an easy way to remember this is to associate the two r's: reduction, right).

In the reaction between $H_2(g)$ and $O_2(g)$ to produce $H_2O(l)$, the equation is

$$H_2(g) + \tfrac{1}{2}O_2(g) \longrightarrow H_2O(l)$$

For a galvanic cell we need to make electrical contact with the chemical species involved. Platinum metal electrodes are usually used for this purpose, because they are chemically inert and they can be prepared with a catalytic surface that aids in attaining nearly reversible conditions. We also need ions present in the liquid water to conduct the electrical current. We can use "inert" ions, such as Na^+ and Cl^-, or we can use involved ions such as H^+ or OH^-. One such arrangement, which uses HCl as the electrolyte, is shown in Fig. 4.8. $H_2(g)$ is provided through a gas bubbler at one Pt electrode, and $O_2(g)$ is provided at a second Pt electrode. The electrodes, in contact with the bubbling gases, are placed in an aqueous solution of HCl. We use a porous barrier or a salt bridge (permeable to diffusion by aqueous ions) between the two electrode compartments to prevent mixing. The reaction to form water is both highly spontaneous ($\Delta G < 0$) and exothermic ($\Delta H < 0$) under normal conditions.

In the galvanic cell shown in Fig. 4.8 the oxidation takes place at the left electrode.

Oxidation: $H_2(g, P_{H_2}) \longrightarrow 2\,H^+(c_{H^+}) + 2\,e^-$

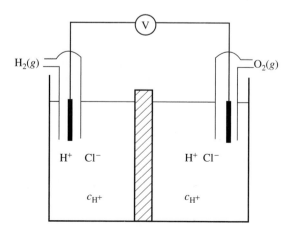

Fig. 4.8 A galvanic cell in which $H_2(g)$ and $O_2(g)$ form water by oxidizing the H_2 at one electrode and reducing the O_2 at the other electrode. The electrodes are inert conductors such as platinum. They allow transfer of the electrons through the external wires and the voltmeter. The electrical connection in the solution is provided by the ions and the semipermeable porous barrier. The voltage measures the difference in chemical potentials of products and reactants—the ΔG of the reaction.

Thus, the left terminal is negative; electrons are being produced. The reduction takes place at the right electrode.

Reduction: $2e^- + \frac{1}{2}O_2(g, P_{O_2}) + 2H^+(c_{H^+}) \longrightarrow H_2O$

Thus the right terminal is positive; electrons are being removed. In addition to preventing diffusive mixing, the porous barrier maintains a path for electrical conduction (by ion diffusion). Such cells are often described using an abbreviated notation.

$$Pt \,|\, H_2(g, P_{H_2}) \,|\, H^+(c_{H^+}) \,\|\, H^+(c_{H^+}) \,|\, O_2(g, P_{O_2}) \,|\, Pt$$

The cell reaction for this cell is the sum of the two half reactions written above, balanced so that the electrons cancel in the cell reaction.

$$H_2(g) + \frac{1}{2}O_2(g) \longrightarrow H_2O(l)$$

Note that H^+ does not appear in the cell reaction for this cell.

To determine the free energy change for the reaction at the gas pressures provided, P_{H_2} and P_{O_2}, we use the measured reversible voltage of the cell in Eqs. (4.39) through (4.41) with $n = 2$. It is easy to measure the reversible voltage of the cell accurately with a potentiometer or electronic voltmeter (represented by V in the diagram). The pressure (concentration) dependence of the reversible voltage can be measured, and by extrapolation the reversible voltage for standard conditions can be determined. This provides the standard free energy change for the reaction.

Standard Electrode Potentials

Reversible voltages have been determined for many reactions at standard conditions. Each reaction of course includes both an oxidation and a reduction half-cell reaction. If we *define* the voltage of one half-cell reaction as zero, all others can be obtained relative to this arbitrary choice. Chemists have chosen the reduction of H^+ to H_2 gas under standard conditions to have a standard electrode potential of zero:

$$H^+(a = 1) + e^- \longrightarrow \frac{1}{2}H_2(g, P = 1 \text{ atm}) \qquad \mathscr{E}^0 = 0.000 \text{ V}$$

With this choice the standard reduction potentials given in Table 4.4 were obtained.

To calculate the standard potential (and therefore the standard free energy) for a chemical reaction from Table 4.4, it is necessary to combine the appropriate potentials for the half-cell reactions. The value of $\mathscr{E}^0_{cell}$ is $\mathscr{E}^0$ for the reduction half-cell reaction minus $\mathscr{E}^0$ for the oxidation half-cell reaction. By convention for galvanic cells, this is equal to $\mathscr{E}^0$ for the right-hand side of the cell minus $\mathscr{E}^0$ for the left-hand side of the cell. For example, let us consider the reaction

$$Zn(s) + Cu^{2+}(a = 1) \longrightarrow Zn^{2+}(a = 1) + Cu(s)$$

The half-cell reactions and their potentials are

$$Cu^{2+}(a = 1) + 2e^- \longrightarrow Cu(s) \qquad \mathscr{E}^0 = +0.337 \text{ V}$$

$$Zn(s) \longrightarrow Zn^{2+}(a = 1) + 2e^- \qquad \mathscr{E}^0 = -(-0.763 \text{ V})$$

$$\overline{\qquad\qquad\qquad\qquad\qquad\qquad \mathscr{E}^0_{cell} = +1.100 \text{ V}}$$

Table 4.4 Standard reduction electrode potentials at 25°C

Oxidant/Reductant	Electrode reaction	$\mathscr{E}^0$ (V)*	$\mathscr{E}^{0\prime}$ (V)[†] (pH 7)
Li^+/Li	$Li^+ + e^- \rightarrow Li$	-3.045	
Na^+/Na	$Na^+ + e^- \rightarrow Na$	-2.714	
Mg^{2+}/Mg	$Mg^{2+} + 2\,e^- \rightarrow Mg$	-2.363	
$OH^-/H_2/Pt$	$2\,H_2O + 2\,e^- \rightarrow H_2 + 2\,OH^-$	-0.8281	
Zn^{2+}/Zn	$Zn^{2+} + 2\,e^- \rightarrow Zn$	-0.7628	
Acetate/acetaldehyde	$OAc^- + 3\,H^+ + 2\,e^- \rightarrow CH_3CHO + H_2O$		-0.581
Fe^{2+}/Fe	$Fe^{2+} + 2\,e^- \rightarrow Fe$	-0.4402	
Gluconate/glucose	$C_6H_{11}O_7^- + 3\,H^+ + 2\,e^- \rightarrow C_6H_{12}O_6 + H_2O$		-0.44
Spinach ferredoxin	$Fd[Fe(III)] + e^- \rightarrow Fd[Fe(II)]$		-0.432
CO_2/formate	$CO_2 + 2\,H^+ + 2\,e^- \rightarrow HCO_2^- + H^+$	-0.20	-0.42
$NAD^+/NADH$[‡]	$NAD^+ + H^+ + 2\,e^- \rightarrow NADH$	-0.105	-0.320
Fe^{3+}/Fe	$Fe^{3+} + 3\,e^- \rightarrow Fe$	-0.036	
$H^+/H_2/Pt$	$2\,H^+ + 2\,e^- \rightarrow H_2$	0	-0.421
Acetoacetate/ β-hydroxybutyrate	$CH_3COCH_2CO_2^- + 2\,H^+ + 2\,e^- \rightarrow$ $CH_3CHOHCH_2CO_2^-$		-0.346
Mn hematoporphyrin IX	$He[Mn(III)] + e^- \rightarrow He[Mn(II)]$		-0.342
$NADP^+/NADPH$	$NADP^+ + H^+ + 2\,e^- \rightarrow NADPH$		-0.324
Horseradish peroxidase	$HRP[Fe(III)] + e^- \rightarrow HRP[Fe(II)]$		-0.271
$FAD/FADH_2$§	$FAD + 2\,H^+ + 2\,e^- \rightarrow FADH_2$		-0.219
Acetaldehyde/ethanol	$CH_3CHO + 2\,H^+ + 2\,e^- \rightarrow CH_3CH_2OH$		-0.197
Pyruvate/lactate	$CH_3COCO_2^- + 2\,H^+ + 2\,e^- \rightarrow$ $CH_3CHOHCO_2^-$		-0.18
Oxaloacetate/malate	$^-O_2CCOCH_2CO_2^- + 2\,H^+ + 2\,e^- \rightarrow$ $^-O_2CCHOHCH_2CO_2^-$		-0.166
Fumarate/succinate	$^-O_2CCH{=}CHCO_2^- + 2\,H^+ + 2\,e^- \rightarrow$ $^-O_2CCH_2CH_2CO_2^-$		$+0.031$
Myoglobin	$Mb[Fe(III)] + e^- \rightarrow Mb[Fe(II)]$		$+0.046$
Dehydroascorbate/ ascorbate	$C_6H_7O_7^- + 2\,H^+ + 2\,e^- \rightarrow C_6H_9O_7^-$		$+0.058$
Ubiquinone	$UQ + 2\,H^+ + 2\,e^- \rightarrow UQH_2$		$+0.10$
$AgCl/Ag/Cl^-$	$AgCl + e^- \rightarrow Ag + Cl^-$	$+0.2223$	
Calomel	$\frac{1}{2}\,Hg_2Cl_2 + e^- \rightarrow Hg + Cl^-$	$+0.268$	
Cytochrome c	$Cyt[Fe(III)] + e^- \rightarrow Cyt[Fe(II)]$		$+0.254$
Cu^{2+}/Cu	$Cu^{2+} + 2\,e^- \rightarrow Cu$	$+0.337$	
$I_2/I^-/Pt$	$I_2 + 2\,e^- \rightarrow 2\,I^-$	$+0.5355$	
$O_2/H_2O_2/Pt$	$O_2 + 2\,H^+ + 2\,e^- \rightarrow H_2O_2$	$+0.69$	$+0.295$
$Fe^{3+}/Fe^{2+}/Pt$	$Fe^{3+} + e^- \rightarrow Fe^{2+}$	$+0.771$	
Ag^+/Ag	$Ag^+ + e^- \rightarrow Ag$	$+0.799$	
$NO_3^-/NO_2^-/Pt$	$NO_3^- + 2\,H^+ + 2\,e^- \rightarrow NO_2^- + H_2O$	$+0.94$	$+0.421$
$Br_2/Br^-/Pt$	$Br_2 + 2\,e^- \rightarrow 2\,Br^-$	$+1.087$	
$O_2/H_2O/Pt$	$O_2 + 4\,H^+ + 4\,e^- \rightarrow 2\,H_2O$	$+1.229$	$+0.816$
$Cl_2/Cl^-/Pt$	$Cl_2 + 2\,e^- \rightarrow 2\,Cl^-$	$+1.359$	
$Mn^{3+}/Mn^{2+}/Pt$	$Mn^{3+} + e^- \rightarrow Mn^{2+}$	$+1.4$	
$Ce^{4+}/Ce^{3+}/Pt$	$Ce^{4+} + e^- \rightarrow Ce^{3+}$	$+1.61$	
$F_2/F^-/Pt$	$F_2 + 2\,e^- \rightarrow 2\,F^-$	$+2.87$	

* $\mathscr{E}^0$ refers to the solute standard state with unit activity for all species.

[†] $\mathscr{E}^{0\prime}$ refers to the biochemist's standard state with pH 7.

[‡] NAD^+ is nicotinamide adenine dinucleotide.

§ FAD is flavin adenine dinucleotide.

The standard reversible voltage is +1.100 V and the standard free energy is obtained from Eqs. (4.39) through (4.41).

$$\Delta G^0 = -2.20 \text{ eV} = -212 \text{ kJ}$$

Because Table 4.4 gives reduction potentials, the sign of the potential must be changed for the oxidation half-cell reaction. However, the voltages must *not* be multiplied by 2 even though there are two electrons involved. The voltage of the cell is independent of the number of electrons involved. The free energy does depend on the number of electrons, and therefore n appears explicitly in Eqs. (4.39) through (4.41). Free energy is extensive, but voltage is intensive.

If the reaction as written does not occur spontaneously, the calculated voltage is found to be negative and the standard free energy is positive.

Concentration Dependence of $\mathscr{E}$

Table 4.4 is analogous to a table of standard free energies. The standard free energies of a great many chemical reactions can be obtained from the table. What if we are interested in the free energies at arbitrary concentrations? We know the relation between standard free energy and free energy:

$$\Delta G = \Delta G^0 + RT \ln Q \qquad (4.4)$$

Substituting Eq. (4.41) for ΔG and ΔG^0, we obtain the *Nernst equation:*

$$\mathscr{E} = \mathscr{E}^0 - \frac{RT}{nF} \ln Q \qquad (4.44)$$

where $\mathscr{E}$ = cell voltage for arbitrary concentrations specified by Q
$\mathscr{E}^0$ = standard cell voltage
R = gas constant in joules = $8.314 \text{ J K}^{-1} \text{ mol}^{-1}$
n = number of moles of electrons involved in the reaction

For the reaction

$$a\text{A} + b\text{B} \longrightarrow c\text{C} + d\text{D}$$

$$Q = \frac{(a_\text{C})^c (a_\text{D})^d}{(a_\text{A})^a (a_\text{B})^b}$$

At 25°C the Nernst equation can be written explicitly as

$$\mathscr{E} = \mathscr{E}^0 - \frac{0.0591}{n} \log Q \qquad (4.45)$$

The Nernst equation can be used to calculate the voltage and free energy for any concentrations of reactants and products in a cell. At equilibrium $\Delta G = 0$; therefore, $\mathscr{E} = 0$, also. Thus the standard potential gives the equilibrium constant for the reaction in the cell:

$$\mathscr{E}^0 = \frac{RT}{nF} \ln K \qquad (4.46)$$

We can solve for K, and at 25°C write

$$K = 10^{n \mathscr{E}^0/0.0591} \qquad (4.47)$$

ELECTRON TRANSFER AND BIOLOGICAL REDOX REACTIONS

Oxidation-reduction reactions are essential for energy storage and conversion in biological organisms. For example, the pyruvate produced as a product of glycolysis (Fig. 4.7) undergoes oxidative decarboxylation to form acetyl coenzyme A, which then enters the *citric acid cycle*. In the citric acid cycle the acetate is transferred to oxaloacetate to form the six-carbon molecule, citrate. Citrate then undergoes a series of at least eight reactions that involve progressive oxidation of two of the carbon atoms to CO_2 and return of the remaining four-carbon portion as oxaloacetate for re-entry into the cycle. Coupled to the oxidation of the acetate to CO_2 is the reduction of 3 mol of NAD^+ to NADH and the production of 1 mol of reduced flavin adenine dinucleotide, $FADH_2$. This reducing power generated in the citric acid cycle is subsequently used for biosynthetic reactions and for the formation of ATP by oxidative phosphorylation (respiration) in mitochondria.

The step that completes the citric acid cycle is the oxidation of malate to oxaloacetate coupled with the reduction of NAD^+ to NADH by the enzyme malate dehydrogenase.

To calculate the $\Delta G^{0\prime}$ associated with this reaction, we can use $\mathscr{E}^{0\prime}$ values for the corresponding redox half-reactions listed in Table 4.4.

$$\mathscr{E}^{0\prime} \text{ (volts)}$$

$$NAD^+ + H^+ + 2\,e^- \longrightarrow NADH \qquad -0.320$$

$$\text{oxaloacetate} + 2\,H^+ + 2\,e^- \longrightarrow \text{malate} \qquad -0.166$$

$$\mathscr{E}^{0\prime} = -0.320 - (-0.166) = -0.154 \text{ V}$$
$$\Delta G^{0\prime} = -(2)(96.485)(-0.154) = 29.7 \text{ kJ}$$

This step in the cycle is not spontaneous under standard conditions. However, it will become spontaneous under conditions where the demand for NADH is high; that is, when $[NADH]/[NAD^+]$ is low. Under these conditions the malate will undergo conversion to oxaloacetate, which serves to turn on the citric acid cycle. This is an example of metabolic regulation whereby supply and demand are kept in reasonable balance.

A principal source of cellular ATP is the process of *oxidative phosphorylation* (respiratory electron transport) carried out in mitochondria.*

This is a series of electron transfer reactions catalyzed by three membrane-associated enzyme complexes whereby NADH is oxidized, ultimately by O_2. Each of the complexes

---

* Mitochondria are present in all organisms that have differentiated cells (eukaryotes). They are small (1 μm) membrane-surrounded structures in which oxidative metabolism occurs.

contributes to the formation of a transmembrane proton gradient which is coupled to the formation of ATP. The overall redox reaction is

$$2\,NADH + 2\,H^+ + O_2(g) \longrightarrow 2\,NAD^+ + 2\,H_2O$$

It is useful to analyze the thermodynamics of this process in terms of each of the three enzyme complexes.

1) *NADH-Q reductase*

NADH, which is a soluble reductant produced in the citric acid cycle and elsewhere, is imported into the mitochondria. It undergoes oxidation at the enzyme complex NADH-Q reductase, which contains several reducible groups. Initially the NADH reduces the flavin, FMN, which in turn reduces an iron-sulfur (FeS) center. Finally the electrons are passed on to reduce ubiquinone, UQ. We can sketch the coupled oxidation-reduction reactions catalyzed by this complex as follows

NADH-Q reductase

The UQH_2 is then released to go on to the next complex. In the course of this process H^+ ions are discharged outside the mitochondrial membranes into an intermembrane space and H^+ is taken up from the internal space. This is the origin of one contribution to the pH gradient that results in ATP formation.

The net reaction for this portion of the electron transfer (per two electrons) is

$$NADH + H^+ + UQ \longrightarrow NAD^+ + UQH_2$$

$\Delta \mathscr{E}^{0\prime} = 0.10 - (-0.32) = 0.42\ V$
$\Delta G^{0\prime} = -(2)(96.485)(0.42) = -81\ kJ$

2) *Cytochrome reductase*

UQH_2 diffuses to a second mitochondrial membrane-bound complex, cytochrome reductase. UQH_2 releases its electrons to an iron-sulfur center in the enzyme complex, which passes them one at a time to water soluble cytochrome *c*. For the overall reaction

$$QH_2 + 2\,Cyt\ c_{ox} \longrightarrow Q + 2\,H^+ + 2\,Cyt\ c_{red}$$

$\mathscr{E}^{0\prime} = 0.254 - 0.10 = 0.154\ V$
$\Delta G^{0\prime} = -(2)(96.485)(0.154) = -29.7\ kJ$

3) *Cytochrome c oxidase*

The third mitochondrial membrane complex, cytochrome *c* oxidase, receives electrons one at a time from Cyt c_{red}, and transmits them to molecular oxygen, which requires four electrons overall to become reduced to two water molecules. The cytochrome *c* oxidase complex accomplishes this one-electron to four-electron transfer through the mediation of two heme centers. Each of the hemes is associated with a copper ion near the heme iron. The net reaction (per two electrons transferred) is

$$2\,\mathrm{Cyt}\;c_{\mathrm{red}} \;+\; \tfrac{1}{2}\mathrm{O}_2 \;+\; 2\,\mathrm{H}^+ \;\longrightarrow\; 2\,\mathrm{Cyt}\;c_{\mathrm{ox}} \;+\; \mathrm{H}_2\mathrm{O}$$

$$\mathscr{E}^{0\prime} = 0.816 - 0.254 = 0.562\;\mathrm{V}$$
$$\Delta G^{0\prime} = -(2)(96.485)(0.562) = -108.5\;\mathrm{kJ}$$

For each of these three stages of mitochondrial electron transport there is a strongly negative value of $\Delta G^{0\prime}$. This free energy is not dissipated, however. A portion of it is retained as chemical potential associated with a transmembrane proton gradient, which in turn provides the proton motive force that drives the formation of ATP. We shall examine the mechanism and the thermodynamics of this coupling in Chapter 5.

Activity Coefficients of Ions

We have usually set activity coefficients equal to 1 in our thermodynamic calculations. For small uncharged molecules and for univalent ions in dilute aqueous solutions, this approximation may not be too bad, but for polymers or multivalent ions such as Mg^{2+} or PO_4^{3-}, activity coefficients may be very different from 1, even at millimolar concentrations. Figure 4.9 illustrates some measured activity coefficients for a 1–1 electrolyte (HC1), a 1–2 electrolyte (H_2SO_4), and a 2–2 electrolyte ($ZnSO_4$). The activity coefficients are much less than 1 for 0.1 M solutions; at concentrations above 1 M the activity coefficients for some electrolytes increase and become even greater than 1.

Previously in this chapter we mentioned the necessity of using mean ionic activity coefficients. The definitions for different electrolytes is best illustrated by examples.

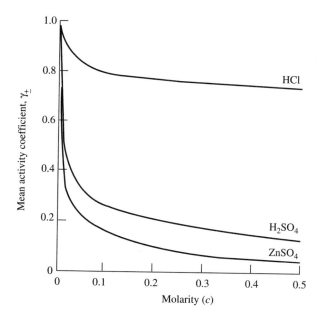

Fig. 4.9 Mean activity coefficient of various electrolytes at 25°C.

For HCl, $ZnSO_4$, or any 1–1 or 2–2 electrolyte, the mean ionic activity coefficient is

$$\gamma_\pm = (\gamma_+\gamma_-)^{1/2} \qquad \text{1–1 electrolyte} \qquad (4.48)$$

For H_2SO_4 or any 1–2 electrolyte,

$$\gamma_\pm = (\gamma_+^2\gamma_-)^{1/3} \qquad \text{1–2 electrolyte} \qquad (4.49)$$

For $LaCl_3$ or any 1–3 electrolyte,

$$\gamma_\pm = (\gamma_+\gamma_-^3)^{1/4} \qquad \text{1–3 electrolyte} \qquad (4.50)$$

The reader can generalize these equations to any type of salt.

To understand why activity coefficients are so different from 1 for electrolytes, and why the ions with the larger charges have the smaller activity coefficients, we need to consider the interactions between the solutes. Activities are effective concentrations. In an HCl solution the effective concentration of the H^+ ions is reduced by the surrounding Cl^- ions; the effective concentration of the Cl^- ions is reduced by the surrounding H^+ ions. Figure 4.9 shows that this decrease in effective concentrations of the ions is greater in H_2SO_4 solution, which has a doubly charged ion, SO_4^{2-}. Debye and Hückel were able to calculate activity coefficients for ions using Coulomb's law for the interaction energy between charged particles. Their result only applies in very dilute solution (less than 0.01 M), but it can provide a qualitative understanding of activity coefficients. Although only mean ionic activity coefficients can be measured because all solutions must be electrically neutral, single ion activity coefficients can be calculated with the Debye-Hückel theory. For ions in water at 25°C

$$\log \gamma_i = -0.509\, Z_i^2 \sqrt{\mu} \qquad (4.51)$$

Z_i = charge on ion ($\pm 1, \pm 2, \pm 3$)
μ = ionic strength = $\frac{1}{2}\Sigma_i\, c_i Z_i^2$

The sum in the ionic strength equation is over all ions in the solution. Ionic strength has been found to be a good measure of how different salts affect equilibria and reaction rates. The ionic strength rather than the concentration of the salt characterizes the effect. Note that 1 M NaCl, 0.33 M $MgCl_2$, and 0.25 M $MgSO_4$ all have ionic strength equal to 1.

Example 4.12   What are the concentrations of all species in a 0.200 M NaOAc solution if we take into account the mean ionic activity coefficients for the ions in the solution? The mean ionic activity coefficient for 1–1 electrolytes in this solution is $\gamma_\pm = 0.755$; we assume the activity coefficient of the uncharged HAc is $\gamma = 1$.

Solution   We have previously considered this problem in Example 4.5 with the assumption that all activity coefficients are unity. The mass balance

$$[Na^+] = 0.200 \qquad (i)$$

$$[HOAc] + [OAc^-] = 0.200 \qquad (ii)$$

and the charge balance

$$[H^+] + [Na^+] = [OH^-] + [OAc^-] \qquad (iii)$$

remain the same. The equilibrium constants K_{HOAc} and K_{H_2O} are

$$K_{HOAc} = \frac{(a_{H^+})(a_{OAc^-})}{(a_{HOAc})}$$

$$= \frac{(c_{H^+})(c_{OAc^-})}{(c_{HOAc})} \frac{\gamma_+\gamma_-}{\gamma_{HOAc}}$$

$$= \frac{[H^+][OAc^-]}{[HOAc]} \frac{\gamma_\pm^2}{\gamma_{HOAc}}$$

$$= 1.80 \times 10^{-5}$$

$$K_{H_2O} = \frac{(a_{H^+})(a_{OH^-})}{(a_{H_2O})}$$

$$= [H^+][OH^-]\gamma_\pm^2$$

$$= 1.00 \times 10^{-14}$$

Substituting in $\gamma_\pm = 0.755$ we obtain

$$\frac{[H^+][OAc^-]}{[HOAc]} = \frac{1.80 \times 10^{-5}}{(0.755)^2} = 3.16 \times 10^{-5} \tag{iv}$$

$$[H^+][OH^-] = \frac{1.00 \times 10^{-14}}{(0.755)^2} = 1.75 \times 10^{-14} \tag{v}$$

Solving Eqs. (i) through (v) the same way as described in Example 4.5 gives

$$[Na^+] = 0.200\ M$$

$$[OAc^-] = 0.200\ M$$

$$[OH^-] = [HOAc] = 1.05 \times 10^{-5}\ M$$

$$[H^+] = 1.67 \times 10^{-9}\ M;\ pH = 8.78$$

The only change introduced by using activity coefficients is in $[H^+]$; previously (Example 4.5) we obtained $[H^+] = 9.50 \times 10^{-10}$; pH = 9.02.

Ionic Effect on Protein–Nucleic Acid Interactions

Macromolecules often carry many charged groups. A double-stranded DNA, for example, has two phosphate groups every 3.4 Å along the molecule, with each phosphate carrying a negative charge. In aqueous solution the negatively charged phosphate groups of the polyelectrolyte interact with positively charged small ions that are present. We call these small ions counterions. For simplicity, we consider a DNA solution containing only Na^+ as the counterions.

The Na^+ counterions interact with DNA in two ways. Because of the very large negative charge density of the DNA there is a direct condensation of a fraction of the counterions to the polyelectrolyte to reduce its charge density. The remaining unneutralized negative charges are screened by a mobile ion atmosphere (Manning, 1979). We let ψ be the effective fraction of each phosphate charge neutralized by Na^+ ions condensed on the DNA. When a protein L binds to a particular site of the DNA, a number of Na^+ originally bound to the DNA are displaced. If L interacts with n phosphate groups of the DNA, $n\psi$ of Na^+ originally condensed on the DNA are released. We express the reaction as

$$P + L \overset{K}{\rightleftharpoons} PL + n\psi Na^+$$

The equilibrium constant K, which is independent of the Na^+ concentration in the solution, is

$$K = \frac{(a_{PL})(a_{Na^+})^{n\psi}}{(a_P)(a_L)}$$

$$K \cong \frac{[PL][Na^+]^{n\psi}}{[P][L]}$$

where [L] = concentration of protein, M
[P] = concentration of DNA, mol phosphate L^{-1}
[PL] = concentration of complex, M

The concentration quotient

$$\frac{[PL]}{[P][L]} \equiv K_{obs}$$

is usually measured in media containing different amounts of Na^+. We note that

$$K_{obs} = K[Na^+]^{-n\psi}$$

Clearly, the observed apparent equilibrium constant K_{obs} will be strongly dependent on $[Na^+]$ if $n\psi$ is large. For double-stranded DNA, the value of ψ is near 0.88. Therefore, from a plot of $\log K_{obs}$ versus $\log [Na^+]$, n can be calculated from the slope; this tells us how many phosphate groups on the DNA the protein L interacts with (see Problem 29 and Record, 1981).

PARTIAL MOLAL QUANTITIES

We introduced partial derivatives in Chapter 3 to discuss changes in G, H, E, S, A with variables T, P, and V. In this chapter we have considered the change in concentration of a chemical and its effect on G, H, and S. This change can be represented as a partial derivative, the partial molal quantity. The partial molal free energy, or chemical potential, for substance A in a system containing other chemicals is

$$\mu_A \equiv \overline{G}_A \equiv \left(\frac{\partial G}{\partial n_A} \right)_{T, P, n_j \neq n_A} \tag{4.52}$$

The T, P, and n_j outside the parentheses tells us that T, P, and the number of moles of all *other* chemicals present in the system are constant; so the partial molal free energy, $\overline{G}_A$, describes the way in which the free energy of an open system changes when the number of moles of A changes, but all other variables are held constant. We see that this becomes equal to the free energy per mole of A if the system consists of pure A. In general, however, the free energy change will depend on what else is in the system, and we must specify what the other components are by specifying the values of n_j. We should remember that $\overline{G}_A$ is related to the activity; $\overline{G}_A = \overline{G}_A^0 + RT \ln a_A$. Some examples should help clarify the ideas. Consider three systems: one is an empty beaker, one is a liter of H_2O, and one is a liter of $1\ M$ KNO_3. We add a small amount of NaCl to each system and ask how the free energy of each system changes. The free energy change will surely be different for each system.

The free energy change for the empty beaker will depend on the free energy per mole of pure sodium chloride:

$$\left(\frac{\partial G}{\partial n_{NaCl}}\right)_{25°C,\ 1\ atm} = \Delta \overline{G}_f^0(NaCl,\ s) = -384.14\ kJ\ mol^{-1}$$

For the liter of H_2O the free energy will depend on the final concentration, c, of NaCl and the standard free energy per mole corresponding to an infinitely dilute aqueous solution extrapolated to $1\ M$ $[\Delta G_f^0(NaCl,\ aq) = -393.13\ kJ\ mol^{-1}]$:

$$\left(\frac{\partial G}{\partial n_{NaCl}}\right)_{25°C,\ 1\ atm,\ dilute\ aq.\ soln.} = -393.13\ kJ\ mol^{-1} + RT \ln a_{NaCl}$$

$$= -393.13\ kJ\ mol^{-1} + RT \ln (a_{Na^+} \cdot a_{Cl^-})$$

$$= -393.13\ kJ\ mol^{-1} + RT \ln (\gamma_+ c \cdot \gamma_- c)$$

$$= -393.13\ kJ\ mol^{-1} + RT \ln (\gamma_\pm)^2 + RT \ln c^2$$

For very dilute NaCl solutions, $\gamma_\pm$ is near 1 and

$$\left(\frac{\partial G}{\partial n_{NaCl}}\right)_{25°C,\ 1\ atm,\ very\ dilute\ aq.\ soln.} = -393.13\ kJ\ mol^{-1} + RT \ln c^2$$

For the $1\ M$ KNO_3 solution, although the NaCl may be dilute, the presence of the KNO_3 will affect the activity of the NaCl. We can no longer calculate the partial molal free energy simply from tables; we must measure the NaCl activity in the KNO_3 solution:

$$\left(\frac{\partial G}{\partial n_{NaCl}}\right)_{25°C,\ 1\ atm,\ 1\ M\ KNO_3} = -393.13\ kJ\ mol^{-1} + RT \ln a_{NaCl}$$

The mean ionic activity coefficient for K^+ and NO_3^- in $1\ M$ KNO_3 is 0.443. If we assume that the value applies also to small added amounts of Na^+ and Cl^-, then

$$\left(\frac{\partial G}{\partial n_{NaCl}}\right)_{25°C,\ 1\ atm,\ 1\ M\ KNO_3} = -393.13\ kJ\ mol^{-1} + RT \ln a$$

$$= -393.13\ kJ\ mol^{-1} + RT \ln (\gamma_\pm)^2 + RT \ln c^2$$

$$= -393.13\ kJ\ mol^{-1} - 4.04\ kJ\ mol^{-1} + RT \ln c^2$$

$$= -397.17\ kJ\ mol^{-1} + RT \ln c^2$$

Whenever we talk about the free energy per mole of a chemical in a mixture, we mean the partial molal free energy. For accurate work we must determine the free energy of the chemical in the system at the concentrations of all substances in the mixture. Often, however, we approximate the standard free energy in solution by the molar free energy of the pure substance. In the Appendix values are listed for some pure compounds and for the infinitely dilute aqueous solution extrapolated to 1 M concentration (labeled 1 M activity); one can see that the values differ by less than 5%.

The introduction of the partial molal free energy allows us to examine more rigorously the important relation between the standard free energy and the equilibrium constant. Consider a closed system consisting of four species A, B, C, and D that are undergoing a reversible chemical reaction

$$aA + bB \rightleftharpoons cC + dD$$

The free energy G of the system is a function of the temperature T, pressure P, the number of moles n_A, n_B, n_C, and n_D of A, B, C, and D, respectively:

$$G = G(T, P, n_A, n_B, n_C, n_D)$$

Applying the standard rules of differentiation gives

$$dG = \left(\frac{\partial G}{\partial T}\right)_{P, n_A \cdots n_D} dT + \left(\frac{\partial G}{\partial P}\right)_{T, n_A \cdots n_D} dP + \left(\frac{\partial G}{\partial n_A}\right)_{T, P, n_j \neq A} dn_A$$

$$+ \left(\frac{\partial G}{\partial n_B}\right)_{T, P, n_j \neq B} dn_B + \left(\frac{\partial G}{\partial n_C}\right)_{T, P, n_j \neq C} dn_C + \left(\frac{\partial G}{\partial n_D}\right)_{T, P, n_j \neq D} dn_D$$

$$= -S\,dT + V\,dP + \mu_A\,dn_A + \mu_B\,dn_B + \mu_C\,dn_C + \mu_D\,dn_D \tag{4.53}$$

In the last step, we have substituted the various partial derivatives by their equivalents according to Eqs. (3.48), (3.49), and (3.52).

From the stoichiometry of the reaction, it is apparent that

$$\frac{dn_A}{a} = \frac{dn_B}{b} = -\frac{dn_C}{c} = -\frac{dn_D}{d}$$

Defining these ratios as $-d\alpha$ and substituting $dn_A = -ad\alpha$, $dn_B = -bd\alpha$, etc. into Eq. (4.53)

$$dG = -S\,dT + V\,dP + (c\mu_C + d\mu_D - a\mu_A - b\mu_B)\,d\alpha$$

$$= -S\,dT + V\,dP + [c(\mu_C^0 + RT\ln a_C) + d(\mu_D^0 + RT\ln a_D)$$

$$- a(\mu_A^0 + RT\ln a_A) - b(\mu_B^0 + RT\ln a_B)]\,d\alpha$$

$$= -S\,dT + V\,dP + [(c\mu_C^0 + d\mu_D^0 - a\mu_A^0 - b\mu_B^0)$$

$$+ RT(c\ln a_C + d\ln a_D - a\ln a_A - b\ln a_B)]\,d\alpha$$

$$= -S\,dT + V\,dP + \left[(c\mu_C^0 + d\mu_D^0 - a\mu_A^0 - b\mu_B^0) + RT\ln\left(\frac{a_C^c a_D^d}{a_A^a a_B^b}\right)\right]d\alpha$$

Here $d\alpha$ is a measure of the extent of the reaction.

If the temperature and pressure are constant, $dT = 0$, $dP = 0$, and we define

$$\Delta\mu^0 = c\mu_C^0 + d\mu_D^0 - a\mu_A^0 - b\mu_B^0$$

Then

$$dG = \left[\Delta\mu^0 + RT\ln\left(\frac{a_C^c a_D^d}{a_A^a a_B^b}\right)\right]d\alpha$$

If, furthermore, the system is at equilibrium, then

$$dG = 0 \qquad \text{[constant } T, P; \text{ at equilibrium; see Eq. (3.26)]}$$

and therefore

$$\Delta\mu^0 + RT \ln \left(\frac{a_C^c a_D^d}{a_A^a a_B^b} \right) = 0$$

or

$$-RT \ln \left(\frac{a_C^c a_D^d}{a_A^a a_B^b} \right)_{eq} = \Delta\mu^0$$

$$\equiv \Delta G^0$$

(4.54)

The subscript "eq" specifies that values of the activities in the quotient above are those when the system is at equilibrium.

Equation (4.54) is, of course, the same as Eq. (4.7). This more rigorous derivation emphasizes, however, that the standard free energy change ΔG^0 refers to the change in free energy when the reactants in their respective standard states go to products in their respective standard states. It should also be apparent from Eq. (4.54) that the numerical value of the equilibrium constant is dependent on what standard states are chosen for the reactants and products.

Most of what we have said about partial molal free energy applies equally to the other extensive thermodynamic variables E, H, S, V, and A. Partial molal volumes are interesting to discuss because they illustrate the large difference that may occur between the volume and the partial molal volume. The volume per mole of solid $MgCl_2$ is about 40 mL mol^{-1}. However, the partial molal volume of $MgCl_2$ in dilute aqueous solution is negative! Adding a small amount of $MgCl_2$ to water causes the volume of the solution to decrease, because the Mg^{2+} ion strongly binds water molecules to it.

This decrease in solution volume upon adding ionized solutes is known as electrostriction. It results from the electrostatic charge on small ions which can organize the surrounding water molecules into a more compact structure than they have in the absence of the ion. This is seen in the case of such molecules as the amino acid glycine, which occurs as a zwitterion (double ion) in dilute aqueous solution. Its partial specific volume is

$$\overline{V}(H_3N^+CH_2COO^-) = 43.5 \text{ mL mol}^{-1}$$

The molecule glycolamide, $HOCH_2CONH_2$, has the same empirical formula but it is not ionic in aqueous solutions.

$$\overline{V}(HOCH_2CONH_2) = 56.3 \text{ mL mol}^{-1}$$

The partial molal volume is significantly smaller for the zwitterionic molecule. For polyelectrolytes, such as the sodium salts of nucleic acids, there is pronounced electrostriction upon forming aqueous solutions. Such information is useful in understanding solute-solvent interactions at the molecular level.

A thermodynamic property of any mixture is defined as the sum of the number of moles of each component times the partial molal property of the component. For volume the equation is

$$V \equiv n_1 \left(\frac{\partial V}{\partial n_1} \right)_{T, P, n_j} + n_2 \left(\frac{\partial V}{\partial n_2} \right)_{T, P, n_j} + \cdots$$

(4.55)

Similarly, for free energy,

$$G = n_1\mu_1 + n_2\mu_2 + \cdots \tag{4.56}$$

SUMMARY

Chemical Reactions

$$aA + bB \longrightarrow cC + dD$$

$$\Delta G = \Delta G^0 + RT \ln Q \tag{4.4}$$

ΔG = free energy change for the reaction at temperature T and concentrations specified by Q

$\Delta G^0 = c\bar{G}_C^0 + d\bar{G}_D^0 - a\bar{G}_A^0 - b\bar{G}_B^0$

= standard free energy change for the reaction at 1 atm pressure and standard conditions

$$Q = \frac{(a_C)^c(a_D)^d}{(a_B)^b(a_A)^a}$$

a_A, a_B, a_C, a_D = activities of reactants and products at the concentrations specified

$$\Delta G^0 = -RT \ln K \tag{4.7}$$

K = equilibrium constant

$$= \frac{(a_C^{eq})^c(a_D^{eq})^d}{(a_B^{eq})^b(a_A^{eq})^a}$$

a_A^{eq}, etc. = activities of reactants and products at equilibrium

$$R = 8.3144 \text{ kJ}^{-1} \text{ mol}^{-1}$$

$$2.3026RT = 5708 \text{ J at 298 K}$$

$$K = e^{-\Delta G^0/RT} = 10^{-\Delta G^0/2.303RT} \tag{4.9}$$

$$K = e^{\Delta S^0/R}e^{-\Delta H^0/RT} \tag{4.10}$$

Activities

$$\bar{G}_A \equiv \bar{G}_A^0 + RT \ln a_A \tag{4.13}$$

$\bar{G}_A$ = free energy per mole of A at temperature T and specified concentration

$\bar{G}_A^0$ = standard free energy per mole of A at temperature T, 1 atm pressure, and standard concentration

a_A = activity of A at temperature T and specified concentration

$$\bar{G}_A \equiv \mu_A \equiv \left(\frac{\partial G}{\partial n_A}\right)_{T,\,P,\,n_j \neq n_A} \tag{4.14}$$

μ_A = chemical potential

$\left(\dfrac{\partial G}{\partial n_A}\right)_{T,\,P,\,n_j \neq A}$ = partial molal free energy

$$a_A = \gamma_A \cdot \text{(concentration of A)} \tag{4.18}$$

γ_A = activity coefficient of A

Standard States

Ideal gases: Standard state is gas at 1 atm pressure:

$$a_A = P_A \quad \text{in atm} \tag{4.16}$$

Pure solids or liquids: Standard state is pure solid or liquid at 1 atm pressure:

$$a_A = 1 \quad \text{at 1 atm pressure}$$

Solvent in solution: Standard state is pure liquid at 1 atm pressure:

$$a_A = \gamma_A X_A \quad (\gamma_A \longrightarrow 1 \quad \text{as} \quad X_A \longrightarrow 1) \tag{4.23}$$

Solutes in solution: Standard state is solute in solution with the properties of an infinitely dilute solution but at a concentration of 1 [either molar (c) or molal (m) concentrations can be used]:

$$a_A = \gamma_A c_A \quad (\gamma_A \longrightarrow 1 \quad \text{as} \quad c_A \longrightarrow 0) \tag{4.24}$$

$$a_A = \gamma_A m_A \quad (\gamma_A \longrightarrow 1 \quad \text{as} \quad m_A \longrightarrow 0) \tag{4.28}$$

Temperature dependence of equilibrium constant:

$$\ln \frac{K_2}{K_1} = -\frac{\Delta H^0}{R}\left(\frac{1}{T_2} - \frac{1}{T_1}\right) \tag{4.35}$$

Galvanic Cells

$$\Delta G = -n\mathscr{E} \text{ (in eV)} = -96.485\, n\mathscr{E} \text{ (in kJ)} = -n\mathscr{E}F \text{ (in J)} \tag{4.39--4.41}$$

$F = 96{,}485 \text{ C mol}^{-1}$

$$\Delta H\,(\text{kJ}) = 96.485n\left[-\mathscr{E} + T\left(\frac{\partial \mathscr{E}}{\partial T}\right)_P\right] \tag{4.43}$$

$$\Delta S\,(\text{J K}^{-1}) = 96{,}485n\left(\frac{\partial \mathscr{E}}{\partial T}\right)_P \tag{4.42}$$

Nernst equation:

$$\mathscr{E} = \mathscr{E}^0 - \frac{RT}{nF} \ln Q \qquad (4.44)$$

$$\mathscr{E}^0 = \frac{RT}{nF} \ln K \qquad (4.46)$$

Debye-Hückel limiting law equation at 25°C in H_2O solutions:

$$\log \gamma_i = -0.509 \, Z_i^2 \sqrt{\mu} \qquad (4.51)$$

γ_i = activity coefficient of ion of charge Z_i

Ionic strength:

$$\mu = \tfrac{1}{2} \sum_i c_i Z_i^2 \qquad (4.51)$$

c_i = molarity of ion with charge Z_i

MATHEMATICS NEEDED FOR CHAPTER 4

Students need to be able to solve quadratic equations. For the equation

$$ax^2 + bx + c = 0$$

the solutions are

$$x = \frac{-b \pm \sqrt{b^2 - 4ac}}{2a}$$

For equilibrium problems only one of the solutions is reasonable; one of the values of x usually turns out to be negative.

Review partial derivatives in Chapter 3.

REFERENCES

The Biochemistry texts listed in Chapter 1 describe metabolism in depth.

SUGGESTED READINGS

BRESLAUER, K. J., R. FRANK, H. BLOCKER, and L. MARKY, 1986. Predicting DNA Duplex Stability from the Base Sequence, *Proc. Natl. Acad. Sci. USA 83,* 3746–3750 .

MANNING, G. S., 1979. Counterion Binding in Polyelectrolyte Theory, *Acc. Chem. Res. 12,* 443–447.

MOSER, C. C., J. M. KESKE, K. WARNKE, R. S. FARID, and P. L. DUTTON, 1992. Nature of Biological Electron Transfer, *Nature 355,* 796–802.

RECORD, M. T., JR., S. J. MAZUR, P. MELANÇON, J.-H. ROE, S. L. SHANER, and L. UNGER, 1981. Double Helical DNA: Conformation, Physical Properties and Interactions with Ligands, *Annu. Rev. Biochem. 50,* 997–1024.

TURNER, D. T., N. SUGIMOTO, and S. M. FREIER, 1988. RNA Structure Prediction, *Annu. Rev. Biophys. Biophys. Chem. 17,* 167–193.

PROBLEMS

1. A key step in the biosynthesis of triglycerides (fats) is the conversion of glycerol to glycerol-1-phosphate by ATP:

$$\text{glycerol} + \text{ATP} \xrightarrow[\alpha\text{-glycerol kinase}]{\text{Mg}^{2+}} \text{glycerol-1-phosphate} + \text{ADP}$$

At a steady state in the living cell, $(\text{ATP}) = 10^{-3}\ M$ and $(\text{ADP}) = 10^{-4}\ M$. The maximum (equilibrium) value of the ratio (glycerol-1-P)/(glycerol) is observed to be 770 at 25°C and pH 7.
   (a) Calculate K and $\Delta G^{0\prime}_{298}$ for the reaction.
   (b) Using the value of $\Delta G^{0\prime}_{298}$ for the reaction $\text{ADP} + \text{P}_i \to \text{ATP} + \text{H}_2\text{O}$, together with the answer to part (a), calculate $\Delta G^{0\prime}_{298}$ and K for the reaction

$$\text{glycerol} + \text{phosphate} \longrightarrow \text{glycerol-1-phosphate} + \text{H}_2\text{O}$$

2. (a) In the frog muscle rectus abdominis, the concentrations of ATP, ADP, and phosphate are $1.25 \times 10^{-3}\ M$, $0.50 \times 10^{-3}\ M$, and $2.5 \times 10^{-3}\ M$, respectively. Calculate the free energy change $\Delta G^{0\prime}$ for the hydrolysis of ATP in muscle. Take the temperature and pH of the muscle as 25°C and 7, respectively.
   (b) For the muscle described, what is the maximum amount of mechanical work it can do per mole of ATP hydrolyzed?
   (c) In muscle, an enzyme creatine phosphokinase catalyzes the following reaction:

$$\text{phosphocreatine} + \text{ADP} \longrightarrow \text{creatine} + \text{ATP}$$

The standard free energy of hydrolysis of phosphocreatine at 25°C and pH 7 is $-43.1\ \text{kJ mol}^{-1}$:

$$\text{phosphocreatine} + \text{H}_2\text{O} \longrightarrow \text{creatine} + \text{phosphate}$$

Calculate the equilibrium constant of the creatine phosphokinase reaction.

3. An important step in the glycolytic path is the phosphorylation of glucose by ATP, catalyzed by the enzyme hexokinase and Mg^{2+}:

$$\text{glucose} + \text{ATP} \xrightarrow[\text{hexokinase}]{\text{Mg}^{2+}} \text{glucose-6-P} + \text{ADP}$$

In the absence of ATP, glucose-6-P is unstable at pH 7, and in presence of the enzyme glucose-6-phosphatase, it hydrolyzes to give glucose:

$$\text{glucose-6-P} + \text{H}_2\text{O} \xrightarrow[\text{G-6-phosphatase}]{} \text{glucose} + \text{phosphate}$$

   (a) Using data from Table 4.2, calculate $\Delta G^{0\prime}$ at pH 7 for the hydrolysis of glucose-6-P at 298 K.
   (b) If the phosphorylation of glucose is allowed to proceed to equilibrium in the presence of equal concentrations of ADP and ATP, what is the ratio (glucose-6-P)/(glucose) at equilibrium? Assume a large excess of ATP and ADP; that is $(\text{ATP}) = (\text{ADP}) \gg [(\text{glucose}) + (\text{glucose-6-P})]$.
   (c) In the absence of ATP (and ADP), calculate the ratio (glucose-6-P)/(glucose) at pH 7, if phosphate $= 10^{-2}\ M$.

4. In the red blood cell, glucose is transported into the cell against its concentration gradient. The energy for this transport is supplied by the hydrolysis of ATP:

$$\text{ATP} + \text{H}_2\text{O} \longrightarrow \text{ADP} + \text{P}_i \qquad \Delta \bar{G}^0 = -31.0\ \text{kJ}$$

Assume that the overall transport reaction is 100% efficient and given by

$$ATP + H_2O + glucose\,(out) \longrightarrow 2\,glucose\,(in) + ADP + P_i$$

(a) At 25°C, under conditions where [ATP], [ADP], and [P_i] are held constant at $1 \times 10^{-2}\,M$ by cell metabolism, what is the maximum value of

$$\frac{[glucose\,(in)]}{[glucose\,(out)]}\quad ?$$

Assume all activity coefficients are equal to 1.

(b) If the stoichiometry of transport were *one* mole of glucose transported per mole of ATP hydrolyzed, what would be the maximum concentration gradient of glucose under the same conditions as part (a)?

(c) In an actual cell, the glucose inside the cell may have an activity coefficient much less than 1 due to nonideal behavior. Would this increase or decrease the maximum *concentration* gradient obtainable? (Assume that all other activity coefficients = 1.)

5. The following reactions can be coupled to give alanine and oxaloacetate:

$$glutamate + pyruvate \rightleftharpoons ketoglutarate + alanine$$

$$\Delta G^{0\prime}_{303} = -1004\,J\,mol^{-1}$$

$$glutamate + oxaloacetate \rightleftharpoons ketoglutarate + aspartate$$

$$\Delta G^{0\prime}_{303} = -4812\,J\,mol^{-1}$$

(a) Write the form of the equilibrium constant for the reaction

$$pyruvate + aspartate \rightleftharpoons alanine + oxaloacetate$$

and calculate the numerical value of the equilibrium constant at 30°C.

(b) In the cytoplasm of a certain cell, the components are at the following concentrations: pyruvate = $10^{-2}\,M$, aspartate = $10^{-2}\,M$, alanine = $10^{-4}\,M$, and oxaloacetate = $10^{-5}\,M$. Calculate the free energy change for the reaction of part (a) under these conditions. What conclusion can you reach about the direction of this reaction under cytoplasmic conditions?

6. (a) Use the following equilibrium constants to calculate ΔH^0 from the slope of $\ln K$ vs. $(1/T)$ for the reaction 3-phosphoglycerate to 2-phosphoglycerate at pH 7.

$T(°C)$	0	20	30	45
K	0.1535	0.1558	0.1569	0.1584

(b) Calculate ΔG^0 at 25°C.

(c) What is the concentration of each isomer of phosphoglycerate if 0.150 M phosphoglycerate reaches equilibrium at 25°C?

7. Oxides of sulfur are important in pollution, arising particularly from burning coal. Use the thermodynamic data at 25°C given below to answer the following questions.

(a) In air the oxidation of SO_2 can occur: $\frac{1}{2}O_2(g) + SO_2(g) \rightarrow SO_3(g)$. Calculate the standard free energy for this reaction at 25°C.

(b) Find the equilibrium ratio of partial pressures of $SO_3(g)$ to $SO_2(g)$ in air at 25°C; the partial pressure of $O_2(g)$ is 0.21 atm.

(c) $SO_3(g)$ can react with $H_2O(g)$ to form sulfuric acid, $H_2SO_4(g)$. Air which is in equilibrium with liquid water at 25°C has a partial pressure of $H_2O(g)$ of 0.031 atm. Find the equilibrium ratio of partial pressures of $H_2SO_4(g)$ to $SO_3(g)$ in air at 25°C.

Compound	ΔH_f^0 (kJ mol^{-1})	S^0 (J K^{-1} mol^{-1})
$O_2(g)$	0	205.1
$H_2O(g)$	−241.8	188.7
$SO_2(g)$	−296.8	248.2
$SO_3(g)$	−395.7	256.8
$H_2SO_4(g)$	−814.0	156.9

8. An important metabolic step is the conversion of fumarate to malate. In aqueous solution an enzyme (fumarase) allows equilibrium to be attained.

$$\text{fumarate} + H_2O \rightleftarrows \text{malate}$$

at 25°C the equilibrium constant $K = (a_M/a_F) = 4.0$. The activity of malate is a_M and the activity of fumarate is a_F, defined on the molarity concentration scale ($a = c$ in dilute solution).
(a) What is the standard free energy change for the reaction at 25°C?
(b) What is the free energy change for the reaction at equilibrium?
(c) What is the free energy change when 1 mol of 0.100 M fumarate is converted to 1 mol of 0.100 M malate?
(d) What is the free energy change when 2 mol of 0.100 M fumarate are converted to 2 mol of 0.100 M malate?
(e) If $K = 8.0$ at 35°C, calculate the standard enthalpy change for the reaction; assume that the enthalpy is independent of temperature.
(f) Calculate the standard entropy change for the reaction; assume that ΔS^0 is independent of temperature.

9. Use the free energy of hydrolysis of ATP under standard conditions at 25°C, 1 atm, to answer the following questions.
(a) Calculate the $\Delta\bar{G}$ for the reaction when $[ATP] = 10^{-2}$ M, $[ADP] = 10^{-4}$ M, and $[\text{phosphate}] = 10^{-1}$ M.
(b) Calculate the maximum available work under the conditions of part (a) when 1 mol of ATP is hydrolyzed. This work can be used, for example, to contract a muscle and raise a weight.
(c) Calculate $\Delta\bar{G}$ and the maximum available work if $[ATP] = 10^{-7}$ M, $[ADP] = 10^{-1}$ M, and $[\text{phosphate}] = 2.5 \times 10^{-1}$ M.

10. What is the equilibrium pressure of $CO(g)$ in equilibrium with the $CO_2(g)$ and $O_2(g)$ in the atmosphere at 25°C? The partial pressure of $O_2(g)$ is 0.2 atm and the partial pressure of $CO_2(g)$ is 3×10^{-4} atm. CO is extremely poisonous because it forms a very strong complex with hemoglobin. Should you worry?

11. For a reaction, $aA + bB \rightarrow cC + dD$,

$$\Delta G_T^0 = c\mu_C^0 + d\mu_D^0 - a\mu_A^0 - b\mu_B^0$$

and

$$\Delta G_T^0 = -RT \ln K \qquad (4.7)$$

Combining these two expressions and dividing each term by RT gives

$$\ln K = -\frac{1}{R}\left(\frac{c\mu_C^0}{T} + \frac{d\mu_D^0}{T} - \frac{a\mu_A^0}{T} - \frac{b\mu_B^0}{T}\right)$$

(a) Show that

$$\frac{d}{dT}\left(\frac{\mu_i^0}{T}\right) = -\frac{\bar{H}_i^0}{T^2}$$

where $\bar{H}_i^0$ is the standard partial molal enthalpy of species i. (Note that at constant T, $\Delta G^0 = \Delta H^0 - T\,\Delta S^0$, and $\mu_i^0 = \bar{H}_i^0 - T\bar{S}_i^0$.)

(b) From the result obtained in part (a) and the relation between $\ln K$ and the chemical potentials, show that $d\ln K/d(1/T) = -\Delta H^0/R$ even if ΔH^0 is temperature dependent.

12. (a) From the ionization constant, calculate the standard free energy change for the ionization of acetic acid in water at 25°C.

   (b) What is the free energy change at equilibrium for the ionization of acetic acid in water at 25°C?

   (c) What is the free energy change for the following reaction in water at 25°C? (Assume that activity coefficients are all equal to 1.)

$$\text{H}^+ (\text{conc.} = 10^{-4}\ M) + \text{OAc}^- (\text{conc.} = 10^{-2}\ M) \longrightarrow \text{HOAc}(1\ M)$$

   (d) What is the free energy change for the following reaction in water at 25°C? (Assume that activity coefficients are all equal to 1.)

$$\text{H}^+ (\text{conc.} = 10^{-4}\ M) + \text{OAc}^- (\text{conc.} = 10^{-2}\ M) \longrightarrow \text{HOAc}(10^{-5}\ M)$$

   (e) What is the free energy change for transferring 1 mol of acetic acid from an aqueous solution of 1 M concentration to an aqueous solution of 10^{-5} M? (Assume that activity coefficients are all equal to 1.)

13. You want to make a pH 7.0 buffer using NaOH and phosphoric acid. The sum of the concentrations of all phosphoric acid species is 0.100 M. The equilibrium constants for concentrations given in mol L^{-1} for the following equilibria are

$$\text{H}_3\text{PO}_4 \rightleftharpoons \text{H}^+ + \text{H}_2\text{PO}_4^- \quad K_1 = 7.1 \times 10^{-3}$$

$$\text{H}_2\text{PO}_4^- \rightleftharpoons \text{H}^+ + \text{HPO}_4^{2-} \quad K_2 = 6.2 \times 10^{-8}$$

$$\text{HPO}_4^{2-} \rightleftharpoons \text{H}^+ + \text{PO}_4^{3-} \quad K_3 = 4.5 \times 10^{-13}$$

   (a) Write the equation which specifies that the solution is electrically neutral. Use $[\text{PO}_4^{3-}]$, $[\text{HPO}_4^{2-}]$, and so on, for the concentrations.

   (b) Calculate the concentrations of all species in the buffer.

   (c) Use data in Table 4.1 to calculate K_2 at 37°C.

14. You made a pH 9.0 buffer solution at 25°C by mixing NaOH and histidine (HisH) to give a solution which is 0.200 M in total concentration of histidine.

   (a) Calculate the concentrations of all species at 25°C.

   (b) What is the pH of the buffer at 40°C? The two most important ΔH^0 values for histidine are given in Table 4.1; assume that $\Delta H^0 = 0$ for the first ionization. You can ignore the volume change of the solution.

   (c) Calculate the concentrations of all species at 40°C.

15. An important reaction in visual excitation is the activation of an enzyme that catalyzes the hydrolysis of guanosine triphosphate.

$$\text{GTP} + H_2O \longrightarrow \text{GDP} + P_i \qquad K'_{298} = 1.9 \times 10^5$$

   (a) Typical concentrations of GTP, GDP, and P_i in the retinal rod cell are 50 mM, 5 mM, and 15 mM, respectively. What is $\Delta G'_{298}$ for the reaction above in the rod cell?
   (b) If the GTP concentration were suddenly doubled, by how much would $\Delta G'_{298}$ change?
   (c) The solution described in (a) is allowed to come to equilibrium. What are the final concentrations of GTP, GDP, and P_i?

16. In general, native proteins are in equilibrium with denatured forms:

$$\text{protein (native)} \rightleftharpoons \text{protein (denatured)}$$

For ribonuclease (a protein), the following concentration data for the two forms were experimentally determined for a total protein concentration of 1×10^{-3} mol L^{-1}:

Temperature (°C)	Native	Denatured
50	9.97×10^{-4} mol L^{-1}	2.57×10^{-6} mol L^{-1}
100	8.6×10^{-4} mol L^{-1}	1.4×10^{-4} mol L^{-1}

Determine $\Delta \bar{H}^0$ for the denaturation reaction, assuming it to be independent of temperature.

17. A single-stranded oligonucleotide which has complementary ends can form a base-paired hairpin loop.

   (a) At 25°C the equilibrium constant $K_1 = 0.86$. What are the concentrations of loop and single strand at equilibrium if the initial concentration of single strands is 1.00 mM? Will increasing the initial concentration of oligonucleotide increase or decrease the fraction of hairpin loop? Explain.
   (b) At 37°C the equilibrium constant $K_1 = 0.51$. Calculate ΔH^0, ΔS^0, ΔG^0 at 37°C.
   (c) As the concentration of oligonucleotide increases, another reaction becomes possible. A double-stranded molecule with an internal loop can form

at 25°C $K_2 = 1.00 \times 10^{-2}$ M^{-1} (and $K_1 = 0.86$). Calculate the concentrations of all three species (single strands, SS; double strand, DS; and hairpin loop, H) at equilibrium in a solution of initial concentration of single strands = 0.100 M.

18. Cytochromes are iron-heme proteins in which a porphyrin ring is coordinated through its central nitrogens to an iron atom that can undergo a one-electron oxidation-reduction reaction. Cytochrome f is an example of this class of molecules, and it operates as a redox agent in chloroplast photosynthesis. The standard reduction potential, $\mathscr{E}^{0\prime}$, of cytochrome f at pH 7 can be determined by coupling it to an agent of known $\mathscr{E}^{0\prime}$, such as ferricyanide/ferrocyanide:

$$\mathrm{Fe(CN)_6^{3-}} + e^- \longrightarrow \mathrm{Fe(CN)_6^{4-}} \qquad \mathscr{E}^{0\prime} = 0.440\ \mathrm{V}$$

In a typical experiment, carried out spectrophotometrically, a solution at 25°C and pH 7 containing a ratio

$$[\mathrm{Fe(CN)_6^{4-}}]/[\mathrm{Fe(CN)_6^{3-}}] = 2.0$$

is found to have a ratio [Cyt f_{reduced}] / Cyt f_{oxidized}] = 0.10 at equilibrium.
   (a) Calculate $\mathscr{E}^{0\prime}$ (reduction) for cytochrome f.
   (b) On the basis of the standard reduction potential, $\mathscr{E}^{0\prime}$, for the reduction of O_2 to H_2O at pH 7 and 25°C, is oxidized cytochrome f a good enough oxidant to cause the formation of O_2 from H_2O at pH 7?

19. Consider the following reaction, in which two electrons are transferred:

$$2\ \text{cytochrome } c \text{ (ferrous)} + \text{pyruvate} + 2\ \mathrm{H^+} \longrightarrow 2\ \text{cytochrome } c \text{ (ferric)} + \text{lactate}$$

   (a) What is $\mathscr{E}^{0\prime}$ for this reaction at pH 7 and 25°C?
   (b) Calculate the equilibrium constant for the reaction at pH 7 and 25°C.
   (c) Calculate the standard free energy change for the reaction at pH 7 and 25°C.
   (d) Calculate the free energy change (at pH 7 and 25°C) if the lactate concentration is 5 times the pyruvate concentration and the cytochrome c (ferric) is 10 times the cytochrome c (ferrous).

20. The cell $\mathrm{Ag}(s)$, $\mathrm{AgI}(s)\,|\,\mathrm{KI}(10^{-2}\ M)\,|\,|\,\mathrm{KCl}(10^{-3}\ M)\,|\,\mathrm{Cl_2}(g,\ 1\ \mathrm{atm})$, $\mathrm{Pt}(s)$ has the voltage 1.5702 V at 298 K.
   (a) Write the cell reaction.
   (b) What is ΔG_{298}?
   (c) What is ΔG^0_{298}?
   (d) Calculate the standard reduction potential for the half-cell on the left.
   (e) Calculate the solubility product of AgI; $K_{\mathrm{AgI}} = (a_{\mathrm{Ag^+}})(a_{\mathrm{I^-}})$.
   (f) The cell has a potential of 1.5797 V at 288 K. Estimate ΔS^0_{298} for the reaction.

21. Ferredoxins (Fd) are iron- and sulfur-containing proteins that undergo redox reactions in a variety of microorganisms. A particular ferredoxin is oxidized in a one-electron reaction, independent of pH, according to the equation

$$\mathrm{Fd_{red}} \rightleftharpoons \mathrm{Fd_{ox}} + e^-$$

To determine the standard emf of $\mathrm{Fd_{red}}/\mathrm{Fd_{ox}}$, a known amount was placed in a buffer at pH 7.0 and bubbled with H_2 at 1 atm pressure. (Finely divided platinum catalyst was present to ensure reversibility.) At equilibrium, the ferredoxin was found spectrophotometrically to be exactly $\frac{1}{3}$ in the reduced form and $\frac{2}{3}$ in the oxidized form.

(a) Calculate K', the equilibrium constant, for the system

$$\tfrac{1}{2} H_2 + Fd_{ox} \rightleftharpoons H^+ + Fd_{red}$$

(b) Calculate $\mathscr{E}^{0\prime}$ for the Fd_{red}/Fd_{ox} half-reaction at 25°C.

22. The conversion of β-hydroxybutyrate, β-HB$^-$, to acetoacetate, AAC$^-$, is an important biochemical redox reaction that uses molecular oxygen as the ultimate oxidizing agent:

$$\text{β-HB}^- + \tfrac{1}{2} O_2(g) \longrightarrow \text{AAC}^- + H_2O$$

(a) Using the standard reduction potentials given in Table 4.4, calculate $\Delta G^{0\prime}$ and the equilibrium constant for this system at pH 7 and 25°C.
(b) In a solution at pH 7 and 25°C saturated at 1 atm with respect to dissolved air (which is 20% oxygen), what is the ratio of $[ACC^-]$ to $[\text{β-HB}^-]$ at equilibrium?

23. Consider the reaction

$$CH_3CH_2OH(aq) + \tfrac{1}{2} O_2(g) \longrightarrow CH_3\overset{\displaystyle O \atop \displaystyle \|}{C}H(aq) + H_2O$$

$$\text{ethanol} \qquad\qquad\qquad \text{acetaldehyde}$$

(a) Calculate $\mathscr{E}^{0\prime}$ for this reaction at 25°C.
(b) Calculate the standard free energy (in kJ) for the reaction at 25°C.
(c) Calculate the equilibrium constant at 25°C for the reaction.
(d) Calculate $\mathscr{E}$ for the reaction at 25°C when: $a(\text{ethanol}) = 0.1$, $P_{O_2} = 4$ atm, $a(\text{acetaldehyde}) = 1$, and $a(H_2O) = 1$.
(e) Calculate $\Delta \bar{G}$ for the reaction in part (d).

24. Magnesium ion and other divalent ions form complexes with adenosine triphosphate, ATP.

$$ATP + Mg^{2+} \longrightarrow \text{complex}$$

(a) Describe an electrochemical cell that would allow you to measure the activity of Mg^{2+} at any concentration in a 0.100 M ATP solution.
(b) Describe how you could measure the thermodynamic equilibrium constant for binding Mg^{2+} by ATP with an electrochemical cell.

25. Photosystem 1, in higher plants, converts light into chemical energy. Energy, in the form of photons, is absorbed by a chlorophyll complex, P700, which donates an electron to A. The electron is passed down an electron transport chain, at the end of which NADP$^+$ is reduced. The reduction potentials of P700$^+$, A, and NADP$^+$, at pH 7.0, 25°C, are 0.490 V, −0.900 V, and −0.350 V, respectively.
(a) Calculate, at pH 7.0, 25°C, $\mathscr{E}^{0\prime}$ of the reaction

$$P700 + A \longrightarrow P700^+ + A^-$$

(b) What is $\Delta G^{0\prime}$, in kJ, for the same reaction?
(c) At pH 7.0, 25°C, find $\Delta G^{0\prime}$, in kJ, for the reaction

$$NADP^+ + H_2(g) \longrightarrow NADPH + H^+$$

26. Certain dyes can exist in oxidized or reduced form in solution. The half-reaction for one such dye, methylene blue (MB), is:

$$MB(ox) + 2H^+ + 2e^- \rightleftarrows MB(red) \qquad \mathcal{E}^0 = +0.4 \text{ V}$$

(blue)   (colorless)

As indicated, the oxidized form is blue and the reduced form is colorless. From the color of the solution, the relative amounts of the two forms can be estimated.

(a) Write the equation for the half-cell reduction potential of methylene blue in terms of $[MB(ox)]$, $[MB(red)]$, $[H^+]$, and $\mathcal{E}^{0\prime}$.

(b) A very small amount of MB(ox) is added to a solution containing an unknown substance. The pH = 7.0 ($[H^+] = 1 \times 10^{-7}$ M). From the color of the solution, it was estimated that the ratio of concentrations $[MB(red)]/[MB(ox)] = 1 \times 10^{-3}$ at equilibrium. Assuming that all activity coefficients are equal to 1, determine the half-cell potential of the unknown substance in solution.

27. In the presence of oxygen most living cells make ATP by oxidative phosphorylation, which takes place in mitochondria. One of the major substrates that is oxidized is NADH. The overall reaction for this energy-yielding process is

$$NADH + H^+ + \tfrac{1}{2}O_2 \longrightarrow NAD^+ + H_2O$$

(a) Use Table 4.4 to calculate the free energy change of the reaction above as it takes place in mitochondria at pH 7 and 25°C. The concentrations of reactants and products are $[NADH] = 1$ mM, $[NAD^+] = 2$ mM, $P_{O_2} = 0.1$ atm.

(b) Given that 3 mol of ATP is made from P_i and ADP for every mole of NADH oxidized, what fraction of the free energy change of part (a) is used in making ATP in mitochondria during oxidative phosphorylation? $[ADP] = 1$ mM, $[ATP] = 3$ mM, $P_i = 10$ mM.

28. Consider the following half-cell reactions and their standard reduction potentials at 298 K and pH 7.0 in aqueous solution:

$$2H^+ + \tfrac{1}{2}O_2 + 2e^- \longrightarrow H_2O \qquad \mathcal{E}^0 = +0.816 \text{ V}$$

$$2H^+ + \text{cystine} + 2e^- \longrightarrow 2\text{cysteine} \qquad \mathcal{E}^0 = -0.34 \text{ V}$$

Although not required for the problem, the structures of cystine and cysteine are as follows:

Cystine:
CH$_2$—S—S—CH$_2$
| |
CH—NH$_2$ CH—NH$_2$
| |
CO$_2$H   CO$_2$H

Cysteine:
CH$_2$—SH
|
CH—NH$_2$
|
CO$_2$H

(a) If you prepare a 0.010 M solution of cysteine at pH 7.0 and let it stand in contact with air at 298 K, what will be the ratio of [cystine]/[cysteine] at equilibrium? The partial pressure of oxygen in the air is 0.20 atm. The activity coefficients may be taken as unity.

(b) What is $\Delta \bar{G}$ for the reaction when the activities of the reactants and products are the equilibrium values?

29. The binding of oligomers (lys)$_n$ of the amino acid lysine, where n indicates the number of lysines in each oligomer, to a synthetic double-stranded RNA poly (rA·rU) has been studied by S. A. Latt and H. A. Sober [*Biochemistry* 6, 3293–3306 (1967)]. The dependence of the apparent binding constant K_{obs} on [NaCl] in the medium is tabulated on the following page:

[NaCl], M	0.06	0.10	0.25	0.39
$\log K_{obs}$, $n = 4$	4.62	—	2.39	1.60
$\log K_{obs}$, $n = 5$	—	4.71	2.79	1.88

Plot $\log K_{obs}$ against $\log$ [Na^+] for each of the oligolysines and interpret the data. The value of ψ for poly($rA \cdot rU$) is close to that of duplex DNA.

30. Self-complementary oligonucleotides can form double-stranded helices in aqueous solution stabilized by Watson-Crick base pairs. For example:

$$2 \text{ (5'–CGCGATATCGCG–3')} \quad \rightleftharpoons \quad \begin{matrix} \text{5'–CGCGATATCGCG–3'} \\ \cdots\cdots\cdots\cdots\cdots \\ \text{3'–GCGCTATAGCGC–5'} \end{matrix}$$

Derive an expression for the equilibrium constant K as a function of c, the initial concentration of single strands and f, the fraction of single strands that are in the double-stranded helix at equilibrium. The equilibrium is

$$2\,S \quad \rightleftharpoons \quad D$$

$$K = \frac{[D]}{[S]^2}$$

Note that

$$f = \frac{2[D]}{2[D] + [S]} = \frac{2[D]}{c}$$

31. (a) Use nearest-neighbor values (Table 4.3) for the thermodynamics of double strand formation to calculate $\Delta G^0(25°C)$, ΔH^0, ΔS^0 for

$$\begin{matrix} \text{5'–GGGCCC–3'} \\ \cdots\cdots\cdots \\ \text{3'–CCCGGG–5'} \end{matrix} \quad \text{and} \quad \begin{matrix} \text{5'–GGTTCC–3'} \\ \cdots\cdots\cdots \\ \text{3'–CCAAGG–5'} \end{matrix}$$

(b) The melting temperature, T_m, is defined as the temperature where f, the fraction of single strands in the helix at equilibrium, is $\frac{1}{2}$. Calculate the melting temperature for each double helix above at 1.0×10^{-4} M initial concentrations of *each* single strand (the initial concentration of strands, c, is 1.0×10^{-4} M for the helix on the left, and 2.0×10^{-4} M for the helix on the right). Note that the double helix on the right is not self-complementary; its equilibrium constant in terms of f and c is slightly different from that in Problem 30.

32. Genomic DNA sequences are detected by hybridization with a radioactively labeled probe which is complementary to the target sequence and forms a double strand.
(a) For the equilibrium

$$\text{Probe } + \text{ Target } \longrightarrow \text{ Double Strand}$$

Express K, the equilibrium constant for the association of the probe with the target, in terms of c_P, the initial probe concentration, c_T, the initial concentration of the target sequence, and f, the fraction of target hydbridized to the probe to form a double strand. Assume that the probe is present in a large excess.

(b) The target sequence is

$$5'-GGGGAATCA-3'$$

Calculate ΔG^0 (use Table 4.3) and K at 25°C for the formation of a double strand with the above sequence at pH 7.0, 1 M NaCl.

(c) Find the "melting temperature of the probe from the target," defined as the temperature at which half of the target is hybridized to the probe. Assume that $c_P = 1.00 \times 10^{-4}$ M and $c_T = 1.00 \times 10^{-8}$ M.

(d) Does the "melting temperature" as defined in part (c) go up, down, or remain the same when the following changes are made? Give reasons for your answers.

    (1) The probe concentration is doubled.

    (2) The target concentration is doubled.

    (3) The probe contains a single base mismatch.

    (4) The salt concentration is decreased.

5

Free Energy and Physical Equilibria

CONCEPTS

Biological organisms are highly inhomogeneous; different regions of each cell have different concentrations of molecules and different biological functions. The storage and expression of genes occur in the nucleus; protein synthesis occurs in the cytoplasm. ATP to drive many of these processes comes from oxidative phosphorylation in mitochondria. Transport of O_2 and essential nutrients into cells and the removal of CO_2 and waste products is required for all these processes. In this chapter we will focus on the physical chemistry underlying the equilibrium and transport of molecules in different regions of a system. Examples include cell compartments separated by membranes, the inside and outside of a membrane, a liquid or solid in contact with a gas, two immiscible liquids, and so forth. In such systems the chemical potential (partial molal free energy) is the thermodynamic property that tells us whether two or more compartments are in equilibrium with one another (chemical potential is the same in all compartments at equilibrium), or whether they are not in equilibrium (unequal chemical potentials).

If a species has different chemical potentials in two phases (or compartments) in contact with one another, it will move to the phase with the lower chemical potential. This will continue until it reaches equilibrium—its chemical potential becomes the same in all phases. This simple idea allows us to understand active and passive transport in cells, the equilibrium concentrations of molecules separated by semipermeable membranes, and the equilibria of molecules between solids, liquids, and gases that characterize freezing points, solubilities, boiling points, and osmotic pressure.

APPLICATIONS

Membranes and Transport

Living cells are separated from their surroundings by membranes and contain a variety of subcellular particles—organelles—also enclosed by membranes. Most biological membranes consist of a lipid bilayer which contains proteins and other molecules that serve as recognition sites, signal transmitters, or ports of entrance and exit. Membranes are so thin, having thicknesses of only one or two molecules, that they are often considered to be two-dimensional phases. The thermodynamic properties of membranes are then described in terms of surface properties, such as the surface chemical potential and the surface tension or pressure. Membranes not only separate the contents of a compartment from its surroundings, they permit the controlled transport of molecules and signals between the inside and outside. Communication between the inside and outside of a cell includes the exchange of metabolites and electrical signals, the flow of heat, and changes in shape. These processes depend on the differences in temperature, pressure, and electrochemical potential on both sides of the membrane. Temperature differences cause heat flow, pressure differences cause changes in shape, and electrochemical potential differences cause molecular transport and electrical signals.

Ligand Binding

Noncovalent interactions that bind ligands like O_2 to hemoglobin, substrates to enzymes, and complementary strands of DNA or RNA to one another are essential dynamic processes in living cells. Equilibrium dialysis provides a method of exploring the binding between macromolecules and small ligand molecules. In equilibrium dialysis a semipermeable membrane allows a ligand to reach equilibrium between two phases, one of which contains a macromolecule. The difference in concentrations of the ligand on both sides of the membrane depends on the interaction of ligand and macromolecule. This provides a very useful and easy method for studying equilibrium binding constants.

Colligative Properties

The chemical potential must be the same for each component present in two or more phases at equilibrium with one another. The component can be the solvent or each of the solutes in a solution. The phases are solids, liquids, gases, or solution compartments separated by a semipermeable membrane. Any change in a property such as temperature, pressure, or activity in one phase that results in a change in chemical potential must be accompanied by an equal change in chemical potential in the other phases for the system to remain in equilibrium. This fundamental fact allows us to understand the *colligative properties,* such as the freezing-point lowering, the boiling-point elevation, the vapor-pressure lowering, and the increase of osmotic pressure when a solute is dissolved in a solvent. These play essential roles in biological cells. Osmotic pressure that results from the activities of components present in the cytoplasm needs to be balanced by a suitable external pressure lest the cell rupture and burst. Colligative properties are used to determine the concentrations and molecular weights of solutes in solution; they can be used to measure association and dissociation equilibrium constants of biopolymers.

PHASE EQUILIBRIA

A very important process is the transfer of a chemical from one phase to another. The evaporation of liquid water into the vapor phase is a familiar example; the heat removed from our bodies by evaporation of sweat allows us to survive in hot climates. Another example is the transport of ions from inside a cell to outside, which is vital to nerve conduction. We can consider the inside and the outside of the cell as two different phases.

We are familiar with the saying "Oil and water do not mix." It describes a commonly observed behavior that applies equally to salad dressings made from oil and vinegar and to crude petroleum leaking into the ocean from a ruptured tanker ship. In fact, oil and water are somewhat soluble in each other, but what one observes experimentally is the presence of two (or more) phases in equilibrium with one another. The phase with the lower density rises to the top in a gravitational field (oil in the salad dressing is less dense than vinegar). In the absence of gravity, as in a satellite or space station, the two phases remain intermixed. Even on earth if the droplets of each phase are very small, their very slow sedimentation rate (see Chapter 6) in the earth's gravitational field means that they will remain closely mixed for a long time. Such mixtures are called emulsions; mayonnaise, milk, and paint are common examples.

Substances which are immiscible—not completely soluble in each other—usually have obviously different kinds of chemical groups. Alkanes, such as n-hexane, are immiscible with water, forming two separate phases when the liquids are mixed. However, n-hexane is completely miscible with benzene, and water is completely miscible with methanol. Here the distinction between the hydrophobic hydrocarbons, on the one hand, and relatively polar groups that can form strong hydrogen bonds, on the other, serves to define two classes. Substances like dioxane and, in fact, methanol that are completely miscible with both water and hydrocarbons are called *amphipathic* or *amphiphilic*.

To examine the thermodynamics of immiscible systems, let us focus on the specific example shown in Fig. 5.1. Water and n-hexane are placed in a bottle that has sufficient

Fig. 5.1 Distinct phases occupy separate regions of space. The two liquid phases and one vapor phase are all in equilibrium with one another.

space to provide for a vapor phase in addition to the two liquid phases. The system is allowed to come to equilibrium across the interfaces between the different phases. Shaking the mixture hastens the approach to equilibrium, because it increases the surface areas of contact but it does not affect the final equilibrium properties. If we chemically analyze each of the three phases, we find that both compounds are present in each phase. The "liquid water" phase is actually an aqueous solution that is saturated with respect to n-hexane. At 25°C such a saturated aqueous solution contains the mole fraction $X_{C_6H_{14}} = 2.1 \times 10^{-6}$ of hexane. This corresponds to a concentration of 3.8×10^{-8} M, which is indeed small. The significant point is that it is not zero. Similarly, the liquid hydrocarbon phase is a saturated solution that contains a small amount of dissolved water. The vapor phase at 25°C consists of a mixture of the two gases at partial pressures $P_{H_2O} = 25.76$ Torr and $P_{C_6H_{14}} = 150.8$ Torr. These are the equilibrium vapor pressures of water and hexane at 25°C; the solubilities of the solutes are so small that the vapor pressures of the saturated solutions are very close to those of the pure solvents. What determines all of these values in the equilibrium system is the chemical potential. Note that although the gas phase is not in direct contact with the aqueous phase, it is still in equilibrium with it.

For a system like that described in Fig. 5.1 to be in equilibrium, each component must have the same chemical potential in each of the three phases. If the concentration of water in the hexane phase is so low that the chemical potential, μ, of the dissolved water is less than that of the water in the aqueous phase, then water will be transferred from the aqueous phase until equilibrium is reached. At that point $\Delta\mu$ for the transfer is zero, and no further change occurs; but a small amount of water is dissolved in the hexane. Once equilibrium is reached, consider the process of moving a small amount (n moles) of hexane from the hexane solution to the aqueous solution.

$$n\ C_6H_{14}(\text{hexane soln.}) \longrightarrow n\ C_6H_{14}(\text{aq. soln.})$$

$$\Delta G(\text{eq}) = 0 = n\bar{G}_{C_6H_{14}}(\text{aq. soln.}) - n\bar{G}_{C_6H_{14}}(\text{hexane soln.})$$

Therefore, at equilibrium

$$\bar{G}_{C_6H_{14}}(\text{aq. soln.}) = \bar{G}_{C_6H_{14}}(\text{hexane soln.})$$

It follows that this is also true for the water in the system and for the components in the vapor phase. At equilibrium,

$$\bar{G}_{C_6H_{14}}(\text{hexane soln.}) = \bar{G}_{C_6H_{14}}(\text{aq. soln.}) = \bar{G}_{C_6H_{14}}(\text{vapor})$$

$$\bar{G}_{H_2O}(\text{hexane soln.}) = \bar{G}_{H_2O}(\text{aq. soln.}) = \bar{G}_{H_2O}(\text{vapor})$$

The first equation states that the free energy per mole of hexane in the hexane solution is equal to the free energy per mole of hexane in the aqueous solution, which is equal to the free energy per mole of hexane in the gas phase. The second equation states the same result for water. The $\bar{G}$ is the partial molal free energy or chemical potential, μ:

$$\bar{G}_A = \left(\frac{\partial G}{\partial n_A}\right)_{T, P, n_j \neq n_A} = \mu_A \qquad (3.53)$$

We have presented criteria for a special case. We can generalize this for phase equilibrium, with T and P constant.

Species A:

$$\mu_A(\text{phase 1}) = \mu_A(\text{phase 2}) = \mu_A(\text{phase 3}) = \ldots \qquad (5.1)$$

Species B:

$$\mu_B(\text{phase 1}) = \mu_B(\text{phase 2}) = \mu_B(\text{phase 3}) = \ldots \qquad (5.1)$$

The chemical potential depends on the activity of each species in each phase.

$$\mu_A = \mu_A^0 + RT \ln a_A = \mu_A^0 + RT \ln \gamma_A c_A \qquad (4.13)$$

Therefore, if the same standard state is used for a component in two different phases, its activity in the two phases will be the same at equilibrium. For phase equilibrium (same standard state in both phases),

$$a_A(\text{phase 1}) = a_A(\text{phase 2}) \qquad (5.2)$$

For the hexane and water mixture, for example, we can choose pure liquid hexane as the standard state in both phases. The activities of the hexane are equal in the two phases at equilibrium, but the concentrations are very different. In the aqueous phase the mole fraction of hexane was found experimentally to be 2.1×10^{-6} at 25°C; in the hexane phase the mole fraction of hexane is nearly 1 (not much water dissolves in hexane).

For some phase equilibria it is reasonable to assume that the activity coefficients are the same for a component in the two phases. This was clearly not correct for the water-hexane mixture. However, in equilibrium dialysis we start with the same solution on both sides of a semipermeable membrane which is permeable to all the species. Clearly there are equal concentrations, activities, activity coefficients, and chemical potentials of all species in the two phases separated by the membrane. Now a macromolecule is added to one side of the membrane. The macromolecule is too big to pass through the membrane; however, it can preferentially bind some of the species. We can determine the amount of binding by assuming that the free (unbound) concentration of a species on the side of the macromolecule is the same as the concentration on the other side of the membrane. We use the reasonable approximation that, except for the binding, the activity coefficients of molecules in the solution are not changed greatly by addition of the macromolecule. For phase equilibrium with activity coefficients equal in both phases (such as in equilibrium dialysis),

$$c_A(\text{phase 1}) = c_A(\text{phase 2}) \qquad (5.3)$$

In summary, if equilibrium is attained for a chemical component among two or more phases at the same T and P, (1) its chemical potential is the same in all phases; (2) if the same standard state is used, its activity is the same in all phases; and (3) if activity coefficients are equal, its concentration is the same in all phases.

Free Energies of Transfer Between Phases

Transfer between two different solvents

The equality of chemical potential for a solute in equilibrium in different phases leads to some very useful but surprising conclusions. Consider a solid solute, A, added in excess to a liquid, 1, to form a saturated solution. In a separate container, solute A is added to a different liquid, 2, to form a saturated solution in the second liquid. A saturated solution is one in which the solution has come to equilibrium with a solid phase. We know that for the solute A:

$$\mu_A(\text{solid}) = \mu_A(\text{saturated soln. 1}) = \mu_A(\text{saturated soln. 2})$$

Thus, for two saturated solutions of a solute at the same T and P, but in different solvents, the chemical potential of the solute is the same in the solutions *even if they are not in contact with one another.* The presence of the solid solute in the saturated solutions is the common feature that links the solute chemical potentials. This is extremely useful. By comparing the activities (or concentrations, if the solutions are sufficiently dilute) of the solute in saturated solutions in different solvents, we can readily determine the standard free energy change, $\Delta\mu_A^0$, for transferring the solute from one solvent to the other. The standard chemical potential of A in solution 1 is $\mu_A^0(\text{soln. 1})$, the solute chemical potential of A in a solution in solvent 1 that has $a_A = 1$. Because we are using solute standard states, $\mu_A^0(\text{soln. 1})$ will not be the same as $\mu_A^0(\text{soln. 2})$. This is different from our discussion on the previous page of the distribution of hexane between hexane and water. There we used a solvent standard state—pure hexane was the standard state for both phases. Relatively hydrophobic solutes, like the fatty acid compounds $C_nH_{2n+1}COOH$ with $n \geq 5$, are more soluble in a hydrocarbon like liquid *n*-heptane than in a dilute aqueous buffer. For palmitic acid, A = $C_{15}H_{31}COOH$, solubility measurements lead to the result that $\Delta\mu_A^0 = \mu_A^0(\text{heptane soln.}) - \mu_A^0(\text{aq. soln.}) = -38$ kJ mol^{-1}. For the series of fatty acids compounds having a number of carbon atoms n_C, where $n_C = 8$ (octanoic), $n_C = 10$ (decanoic), $n_C = 12$ (dodecanoic), etc. to $n_C = 22$ (behenic), the solubility data can be used to generate the relation (in kJ mol^{-1}).

$$\mu_A^0(\text{heptane soln.}) - \mu_A^0(\text{aq. soln.}) = 17.82 - 3.45 \, n_C \qquad (5.4)$$

This equation gives you the standard free energy change in transferring one mole of fatty acid from liquid water to liquid heptane. The positive value for the first term on the right side of Eq. (5.4) reflects the preference of the polar COOH group for water as solvent. The addition of CH_2 groups to increase the length of the hydrocarbon part of the molecule then results in a decrease of the strength of the interaction with water relative to that with heptane by 3.45 kJ mol^{-1} per CH_2 group. For $n_C \geq 6$ the fatty acids are more soluble in hydrocarbons, whereas the shorter chain carboxylic acids are more soluble in water. Figure 5.2 shows that the experimental data follow closely a linear relation between $\Delta\mu_A^0$ and n_C.

Another useful application of this approach is to determine the relative hydrophobicities of the side-chains of the amino acids that are the constituents of proteins. Many structures of water-soluble proteins are now known in detail, based on x-ray crystallography (Chapter 12) or high-resolution nuclear magnetic resonance (Chapter 10). An almost uniform characteristic of these structures is that charged or polar amino acid side chains are exposed to the aqueous medium at the surface of the protein molecules, whereas nonpolar hydrocarbon side chains are buried in the interior. The free energy of the aqueous protein solution has a lower free energy this way. For proteins that are embedded in lipid membranes, the hydrocarbon side chains are on the outside of the protein and the polar groups are hidden inside.

Amino acid side chains can be characterized as polar—hydrophilic, or as hydrophobic with properties of hydrocarbons. A quantitative evaluation is provided by a *hydropathy index* based on measurements of the solubility of the different amino acid side chains in polar versus nonpolar solvents. One such index is presented in Table 5.1.

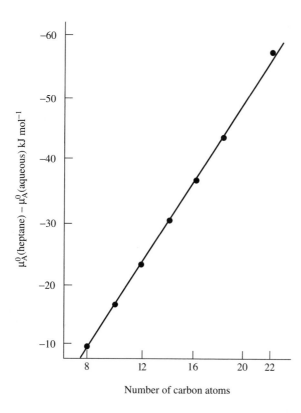

Fig. 5.2 The standard free energies of transfer from aqueous solution to heptane solution of aliphatic carboxylic acids as a function of number of carbons in the aliphatic chain. The negative values of the free energy differences mean that the fatty acids prefer to be in the nonpolar hydrocarbon solution rather than in the aqueous solution. The data were measured at 23 to 25°C; a least-squares fit to the data provides the parameters for [Eq. (5.4)]. Data are from Fig. 3-1 of C. Tanford, *The Hydrophobic Effect,* 2nd ed., 1980, John Wiley, New York.

Table 5.1   A hydropathy index for the side chains of the amino acids in proteins*

Side chain	Hydropathy index	Side chain	Hydropathy index
Isoleucine	4.5	Tryptophan	−0.9
Valine	4.2	Tyrosine	−1.3
Leucine	3.8	Proline	−1.6
Phenylalanine	2.8	Histidine	−3.2
Cysteine/cystine	2.5	Glutamic acid	−3.5
Methionine	1.9	Glutamine	−3.5
Alanine	1.8	Aspartic acid	−3.5
Glycine	−0.4	Asparagine	−3.5
Threonine	−0.7	Lysine	−3.9
Serine	−0.8	Arginine	−4.5

* The hydropathy index is the relative hydrophobicity-hydrophilicity of the side-chain groups in proteins at neutral pH. Positive values characterize hydrophobic groups which tend to be buried in water-soluble proteins; negative values characterize hydrophilic groups which appear on the surface of the proteins. The relative values were deduced from free energies of transfer of side-chain molecules from water to organic solvents or from water to the gas phase. See J. Kyte and R. F. Doolittle, *J. Mol. Biol. 157,* 105–132 (1982).

Transfer between the same solvents

There is a change in free energy when a molecule is transferred from a solution at one concentration to a different concentration in the same solvent. Examples include transfer of molecules from inside a cell to outside, or transfer across a dialysis membrane. We can even treat the diffusion of a molecule from high to low concentration in a poorly mixed solution by the same method. For these transfers the standard free energies are equal because the solvents are the same, and we keep T and P constant. As the concentrations in the different phases do not correspond to equilibrium, and the standard free energy difference $\Delta\mu_A^0$ is zero, the free energy per mole for transferring A from one phase to another is

$$\Delta\bar{G} = \mu_A(\text{phase 2}) - \mu_A(\text{phase 1})$$

$$= RT\ln\frac{a_A(\text{phase 2})}{a_A(\text{phase 1})} \tag{5.5}$$

If the activity coefficients for A are the same in the two phases, the free energy of transfer per mole is

$$\Delta\bar{G} = RT\ln\frac{c_A(\text{phase 2})}{c_A(\text{phase 1})} \tag{5.6}$$

These equations apply to uncharged molecules and to molecules in the absence of an electric field. If there is a voltage difference of V volts between the two phases, an additional term is required for charged molecules. The associated free energy change is equal to the electrical work [Eqs. (4.38) and (4.39)] of transferring an ion of charge Z against a voltage difference V. The corresponding free energy for 1 mol of ions is N_0ZV (where N_0 is Avogadro's number and the units are electron volts, eV), or FZV (where $F = 96{,}485$ J (eV)$^{-1}$, the Faraday conversion factor, and the units are joules). The total free energy of transfer for charged species in the presence of a field is therefore

$$\Delta\bar{G} = \mu_A(\text{phase 2}) - \mu_A(\text{phase 1}) + FZV$$

$$= RT\ln\frac{a_A(\text{phase 2})}{a_A(\text{phase 1})} + FZV \tag{5.7}$$

where Z = charge on ion, $\pm 1, \pm 2, \pm 3, \ldots$
$\quad\quad\quad F$ = Faraday = $96{,}485$ J (eV)$^{-1}$
$\quad\quad\quad V \equiv \phi_2 - \phi_1$ is the potential differerence in volts between the two phases; ϕ_1 and ϕ_2 are the electric potential of phase 1 and phase 2, respectively. The sign of V is positive when ϕ_2 is electrically positive relative to ϕ_1.

These equations will be useful in understanding active transport in biological cells. The transport of ions and metabolites across membranes is characterized by Eqs. (5.4) through (5.7). Equation (5.7) can be rearranged in the form

$$\Delta\bar{G} = (\mu_A + ZF\phi)(\text{phase 2}) - (\mu_A + ZF\phi)(\text{phase 1})$$

It is convenient to define

$$\mu_{A,\,\text{total}} \equiv \mu_A + ZF\phi \tag{5.8}$$

as the total chemical potential of A in the presence of the electric potential ϕ. The condition for equilibrium between the two phases in the presence of a field is then, at constant T and P,

$$\mu_{A,\,total}(\text{phase 2}) \;=\; \mu_{A,\,total}(\text{phase 1})$$

This is the first example we have given of an external field affecting the free energy of a species. We will discuss the effects of other external fields, such as centrifugal fields and gravitational fields, in the next chapter. The important idea to remember is that the free energy depends on the interactions of molecules with other molecules, and with their surroundings. If the surroundings include electric fields, or other fields, we must consider the *total* chemical potential in calculating the free energy.

Equilibrium Dialysis and Scatchard Plots

The concept that two species in equilibrium across a membrane must have the same chemical potential is crucial to the study of binding of small molecules to macromolecules by equilibrium dialysis. The binding of inhibitors to enzymes, of hormones to proteins, and of antibiotics to DNA have all been studied by this method, which is illustrated in Fig. 5.3. A macromolecule is placed on the inside of a dialysis membrane. The semipermeable dialysis membrane allows water and other small molecules to pass through freely but prevents passage of the macromolecule. Therefore, equilibrium can be reached for all species except the macromolecule. At equilibrium if the macromolecule binds one of the small-molecule species, we expect that the concentration of this species will be higher inside the dialysis chamber than outside. If the macromolecule has no effect on a small-molecule species, we

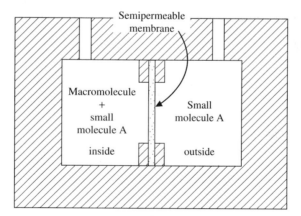

Fig. 5.3 An equilibrium dialysis cell in which two compartments are separated by a semipermeable membrane. A solution containing a macromolecule is placed on one side of the membrane; a solution containing a small molecule A is placed on the other side. The solvent plus all other small molecules can pass through the membrane and can attain thermodynamic equilibrium. The binding of small molecule A to the macromolecule can be measured from the difference in concentration of A on both sides of the membrane.

expect the concentration of the small molecule to be the same on both sides of the membrane at equilibrium. Thus, by simply measuring concentrations of molecules in the solutions on either side of the membrane, we can determine quantitative binding to a macromolecule.

The chemical potential of each species at equilibrium is the same on both sides of the membrane.

$$\mu_A(\text{inside}) = \mu_A(\text{outside})$$

It is convenient to use the same standard state for small molecule A in both phases; therefore,

$$a_A(\text{inside}) = a_A(\text{outside}) \tag{5.2}$$

The activity coefficient of A inside will usually be smaller than A outside; consequently, the concentration of A inside will be greater than that outside. Thermodynamically, we can say only that the activity of A is lowered in the presence of the macromolecule, but a reasonable interpretation of this is that A is bound to the macromolecule. To calculate the amount bound, we assume that the *free* molecule A *that is not bound to the macromolecule* has the same activity coefficient as A outside. The only difference between inside and outside is the presence of the macromolecule inside; it binds some A molecules but has no effect on the free A molecules. Then we can write at equilibrium

$$c_A(\text{free inside}) = c_A(\text{outside})$$

But the total concentration of A inside is just the sum of the free and bound A:

$$c_A(\text{inside}) = c_A(\text{free inside}) + c_A(\text{bound inside})$$

or

$$c_A(\text{bound inside}) = c_A(\text{inside}) - c_A(\text{free inside})$$

Therefore, from the equilibrium dialysis,

$$c_A(\text{bound inside}) = c_A(\text{inside}) - c_A(\text{outside}) \tag{5.9}$$

or

$$c_A(\text{bound to macromolecule}) = c_A(\text{macromolecule side}) - c_A(\text{solvent side})$$

We have obtained the concentration of A bound to a macromolecule in terms of the difference of two measurable quantities: the total concentration of A on each side of the dialysis membrane.

The equilibrium constant for binding molecule A to the macromolecule can be calculated from the total concentration of macromolecule, c_M, and the concentrations of free A and bound A at equilibrium. Consider the equilibrium

$$P + A \underset{}{\overset{K}{\rightleftharpoons}} P \cdot A$$

$$K = \frac{[P \cdot A]}{[P][A]}$$

where P is a macromolecule with a single binding site for A. If we assume activities can be replaced by concentrations,

$$[\text{P} \cdot \text{A}] = \text{concentration of complex at equilibrium}$$
$$= c_A(\text{bound}) = c_A(\text{inside}) - c_A(\text{outside})$$
$$[\text{P}] = \text{concentration of free macromolecule at equilibrium}$$
$$= c_M - c_A(\text{bound})$$
$$[\text{A}] = \text{concentration of free A at equilibrium}$$
$$= c_A(\text{outside})$$

We thus have the equilibrium constant K in terms of measurable quantities.

$$K = \frac{[c_A(\text{bound})]}{[c_M - c_A(\text{bound})][c_A(\text{outside})]}$$

The Scatchard Equation

The equation for K can be simplified and generalized by introducing the number, v, of small molecules bound per macromolecule at equilibrium.

$$v = \frac{[c_A(\text{bound})]}{c_M} \tag{5.10}$$

where
$$v = \text{average number of A's bound per macromolecule}$$
$$c_A(\text{bound}) = \text{concentration of A bound}$$
$$c_M = \text{concentration of macromolecule}$$

Writing the expression for K in terms of v (divide the numerator and denominator by c_M) we obtain

$$K = \frac{[v]}{[1 - v][c_A(\text{outside})]} = \frac{v}{(1 - v)[A]}$$

or

$$\frac{v}{[A]} = K \cdot (1 - v)$$

The equilibrium constant was written with the assumption that only one molecule of A was bound per macromolecule; this means that v can vary only from 0 (no A bound) to 1 (each macromolecule has bound an A). However, many macromolecules have more than one binding site for a ligand. Now v can vary from 0 to N, the number of binding sites on each macromolecule. If the sites are identical and independent it is very easy to generalize the equation. If there are N *identical and independent binding sites* per macromolecule this means that the N sites have the same binding equilibrium constant, K, and that binding at one site does not change the binding at another site. Then we can replace v in the equation for the equilibrium constant above by v/N.

$$\frac{v/N}{[A]} = K \cdot (1 - v/N) \tag{5.11}$$

By rearranging, we obtain the *Scatchard equation:*

$$\frac{v}{[A]} = K \cdot (N - v) \tag{5.12}$$

The Scatchard equation (5.12) is often used to study binding to a macromolecule. The binding can be measured by any method, but equilibrium dialysis is often convenient. The value of [A] which we designated as c_A (free) is the concentration on the solvent side of the dialysis membrane at equilibrium. The value of v comes from the concentration of A on the macromolecule side at equilibrium [see Eqs. (5.9) and (5.10)]. A plot of $v/[A]$ versus v should give a straight line with slope of minus K, y intercept of NK, and x intercept of N, as shown in Fig. 5.4. If a Scatchard plot does not give a straight line, this indicates that the binding sites are not identical or not independent. One often finds a straight line for low v but curvature for higher number of A's bound. The first few molecules bound may not change the properties of the macromolecule very much and not interact with each other, but as more molecules are bound, interactions between them become more likely. The consequences of these interactions between adjacent binding sites are discussed more fully in the next section.

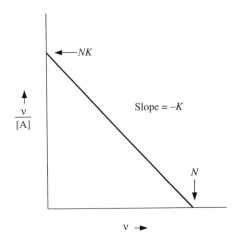

Fig. 5.4 Scatchard plot of the binding of substrate A to a macromolecule containing N binding sites, where v is the average number of A molecules bound per macromolecule, and K is the equilibrium binding constant.

Example 5.1 The formyltetrahydrofolate synthetases can utilize the energy of hydrolysis of ATP (to ADP and phosphate) for the formation of a carbon-nitrogen bond between (*l*)-tetrahydrofolate and formate. Binding of the Mg complex of ATP to such an enzyme from *C. cylindrosporum* has been measured; the data are plotted in Scatchard form in Fig. 5.5.

The data fit the identical-and-independent-sites model well. The slope of the plot gives an equilibrium constant K of 1.37×10^4 M^{-1}. The intercept gives $N = 4.2$. Since the enzyme is known to have four identical subunits (each has a molecular weight of 60,000), the binding results are consistent with one Mg-ATP binding site per subunit.

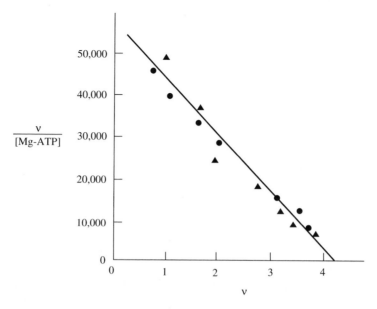

Fig. 5.5 Binding of Mg-ATP by tetrahydrofolate synthetase from *C. cylindro-sporum*. The circles and triangles represent two sets of data obtained by two different techniques. Measurements were carried out at 23°C and a pH of 8.0. [Data from N. P. Curthoys and J. C. Rabinowitz, *J. Biol. Chem. 246*, 6942 (1971).]

In the discussion leading to Eq. (5.12) we defined the quantities ν and N as follows:

$$\nu = \text{number of bound A } per\ macromolecule$$

$$N = \text{number of binding sites } per\ macromolecule$$

A macromolecule is made of monomer units, such as the peptide units in a protein. A DNA molecule is a polynucleotide consisting of nucleotides as its monomer building blocks. Sometimes it is more convenient to define the binding parameters on a *per monomer unit basis* as follows:

$$r = \text{number of bound A } per\ monomer\ unit\ of\ the\ macromolecule$$

$$n = \text{number of binding sites } per\ monomer\ unit\ of\ the\ macromolecule$$

The Scatchard equation is unchanged by these new definitions.

$$\frac{r}{[A]} = K(n - r) \tag{5.13}$$

Thus, the Scatchard plot can also be presented with r and n as the parameters.

In Fig. 5.6, data are plotted for the binding of the trypanocide drug ethidium (ethidium is sometimes used to treat animals infected by parasitic trypanosomes) to a double-stranded DNA. Note that the data are represented reasonably well by the identical-and-independent-sites model. The r intercept gives $n \approx 0.23$, or approximately one binding site per four nucleotides (two base pairs). This is consistent with the fact that ethidium binds to

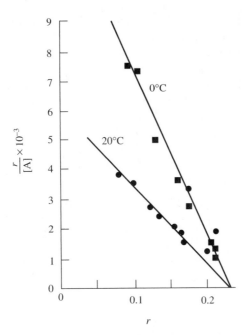

Fig. 5.6 Scatchard plot of the binding of ethidium to DNA in a medium containing 3 *M* CsCl, 0.01 *M* Na$_3$EDTA.

DNA by *intercalation*. Planar, aromatic molecules like ethidium slide between two adjacent base pairs in DNA so that the hydrophobic parts are stacked on the base pairs; this is intercalation. Note also that at the lower temperature the line is steeper, indicating a greater binding constant. This temperature dependence of the binding constant yields a negative ΔH^0 (calculated to be ~ −25 kJ mol^{-1}) for the binding of the drug. The fairly strong interaction ($K = 2.66 \times 10^4$ M^{-1} and 5.60×10^4 M^{-1} at 20°C and 0°C, respectively) between the drug and DNA is evidence that the target of the drug *in vivo* is the DNA of the trypanosome.

Cooperative Binding and Anticooperative Binding

In the previous section, a macromolecule with many binding sites was assumed to have identical and noninteracting sites. Frequently, however, although the sites are identical there is significant interaction among them. The binding of a small molecule to one site affects the binding of small molecules to other sites. The first ligand bound affects the binding of the next one, which affects the binding of the next one, and so on. In *cooperative binding* the first ligand bound makes it easier for the next one to be bound. This may be caused by a conformational change in a multisubunit protein which makes it easier for successive ligands to be bound. The limiting case of cooperative binding is *all-or-none binding*. In all-or-none binding the first ligand increases the binding of the other ligands so much that all *N* ligands are bound at once. As the concentration of ligand is increased, the number of ligands bound increases sharply to the maximum possible. The macromolecule has either no ligands bound or *N* ligands bound.

In *anticooperative binding* each succeeding ligand is bound less strongly than the previous one. An example of anticooperative binding occurs, for example, in the titration of a protein which has 10 carboxyl groups from 10 glutamic acid side chains. The carboxylate groups with pK's near 4 have the same intrinsic binding constant for hydrogen ions; however, they can interact strongly (by Coulomb's law). As the pH is lowered towards 4 the first proton is attracted by 10 negative charges; it is bound strongly. The next proton is not attracted as strongly, and so forth. Each proton bound makes it harder to bind the next proton. For some macromolecules and ligands the anticooperativity may be so strong that the binding of a ligand prevents the binding of another ligand at a neighboring site. This is *excluded-site binding*—the binding at one site excludes binding at another site. An example of excluded-site binding is the binding of the antibiotic drug actinomycin to DNA. An actinomycin molecule consists of a heterocyclic ring phenoxazone, with two cyclic pentapeptides attached to the ring. It bonds to double-stranded DNA by intercalating the phenoxazone ring between two adjacent base pairs and tucking the peptides into the narrow groove of the DNA helix. One guanine-cytosine (G·C) base pair must be present at each binding site. A Scatchard plot from data for the binding of actinomycin to DNA is shown in Fig. 5.7. Note that the Scatchard plot is curved; this indicates interaction between sites, or it indicates that the sites have different intrinsic binding constants. The actinomycin molecule with its two pentapeptides is so big that each molecule that binds prevents other molecules from binding nearby. An excluded-site model is appropriate, and the solid curve in the figure shows that the model fits the data well.

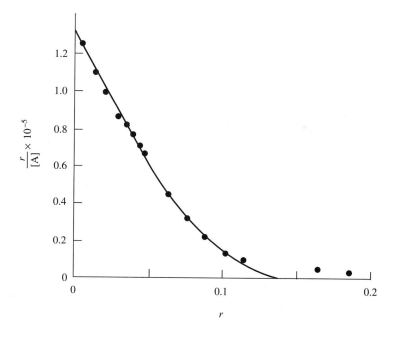

Fig. 5.7 Binding of actinomycin C_3 to calf thymus DNA. The solid curve shown, which fits the data well for $r \leq 0.11$, is obtained by using the excluded-site model. The parameters for the curve are $K = 10.6 \times 10^5$ and that two bound actinomycin molecules must be separated by at least six base pairs. It has also been assumed that every G·C is a site of binding and that the G + C pairs (41% of the base pairs are G·C in calf thymus DNA) are randomly distributed in the DNA. [Data are from W. Müller and D. M. Crothers, *J. Mol. Biol. 35*, 251 (1968).]

A useful way to plot binding data and to learn about the cooperativity (or anticooperativity) of binding is to plot fraction of sites bound versus concentration of free ligand. A vital example of cooperative binding is the binding of oxygen by hemoglobin. Hemoglobin has four heme binding groups which bind four oxygen molecules cooperatively. The binding is cooperative so that hemoglobin releases most of its bound oxygen at the low oxygen pressure in the tissues, but binds the maximum amount of oxygen in the lungs. Figure 5.8 compares the binding of oxygen to myoglobin with only one heme binding site and to hemoglobin with four interacting binding sites. The cooperative binding curve for hemoglobin shows a characteristic *sigmoidal* shape; sigmoidal means shaped like the letter S. Myoglobin, which has only one binding site so it cannot bind cooperatively, has a binding curve characteristic of a molecule with any number of identical and noninteracting sites. That is, the data for myoglobin is represented by the Scatchard equation; a linear Scatchard plot is obtained.

A *Hill plot* is a plot to quantitatively assess the cooperativity of binding. The Scatchard equation was derived for no interaction among sites. In the form of Eq. (5.11) it can be rewritten as

$$\frac{v/N}{(1 - v/N)} = \frac{f}{(1 - f)} = K \cdot [A] \tag{5.11}$$

where $f = v/N$ is the fraction of sites bound. This equation represents the binding curve for myoglobin or for any number of noninteracting and identical sites. For cooperative binding the form of the equation becomes

$$\frac{f}{(1 - f)} = K \cdot [A]^n \tag{5.14}$$

n = Hill coefficient
K = a constant, not the binding constant for one ligand

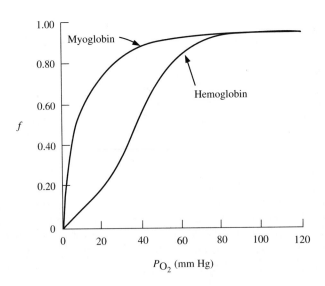

f

P_{O_2} (mm Hg)

Fig. 5.8 Fraction of hemes of myoglobin or hemoglobin occupied by oxygen as a function of the pressure of oxygen. Myoglobin has only one heme group and therefore cannot show cooperativity. Hemoglobin has four hemes; the binding is cooperative.

The Hill coefficient, n, varies from 1 for no cooperativity to N for all-or-none binding of N ligands. The way to determine the Hill coefficient or cooperativity coefficient is to plot the logarithm of Eq. (5.14).

$$\log \frac{f}{(1-f)} = n \log [A] + \log K \qquad (5.15)$$

Clearly the slope of the plot gives n, the cooperativity coefficient. This is illustrated in Fig. 5.9 for myoglobin and hemoglobin. For myoglobin the cooperativity coefficient is 1, as it must be for a single site. The slope is 2.8 for the cooperative binding of oxygen by hemoglobin, which is greater than 1 but less than the all-or-none value of 4.

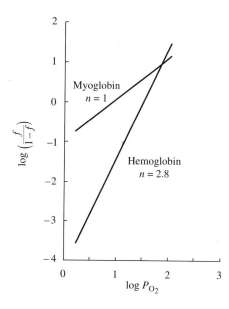

Fig. 5.9 Hill plots [Eq. (5.15)] for the binding of oxygen by myoglobin and by hemoglobin. The slope of the line is the Hill cooperativity coefficient. For no cooperativity the slope is 1; for maximum (all-or-none) cooperativity the slope is N, the total number of binding sites. Hemoglobin has a Hill coefficient of 2.8 for 4 binding sites. The binding is cooperative, but not all-or-none.

Donnan Effect and Donnan Potential

We described equilibrium dialysis as an easy way to measure binding of a small molecule ligand by a macromolecule. The method becomes more complicated if the macromolecule and ligand are charged. The requirement that the solutions on each side of the dialysis membrane must be electrically neutral means that there can be an apparent increase in binding of a ligand with the opposite charge to that of the macromolecule, and a decrease in binding of a ligand with the same charge. These effects depend on the net charge on the macromolecule and are not caused by binding at specific sites. The effect of the net charge of the macromolecule on the apparent binding of a ligand can be minimized by using high concentrations of a salt not involved in binding. Thus, the charged macromolecule and ligands are not the main contributors to the total concentration of ions in the solutions.

When equilibrium (except for the macromolecule) is reached for charged species, a voltage is developed across the membrane. The asymmetric distribution of ions caused by the charged macromolecule is called the *Donnan effect* and the transmembrane potential is called the *Donnan potential*. We can use thermodynamics to quantitatively calculate these effects. For simplicity, we consider the dialysis of a macromolecule with a net charge of Z_M and concentration c_M, against a NaCl solution. The requirement of electrical neutrality outside the membrane means that

$$c_{Na^+}(\text{outside}) = c_{Cl^-}(\text{outside}) = c \qquad (5.16)$$

where c is the concentration of NaCl outside the membrane at equilibrium. At equilibrium, because the membrane is permeable to NaCl,

$$\mu_{NaCl}(\text{inside}) = \mu_{NaCl}(\text{outside})$$

But

$$\mu_{NaCl} = \mu_{NaCl}^0 + RT \ln (a_{Na^+})(a_{Cl^-}) \qquad (4.25)$$

We make the simplifying assumption that the activity coefficients are approximately equal inside and outside the membrane; thus

$$c_{Na^+}(\text{inside}) \cdot c_{Cl^-}(\text{inside}) = c_{Na^+}(\text{outside}) \cdot c_{Cl^-}(\text{outside}) \qquad (5.17)$$

From Eq. (5.16) we see that

$$c_{Cl^-}(\text{inside}) = \frac{c^2}{c_{Na^+}(\text{inside})} \qquad (5.18)$$

But for electrical neutrality inside we have

$$c_{Cl^-}(\text{inside}) = Z_M c_M + c_{Na^+}(\text{inside}) \qquad (5.19)$$

We can equate Eqs. (5.18) and (5.19) and solve the quadratic equation for $c_{Na^+}(\text{inside})$. The positive solution of the quadratic equation is

$$c_{Na^+}(\text{inside}) = \frac{-Z_M c_M + \sqrt{(Z_M c_M)^2 + 4c^2}}{2}$$

The ratio r of $c_{Na^+}(\text{inside})$ to $c_{Na^+}(\text{outside})$ is

$$r = \frac{c_{Na^+}(\text{inside})}{c}$$

$$= \frac{-Z_M c_M}{2c} + \sqrt{\left(\frac{Z_M c_M}{2c}\right)^2 + 1} \qquad (5.20)$$

From Eq. (5.18) we can also calculate the ratio of Cl$^-$ inside to outside

$$\frac{c_{Cl^-}(\text{inside})}{c_{Cl^-}(\text{outside})} = \frac{c}{c_{Na^+}(\text{inside})} = \frac{1}{r}$$

The equations above show that for a positively charged macromolecule the concentration of positive ions inside will be less than that outside (r less than 1), and the concentration of negative ions inside will be greater than outside. For a negatively charged macromolecule, r is greater than 1 and the concentration of positive ions is greater inside than outside. For example, for a macromolecule with a net positive charge of 10 ($Z_M = 10.0$) at a concentration of 1 mM ($c_M = 1.00 \times 10^{-3}$) dialyzed against 0.10 M NaCl ($c = 0.100$), the value of $r = 0.95$ from Eq. (5.20). This means that the ratio of Na$^+$ inside to outside is 0.95, and the ratio of Cl$^-$ inside to outside is $1/r = 1.053$. Increasing the positive charge or concentration of a macromolecule will make r decrease; increasing the concentration of NaCl will make r approach 1.

The asymmetry in Na$^+$ or Cl$^-$ concentration on the two sides of the membrane is surprising because the membrane is permeable to Na$^+$ or Cl$^-$ and we might think that the relations μ_{Na^+}(inside) $= \mu_{Na^+}$(outside) and μ_{Cl^-}(inside) $= \mu_{Cl^-}$(outside) would predict equal concentrations of Na$^+$ and Cl$^-$ on the two sides of the membrane. The reason that this is not so lies in the presence of a potential difference V, the Donnan potential, across the membrane. From Eq. (5.7), and the requirement $\Delta \overline{G} = 0$ when the inside and outside are in equilibrium with each other, we obtain

$$RT \ln \frac{a_{Na^+}(\text{inside})}{a_{Na^+}(\text{outside})} + ZFV = 0$$

or

$$V = \frac{-RT}{ZF} \ln \frac{a_{Na^+}(\text{inside})}{a_{Na^+}(\text{outside})}$$

$$\approx -\frac{RT}{ZF} \ln \frac{c_{Na^+}(\text{inside})}{c_{Na^+}(\text{outside})} \tag{5.21}$$

$$V \approx -\frac{RT}{ZF} \ln r$$

From the example above with $r = 0.95$, the charge $Z = 1$ on Na$^+$, $T = 298$ K, $R = 8.314$ J K^{-1}, $F = 96,485$ J (eV)$^{-1}$, we can calculate a transmembrane voltage of +1.3 mV. The plus sign means that the electrical potential is higher outside than inside the membrane because the concentration of Na$^+$ ions is higher outside. We, of course, will calculate the same value for the voltage if we use the ratio of Cl$^-$ concentration inside and outside.

SURFACES, MEMBRANES, AND SURFACE TENSION

When we think of phases we think of the usual solid, liquid, or gaseous states. However, whenever there are two phases in contact, there is also a surface, or interface, between them. This surface has properties different from those in the two bulk phases and therefore will have different behavior. The surface has thermodynamic properties specified by its free energy, enthalpy, and so on, just as bulk phases have. However, there are differences. A surface is two-dimensional instead of three-dimensional. This means that concentration units for a surface are mol cm^{-2} instead of mol cm^{-3}. Furthermore, because the surface properties are determined relative to the bulk phase, the surface concentration is treated as an excess or a deficiency and can be either positive or negative. The surface pressure has dimensions of force per unit length instead of force per unit area; the surface free energy, surface enthalpy, and so on have units of J m^{-2}.

Usually, we ignore the surface contributions to the total thermodynamic properties of a system. However, we can focus attention on the properties of the surface if we choose to. Sometimes the surface area is so large that we cannot ignore its properties. This occurs for a system divided into very small pieces where a large fraction of the molecules are at the surface. One gram of finely divided carbon black can have a surface area of 500 m^2!

Biological cells are small volumes surrounded by membranes. (Membranes are present inside the cells, also.) The membranes are a different phase from the rest of the cell, and for some purposes they can be thought of as a surface phase. The membranes, of course, have a finite thickness and definite volumes, but it may be more useful to think about their concentrations in molecules m^{-2} instead of molecules m^{-3}.

Surface Tension

Because of intermolecular attractions, the molecules at the surface of a liquid are attracted inward. This creates a force in the surface which tends to minimize the surface area. If the surface is stretched, the free energy of the system is increased. The *free energy per unit surface area,* or the force per unit length on the surface, is called the *surface tension.* Note that the units for energy per unit area and force per unit length are identical. The SI units are millinewtons per meter (mN m^{-1}), which are equal to mJ m^{-2}. Data in Table 5.2 illustrate the range of surface tensions that can occur. What happens when a substance is dissolved in the liquid, or added to the surface? The surface tension either decreases or does not change very much. It never increases greatly. There is a thermodynamic reason for this, but before stating it let us consider what would happen if we could greatly increase the surface tension of water. Water drops from your faucet could become the size of basketballs, because the size of the drop is directly proportional to the surface tension. You might have to cut water with a knife and chew it well before swallowing. Water would not wet anything, so processes depending on capillary action would not work.

Table 5.2   Surface tensions of pure liquids in air

Substance	Surface tension (mN m^{-1})	Temperature (°C)
Platinum	1819	2000
Mercury	487	15
Water	71.97	25
	58.85	100
Benzene	28.9	20
Acetone	23.7	20
n-Hexane	18.4	20
Neon	5.2	−247

Any substance that tends to raise the surface tension of a liquid raises the free energy of the surface. The substance will therefore not concentrate at the surface. Thus we are saved from these possible catastrophes because the surface tension, or the surface free energy, is just $(\partial G/\partial A)_{T, P}$, where A is the surface area. However, substances that lower the surface tension also lower the free energy of the surface; they will preferentially migrate to the surface. Thus substances that lower the surface tension will concentrate at the surface and can give large decreases in surface tension, but substances that raise the surface tension will avoid surfaces and give only small increase in surface tension. The quantitative expression for this is called the *Gibbs adsorption isotherm:*

$$\Gamma = -\frac{1}{RT}\frac{d\gamma}{d\ln a} \cong -\frac{1}{RT}\frac{d\gamma}{d\ln c} \tag{5.22}$$

where Γ = adsorption (excess concentration) of solute at surface, mol m^{-2}
γ = surface tension, N m^{-1}
R = gas constant = 8.314 J K^{-1} mol^{-1}
a = activity of solute in bulk solution
c = concentration of solute in bulk solution (any units can be used)

The sign of the excess surface concentration, Γ, is opposite to the sign of the change of the surface tension with concentration (or activity) of solute in the solution. Figure 5.10 shows the change in surface tension of water when LiCl or ethanol is added. Ionizing salts are almost the only solutes that raise the surface tension of water.

Why do some molecules preferentially absorb at a surface separating the two phases and thus reduce the surface tension? The molecule may consist of two parts or regions, one of which interacts primarily with one phase and the other with the second phase. As mentioned in Chapter 3, such molecules are said to be amphiphilic, and they tend to locate preferentially at the interface. This is what happens to detergent molecules that have a hydrocarbon tail and a polar head; they are called *surface-active molecules,* or *surfactants.*

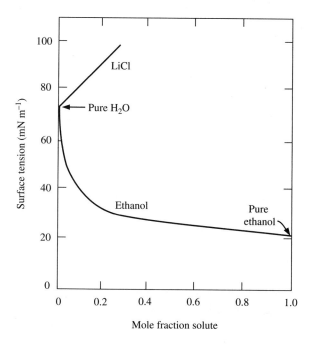

Fig. 5.10 Effect of solute concentration on surface tension in aqueous solution at 20°C.

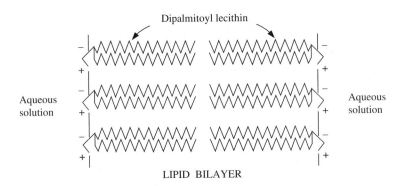

$$Na^+ {}^-O_3SO(CH_2)_{11}CH_3 \qquad \text{Sodium dodecylsulfate}$$

$$Na^+ {}^-O_2C(CH_2)_{14}CH_3 \qquad \text{Sodium palmitate}$$

$$(CH_3)_3\overset{+}{N}(CH_2)_2O\overset{-}{P}O_3CH_2 \qquad \text{Dipalmitoyl lecithin}$$
$$|$$
$$CHO_2C(CH_2)_{14}CH_3$$
$$|$$
$$CH_2O_2C(CH_2)_{14}CH_3$$

Air

Sodium palmitate

Water phase

SURFACE TENSION LOWERING

Water phase

Sodium dodecyl-sulfate

Dirt phase

DETERGENT ACTION

Dipalmitoyl lecithin

Aqueous solution

Aqueous solution

LIPID BILAYER

Fig. 5.11 Some common surface-active molecules and their orientations at interfaces.

Sodium dodecylsulfate (see Fig. 5.11), a principal component of many detergents, orients at the surface of dirt particles and makes them soluble in water. The hydrocarbon tail is attracted to the oily dirt "phase," and the polar sulfate head faces the water. The hydrocarbon part is hydrophobic; the polar group is hydrophilic, which is typical of the characteristics of amphiphilic molecules.

There are many naturally occurring detergents in plants and animals. Dipalmitoyl lecithin forms a layer (see Fig. 5.11) that lowers the surface tension at the surface of the alveoli* in the lung and allows one to breathe. A large surface area is necessary for the efficient exchange of gases in the lung. Dipalmitoyl lecithin lowers the surface tension of water to nearly zero and allows large aqueous surface areas to exist in the lung. Premature babies lack this vital surfactant and have great difficulty breathing effectively.

The surface-tension decrease caused by surface-active molecules is easily measured with a Langmuir film balance, as illustrated in Fig. 5.12. One adds some molecules to the surface of water and uses the movable barrier to compress them. The surface pressure necessary to maintain the film in the compressed state is just the difference between the surface tension γ_0 of the water and the surface tension γ of the coated water.

$$P_{surface} \ (mN \ m^{-1}) = \gamma_0 - \gamma$$

The surface concentration in mol m^{-2} depends on the surface pressure. The experiment is the two-dimensional analog of a pressure versus volume experiment in three dimensions. It is traditional to plot surface pressure versus area per molecule in Å^2 molecule^{-1}. The area per molecule is just the reciprocal of the surface concentration. Some representative data are given in Fig. 5.13. Note that the high surface pressures result from the low surface tensions produced by the surface-active molecules. The maximum possible surface pressure, equal to the surface tension of pure water, corresponds to zero surface tension of the film. This is implicit in the preceding equation.

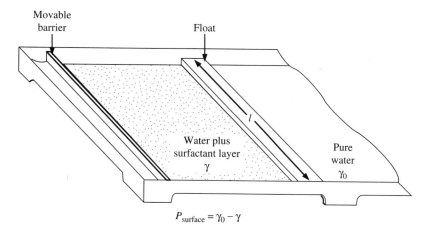

Fig. 5.12 Surface tension can be measured using a Langmuir film balance. The force per unit length, F/l, required to compress a surface containing a surfactant layer is known as the surface pressure, $P_{surface}$. The float is connected to a torsion wire (not shown) to measure the force.

* The alveoli are the smallest air compartments found in the lung.

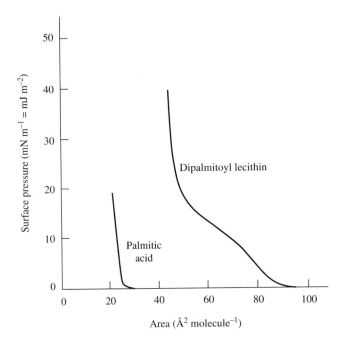

Fig. 5.13 Surface pressure versus area curves for two amphiphilic molecules (surfactants) at an air-water interface. The molecular areas correspond to average cross sections of the molecules projected on the surface. The steeply rising portions of the curves occur when the molecules in the surface layer have been compressed into a compact arrangement.

Vapor Pressure and Surface Tension

If small droplets and a much larger drop of water are placed in a closed container saturated with water vapor, the small droplets eventually disappear and the large one becomes larger. This phenomenon is a result of surface tension; the small droplets have a higher vapor pressure than the large one. When water or any other substance is finely divided, the surface effect becomes significant and we need to include the term $\gamma \, dA$ in the expression for the change in free energy. We replace

$$dG = -S \, dT + V \, dP + \mu_{H_2O} \, dn_{H_2O} \tag{5.23}$$

for bulk water by

$$dG = -S \, dT + V \, dP + \gamma \, dA + \mu_{H_2O} \, dn_{H_2O} \tag{5.24}$$

for the small droplets, where $\gamma \, dA$ is the change in free energy when the surface area is changed. The free energy per mol of bulk water (water in the interior of the liquid) is μ_{H_2O}. Equation (5.24) shows that the free energy of a liquid is increased by increasing its surface area, thus a liquid will minimize its free energy by decreasing its surface area. This means a liquid will form a sphere if there are no other forces acting on it, such as gravity or other surfaces. Combining two spheres into one lowers the total surface area and therefore the total free energy. Thus small spheres will tend to combine to form larger spheres. Mercury with its higher surface tension illustrates this better than water. We usually ignore the effect of surface area and the contribution of surface free energy to the total free energy; the effects are too small. However, significant effects can occur as shown by the higher vapor pressure of small—micron-sized—droplets. We can combine the terms $\gamma \, dA + \mu_{H_2O} \, dn_{H_2O}$

of Eq. (5.24) to obtain the *total chemical potential* of the liquid water; this takes into account the effect of the surface on the water molecules. Water in small droplets (with a high surface to volume ratio) has a higher total chemical potential and higher vapor pressure than bulk water and can therefore transfer spontaneously into a larger drop.

Total Chemical Potential

It is useful to summarize our discussion of total chemical potential. Consider a column of solution of a macromolecule in a gravitational field. If the gravitational acceleration g is very large, the macromolecule will sediment, and its concentration at a particular location within the column of solution will depend on the height h of its location.

The free energy change for the transfer of dn moles of macromolecule from height h_1 to h_2 is

$$[\mu(\text{at } h_2) - \mu(\text{at } h_1)] \, dn \tag{5.25}$$

The gravitational force acting on the macromolecule depends on its molecular weight and on the buoyant effect of the solvent. The effective molecular weight is the weight minus the weight of solvent displaced.

$$M_{\text{eff}} = M \cdot (1 - \bar{v}\rho)$$

where M = molecular weight of macromolecule
$\bar{v}$ = volume per gram of macromolecule
ρ = density of solvent

The gravitational force acting on dn moles of macromolecule of effective molecular weight M_{eff} is $gM_{\text{eff}} \, dn$, and the work required for this transfer is $gM_{\text{eff}} \, dn \cdot (h_2 - h_1)$ [Eq. (2.7)]. From the relation between the free energy change and work [$\Delta G = w_{\text{max, useful}}$, Eq. (4.37)], the effect of the gravitational field is to add an amount $gM_{\text{eff}} \cdot (h_2 - h_1) \, dn$ for this transfer. The total free energy change at constant T and P is therefore

$$
\begin{aligned}
dG &= [\mu(\text{at } h_2) - \mu(\text{at } h_1)] \, dn + gM_{\text{eff}} \cdot (h_2 - h_1) \, dn \\
&= [(\mu + gM_{\text{eff}}h)(\text{at } h_2) - (\mu + gM_{\text{eff}}h)(\text{at } h_1)] \, dn
\end{aligned}
\tag{5.26}
$$

If equilibrium is established, then $dG = 0$ and

$$\mu + gM_{\text{eff}}h = \text{constant} \tag{5.27}$$

We define the total chemical potential for a solute molecule

$$\mu_{\text{total}} = \mu + gM_{\text{eff}}h \tag{5.28}$$

in a gravitational field; the condition for equilibrium at constant T and P in a gravitational field is that the total chemical potential μ_{total} is constant everywhere.

Similarly, if a solution of macromolecules is spun in a centrifuge, at equilibrium the total chemical potential

$$\mu_{\text{total}} = \mu - \tfrac{1}{2} M_{\text{eff}}\omega^2 x^2 \tag{5.29}$$

where ω is the angular velocity and x is the distance from the center of rotation. Note that in Eq. (5.29), the acceleration $\omega^2 x$ in a centrifuge replaces g in the gravitational case. The change of the sign is a result of a difference in conventions. In a gravitational field h is taken to be increasing in the opposite direction of acceleration, whereas in a centrifuge x is taken to be decreasing in the direction of acceleration.

The concept of the total chemical potential is a very useful one. In Table 5.3 we list the cases that we have considered.

Table 5.3 Total chemical potential in different fields

Field	Total chemical potential
None	μ
Electric	$\mu + ZF\phi$
Gravitational	$\mu + M_{eff}gh$
Centrifugal	$\mu - \frac{1}{2}M_{eff}\omega^2 x^2$

Biological Membranes

The presence of molecules at a surface markedly changes other surface properties besides the surface tension. The rate of transport of water, ions, or other solutes across the surface is strongly affected. The composition and orientation of the molecules in biological membranes determine their function. Along with the diversity in composition comes a richness of properties that correlates with the variety of membrane functions. Usually, membranes contain an assortment of lipid molecules with diverse chemical structures, together with proteins and sometimes polysaccharides. The lipids are typically fatty acid esters that differ in the length of the fatty acid chain, the degree of unsaturation, the charge or polarity of the esterifying group, and the number of fatty acids esterified per molecule. The proteins may be intrinsic or integral, in the sense that they are incorporated directly into the membrane structure, or extrinsic, if they are attached to the membrane surface or interact strongly with it. Other components, such as cholesterol, may also be present.

The variety of membrane composition results in a range of physical properties. For example, the mobility of the molecules in two dimensions in a membrane may correspond to that of a typical liquid or to that of a solid. In fact, it is relatively common to encounter phase transitions in membranes or artificial surface films, and these transitions are closely analogous to their three-dimensional counterparts. The transition temperature or melting point, for example, is sensitive to the lipid composition; and the thermodynamic analysis of such two-dimensional solutions can be carried out just as for three-dimensional systems.

The relation between the intrinsic proteins and the lipids is more complex and is not yet well understood. Many proteins are amphiphilic and interact both with the hydrophobic lipids and with the polar aqueous interface. These hydrophobic and hydrophilic sites are located in different regions of the protein molecule, resulting in a strong orientation with respect to the membrane surface. Apart from this orientation, the protein may behave much like a solute in the lipid solvent; essentially a two-dimensional solution is formed. There is even a two-dimensional analog of precipitation, in which the solution becomes saturated with respect to a protein constituent and the protein separates as a distinct phase. Electron microscopy of membrane surfaces provides a useful method for detecting and characterizing the phase changes.

ACTIVE AND PASSIVE TRANSPORT

Frogs, seaweed, and other organisms that live in contact with water have semipermeable skins. Water and some ions and small molecules pass through the skins; macromolecules generally do not. The frog or the seaweed can selectively concentrate certain molecules inside itself and selectively exclude or excrete other molecules. How do they do it? The thermodynamic question is: If a molecule can easily pass through the skin, how can the inside concentration be maintained at a value that is different from the outside concentration? The answer is to ensure that the free energy change for transporting the molecule inside is negative. For example, the presence of a protein inside seaweed which strongly binds iodide ion ensures that the iodide concentration in the seaweed is always higher than in the seawater. If the concentration of *free* iodide is the same inside and outside, the bound iodide would account for the concentrating effect of the seaweed. This effect, known as *passive transport,* does not depend on whether the seaweed is alive or dead; that is, metabolism is not involved in the concentrating effect. The presence of the specific binding protein lowers the chemical potential of the iodide, which therefore concentrates in the seaweed. In equilibrium dialysis we used the transport of small molecules through a dialysis membrane to measure binding to a macromolecule on one side of the membrane. This is clearly passive transport.

There is another kind of transport, known as *active transport,* which is closely dependent on active cellular metabolism. Active transport is defined as the transport of a substance from a lower to a higher chemical potential. Because the total free energy change of the process must be negative, active transport must be tied to a chemical reaction that has a negative free energy change. In a biological system this means that metabolism must be occurring. Therefore, an experimental test of whether active transport is involved is to poison the metabolic activity and see if the transport also stops. A well-known example of active transport is the concentration of K^+ ions and the removal of Na^+ ions that occurs inside the cells of animals. A protein complex called the sodium-potassium pump uses the free energy of hydrolysis of ATP to pump Na^+ ions out of the cell and to pump K^+ ions into the cell. The net reaction for the active transport is thought to be

$$\left.\begin{array}{c} 3\,Na^+(\text{inside}) \\ + \\ 2\,K^+(\text{outside}) \end{array}\right\} + ATP \longrightarrow ADP + \text{phosphate} + \left\{\begin{array}{c} 3\,Na^+(\text{outside}) \\ + \\ 2\,K^+(\text{inside}) \end{array}\right.$$

An illustration involving representative concentration differences across a typical animal cell membrane is shown in Fig. 5.14. There is also a voltage difference of about -70 millivolts across the cell membrane; the inside is negative relative to the outside as shown.

A mechanism for the reaction is given in Fig. 5.15. The phosphorylation of the sodium-potassium pump by ATP causes a conformational change in the protein which exchanges bound Na^+ ions for K^+ ions outside the cell. Subsequent hydrolysis of the phosphate group from the sodium-potassium pump causes a conformational change, which exchanges bound K^+ ions for Na^+ ions inside the cell. The molecular machine is now ready for another cycle. The details of the mechanism are not known, but it is important to make sure that the process is consistent with the laws of thermodynamics. If it is not, we know that we have left something out.

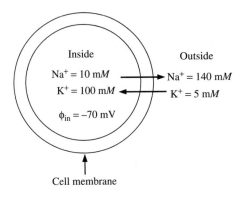

Fig. 5.14 Ion and voltage gradients occur typically across cell membranes in plants and animals. These are maintained by active transport.

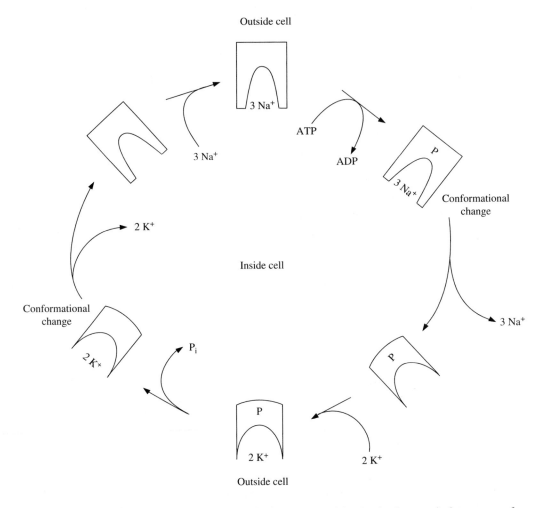

Fig. 5.15 The Na⁺-K⁺ pump is a membrane-bound protein which uses the decrease in free energy of hydrolysis of ATP to actively pump Na⁺ outside the cell and to actively pump K⁺ inside the cell. For each ATP hydrolyzed 3 Na⁺ are pumped out and 2 K⁺ are pumped in.

The free energy for the net reaction must be negative. This means that the positive free energy of actively transporting ions against concentration and voltage gradients must be balanced by the negative free energy of ATP hydrolysis.

Let us divide the process into three parts: the transport of Na^+ ions out, the transport of K^+ ions in, and the hydrolysis of ATP. The free energy per mole for the transport of Na^+ ions is

$$\Delta \bar{G}(\text{transport of } Na^+ \text{ out}) = RT \ln \frac{a_{Na^+}(\text{outside})}{a_{Na^+}(\text{inside})} + ZFV \qquad (5.7)$$

We will replace ratios of activities by ratios of concentrations and use a temperature of 37°C. For an answer in kJ

$$\Delta \bar{G}(\text{transport of } Na^+ \text{ out}) = (0.008314)(310) \ln \frac{140}{10} + (1)(96.485)(+0.07)$$

$$\Delta \bar{G}(\text{transport of } Na^+ \text{ out}) = +13.6 \text{ kJ mol}^{-1}$$

The last term has $+0.07$ because the voltage is $\phi_{out} - \phi_{in} = 0.0 - (-0.07)$. The free energy per mole for the transport of K^+ ions is

$$\Delta \bar{G}(\text{transport of } K^+ \text{ in}) = (0.008314)(310) \ln \frac{100}{5} + (1)(96.485)(-0.07)$$

$$\Delta \bar{G}(\text{transport of } K^+ \text{ in}) = +1.0 \text{ kJ mol}^{-1}$$

Note that the difference in free energy for moving Na^+ ions out and K^+ ions in is that the electrical potential difference of -70 mV favors transport of positive ions in, thus the favorable effect of the electric field nearly cancels the effect of the concentration gradient. The total free energy cost of transporting 3 moles of Na^+ ions out and 2 moles of K^+ ions in is

$$\Delta G(\text{total ion transport}) = 3\Delta \bar{G}(\text{transport of } Na^+ \text{ out}) + 2\Delta \bar{G}(\text{transport of } K^+ \text{ in})$$

$$\Delta G(\text{total ion transport}) = +42.8 \text{ kJ} \qquad (5.30)$$

The free energy for the hydrolysis of ATP (ATP $\rightarrow$ ADP + P) is

$$\Delta G = \Delta G^0 + RT \ln Q \qquad (4.4)$$

$$\Delta G = \Delta G^0 + RT \ln \frac{[ADP][P]}{[ATP]}$$

ΔG^0 is -31.0 kJ for ATP hydrolysis at 298 K and -31.3 at 310 K. The ratio of ATP to ADP in cells is of order 100 or larger. The concentration of phosphate varies, but estimates of Q range from 10^{-2} to 10^{-3}. This gives

$$\Delta G(\text{ATP hydrolysis}) = -31.3 + (0.008314)(310) \ln Q$$

$$\Delta G(\text{ATP hydrolysis}) = -31.3 - 11.8 \text{ or } = -31.3 - 17.8$$

$$\Delta G(\text{ATP hydrolysis}) = -43.1 \text{ to } -49.1 \text{ kJ} \qquad (5.31)$$

Clearly the free energy of ATP hydrolysis is sufficient to account for the active transport of the ions; the negative free energy values of Eq. (5.31) are larger than the positive value of Eq. (5.30). The free energy calculation surely does not prove the mechanism. It only proves that the mechanism is a possible one; it does not violate thermodynamic principles. This is a very important test, however, because if the calculation did not give a net negative free energy, the mechanism would have been immediately disproven.

COLLIGATIVE PROPERTIES

The boiling point and freezing point of a liquid depend on the pressure and on the concentration of any solutes dissolved in the liquid. The solubility of gases and solids in a solvent depends on the temperature and pressure. These properties involve phase equilibria between gas and liquid, or between solid and liquid. They can be understood by a common thermodynamic analysis.

The osmotic pressure of a solution—the pressure that must be applied to a solution to keep it in equilibrium with pure solvent—depends on the concentrations of the solutes dissolved. If a solution is separated from a solvent by a semipermeable membrane, solvent will flow through the membrane into the solution as long as equilibrium is not attained. This effect is vital in transporting water from the roots up to the leaves of trees. It will also cause rupture of red blood cells in contact with pure water or with a solution that is too dilute—is hypotonic.

The properties described above are called *colligative* because they depend on the *collection* of molecules. In dilute solution they depend only on the number of solute molecules, not on the kinds of solute molecules. This means that colligative properties can be used to count the number of molecules present in a solution. We can easily prepare a solution containing a known weight of solute molecules, thus we can use a colligative property to determine their molecular weight.

$$\text{Molecular weight} = \frac{\text{weight of molecules}}{\text{number of molecules}}$$

Boiling Point and Freezing Point of a Pure Component

The history of mountaineering and the thermodynamics of gas-liquid equilibria have always been closely linked. The reason is that the boiling point of water, or any liquid, depends on the pressure of its vapor in equilibrium with it. In the 1850s, intrepid mountaineers on first ascents carried 760-mm-long mercury manometers to record the pressure and thus the altitude at the peak. The scientist mountaineer carried a thermometer to measure the boiling point of water at the peak and calculated the pressure from the measured boiling point. Present-day mountaineers who climb above 15,000 ft routinely carry a pressure cooker for meals and an aneroid barometer to measure altitudes.

It is straightforward to derive the equation relating the boiling point of a pure liquid to the pressure of its gas phase in equilibrium. Consider the reaction

$$A(\text{pure liquid}) \longrightarrow A(\text{gas})$$

The equilibrium constant is

$$K = \frac{a_A(\text{gas})}{a_A(\text{pure liquid})}$$

Gases are always much more nearly ideal than liquids because the interactions of the molecules in the gas phase are always less than the same molecules in the liquid. In liquids molecules are in direct contact; in gases (except for very high pressures) they are often far apart. Therefore it is appropriate to approximate the gas as an ideal gas, and to replace the activity of the gas by the pressure of the gas in atmospheres.

$$a_A(\text{gas}) = P(\text{atm})$$

We can ignore the effect of pressure on the activity of the liquid,

$$a_A(\text{pure liquid}) = 1$$

Therefore, the equilibrium constant is just the equilibrium gas pressure,

$$K = P$$

The temperature dependence of the equilibrium constant [Eq. (4.35)] gives

$$\ln \frac{P_2}{P_1} = \frac{-\Delta \bar{H}_{vap}}{R}\left(\frac{1}{T_2} - \frac{1}{T_1}\right) \tag{5.32}$$

where $\Delta \bar{H}_{vap} = \bar{H}_{gas} - \bar{H}_{liquid} =$ molar enthalpy (heat) of vaporization. This equation is known as the *Clausius-Clapeyron equation.* Let us consider its uses.

The pressures P_2 and P_1 are the equilibrium vapor pressures of a liquid at temperatures T_2 and T_1, respectively. The normal boiling point, T_b, is defined as the temperature where the vapor pressure is 1 atm. Therefore, for a pure liquid whose molar enthalpy of vaporization and normal boiling point are known, one can use Eq. (5.32) to obtain the equilibrium vapor pressure at other temperatures.

Example 5.2   At 20,320 ft altitude at the summit of Mt. McKinley, pure water boils at only 75°C. We note in passing that "hot" tea will be tepid and weak at this altitude. What is the atmospheric pressure?

Solution   The enthalpy of vaporization of water depends on temperature; for Eq. (5.32) we use an approximate value of $\Delta \bar{H}_{vap} = 42{,}000$ J mol^{-1}. $P_1 = 1$ atm, and $T_1 = 373$ K (the normal boiling point of water).

$$\ln P = \frac{-42{,}000}{8.314}\left(\frac{1}{348} - \frac{1}{373}\right)$$

$$P = 0.38 \text{ atm}$$

This is the pressure of the water in equilibrium with the boiling liquid. Liquid boils when the vapor pressure of the liquid equals that of the surrounding atmosphere; therefore, the atmospheric pressure is 0.38 atm. The effect of this reduced pressure on the equilibrium binding of oxygen to hemoglobin in the lungs is another problem in phase equilibria that concerns mountaineers.

The Clausius-Clapeyron equation can be used to calculate a normal boiling point temperature from two measured vapor pressures at different temperatures. First, the enthalpy of vaporization is calculated; then the temperature corresponding to a vapor pressure of 1 atm is calculated. An easy way to apply the Clausius-Clapeyron equation is to plot ln P versus $1/T$. The slope is equal to $-\Delta \bar{H}_{vap}/R$. The vapor pressure at any temperature, or the boiling point at any pressure, can be read from the graph.

In using Eq. (5.32) we should remember its limitations. The main one is the assumption that $\Delta \bar{H}_{vap}$ is independent of temperature. Over a wide-enough temperature range ln P versus $1/T$ will be a curve; the slope at any point will give $\Delta \bar{H}_{vap}$ at that temperature. The assumption of gas ideality and the neglect of pressure dependence of the liquid activity are usually valid. We can increase the activity and therefore increase the vapor pressure of a liquid by applying an external pressure. However, this effect is small compared to the large effect of temperature on the vapor pressure.

For the effect of pressure on melting-point temperature, we need to use a somewhat different and even more general approach. Consider a solid and a liquid in equilibrium at temperature T and pressure P. Then

$$\bar{G}_s = \bar{G}_l$$

at T and P. Now increase the temperature by an amount dT and the pressure by an amount dP so as to maintain the solid and liquid in equilibrium. Under the new conditions,

$$\bar{G}_s + d\bar{G}_s = \bar{G}_l + d\bar{G}_l$$

at $T + dT$, $P + dP$. Subtracting the first equation from the second, we obtain

$$d\bar{G}_s = d\bar{G}_l$$

This states that, if we change the pressure of a solid substance in equilibrium with its liquid, we must also change the temperature so as to keep the increment in free energy the same for the two phases. Otherwise, the one with higher free energy will be converted into the one with lower free energy.

Using

$$d\bar{G} = -\bar{S}\,dT + \bar{V}\,dP \qquad (3.47)$$

we obtain

$$-\bar{S}_s\,dT + \bar{V}_s\,dP = -\bar{S}_l\,dT + \bar{V}_l\,dP$$

which can be rearranged to give

$$\frac{dT}{dP} = \frac{\Delta \bar{V}_{fus}}{\Delta \bar{S}_{fus}}$$

$$\Delta \bar{V}_{fus} = \bar{V}_l - \bar{V}_s \qquad \Delta \bar{S}_{fus} = \bar{S}_l - \bar{S}_s$$

Because the process is reversible,

$$\Delta \bar{S}_{fus} = \frac{\Delta \bar{H}_{fus}}{T}$$

and

$$\frac{dT}{dP} = \frac{T \Delta \overline{V}_{fus}}{\Delta \overline{H}_{fus}} \tag{5.33}$$

This is an example of a general expression that can be applied to equilibrium between any two phases. For example, Eq. (5.32) can be derived as a special case for liquid-vapor equilibrium.

The effect of external pressure on melting points is not large, so we can write Eq. (5.33) as

$$\frac{T_2 - T_1}{P_2 - P_1} = T \frac{\Delta \overline{V}_{fus}}{\Delta \overline{H}_{fus}} \tag{5.34}$$

where T = average of T_2 and T_1
$\Delta \overline{V}_{fus} = \overline{V}(\text{liquid}) - \overline{V}(\text{solid})$ = molar volume change of fusion
$\Delta \overline{H}_{fus} = \overline{H}(\text{liquid}) - \overline{H}(\text{solid})$ = molar enthalpy (heat) of fusion

We must be careful with units in this equation. If P is in atm, $\Delta \overline{V}_{fus}$ in $m^3 \, mol^{-1}$, and $\Delta \overline{H}_{fus}$ in $J \, mol^{-1}$, the conversion factor of $9.869 \times 10^{-6} \, m^3 \, atm \, J^{-1}$ must be used. The effect of pressure on a melting point depends on the signs and magnitudes of two thermodynamic variables: $\Delta \overline{H}_{fus}$ and $\Delta \overline{V}_{fus}$. The heat of fusion is always positive; heat is always absorbed on melting. But the volume of fusion can be positive or negative, although it is usually positive. H_2O is an exception. The volume of water decreases on melting.

Ideal Solutions

The vapor pressure, boiling point, freezing point, and osmotic pressure of a solution are known as colligative properties. For sufficiently dilute solutions, these properties are linearly dependent on the concentration of solutes present. Furthermore, they are all closely related to each other. If one is measured, the others can be calculated. For example, if the osmotic pressure of a solution is measured, the vapor pressure of the solvent is immediately known. For dilute or ideal solutions, from the measured boiling-point rise, the freezing-point lowering can be calculated. We find that in ideal solutions the colligative properties depend only on the number of solute molecules. So if we measure the molar concentration of the solute in solution, we can calculate any colligative properties. The effective link between the measured colligative properties is the activity of the solvent. For a real solution we can obtain the activity of the solvent in the solution from one measured colligative property. Then any other colligative property at the same temperature can be calculated.

In this section we will consider only the empirical results found for ideal solutions. The general thermodynamic equations relating the activity of the solvent to colligative properties will be discussed in a later section.

Raoult's law

Adding a solute to a solvent decreases the vapor pressure of the solvent. Raoult found experimentally that the vapor pressure of the solvent in a solution is directly proportional to its mole fraction. *Raoult's law* is stated:

$$P_A = X_A P_A^0 \tag{5.35}$$

where P_A = vapor pressure of component A in solution
X_A = mole fraction of A
P_A^0 = vapor pressure of pure A

Raoult's law is like the ideal gas law, in that it applies approximately to many solutions and it applies exactly to all very dilute solutions. Therefore, for a dilute solution you can always calculate the vapor pressure of the solvent simply by knowing its mole fraction, X_A, and the vapor pressure of pure solvent at the same temperature. If Raoult's law applies, the solution is defined to be ideal.

Henry's law

The solutes in the solution may also be volatile. What about their vapor pressures? Henry found experimentally that the solute vapor pressure is proportional to the solute mole fraction, but the proportionality constant is not, in general, the vapor pressure of the pure solute. *Henry's law* is stated:

$$P_B = k_B X_B \tag{5.36}$$

where P_B = vapor pressure of component B
X_B = mole fraction of B
k_B = Henry's law constant for component B in solvent A

You can think of Henry's law as specifying the pressure dependence of the solubility of a gas in a solution. It says that the gas solubility should be directly proportional to pressure:

$$X_B = \frac{P_B}{k_B} \tag{5.37}$$

The solubility will also depend on temperature and the nature of the solvent. These must be specified for each value of the Henry's law constant k_B. Table 5.4 gives Henry's law constants for various gases in water in order of decreasing vapor pressure of the gas and therefore in order of increasing solubility of the gas. Using Eq. (5.37), you will obtain the

Table 5.4   Henry's law constants
$(k_B = P_B/X_B)$ for aqueous
solutions at 25°C

Gas	k_B (atm)
He	131×10^3
N_2	86×10^3
CO	57×10^3
O_2	43×10^3
CH_4	41×10^3
Ar	40×10^3
CO_2	1.6×10^3
C_2H_2	1.3×10^3

solubility as mole fraction, but because these are dilute aqueous solutions, it is easy to see that the molarity is just 55.6 (= 1000/18) times the mole fraction; for dilute aqueous solutions: $c = 55.6X$.

Example 5.3 Divers get the bends when bubbles of N_2 gas form in their bloodstream if they rise too rapidly from a deep dive. Calculate the solubility (mol L^{-1}) of N_2 in water (this is roughly equal to the solubility in blood serum) at a depth of 300 ft below the surface of the ocean and at sea level.

Solution   First, we need to find the pressure at a depth of 300 ft. We know that a column of mercury 760 mm (= 2.49 ft) high exerts a pressure of 1 atm. An equal column of seawater will exert a pressure proportional to the ratio of the density of seawater (1.01) to mercury (13.6); therefore, the pressure 300 ft deep in the ocean is

$$P = \frac{300 \text{ ft}}{2.49 \text{ ft atm}^{-1}} \left(\frac{1.01}{13.6}\right) + 1 \text{ atm}$$

$$= (8.9 + 1)\text{atm} = 9.9 \text{ atm}$$

The solubilities are obtained from Eq. (5.37) and $c = 55.6X$.

$$c_{N_2} = \frac{55.6 P_{N_2}}{k_{N_2}} = \frac{55.6 P_{N_2}}{86 \times 10^3}$$

$$= 0.65 \times 10^{-3} \text{ mol } L^{-1} \text{ at 1 atm}$$

$$= 6.4 \times 10^{-3} \text{ mol } L^{-1} \text{ at 9.9 atm}$$

The diver should rise slowly to allow plenty of time for the extra dissolved nitrogen in his blood to equilibrate with the decreasing external pressure.

Boiling-point rise and freezing-point lowering

When a small amount of solute is added to a solvent, the boiling point of the solution is raised and the freezing point of the solution is lowered relative to that of the solvent. The change in the boiling point and freezing point is found to be directly proportional to the concentration of the solute in dilute solution.

$$T_0 - T_f = K_f m \tag{5.38}$$

$$T_b - T_0 = K_b m \tag{5.39}$$

where T_0 = freezing point or boiling point of pure solvent (it depends on pressure)
T_f, T_b = freezing point and boiling point of solution
K_f, K_b = constants that depend only on the solvent
(K_f = 1.86 and K_b = 0.51 for water)
m = molality of solution = number of moles of solute in 1000 g of solvent

Because K_f and K_b depend only on the solvent, Eqs. (5.38) and (5.39) can be used to measure total concentration of solutes in complicated mixtures of solutes in solution. For example, the freezing point of seawater immediately gives a measure of the total number of moles of ions and other solutes of all kinds present in the seawater. In using Eqs. (5.38) and (5.39), we must remember that the solutions should be dilute and that only the pure solvent must freeze or evaporate. The solutes should not be volatile at the boiling point or soluble in the solid solvent at the freezing point for the simple equations to apply.

We have been using three different concentration scales, so it is well to know the relations among them. From the definition of mole fraction, X, molality, m, and molarity, c, we can convert from one to the other. Here we will give the relations that apply to dilute solutions (m or $c < 0.1$):

$$X_B = \frac{M_A m}{1000} \tag{5.40}$$

$$c = \rho_0 m \tag{5.41}$$

where X_B = mole fraction of solute
M_A = molecular weight of solvent
m = molality of solute
c = molarity of solute
ρ_0 = density of solvent

We note that for dilute aqueous solution,

$$X_B = \frac{m}{55.6} \tag{5.42}$$

$$c = m$$

Osmotic pressure

It has happened that a hospital patient died because pure water was accidentally injected into the veins. The osmotic pressure of the solution inside the blood cells is high enough to cause them to break under these circumstances. Osmotic pressure is defined as the pressure that must be applied to a solution to keep its solvent in equilibrium with pure solvent at the same temperature. If a solution is separated from a solvent by a semipermeable membrane (such as the membrane of a red blood cell), solvent will flow into the solution side until equilibrium is reached. At constant temperature, equilibrium can be attained by increasing the pressure on the solution or by diluting the solution until it is essentially pure solvent. A red blood cell placed in pure water or in a dilute salt solution will burst before equilibrium is reached.

Osmotic pressure is a colligative property closely related to the vapor pressure, freezing point, and boiling point. The activity of the solvent is the parameter that is most easily related to the osmotic pressure.

Adding a solute to a solution lowers the activity of the solvent; by raising the pressure on the solution, the activity of the solvent is increased. Van't Hoff found empirically that for dilute solutions the osmotic pressure, π, is directly proportional to the concentration of the solute. Because of the importance of osmotic pressure in biology, we will derive the relevant equations.

Figure 5.16 shows the condition at equilibrium for a solution in contact with pure solvent through a semipermeable membrane that allows only solvent to pass. The equilibrium

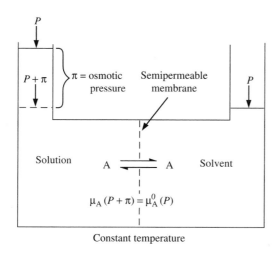

Fig. 5.16 Osmometer showing conditions at equilibrium. A is a solvent molecule.

occurs in an osmometer in which the extra pressure, π, necessary for equilibrium is produced by the hydrostatic pressure of a water column on the solution side. At equilibrium, there will be no net flow of solvent across the membrane, so the chemical potential of the pure solvent at pressure P must be equal to the chemical potential of the solvent in solution at pressure $P + \pi$:

$$\mu_A(\text{solution}, P + \pi) = \mu_A(\text{solvent}, P)$$

The temperature is constant throughout. If we let $P = 1$ atm, then at equilibrium,

$$\mu_A(\text{solution}, 1 + \pi) = \mu_A^0(\text{solvent})$$

Because of the presence of the solute in solution, the chemical potential of A at 1 atm pressure is decreased by an amount that can be calculated using Eq. (4.13):

$$\Delta\mu_{\text{osm}} = \mu_A(\text{solution}, 1\ \text{atm}) - \mu_A^0(\text{solvent}) = RT \ln a_A$$

To compensate for this change and restore equilibrium with the solvent, we need to increase the chemical potential of A by an equal amount by applying external pressure π to the solution. Remembering that

$$\bar{G}_2 - \bar{G}_1 = \int_{P_1}^{P_2} \bar{V}\, dP = \bar{V} \cdot (P_2 - P_1) = \Delta\mu \tag{3.33}$$

we obtain

$$\Delta\mu_{\text{osm}} = RT \ln a_A = \int_{1+\pi}^{1} \bar{V}_A\, dP = -\bar{V}_A \pi$$

and

$$\ln a_A = \frac{-\pi\bar{V}_A}{RT} \tag{5.43}$$

For dilute solutions, we can use the following approximation for $\ln a_A$:

$$\ln a_A = \ln X_A = \ln (1 - X_B) \cong -X_B = -\frac{n_B}{n_A + n_B} \cong -\frac{n_B}{n_A}$$

Then

$$\pi = \frac{RT}{n_A \overline{V}_A} n_B \qquad (5.44)$$

Furthermore, in dilute solutions, $n_A \overline{V}_A$ is essentially equal to the volume of solution (the volume of solute is negligible). Therefore, for dilute solutions,

$$\pi V = n_B RT \qquad (5.45a)$$

$$\pi = cRT \qquad (5.45b)$$

where π = osmotic pressure, atm
  V = volume of solution, L
  n_B = number of moles of solute in solution
  R = gas constant = 0.08205 L atm K^{-1} mol^{-1}
  c = concentration of solute, mol L^{-1}

Equation (5.45b) is the relation that was first found empirically by van't Hoff. These equations state that the osmotic pressure of a solution is directly proportional to the concentration of solute in dilute solution. Equation (5.45) is written to look like the ideal gas equation, to make it easy to remember. However, π is not the pressure of the solute, but rather the pressure that must be applied to the solution to keep solvent from flowing in through a semipermeable membrane.

To gain an appreciation for the magnitude of osmotic pressures, note that according to Eq. (5.45) a solution with $c = 0.5$ mol L^{-1} at 310 K (37°C) will have an osmotic pressure of 12.7 atm. This is typical of the osmotic concentration of the cytoplasm inside animal cells. When such cells come into contact with pure water, the osmotic pressure may rupture the cell membranes. This happens in the hemolysis of red blood cells when the blood serum is diluted with pure water.

MOLECULAR-WEIGHT DETERMINATION

The molecular weight is an important property of any molecule, but is particularly important for proteins, nucleic acids, and other macromolecules. Just learning the rough size of the molecule may be sufficient for some purposes (M = 1000?, 10,000?, 1,000,000?), but usually we need more accurate molecular weights. The colligative properties that we have been discussing in the previous section are all used to measure molecular weights. Osmotic pressure is the most useful for macromolecules.

Vapor-pressure lowering, freezing-point lowering, boiling-point rise, and osmotic pressure are all related to the mole fraction of solvent in dilute solution. This means that they all depend on the number of solute molecules (or ions, in the case of electrolytes) added to the solvent. If we divide the weight of solute by the number of solute molecules, we obtain the molecular weight.

Vapor-Pressure Lowering

In dilute solution we can write Raoult's law,

$$X_A = \frac{P_A}{P_A^0} \tag{5.35}$$

as

$$X_B = 1 - X_A = \frac{P_A^0 - P_A}{P_A^0} = \frac{n_B}{n_A + n_B} \cong \frac{n_B}{n_A}$$

where
X_B = mole fraction of solute
P_A^0 = vapor pressure of pure solvent
P_A = vapor pressure of solvent in solution
$P_A^0 - P_A$ = vapor pressure lowering of solvent
n_A, n_B = number of moles of solvent and solute, respectively

The number of moles of solute, n_B, is equal to wt_B/M_B; therefore,

$$M_B = \frac{wt_B}{n_A} \cdot \frac{P_A^0}{P_A^0 - P_A} \tag{5.46}$$

where M_B = molecular weight of solute
wt_B = weight of solute

For a solution of known weight concentration, vapor-pressure measurements provide the molecular weight of the solute. The solution should be dilute and ideal, so measurements are made as a function of concentration and extrapolated to zero concentration. Inaccuracies in measuring vapor pressures put an upper limit on the molecular weight that can be measured. For very high molecular weights, the vapor-pressure lowering is so small that slight temperature differences between solvent and solution can ruin the results.

A more significant problem for high-molecular-weight solutes is the effect of impurities. Because the vapor pressure depends on the mole fraction of solvent, each solute molecule contributes equally to decreasing the vapor pressure. In dilute solution a large molecule has the same effect on vapor pressure as a small molecule. On a weight basis, the small molecule produces a much larger effect. One milligram of solute with a molecular weight of 100 lowers the vapor pressure of solution as much as 1 g of solute with a 100,000 molecular weight. It is clear that vapor-pressure lowering cannot be used to measure the molecular weight of a protein dissolved in a buffer solution; the buffer ions would dominate the vapor-pressure lowering.

Boiling Points and Freezing Points

The change of the boiling point or freezing point of a dilute solution is inversely proportional to the molecular weight of the solute.

Equations (5.38) and (5.39) can be written as:

$$\text{Boiling point:} \qquad M_B = \frac{\text{wt}_B}{\text{kg solvent}} \cdot \frac{K_b}{\Delta T_b} \qquad\qquad (5.47)$$

$$\text{Freezing point:} \qquad M_B = \frac{\text{wt}_B}{\text{kg solvent}} \cdot \frac{K_f}{\Delta T_f} \qquad\qquad (5.48)$$

where
M_B = molecular weight of solute
ΔT_b = increase in boiling-point temperature
ΔT_f = decrease in freezing-point temperature
$\dfrac{\text{wt}_B}{\text{kg solvent}}$ = weight of solute per 1000 g of solvent

These methods have the same limitations as do vapor-pressure measurements. Molecular weights should be no higher than about 10,000 and impurities must be absent. In addition, the measurements are limited to temperatures near the boiling point or freezing point of the solvent. These methods have limited applicability to purely biological systems, but in special cases they may be quite useful. A study of the molal concentration of total solute particles in different cells of the kidney was made by putting a kidney slice on the cold stage of a microscope. The freezing point determined for each separate cell immediately gave the total molality of its contents.

Osmotic Pressure

Osmotic-pressure measurement is the most useful colligative method for determining molecular weights. It is accurate for molecular weights approaching 1 million, and small molecule impurities, buffers, or salts do not interfere if suitable membranes are used. The molecular weight of only the solutes that do not pass the semipermeable membrane is obtained.
    Equation (5.45) can be rewritten as

$$\pi = \frac{w}{M} RT \qquad\qquad (5.49)$$

where π = osmotic pressure, atm
w = concentration of solution, g L^{-1}
R = gas constant = 0.08205 L atm K^{-1} mol^{-1}
M = molecular weight
T = absolute temperature

Usually, a series of concentrations is measured, and extrapolation is made to zero concentration. A plot of π/w versus w gives RT/M as the intercept; the slope characterizes deviations from ideality (see Fig. 5.17). A positive slope is usually found for these plots; it can be accounted for in terms of the activity coefficient of the solute. A negative slope indicates an increase in molecular weight with increasing concentration. This will occur if the solute undergoes aggregation. If we assume that the solution is ideal, then π/w can be equated to RT/M at any concentration; for real solutions this relation may begin to fail for concentrations greater than 0.01 M. Osmotic pressure, as well as the other colligative methods, can be used to study aggregation in solution at higher concentrations.

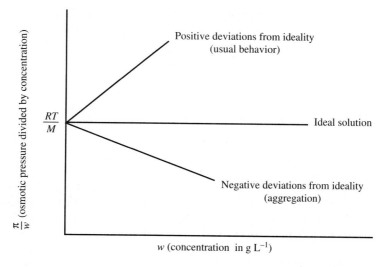

Fig. 5.17 Dependence of osmotic pressure on weight concentration.

Number-Average Molecular Weights

For aggregating molecules (or synthetic polymers or mixtures of molecules), the methods that we have been discussing will give average molecular weights. All colligative methods give a number-average molecular weight. This is defined as

$$M_n = \frac{\sum_i n_i M_i}{\sum_i n_i} \tag{5.50}$$

where M_n = number-average molecular weight
n_i = number of molecules with molecular weight M_i

The sums are over all solute species. For example, if we use vapor-pressure lowering or some other colligative method to measure the molecular weight of sodium chloride in water, we will obtain the arithmetic mean of the molecular weights of Na^+ and Cl^-:

$$M_n = \frac{M_{Na^+} + M_{Cl^-}}{2} = \frac{23 + 35.5}{2} = 29.25$$

Equation (5.50) can also be written in terms of the weight of each species present in solution:

$$M_n = \frac{\sum_i wt_i}{\sum_i \frac{wt_i}{M_i}} \tag{5.51}$$

where wt_i is the weight of molecules with molecular weight M_i. This equation illustrates that for equal weights of various species the lower-molecular-weight species dominate the number average. This is why impurities can be so significant in molecular-weight determinations of macromolecules.

Free Energy and Physical Equilibria   **235**

Example 5.4 How will 1.0 wt % impurity affect the determination of the number-average molecular weight of a macromolecule whose true molecular weight is 100,000? Assume that the impurity has a molecular weight of 100.

Solution

$$M_n = \frac{0.99 + 0.01}{\dfrac{0.99}{100,000} + \dfrac{0.010}{100}} = \frac{1.00}{0.000110}$$

$$= 9.1 \times 10^3$$

Although the sample is 99% pure, the molecular weight obtained is more than ten-fold too small. Osmotic-pressure measurements that allow the small-molecule impurities to equilibrate and not contribute to the sums in Eqs. (5.50) and (5.51) are obviously preferable.

Weight-Average Molecular Weights

Some molecular-weight methods such as equilibrium ultracentrifuge measurements give weight-average molecular weights

$$M_w = \frac{\sum\limits_i \text{wt}_i M_i}{\sum\limits_i \text{wt}_i} \tag{5.52}$$

where M_w = weight-average molecular weight
wt_i = weight of all materials with molecular weight M_i

This can also be written as

$$M_w = \frac{\sum\limits_i n_i M_i^2}{\sum\limits_i n_i M_i} \tag{5.53}$$

This average depends more on the high-molecular-weight species than does the number-average weight.

Example 5.5 How will 1.0 wt % impurity affect the determination of the weight-average molecular weight of a macromolecule whose molecular weight is 100,000? Assume that the impurity has a molecular weight of 100.

Solution

$$M_n = \frac{(0.99)(100,000) + (0.010)(100)}{0.99 + 0.010} = \frac{99,001}{1.00}$$

$$= 0.99 \times 10^5$$

The weight average is almost unaffected by 1% impurity.

A good way of determining whether a sample is pure or a mixture is to measure both the number-average and the weight-average molecular weight. If the two averages agree, the sample is probably a single-molecular-weight species. If the averages disagree, either the sample is a mixture or one of the measurements is in error.

ACTIVITY OF THE SOLVENT AND COLLIGATIVE PROPERTIES

Thermodynamics can be used to relate the empirical Raoult's law and Henry's law to other properties of solutions, such as freezing-point lowering, boiling-point rise, or osmotic pressure. The activity of the solvent is the link that connects these seemingly unrelated colligative properties. The general definition of activity is given by

$$\mu = \mu^0 + RT \ln a \tag{4.13}$$

We use the solvent standard state for the activity of the solvent, so the chemical potential of the standard state, μ^0, is chosen to be that of the pure solvent.

Vapor Pressure

For equilibrium between any two phases, there is equality of the chemical potentials of each component in the two phases. For example,

$$\mu_A(\text{solution}) = \mu_A(\text{gas}, P_A)$$

$$\mu_A^0(\text{pure solvent}) = \mu_A^0(\text{gas}, P_A^0)$$

where P_A = vapor pressure of A for solution
P_A^0 = vapor pressure of pure solvent A

For an ideal gas, we can use Eq. (3.35) to obtain

$$\mu_A(\text{gas}, P_A) - \mu_A^0(\text{gas}, P_A^0) = RT \ln \frac{P_A}{P_A^0}$$

Any nonideality in the gas at pressures near 1 atm can be ignored, because it will be small compared to nonidealities in solution. Therefore,

$$\mu_A(\text{solution}) - \mu_A^0(\text{pure solvent}) = RT \ln \frac{P_A}{P_A^0}$$

We can now relate vapor pressures to the activity of the solvent. From Eq. (4.13),

$$\mu_A(\text{solution}) - \mu_A^0(\text{pure solvent}) = RT \ln a_A$$

Therefore, at equilibrium with gas, from Eq. (4.16),

$$RT \ln \left(\frac{P_A}{P_A^0} \right) = RT \ln a_A$$

$$a_A = \frac{P_A}{P_A^0} \tag{5.54}$$

where a_A = activity of solvent

P_A = vapor pressure of solvent in solution

P_A^0 = vapor pressure of pure liquid solvent

All of these must be measured at the same temperature. Vapor-pressure measurements thus provide a simple method for measuring the activity of the solvent.

Because the activity of the solvent can be written

$$a_A = \gamma_A X_A$$

it is clear that Eq. (5.54) reduces to Raoult's law for ideal solutions, where $\gamma_A \equiv 1$. It can be shown that when Raoult's law applies to the solvent in a solution, Henry's law must apply to the solute.

Boiling Point

Another method of determining the activity of the solvent in a solution is to measure the boiling-point rise. The *normal boiling point* of a solution is defined as the temperature where the vapor pressure of the solvent equals 1 atm. When a nonvolatile solute is added to a solvent, the vapor pressure of the solution decreases, and therefore the normal boiling point is raised. How much higher must the temperature of the solution be so that its vapor pressure is 1 atm, the same as that of the pure solvent at the normal boiling point, T_0? Because the vapor pressure is decreased in the solution, we need to raise the temperature from T_0 to T_b, which we choose so that $P_A = 1$ atm. The vapor pressure of the pure solvent at temperature T_b would be $P_A/a_A = (1\ \text{atm})/a_A$. Now we can use Eq. (5.32), which gives the vapor pressure as a function of temperature. Let $P_1 = 1$ atm at $T_1 = T_0$ and $P_2 = (1\ \text{atm})/a_A$ at $T_2 = T_b$. After rearranging terms, we obtain

$$\ln a_A = \left(\frac{\Delta \bar{H}_{\text{vap}}}{R}\right) \cdot \left(\frac{1}{T_b} - \frac{1}{T_0}\right) \tag{5.55}$$

where a_A = activity of solvent in solution; if the solution is ideal, $a_A = X_A$

$\Delta \bar{H}_{\text{vap}}$ = enthalpy of vaporization of solvent

T_b = boiling point of solution

T_0 = boiling point of pure solvent

Equation (5.55) looks different from the familiar equation (5.39) relating boiling point and concentration in dilute solution, but Eq. (5.55) reduces to the familiar result as follows. For dilute (two component) solutions we have the relations

$$\frac{1}{T_b} - \frac{1}{T_0} = \frac{T_0 - T_b}{T_b T_0} \cong \frac{T_0 - T_b}{T_0^2}$$

$$\ln a_A \cong -\frac{n_B}{n_A}$$

$$\ln a_A \cong -\frac{m M_A}{1000} \tag{5.56}$$

where a_A = activity of solvent
m = molality of solute (moles of solute per 1000 g of solvent)
M_A = molecular weight of solvent

Therefore, from Eqs. (5.55) and (5.56),

$$T_b - T_0 = \frac{RT_0^2 M_A}{1000\, \Delta\bar{H}_{\text{vap}}}\, m = K_b m \tag{5.57}$$

The boiling-point rise is directly proportional to the molality, just as in Eq. (5.39). The proportionality constant depends only on the properties of the solvent. In fact, Eq. (5.57) allows us to calculate values of K_b from known thermodynamic properties of the solvent. For water, we can use $T_0 = 373$ K, $M_A = 18.0$, and $\Delta\bar{H}_{\text{vap}} = 40.66$ kJ mol^{-1} at 373 K to calculate a value of $K_b = 0.51$ K m^{-1} (m = molal). This agrees well with the experimental value of 0.52 K m^{-1}, obtained by extrapolating to infinite dilution.

Freezing Point

Adding a solute to a solution decreases the activity of the solvent and, thus, lowers the melting point. The temperature of the solution must be lowered to keep it in equilibrium with the pure solid solvent. The equation is

$$\ln a_A = \frac{\Delta\bar{H}_{\text{fus}}}{R} \cdot \left(\frac{1}{T_0} - \frac{1}{T_f}\right) \tag{5.58}$$

where a_A = activity of pure solvent; if the solution is ideal, $a_A = X_A$
$\Delta\bar{H}_{\text{fus}}$ = enthalpy of fusion per mole
T_f = freezing point of solution
T_0 = freezing point of pure solvent

For a dilute solution,

$$T_0 - T_f = \frac{RT_0^2 M_A}{1000\, \Delta\bar{H}_{\text{fus}}}\, m \tag{5.59}$$

$$\Delta T_f = K_f m \tag{5.39}$$

Again, for water at $T_0 = 273$ K, $\Delta\bar{H}_{\text{fus}} = 6.007$ kJ mol^{-1}, and we can calculate $K_f = 1.875$ K m^{-1}, compared with the experimental value of 1.852 K m^{-1}.

When the solid in equilibrium with a solution is pure solute instead of pure solvent, we think of the equilibrium as a solubility. We can then derive an equation for the temperature dependence of solubility. For an ideal solution,

$$\ln \frac{X_B(T_2)}{X_B(T_1)} = \frac{-\Delta\bar{H}_{\text{sat}}}{R} \cdot \left(\frac{1}{T_2} - \frac{1}{T_1}\right) \tag{5.60}$$

where $X_B(T_2)$, $X_B(T_1)$ = solubility of solute at temperatures T_2, T_1
$\Delta\bar{H}_{\text{sat}} = \bar{H}_B(\text{solution}) - \bar{H}_B(\text{solid})$ = molar enthalpy (heat) of solution
of solute in saturated solution

This equation is most useful for slightly soluble solutes, because then the solutes will be dilute enough for the assumption of ideality to be valid. For dilute solutions the mole fraction is proportional to molarity and molality, so the ratio of solubilities in any convenient units can be used in Eq. (5.60). Usually, we expect solubility to increase with increasing temperature. However, we now see that the change of solubility with temperature actually depends on the sign of the heat of solution. Many ionic salts decrease in solubility with increasing temperature.

Osmotic Pressure

As derived earlier, the osmotic pressure of a solution is a measure of the activity of the solvent:

$$\ln a_A = \frac{-\pi \overline{V}_A}{RT} \tag{5.43}$$

Both vapor pressure and osmotic pressure can be measured at any temperature. They are therefore more useful methods than boiling-point or freezing-point measurements.

PHASE RULE

We have considered equilibria between various phases, but not the number of phases that may coexist or the constraints that apply to the thermodynamics of systems of more than one phase. Willard Gibbs deduced the thermodynamic implications of multiple phases in equilibrium and obtained the *phase rule*. This rule answers the questions in the first sentence above. The maximum number of phases that can coexist at equilibrium in the absence of external fields is equal to the number of components plus two. Pure water (one component) can have up to three phases in equilibrium. This can only occur at one temperature and pressure; it is called the *triple point*. For example, the triple point for the vapor, the liquid, and the solid of pure water occurs at $T = 0.0075°C$ and $P = 4.58$ Torr. In addition to the common form of solid water, ice, five other solid phases of water with different crystal structures have been detected at high pressure. The phase rule states that no more than three of these can coexist in equilibrium. If liquid and gaseous water are also present, only the familiar form of ice will be stable.

For a two-component system (NaCl in water) four phases can be in equilibrium. For example, at the freezing point of a saturated NaCl solution we could have solid NaCl, solid H_2O, the liquid solution, and H_2O vapor all in equilibrium, but again only at a unique value of T and P. A system that has the maximum number of coexistent phases is said to have no degrees of freedom; all the properties of the system are defined. The degrees of freedom of the systems that we normally consider are T and P and concentrations of the components. Every time a new phase appears in a system at equilibrium the number of degrees of freedom is reduced by one. We can adjust the T and P of pure gaseous water (one component, one phase) to any values that we choose over a wide range. But if we have liquid water in equilibrium with gaseous water (one component, two phases), the T and P are no longer independent. For each temperature we choose, only one pressure (the vapor pressure) will keep the liquid and gas in equilibrium. For equilibrium of solid, liquid, and gaseous water we cannot choose either T or P; the triple point is fixed by nature. The Gibbs phase rule that summarizes the discussion above is

$$F = C - P + 2 \qquad (5.61)$$

where F = number of degrees of freedom of the system
C = number of components of the system
P = number of phases at equilibrium in the system

To find the maximum number of phases for a given number of components, just set F equal to zero. The value of 2 in Eq. (5.61) occurs because we consider only two variables (T and P) besides concentrations. For each additional variable other than T and P that we consider (electric fields, magnetic fields, etc.), the number 2 in Eq. (5.61) is increased by 1.

SUMMARY

Phase Equilibrium

Equilibrium conditions:

$$T(\text{phase 1}) = T(\text{phase 2})$$
$$P(\text{phase 1}) = P(\text{phase 2}) \qquad (5.1)$$
$$\mu_A(\text{phase 1}) = \mu_A(\text{phase 2})$$

The last equation is modified to

$$\mu_{A,\,total}(\text{phase 1}) = \mu_{A,\,total}(\text{phase 2})$$

when effects such as surface tension, and electrical, gravitational, and centrifugal fields are included. See Table 5.3 for the relations between $\mu_{A,\,total}$ and μ_A for the various cases.

Free energy of transport of n moles of an uncharged molecule from phase 1 to phase 2:

$$\Delta G = nRT \ln \frac{a_A(\text{phase 2})}{a_A(\text{phase 1})} \cong nRT \ln \frac{c_A(\text{phase 2})}{c_A(\text{phase 1})} \qquad (5.5),\ (5.6)$$

Free energy of transport of n moles of an ion with charge Z from phase 1 to phase 2:

$$\Delta G = n\left[RT \ln \frac{a_A(\text{phase 2})}{a_A(\text{phase 1})} + FZV \right] \qquad (5.7)$$

F = Faraday = 96,485 J (eV)$^{-1}$
V = potential differerence between phases

Gibbs' phase rule (for T and P as only external variables):

$$F = C - P + 2 \qquad (5.61)$$

F = number of degrees of freedom
C = number of components
P = number of phases at equilibrium

Scatchard equation for binding to a macromolecule:

$$\frac{\text{v}}{[\text{A}]} = K \cdot (N - \text{v}) \tag{5.12}$$

v = average number of A's bound to macromolecule
N = number of binding sites on macromolecule

Scatchard equation for binding per monomer unit of a macromolecule:

$$\frac{r}{[\text{A}]} = K \cdot (n - r) \tag{5.13}$$

r = number of bound A per monomer unit of the macromolecule
n = number of binding sites per monomer unit of the macromolecule

Hill equation to assess the cooperativity of binding:

$$\frac{f}{(1 - f)} = K \cdot [\text{A}]^n \tag{5.14}$$

f = fraction of sites bound
n = Hill coefficient; it varies from 1 for no cooperativity to N for all-or-none binding of N ligands
K = a constant, not the binding constant for one ligand

Solutions

Gibbs' adsorption isotherm for adsorption at a surface of a solution:

$$\Gamma = -\frac{1}{RT}\frac{\partial \gamma}{\partial \ln c} \tag{5.22}$$

Γ = adsorption (excess concentration) of solute at surface, mol m^{-2}
γ = surface tension, N m^{-1} = J m^{-2}
R = gas constant = 8.314 J K^{-1} mol^{-1}
c = concentration of solute in bulk solution

Raoult's law for solvent:

$$P_A = X_A P_A^0 \tag{5.35}$$

P_A = vapor pressure of solvent in solution with mole fraction X_A

$$X_A = \frac{n_A}{n_A + n_B}$$

P_A^0 = vapor pressure of pure solvent

Henry's law for solute:

$$P_B = k_B X_B \tag{5.36}$$

k_B = Henry's law constant (from tables)

Clausius-Clapeyron equation for the change of the boiling point with pressure and the change of the vapor pressure with temperature:

$$\ln \frac{P_2}{P_1} = \frac{-\Delta \bar{H}_{vap}}{R} \cdot \left(\frac{1}{T_2} - \frac{1}{T_1} \right)$$ (5.32)

$\Delta \bar{H}_{vap}$ = molar enthalpy of vaporization

Pressure dependence of the freezing point:

$$\frac{T_2 - T_1}{P_2 - P_1} = T \frac{\Delta \bar{V}_{fus}}{9.869 \text{ cm}^3 \text{ atm J}^{-1} \, \Delta \bar{H}_{fus}}$$ (5.34)

$\Delta \bar{V}_{fus}$ = molar volume (cm^3 mol^{-1}) of fusion
$\Delta \bar{H}_{fus}$ = molar heat (J mol^{-1}) of fusion
T = average of T_2 and T_1

P must be in atmospheres to be consistent with the unit conversion factor.

Concentration units:

mole fraction: $X_B = \dfrac{n_B}{n_A + n_B}$

molality: m = moles of solute per 1000 g of solvent
molarity: c = moles of solute per liter of solution

Dilute solutions:

$$X_B = \frac{M_A m}{1000}$$ (5.40)

$$c = \rho_0 m$$ (5.41)

M_A = molecular weight of solvent of density ρ_0

Boiling-point rise:

$$T_b - T_0 = K_b m = \frac{M_A R T_0^2}{1000 \, \Delta \bar{H}_{vap}} m$$ (5.39), (5.57)

For H_2O, $K_b = 0.51$.

Freezing-point lowering:

$$T_0 - T_f = K_f m = \frac{M_A R T_0^2}{1000 \, \Delta \bar{H}_{fus}} m$$ (5.38), (5.59)

For H_2O, $K_f = 1.86$.

Osmotic pressure (in atm):

$$\pi = \frac{n_B RT}{V} = cRT \qquad \text{(5.45a), (5.45b)}$$

$R = 0.08205$ L atm K^{-1} mol^{-1}
$V =$ volume of solution containing n_B moles solute, L
$c =$ concentration, mol L^{-1}

Number-average molecular weight:

$$M_n = \frac{\sum\limits_{i} n_i M_i}{\sum\limits_{i} n_i} \qquad \text{(5.50)}$$

Weight-average molecular weight:

$$M_w = \frac{\sum\limits_{i} \text{wt}_i M_i}{\sum\limits_{i} \text{wt}_i} \qquad \text{(5.52)}$$

$n_i =$ number of molecules with molecular weight M_i
$\text{wt}_i =$ weight of all materials with molecular weight M_i

Number-average molecular weights for solutes are obtained from colligative properties in dilute solution.

Osmotic-pressure molecular weight:

$$M = \frac{wRT}{\pi} \qquad \text{(5.49)}$$

$w =$ concentration of solution, g L^{-1}

Vapor-pressure-lowering molecular weight:

$$M = \frac{\text{wt of solute}}{\text{moles of solvent}} \cdot \frac{P_A^0}{P_A^0 - P_A} \qquad \text{(5.46)}$$

$P_A =$ vapor pressure of solution
$P_A^0 =$ vapor pressure of pure solvent

Boiling-point and freezing-point molecular weights:

$$M = \frac{\text{wt of solute}}{\text{kg solvent}} \cdot \frac{K_b}{\Delta T_b} \qquad \text{(5.47)}$$

$$M = \frac{\text{wt of solute}}{\text{kg solvent}} \cdot \frac{K_f}{\Delta T_f} \qquad \text{(5.48)}$$

$\Delta T_b, \Delta T_f =$ change in boiling point or freezing point
$K_b, K_f =$ constants from tables

The activity of the solvent is obtained from colligative properties.

Osmotic pressure, π:

$$\ln a_A = \frac{-\pi \overline{V}_A}{RT} \tag{5.43}$$

$\overline{V}_A$ = molar volume of solvent

Vapor pressure:

$$a_A = \frac{P_A}{P_A^0} \tag{5.54}$$

Boiling point and freezing point:

$$\ln a_A = \frac{\Delta \overline{H}_{vap}}{R} \cdot \left(\frac{1}{T_b} - \frac{1}{T_0} \right) \tag{5.55}$$

$$\ln a_A = \frac{\Delta \overline{H}_{fus}}{R} \cdot \left(\frac{1}{T_0} - \frac{1}{T_f} \right) \tag{5.58}$$

$\Delta \overline{H}_{vap}, \Delta \overline{H}_{fus}$ = molar heats of vaporization, fusion

REFERENCES

Branton, D., and D. Deamer, 1972. *Membrane Structure,* Springer-Verlag, New York.
Nicholls, D. G., 1982. *Bioenergetics, An Introduction to Chemiosmotic Theory,* Academic Press, New York.
Tanford, C., 1980. *The Hydrophobic Effect: Formation of Micelles and Biological Membranes,* 2nd ed., John Wiley, New York.

SUGGESTED READINGS

Dickerson, R. E., and I. Geis, 1983. *Hemoglobin: Structure, Function, Evolution, and Pathology,* Benjamin/Cummings, New York.
Linder, M. E., and A. G. Gilman, 1992. G Proteins, *Sci. Am. 267* (July), 56–65. G proteins are membrane-bound proteins which cause cellular responses to chemical messengers.
Singer, S. J., and G. L. Nicholson, 1972. The Fluid Mosaic Model of the Structure of Cell Membranes, *Science 175,* 720–731.
Tosteson, D. C., 1981. Lithium and Mania, *Sci. Am. 244* (April), 164–174.
Ward, P., L. Greenwald, and O. E. Greenwald, 1980. The Buoyancy of the Chambered Nautilus, *Sci. Am. 243* (October), 190–203.
Wolfenden, R., 1983. Waterlogged Molecules, *Science 222,* 1087–1093.

PROBLEMS

1. (a) How many kg of water can be evaporated from a 1 square meter surface of water in a lake on a hot, clear summer day, if it is assumed that the limiting factor is the sun's heat of 4 J min^{-1} cm^{-2} for an 8-hour day. Assume that the air and water temperatures remain constant at 40°C.

   (b) How many liters of air are needed to hold this much water vapor at 40°C?

   (c) When the sun goes down the air temperature drops to 20°C. Assuming that the excess water vapor is precipitated as rain, calculate the weight of the rain and the amount of heat released when the water condenses. When it starts raining do you expect the temperature of the air to rise or fall? Explain.

   (d) Calculate the vapor pressure of liquid water at 200°C. Assume that the heat of vaporization does not change from its value at 100°C.

   (Table 2.2 is useful for this problem.)

2. The vapor pressure of pure, solid pyrene at 25°C is P_1 atm. The solubility at 25°C of pyrene in water is 1.0×10^{-4} M; the solubility of pyrene in ethanol at 25°C is greater than 10^{-4} M. Neither water nor ethanol dissolves significantly in pyrene.

   (a) What is the vapor pressure of pyrene above a saturated solution of pyrene in water at 25°C?

   (b) If ethanol is added to the saturated aqueous solution, more pyrene dissolves. When equilibrium with the solid pyrene is reached again, will the vapor pressure of the pyrene above the new solution be greater than, less than, or the same as in part (a)?

   (c) The solubility of pyrene at 25°C is 1.1×10^{-3} M in an aqueous solution of 0.10 M of cytosine. This increase in solubility is due to the formation of a complex between pyrene and cytosine. Calculate the values of the equilibrium constants for the following reactions at 25°C, where P = pyrene and C = cytosine. Use concentration in M instead of activities.

   $$P(\text{solution}) + C(\text{solution}) = P \cdot C(\text{solution}) \quad (1)$$
   $$P(\text{solid}) + C(\text{solution}) = P \cdot C(\text{solution}) \quad (2)$$

3. (a) The concentration of creatine in blood is 20 mg L^{-1} and in urine is 750 mg L^{-1}. Calculate the free energy per mole required for its transfer from blood to urine at 37°C.

   (b) To determine the molecular weight of creatine, a sample was purified and used to prepare a solution containing 0.1 g L^{-1}. This solution exhibited an osmotic pressure against pure water of 13.0 Torr at 25°C. What is the molecular weight of creatine?

   (c) Using your answer to part (b), calculate the vapor-pressure lowering of water above a solution containing 0.1 g of creatine per liter at 25°C. The vapor pressure of pure water at 25°C is 23.8 Torr.

4. Myoglobin is a skeletal muscle protein that binds oxygen. The standard free energy for the reaction

   $$\text{myoglobin} + O_2 \longrightarrow \text{oxymyoglobin}$$

   is $\Delta G^{0'} = -30.0$ kJ mol^{-1} at 298 K and pH 7. The standard state of O_2 is the dilute solution, molarity scale; therefore the concentration of O_2 must be in M.

   What is the ratio (oxymyoglobin)/(myoglobin) in an aqueous solution at equilibrium with a partial pressure of oxygen $P(O_2) = 30$ Torr? Assume ideal behavior of O_2 gas, activity coefficient for the protein in solution = 1, and that Henry's law holds for O_2 dissolved in water.

5. According to the chemiosmotic theory, an electrochemical proton gradient is used to synthesize ATP in mitochondria. The enzyme that does this is located on the inside of the mitochondrial membrane. The oxidation of carbohydrates and fats is used to pump protons outside the

mitochondrial membrane until the steady state membrane potential is -140 mV and the pH gradient is 1.5 pH units. Inside the mitochondrion pH = 7.0, [ATP] = 1 mM, [P$_i$] = 2.5 mM, [ADP] = 1 mM, and $T = 298$ K.

(a) How much free energy is required to synthesize ATP inside the mitochondria?

(b) How much free energy is made available by moving one mol of protons from the outside to the inside? Is this enough to drive ATP synthesis?

(c) How many protons must be translocated per ATP synthesized?

6. In living biological cells the concentration of sodium ions inside the cell is kept at a lower concentration than the concentration outside the cell, because sodium ions are actively transported from the cell.

(a) Consider the following process at 37°C and 1 atm.

$$1 \text{ mol NaCl}(0.05 \ M \text{ inside}) \longrightarrow 1 \text{ mol NaCl}(0.20 \ M \text{ outside})$$

Write an expression for the free energy change for this process in terms of activities. Define all symbols used.

(b) Calculate $\Delta \bar{G}$ for the process. You may approximate the activities by concentrations in M. Will the process proceed spontaneously?

(c) Repeat the calculation in part (b), except move 3 mol of NaCl from inside to outside.

(d) Calculate $\Delta \bar{G}$ for the process if the activity of NaCl inside is equal to that outside.

(e) Calculate $\Delta \bar{G}$ for the process at equilibrium.

(f) The standard free energy for hydrolysis of ATP to ADP (ATP + H$_2$O → ADP + phosphate) in solution is $\Delta G^0 = -31.0$ kJ mol^{-1} at 37°C, 1 atm. The free energy of this reaction can be used to power the sodium-ion pump. For a ratio of ATP to ADP of 10, what must be the concentration of phosphate to obtain -40 kJ mol^{-1} for the hydrolysis? Assume that activity coefficients are 1 for the calculation.

(g) If the ratio of ATP to ADP is 10, what is the concentration of phosphate at equilibrium? Assume that $\gamma = 1$ for the calculation. What do you conclude from your answer?

7. The binding of a ligand A to a macromolecule was studied by equilibrium dialysis. In each measurement a $1.00 \times 10^{-6} \ M$ solution of the macromolecule was dialyzed against an excess amount of a solution containing A. After equilibrium was reached, the total concentration of A on each side of the dialysis membrane was measured. The following data were obtained:

Total concentration of A (M)

The side without macromolecules	The side with $1.00 \times 10^{-6} \ M$ macromolecules
0.51×10^{-5}	0.67×10^{-5}
1.02×10^{-5}	1.28×10^{-5}
2.01×10^{-5}	2.34×10^{-5}
5.22×10^{-5}	5.62×10^{-5}
$10.5 \ \ \times 10^{-5}$	10.91×10^{-5}
$20.0 \ \ \times 10^{-5}$	20.47×10^{-5}

From these data, calculate v as a function of [A]. Make a Scatchard plot, and evaluate the intrinsic equilibrium constant K and the total number of sites per macromolecule, if the independent-and-identical-sites model appears to be applicable.

8. Plot $v/[A]$ versus v for the following cases:

(a) There is a total of 10 identical and independent sites per polymer, and the intrinsic binding constant is $K = 5 \times 10^5 \ M^{-1}$.

(b) There is a total of 10 independent sites per polymer. Nine of the 10 sites are identical, with $K = 5 \times 10^5 \ M^{-1}$. The binding constant for the tenth site is $5 \times 10^6 \ M^{-1}$. For more than one type of site, the Scatchard equation can be generalized to

$$\frac{v}{[A]} = \sum_i \frac{N_i K_i}{1 + K_i[A]}$$

[Note that the deviation from linearity is small for part (b). Because experimental error in such measurements is usually appreciable, one should be cautious in concluding from a linear Scatchard plot that the sites are identical.]

(c) Repeat the calculation and plot of part (b) for five identical sites of each type.

9. The enzyme tetrahydrofolate synthetase has four identical and independent binding sites for its substrate ATP. For an equilibrium dialysis experiment a solution of the enzyme was prepared. First the osmotic pressure of this solution was measured to be 2.4×10^{-3} atm at 20°C, using an osmometer. Then the enzyme solution was placed in a dialysis bag and the binding of ATP to the enzyme was measured by equilibrium dialysis at the same temperature. In one run, after the binding equilibria was established, the concentration of free ATP outside the bag was found to be $1.0 \times 10^{-4} \ M$, and the total ATP concentration inside the bag was found to be $3.0 \times 10^{-4} \ M$.

(a) On the basis of the information above, calculate K, the equilibrium constant for the binding of ATP to the enzyme at 20°C.

(b) The equilibrium dialysis experiment was repeated for a series of outside ATP concentrations at two different temperatures (20°C and 37°C). The following Scatchard plots were obtained:

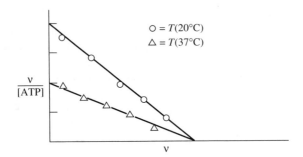

Using this information, calculate ΔH^0 and ΔS^0 for the binding of ATP to its site on the enzyme. (Assume that ΔH^0 and ΔS^0 are independent of temperature.)

10. (a) Using the Henry's law constants in Table 5.4, calculate the solubility (in M) of each gas in water at 25°C if $P_{O_2} = 0.2$ atm, $P_{N_2} = 0.75$ atm, and $P_{CO_2} = 0.05$ atm.

(b) What will be the vapor pressure at 25°C of the water in this solution if Raoult's law holds? The vapor pressure of pure water at 25°C is 23.756 Torr.

11. (a) Assume that Raoult's law holds and calculate the boiling point of a 2 M solution of urea in water.

(b) Urea actually forms complexes in solution in which two or more molecules hydrogen bond to form dimers and polymers. Will this effect tend to raise or lower the boiling point?

12. When cells of the skeletal vacuole of a frog were placed in a series of aqueous NaCl solutions of different concentrations at 25°C, it was observed microscopically that the cells remained unchanged in 0.7 wt % NaCl solution, shrank in more concentrated solutions, and swelled in more dilute solutions. Water freezes from the 0.7% salt solution at −0.406°C.
(a) What is the osmotic pressure of the cell cytoplasm relative to that of pure water at 25°C?
(b) Suppose that sucrose (mol wt 342) was used instead of NaCl (mol wt 58.5) to make the isoosmotic solution. Estimate the concentration (wt %) of sucrose that would be sufficient to balance the osmotic pressure of the cytoplasm of the cell. Assume that sucrose solutions behave ideally.

13. On the planet Taurus II, ammonia plays a role similar to that of water on the earth. Ammonia has the following properties: Normal (1 atm) boiling point is −33.4°C, where its heat of vaporization is 1368 kJ kg^{-1}. Normal freezing point is −77.7°C.
(a) Estimate the temperature at which the vapor pressure of NH_3(liquid) is 60 Torr.
(b) Calculate the entropy of vaporization of NH_3 at its normal boiling point. Compare this value with that predicted by Trouton's rule ($\Delta S_{vap} \cong 88$ J K^{-1}) for typical liquids. Give a likely reason for any discrepancy.
(c) Ammonia undergoes self-dissociation according to the reaction

$$2\ NH_3(l) \ \rightleftarrows \ NH_4^+(am) \ + \ NH_2^-(am)$$

where (am) refers to the "ammoniated" ions in solution. At −50°C, the equilibrium constant for this reaction is

$$K \ = \ (NH_4^+)(NH_2^-) \ = \ 10^{-30}$$

Calculate ΔG_{223}^0 for the self-dissociation. What is the standard state for which $a_{NH_3} = 1$?
(d) When 1 mol of ammonium chloride, NH_4Cl, is dissolved in 1 kg of liquid ammonia, the boiling point at 760 Torr is observed to occur at −32.7°C. What conclusions can you reach about the nature of this solution? State the evidence for your answer.

14. One of the important factors responsible for geologic evolution (erosion in particular) is the pressure developed by the freezing of water trapped in enclosed spaces in rock formations. Estimate the maximum pressure that can be developed by water freezing to ice on a cold winter's night ($T = -10$°C). The densities of ice and water at −10°C may be taken as 0.9 g cm^{-3} and 1.0 g cm^{-3}, respectively.

15. Proteins may be hydrolyzed to amino acids by heating in very dilute aqueous NaOH solution. The protein itself is also very dilute. The reaction is carried out at 120°C by heating the solution in a sealed tube which was evacuated prior to sealing. What internal pressure must the tube withstand at this temperature? The normal boiling point and $\Delta \bar{H}_{vap}$ of water may be taken as 100°C and 40.66 kJ mol^{-1}, respectively.

16. Indicate whether each of the following statements is true or false as it is written. Alter each false statement so as to make it correct.

(a) If two aqueous solutions containing different nonvolatile solutes exhibit exactly the same vapor pressure at the same temperature, the activities of water are identical in the two solutions.

(b) The most important reason that foods cook more rapidly in a pressure cooker is that equilibrium is shifted in the direction of products at the higher pressure.

(c) If solutions of a single solute are prepared at equal molalities but in different solvents, the freezing-point lowering will be the same for the different solutions, assuming that each behaves ideally.

(d) If solutions of a single solute are prepared at equal mole fractions but in different solvents, the fractional vapor-pressure lowering will be the same for the different solutions, assuming that each behaves ideally.

(e) A saturated solution of sodium chloride in liquid water has one degree of freedom.

(f) If two liquids, such as benzene and water, are not completely miscible in one another, the mixture can never be at equilibrium.

17. A solution of the sugar mannitol (mol wt 182.2) is prepared by adding 54.66 g of mannitol to 1000 g of water. The vapor pressure of pure liquid water is 17.54 Torr at 20°C. Mannitol is nonvolatile and does not ionize in aqueous solutions.

(a) Assuming that aqueous mannitol solutions behave ideally, calculate the vapor-pressure lowering (the difference between the vapor pressure of pure water and that of the solution) for the above solution at 20°C.

(b) The *observed* vapor-pressure lowering of the mannitol solution above is 0.0930 Torr. Calculate the activity coefficient (based on mole fraction) of water in this solution.

(c) Calculate the osmotic pressure of the mannitol solution of part (b) when it is measured against pure water, and compare it with the osmotic pressure of the ideal solution.

18. (a) Assuming that glucose and water form an ideal solution, what is the equilibrium vapor pressure at 20°C of a solution of 1.00 g of glucose (mol wt 180) in 100 g of water? The vapor pressure of pure water is 17.54 Torr at 20°C.

(b) What is the osmotic pressure, in Torr, of the solution in part (a) versus pure water?

(c) What is the activity of water as a solvent in such a solution?

(d) What would be the osmotic pressure, versus pure water, of a solution containing both 1.00 g of glucose and 1.00 g of sucrose (mol wt 342) in 100 g of water at 20°C?

19. Workers in underwater caissons or diving suits necessarily breathe air at greater than normal pressures. If they are returned to the surface too rapidly, N_2, dissolved in their blood at the previously higher pressure, comes out of solution and may cause emboli (gas bubbles in the bloodstream), bends, and decompression sickness. Blood at 37°C and 1 atm pressure dissolves 1.3 mL of N_2 gas (measured for pure N_2 at 37°C and 1 atm) in 100 mL of blood. Calculate the volume of N_2 likely to be liberated from the blood of a caisson worker returned to 1 atm pressure after prolonged exposure to air pressure at 300 m of water (below the surface). The total blood volume of the average adult is 3.2 L; density of mercury is 13.6 g mL^{-1}; air contains 78 vol % N_2.

20. Hexachlorobenzene, C_6Cl_6, is a solid compound that is somewhat volatile, but not very soluble in water; water is not soluble in C_6Cl_6. Solid C_6Cl_6 and liquid water are allowed to reach equilibrium at 20°C to form a solid phase, an aqueous phase, and a gaseous phase. The vapor pressure of pure solid C_6Cl_6 is 1.0×10^{-2} Torr at 20°C and 1.0×10^{-1} Torr at 40°C.

(a) A sample of the equilibrium aqueous phase was removed and the equilibrium concentration of C_6Cl_6 was found to be mole fraction, $X = 1.00 \times 10^{-5}$ and molarity $c = 55.6 \times 10^{-5}$. What are the vapor pressures of H_2O and C_6Cl_6 for this phase at 20°C?

(b) C_6Cl_6 binds to proteins. A protein is added to the aqueous solution and equilibrium with solid C_6Cl_6 is established. State whether each of the following increases, decreases, or remains unchanged compared to (a):

(1) The vapor pressure of C_6Cl_6 for the aqueous solution.

(2) The vapor pressure of water for the aqueous solution.

(3) The concentration of dissolved (free) C_6Cl_6 in the aqueous solution.

(c) What is the mole fraction of C_6Cl_6 in the gas phase at equilibrium at 20°C?

(d) Calculate the heat of vaporization of solid C_6Cl_6.

21. The solubilities of two amino acids in two solvents at 25°C are given below; they are the concentrations present in saturated solutions.

	In water	In ethanol
Glycine	3.09 M	$4.04 \times 10^{-4}\ M$
Valine	0.60 M	$1.32 \times 10^{-3}\ M$

(a) Calculate the standard free energy of transfer, $\Delta\mu^0$, of 1 mol of glycine from the solid to the aqueous solution at 25°C. The standard state for the solid is the pure solid; the standard state in the solution corresponds to $a = 1$ extrapolated from a dilute solution; the molarity concentration scale is used. You may consider the solutions to be ideal.

(b) Calculate the standard free energy of transfer, $\Delta\mu^0$, of 1 mol of glycine from ethanol to the aqueous solution at 25°C.

(c) Assume that the effects of the backbone and side chain are simply additive (glycine essentially has no side chain), and calculate the standard free energy of transfer of 1 mol of the $(CH_3)_2CH$— side chain of valine from water to ethanol at 25°C.

(d) Ethanol can be considered to mimic the interior of a protein. Will the mutation of a glycine to a valine in the interior of a protein favor the folding of the protein if interaction with the solvent is the dominant effect? Explain.

(e) Assume that the protein folding equilibrium is a two-state transition: folded or unfolded. Calculate the change in equilibrium constant due to the mutation from glycine to valine for folding at 25°C, if interaction with the solvent is the dominant effect.

22. The human red-blood-cell membrane is freely permeable to water but not at all to sucrose. The membrane forms a completely closed bag, and it is found by trial and error that adding the cells to a solution of a particular concentration of sucrose results in neither swelling nor shrinking of the cells. In a separate experiment, the freezing point of that particular sucrose solution is found to be −0.56°C.

(a) If the exterior of the cell contains only KCl solution, estimate the KCl concentration. Assume that the membrane is also impermeable to KCl.

(b) If these cells are suspended in distilled water at 0°C, what would be the internal hydrostatic pressure at equilibrium, assuming that the cell did not change volume?

23. (a) What is the osmotic pressure in atm of a 0.1 M solution of urea in ethanol compared with that of pure ethanol at 27°C? State what assumptions you made in the calculation.

(b) What is the osmotic pressure in atm of a 0.1 M solution of NaCl in water compared with that of pure water at 27°C? State what assumptions you made in the calculation.

(c) An aqueous solution of osmotic pressure 5 atm is separated from an aqueous solution of osmotic pressure 2 atm. Which way will the solvent flow? How large a pressure must be applied to prevent this solvent flow?

(d) What is the free energy change for transferring 1 mol of water from a solution of osmotic pressure 5 atm to a solution of osmotic pressure 2 atm?

24. A protein solution containing 0.6 g of protein per 100 mL of solution has an osmotic pressure of 22 mm of water at 25°C and at a pH where the protein has no net charge (the isoelectric point). What is the molecular weight of the protein?

25. A climber has carefully measured the boiling point of water at the top of a mountain and found it to be 95°C. Calculate the atmospheric pressure at the mountaintop. The boiling point of water at 1 atm pressure is 100°C, and the heat absorbed in vaporizing 1 mol of water at constant pressure is 40.66 kJ mol^{-1}.

26. A cell membrane at 37°C is found to be permeable to Ca^{2+}, and analysis shows the inside concentration to be 0.100 M and the outside concentration to be 0.001 M in Ca^{2+}.

(a) What potential difference in volts would have to exist across the membrane for Ca^{2+} to be in equilibrium at the stated concentrations? Assume that activity coefficients are equal to 1. Give the sign of the potential inside with respect to that outside.

(b) If the measured inside potential is +10 mV with respect to the outside, what is the minimum (reversible) work required to transfer 1 mol of Ca^{2+} from outside to inside under these conditions?

27. The water in freshwater lakes is supplied by transfer from the oceans. Consider the oceans to be a 0.5 M NaCl solution and freshwater lakes to be a 0.005 M solution of $MgCl_2$. For simplicity, consider the salts to be completely dissociated, and the solutions sufficiently dilute so that all the equations for dilute solutions apply:

(a) Calculate the osmotic pressure of the ocean water and lake water at 25°C.

(b) Give a simple relation between the osmotic pressure and the freezing point depression of a solution by which you could calculate ΔT_f without knowing the freezing-point depression constant for the solvent.

(c) How much free energy is required to transfer a mole of pure water from the ocean to the lake at 25°C? (In the actual process, the sun supplies this energy.)

(d) Which solution, the ocean or the lake, has the highest vapor pressure of water?

(e) The observed water vapor pressure at 100°C for 0.5 M NaCl is 747.7 Torr. What is the activity of water at this temperature? The vapor pressure of pure water at 100°C is 760 Torr.

28. For each of the following processes state whether ΔG, ΔH, and ΔS of the system increase, decrease, or do not change. If there is not enough information given, state what experimental measurements need to be made to allow you to deduce the direction of the change.
   (a) A system contains a 1 M sucrose solution and pure water separated by a membrane permeable to water only. A small amount of water is transferred through the membrane from the pure water to the sucrose solution at constant temperature and pressure.
   (b) Liquid water at 100°C, 1 atm pressure is evaporated to water vapor at 100°C, 1 atm pressure. The system is the water.
   (c) A system contains pure solid water at 0°C suspended in a 0.01 M KCl solution at the same temperature. The pressure on the system is 1 atm. Some of the liquid water is frozen to ice at constant temperature and pressure. The system is the solid water plus the KCl solution.
   (d) Water vapor at 1 atm pressure and 25°C is converted to liquid water at the same temperature and pressure along a reversible path. The system is the water.

29. For the purpose of this problem a red blood cell will be considered to be a 0.15 M solution of NaCl (the cytoplasm) surrounded by a membrane permeable to water only. NaCl is completely dissociated in water. Assume that all solutions behave ideally.
   (a) If the cell is suspended in an aqueous solution of sucrose at 25°C, what must be the molar concentration of sucrose to prevent the entry or exit of water from the cytoplasm?
   (b) If the cell is suspended in a 0.10 M sucrose solution in an open container at 25°C, what will be the pressure within the cell at equilibrium? Assume that the membrane is rigid and the movement of water does not dilute the solutions. Take the partial molar volume of water to be 0.018 L mol^{-1}, and, if you wish to use them, the mole fractions of water in 0.15 M NaCl and 0.1 M sucrose are 0.995 and 0.998, respectively.

30. A spherical cell 1 μm in diameter divides into two spherical cells.
   (a) Calculate the work necessary to increase the surface area if its surface tension is 12.3×10^{-3} N m^{-1}. Assume that the volume of the daughter cells taken together is the same as that of the parent.
   (b) If the hydrolysis of ATP in this cell yields 25 kJ mol^{-1} of free energy, how many molecules of ATP are needed for the division above?

31. A 1.0 M sucrose solution and a 0.5 M sucrose solution (both in water) in separate open containers are placed next to each other in a sealed vessel at constant temperature. What, if anything, will occur? Explain your answer briefly.

32. The measured osmotic pressure of seawater is 25 atm at 273 K.
   (a) What is the activity of water in seawater at 273 K? Take the partial molar volume of water in seawater to be 0.018 M^{-1}.
   (b) The vapor pressure of pure water at 273 K is 4.6 Torr. What is the vapor pressure of water in seawater at this temperature?
   (c) What temperature would you expect to find for seawater in equilibrium with the polar ice caps? The melting point of pure water at 1 atm pressure is 0°C, and $\Delta \bar{H}_{fus}$ may be taken as 5.86 kJ mol^{-1}, independent of temperature.

33. The following device is a candidate for a perpetual motion machine.

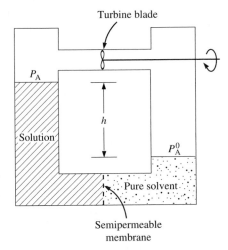

Turbine blade

P_A

h

Solution

P_A^0

Pure solvent

Semipermeable
membrane

(1) All parts of the machine are at the *same* temperature T, equal to that of the heat bath surrounding the machine.

(2) The compartment at the left contains a solution, separated from the compartment at the right (containing pure solvent) by a semipermeable membrane, permitting solvent transfer only. A pressure difference $\Delta P = \rho g h$ balances the osmotic pressure difference π across the membrane.

(3) The vapor pressure P_A^0 above pure solvent is higher than the pressure P_A above the solution (Raoult's law). There is therefore a pressure difference $\Delta P = P_A^0 - P_A$ in the tube, and the flow of vapor in the upper connecting tube, from right to left, drives a turbine blade.

(4) Vapor moving through the connecting tube condenses at the surface of the solution. It tends to dilute the solution, therefore reducing its osmotic pressure. This means that solvent will flow back through the semipermeable membrane, restoring the original concentration of the solution.

(5) The solvent is therefore set into perpetual motion, flowing as vapor from right to left in the upper part, and from left to right in the lower part, yielding energy.

Analyze the operation of this machine carefully. First, decide whether this machine works or fails. Then defend your answer as quantitatively and clearly as possible.

6

Molecular Motion and Transport Properties

CONCEPTS

We have learned how the thermodynamic properties of molecules—energy and entropy—are related to the energy per molecule, to molecular interactions, and to the randomness of the collection of molecules (Chapters 2 and 3). Now we will think about properties that depend directly on how fast molecules travel.

The random motion of molecules is dependent on the temperature. At 0 K molecular motion ceases except for quantum mechanical zero-point vibration. As the temperature is raised molecules increasingly vibrate, rotate, and translate—they move. It is easiest to analyze their motion in the gas phase where molecular interactions are least important. For ideal gases we find that the velocity of a molecule just depends on the temperature and the molecular weight of the molecule. Of course all molecules in a gas do not have the same velocity. A molecule may move forward, collide, change direction, speed up, slow down, and so forth. We therefore can calculate only the probability that a molecule will have a particular velocity, or we can calculate the average velocity for all the molecules. The *Maxwell-Boltzmann theory* allows us to understand molecular velocities and kinetic energy.

In a liquid the random motion of molecules depends on the temperature, but because collisions are so important in liquids, molecular motion also depends on the sizes and shapes of molecules. Big molecules will move slower than small ones, and for two molecules with the same volume, spherical molecules will move faster than others. The random motion of molecules is *diffusion;* by measuring the diffusion we learn about size and shape.

If a force is applied to a molecule its motion is no longer random; it is directed. On earth's surface all molecules experience the force of gravity pulling down on them. They will tend to sediment to the bottom of a container unless their random motion—their diffusion—keeps the solution mixed. Light molecules in solution will remain uniformly mixed;

heavy molecules will sediment to the bottom. Earth's gravitational field is not very strong, so only large particles, such as viruses or chromosomes, will sediment significantly. By using a centrifugal field, which can be many orders of magnitude larger than earth's field, any molecule can be sedimented. *Sedimentation velocity* is the motion of a particle in a gravitational or centrifugal field; its rate depends on the mass plus the size and shape of the particle. *Sedimentation equilibrium* refers to the distribution of molecules in a gravitational or centrifugal field after equilibrium has been reached. The balance between the sedimenting force in one direction and the randomizing effect of diffusion means that a concentration gradient is formed; there is a higher concentration near the bottom of the container. The concentration gradient at sedimentation equilibrium depends on the molecular weight of the molecules.

The *viscosity* of a fluid is its resistance to flow. Flow means that one part of a fluid is moving faster than another part. For example in flowing through a tube, the part of the fluid at the wall is stationary, whereas the part in the middle of the tube is flowing most rapidly. The molecules in different parts of the fluid have different average velocities in the direction of flow. For a solvent the viscosity depends on the interactions between the solvent molecules. For example the three hydrogen bond donors and acceptors in glycerol, [$HOCH_2CH(OH)CH_2OH$], make it more than 1000 times more viscous than water. When a macromolecule is added to a solvent, the viscosity increases. The amount of increase depends on the concentration, plus the size and shape of the molecule.

Electrophoresis is the motion of charged molecules in an electric field. It depends on the charge on the molecule, plus its size and shape. For proteins and nucleic acids with many ionizable groups, the charge on the molecule can be manipulated by changing the pH of the solution. In aqueous solutions it is very difficult to quantitate the effective electric field and effective charge on a macromolecule because of all the other ions present. Thus, electrophoresis is used to measure relative sizes and shapes of molecules rather than absolute values.

All the properties we have described depend on the sizes and shapes of macromolecules; the properties are called *transport properties*. Diffusion and sedimentation measure mass transport; viscosity measures momentum transport; and electrophoresis measures charge transport. We care about transport properties because they tell us what the molecules look like in solution. Are they big or small, compact or extended, rigid or flexible? Furthermore, diffusion and flow are important methods of transport of molecules in biological cells, in animals, and in plants. We need to know how fast molecules can move from one place to another in a cell, for example, and how much their speed changes when the temperature or molecular size changes.

APPLICATIONS

Transport properties are used to analyze, separate, and purify cellular particles, proteins, and nucleic acids. Sedimentation provides fractionation based on differences in sedimentation coefficients, which depend on the mass of the particle, its shape, and its buoyancy. Gel electrophoresis is used to separate native proteins that differ by only one charge, or denatured proteins that differ by one peptide unit. Two-dimensional gel electrophoresis has the ability to show nearly every protein present in a human cell. The changes in the pattern of proteins produced during differentiation of cells can be followed. Gel electrophoresis can separate nucleic acids that differ by one nucleotide, and thus determine their sequence, or separate DNA fragments millions of base pairs in length. Both of these abilities are crucial for the Human Genome Project, whose goal is to obtain the complete sequence of all human chromosomes.

The absolute molecular weights of macromolecules can be determined by a combination of sedimentation and diffusion, or by sedimentation equilibrium. Relative molecular weights can be determined easily by gel electrophoresis. The molecular weights of macromolecules are a key to their identification. The interactions of macromolecules to form higher order structures can be characterized by the molecular weight of the complex.

The shapes of macromolecules and how they change in different environments can be followed by any of the transport properties. Thus all sorts of reactions of macromolecules are studied by measurements of diffusion, sedimentation, viscosity, or electrophoresis.

KINETIC THEORY

We consider gases first because they are the simplest to understand, but the concepts and methods we learn are applicable to more complicated problems like liquids and solutions. In a gas at ordinary pressures, the molecules move freely in a volume much greater than their size. While the molecules in 1 g of liquid water or ice are crowded into a volume of approximately 1 cm^3, when vaporized at 100°C, the same number of molecules occupy a volume of 1700 cm^3 at 1 atm. The molecules in a dilute gas are usually far apart from one another, and consequently, they interact only weakly. Therefore, as an approximation we can assume that the molecules are moving about more or less randomly in the gas phase; the motion of one molecule is little affected by the others unless it encounters another molecule in a direct collision.

For the time being we will not consider molecular collisions. We can assume that the molecules are vanishingly small, or that the gas is so dilute, that collisions between molecules are highly improbable. Molecules in such a gas are nevertheless colliding with the walls of the container, and thereby exerting pressure on the walls.

Velocities of Molecules, Translational Kinetic Energy, and Temperature

We will relate the pressure of a gas to the collisions of the gas molecules on the walls of the container. This will provide a relation between the pressure-volume product of the gas and the velocities of the gas molecules. Comparing this expression with the ideal gas equation, we learn how the translational kinetic energy and the average velocity of the molecules depend on temperature. This is a very important result which can be applied to liquids and solids. In condensed phases the molecules are continually colliding and interacting. There is constant interchange of kinetic and potential energy. However, if we consider the motion of each atom in each of the molecules in the sample, the average translational kinetic energy of all the atoms depends only on the temperature of the sample.

To relate molecular velocities with kinetic energy and with pressure we need to use definitions from elementary physics. These are:

Translational kinetic energy (U) equals one-half mass (m) times velocity (u) squared.

$$U = \tfrac{1}{2} mu^2 \tag{6.1}$$

Pressure (P) equals force (f) per unit area (A).

$$P = \frac{f}{A} \tag{6.2}$$

Newton's law [force equals mass times acceleration ($\boldsymbol{a}$)] in the form of force equals the derivative of the momentum ($m\boldsymbol{u}$) with respect to time (t).

$$\boldsymbol{f} = m\boldsymbol{a} = \frac{d(m\boldsymbol{u})}{dt} \tag{6.3}$$

Force, acceleration, and velocity are all vectors; they have both magnitude and direction. We therefore use a **boldface** notation for them. Usually we will be referring to the magnitude of the vectors; then they will be written normally. The word speed will be used for the magnitude of the velocity; speed is always positive.

We can now calculate the pressure exerted by one molecule of mass m and velocity $\boldsymbol{u}$ on the walls of a cubic container of length l; the area of each wall is $A = l^2$ (Fig. 6.1). The velocity $\boldsymbol{u}$ can be resolved into its three Cartesian components u_x, u_y, and u_z along the three axes. Let us first focus our attention on the collisions between this molecule and face A. If the x component of the velocity is $\pm u_x$, then in 1 s the molecule travels a distance of u_x meters. If u_x is much greater than l, the dimension of the container, the molecule will pass many times in 1 s between the two walls perpendicular to the x axis. Because each round trip spans a distance of $2l$, the number of times the molecule hits face A per second is $u_x/2l$. Now consider the momentum change per collision. Each time the molecule approaches

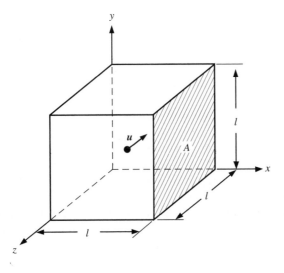

Fig. 6.1 Molecule of mass m and velocity $\boldsymbol{u}$ in a cubic container of side l. To calculate the pressure we need to know the number of collisions between the molecule and the crosshatched face labeled A. The velocity is shown as a vector; it has components along three perpendicular directions. To calculate the pressure we need to use the x-component of the velocity, u_x.

wall A it has a perpendicular velocity component u_x and a corresponding momentum mu_x. As it departs in the opposite direction after a collision, the velocity component perpendicular to the wall changes to $-u_x$, with a corresponding momentum $-mu_x$. Thus the momentum change along the x direction is $2mu_x$ per collision.

The total momentum change along the x axis per second, which according to Newton's second law [Eq. (6.3)] is the force exerted by the molecule along the x axis, is then

$$f_x = \text{(momentum change per collision)} \times \text{(number of collisions per second)}$$

$$= 2mu_x \frac{u_x}{2l} \tag{6.4}$$

$$= \frac{mu_x^2}{l}$$

The pressure caused by one molecule is then from Eq. (6.2) equal to f_x/l^2.

$$P = \frac{mu_x^2}{l^3}$$

If there are N identical molecules in the container, they will each have the same mass m, but each one will have a different velocity, u_i. The total force in the direction perpendicular to the wall A exerted by all the molecules is

$$F_x = \frac{mu_{x1}^2}{l} + \frac{mu_{x2}^2}{l} + \ldots + \frac{mu_{xN}^2}{l}$$

$$= \frac{m}{l} \sum_{i=1}^{N} u_{xi}^2$$

The force per unit area, in the direction normal to wall A, is by definition the pressure P. Thus,

$$P = \frac{F_x}{l^2}$$

$$= \frac{m}{l^3} \sum_{i=1}^{N} u_{xi}^2 \tag{6.5}$$

$$= \frac{m}{V} \sum_{i=1}^{N} u_{xi}^2$$

where $V = l^3$ is the volume of the container.

We define the mean-square velocity; or mean-square speed (the mean of the velocities squared) of the molecules in the x direction as

$$\langle u_x^2 \rangle = \frac{1}{N} \sum_{i=1}^{N} u_{xi}^2 \tag{6.6}$$

Equation (6.5) then becomes

$$P = \frac{Nm}{V} \langle u_x^2 \rangle \tag{6.7}$$

We now have an expression for pressure in terms of the mean-square speed in the x direction. But there is nothing special about the x direction. We need an expression in terms of molecular velocities, not just one component of the velocity. The square of the magnitude of any vector is the sum of the squares of its components, so for each velocity u_i

$$u_i^2 = u_{xi}^2 + u_{yi}^2 + u_{zi}^2$$

By adding all these equations and dividing each side by N, we obtain

$$\frac{1}{N} \sum_{i=1}^{N} u_i^2 = \frac{1}{N} \sum_{i=1}^{N} u_{xi}^2 + \frac{1}{N} \sum_{i=1}^{N} u_{yi}^2 + \frac{1}{N} \sum_{i=1}^{N} u_{zi}^2 \tag{6.8}$$

We define the mean-square speed, $\langle u^2 \rangle$, as

$$\langle u^2 \rangle = \frac{1}{N} \sum_{i=1}^{N} u_i^2$$

Therefore, Eq. (6.8) becomes

$$\langle u^2 \rangle = \langle u_x^2 \rangle + \langle u_y^2 \rangle + \langle u_z^2 \rangle \tag{6.9}$$

Since the molecules are moving about in space randomly, we expect no bias for the x, y, or z directions. Therefore,

$$\langle u_x^2 \rangle = \langle u_y^2 \rangle = \langle u_z^2 \rangle$$

and

$$\langle u_x^2 \rangle = \tfrac{1}{3} \langle u^2 \rangle$$

$$\left. \begin{array}{l} \\ \\ \end{array} \right\} \tag{6.10}$$

Substituting Eq. (6.10) into (6.7), we obtain

$$P = \frac{1}{3} \frac{Nm}{V} \langle u^2 \rangle$$

or

$$PV = \tfrac{1}{3} Nm \langle u^2 \rangle \tag{6.11}$$

We thus have related the pressure of a dilute (ideal) gas to the masses and average speeds of the molecules.

According to the ideal gas law,

$$PV = nRT = \frac{N}{N_0} RT \tag{6.12}$$

where $n = N/N_0$, N_0 is Avogadro's number, and n is the number of moles of gas. Combining Eqs. (6.11) and (6.12) gives

$$RT = \tfrac{1}{3} N_0 m \langle u^2 \rangle$$

$$= \tfrac{2}{3} \left(\tfrac{1}{2} N_0 m \langle u^2 \rangle \right) \tag{6.13}$$

Note that $\frac{1}{2} m \langle u^2 \rangle$ is the average translational kinetic energy per molecule [Eq. (6.1)], and that $\frac{1}{2} N_0 m \langle u^2 \rangle$ is the translational kinetic energy $\bar{U}_{tr}$ of 1 mol of gas. Therefore,

$$RT = \tfrac{2}{3} \bar{U}_{tr}$$

and

$$\bar{U}_{tr} = \tfrac{3}{2} RT \qquad (6.14)$$

Note that $\bar{U}_{tr}$ is a function of T only. Also, since $N_0 m$ is the molecular weight, M, Eq. (6.13) gives

$$\langle u^2 \rangle = 3 \frac{RT}{M} \qquad (6.15)$$

Equation (6.15) is an important result. It states that for a gas consisting of molecules with molecular weight M, the mean-square speed is a function of T only. In deriving Eq. (6.15), we have neglected molecular collisions; the same result is obtained without this omission.

The root-mean-square speed (the square root of the mean-square velocities)

$$\langle u^2 \rangle^{1/2} = \left(\frac{3RT}{M} \right)^{1/2} \qquad (6.16)$$

is one measure of the average speed of the molecules.

Example 6.1   Calculate the root-mean-square speed of oxygen and geraniol vapor at 20°C. Geraniol, an alcohol responsible for the fragrance of roses, has a molecular weight of 154.2.

Solution   The molecular weight of O_2 is 32.0. To obtain reasonable units (m s^{-1}) for the velocity, we must express the gas constant R as 8.314 J K^{-1} mol^{-1} = 8,314 g m^2 s^{-2} K^{-1} mol^{-1}.

From Eq. (6.16),

$$\langle u^2 \rangle^{1/2} = \left(\frac{3RT}{M} \right)^{1/2}$$

$$= \left[3(8{,}314 \text{ g m}^2 \text{ s}^{-2} \text{ K}^{-1} \text{ mol}^{-1}) \ \frac{(293 \text{ K})}{32 \text{ g mol}^{-1}} \right]^{1/2}$$

$$= (2.284 \times 10^5 \text{ m}^2 \text{ s}^{-2})^{1/2}$$

$$= 478 \text{ m s}^{-1}$$

Similarly, for geraniol,

$$\langle u^2 \rangle^{1/2} = 218 \text{ m s}^{-1}$$

This example shows that the speeds are rather high—several tenths of kilometers per second. The values that we have calculated for oxygen and geraniol are of the same order of magnitude as the speed of sound in air (~330 m s^{-1}).

The average translational kinetic energy of each atom in a molecule in a solid, liquid, or gas depends only on the absolute temperature. The equation is the same one we derived for an ideal gas [Eq. (6.14)]. For an average atom in a molecule in any sample (averaged over all the atoms in the sample), the average translational kinetic energy is

$$\langle U_{tr} \rangle = \tfrac{3}{2} kT \tag{6.14}$$

k = Boltzmann constant = R/N_0

  = 1.380×10^{-23} J K^{-1} molecule^{-1}

Furthermore, the mean-square speed of each atom is the same as for an ideal gas as given in Eq. (6.15). The fact that collisions, interactions, and even bonding the atoms into molecules does not change these averages is very surprising, and very useful. As long as the molecular and atomic motions can be treated classically, the simple equations for average energies and average speeds are correct. They are applicable to the motions and interactions of proteins and nucleic acids in solution, for example.

Maxwell-Boltzmann Distribution of Velocities

We have found that the mean-square speed of all the molecules in a gas is simply related to the absolute temperature. The squares of the speeds—equal to the squares of the velocities—are pertinent because kinetic energy depends on velocity squared. Other measurable properties such as the rate of molecular collisions, which is crucial for kinetics, depend on the mean speed $\langle u \rangle$. It is easy to see that unless all the speeds are equal, the root-mean-square speed, $\langle u^2 \rangle^{1/2}$, can never equal the mean speed $\langle u \rangle$. For any distribution of speeds the root-mean-square speed, $\langle u^2 \rangle^{1/2}$, must be greater than the mean speed $\langle u \rangle$. Similarly, the root-mean-cube speed, $\langle u^3 \rangle^{1/3}$ will be greater than $\langle u^2 \rangle^{1/2}$ and $\langle u \rangle$. If we knew the molecular speeds, u_i, for all the molecules at any one time we could calculate any average speed such as the mean, the median, the most probable, the root-mean-square, etc. Ludwig Boltzmann of entropy fame, and James Clerk Maxwell (best known for his electromagnetic equations) independently derived equations for the probability of finding a molecule with any chosen speed.

To help understand their derivation, let us think about a large number, such as Avogadro's number, of (dilute) gas molecules in a container. In a thought experiment we can instantaneously measure the speed, and thus the translational kinetic energy, of each molecule at any time. For example, we can count the number of molecules that have nearly zero speed. We have to specify some range of speeds because there is always some uncertainty in any measurement. So we count the number of molecules with speed between zero m s^{-1} and 0.1 m s^{-1}, and between 100 m s^{-1} and 100.1 m s^{-1}, and so on. These numbers will be independent of time as long as the conditions, such as temperature, number of molecules, etc., do not change. Instead of the numbers of molecules within each range of speeds it is more useful to know the fraction of molecules within each range of speeds. It does not matter whether the container has one mol of O_2 molecules or 10 mols; the distribution of speeds will be the same for the dilute gas molecules. Because the molecules are identical, the fraction of molecules within a certain range of speeds is equal to the probability that any one molecule will be in this range. For example, we can say that 10% of the molecules have speed less than 50 m s^{-1}, or equally correctly say that the probability of finding a molecule with speed less than 50 m s^{-1} is 0.10.

Let us see how Maxwell and Boltzmann derived the speed distribution for molecules in dilute gases. The key idea is that statistics and probability work best for large numbers, such as the number of molecules in a mole. Although there is a very wide range of speeds and energies for the large number of molecules, the most probable distribution of speeds is the only one we need consider. Boltzmann derived the general result that the probability of finding a molecule (in a dilute gas) with energy E_i is proportional to $e^{-E_i/kT}$ with k Boltzmann's constant and T the absolute temperature.

$$P_i \propto g_i \, e^{-E_i/kT}$$

P_i = probability of finding a molecule with energy E_i
g_i = degeneracy—the number of different states of a molecule corresponding to energy E_i

We see that the probability of a molecule having energy E_i depends on the exponential of E_i/kT as well as on the degeneracy g_i. The magnitude of the degeneracy depends on the type of energy. For electronic energy levels the degeneracies are small integers, such as 1, 2, or 3 as described for the hydrogen atom in Chapter 9. For translational energies the degeneracies are very large and increase rapidly with energy.

The ratio of the number of molecules in two different energy levels according to Boltzmann is

$$\frac{P_j}{P_i} = \frac{N_j}{N_i} = \frac{g_j \, e^{-E_j/kT}}{g_i \, e^{-E_i/kT}} = \frac{g_j}{g_i} \, e^{-(E_j - E_i)/kT} \tag{6.17}$$

$\dfrac{N_j}{N_i}$ = the ratio of the number of molecules with energies E_j and E_i at equilibrium

A probability, like the fraction of the number of molecules, must be less than 1. Thus, the Boltzmann most-probable energy distribution (the distribution at equilibrium) is

$$P_j = \frac{N_j}{N} = \frac{g_j \, e^{-E_j/kT}}{\displaystyle\sum_i g_i \, e^{-E_i/kT}} \tag{6.18}$$

P_j = the probability of finding a molecule with energy E_j
N_j/N = the fraction of molecules with energy E_j

By dividing by the sum over all the exponential terms (which are each proportional to the probability of finding a molecule with energy E_i) we make sure that the sum over all probabilities is equal to 1.

$$\sum_1 P_i = 1$$

The Boltzmann equation is very important in all fields of science. It applies to any type of energy, not just translational kinetic energy. It applies to vibrational, rotational, and electronic energies, for example. It applies to quantized energy states (as required to treat electronic energies) as well as energies that can be treated classically, such as translational energies. For interacting molecules, as found in liquids, solids, and high pressure gases, the same equation applies except that now the energies E_i refer to the entire collection of molecules. We can, in principle, calculate the probability of finding a protein in a particular folded state in solution by calculating the energies of possible folded and unfolded states of the protein in the solvent. We will discuss this further in Chapter 9.

Maxwell derived a special case of the Boltzmann equation [Eq. (6.18)] specifically for translational kinetic energies with E_i replaced by $(1/2)\, mu_i^2$. Furthermore, he considered a continuous distribution of energies and speeds. Although all energies are quantized, as we will describe in Chapter 9, the translational energy levels are so closely spaced that they can be treated as if they were continuous. For a continuous distribution of energies the sum over energies in Eq. (6.18) can be replaced by an integral from zero to infinity. The Maxwell-Boltzmann distribution of speeds is obtained by replacing E by $(1/2)\, mu^2$ in the exponential and using the degeneracies pertinent to translational energies in three dimensions.

$$P(u)\, du = F(u)\, du = \frac{u^2\, e^{-mu^2/2kT}\, du}{\displaystyle\int_0^\infty u^2\, e^{-mu^2/2kT}\, du} \qquad (6.19)$$

$P(u)\, du =$ the probability of finding a molecule with speed between u and $u + du$
$F(u)\, du =$ the fraction of molecules with speed between u and $u + du$

For both Eqs. (6.18) and (6.19) it is obvious that the sum over all energies, and the integral over all speeds is equal to 1.

$$\sum_i P_i = \int_0^\infty P(u)\, du = 1$$

From Eq. (6.19) and a table of integrals that tells us that

$$\int_0^\infty x^2\, e^{-ax^2}\, dx = \frac{1}{4\pi}\left(\frac{\pi}{a}\right)^{3/2}$$

and noting that a is $(m/2kT)$ in Eq. (6.19), we obtain the Maxwell-Boltzmann equation for the probability of finding a molecule in a dilute gas with speed between u and $u + du$.

$$P(u)\, du = F(u)\, du = 4\pi\left(\frac{m}{2\pi kT}\right)^{3/2} u^2\, e^{-mu^2/2kT}\, du \qquad (6.20)$$

We notice that the probability only depends on the mass m of the molecule and the absolute temperature. A plot of $P(u)\, du$ from Eq. (6.20) versus u is shown in Fig. 6.2. You can see that as the temperature increases, the distribution of speeds broadens and shifts to higher speeds. The molecules move faster and have a wider spread of speeds. Decreasing the mass of each molecule has a similar effect.

The Maxwell-Boltzmann distribution [Eq. (6.20)] and the curves shown in Fig. 6.2 give us more information about the speeds of molecules in dilute (ideal) gases than we can measure. We believe the predictions because using the equation we can derive properties that can be tested. Obviously we can calculate any average speed we need. The mean speed $\langle u \rangle$ and the mean-square speed $\langle u^2 \rangle$ are obtained from

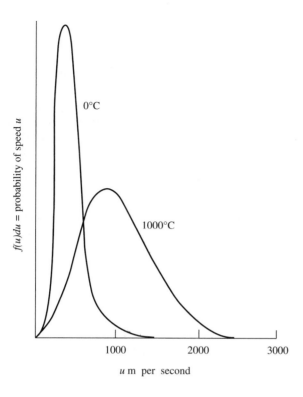

Fig. 6.2 Maxwell-Boltzmann distribution of speeds at two temperatures. Notice that as the temperature increases the most probable speed increases (the maximum in the curve) but the spread of the speeds increases more markedly. This means that the fraction of molecules with the highest speeds increases much faster with temperature than the average speeds. This is one reason why temperature has such a large effect on reaction rates.

$$\langle u \rangle = \int_0^\infty u\, P(u)\, du \tag{6.21}$$

$$\langle u^2 \rangle = \int_0^\infty u^2\, P(u)\, du \tag{6.22}$$

Substitution of $P(u)$ from Eq. (6.20) into these integrals and the use of a table of integrals gives the results

$$\langle u \rangle = \left(\frac{8kT}{\pi m} \right)^{1/2} \tag{6.23}$$

$$\langle u^2 \rangle = \frac{3kT}{m} \tag{6.24}$$

We had obtained Eq. (6.24) earlier with k/m replaced by the equivalent R/M in Eq. (6.16).

We have described molecular velocities in the gas phase in great detail because they provide a simple example of the ability to quantitatively relate molecular properties to macroscopic properties. In the rest of this chapter and in succeeding chapters, we will treat more complex systems, and more relevant systems to biochemical applications. However, the ideas introduced here—such as the Boltzmann distribution of energies—can be applied directly.

Molecular Collisions

The speeds of molecules in the gas phase are very large. Figure 6.2 shows ranges of hundreds to thousands of m s^{-1}; Example 6.1 gives a root-mean-square speed of 218 m s^{-1} at room temperature for the fragrance of roses. You might think that if a bottle of perfume is opened in the center of a room in still air, the fragrance would reach the corners in a fraction of a second. This does not happen; a much longer time is required. To see why this is so, we must consider molecular collisions, mean free paths, and diffusion. Although the molecules have high speeds, they do not travel far before colliding and changing their direction. They move fast, but they move randomly and do not quickly get very far from where they started.

Let us think about the number of collisions per second, z, that one molecule undergoes in a gas containing N molecules in a volume V. What will the number of collisions depend on? The larger the concentration of molecules the more collisions will occur, so z will be proportional to N/V. Surely the faster the molecules are moving the more collisions will occur. This means z will depend on the speeds of the molecules. The number of collisions must depend on the sizes of the molecules. When two molecules collide, the size of each molecule is important. We will consider the simplest case of two identical, spherical molecules each with diameter σ (sigma) The number of collisions per second for one molecule, z, is thus proportional to σ^2. Putting all this together, and using the mean speed, $\langle u \rangle$, as a measure of the molecular speeds, we obtain the result that

$$z \text{ is proportional to } \frac{N}{V}\sigma^2\langle u \rangle$$

To make sure that our reasoning makes sense we check that the units of the expression above are correct. The units of z are s^{-1}, so the product of $(N/V)\sigma^2\langle u \rangle$ must also equal s^{-1}, which it does. A detailed derivation shows that the mean speed is the correct one to use and provides the proportionality constant. The number of collisions per second for one molecule is

$$z = \sqrt{2}\,\pi\,\frac{N}{V}\,\sigma^2\langle u \rangle \tag{6.25}$$

Using Eq. (6.23) for $\langle u \rangle$ we obtain

$$z = 4\sqrt{\pi}\,\frac{N}{V}\,\sigma^2\left(\frac{RT}{M}\right)^{1/2} \tag{6.26}$$

Since there are N/V molecules per unit volume, there are Nz/V molecules per unit volume undergoing collisions per unit time. The total number of collisions per unit volume per unit time, Z, by all the molecules present is then

$$Z = \left(\frac{N}{V}\right)\frac{z}{2} \tag{6.27}$$

The factor 2 results from the fact that each collision involves two molecules. Combining Eqs. (6.27) and (6.26) we obtain for the total number of collisions per volume per second

$$Z = 2\sqrt{\pi}\left(\frac{N}{V}\right)^2 \sigma^2 \left(\frac{RT}{M}\right)^{1/2} \tag{6.28}$$

The total number of molecular collisions in a container is seen to be proportional to the square of the concentration, the square of the molecular diameter, the square root of the absolute temperature, and to be inversely proportional to the square root of the molecular weight.

Mean Free Path

The mean free path, l, is defined as the average distance a molecule travels between two successive collisions with other molecules. Because $\langle u \rangle$ is the average distance a molecule travels per unit time and z is the number of intermolecular collisions it encounters during that time, the mean free path is given by

$$l = \frac{\langle u \rangle}{z} = \frac{1}{\sqrt{2}\pi(N/V)\sigma^2} \tag{6.29}$$

using Eq. (6.25).

Example 6.2 For N_2 gas at 1 atm and 25°C, calculate (a) the number of collisions each N_2 molecule encounters in 1 s; (b) the total number of collisions in a volume of 1 m³ in 1 s; (c) the mean free path of an N_2 molecule. The diameter of an N_2 molecule can be taken as 3.74 Å (3.74×10^{-10} m).

Solution For N_2 gas at 1 atm and 25°C, the number of moles per unit volume is P/RT, and the number of molecules per unit volume is

$$\frac{N}{V} = \frac{N_0 P}{RT}$$

$$= \frac{(6.023 \times 10^{23} \text{ molecules mol}^{-1})(1 \text{ atm})}{(0.08205 \text{ L atm K}^{-1} \text{ mol}^{-1})(298 \text{ K})}$$

$$= 2.46 \times 10^{22} \text{ molecules L}^{-1}$$

$$= 2.46 \times 10^{25} \text{ molecules m}^{-3}$$

From Eq. (6.26), the number of collisions each molecule encounters per unit time is

$$z = 4\sqrt{\pi}(2.46 \times 10^{25} \text{ m}^{-3})(3.74 \times 10^{-10} \text{ m})^2$$

$$\times \left[(8.31) \text{ J K}^{-1} \text{ mol}^{-1})\left(\frac{298 \text{ K}}{0.028 \text{ kg mol}^{-1}}\right)\right]^{1/2}$$

$$= 7.27 \times 10^9 \text{ (J m}^{-2} \text{ kg}^{-1})^{1/2}$$

but 1 J = 1 kg m² s⁻², therefore

$$z = 7.27 \times 10^9 \text{ s}^{-1}$$

(b) From Eq. (6.27), the total number of collisions is

$$Z = \left(\frac{N}{V}\right)\frac{z}{2}$$

$$= \frac{(2.46 \times 10^{25} \text{ m}^{-3})(7.26 \times 10^9 \text{ s}^{-1})}{2}$$

$$= 8.94 \times 10^{34} \text{ collisions m}^{-3} \text{ s}^{-1}$$

(c) From Eq. (6.29), the mean free path is

$$l = \frac{1}{\sqrt{2}\pi(2.46 \times 10^{25} \text{ m}^{-3})(3.74 \times 10^{-10} \text{ m})^2}$$

$$= 6.53 \times 10^{-8} \text{ m}$$

Thus at 1 atm and 25°C a typical N_2 molecule goes a distance almost 200 times its diameter between collisions.

DIFFUSION

The Random Walk and Diffusion in a Gas

A molecule in a gas may have a high speed (hundreds of meters per second) but, because its movement is random, after 1 s it does not end up very far from where it started. This is illustrated by the random walk in Fig. 6.3; clearly the total path taken by the molecule is large compared to the distance from the beginning to the end of the path. This straight-line distance is called the displacement, d. Let us consider the average displacement per second of a molecule. In 1 s the molecule undergoes z collisions with other molecules. The distance it travels between two successive collisions is, on the average, the mean free path, l. Since molecules in a gas at equilibrium are moving randomly, the molecules colliding with our selected molecule come from random directions. Therefore, after each collision, the molecule will change its direction in a random manner (Fig. 6.3). This type of motion is called

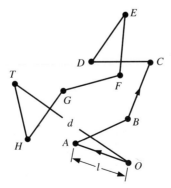

Fig. 6.3 Random path of a molecule. At time zero, its position is at O, and collisions are encountered at points A, $B, \ldots, H$. At time t, its position is at T. The average distance it travels between two successive collisions is the mean free path, l. In our illustration, the distances $\overline{OA}$, $\overline{AB}$, $\overline{BC}$, etc., are depicted as equal to l. The displacement, d, that occurs during t is the distance $\overline{OT}$.

Brownian motion, because it was first observed by Brown in pollen grains suspended in water. The random path of a pollen grain caused by collisions with solvent is completely analogous to the random path of a gas molecule caused by collisions with other gas molecules.

The problem of the random path occurs in many different areas of science. The shape of a flexible polymer such as polyethylene or a random polypeptide (a denatured protein), for example, can be calculated from a random path. The question is: What is the average distance between the beginning and end of a path of N random steps of length l? We can deduce the answer easily for a random path in one dimension; a detailed derivation is given in Chapter 11. Think of a person who takes a step randomly either to the right or left along a line; he flips a coin to decide whether to step right or left. How far does he get after N steps of unit length? The table below gives the answers for a small number of steps. In the table, r means a step right and l means a step left.

Table 6.1 A random walk in one dimension

No. of steps N	Possible paths	Possible locations, d_i	Mean displacement $\langle d \rangle$	Mean-square displacement $\langle d^2 \rangle$
1	r	+1	0	1
	l	−1		
2	rr	+2	0	2
	rl, lr	0, 0		
	ll	−2		
3	rrr	+3	0	3
	rrl, rlr, lrr	+1, +1, +1		
	rll, lrl, llr	−1, −1, −1		
	lll	−3		

In the table we calculated the means by averaging over all possible paths for a given number of steps, such as the eight possibilities for three steps. The mean displacement

$$\langle d \rangle = \frac{\sum_i^{paths} d_i}{no.\ of\ paths} \tag{6.30}$$

and the mean-square displacement

$$\langle d^2 \rangle = \frac{\sum_i^{paths} d_i^2}{no.\ of\ paths} \tag{6.31}$$

are calculated for the possible locations from the origin after 1, 2, and 3 steps. The mean displacement is always zero because there is equal probability of moving left or right. We find that the mean-square displacement is always equal to the number of steps. If we do the calculation in three dimensions—to produce what is sometimes called a random flight—we get exactly the same result. So for N random steps in any direction of length l the mean-square displacement is

$$\langle d^2 \rangle = N \, l^2 \qquad (6.32)$$

and the root-mean-square displacement is

$$\langle d^2 \rangle^{1/2} = \sqrt{N} \, l \qquad (6.33)$$

We can use this equation to calculate the average end-to-end distance in a random polymer of N monomer units. A real polymer (for example, polyethylene) is not random, so the measured average end-to-end distance is slightly larger, but the square root dependence on the number of monomer units is correct.

For the random diffusion of a molecule in a gas, the random walk equation equivalent to Eq. (6.32) is

$$\langle d^2 \rangle = z \, l^2 \qquad (6.34)$$

Here $\langle d^2 \rangle$ is the mean-square displacement of the molecule per second, z is the number of collisions per second, and l is the mean free path. Using Eq. (6.29) for the mean free path in a dilute gas we obtain

$$\langle d^2 \rangle = \frac{\langle u \rangle^2}{z} \qquad (6.35)$$

The square root of $\langle d^2 \rangle$ is the root-mean-square displacement,

$$\langle d^2 \rangle^{1/2} = \sqrt{z} \, l \qquad (6.36\text{a})$$

$$= \frac{\langle u \rangle}{\sqrt{z}} \qquad (6.36\text{b})$$

which is a measure of the average displacement per unit time.

We mentioned earlier that if a bottle of perfume is opened in the middle of a windless room, the scent takes much longer to reach the other parts of the room than one might naively anticipate from the high value of $\langle u \rangle$. The explanation lies in Eq. (6.36). While $\langle u \rangle$ is large, any molecule encounters many collisions per unit time, and each collision changes its direction. Thus, the root-mean-square displacement per unit time is $\langle u \rangle / \sqrt{z}$. At atmospheric pressure, $\sqrt{z}$ is a large number ($\sim 10^{10}$ s^{-1}), as we found in Example 6.2.

Example 6.3   Consider a molecule of N_2 gas at 1 atm and 25°C.

(a) Estimate how far it will move in 1 s away from where it started.
(b) What is the total distance traveled along the zigzag path in 1 s?

Solution

(a) The root-mean-square displacement $\langle d^2 \rangle^{1/2}$ is a measure of its displacement. From Eq. (6.36a) and using the results of Example 6.2,

$$\langle d^2 \rangle^{1/2} = \sqrt{z}\, l$$

$$= (7.27 \times 10^9)^{1/2}(6.53 \times 10^{-8}\,\text{m})$$

$$= 5.57 \times 10^{-3}\,\text{m} = 0.557\,\text{cm}$$

(b) The total distance traveled per second is

$$zl = (7.27 \times 10^9)(6.53 \times 10^{-8}\,\text{m})$$

$$= 475\,\text{m}$$

Note that these distances differ by a factor of almost 10^5 for this example.

Diffusion Coefficient and Fick's First Law

The diffusion of the molecules which produce the scent of a perfume is an example of diffusion in the gas phase. We can calculate the rate of diffusion in a dilute gas from the mean velocity and the number of collisions per second of the molecules. In a liquid there are many more collisions per second than in a gas, and it is more difficult to calculate their number. Instead of trying to calculate the number of collisions, we consider the molecule to move in a fluid continuum. The molecule is like a submarine in the ocean. The rate of diffusion depends on the size and shape of the molecule, and on the properties—such as viscosity—of the solvent.

We must first define the measurable properties that characterize diffusion. Diffusion occurs whenever there is a concentration difference in a container; diffusion will eventually make the concentration uniform. Diffusion is a slow process compared to mixing caused by stirring, for example, but over small distances—as occur in biological cells—diffusion can be an effective method for transporting molecules.

Consider a 1-cm^2 cross-sectional area in the yz plane (Fig. 6.4). We define the flux J_x as the net amount of substance that diffuses through this unit area, per unit time, in the x direction. Units of flux, for example, would be mol cm^{-2} s^{-1}. If there is no concentration gradient in the x direction, you expect equal numbers of molecules crossing this area from the

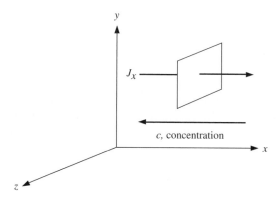

Fig. 6.4 Flux J_x is the net diffusional transport of material per unit time (mol cm^2 s^{-1}), in the x direction, across a unit cross-sectional area perpendicular to x. If the concentration c decreases with x, the flux J_x is in the direction opposite to the concentration gradient.

left and from the right. Therefore, the net flux J_x, is zero. Suppose that there is a concentration gradient in the x direction, say $dc/dx < 0$. The concentration decreases with x; therefore, there are more molecules per unit volume to the left of the area than to the right. You expect, then, that more molecules per unit time will diffuse across the area from the left than from the right. In other words, there will be a net transport of material in the direction opposite to the concentration gradient, as shown in Fig. 6.4. The steeper the concentration gradient, the larger the net flux. These considerations lead to the equation for *Fick's first law:*

$$J_x = -D\frac{dc}{dx} \tag{6.37}$$

where D, the proportionality constant, is called the *diffusion coefficient.* The negative sign indicates that the net transport by diffusion is in a direction opposite to the concentration gradient. The units of D are $cm^2\,s^{-1}$; the concentration gradient is in $mol\,cm^{-4}$.

As diffusion occurs the solution becomes more uniform; the concentration gradient dc/dx, decreases. As time goes on, the system gradually approaches homogeneity, and dc/dx approaches zero. Thus, Eq. (6.37) is given in terms of the instantaneous flux at any time t. Equation (6.37), Fick's first law of diffusion, has been shown experimentally to be correct.

Fick's Second Law

It is also possible to describe how the concentration in a gradient changes with time. In a uniform concentration gradient, where $dc/dx = $ constant for all values of x, Eq. (6.37) tells us that the flux J_x is the same at all positions and, therefore, c will not change with time. (The flux into every volume element from one side is exactly equal to that going out the other side.) This describes a steady-state flow of material by diffusion.

However, if the concentration gradient is not the same everywhere, the concentration will change with time. The change of concentration with time $(\partial c/\partial t)$ at position x is proportional to the gradient in the flux $(\partial J/\partial x)$ at that point. Intuitively, if more material is diffusing into the volume element from the left than is diffusing out to the right, then $(\partial J/\partial x) < 0$, and the concentration inside should increase with time, $(\partial c/\partial t) > 0$. This logic leads to *Fick's second law.*

$$\left(\frac{\partial c}{\partial t}\right)_x = D\left(\frac{\partial^2 c}{\partial x^2}\right)_t \tag{6.38}$$

We use the notation of partial derivatives because the concentration depends on both time t and distance x. Fick's second law states that the change in concentration with time at any position x, $(\partial c/\partial t)_x$, is proportional to the second derivative of the concentration with respect to x at time t, $(\partial^2 c/\partial x^2)_t$. The proportionality constant is the diffusion coefficient D, assumed to be independent of concentration. An illustration of the application of Fick's second law is shown in Fig. 6.5.

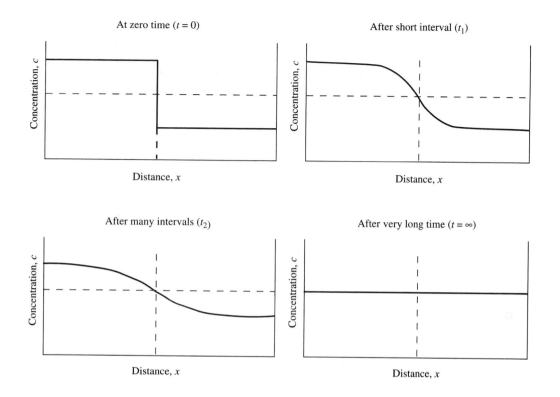

At zero time ($t = 0$)

After short interval (t_1)

After many intervals (t_2)

After very long time ($t = \infty$)

Fig. 6.5 Diffusion of a material with a concentration gradient that is initially a step. Profiles are shown at time zero (t_0) and at successive times, t_1, t_2, and t_∞. The exact behavior is described by Fick's second law [Eq. (6.38)].

Determination of the Diffusion Coefficient

In principle, D can be determined by the use of either Fick's first law [Eq. (6.37)] or Fick's second law [Eq. (6.38)]. A simple way to measure a diffusion coefficient is to use Fick's first law and to measure the amount of material which is transferred through unit area per unit time—the flux, J. We use a porous glass disk of thickness Δx to separate two solutions of two different concentrations as shown in Fig. 6.6. The rate of transfer of material (mols s^{-1} or grams s^{-1}) across the disk can be measured using a radioactive label, for example. The effective area of the porous disk is determined by calibrating the disk with a substance of known diffusion coefficient. Dividing the rate of transfer of material by the effective area of the porous disk gives the flux, J. The diffusion coefficient is obtained from Fick's first law [Eq. (6.37)].

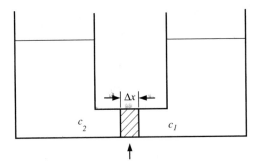

Fig. 6.6 Measurement of a diffusion coefficient D by measuring the diffusion through a porous glass disk. The disk's effective porous area can be calibrated using a molecule with a known diffusion coefficient. The two solutions are stirred well and the concentrations are kept essentially constant during the experiment. The flux (amount transferred per unit area per unit time) is measured. D is calculated from Fick's first law, Eq. (6.39).

Porous glass disk with effective area A and thickness Δx

$$D = -J\left(\frac{\Delta x}{c_2 - c_1}\right) \qquad (6.39)$$

J = flux with units of mols cm^{-2} s^{-1} or grams cm^{-2} s^{-1}
c_2, c_1 = concentrations with units of mols cm^{-3} or grams cm^{-3}
Δx = thickness of porous glass disk in cm
D = diffusion coefficient with units of cm^2 s^{-1}

The concentrations are kept constant during the experiment by using large volumes on each side and by stirring near the porous disk. Although Fick's second law [Eq. (6.38)] was written for a concentration-independent diffusion coefficient, in general D will vary with concentration. In the experiment shown in Fig. 6.6 the measured value of D corresponds to the average of the two concentrations. By changing these concentrations, we can obtain D at any desired concentration, and we can extrapolate D to zero concentration.

To use Fick's second law [Eq. (6.38)] to measure a diffusion coefficient we measure the concentration c or the concentration gradient dc/dx as a function of time. For example, the experiment shown in Fig. 6.5 can provide a diffusion coefficient. A step concentration gradient is prepared at the beginning of the experiment; then the diffusion of the gradient is followed as a function of time. From measured values of the concentration c as a function of position x and time t, the derivatives needed to evaluate D from Eq. (6.38) can be obtained. An integrated form of Eq. (6.38) is most useful. The concentration is usually measured optically using the absorbance or refractive index of the solution. Thus, pictures are taken of the concentration gradient at different times and analyzed in terms of Fick's second law.

Relation Between the Diffusion Coefficient and the Mean-Square Displacement

In our discussion on diffusion in a gas, we have used the quantity $\langle d^2 \rangle$, which is the average of the square of the displacement of a molecule in 1 s. Because diffusion is due to the random motion of molecules, we expect that the larger the value of $\langle d^2 \rangle$, the larger the diffusion coefficient D. That is, the faster the molecules can move from one point to another, the greater the flux, J, for a given concentration gradient.

To derive the relationship between a diffusion coefficient and the mean-square displacement, we consider a very thin layer of a solution containing w grams of solute per unit area sandwiched between layers of solvent at time $t = 0$, as illustrated in Fig. 6.7a.

For this case, Eq. (6.38) can be solved to give c in g cm^{-3} at any position x and time t,

$$c = \frac{w}{(4\pi Dt)^{1/2}} e^{-x^2/4Dt} \qquad (6.40)$$

This is plotted in Fig. 6.7b. We can consider the concentration profile as a distribution in x. That is, during time t some solute molecules have moved far away from their original positions ($x = 0$), but others have hardly moved away at all. The concentration profile is that of a Gaussian distribution in x at any time. The mean-square displacement $\langle x^2 \rangle$ can be evaluated directly from this distribution using Eq. (6.22) to give

$$\langle x^2 \rangle = 2Dt$$

$$D = \frac{\langle x^2 \rangle}{2t} \qquad (6.41)$$

The diffusion coefficient is one-half of the mean-square displacement per unit time. This relation was first derived by Einstein in 1905.

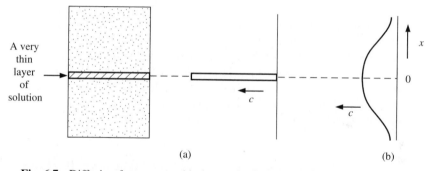

(a)   (b)

Fig. 6.7   Diffusion from a very thin layer of solution: (a) initial state; (b) concentration profile after time t. The concentration is a Gaussian distribution in x at any time t [Eq. (6.40)].

Example 6.4   The diffusion coefficient of the oxygen storage protein myoglobin is $D_{20, w} = 11.3 \times 10^{-7}$ cm^2 s^{-1}, at 20°C in water. Estimate a mean value for the time required for a myoglobin molecule to diffuse a distance of 10 μm, which is the order of the size of a cell.

Solution   The mean-square displacement, $\langle x^2 \rangle$, is a measure of the displacement of an average molecule. From Eq. (6.41),

$$t = \frac{\langle x^2 \rangle}{2D} = \frac{(10 \times 10^{-4} \text{ cm})^2}{2(11.3 \times 10^{-7} \text{ cm}^2 \text{ s}^{-1})} = 0.44 \text{ s}$$

Although the diffusion coefficient of myoglobin in a cell is smaller than in water because of the higher viscosity of cytoplasm, nevertheless the diffusion of even relatively large molecules such as myoglobin (mol wt 17,000) occurs rapidly across dimensions comparable to the size of a cell.

For the Gaussian distribution [Eq. (6.40)] shown in Fig. 6.7, the diffusion coefficient D can be measured from the width of the distribution at half maximum. The concentration c is maximum at $x = 0$. For c equal to half its maximum value, Eq. (6.40) shows that $x = \pm 2 (\ln 2 \, Dt)^{1/2}$; the full width at half maximum is $4 (\ln 2 \, Dt)^{1/2}$. Thus the spread with time t of the initially sharp distribution is a direct measure of the diffusion coefficient D. Note the appearance of the square root of time $\sqrt{t}$, which is related to the square root of the number of steps, or of collisions, expected for random walks.

A diffusion coefficient can be measured from the disappearance of any concentration gradient caused by diffusion. One way to create a concentration gradient, and to measure the concentration of the diffusing species as a function of time, is by photobleaching (Axelrod et al., 1976). Proteins embedded in a cell membrane are often able to diffuse in the plane of the membrane. A laser is focused into a small spot on the membrane of a living cell and an intense laser pulse is used to photochemically destroy absorbing groups in the proteins in the spot. The laser (with much lower intensity) can now be used to measure the return by diffusion of neighboring protein molecules with unchanged absorbing groups. Thus a hole is produced in the apparent color of a membrane, and the time that it takes for the hole to disappear provides a diffusion coefficient for the absorbing species. By making the hole, or spot, a few microns in size the diffusion experiment need only take a few seconds. See Example 6.4 for the time it takes for myoglobin to diffuse 10 μm.

Fluorescence is a more sensitive method for detecting membrane proteins than is absorption, so fluorescence is usually used to follow the motion of the molecules. The terms fluorescence photobleaching recovery (FPR) and fluorescence recovery after photobleaching (FRAP) have been coined for this method.

Determination of the Diffusion Coefficient by Laser Light Scattering

The methods described above start with a nonequilibrium system, and the diffusion coefficient is evaluated from the way the system moves toward equilibrium. But since the diffusion coefficient D is related to the random motion of the molecules, which occurs in equilibrium as well as nonequilibrium systems, one should be able to measure D for a system at equilibrium by monitoring the random motion of molecules directly. One such method is laser light scattering, sometimes called quasi-elastic light scattering.

There is random motion of molecules in a homogeneous solution. If a beam of monochromatic light of frequency ν_0 passes through the solution, the scattered light is no longer monochromatic. Some molecules will be moving toward the light and some away; this causes a Doppler broadening of the scattered light. The intensity of the scattered light, when plotted as a function of frequency, has a maximum at ν_0, and its width at half-height is directly proportional to D (for molecules whose characteristic dimensions are small compared with the wavelength of the light). For a typical protein molecule with a D of 10^{-6} cm^2 s^{-1}, the width of the spectrum of the scattered light is of the order of 10^4 hertz

(Hz). This is still an extremely sharp line, considering the fact that v_0 is of the order of 10^{15} Hz. The availability of highly monochromatic light from laser sources has made it possible to measure diffusion coefficients of macromolecules by this method. It provides a rapid method of measuring diffusion coefficients without the need to wait for the return of a concentration gradient to a uniform concentration.

Some representative results are given in Table 6.2. For small rigid proteins the precision of the measurements is about ±3%, and the agreement with literature values is quite good. The value for tobacco mosaic virus is less reliable because the virus is not small compared with the wavelength of light. Nevertheless, for suitable systems, the light-scattering method is rapid and convenient.

Table 6.2 Diffusion coefficients for macromolecules measured using laser light-scattering (LS) or conventional (Lit) methods

	pH	Salt	$D \times 10^7$ (cm^2 s^{-1})	
			LS	Lit
Bovine serum albumin	6.8	0.5 M KCl	6.7 ± 0.1	6.7
Ovalbumin	6.8	0.5 M KCl	7.1 ± 0.2	8.3
Lysozyme	5.6	—	11.5 ± 0.3	11.6
Tobacco mosaic virus	7.2	0.01 M Na phosphate	0.40 ± 0.02	0.3

SOURCE: S. B. Dubin, J. H. Lunacek, and G. B. Benedek, *Proc. Natl. Acad. Sci. USA 57,* 1164–1171 (1967).

Diffusion Coefficient and Molecular Parameters

An important application of transport measurements is to provide information on the size and shape of macromolecules. According to Newton, a force acting on any object causes an acceleration. However, in a viscous medium there is a counterforce, called a frictional force, which is proportional to the velocity of the object.

$$\text{frictional force} = fu$$

where f is called the *frictional coefficient.* Because the frictional force increases with u, a velocity is soon reached at which the frictional force balances the driving force, and the acceleration becomes zero. This velocity is called the terminal velocity. In a viscous medium the terminal velocity is quickly reached, so it is the frictional coefficient which characterizes how fast a molecule will move under the forces acting in diffusion, sedimentation, electrophoresis, and other transport measurements. Although, of course, Newton's law— force equals mass times acceleration—is acting on each of the particles in the solution, the observed effect is that: driving force equals frictional coefficient f times terminal velocity u. For a particle in a viscous medium

$$\text{force} = fu \tag{6.42}$$

The units of f are (force/velocity) equal to (mass acceleration/velocity); we will use g s^{-1}. The frictional coefficient of a particle, such as a protein or nucleic acid, is the molecular property that characterizes how fast the particle moves under a driving force. A larger frictional coefficient means a smaller velocity. By measuring the frictional coefficient we can learn about sizes and shapes of molecules, and about their interactions with the surrounding medium.

In a diffusion experiment the driving force is the average of the random forces which comes from the thermal kinetic energy of the surrounding molecules. A measure of the average velocity of motion is the diffusion coefficient. The basis for relating the diffusion coefficient to molecular parameters is the equation

$$D = \frac{kT}{f} \tag{6.43}$$

where k, the Boltzmann constant, is equal to the gas constant R divided by Avogadro's number, N_0. This equation was first derived by Einstein as part of the work for his Ph.D. thesis. This equation is very important to us because it relates a macroscopic measurement, the diffusion coefficient, to a molecular property—the frictional coefficient. The frictional coefficient can be related to the size and shape of a molecule. For a sphere, or spherical macromolecule, of radius r, Stokes found that

$$f = 6\pi\eta r \tag{6.44}$$

where η is the viscosity coefficient, a measure of how viscous the surrounding medium is. We shall discuss viscosity in more detail in a later section.

If a macromolecule of known volume is spherical and unsolvated we can calculate its radius, and its frictional coefficient from the Stokes equation (6.44). This calculated frictional coefficient can be compared with the measured frictional coefficient determined from the diffusion coefficient [Eq. (6.43)]. The calculated f is usually smaller than the measured f. The explanation for this is that the macromolecule is either solvated (which increases its radius), or nonspherical, or both.

In the following sections we will discuss these effects quantitatively, and learn that both increase the frictional coefficient.

Solvation

Some solvent is usually associated with a macromolecule in a solution. This solvation increases the effective or hydrodynamic volume of the macromolecule and, therefore, its frictional coefficient. We can measure the partial specific volume of a macromolecule in solution. It is the change in volume of a solution when g_2 grams of solute are added.

$$\bar{v}_2 \equiv \left(\frac{\partial V}{\partial g_2}\right)_{T, P, g_1} \tag{6.45}$$

The partial specific volume can be measured just as implied by Eq. (6.45). The volume of a solvent is measured, then a known weight of solute is added. The partial specific volume $\bar{v}_2$ is the change in volume of the solution Δv divided by the added weight Δg. The partial specific volume can depend on the concentration of the solute, the composition of the solvent, and temperature and pressure. We specify that these parameters are held constant by the subscripts g_1, T, P in the definition of $\bar{v}_2$. For an impermeable object, such as a glass bead, the increase in the volume of the solution is just equal to the volume of the bead.

Therefore, the partial specific volume $\bar{v}_2$, is equal to the specific volume, v_2 (volume per gram), of the bead. It will be independent of solvent composition and of solute concentration.

For a macromolecule the partial specific volume may be very dependent on such variables as pH, salt concentration, and macromolecule concentration. For some solutes the partial specific volume can even be negative! Addition of $MgSO_4$ to water *decreases* the volume of the solution at concentrations below about 0.07 M. In general when solute is added to a solution, the volume is increased due to the volume of the added solute, but is decreased by solvent bound to the solute if the bound solvent has a smaller volume than the free solvent. The Mg^{2+} cation binds water strongly and the bound water has a smaller volume than the free solvent. This is why the partial specific volume—the change in volume of the solution when the salt is added—of multivalent cation salts can be negative in dilute solution. The volume of each solvated ion is positive, but the volume of the solution is decreased by addition of salt.

The frictional coefficient, and thus the hydrodynamic properties, of a particle depends on its volume in solution. The volume of an unsolvated molecule is

$$ V = \frac{M}{N_0} v_2 = \frac{M}{N_0} \bar{v}_2 $$

where M is the molecular weight of the molecule and N_0 is Avogadro's number. For a solvated molecule which binds δ_1 grams of solvent per gram of molecule, the solvated volume V_s is

$$ V_s = \frac{M}{N_0} (\bar{v}_2 + \delta_1 v_1^0) \tag{6.46} $$

where $\bar{v}_2$ = partial specific volume ($cm^3\ g^{-1}$)
  δ_1 = hydration = weight of solvent bound per weight of molecule (unitless)
  v_1^0 = specific volume of the free solvent; it is the reciprocal of the density of the solvent ($cm^3\ g^{-1}$)

The partial specific volume, $\bar{v}_2$, and the hydration, δ_1, have been measured for many proteins and nucleic acids in aqueous solution. For most proteins $\bar{v}_2$ is 0.7 to 0.75 $cm^3\ g^{-1}$ and for double stranded DNA $\bar{v}_2$ is 0.4 to 0.5 $cm^3\ g^{-1}$. Values of hydration, δ_1, range from about 0.2 to 0.6 (g of water per gram of protein) for proteins, and from about 0.5 to 0.7 for DNA.

Shape Factor

We can now understand the effect of solvation and shape on the frictional coefficient. For a spherical macromolecule we can calculate the frictional coefficient from the Stokes equation (6.44), if we know the radius. But the radius of a sphere is related to its volume by $r = (3\ \text{volume}/4\pi)^{1/3}$. Therefore, from Eq. (6.46) we can calculate the radius of a *solvated* sphere.

$$ r = \left[\frac{3M(\bar{v}_2 + \delta_1 v_1^0)}{4\pi N_0} \right]^{1/3} \tag{6.47} $$

The frictional coefficient is directly proportional to the radius of a sphere. Therefore we can calculate that for typical values of $\bar{v}_2$ and δ_1, hydration can increase the frictional coefficient of a spherical macromolecule by 10 to 20%.

For a nonspherical macromolecule the frictional coefficient is greater than that for a sphere of the same size. It seems reasonable that any deviation from a spherical shape for the same size particle will increase its frictional coefficient. Computer calculations can be done to calculate the frictional coefficient for any shape particle, and tables exist for frictional coefficients of simple shapes. Figure 6.8 illustrates the increase in frictional coefficient as a sphere is changed to an ellipsoid of revolution with the same volume. An oblate

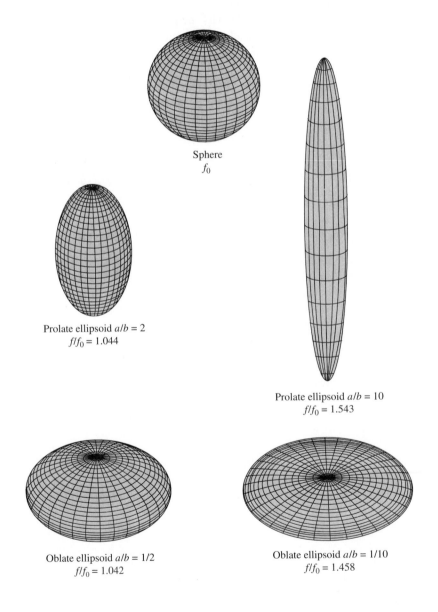

Sphere
f_0

Prolate ellipsoid $a/b = 2$
$f/f_0 = 1.044$

Prolate ellipsoid $a/b = 10$
$f/f_0 = 1.543$

Oblate ellipsoid $a/b = 1/2$
$f/f_0 = 1.042$

Oblate ellipsoid $a/b = 1/10$
$f/f_0 = 1.458$

ellipsoid of revolution is generated by rotating an ellipse about its minor axis; a prolate ellipsoid of revolution is generated by rotating an ellipse about its major axis. Figure 6.8 shows that as the axial ratio increases, the frictional coefficient increases faster for a prolate ellipsoid than for an oblate ellipsoid. Proteins and nucleic acids are not spheres or ellipsoids, but Fig. 6.8 shows that for small deviations from spheres, the frictional coefficient calculated from Stokes' equation [Eq. (6.44)] is a reasonable approximation.

Example 6.5   The diffusion coefficient D and the partial specific volume of an enzyme (ribonuclease) from bovine pancreas, which digests RNA, have been measured in a dilute buffer at 20°C. The values are:

$$D = 13.1 \times 10^{-7} \text{ cm}^2 \text{ s}^{-1}$$
$$\bar{v}_2 = 0.707 \text{ cm}^3 \text{ g}^{-1}$$

The molecular weight of the protein is 13,690.

(a) Calculate the frictional coefficient f.
(b) Assuming that the protein molecule is an unhydrated sphere ($\bar{v}_2 = v_2$), calculate its volume, radius r, and the frictional coefficient f_0 corresponding to this unhydrated sphere.
(c) From the ratio f/f_0, estimate the extent of hydration if the difference from unity is due entirely to hydration. That is, we assume that the molecule is spherical.

Solution

(a) From Eq. (6.43),

$$f = \frac{kT}{D}$$

$$k = 1.380 \times 10^{-16} \text{ erg K}^{-1} \text{ molecule}^{-1}$$
$$= 1.380 \times 10^{-16} \text{ g cm}^2 \text{ s}^{-2} \text{ K}^{-1} \text{ molecule}^{-1}$$

Therefore,

$$f = \frac{(1.380 \times 10^{-16} \text{ g cm}^2 \text{ s}^{-2} \text{ K}^{-1})(298 \text{ K})}{13.1 \times 10^{-7} \text{ cm}^2 \text{ s}^{-1}}$$

$$= 3.09 \times 10^{-8} \text{ g s}^{-1}$$

Fig. 6.8   The frictional coefficients of ellipsoids of revolution relative to a sphere of the same volume. Prolate ellipsoids are shaped like footballs; oblate ellipsoids are shaped like pancakes. The axis of revolution is designated by a for each molecule. Note that for slight deviations from a sphere—axial ratios of less than 2 to 1—the increase of f is less than 5%. The frictional coefficient increases faster for a prolate ellipsoid than for an oblate ellipsoid as the axial ratio increases.

(b) The mass of a ribonuclease molecule is $(13{,}690 \text{ g}/6.023 \times 10^{23})$, or 2.27×10^{-20} g. The volume is the mass times the partial specific volume $\bar{v}_2$:

$$V = (2.27 \times 10^{-20} \text{ g})(0.707 \text{ cm}^3 \text{ g}^{-1})$$

$$= 1.61 \times 10^{-20} \text{ cm}^3$$

Since

$$\tfrac{4}{3}\pi r^3 = 1.61 \times 10^{-20} \text{ cm}^3$$

$$r = 1.57 \times 10^{-7} \text{ cm}$$

$$= 15.7 \text{ Å}$$

$$f_0 = 6\pi\eta r$$

The viscosity coefficient for a dilute aqueous buffer is approximately equal to that of water, which is, at 20°C, 1.00 centipoise (cP) or $1.00 \times 10^{-2} \text{ g s}^{-1} \text{cm}^{-1}$; this is equal to 1.00 mPa s (see Table 2.2). Thus

$$f_0 = 6\pi(1.00 \times 10^{-2} \text{ g s}^{-1} \text{ cm}^{-1})(1.57 \times 10^{-7} \text{ cm})$$

$$= 2.95 \times 10^{-8} \text{ g s}^{-1}$$

(c) $f/f_0 = (3.09 \times 10^{-8})/(2.95 \times 10^{-8}) = 1.05$. From Eqs. (6.44) and (6.47) we can derive

$$\frac{\bar{v}_2 + \delta_1 v_1^0}{\bar{v}_2} = \left(\frac{f}{f_0}\right)^3$$

$$= 1.15$$

Thus $\delta_1 v_1^0 = 1.15 \bar{v}_2 - \bar{v}_2 = 0.15 \bar{v}_2 = 0.105 \text{ cm}^3 \text{ g}^{-1}$. The specific volume of bulk water, v_1^0, is $1.00 \text{ cm}^3 \text{ g}^{-1}$. Thus

$$\delta_1 = 0.105$$

That is, the amount of hydration is 0.105 g of water per gram of protein.

Note that in Example 6.5 the calculated ratio of $f/f_0 = 1.05$ could have been caused by a nonspherical shape. A measured frictional coefficient larger than that calculated for an unhydrated sphere can be caused either by hydration, or by a shape that is nonspherical.

Diffusion Coefficients of Random Coils

The concepts underlying our discussions in the preceding section are useful for macromolecules that are reasonably rigid. The situation is different when very flexible macromolecules are involved. Consider, for example, a large DNA molecule. Such a molecule, with its length much greater than its diameter, resembles somewhat a loose ball of thread in solution. As the macromolecule moves through solution, two limiting possibilities arise with regard to the solvent molecules within the domain of the coiled macromolecules. They can move freely, independent of the motion of the macromolecule, or they can be "trapped" by the macromolecule and move with it. These two limiting possibilities are called "free draining" and "nondraining," respectively. For DNA molecules in aqueous media, the "nondraining" model describes their hydrodynamic properties adequately. The amount of solvent hydrodynamically associated with a DNA molecule is large. Therefore, as an approximation the DNA molecule can be considered to be a highly hydrated sphere. The radius of this sphere is expected to be directly proportional to the average three-dimensional size of the molecule.

The root-mean-square end-to-end distance in a random polymer of N units is given by Eq. (6.33).

$$\langle d^2 \rangle^{1/2} = \sqrt{N}\, l \tag{6.33}$$

The root-mean-square radius of a random polymer is also proportional to the square root of the number of monomer units. Thus the root-mean-square radius is proportional to the square root of the molecular weight of a macromolecule if the macromolecule is a very flexible coil. DNA molecules with molecular weights, M, in the millions act as flexible coils; therefore their frictional coefficients are expected to be proportional to $M^{1/2}$, and their diffusion coefficients are expected to be proportional to $M^{-1/2}$. This is found to be approximately true.

SEDIMENTATION

Sedimentation is used to separate, purify, and analyze all sorts of cellular species ranging from individual proteins and nucleic acids to viruses, chromosomes, mitochondria, and so forth. Let us first consider sedimentation in a gravitational field. The free fall of a particle of mass m and partial specific volume $\bar{v}_2$ through a viscous medium of density ρ is illustrated in Fig. 6.9. The driving force that causes the particle to sink is the difference between the gravitational force and the buoyant force:

$$\text{driving force} = mg - m\bar{v}_2\rho g = m \cdot (1 - \bar{v}_2\rho)g$$

where g is the gravitational acceleration.

At time zero, when the particle is released, its velocity is zero. As it falls, it picks up speed. Let u be its velocity at time t. As the particle moves through the medium with a velocity u, it experiences a frictional force proportional to u and the frictional coefficient, f. From Newton's law we obtain

$$m \cdot (1 - \bar{v}_2\rho)g - fu = m\frac{du}{dt}$$

where du/dt is the acceleration.

Since the frictional force increases with u, a velocity is soon reached at which the frictional force balances the driving force, and the acceleration becomes zero. This velocity is called the terminal velocity, u_t:

$$m \cdot (1 - \bar{v}_2\rho)g - fu_t = 0$$

or

$$u_t = \frac{m \cdot (1 - \bar{v}_2\rho)g}{f} \tag{6.48}$$

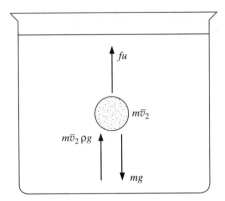

Fig. 6.9 Free fall of a particle, of mass m and specific volume $\bar{v}_2$, through a viscous medium of density ρ. The forces acting on the particle are, in one direction, the gravitational force mg, and in the other direction, the buoyant force $m\bar{v}_2\rho g$ and the frictional force fu, where u is the velocity of the particle and f is the frictional coefficient.

In a centrifugal field, the centrifugal acceleration is $\omega^2 x$, where ω is the angular velocity of the centrifuge in units of rad s^{-1}, and x is the distance from the center of rotation. By analogy to the free fall of a sphere in a gravitational field, a molecule (of mass m and partial specific volume $\bar{v}_2$) sedimenting through a viscous medium in a centrifugal field also reaches a terminal velocity u_t:

$$u_t = \frac{m \cdot (1 - \bar{v}_2\rho)}{f} \omega^2 x \qquad (6.49)$$

The centrifugal acceleration $\omega^2 x$ now replaces the gravitational acceleration g in Eq. (6.48).

The quantity $u_t/\omega^2 x$, which is the velocity per unit acceleration, is called the *sedimentation coefficient*. The symbol s is usually used for the sedimentation coefficient. The dimension of s is (cm sec^{-1})/(cm sec^{-2}) or sec. (Here we will abbreviate seconds as sec to avoid confusion with the sedimentation coefficient.) A convenient unit for s is the *svedberg*, named in honor of T. Svedberg, who pioneered much research on sedimentation in an ultracentrifuge. One svedberg, or 1 S, is defined as 10^{-13} sec. From the definition of s and Eq. (6.49), we have

$$s = \frac{m \cdot (1 - \bar{v}_2\rho)}{f} \qquad (6.50)$$

Determination of the Sedimentation Coefficient

Two methods can be used to measure the sedimentation coefficients of macromolecules. In one, a homogeneous solution is spun in an ultracentrifuge. As the macromolecules move down the centrifugal field, a solution-solvent boundary is generated. By following the movement of the boundary with time, the sedimentation coefficient can be calculated. This method is called *boundary sedimentation* and is illustrated in Fig. 6.10.

The generation of a boundary means that a concentration gradient is also generated, which, according to Eq. (6.37), results in diffusional transport of solute molecules. Thus, transport by sedimentation is coupled with transport by diffusion. If the macromolecules are very large or the centrifugal acceleration is very high, transport by sedimentation is much greater than transport by diffusion. Under such conditions, the boundary is very sharp. If transport by diffusion is significant, the boundary broadens as it moves downfield. As an approximation, we can assume that the diffusion process does not affect the rate of

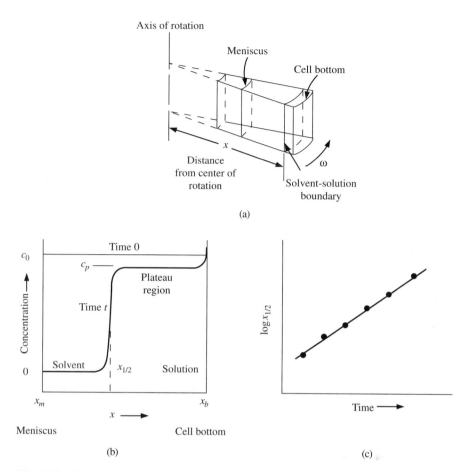

Fig. 6.10 Boundary sedimentation. (a) The compartment of the centrifuge cell containing the solution is sector-shaped, with the center of the sector at the axis of rotation. The dimensions of the cell are exaggerated for clarity. (b) Concentration as a function of the distance from the center of rotation. At time 0, the concentration (c_0) is uniform across the cell. At time t, owing to sedimentation of the macromolecular solute molecules, a sharp boundary has been generated, with solvent to its left and solution to its right. The concentration in the plateau region, c_p, is independent of position; c_p is lower than c_0 because of the sector shape of the cell compartment and because the centrifugal field increases with increasing x. It can be shown that $c_p/c_0 = (x_m/x_{1/2})^2$, where x_m is the position of the meniscus and $x_{1/2}$ is the position of the boundary. (c) Plot of $\log x_{1/2}$ versus t gives a straight line. The sedimentation coefficient can be obtained from the slope of this line [Eq. (6.51)].

movement of the boundary. It is usually sufficiently accurate to calculate s from the position of the midpoint of the boundary, $x_{1/2}$:

$$s = \frac{u_t}{\omega^2 x} = \frac{dx_{1/2}/dt}{\omega^2 x_{1/2}} = \frac{1}{\omega^2}\frac{d \ln x_{1/2}}{dt} = \frac{2.303}{\omega^2}\frac{d \log x_{1/2}}{dt} \qquad (6.51)$$

For a more rigorous discussion of the calculation of s, a more advanced text should be consulted. Several references are listed at the end of this chapter.

Example 6.6 *Escherichia coli* DNA ligase, an enzyme that catalyzes the formation of a phosphodiester bond from a pair of 3'-hydroxyl and 5'-phosphoryl groups in a double-stranded DNA, has a molecular weight of 74,000 and a $\bar{v}_2$ of 0.737 cm^3 g^{-1} at 20°C. Boundary sedimentation in a dilute aqueous buffer (0.02 M potassium phosphate, 0.01 M NH$_4$Cl, and 0.2 M KCl, pH 6.5) at 20.6°C and a speed of rotation of 56,050 rpm gave the following results (data courtesy of P. Modrich):

Time (min)	$x_{1/2}$ (cm)	log $x_{1/2}$
0	5.9110	0.7717
20	6.0217	0.7797
40	6.1141	0.7863
60	6.2068	0.7929
80	6.3040	0.7996
100	6.4047	0.8065
120	6.5133	0.8138
140	6.6141	0.8205

(a) Calculate s.
(b) Calculate the frictional factor f in the dilute aqueous buffer; the density of the buffer at 20.6°C is 1.010 g cm^{-3}.

Solution

(a) A plot of log $x_{1/2}$ versus t gives a straight line with a slope of 3.42×10^{-4} min^{-1}. The angular velocity ω is

$$\omega = (56{,}050 \text{ revolutions min}^{-1})(2\pi \text{ rad revolution}^{-1}) \, (1 \text{ min/60 sec})$$
$$= 5.870 \times 10^3 \text{ rad sec}^{-1}$$

(note that the unit radian is defined as an arc length divided by a radius and is therefore dimensionless)

From Eq. (6.51),

$$s = \frac{2.303}{\omega^2}(3.42 \times 10^{-4} \text{ min}^{-1})$$

$$= \frac{2.303(3.42 \times 10^{-4} \text{ min}^{-1})}{(5.870 \times 10^3 \text{ sec}^{-1})^2(60 \text{ sec min}^{-1})}$$

$$= 3.81 \times 10^{-13} \text{ sec}$$

$$= 3.81 \text{ S}$$

(b) From Eq. (6.50),

$$f = \frac{m \cdot (1 - \bar{v}_2\rho)}{s}$$

$$m = \frac{74{,}000}{6.023 \times 10^{23}} = 1.23 \times 10^{-19} \text{ g}$$

$$f = \frac{(1.23 \times 10^{-19} \text{ g})(1 - 0.737 \times 1.010)}{3.81 \times 10^{-13} \text{ sec}}$$

$$= 8.24 \times 10^{-8} \text{ g sec}^{-1}$$

A second method of sedimentation is called *zone sedimentation*. A thin layer of a solution of macromolecules is placed on top of a solvent at the beginning of centrifugation. As centrifugation continues, the macromolecules sediment through the solvent as a zone or band. However, because the density of the macromolecular solution will be higher than the solvent, an instability will occur. The sedimenting zone will be denser than the solvent below it and will tend to mix. To avoid such instability, a solvent gradient is usually imposed so that the net density gradient in the direction of the centrifugal field is always positive. For example, a preformed sucrose gradient can be used, as illustrated in Fig. 6.11. Alternatively, if a concentrated salt solution is used as the medium for sedimentation measurements, the density gradient generated by the sedimentation of the salt is sufficient to provide the stabilization.

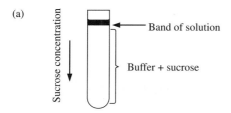

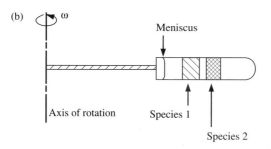

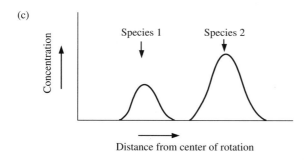

Fig. 6.11 Zone sedimentation through a preformed density gradient. (a) At time 0, a layer of solution is placed on a preformed gradient. In this illustration, a sucrose gradient (for example, a linear gradient from 5 to 20% sucrose) is employed. The buffer concentration is constant along the tube. (b) After spinning the tube in a centrifuge at an angular speed ω for a certain time, t, the macromolecular species have sedimented. In this example, the original solution contained two macromolecular species with different sedimentation coefficients. (c) The concentration profile after time t. A typical measurement technique involves the use of radioactively labeled macromolecules. The centrifuge is stopped after time t, a hole is punched through the bottom of the tube, and fractions are collected. The radioactivity of each fraction is then determined.

Standard Sedimentation Coefficient

To facilitate comparison of sedimentation coefficients measured in solvents with different values of solvent density, ρ, and solvent viscosity, η, it is common practice to standardize the measured s. From Eq. (6.50),

$$s = \frac{m \cdot (1 - \bar{v}_2 \rho)}{f}$$

we know that s is directly proportional to the buoyancy factor and inversely proportional to the frictional coefficient, f. The frictional coefficient is directly proportional to the solvent viscosity, and the buoyancy factor depends on the solvent density. Therefore we can calculate a standard value of a sedimentation coefficient, chosen to be the value at 20°C in water, $s_{20, w}$, by multiplying the measured s by ratios of viscosities and buoyancy factors.

$$s_{20, w} = s \cdot \left(\frac{\eta}{\eta_{20, w}}\right) \frac{(1 - \bar{v}_2 \rho)_{20, w}}{(1 - \bar{v}_2 \rho)} \tag{6.52}$$

To use Eq. (6.52) to convert the measured s to $s_{20, w}$, we must measure the density and viscosity of the solvent we use for the sedimentation experiments. For the calculation to be valid for different solvents, it is necessary that the solvent not change the shape or solvation of the macromolecule significantly. In Eq. (6.52) we assume that the only effect on the frictional coefficient is through the solvent viscosity. If two different solvents give very different values of $s_{20, w}$, we know that the macromolecule has different conformations in the two solvents.

Example 6.7   Convert the value of s obtained in Example 6.6(a) to $s_{20, w}$. At 20°C the ratio of the viscosity of the buffer, η_b, to that of water, η_w, has been measured to be $\eta_b / \eta_w = 1.003$.

Solution   For a dilute buffer, it is sufficiently accurate to assume that its temperature dependence of viscosity is the same as that of water. Thus

$$\frac{\eta_b}{\eta_w} = 1.003$$

From a handbook we obtain $\eta_{20.6, w} = 0.9906$ cP and $\eta_{20, w} = 1.0050$ cP. Therefore,

$$\frac{\eta_{20.6, w}}{\eta_{20, w}} = \frac{1.003 \times 0.9906}{1.0050} = 0.989$$

We assume for the protein that $\bar{v}_2$ at 20.6°C is the same as $\bar{v}_2$ at 20°C. The density of water at 20°C is 0.9982 g cm^{-3} from Table 2.2. Thus

$$\frac{(1 - \bar{v}_2 \rho)_{20, w}}{(1 - \bar{v}_2 \rho)} = \frac{1 - 0.737 \times 0.9982}{1 - 0.737 \times 1.010} = 1.034$$

Substituting these values into Eq. (6.52), we obtain

$$s_{20, w} = 3.81 \text{ S} \times 0.989 \times 1.034 = 3.90 \text{ S}$$

Determination of Molecular Weights from Sedimentation and Diffusion

Equation (6.50) states that the sedimentation coefficient s depends on the mass m, the frictional coefficient f, and the buoyancy factor $(1 - \bar{v}_2 \rho)$ of a molecule. But f can be obtained from a measured diffusion coefficient.

$$f = \frac{kT}{D} \tag{6.43}$$

Combining Eqs. (6.50) and (6.43), and replacing the molecular mass by the mass per mol of molecules ($M = mN_0$) we obtain

$$M = \frac{RTs}{D \cdot (1 - \bar{v}_2 \rho)} \tag{6.53}$$

Thus the molecular weight of a molecule can be obtained by measuring its sedimentation coefficient s, its diffusion constant D, and its partial specific volume $\bar{v}_2$. This provides an absolute method for determining molecular weights.

> **Example 6.8**   The following data have been obtained for ribosomes from a paramecium: $s_{20, w} = 82.6$, $D_{20, w} = 1.52 \times 10^{-7} \, \text{cm}^2 \, \text{sec}^{-1}$, $\bar{v}_{20} = 0.61 \, \text{cm}^3 \, \text{g}^{-1}$ [from A. H. Reisner, J. Rowe, and H. M. Macindoe, *J. Mol. Biol. 32*, 587 (1968)]. Calculate the molecular weight of the ribosomes.
>
> **Solution**   From Eq. (6.53),
>
> $$M = \frac{(8.314 \, \text{J K}^{-1} \, \text{mol}^{-1})(293 \, \text{K})(82.6 \times 10^{-13} \, \text{s})}{(1.52 \times 10^{-7} \, \text{cm}^2 \, \text{s}^{-1})(10^{-2} \, \text{m cm}^{-1})^2 (1 - 0.61 \times 1.00)}$$
>
> $$= 3.4 \times 10^3 \, \text{J s}^2 \, \text{m}^{-2}$$
>
> $$= 3.4 \times 10^3 \, \text{J s}^2 \, \text{m}^{-2} \left(\frac{\text{kg m}^2 \, \text{s}^{-2}}{\text{J}} \right) \left(\frac{10^3 \, \text{g}}{\text{kg}} \right)$$
>
> $$= 3.4 \times 10^6 \, \text{g mol}^{-1}$$

In a given medium for a family of macromolecules of the same $\bar{v}_2$, we expect s to be proportional to M/f. We discussed previously that, for random coils in the nondraining limit, f is proportional to $M/M^{1/2}$ or $M^{1/2}$. Experimentally, for large DNA molecules s is found to be proportional to $M^{0.44}$, close to our expectation. For spherical, unsolvated molecules of the same partial specific volume, f is proportional to r and therefore to $M^{1/3}$, and s is expected to be proportional to $M^{2/3}$.

Determination of Molecular Weights from Sedimentation Equilibrium

Suppose that a homogeneous two-component solution (a solute plus a solvent) is spun in an ultracentrifuge. Because of sedimentation, a concentration gradient is generated. Diffusion then sets in. Since transport by sedimentation and by diffusion go in opposite directions, an equilibrium concentration gradient can be generated by centrifugation, in which transport by sedimentation exactly balances transport by diffusion. Similar concentration gradients develop in the earth's atmosphere and in the oceans, although convective currents can disrupt the balance between gravitational and diffusive transport.

The concentration at equilibrium in a gravitational field or a centrifugal field does not depend on frictional coefficients, sedimentation coefficients, or diffusion coefficients of the molecules. At equilibrium, thermodynamics, not kinetics, is controlling. We will describe the equilibrium concentration gradient for an ideal solution or for real solutions at low concentrations. A rigorous thermodynamic derivation is given, for example, in Cantor and Schimmel (1980).

The concentration of molecules at equilibrium in the earth's gravitational field will be higher at the bottom of a container than at the top because of their lower potential energy in the field at the bottom. For a molecule in an external field the total chemical potential (see Table 5.3) determines the concentration at equilibrium. For a quantitative statement we use the Boltzmann equation [Eq. (6.17)] for the probability of finding a molecule with energy E_i. The ratio of the number of molecules with energy E_j to the number with energy E_i (when the degeneracy factors are equal) is

$$\frac{P_j}{P_i} = \frac{e^{-E_j/kT}}{e^{-E_i/kT}} \tag{6.54}$$

The ratio of probabilities is equal to the ratio of concentrations. We can now write the equation as

$$\frac{c_j}{c_i} = e^{-(E_j - E_j)/kT} \tag{6.55a}$$

For energies per mol instead of energies per molecule

$$\frac{c_j}{c_i} = e^{-(\bar{E}_j - \bar{E}_j)/RT} \tag{6.55b}$$

For a molecule in a solvent of density ρ, the energy for a mole of molecules in a gravitational field with acceleration g is from Table 5.3

$$E = M \cdot (1 - \bar{v}_2\rho)\, g\, x$$

Substituting this expression into Eq. (6.55b) we obtain

$$\frac{c_2}{c_1} = e^{-M \cdot (1 - \bar{v}_2\rho)\, g \cdot (x_2 - x_1)/RT} \tag{6.56}$$

Here x is measured from the bottom of the container. Equation (6.56) states that at equilibrium the concentration of any molecule will vary exponentially with position above the bottom of the container. In the laboratory for distances of a few cm the effect is large only for high molecular weights (M greater than 10^8). To obtain equilibrium it is important not to stir the solution, nor to allow temperature gradients to cause convective mixing.

For large distances the exponential decrease of concentration with height above the earth's surface is important for all molecules. The change of pressure with altitude is approximately characterized by Eq. (6.56). With the buoyancy factor set equal to 1 and using the molecular weight of oxygen gas, we calculate reasonable values for the oxygen pressure as a function of altitude. Of course we can also take into account the change of temperature with altitude, and even the change of g with altitude.

To determine molecular weights of most molecules we need stronger fields than gravity to establish measurable concentration gradients, so we use centrifuges. The centrifugal acceleration for a centrifuge spinning at ω radians s^{-1} ($\omega = 2\pi$ revolutions per second) is $\omega^2 x$, where x is measured from the center of rotation. The force acting towards the bottom of the cell is $m \cdot (1 - \bar{v}_2\rho)\omega^2 x$. The potential energy decreases with increasing x; the energy is the negative integral of the force with respect to x. Thus, the energy for a mol of molecules in a centrifugal field with acceleration $\omega^2 x$ is (see Table 5.3)

$$E = -M \cdot (1 - \bar{v}_2\rho)\omega^2 x^2/2$$

Substituting this expression into Eq. (6.55b) we obtain

$$\frac{c_2}{c_1} = e^{+M \cdot (1 - \bar{v}_2\rho)\omega^2 \cdot (x_2^2 - x_1^2)/2RT} \tag{6.57}$$

The difference in sign between Eqs. (6.56) and (6.57) is due to the different convention for measuring x. In both cases the concentration increases toward the bottom of the cell, where the bottom in a centrifuge cell is farthest from the center of rotation. By choosing the speed of rotation ω, we can sediment molecules of any molecular weight. Note that the buoyancy factor can be positive or negative; this means that some molecules sink, but others, such as lipoproteins, float.

To determine a molecular weight from a sedimentation equilibrium experiment we take the logarithm of Eq. (6.57).

$$\ln\left(\frac{c_2}{c_1}\right) = \frac{M \cdot (1 - \bar{v}_2\rho)\omega^2 \cdot (x_2^2 - x_1^2)}{2RT} \tag{6.58}$$

We see that a plot of $\ln c$ vs. x^2 is a straight line with slope equal to

$$\frac{M \cdot (1 - \bar{v}_2\rho)\omega^2}{2RT}$$

Thus, to obtain an absolute molecular weight we use

$$M = \frac{2RT}{(1 - \bar{v}_2\rho)\omega^2} \cdot (\text{slope of } \ln c \text{ vs. } x^2)$$

Absolute molecular weights are emphasized because with gel electrophoresis methods (see following sections) we can only measure relative molecular weights. The most accurate molecular weights are obtained from the known amino acid sequence of a protein, or the nucleotide sequence of a nucleic acid. These provide good standards. However, for proteins containing more than one polypeptide, or if association or dissociation of macromolecules occurs, then sedimentation-diffusion or sedimentation equilibrium can provide the molecular weights of the complexes.

Density Gradient Centrifugation

A second application of equilibrium centrifugation involves the spinning of a concentrated salt solution, such as CsCl, at high speed. A concentration gradient is generated, which results in a density gradient, because the density of a CsCl solution increases with increasing concentration. If a macromolecular species is also present in the solution, it will form a band at a point in the salt gradient where the macromolecules are "buoyant." The buoyancy term $(1 - \bar{v}_2\rho)$ is zero at the value of solvent density ρ at this position in the density gradient. The macromolecules above this point sink; those below it float. This means that macromolecules that were uniformly distributed throughout the cell before sedimentation concentrate at this position when the density gradient forms. This is illustrated in Fig. 6.12.

(a)

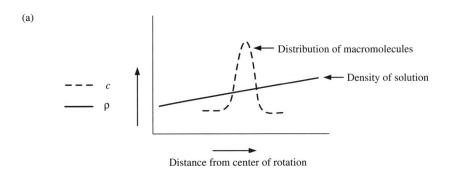

(b)

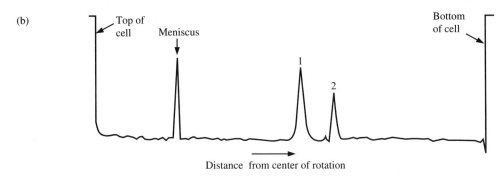

Fig. 6.12 Density-gradient centrifugation. (a) A macromolecular species in a concentrated salt solution of an appropriate density is spun in an ultracentrifuge. The solution was initially homogeneous. After a certain time, equilibrium is reached. The concentration of the salt, and consequently, the density of the solution, increases with increasing distance from the center of rotation. The macromolecular species forms a band at a position at which the solvated molecules are buoyant. (b) An actual tracing of two DNA species in a CsCl solution. The initial homogenous solution has a density of 1.739 g cm^{-3}. After 17 hr at 44,770 revolutions min^{-1} and 25°C the DNA species form two sharp bands. Species 1 is a bacterial virus DNA with a molecular weight 20×10^6. Species 2 is the same DNA except that it contains a heavier isotope of nitrogen (^{15}N rather than the usual ^{14}N). The substitution of ^{14}N by ^{15}N increases the buoyant density of this DNA by 0.012 g cm^{-3}.

The higher the molecular weight of the macromolecular species, the sharper will be the band that forms in the density gradient. Many biological macromolecules have "buoyant densities" sufficiently different that they can be resolved by the density-gradient centrifugation method.

The original experiment that showed that DNA replicated by making a new complementary strand for each original strand in the parent DNA was done by using density gradient centrifugation. Bacteria with DNA containing ^{14}N nitrogen were grown in a medium containing ^{15}N nitrogen. The DNA was isolated as a function of time and analyzed in a CsCl density gradient as shown in Fig. (6.12). The original DNA had both strands labeled with ^{14}N; this DNA is designated as ^{14}N:^{14}N. After one generation DNA appears with one ^{14}N strand and one ^{15}N strand (^{14}N:^{15}N). It is not until the second generation that ^{15}N:^{15}N DNA appears.

VISCOSITY

When an external force F is applied to a solid particle to make it move through a liquid with a velocity u_t, the molecules of the liquid at the interface with the particle move at the same velocity u_t because of the attractive forces between the two. Far away from the particle, the liquid remains stationary. Thus the movement of the particle through the liquid generates a velocity gradient in the medium.

Whenever a velocity gradient is maintained in a liquid, momentum is constantly transferred in a direction opposite to the direction of the velocity gradient. This is illustrated in Fig. 6.13. Here the liquid is moving in the direction x. The x components of the average velocities of molecules in different layers are represented by arrows of different lengths. The velocity gradient is in the y direction; that is, the x component of the velocity, u_x, increases with increasing y. The molecules are also moving in the y and z directions.

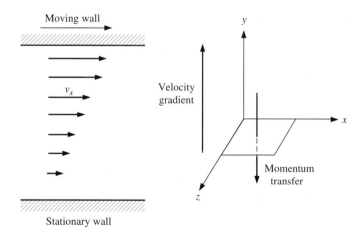

Fig. 6.13 Uniform velocity gradient produced in a fluid placed between a moving wall and a parallel stationary wall. The velocity gradient is a vector in the direction of increasing velocity, y, and is perpendicular to the direction of flow, x. Momentum transfer occurs in the direction $-y$, opposite to the velocity gradient.

However, because there is no net flow in the y and z directions, the movement of the molecules in these directions is random. Now consider a unit cross-sectional area in the xz plane. In 1 s, a certain number of molecules will move across this area from below, and an equal number of molecules will move across this area from above. But since molecules from below have lower u_x and consequently lower momentum in the x direction, there is a net transfer of momentum in the x direction from above to below, that is, in a direction opposite to the velocity gradient. The steeper the velocity gradient, the larger the net momentum transfer. Mathematically, we write

$$J_{mu} \propto -\frac{du_x}{dy}$$

or

$$J_{mu} = -\eta \frac{du_x}{dy} \tag{6.59}$$

where J_{mu} is the rate of momentum transfer per unit time per unit cross-sectional area.

The quantity η (eta) in Eq. (6.59) is called the *viscosity coefficient* or viscosity. Its SI units are kg m^{-1} s^{-1} ≡ Pa s, but viscosities are often given in poise [1 poise (P) = 1 g cm^{-1} s^{-1}; P = 10^{-1} Pa s; 1 cP = 1 mPa s]. If η is a constant independent of du_x/dy, the fluid is called a *Newtonian fluid*. If η is itself dependent on du_x/dy, the fluid is *non-Newtonian*.

Thus, when a particle moves through a stationary fluid under an external force F, a velocity gradient is generated and the velocity gradient, in turn, imposes a viscous drag F' on the particle. The final velocity gradient is such that F and F' balance each other, and the particle moves at the terminal velocity u_t.

Measurement of Viscosity

In our discussion of the free fall of a particle through a viscous medium, we obtained

$$u_t = \frac{m \cdot (1 - \bar{v}\rho)g}{f} \tag{6.48}$$

where we have dropped the subscript 2 on $\bar{v}$ for simplicity. Substituting the Stokes equation for a sphere, $f = 6\pi\eta r$, we obtain, upon rearranging,

$$\eta = \frac{m \cdot (1 - \bar{v}\rho)g}{6\pi r u_t} \tag{6.60}$$

Thus η for a fluid can be determined by measuring u_t for a sphere falling through it. Note also that from measurements using the same particle, the relative viscosities of two liquids η_2 and η_1 can be calculated from

$$\frac{\eta_2}{\eta_1} = \left(\frac{1 - \bar{v}\rho_2}{1 - \bar{v}\rho_1}\right)\frac{u_{t1}}{u_{t2}} \tag{6.61}$$

where the subscripts 1 and 2 refer to the quantities for liquids 1 and 2, respectively.

A more convenient method is to measure the volume rate of flow through a capillary. A simple viscometer, called an Ostwald viscometer, is shown in Fig. 6.14. Here the time t

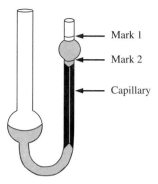

Mark 1

Mark 2

Capillary

Fig. 6.14 Simple Ostwald viscometer. Liquid is drawn up initially into the arm on the right to above mark 1. Upon release, it begins to flow back under the influence of gravity but restricted by viscosity primarily in the capillary region. The elapsed time, t, is measured from when the meniscus passes mark 1 until it passes mark 2.

required for the liquid level to drop from mark 1 to mark 2 is measured. The relative viscosities of two liquids are measured from the ratio of their flow times, t_1 and t_2, and the ratio of their densities, ρ_1 and ρ_2.

$$\frac{\eta_2}{\eta_1} = \frac{\rho_2 t_2}{\rho_1 t_1} \tag{6.62}$$

Viscosities of Solutions

Adding macromolecules to a solvent increases the viscosity of the solution. The viscosity of a solution of macromolecules depends on concentration, and the size and shape of the macromolecule. The *specific viscosity* of a solution is defined as

$$\eta_{sp} \equiv \frac{\eta' - \eta}{\eta} \tag{6.63}$$

where η is the viscosity of the solvent and η' is the viscosity of the solution. The specific viscosity is unitless. To separate the effect of concentration on viscosity from that of molecular size and shape, the *intrinsic viscosity* is defined as the limit of the specific viscosity divided by concentration, as the concentration approaches zero.

$$[\eta] \equiv \lim_{c \to 0} \frac{\eta_{sp}}{c} \tag{6.64}$$

The units of $[\eta]$ are the reciprocal of concentration c. We will use c in g cm^{-3}, thus the units of $[\eta]$ are g^{-1} cm^3. The intrinsic viscosity can be related to molecular properties by

$$[\eta] = \nu \cdot (\bar{v}_2 + \delta_1 v_1^0) \tag{6.65}$$

where ν (nu) is a unitless shape factor which equals 2.5 for spheres, and increases rapidly for nonspherical shapes. Prolate ellipsoids have higher values than oblate ellipsoids; for an axial ratio of 10 the value of ν is about 15 for prolate ellipsoids and about 8 for oblate ellipsoids (see Fig. 6.15). The intrinsic viscosity depends on the volume of the particle, so it depends on its partial specific volume $\bar{v}_2$ as well as its hydration δ_1 and the specific volume of the solvent v_1^0.

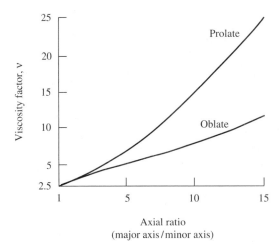

Fig. 6.15 Factor for the viscosity increment of ellipsoids, for relatively low values of the axial ratio (major axis/minor axis).

Measurement of the viscosity of a solution of macromolecules is a convenient method to study denaturation of a protein, or the assembly of several molecules into a large particle. The change of viscosity as a function of time, or pH, or temperature, is a measure of the change in shape or volume of the molecules.

ELECTROPHORESIS

Biological macromolecules are usually charged, therefore they will move in the presence of an electric field. Obviously, molecules with a net positive charge will move toward the negative electrode and molecules with a net negative charge will move away from the negative electrode. The velocity of motion depends on the magnitude of the electric field E, the net charge of the molecule, and the size and shape of the molecule as characterized by its frictional coefficient, f. The net charge on the molecule is designated by Ze where Z is the number of electronic charges e.

For a particle in a nonconducting solvent the velocity u of migration (m s^{-1}) in an electric field is

$$u = \frac{ZeE}{f} \tag{6.66}$$

where Z is the (unitless) number of charges, $e = 1.6022 \times 10^{-19}$ coulombs, E is the electric field in volts per m, and f is the frictional coefficient in kg s^{-1}. The velocity per unit electric field, u/E, is called the *electrophoretic mobility, μ*. The electrophoretic mobility and Eq. (6.66) can be used to measure the charge of a particle in a nonconducting medium. The charge of the electron was originally measured this way with a charged oil drop in air. However, biological macromolecules are found in aqueous solutions with other ions, buffers, and so on. Therefore, the charged macromolecule is surrounded by an atmosphere of small ions. This ion atmosphere greatly complicates the interpretation of electrophoretic mobility. The shielding effect of the ion atmosphere reduces the electric field experienced by the macromolecule. Furthermore, when the macromolecule moves, it drags its ion atmosphere with it. Therefore, the frictional coefficient is changed. The complications make it very difficult to interpret electrophoretic mobility quantitatively as we have done for the other hydrodynamic measurements. Electrophoresis is a powerful analytical tool, however, and macromolecules with very small differences in their properties can be resolved. One example is the separation of normal hemoglobin and hemoglobin from patients who suffer from sickle-cell anemia. Sickle-cell hemoglobin S differs from normal hemoglobin A by a single change of valine for glutamic acid in each of the two β-chain peptides. As a consequence, the proteins differ by two charges per molecule, which is sufficient to effect a clear separation by electrophoresis.

In the following sections we will describe some applications of electrophoresis to the characterization of macromolecules.

Gel Electrophoresis

Nearly all electrophoresis of macromolecules is done in gels. We want to separate molecules according to charge or size or shape; the gel greatly reduces mixing of the molecules by convection and by diffusion.

A gel is a three-dimensional polymer network dispersed in a solvent. A variety of gels have been used. For example, an agarose gel consists of an aqueous medium and a polysaccharide obtained from agar. An acrylamide gel consists of an aqueous medium and a copolymer of acrylamide (CH_2=CH—CO—NH_2) and N, N'-methylenebisacrylamide (NH_2=CH—CO—NH—CH_2—NH—CO—CH=CH_2). In the latter gel, the water-soluble acrylamide, with its reactive double bond, can polymerize into a linear chain.

$$\text{—}(CH_2-CH-CH_2-CH)\text{—}$$
$$| \qquad \qquad |$$
$$CO \qquad \quad CO$$
$$| \qquad \qquad |$$
$$NH_2 \qquad \quad NH_2$$

Adding the bisacrylamide, which has two double bonds, results in the formation of cross-links between different chains. The degree of cross-linking can be controlled by the ratio of the bis compound to acrylamide. In a typical gel, over 90% of the space is occupied by the aqueous medium, but the presence of the three-dimensional polymer network prevents convectional flow.

The gel contributes additional factors that affect the electrophoretic mobility: (1) the actual path traveled by a macromolecule through the porous gel is likely to be considerably longer than the length of the gel; (2) the gel imposes additional frictional resistance to the macromolecules as they move through pores of comparable dimensions; (3) the macromolecules may interact with the gel network (for example, through electrostatic interactions, if the gel network contains charged groups); and (4) the parts of the gel with pores smaller than a particular macromolecular species are inaccessible to that species. The combination of these factors further improves the resolution of electrophoresis.

DNA Sequencing

Gel electrophoresis in a denaturing solvent provides a rapid, effective method to determine the sequence of a single strand of DNA. Single-stranded nucleic acids usually contain regions of complementary base sequences. As a consequence, intramolecular base pairing can occur. Since such pairing affects the frictional coefficients, the electrophoretic mobility of single-stranded nucleic acids is not a simple function of molecular weight. However, if electrophoresis is carried out in a gel containing a high concentration of a neutral reagent (such as urea) that disrupts base pairing, the mobility of a nucleic acid is determined by the number of nucleotide phosphate groups in the strand. This simple method has been combined with brilliant logic to invent a quick method for determining the base sequence of a nucleic acid. The problem was to determine the sequence of four bases in a single strand of nucleic acid. The solution was to make a break in the strand after only one type of base (such as guanine) but not at every location of this base. If one end of the strand is labeled, for example, with ^{32}P, we end up with strands of different length, each with ^{32}P on one end and guanine on the other. Measuring the chain length by gel electrophoresis gives us the positions of all the guanines. Repeating the procedure for each of the other bases gives the base sequence. This method was invented independently by Maxam and Gilbert [*Proc. Natl. Acad. Sci. USA 74*, 560–564 (1977)] and by Sanger and coworkers [*Proc. Natl. Acad. Sci. USA 74*, 5463–5467 (1977)]. Figure 6.16 shows an example of the procedure.

Sanger's method uses an enzyme, DNA polymerase, to make a complement of the DNA being sequenced. A primer (an oligonucleotide complementary to the DNA sequence) is used to specify where the DNA polymerase starts synthesis. The enzyme adds deoxynucleoside triphosphates (dNTPs) to the primer to form the newly synthesized strand; it is the one actually sequenced. The breaks in the strand, which give the different length fragments, are made by adding a few percent of a dideoxynucleoside triphosphate (ddNTP) to the dNTP mixture. Incorporation of a ddNTP into the new strand stops the polymerization at that point. For example, to obtain the positions of the guanines in the newly synthesized strand, 10% ddGTP is added. Most of the time dGTP, which allows the chain to continue, is incorporated, but 10% of the time ddGTP, which stops the chain, is incorporated. Thus a ladder of chains with increasing lengths is obtained. All strands end in G; all start at the primer. Electrophoresis as shown in Fig. 6.16 gives the chain lengths. Repeating with each of the other bases completes the sequence. By using fluorescent derivatives of the ddNTPs we can do the gel electrophoresis in a single lane instead of four separate lanes. The lane is scanned by a photodetector. Thus, the position of the band gives the chain length; its fluorescence characteristics identify the base. This method allows many different experiments to be done on a single gel and provides a more efficient recording of the sequence data.

The Human Genome Project has as its goal the complete sequence of a human genome: sequences for all 24 human chromosomes—about 3×10^9 base pairs in all. To accomplish this in a reasonable time will require a tremendous increase in the speed of sequencing. Even sequencing at one million base pairs a day, it will take 10 years to finish. Hunkapiller et al. (1991) discuss new methods, and improvements in automation and data acquisition to greatly increase the rate of sequencing.

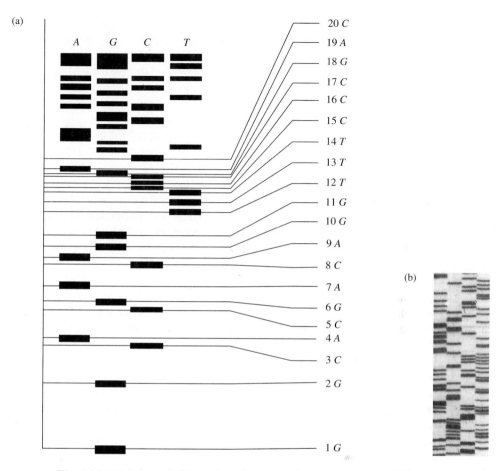

Fig. 6.16 (a) Schematic illustration of a sequencing gel electrophoresis pattern. The sequence of the first 20 bases of a single strand of DNA is shown. The strand is labeled at the 5′-hydroxyl group of the first nucleotide with radioactive ^{32}P. The strand is broken at adenines (A), guanines (G), cytosines (C), or thymines (T) and placed in four lanes at the top of a denaturing gel of 20% acrylamide, 0.67% methylene bisacrylamide, 7 M urea. After electrophoresis (the positive electrode is at the bottom of the gel), autoradiography for 8 hr produced the pattern shown. The sequence is now simply read off. [A. M. Maxam and W. Gilbert, *Proc. Natl. Acad. Sci. USA 74,* 560–564 (1977).] (b) An actual gel pattern.

Double-Stranded DNA

Double-stranded DNA molecules can be separated according to molecular weight by electrophoresis in either polyacrylamide gels or agarose gels. No denaturant, such as urea, is added, so the gels are called native gels; the DNA runs in its native form. DNA has a uniform charge density because each base pair adds two phosphate charges, therefore in free solution the electrophoretic mobility is essentially independent of molecular weight. In gels the DNA molecules have to wander through the pores of the gel and good separation can be obtained in the range of 10 to 100,000 base pairs. The concentration of the agarose, or the cross-linking of the polyacrylamide, is optimized to give the best separation for each size range. Figure 6.17 illustrates the excellent separation of six DNA fragments in the range from 100 to 1,000 base pairs.

DNA Fingerprinting

Although all humans have similar sequences of their DNA—the similarity is what we plan to find through the Human Genome Project—we are all unique in our total DNA sequence. Identical twins start out with identical sequences at conception, but mutations during growth and development will cause slight divergences of their sequences. A DNA sample obtained from skin, blood, semen, hair, etc., left at a crime scene, for example, can be compared with the DNA from a suspect to see if they match.

Only about 5% of human DNA codes for proteins; much of the rest is simple repeating sequences. The number of repetitions of the sequence and the sequences differ from person to person. A region of the DNA where this occurs has been called the hypervariable fingerprint region (Jeffries et al., 1986). Restriction enzymes cut double-stranded DNA at specific sequences (such as CAATTG). They produce a distinct set of different length fragments of the fingerprint DNA; the fragments are separated by native gel electrophoresis (see Fig. 6.17). For identifying individual people Jeffries et al. (1986) used the size range of 2,000 to 20,000 base pairs separated on an agarose gel. The fragments are visualized by adding a radioactive probe oligonucleotide, which is complementary to the repeating sequence and binds to it. Use of a few different restriction enzymes and a few different probes is sufficient to distinguish even between members of a single family, but not identical twins. In criminal cases prosecutors, defense lawyers, and jurors are having to learn about reproducibility of gel electrophoresis patterns, and to judge when similar patterns mean guilt or innocence.

Conformations of Nucleic Acids

We have been describing the separation of nucleic acids based solely on their molecular weights—the number of nucleotides in single strands on denaturing gels, or the number of base pairs in double strands on native gels. But nucleic acids of the same molecular weight will have different electrophoretic mobilities when they have different shapes. An obvious example is linear double-stranded DNA versus circular double-stranded DNA. The compact circular DNA travels faster than the linear molecule. Many natural DNAs occur as circles, or twisted circles (figure eights and so forth). These superhelical DNAs differ only in their topology—the number of twists in the DNA before the ends are joined; they are called topological isomers or topoisomers. The topoisomers can be separated and quantitated by gel electrophoresis. The concentrations of DNA topoisomers are used to assay for topoisomerases, enzymes which wind and unwind the DNA during its replication. These enzymes are targets for anticancer drugs as discussed by Liu (1989), because the most rapidly dividing cells require the largest amounts of these enzymes.

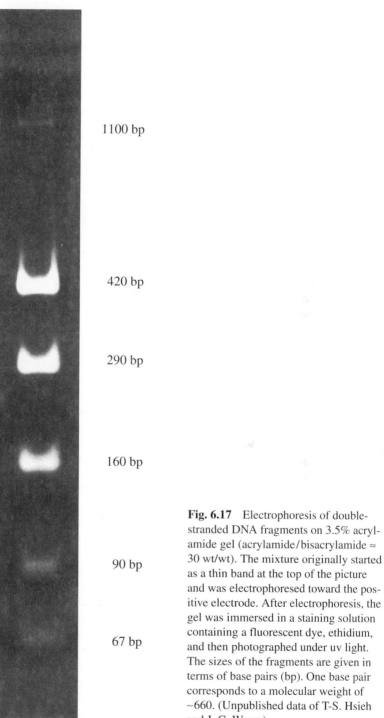

1100 bp

420 bp

290 bp

160 bp

90 bp

67 bp

Fig. 6.17 Electrophoresis of double-stranded DNA fragments on 3.5% acrylamide gel (acrylamide/bisacrylamide ≈ 30 wt/wt). The mixture originally started as a thin band at the top of the picture and was electrophoresed toward the positive electrode. After electrophoresis, the gel was immersed in a staining solution containing a fluorescent dye, ethidium, and then photographed under uv light. The sizes of the fragments are given in terms of base pairs (bp). One base pair corresponds to a molecular weight of ~660. (Unpublished data of T-S. Hsieh and J. C. Wang.)

The DNA double helix is a flexible linear rod that is approximately linear for a few hundred base pairs, but because of its flexibility it acts more like a random coil when it is thousands of base pairs in length. However, certain sequences of DNA, such as repeating A·T pairs, cause the DNA to bend. Many chemicals that react with DNA, including carcinogens (such as *N*-acetylaminofluorene) and anticancer drugs [such as *cis*-diamminedichloroplatinum (II)] also cause the DNA to bend. All these bent DNAs have anomalous electrophoretic mobilities and can be distinguished from linear molecules of the same size; this is reviewed by Leng (1990).

RNA molecules are synthesized as single strands, which then fold into intramolecular double-strand regions with single-stranded loops. The simplest folded structure is called a hairpin—a single stem and loop. More complex structures are formed by further folding and interactions of different kinds of loops and stems. These structures are vital for the correct interactions of RNA molecules with proteins and with other RNAs. One example is transfer RNA, which adds each amino acid to the growing polypeptide chain during protein synthesis. The folding of RNA molecules can be easily monitored by native gel electrophoresis; mutations can be made in the RNA to determine how the sequence determines the folding (Jacques and Susskind, 1991).

Pulsed-Field Gel Electrophoresis

In the usual gel electrophoresis the gel pores act like a sieve and different size DNA molecules pass through with different mobilities. For DNA molecules with more than 50,000 base pairs, separation by the usual gel electrophoresis method becomes poor. The DNA can no longer pass through the pores in its coiled form. To travel through the gel the DNA is distorted by the field and moves along its long axis through the pores. It travels like a snake. Once the DNA is stretched out, its mobility becomes nearly independent of length and no separation according to molecular weight is obtained. The solution to this problem is to make the DNA molecules change direction during electrophoresis. If an electric field is applied perpendicular to the direction of motion of the DNA, the molecule must change shape to allow it to travel in the new direction. The time necessary for this to occur depends on the size of the DNA. By using a pulsed electric field that alternates in direction, Schwartz and Cantor (1984) were able to separate DNA molecules in the million base pair range. Figure 6.18 shows one configuration of a pulsed-field gel electrophoresis (PFGE) apparatus; many other arrangements of electrodes have been used. The key variable for obtaining optimum separation is the length of each pulse; the longer the molecules, the longer the pulses required. Individual DNA molecules can be seen moving through a gel by using acridine-labeled DNA and a fluorescence microscope. Computer modeling of this motion allows calculation of the optimum pulse length to separate a given size range (see Smith et al., 1991).

An example of the effect of pulse length in field-inversion gel electrophoresis (FIGE) is shown in Fig. 6.19. The sign of the electric field is changed periodically to obtain improved separation of the DNA molecules. A field of 0.50 s plus and 0.25 s minus gives good separation in the size range from 10,000 to 50,000 base pairs. A field of 3 s plus and 1 s minus gives good separation in the size range from 50,000 to 200,000 base pairs.

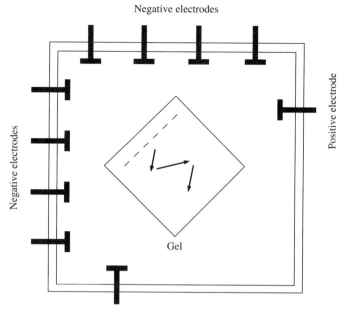

Negative electrodes

Negative electrodes

Positive electrode

Gel

Positive electrode

Fig. 6.18 A pulsed-field gel electrophoresis (PFGE) apparatus in which the DNA molecules move alternately down and to the right in the figure. The gel is actually horizontal and is immersed in a buffer solution. The net motion of the DNA is along the diagonal of the apparatus and is approximately straight in the gel. The angle between alternate electric fields ranges from 100° to 150°; this improves resolution. Each electrical pulse continues from 1 second to several minutes depending on the size of the molecules.

Protein Molecular Weights

Nucleic acids can be separated according to molecular weight by electrophoresis because there is one charged phosphate group for each nucleotide monomer unit. For proteins the number of charges depends on their amino acid compositions and the pH of the buffer. Furthermore, a polypeptide chain of a specific length can fold into different shapes with different frictional coefficients. Therefore, to use electrophoresis to determine protein molecular weights it is necessary to denature the protein and to introduce a charge on each peptide. This is done by adding an anionic detergent, sodium dodecylsulfate (SDS), and 2-mercaptoethanol; the latter disrupts sulfur-sulfur linkages in proteins. The combined action of these reagents causes the proteins to unfold. Furthermore, for most proteins at an SDS concentration greater than 10^{-3} M, a nearly constant amount of SDS is bound per unit weight of protein (approximately 0.5 detergent molecule is bound per amino acid residue). Thus, the charge of the protein-dodecylsulfate complex is due primarily to the charges of the bound dodecylsulfate groups, making the charge per unit weight approximately the same for most proteins.

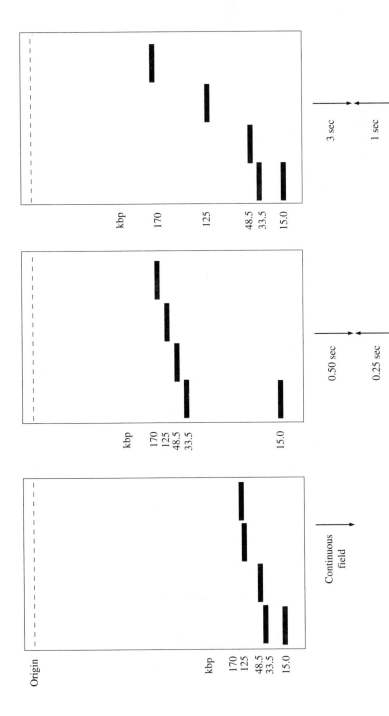

Fig. 6.19 Separation of DNA molecules by field-inversion gel electrophoresis (FIGE). The efficiency of separation is increased by alternately reversing the direction of the field. Three separate experiments are shown on DNA molecules ranging in size from 15.0 kilobase pairs (kbp) to 170 kbp. The DNA molecules are: T4 bacteriophage, 170 kbp; T5 bacteriophage, 125 kbp; lambda bacteriophage, 46.5 kbp; two restriction enzyme fragments of lambda bacteriophage, 15.0 and 33.5 kbp. The molecules start at the top of the figure labeled origin and move down in a 1% agarose gel. In a continuous field applied for 4 hr all sizes move with similar mobilities. In a field of 0.50 s forward and 0.25 s back for 12 hr, the two restriction fragments are well separated. In a field of 3 s forward and 1 s back for 12 hr, the lambda, T4, and T5 DNAs are well separated. The data are from Carle et al., *Science 232*, 65–68 (1986).

Under these conditions the mobilities of the SDS-treated proteins are determined by their molecular weights. If M is the molecular weight of a protein and x is the distance migrated in the gel (proportional to the mobility), the relation

$$\log M = a - bx \qquad (6.67)$$

is usually observed, where a and b are constants for a given gel at a given electric field. A set of proteins of known molecular weight must be used for calibration in the determination of a protein of unknown molecular weight. An example is shown in Fig. 6.20.

We emphasize that Eq. (6.67) is dependent on two factors: a constant amount of bound detergent per unit weight of protein, and charges due to bound detergent dominate those carried by the protein itself. Deviations from this type of relation occur when a protein binds an abnormal amount of dodecylsulfate (such as glycoproteins*) or carries a large number of charges (such as histones).*

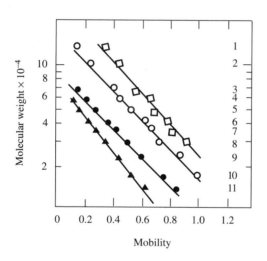

Mobility

Fig. 6.20 The electrophoretic mobilities of proteins in SDS-polyacrylamide gel electrophoresis (SDS-PAGE). The logarithms of the molecular weights are linear in the mobilities of the proteins, as stated in Eq. (6.67). The slope and intercept of the linear function depends on the amount of cross-linking and the concentration of the gel. The acrylamide concentrations illustrated are 15% (▲), 10% (●), 7.5% (O), and 5% (□). The weight ratio of acrylamide to methylenebisacrylamide is 37:1. The numbers 1 to 11 refer to the following proteins: β-galactosidase, phosphorylase a, serum albumin, catalase, fumarase, aldolase, glyceraldehyde-phosphate dehydrogenase, carbonic anhydrase, trypsin, myoglobin, and lysozyme. [From K. Weber, J. R. Pringle, and M. Osborn, *Methods Enzymol. 26*, 3 (1972).]

Protein Charge

For native proteins the mobility depends on the net charge and the frictional coefficient. The net charge depends on the amino acid composition, and the charges of any ligands bound covalently or reversibly. The net charge is always a function of pH; therefore, by judicious choice of pH, different proteins can be separated. All proteins will be positively charged at low enough pH because the carboxyl groups of aspartic and glutamic acid will

---

* A glycoprotein is a protein that has oligosaccharide groups attached. Histones are highly positively charged proteins rich in lysine and arginine residues.

be neutral (COOH), but the amino groups of lysine and arginine will be positive (NH_3^+). At high enough pH all proteins will be negatively charged because now the carboxyls are negative (COO^-) and the aminos are neutral (NH_2). This means that at some pH each protein must be neutral. This pH is the *isoelectric point*—the pH at which the protein has zero mobility.

Isoelectric focusing (IEF) uses the different isoelectric points of proteins to provide an effective separation method. Buffers are used to establish a pH gradient in the gel with high pH at the negative electrode and low pH at the positive electrode. We can start the sample on the high pH side near the negative electrode. The negatively charged proteins will move toward the positive electrode until each one reaches its isoelectric point and stops moving. Similarly, if we start the proteins at low pH near the positive electrode, each will move to its isoelectric point. Combining isoelectric focusing in one direction followed with SDS electrophoresis to separate by molecular weight at right angles to the first separation makes a very powerful analytical method. Figure 6.21 shows an example of the hundreds of different proteins that can be detected from a single cell.

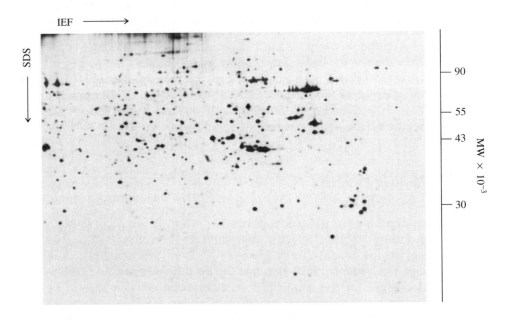

Fig. 6.21 A two-dimensional gel electrophoresis pattern of acidic proteins (isoelectric points are between pH 7.5 and 4.5) separated by isoelectric focusing (IEF) in the first dimension, and SDS-PAGE in the second dimension. The proteins are radioactively labeled by growing cells on ^{3}H or ^{35}S media; the radioactivity is detected by placing an x-ray sensitive film on the gel. [From R. Bravo, Fig. 6, p. 22 in *Two-Dimensional Gel Electrophoresis of Proteins*, edited by J. D. Celis and R. Bravo, Academic Press, 1984.]

Macromolecular Interactions

Gel electrophoresis can be used to study interactions between macromolecules. To determine the interaction between a protein and a nucleic acid, we do gel electrophoresis of the mixture. For example, to learn which restriction fragment of DNA binds to a protein, we compare the gel pattern of the ^{32}P-labeled DNA with and without protein. The fragment that is retarded in the presence of the protein is the one that is bound. Similar methods can be used to study the interaction of two proteins or two nucleic acids.

If the kinetics of the macromolecular reactions is slow enough so that re-equilibration does not occur during electrophoresis, equilibrium constants for reaction can be measured by gel electrophoresis. Mixtures are equilibrated before the mixture is applied to the gel, then electrophoresis is done to separate the individual species. If it is assumed that electrophoresis does not change the equilibrium concentrations, the concentration of each species determined after separation gives the equilibrium constant. For example, the binding of an RNA enzyme—a ribozyme—to its substrate was measured by this method (Pyle et al., 1990). It is well to emphasize that the gel electrophoresis method for measuring equilibrium binding is valid only if the kinetics of dissociation is slow. Otherwise, as the complex is moved away from its substrate it will dissociate; the apparent equilibrium constant will be a complicated function of the rate of separation and the rate of reaction.

SIZE AND SHAPE OF MACROMOLECULES

We have seen that measurements of the transport properties of a macromolecule allow the evaluation of its molecular weight M. Relevant properties for some representative macromolecules and particles are given in Table 6.3. From the diffusion coefficient or the sedimentation coefficient, the frictional coefficient f can be calculated. If the macromolecule is reasonably rigid, we can calculate for comparison the frictional coefficient f_0 for an unsolvated spherical molecule with molecular weight M and partial specific volume $\bar{v}_2$ that are the same as those of the macromolecule of interest. If f_0 is close to 1, we conclude that the macromolecule is approximately spherical and only lightly solvated. If f_0 is much greater than 1, the molecule is either highly solvated or asymmetric or both. Rough estimates of the extent of solvation can usually be made. For most proteins in dilute aqueous buffers, the amount of hydrodynamically associated water is in the range 0.2 to 0.6 g per gram of protein. Measurements, such as that of low-angle x-ray scattering, also provide determinations of the extent of solvation. Once a reasonable estimate or measurement is available for the solvation, the value of f/f_0 permits determination of the asymmetry of the macromolecule. We can calculate the dimensions of a prolate or oblate ellipsoid that fit the values the best. Additional information is provided by the intrinsic viscosity. The shape factor ν contains information that complements that obtained from the ratio f/f_0.

While the transport properties provide important information on the size and shape of a macromolecule, more detailed knowledge on the structural features and functional aspects of the macromolecule can be obtained only by other physical-chemical methods, some of which we will discuss in later chapters. The understanding of how molecules function in biological processes is usually achieved by a combination of many methods.

Table 6.3 Transport and related properties of some proteins

	$s_{20,w} \times 10^{13}$ (s)	$D_{20,w} \times 10^{7}$ (cm^2 g^{-1})	$\bar{v}_{20,w}$ (cm^3 s^{-1})	M	f/f_0
Ferricytochrome c (bovine heart)	1.91	13.20	0.707	12,744	1.077
Ribonuclease (bovine pancreas)	2.00	13.10	0.707	13,690	1.066
Myoglobin (horse heart)	2.04	11.30	0.741	16,951	1.105
Lysozyme (chicken egg white)	1.91	11.20	0.703	14,314	1.210
Chymotrypsinogen (bovine pancreas)	2.54	9.50	0.721	25,666	1.193
Immunoglobulin G (human)	6.6–7.2	4.00	0.739	156,000	1.513
Myosin	6.4	1.1	0.728	570,000	3.6
Tobacco mosaic virus	192	0.44	0.73	40,000,000	2.6

SOURCES: Selected from C. R. Cantor and P. R. Schimmel, *Biophysical Chemistry,* Part II, W. H. Freeman, San Francisco, 1980, and from other sources.

Lastly, since the transport properties are sensitive to the sizes and shapes of molecules, they can be used to study interactions between molecules, and conformational changes of molecules. Some examples are interactions between subunits of a macromolecular structure (such as a multisubunit protein or a ribosome), conformational change of a protein due to the binding of a substrate, circularization of a linear DNA, unfolding of a protein chain, and antibody-antigen interactions.

SUMMARY

Kinetic Theory

Average translational kinetic energy:

$$\langle U_{tr} \rangle = \tfrac{3}{2} kT \tag{6.14}$$

$\langle U_{tr} \rangle$ = the average translational kinetic energy of an atom in a molecule in any medium
k = Boltzmann's constant
T = absolute temperature

Mean-square velocity:

$$\langle u^2 \rangle = \frac{3RT}{M} \tag{6.15}$$

$\langle u^2 \rangle \equiv$ average of the square of the velocities (m s^{-1}) of molecules in a gas of molecular weight M (kg) at T K
$R =$ gas constant 8.314×10^7 erg K^{-1} mol^{-1} = 8.314 J K^{-1} mol^{-1}
1 erg = 1 g cm^2 s^{-2}
1 J = 1 kg m^2 s^{-2}

Boltzmann most-probable distribution of energies:

$$P_i = \frac{N_i}{N} = \frac{g_i \, e^{-E_i/kT}}{\sum_i g_i \, e^{-E_i/kT}} \tag{6.18}$$

$P_i =$ the probability of finding a molecule with energy E_i and degeneracy g_i
$N_i/N =$ the fraction of molecules with energy E_i and degeneracy g_i

The sum is over all possible energies.

Collision frequency:

$$z = 4\sqrt{\pi} \, \frac{N}{V} \, \sigma^2 \cdot \left(\frac{RT}{M}\right)^{1/2} \tag{6.26}$$

$z \equiv$ number of collisions a molecule encounters per second
$\dfrac{N}{V} \equiv$ number of molecules per unit volume
$\sigma \equiv$ diameter of a molecule (each molecule is considered to be a hard sphere)

Total number of collisions Z per unit volume per second:

$$Z = \frac{N}{V} \frac{z}{2} \tag{6.27}$$

Mean free path:

$$l = \frac{1}{\sqrt{2}\pi(N/V)\sigma^2} \tag{6.29}$$

Random walk:

$$\langle d^2 \rangle^{1/2} = \sqrt{N} \, l \tag{6.33}$$

$\langle d^2 \rangle^{1/2} =$ the root-mean-square displacement after N random steps of length l

Diffusion

Fick's first law:

$$J_x = -D \cdot \left(\frac{\partial c}{\partial x}\right)_t \tag{6.37}$$

J_x ≡ net amount of material that diffuses in the x direction per second, across an area 1 cm^2 perpendicular to x

D ≡ diffusion coefficient, which is directly related to the mean-square displacement per unit time; the units are cm^2 s^{-1}

$\left(\frac{\partial c}{\partial x}\right)_t$ = concentration gradient; it is the change of concentration with respect to x at a specified time t

Fick's second law:

$$\left(\frac{\partial c}{\partial t}\right)_x = D \cdot \left(\frac{\partial^2 c}{\partial x^2}\right)_t \tag{6.38}$$

$\left(\frac{\partial c}{\partial t}\right)_x$ = change of concentration versus time at a given position x

$\left(\frac{\partial^2 c}{\partial x^2}\right)_t$ = second derivative of the concentration c with respect to the position x, at a given time t

Diffusion coefficient D and frictional coefficient f:

$$D = \frac{kT}{f} \tag{6.43}$$

k = Boltzmann constant ≡ R/N_0
= 1.380 × 10^{-16} g cm^2 s^{-2} K^{-1} molecule^{-1}
= 1.380 × 10^{-23} J K^{-1} molecule^{-1}

Sedimentation

$$s = \frac{m \cdot (1 - \bar{v}_2 \rho)}{f} \tag{6.50}$$

$s \equiv \dfrac{u_t}{\omega^2 x}$ = velocity of sedimentation per unit centrifugal acceleration

The dimension of s is seconds. A convenient unit for s, the svedberg (S), is equal to 10^{-13} sec (1 S = 10^{-13} sec).

m = mass of the macromolecule
$\bar{v}_2$ = partial specific volume of the macromolecule
ρ = density of the solution
f = frictional coefficient

$$s_{20,\,w} = s\,\frac{\eta}{\eta_{20,\,w}}\,\frac{(1 - \bar{v}_2\rho)_{20,\,w}}{1 - \bar{v}_2\rho} \tag{6.52}$$

$s_{20,\,w}$ ≡ sedimentation coefficient corrected to give the expected value at 20°C in water

The subscript $s_{20,\,w}$ refers to quantities in water at 20°C.

Frictional Coefficient and Molecular Parameters

f can be obtained from either s or D. For a sphere of radius r,

$$f = 6\pi\eta r \tag{6.44}$$

η = viscosity of medium

For an unsolvated, spherical molecule of mass m and partial specific volume $\bar{v}_2$, its radius r is

$$r = \left(\text{volume} \cdot \frac{3}{4\pi}\right)^{1/3}$$

$$= \left(\frac{3m\bar{v}_2}{4\pi}\right)^{1/3}$$

The frictional coefficient f_0 for this unsolvated sphere is

$$f_0 = 6\pi\eta \cdot \left(\frac{3m\bar{v}_2}{4\pi}\right)^{1/3}$$

The deviation of f/f_0 from unity for a macromolecule can be due to either solvation or asymmetric shape.

For a *solvated* spherical molecule of mass m and partial specific volume $\bar{v}_2$, its radius r is

$$r = \left[\frac{3m(\bar{v}_2 + \delta_1 v_1^0)}{4\pi}\right]^{1/3} \tag{6.47}$$

δ_1 ≡ grams of solvent hydrodynamically associated with each gram of macromolecule
v_1^0 ≡ specific volume of the solvent; for dilute aqueous buffer, $v_1^0 = 1\ \mathrm{cm^3\ g^{-1}}$

f/f_0 is also determined by the shape of the molecule; it is 1 for a sphere, and values are tabulated for ellipsoids of revolution; for very flexible coils (random coils), f is expected to be proportional to $M^{1/2}$.

Combination of Diffusion and Sedimentation

$$M = \frac{RTs}{D \cdot (1 - \bar{v}_2\rho)} \tag{6.53}$$

Molecular weight M can be calculated from s, D, and the partial specific volume $\bar{v}_2$. ρ is the density of the solution.

Equilibrium centrifugation:

$$M = \frac{2RT}{\omega^2 \cdot (1 - \bar{v}_2\rho)} \frac{d \ln c}{d(x^2)} \tag{6.58}$$

$\omega \equiv 2\pi\nu = $ angular speed, rad s^{-1} ($\nu = $ revolutions s^{-1})
$c \equiv $ concentration
$x \equiv $ distance from the center of rotation

When centrifuged to equilibrium, the molecular weight of a macromolecule can be calculated from the slope of a plot of $\ln c$ versus x^2.

Viscosity

$$J_{mu} = -\eta \frac{du_x}{dy} \tag{6.59}$$

$J_{mu} \equiv $ rate of transfer of momentum in the x direction per second across a cross-sectional area 1 cm^2 in the xz plane

$\dfrac{du_x}{dy} \equiv $ velocity gradient, or the rate of change of the x component of the velocity with respect to y

$\eta \equiv $ viscosity coefficient

Newtonian fluid: η is independent of du_x/dy.
Non-Newtonian fluid: η is dependent on du_x/dy.
Solutions of macromolecules are often non-Newtonian and viscosity measurements are usually done at several values of du_x/dy and extrapolated to $du_x/dy = 0$.

Specific viscosity:

$$\eta_{sp} = \frac{\eta' - \eta}{\eta} \tag{6.63}$$

$\eta' \equiv $ viscosity coefficient of a solution of a macromolecular species
$\eta \equiv $ viscosity coefficient of solvent

Intrinsic viscosity is the limiting value of η_{sp}/c as c approaches 0, where c is the concentration of macromolecules in g cm^{-3}.

$$[\eta] \equiv \lim_{c \to 0} \frac{\eta_{sp}}{c} \tag{6.64}$$

Relation between $[\eta]$ and molecular parameters:

Rigid molecules:

$$[\eta] = v \cdot (\bar{v}_2 + \delta_1 v_1^0) \qquad (6.65)$$

$v \equiv$ shape factor; $v = 2.5$ for a sphere; values of v for ellipsoids of revolution are given in Fig. 6.15.
$\bar{v}_2 \equiv$ partial specific volume of the macromolecule
$\delta_1 \equiv$ grams of solvent molecules hydrodynamically associated with each gram of macromolecule
$v_1^0 \equiv$ specific volume of the solvent; for a dilute aqueous solution, $v_1^0 = 1 \text{ cm}^3 \text{ g}^{-1}$

Very flexible coils (random coils):

$$[\eta] \text{ approximately} \propto M^{1/2}$$

Electrophoresis

Electrophoretic mobility:

$$\mu = \frac{u}{E}$$

$u =$ velocity of charged particle, cm s^{-1}
$E =$ electric field strength, V cm^{-1}

$$\mu = \frac{ZeE}{f} \qquad (6.66)$$

$Z =$ number of charges on particle
$f =$ frictional coefficient
$e =$ electronic charge

Electrophoretic mobility $\equiv \mu/E$, or velocity per unit electric field.

Gel Electrophoresis

Molecular weight:

$$\log M = a - bx \qquad (6.67)$$

$M =$ molecular weight of protein
$x =$ distance traveled in gel (proportional to electrophoretic mobility)
$a, b =$ parameters established by measuring reference proteins of known molecular weights

REFERENCES

The following books are introductory texts similar in level to this chapter.

EISENBERG, D., and D. CROTHERS, 1979. *Physical Chemistry, with Applications to the Life Sciences,* Benjamin/Cummings, Menlo Park, California.

VAN HOLDE, K. E., 1971. *Physical Biochemistry,* Prentice-Hall, Englewood Cliffs, New Jersey.

The material in this chapter is covered at a more advanced level in

CANTOR, C. R., and SCHIMMEL, P. R., 1980. *Biophysical Chemistry,* Part II: *Techniques for the Study of Biological Structure and Function,* W. H. Freeman, San Francisco.

CELIS, J. D., R. BRAVO, eds., 1984. *Two-Dimensional Gel Electrophoresis of Proteins*, Academic Press, Orlando, Florida.

SUGGESTED READINGS

AXELROD, D., D. E. KOPPEL, J. SCHLESSINGER, E. ELSON, and W. W. WEBB, 1976. Mobility Measurement by Analysis of Fluorescence Photobleaching Recovery Kinetics, *Biophys. J. 16,* 1055–1069.

BAUER, W. R., F. H. C. CRICK, and J. H. WHITE, 1980. Supercoiled DNA, *Sci. Am. 243* (July), 118.

CARLE, G. F., M. FRANK, and M. V. OLSON, 1986. Electrophoretic Separation of Large DNA Molecules by Periodic Inversion of the Electric Field, *Science 232,* 65–68.

HUNKAPILLER, T., R. J. KAISER, B. F. KOOP, and L. HOOD, 1991. Large-Scale and Automated DNA Sequence Determination, *Science 254,* 59–67 (1991).

JACQUES, J.-P., and M. M. SUSSKIND, 1991. Use of Electrophoretic Mobility to Determine the Secondary Structure of a Small Antisense RNA, *Nucleic Acids Res. 19,* 2971–2977.

JEFFREYS, A. J., V. WILSON, and S. L. THEIN, 1986. Individual-Specific "Fingerprints" of Human DNA, *Nature 316,* 76–79.

LENG, M., 1990. DNA Bending Induced by Covalently Bound Drugs: Gel Electrophoresis and Chemical Probe Studies, *Biophys. Chem. 35,* 155–163.

LIU, L., 1989. DNA Topoisomerase Poisons as Anticancer Drugs, *Annu. Rev. Biochem. 58,* 351–375.

PYLE, A. M., J. A. McSWIGGEN, and T. R. CECH, 1990. Direct Measurement of Oligonucleotide Substrate Binding to Wild-Type and Mutant Ribozymes from *Tetrahymena, Proc. Natl. Acad. Sci. USA, 87,* 8187–8191.

RIGHETTI, P. G., 1990. Recent Developments in Electrophoretic Methods, *J. Chromatography 516,* 3–22.

SCHWARTZ, D. C., and C. R. CANTOR, 1984. Separation of Yeast Chromosome-Sized DNAs by Pulsed Field Gradient Gel Electrophoresis, *Cell 37,* 67–75.

SMITH, S. B., C. HELLER, and C. BUSTAMANTE, 1991. Model and Computer Simulations of the Motion of DNA Molecules During Pulse-Field Gel Electrophoresis, *Biochemistry 30,* 5264–5274.

PROBLEMS

1. The collisional diameter σ of a H_2 molecule is about 2.5 Å or 2.5×10^{-8} cm. For H_2 gas at 0°C and 1 atm, calculate
   (a) The root-mean-square velocity.
   (b) The translational kinetic energy of 1 mol of H_2 molecules.
   (c) The number of H_2 molecules in 1 cm^3 of the gas.
   (d) The mean free path.
   (e) The number of collisions each H_2 molecule encounters in 1 s.
   (f) The total number of intermolecular collisions in 1 s in 1 cm^3 of the gas.

2. A proton in a magnetic field can be in only one of two energy levels; each level has a degeneracy of 1. The energy spacing, ΔE, corresponds to a transition frequency, ν, of 100 MHz, using the equation $\Delta E = h\nu$, with h = Planck's constant. Use the Boltzmann distribution to calculate the absolute temperature when:
   (a) The proton is in the lower state only.
   (b) The probability of finding the proton in either level is equal.
   (c) The probability of finding the proton in the lower level is 1.000015 times the probability of finding it in the upper level. (Remember that $e^x = 1 + x + \cdots$ when x is small.)

3. The following data were reported for human immunoglobulin G (IgG) at 20°C in a dilute aqueous buffer,

$$M = 156,000$$

$$D_{20, w} = 4.0 \times 10^{-7} \text{ cm}^2 \text{ s}^{-1}$$

$$\bar{v}_2 = 0.739 \text{ cm}^3 \text{ g}^{-1}$$

   (a) Calculate f, the frictional coefficient.
   (b) Calculate f_0, the frictional coefficient of an unhydrated sphere of the same M and $\bar{v}_2$ as IgG.
   (c) From the ratio f/f_0, estimate the extent of hydration of IgG if it is a spherical molecule.
   (d) If IgG is not significantly hydrated, use Fig. 6.8 to estimate the dimensions of a prolate ellipsoid that best fit the data.

4. Diffusion, sedimentation, and electrophoresis can be used to characterize, identify, or separate macromolecules.
   (a) Describe one experimental method to measure a diffusion coefficient, one to measure a sedimentation coefficient, and one to measure an electrophoretic mobility. For each method state what is measured and how it is used to obtain the desired parameter.
   (b) Describe what you could learn about a protein by applying the three methods.
   (c) Describe what you could learn about a nucleic acid by applying the three methods.

5. The sedimentation coefficient of a certain DNA in 1 M NaCl at 20°C was measured by boundary sedimentation at 24,630 rpm. The following data were recorded:

Time, t (min)	Distance of boundary from center of rotation, x (cm)
16	6.2687
32	6.3507
48	6.4380
64	6.5174
80	6.6047
96	6.6814

   (a) Plot log x versus t. Calculate the sedimentation coefficient s.
   (b) The partial specific volume of the sodium salt of DNA is 0.556 cm^3 g^{-1}. The viscosity and density of 1 M NaCl and the viscosity of water can be found in the International Critical Tables; the values are 1.104 cP, 1.04 g cm^{-3}, and 1.005 cP, respectively. Calculate $s_{20, w}$ of the DNA.

6. For a bacteriophage T7, the following data at zero concentration have been obtained [Dubin et al., *J. Mol. Biol. 54*, 547 (1970)]:

$$s_{20,w}^0 = 453 \text{ S}$$

$$D_{20,w}^0 = 6.03 \times 10^{-8} \text{ cm}^2 \text{ s}^{-1}$$

$$\bar{v}_2 = 0.639 \text{ cm}^3 \text{ g}^{-1}$$

(a) Calculate the molecular weight of the phage.
(b) It can be calculated from phosphorus and nitrogen analyses of the phage that 51.2% (by weight) of the phage is DNA. Calculate the molecular weight of T7 DNA. Each phage contains one DNA molecule.

7. Two *spherical* viruses of molecular weights M_1 and M_2, respectively, happen to have the same partial specific volume $\bar{v}_2$. Neglecting hydration, what are the expected values for the ratios s_2/s_1, D_2/D_1, and $[\eta]_2/[\eta]_1$, where s_1, D_1, and $[\eta]_1$ are the sedimentation coefficient, diffusion coefficient, and the intrinsic viscosity, respectively, of the virus of molecular weight M_1, and s_2, D_2, and $[\eta]_2$ are the corresponding parameters of the virus of molecular weight M_2?

8. For a rodlike particle with length L and diameter d, its hydrodynamic properties are similar to those of a prolate ellipsoid of the same length and volume. Show that

$$\frac{L}{d} = \left(\frac{3}{2}\right)^{1/2} \frac{a}{b}$$

where a and b are the long and short semiaxes of the ellipsoid, respectively.

9. Each question below can be answered by the application of the equations discussed in this chapter.
(a) Some lipoproteins sediment in a centrifugal field and others float. What is the partial specific volume of a lipoprotein that neither sinks nor floats in a solution of density 1.125 g cm^{-3}?
(b) Calculate the sedimentation coefficient in seconds of a parachutist who is falling toward the earth at a velocity of 2 m s^{-1} (about 4.5 miles per hour).
(c) The intrinsic viscosity of a solution of spherical viruses is 1.5 cm^3 g^{-1}. The molecular weight is 5×10^6 g mol^{-1}. What is the volume of each solvated virus particle?
(d) A mutant protein with molecular weight 100,000 differs from the normal protein by the replacement of an alanine amino acid (side chain —CH$_3$) by an aspartic acid (side chain —CH$_2$COO$^-$). This small change has a negligible effect on the molecular weight and frictional coefficient. How would you separate the two proteins from each other?
(e) The rate of a gas phase reaction is directly proportional to the total number of collisions per second which occur in the container. State by what factor the rate of the reaction increases when the concentration of molecules doubles; when the collision cross section (σ^2) doubles; when the absolute temperature doubles; when the molecular weight doubles.

10. Nucleosome particles are composed of DNA wrapped around a "core" of protein. Nucleosome particles were analyzed using ultracentrifugation and dynamic light scattering. In a solution at 20°C the diffusion constant was measured to be 4.37×10^{-7} cm^2 s^{-1}. In the same solution boundary centrifugation was done with angular speed of rotation of $\omega = (2\pi\,18,100)$ radians per minute to obtain the following data for the position of the boundary measured from the center of rotation.

Time (minutes)	Boundary position (cm)
0	4.460
80	4.593
160	4.713
240	4.844

The density of the solvent was 1.02 g cm^{-3} and the partial specific volume of the nucleosome was determined to be 0.66 cm^3 g^{-1}.

(a) What is the molecular weight of a nucleosome particle?

The gel pattern shown below was found when the DNA in nucleosomes was cut with an enzyme, all protein was removed, and then gel electrophoresis of the DNA was done.

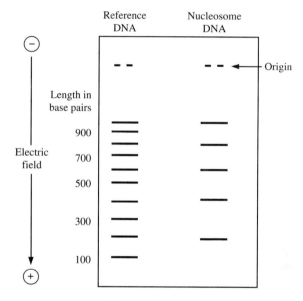

We can assume that the shortest fragment found corresponds to a piece of DNA which had been wrapped around a single nucleosome core, and longer fragments came from DNA associated with 2, 3, 4, etc. nucleosome particles. One DNA base pair has a molecular weight 660 g mol^{-1}.

(b) How many DNA base pairs are wrapped around each protein core?

(c) What is the molecular weight of a single protein in the nucleosome if (as other evidence suggests) this "nucleosome core" is made up of eight protein molecules that bind together, each of about the same molecular weight?

11. Use the data for lysozyme given in Table 6.3 to answer the following questions.

(a) Calculate the frictional coefficient, f, and the frictional coefficient, f_0, if lysozyme were an unhydrated sphere, in g s^{-1}.

(b) Under certain conditions, lysozyme dimerizes. If the ratio $f_{dimer}/f_{monomer} = 1.6$, calculate $s_{20, w}$ for the lysozyme dimer.

12. Consider a mixture of two proteins with molecular weights of 20,000 and 200,000. For simplicity of calculation, both may be approximated as unhydrated spheres with $\bar{v}_2 = 0.740 \ cm^3 \ g^{-1}$.

  (a) Calculate the ratio of the sedimentation coefficients of the proteins in the same medium (density = 1.05 g cm^{-3}) at the same temperature.

  (b) Consider the following experiments. A mixture of the two proteins is placed on top of a centrifuge tube filled with a dilute aqueous buffer containing a linear gradient of sucrose ranging from 5% at the top to 20% at the bottom, and the tube is spun in a centrifuge, as illustrated in the diagram. The top and bottom of the liquid column are 4.0 and 8.0 cm from the axis of rotation, respectively. When the larger protein has sedimented a distance of 3.0 cm, how far has the smaller protein traveled? [Use your answer to part (a) in the calculation and neglect the viscosity and density gradients that are associated with the sucrose concentration gradient.] Based on your calculation, is this an effective method of separation for the two proteins?

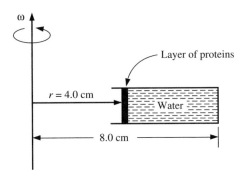

  (c) As the proteins are sedimenting, the initial concentrated thin layer of protein molecules will spread out due to diffusion. If the distance over which the protein molecules spread due to diffusion is much greater than the distance separating the two proteins, this method will not be useful as a separation approach since the protein layers will overlap. To see if this might be a problem, calculate the average distances the two proteins move due to diffusion at 298 K if the sedimentation experiment lasts 12 hr. Take the average viscosity of the medium as 1.5 centipoise. Is the "spreading distance" due to diffusion significant compared to the separation of the proteins due to sedimentation?

  (d) At sedimentation equilibrium, do you expect to find a gradual variation in the concentration of protein along r near the bottom of the tube?

13. Consider a small spherical molecule of radius 2 Å (2×10^{-8} cm).

  (a) At 298 K, what is the average time required for such a molecule to diffuse across a phospholipid bilayer 40 Å thick? The viscosity of the bilayer interior is about 0.10 P.

  (b) At a constant temperature the average time to diffuse across a bilayer hydrophobic interior depends only on the radius of the diffusing molecule. Hence all molecules with the same size will cross in the same time. However, it is well known that for the same concentration gradient the flux of nonpolar molecules across bilayers is much greater than that for polar molecules of the same size. Are these statements contradictory? Explain your answer.

14. The pK_a values of various groups in proteins are tabulated on the next page.

		pK_a
	α-COOH	~2
	α-NH$_3^+$	~9
Side chain	—COOH	~4
Side chain	—NH$_3^+$	~11
	Histidine	~6

A biochemist studying the properties of the enzyme hexokinase isolates and purifies two different mutants (I and II) for this enzyme from bacteria. She then performs isoelectric focusing and runs an SDS-gel for these mutant enzymes as well as the normal enzyme. Her gels after protein staining are shown below:

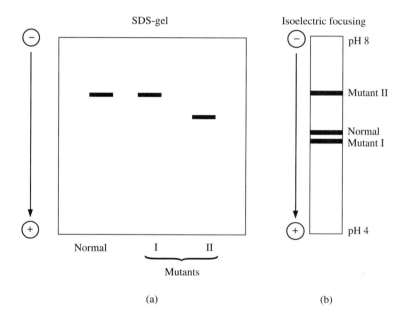

(a)   (b)

(a) On the basis of the gel results, what can you say about the possible kinds of changes in the amino acid composition of these two mutants? Explain briefly.

(b) You have a small unidentified peptide. You determine its molecular weight by SDS-gel electrophoresis to be 3000 and its isoelectric point to be 10. With a specific protease you cleave a *single* peptide bond in your protein. You find that your cleaved peptide still has approximately the same molecular weight on an SDS-gel, but its isoelectric point is reduced to about 8. How would you explain your results?

15. For prolate ellipsoids of large axial ratios, say with $a/b > 20$, it can be shown that the viscosity shape factor $\nu \approx 0.207(a/b)^{1.723}$. A certain rodlike macromolecule is believed to form an end-to-end dimer. Show that the intrinsic viscosity of the dimer is expected to be 3.32 times that of the monomer.

16. The DNA from an animal virus, polyoma, has a sedimentation coefficient $s_{20,\,w}$ of 20 S. Digesting the DNA very briefly with pancreatic DNase I, an enzyme that introduces single-chain breaks into a double-stranded DNA, converts it to a species sedimenting at 16 S. This

reduction in s could be due to either a reduction in molecular weight or a conformational change of the DNA (so that its frictional coefficient is increased). How would you design an experiment to decide between these possibilities?

17. The following data have been obtained for human serum albumin:

$$s_{20, w} = 4.6 \text{ S}$$

$$D_{20, w} = 6.1 \times 10^{-7} \text{ cm}^2 \text{ s}^{-1}$$

$$[\eta] = 4.2 \text{ cm}^3 \text{ g}^{-1}$$

$$\bar{v}_2 = 0.733 \text{ cm}^3 \text{ g}^{-1}$$

Calculate the molecular weight of this protein.

18. The DNA from a simian virus SV40 (simian viruses infect monkeys) is a twisted, double-stranded ring. A restriction enzyme Eco RI cleaves this ring structure at a unique location and converts the DNA to a linear form. The ratio of the sedimentation coefficients of the two forms is measured to be 1.45. Estimate the ratio of the intrinsic viscosities of the two forms.

19. In 6 M guanidine hydrochloride and in the presence of 2-mercaptoethanol, it is generally believed that complete unfolding of proteins occurs. Could you test whether this is true by viscosity measurements? Give a brief and concise discussion. Some experimental data are listed below [taken from C. Tanford, *Adv. Protein Chem.* **23**, 121 (1968)]:

		$[\eta]$ (cm^3 g^{-1})	
			In 2 M guanidine·HCl
	M^*	In dilute	and in the presence of
Protein	molecular weight	aqueous buffer	2-mercaptoethanol
Ribonuclease (cow)	13,690	3.3	16.6
Myoglobin (horse)	17,568	3.1	20.9
Chymotrypsinogen (cow)	25,666	2.5	26.8
Serum albumin (cow)	66,296	3.7	52.2

* These proteins have only one polypeptide chain per molecule; therefore, the molecular weight does not change upon unfolding of the protein.

20. The O_2-carrying protein, hemoglobin, contains a total of four polypeptide chains: two α chains and two β chains per molecule. The hemoglobin of a certain person from Boston, designated hemoglobin M Boston, differs from normal hemoglobin (hemoglobin A) in that a histidine residue in each of the α chains of hemoglobin A is substituted by a tyrosine residue. From the ionization constants given in Table 4.1, do you expect the electrophoretic mobilities at pH 7 of the two proteins, hemoglobin M Boston and hemoglobin A, to differ? Give your reasons. What would be a reasonable pH to use to separate the two by electrophoresis? Which is more negatively charged at this pH?

21. The protein β-lactalbumin (molecular weight = 14,000) has been studied under different solution conditions by dynamic light scattering. At pH 7 and 40°C in a dilute solution the diffusion constant was found to be 14.25×10^{-7} cm^2 s^{-1}.
    (a) If the viscosity of the solvent was 0.0101 P, estimate the diameter of the protein assuming that it is spherical in shape.

(b) At pH <2 this protein becomes inactive and changes spectral characteristics. Under these conditions the diffusion constant was found to be 12.80×10^{-7} cm^2 s^{-1}. Calculate the change in volume of the protein (assume all else remained the same).

(c) At intermediate pH there is an equilibrium between these two forms of the protein. Calculate the ratio of sedimentation coefficients ($s_{\text{high pH}}/s_{\text{low pH}}$) for these two forms. Assume the solvent has a density of 1.04 g cm^{-3}.

22. A protein involved in light harvesting for photosynthesis in bacteria was investigated using sedimentation equilibrium and SDS-gel electrophoresis. Two centrifuge experiments were done, the first in 1 M NaCl solution, and the second in 1 M NaSCN. The samples were spun at 20°C at 21,380 rpm until an equilibrium concentration gradient was established. The partial specific volume of the protein, $\bar{v}_2$, was found to be 0.709 cm^3 g^{-1}; the density, ρ, for the salt solutions was 1.20 g cm^{-3}. Ignore any possible effects of density gradients that are established due to the salt in the solutions. The concentration profile was determined and used to generate the graph below to the left; c = protein concentration at position x in the centrifuge cell. On the right is a drawing of the denaturing SDS-gel.

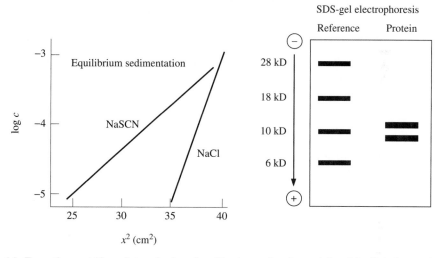

(a) From the centrifuge data calculate the effective molecular weight of the light harvesting protein in each of the two salt solutions.

(b) From the SDS-gel electrophoresis data estimate the molecular weights of the two polypeptides that make up the light harvesting protein.

(c) How can you explain the two sets of data? Give a concise explanation of what each observation indicates and how the observations fit together.

23. The enzyme aspartate transcarbamylase (ATCase) has a molecular weight of 310 kD, and undergoes a change in shape upon binding of substrate or inhibitors. This change has been characterized by ultracentrifugation. In the absence of any ligands, a sedimentation coefficient of $s_{20, w} = 11.70$ S was measured. The solvent density was 1.00 g cm^{-3}, the viscosity was 1.005 cP, and the specific volume of the protein was 0.732 cm^3 g^{-1}.

(a) Find the radius of the protein in Å assuming that it is a sphere.

(b) Upon adding a ligand that bound to the enzyme, the sedimentation coefficient increased by 3.5%. What is the radius of the ligated enzyme?

(c) From x-ray diffraction it was shown that the presence of ligands causes a contraction of the enzyme of about 12 Å along one axis. Why did the radius not decrease by 6 Å? (Explain briefly.)

24. T4 is a large bacterial virus. The virus has a symmetrical (approximately spherical) head group which contains DNA. These head particles were studied and were found to have the following parameters: $s_{20, w} = 1025$ S; $D_{20, w} = 3.60 \times 10^{-8}$ cm^2 sec; $\bar{v}_2 = 0.605$ cm^3 g^{-1} at 20°C. Calculate the following:
    (a) molecular weight of the head group.
    (b) volume of the head group (unhydrated, in cm^3).
    (c) the frictional coefficient of the head group.
    (d) volume of the head group (hydrated, in cm^3).
    (e) hydration parameter δ_1 (g water/g head particle).

25. A small spherical virus has a molecular weight of 1.25×10^6 g mol^{-1} and a diameter of 100 Å; it is not significantly hydrated. Calculate the intrinsic viscosity in units of cm^3 g^{-1} of an aqueous solution of this virus.

26. For each of the following changes state whether the sedimentation coefficient of the particles will increase, decrease, remain the same, or whether it is impossible to tell. Also give an equation or a one- or two-sentence explanation that supports your answer.
    (a) The temperature of the solvent is increased from 20°C to 30°C.
    (b) The long axis of the particle (which is a prolate ellipsoid) is cut in half.
    (c) ^{15}N is substituted for ^{14}N in the particle.
    (d) Provide answers for the diffusion coefficient of the particle in parts (a), (b), and (c).

27. An important step in the blood coagulation process is the enzymatic conversion of the protein prothrombin to the lower-molecular-weight clotting agent, thrombin, via cleavage of a portion of the prothrombin polypeptide.

$$\text{prothrombin} \longrightarrow \text{thrombin} + \text{cleaved peptide}$$

The values of $s_{20, w}$ and $D_{20, w}$ for prothrombin have been found to be 4.85×10^{-13} sec and 6.24×10^{-7} cm^2 sec^{-1}, respectively. The diffusion coefficient for thrombin is 8.76×10^{-7} cm^2 sec^{-1}. Assume that both thrombin and prothrombin are unhydrated spheres with $\bar{v}_2 = 0.70$ cm^3 g^{-1}, and calculate the molecular weights of prothrombin, thrombin, and the cleaved peptide.

28. (a) Gel electrophoresis is used to determine the sequences of DNA molecules. Describe how this is done. In particular describe how the positions of bands on a gel are related to the sequence of the DNA.
    (b) Gel electrophoresis is used to determine molecular weights of proteins. Describe how this is done and mention the limitations to this method.
    (c) Describe one other method that could be used to determine the molecular weight of a protein. State what is measured and how this is related to molecular weight.
    (d) Describe how an equilibrium binding constant can be measured by gel electrophoresis.

29. Three transport properties of proteins are easy to measure: s (sedimentation coefficient), D (diffusion coefficient), μ (electrophoretic mobility). State what happens to each of these three quantities for the following changes. Be quantitative and give equations to justify your answers.
    (a) The molecular weight of the protein doubles, but the frictional coefficient is constant.
    (b) The frictional coefficient of the protein doubles, but the molecular weight is constant.
    (c) The charge on the protein doubles, but nothing else changes.
    (d) The protein changes from a prolate ellipsoid to an oblate ellipsoid of the same axial ratio; nothing else changes.
    (e) The amount of hydration of the protein doubles.

7

Kinetics: Rates of Chemical Reactions

CONCEPTS

A chemical reaction can occur when two molecules collide. However, for a reaction to occur the molecules must have enough energy to break the covalent bonds of the reactants and to form the new bonds of the products. This explains why increasing the concentrations of reactants and increasing the temperature nearly always increase the rate of a chemical reaction. The higher concentrations mean more collisions; the higher temperature provides more energy per collision and also more collisions per second. However, not all reaction rates increase with increasing temperature and concentration. We need to understand how to explain the results no matter what is observed experimentally.

   Chemical kinetics includes two distinct parts. One is the measurement of the rate of a reaction and its dependence on experimental conditions. You measure the dependence on concentrations of reactants and, possibly, of products. (There is often product inhibition of the rate, and there may be catalysis by the product.) You change the solvent environment. Concentrations of molecules not directly involved in the reaction can have a large effect on the rate. Added salts, metal ions, pH, and so forth can act as catalysts or inhibitors of the reaction. The effect of temperature is determined. Finally, the effect of other variables can be tested. The rate may depend on whether the reaction is done in a glass container or a plastic container. It may depend on whether it is done in a dark room or in the sunlight. Joseph Priestley accidentally learned that plants would produce oxygen only in the sunlight, not in the dark. We have since learned much more about photosynthesis, and photochemical reactions in general. The important lesson to learn, however, is that many things may have large effects on the rate of a reaction. A spark introduced into a container of H_2 and O_2 gas can change the rate of formation of H_2O by more than 10 orders of magnitude. A stable gas mixture is converted into a bomb.

Once the effects of many variables on the rate are determined, the next step is to try to explain the effects in terms of what the molecules are doing. The explanation is called a *mechanism* of the reaction. A mechanism is a set of reactions showing which molecules collide and which bonds break or form. Each reaction step involves one or two molecules at a time, never more than three molecules. The mechanism is distinct from the *stoichiometric reaction*. The stoichiometry tells you what the reactants and products are, and the number of moles of reactants that produce the number of moles of products. The stoichiometry tells you nothing about the dependence of rate on concentration; the mechanism is an explanation of the experimental rate data. For a chemical reaction there is only one stoichiometric reaction; however, there can be many mechanisms that are consistent with the rate data. Each mechanism is a hypothesis to explain the kinetic data; the mechanism is used to design further experiments to better understand the results.

An important concept in understanding the mechanism of a reaction and the temperature dependence of a rate is the *transition state*. For a reaction in which one species is converted to another, the structure of the transition state is assumed to be halfway between the two species. It is an unstable structure (unstable means that it exists only for the time of a molecular vibration) which quickly transforms to either stable species. The transition state corresponds to an energy maximum between two stable species which exist in energy minima. The *transition state energy* is the energy difference between the transition state and a reactant; its magnitude characterizes the temperature dependence of the rate of the reaction. A small transition state energy means that little energy is required to transform reactant to product, and the rate does not depend much on temperature. A large transition state energy produces a large temperature dependence of the rate.

APPLICATIONS

Kinetics—the study and understanding of the rate of change of anything—is useful in all areas of medicine, biology, and biochemistry. The effect of food supply, predators, and climate on the number of animals in a population can be treated by the same differential equations used for treating molecular reactions in the atmosphere, or in a single biological cell. Bacterial growth rates, radioactive decay, biosynthesis of deoxynucleotides, viral infectivity, and antibiotic cures all depend on the rates of reactions. We will learn how to measure rates of reaction and to study the variables that affect the rates. For some purposes this may be enough. For example, if we use an enzyme as a catalyst it may be sufficient to assay its activity every day. Its activity may decrease every day, but all we care about is the number of catalytic units we have. When the activity gets too low, we throw out the old solution and make new batch. The rate of denaturation of the enzyme is important to us, but the mechanism may be too complicated to try to understand. We are satisfied with the empirical rate data.

For most reactions we do want to understand what is happening. Why do some reaction rates increase by a factor of 2 for a 10-degree rise in temperature, but some do not change, or decrease? When should an increase in concentration of a salt not directly involved in the reaction increase or decrease the rate? The energy and structure of the transition state is important in answering these questions. You may think it is not possible to generalize the concept of a mechanism and a transition state to the complex rates involved in predator-prey dependence. However, the idea is simply to propose a hypothesis that explains your data. The test of your hypothesis is how well it explains other data taken for different animals under different conditions. If your hypothesis is generally useful, it may be given a name which students studying the subject in a few years will want to memorize.

KINETICS

Chemical kinetics is the study of the rate of reactions. Some reactions, such as that between hydrogen gas and oxygen gas in an undisturbed clean flask, occur so slowly as to be unmeasurable. Radioisotopes of some nuclei have very long lifetimes; the carbon isotope $^{14}_{6}C$ decays so slowly that half of the initial amount is still present after 5770 years. Other radionuclei have half-lives that are orders of magnitude longer than this. Processes such as the growth of bacterial cells are slow but easily measurable. The rate of reaction between H^+ and OH^- to form water is so fast that it requires special techniques, such as temperature-jump kinetics. The frontier is expanding to include still faster processes, and reaction times of fractions of a picosecond (1 ps = 10^{-12} s) are currently being reported using apparatus based on the mode-locked laser. Nuclear reactions involving species with lifetimes shorter than 10^{-20} s are known. Clearly, the methods of observation are very different—to include processes over such an enormous range of time.

The first chemical reaction whose rate was studied quantitatively involved a compound of biological origin. In 1850 L. Wilhelmy reported that the hydrolysis of a solution of sucrose to glucose and fructose occurred at a rate that decreased steadily with time but always remained proportional to the concentration of sucrose remaining in the solution. He was able to follow the reaction indirectly by measuring the change with time of the optical rotation—the rotation of the direction of plane-polarized light passing through the solution. The phenomenon of optical rotation results from the molecular chirality of sugars such as sucrose, glucose, and fructose; chirality (handedness) refers to the fact that these molecules are structurally (and chemically) distinct from their mirror-image molecules (see Chapter 10).

Subsequent pioneering work by Guldberg and Waage (1863) and by van't Hoff (1887) demonstrated that systems at equilibrium are not static but are undergoing transformations between reactants and products in both directions and at equal rates.

The role of catalysts, substances that increase the rates of reactions without themselves being consumed, was recognized early from the influence of hydrogen ion on the rate of sucrose hydrolysis. The early "ferments" used to convert sugars from grain or grapes into beverages are now known to contain enzyme catalysts that greatly speed up the rates of these processes. Biological organisms contain thousands of different enzymes—protein molecules that not only can provide increases of many millionfold in rates of reaction but also can be highly selective in their choices of reactant molecules from the large assortment present in living cells. The dramatic effect of some catalysts is easily seen in such cases as the hydrogen-oxygen gas mixture, where the introduction of a trace of finely divided metal such as platinum leads to a violent explosion. Ostwald pointed out that catalysts that affect the rates of reactions nevertheless have no effect on the position of equilibrium. We can conclude that the hydrogen-oxygen mixture in the absence of a catalyst is not at equilibrium.

The ability of increases in temperature to speed up most chemical reactions was put on a quantitative bases by Arrhenius (1889). Subsequent studies have made important contributions to our understanding of the mechanisms of reactions, which is a primary goal of kinetics. The mixture of hydrogen and oxygen, which is stable indefinitely at room temperature, can be made to explode by heating it to temperatures in excess of 400°C. In other cases, such as the processes that occur in biological cells, an increase in temperature causes the enzyme-catalyzed reactions to cease altogether.

Light or electromagnetic radiation serves as a "reagent" in some biological processes that are essential to our survival. Photosynthesis and vision are just two of the most obvious examples. Other processes that are not chemical in nature, such as the nuclear reactions that provide the energy source of the sun and have been the major source of heat within the earth, can nevertheless be described using the methods of chemical kinetics. Population dynamics, ecological changes and balance, atmospheric pollution, and biological waste disposal are just a few of the relatively new applications of this powerful approach, which has its origins in physical chemistry.

Dynamic processes leading to growth, change, and evolution permeate all of biology. Some of these, such as our visual response or those of the involuntary muscles, are exceedingly fast even though they represent the consequences of an extensive chain of events. On a more leisurely time scale are the processes of cell division, growth, and death of organisms and the associated steps of metabolism. Extremely slow changes are of significance in genetic evolution and in adaptation to the modifications of our environment. The trend from reducing conditions to oxidizing conditions during the time since life first appeared on the earth can be traced in the response of photosynthetic organisms. Initially, there were much better reducing agents than water available in the environment, and there was no need to build up protection against the potentially toxic effects of oxygen. When the reductants became depleted and oxygen became abundant in the atmosphere, the modified characteristics of higher plants developed and new forms of photosynthesis continued to support life. In more recent times, such phenomena as the fluctuations in climatic temperature associated with the ice ages have led to important biological responses that can be investigated from fossil remains.

The methods of chemical kinetics or reaction-rate analysis were developed for the resolution and understanding of the relatively simple systems encountered by chemists. It may be surprising to find that these approaches are also valuable in analyzing the much more complex processes of biology. The reason that the methods do work is often because one or a few steps control the rate of an extensive chain of reactions. All the steps involved in metabolism, cell division and replication, muscular contraction, and so on are subject to the same basic principles as are the elementary reactions of the chemist.

The rate or velocity, v, of a reaction or process describes how fast it is occurring. Usually the velocity is expressed as a change in concentration per unit time,

$$v = \frac{dc}{dt} = \text{rate of reaction}$$

but it may alternatively express the change of a population of cells with time, the increase or decrease in the pressure of a gas with time, or a change in the absorption of light by a colored solution with time. In general, the rate of a process depends in some way on the concentrations or amounts involved; the rate is a function of the concentrations. This relation is known as the *rate law:*

$$v = f(\text{concentrations})$$

It may be simple ($v =$ constant, for example) or complex, but it gives important information about the mechanism of the process. One of the main objectives of research in kinetics is the determination of the rate law for the phenomenon under investigation.

Rate Law

Substances that influence the velocity of a reaction can be grouped into two categories:

1. Those whose concentration changes with time during the course of the reaction:
   Reactants—decrease with time.
   Products—increase with time.
   Intermediates—increase and then decrease during the course of the reaction. An example is substance C in the following chain reaction:

   $$A \longrightarrow C \longrightarrow B$$

2. Those whose concentrations do not change with time:
   Catalysts (both promoters and inhibitors), including enzymes and active surfaces.
   Intermediates in a steady-state process, including reactions under flowing conditions.
   Components that are buffered by means of equilibrium with large reservoirs.
   Solvents and the environment in general.

These concentrations or influences do not change during a single run, but they can be changed from one experiment to the next. The concentrations of these components frequently do influence the rates of reactions.

Order of a Reaction

It is important in kinetics to learn immediately the vocabulary that kineticists use. We need to distinguish the stoichiometric reaction, the order of the reaction, and the mechanism of the reaction. It is essential to understand these terms. The *stoichiometry* of the reaction tells you how many moles of each reactant are needed to form each mole of products. Only ratios of moles are significant. For example,

$$H_2 + \tfrac{1}{2}O_2 = H_2O$$

and

$$2\,H_2 + O_2 = 2\,H_2O$$

are both correct stoichiometric reactions. The *mechanism* of a reaction tells you how the molecules react to form products. The mechanism is, in general, a set of *elementary reactions* consistent with the stoichiometric reaction. For the reaction of H_2 and O_2 in the gas phase, the reaction is thought to be a chain involving H, O, and OH radicals:

$$H_2 \longrightarrow 2\,H$$

$$H + O_2 \longrightarrow OH + O$$

$$OH + H_2 \longrightarrow H_2O + H$$

$$O + H_2 \longrightarrow OH + H$$

Each reaction is an elementary reaction; the four reactions describe the proposed mechanism.

The kinetic *order* of a reaction describes the way in which the velocity of the reaction depends on the concentration. Consider a reaction whose stoichiometry is

$$A + B \longrightarrow P$$

For many such reactions the rate law is of the form

$$v = kc_A^m c_B^n c_P^q \tag{7.1}$$

where the concentrations c_A, c_B, etc., are raised to powers m, n, etc., that are usually integers or zero ($c_A^0 = $ constant), but may be nonintegral as well. The order of the reaction with respect to a particular component A, B, P, etc., is just the exponent of the concentration term. Because the velocity may depend on the concentrations of several species, we need to distinguish between the order with respect to a particular component and the overall order, which is the sum of the exponents of all components. Some representative examples are listed in Table 7.1.

Table 7.1 Rate laws and kinetic order for some reactions

Stoichiometric reaction	Rate law	Kinetic order
Sucrose + H_2O $\xrightarrow{\text{H}^+}$ fructose + glucose	$v = k[\text{sucrose}]$	1
L-Isoleucine $\longrightarrow$ D-isoleucine	$v = k[\text{L-isoleucine}]$	1
$^{14}_{6}C \longrightarrow {}^{14}_{7}N + \beta^-$	$v = k[^{14}_{6}C]$	1
2 Proflavin $\longrightarrow$ proflavin dimer	$v = k[\text{proflavin}]^2$	2
p-Nitrophenyl acetate + 2 OH$^-$ $\xrightarrow{\text{pH 9}}$ p-nitrophenolate$^-$ + acetate$^-$ + H_2O	$v = k[p\text{-nitrophenyl acetate}][\text{OH}^-]$	2 (overall)
Hemoglobin $\cdot$ 3 O_2 + O_2 $\longrightarrow$ Hb $\cdot$ 4 O_2	$v = k[\text{Hb} \cdot 3\,O_2][O_2]$	2 (overall)
$H_2 + I_2 \longrightarrow$ 2 HI	$v = k[H_2][I_2]$	2 (overall)
$H_2 + Br_2 \longrightarrow$ 2 HBr	$v = \dfrac{k[H_2][Br_2]^{1/2}}{k' + [\text{HBr}]/[Br_2]}$	Complex
$CH_3CHO \longrightarrow CH_4 + CO$	$v \cong k[CH_3CHO]^{3/2}$	$\frac{3}{2}$ (approx.)
$C_2H_5OH \xrightarrow[\text{enzymes}]{\text{liver}} CH_3CHO$	$v = $ constant	0

If the concentration of a component is unchanged during the course of the reaction, it is frequently omitted in the rate-law expression. A more complete rate law for the first reaction tabulated in Table 7.1 is

$$v = k'[\text{sucrose}][\text{H}^+][\text{H}_2\text{O}]$$

However, H^+ is a catalyst and its concentration is constant during a run; the concentration of H_2O, the solvent, is also little changed because it is present in vast excess. Therefore, the terms $[\text{H}^+]$ and $[\text{H}_2\text{O}]$ are omitted in the rate law given in the table. When the reaction is carried out in the presence of different concentrations of H^+, or with an added inert solvent, the first-order dependence of the reaction on $[\text{H}^+]$ and on $[\text{H}_2\text{O}]$ is seen.

Note that there is no simple relation between the stoichiometry and the rate law. It is never possible to *deduce* the order of the reaction by inspection of the stoichiometric equation. Kinetic experiments must be done to measure the order of the reaction. The reaction of H_2 and I_2 is first order with respect to each reactant over a wide range of conditions, whereas the similar reaction of $\text{H}_2 + \text{Br}_2$ exhibits a complex rate law that cannot be described by a single "order" under all conditions. Note that in this case the velocity depends on the concentration of a product as well as on reactants. This comes about because of a reverse step in the mechanism that becomes important as the product concentration builds up. Reaction orders may be nonintegral, as in the case of the thermal decomposition of acetaldehyde, and they may be significantly different during the initial stages, when the reaction is getting under way, or at the end, when other complications set in. The significance of a zero-order reaction is that the velocity is constant and independent of the concentration of the reactants. This is characteristic of reactions catalyzed by enzymes, such as liver alcohol dehydrogenase, under the special conditions where the enzyme is saturated with reactants (called *substrates* in enzyme reactions). The role of enzymes and other catalysts, such as H^+ in the sucrose hydrolysis reactions, can be detected by observing that the changes in the catalyst concentration are reflected by changes in the experimental rate constants for the reactions, even though the catalyst concentration does not vary with time during the course of any single experiment. Even these preliminary comments indicate that the rate law contains important information about the mechanism of the reaction.

Experimental Rate Data

The rate law for a reaction must be determined from experimental data. We may simply want to know how the rate depends on concentrations for practical reasons, or we may want to understand the mechanism of the reaction. For example, if we are inactivating a virus by reaction with formaldehyde, it is vital for us to know how the rate depends on the concentrations of virus and formaldehyde. If the rate is first order in virus concentration, the rate will be one-tenth as fast for 10^5 viruses per milliliter as for 10^6 viruses per milliliter. If the rate is second order in virus concentration (unlikely), the rate will be only one-hundredth as fast. Knowledge of the rate law is necessary to determine the time of treatment with formaldehyde needed to ensure that all the viruses are inactivated before they are used to immunize a population. We also must know how temperature, pH, and solvent affect the rate.

There are many convenient ways to obtain the rate data. A usual method is to obtain concentrations of reactants and products at different times during the reaction. For the virus inactivation we would presumably measure live virus versus time by an infectivity assay. In general, any analytical method that determines concentration can be used. If the time required to perform the analysis is long relative to the rate of the reaction, "quenching" or sudden stopping of the reaction is necessary. An enzyme-catalyzed reaction, for example, can be quenched by cooling the reaction mixture quickly, by the addition of an agent that denatures the enzyme, or by the addition of a chelating compound if a multivalent metal ion is necessary for catalytic activity. In a *quenched-flow* apparatus, reactants in two syringes are forced through a mixing chamber to initiate the reaction. The mixture flows through a tube into a second mixing chamber, where it is mixed with a stopping reagent. The reaction time in such an experiment is the time it takes for the solution to flow from the first to the second mixing chamber, and can be as short as several milliseconds.

If there is a physical property that changes significantly as the reaction proceeds, it can usually be used to follow the reaction. If the reactants and products absorb light of a certain wavelength to a different extent, the absorbance at this wavelength can be related to the extent of the reaction (see Chapter 10). The concentration of H^+ can be easily monitored with a glass electrode, but different techniques are needed when the pH changes rapidly with time. Reactions between charged species can often be followed by the electrical conductivity of the solution.

Very fast reactions require special experimental approaches. The *relaxation methods,* such as temperature-jump methods, will be described later in this chapter.

Zero-Order Reactions

A *zero-order reaction* corresponds to the rate law:

$$\frac{dc}{dt} = k_0 \tag{7.2}$$

where c is the concentration of a product. The units of the rate constant, k_0, for a zero-order reaction are obviously concentration per time, such as $M\ s^{-1}$. This expression can readily be integrated by writing it in the differential form with the variables c and t separated on each side of the equality:

$$dc = k_0\ dt$$

Integrating both sides, we obtain

$$\int dc = k_0 \int dt + \text{constant}$$

where the rate constant, k_0, is placed outside the integral sign because it does not depend on time. Once the initial and final conditions are specified, the equation can be written as a definite integral; that is, the constant can be evaluated. If the concentration is c_1 at time t_1 and c_2 at time t_2, then

$$\int_{c_1}^{c_2} dc = k_0 \int_{t_1}^{t_2} dt$$

$$c_2 - c_1 = k_0 \cdot (t_2 - t_1)$$

A different form is commonly used, where c_0 refers to the initial concentration at zero time, and the concentration is c at any later time t. Then

$$c - c_0 = k_0 t \qquad (7.3)$$

This is the equation for a straight line giving the dependence of c on t; the slope is the rate constant, k_0.

This behavior is illustrated by the conversion of ethanol to acetaldehyde by the enzyme liver alcohol dehydrogenase (LADH). The oxidizing agent is nicotinamide adenine dinucleotide (NAD^+) and the reaction can be written

$$CH_3CH_2OH + NAD^+ \xrightarrow{\text{LADH}} CH_3CHO + NADH + H^+$$

In the presence of an excess of alcohol over the enzyme and with the NAD^+ buffered via metabolic reactions that rapidly restore it, the rate of this reaction in the liver is zero-order over most of its course,

$$v = -\frac{d[CH_3CH_2OH]}{dt} = \frac{d[CH_3CHO]}{dt} = k_0 \qquad (7.4)$$

The negative sign is used with the reactant, ethanol, because its concentration decreases with time; the concentration of the product, acetaldehyde, increases with time. This behavior is illustrated in Fig. 7.1.

The reaction cannot be of zero order for all times; because obviously the reactant concentration cannot become less than zero, and the product concentration must also reach a limit. For the oxidation of alcohol by LADH the reaction is zero order only while alcohol is in excess.

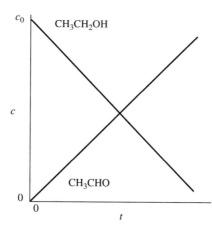

Fig. 7.1 Plot of concentration versus time for a zero-order reaction. The magnitude of the slope of each straight line is equal to the rate constant, k_0. The order must eventually change from being zero order as the concentration of CH_3CH_2OH approaches zero.

First-Order Reactions

A *first-order reaction* corresponds to the rate law:

$$\frac{dc}{dt} = k_1 c \qquad (7.5)$$

The units of k_1 are time^{-1}, such as s^{-1}. There are no concentration units in k_1, so it is clear that we do not need to know absolute concentrations; only relative concentrations are needed. An elementary step in a reaction of the form

$$A \longrightarrow B$$

has a rate law of the form

$$v = -\frac{d[A]}{dt} = \frac{d[B]}{dt} = k_1[A] \tag{7.6}$$

where k_1 is the rate constant for the reaction and [A] and [B] are concentrations. The velocity of the reaction can be expressed in terms of either the rate of disappearance of reactant, $-d[A]/dt$, or the rate of formation of product, $d[B]/dt$. The stoichiometric equation assures us that these two quantities will always be equal to one another. To solve the rate-law expression we choose the form involving the smallest number of variables.

$$-\frac{d[A]}{dt} = k_1[A] \tag{7.7}$$

Here time is one variable and the concentration of A is the other. Dividing both sides by [A], we obtain

$$\frac{d[A]}{[A]} = -k_1 \, dt$$

In this form the variables are *separated* in the sense that the left side depends only on [A] and the right side only on t. Once the variables are separated, the equation can be integrated, separately, on each side.

$$\int \frac{d[A]}{[A]} = \int - k_1 dt = -k_1 \int dt$$

$$\ln [A] = -k_1 t + C \tag{7.8}$$

where C is a constant of integration. This states that for a first-order reaction, the *logarithm* of the concentration will be a linear function of time, as shown in Fig. 7.2. To evaluate the

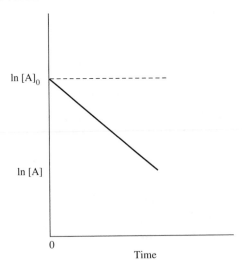

Fig. 7.2 Semilogarithmic straight-line plot for a first-order reaction.

constant C we need to know one specific concentration at a particular time. For example, if $[A]_0$ is the value of the concentration initially when $t = 0$, it, too, must satisfy Eq. (7.8). Substitution gives

$$\ln [A]_0 = C$$

which provides an alternative form of Eq. (7.8),

$$\ln \frac{[A]}{[A]_0} = -k_1 t \qquad (7.9)$$

A more general form involving any two points during the course of a first-order reaction is

$$\ln \frac{[A]_2}{[A]_1} = -k_1(t_2 - t_1) \qquad (7.10)$$

An alternative form of Eq. (7.9) is

$$[A] = [A]_0 \, e^{-k_1 t} = [A]_0 \, 10^{-k_1 t/2.303} \qquad (7.11)$$

This says that the concentration of A decreases exponentially with time for a first-order reaction. It starts at an initial value $[A]_0$, since $e^0 = 1$, and reaches zero only after infinite time! Strictly speaking, the reaction is never "finished." Before worrying about the philosophical implications of this, be assured that the inability to detect any remaining $[A]$ will occur in finite time even using the most sensitive analytical methods.

Let us consider the stability of the antibiotic penicillin as a practical example. Assume that our job is to learn how long penicillin remains active when it is stored at room temperature. We want to determine its activity vs. time. The general structure of penicillin is

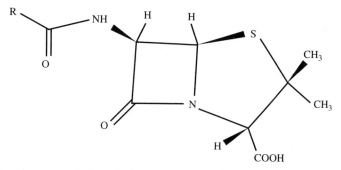

Ultraviolet absorbance, or infrared absorbance, or nuclear magnetic resonance could be used to measure its concentration, but we want to know if it will kill bacteria, therefore we must measure its antibiotic properties. We use a standard assay to count the number of bacterial colonies that survive after treatment with successive dilutions of the penicillin solution under standardized conditions. We thus obtain the number of standard units of penicillin in the sample. The penicillin is in a well-buffered solution at pH 7 kept at 25°C. The mean values for data measured in triplicate are given in Table 7.2.

A plot of penicillin units vs. time is shown in Fig. 7.3. The curve looks qualitatively exponential.

$$[\text{penicillin}] = [\text{penicillin}]_0 \, e^{-kt}$$

Table 7.2   The number of units of penicillin present after storage at 25°C. Penicillin units are proportional to the concentration of penicillin; they are a measure of the antibiotic effectiveness of the penicillin solution.

Time (weeks)	Penicillin units*	ln (penicillin units)
0	10,100	9.220
1.00	8180	9.009
2.00	6900	8.839
3.00	5380	8.590
4.00	4320	8.371
5.00	3870	8.261
6.00	3130	8.049
7.00	2190	7.692
8.00	2000	7.601
9.00	1790	7.490
10.00	1330	7.193
11.00	1040	6.947
12.00	898	6.800
13.00	750	6.620
14.00	572	6.349
15.00	403	5.999
16.00	403	5.999
17.00	314	5.749
18.00	279	5.631
19.00	181	5.198
20.00	167	5.118

* The precision of the data produces only three significant figures for this column.

We note that after three to four weeks the penicillin is only half as active as it was originally. The data clearly are not consistent with zero-order kinetics; the concentration is not linear in time. To test for first-order kinetics the natural logarithm (or the base 10 logarithm) of the penicillin units is plotted vs. time as shown in Fig. 7.4. A linear plot is found; this indicates that the data are consistent with first-order kinetics. A best-least-squares line through the data (a linear regression) has the equation

$$\text{ln (penicillin units)} = 9.237 - 0.2059 \text{ time}$$

Comparison with Eq. (7.9) shows that the slope of the line is the first-order rate constant; $k_1 = 0.2059 \text{ week}^{-1}$. The intercept is ln (penicillin units) extrapolated to zero time; therefore ln $[\text{penicillin}]_0 = 9.237$ and $[\text{penicillin}]_0 = e^{9.237} = 10{,}270$. This is consistent with the measured value of 10,100.

A common parameter used to describe the kinetics of first-order rate processes is the *half-life* $t_{1/2}$. This is simply the time required for half the initial concentration to react; we saw that for the data in Fig. 7.3 this was about 3.5 weeks. For a quantitative determination we put $[A] = (1/2)[A]_0$ into Eq. (7.11).

$$\tfrac{1}{2}[A]_0 = [A]_0 e^{-k_1 t_{1/2}}$$

$$\tfrac{1}{2} = e^{-k_1 t_{1/2}}$$

Taking logarithms of both sides, we have

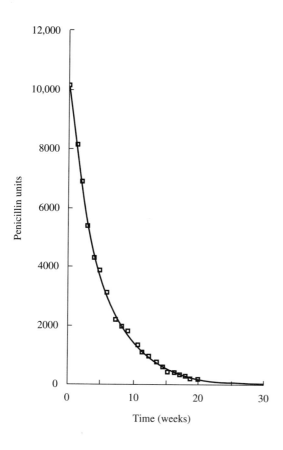

Fig. 7.3 The concentration (in units of antibiotic activity) of penicillin plotted vs. time (in weeks). The data are not consistent with a zero-order reaction; a zero-order reaction would give a linear plot.

$$\ln \tfrac{1}{2} = -k_1 t_{1/2}$$

Since $\ln (1/2) = -0.6931$ and therefore $\ln 2 = +0.6931$, we see that

$$t_{1/2} = \frac{\ln 2}{k_1} = \frac{0.6931}{k_1} \qquad (7.12)$$

For the penicillin data $k_1 = 0.2059$ week^{-1} and the half-life is $t_{1/2} = 3.37$ weeks.

Exercise From the definition of the half-life show that

$$\frac{[A]}{[A]_0} = 2^{-t/t_{1/2}} \qquad (7.13)$$

and that $t_{1/2}$ is independent of the initial concentration of reactant. Just combine Eqs. (7.11) and (7.12) to obtain Eq. (7.13), then show that the time for the concentration to change from $[A]_0$ to $(1/2)[A]_0$ is the same as the time for the concentration to change from $(1/2)[A]_0$ to $(1/4)[A]_0$.

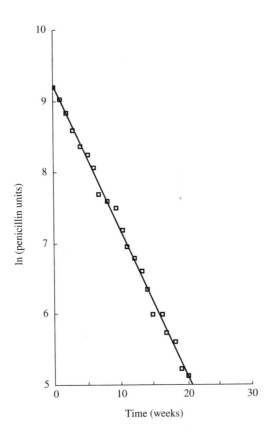

Fig. 7.4 The ln (concentration) of penicillin plotted vs. time. The data are consistent with a first-order kinetics; the slope of the line is equal to the rate constant, k_1, for the reaction.

Another parameter used to characterize first-order kinetic data is the *relaxation time* τ; it is the reciprocal of the rate constant k. From Eq. (7.11)

$$\frac{[A]}{[A]_0} = e^{-t/\tau} \tag{7.14}$$

$$\tau = 1/k_1 \tag{7.15}$$

The relaxation time τ is the time required for the concentration to decrease to e^{-1} ($e^{-1} = 1/e = 0.3679$) of its initial value.

Note that the half-life, the relaxation time, and the first-order rate constant only depend on ratios of concentrations. We do not need to know actual concentrations to characterize the kinetics of first-order reactions. We can measure any property proportional to concentration, such as the optical absorbance or the mutagenic ability. It is only for first-order reactions that the rate constant (with units of time^{-1}) does not contain units of concentration. A familiar example is radioactive decay, which is always first order. It does not matter whether there is a large amount or small amount of a radioactive isotope—the half-life is the same. For all other rate processes the rate constant does contain units of concentration, and half-lives and relaxation times depend on initial concentrations.

We can now characterize the loss of activity of the penicillin at one temperature and one buffer solution by a single number—the half-life of the first-order reaction is 3.37 weeks. It is useful to be able to characterize the data taken over a 20-week period by one number instead of a table of many numbers such as Table 7.2. How can knowledge of kinetics help further? Knowing that most reactions are faster at higher temperatures suggests that the penicillin will remain active longer when stored in a refrigerator at 4°C, or in liquid nitrogen at 77 K. Because it might take years to measure the stability at low temperatures, it will be quicker to measure the rate of decomposition at higher temperatures and use the Arrhenius equation, which we will discuss shortly, to extrapolate to 4°C. We would need to make measurements in frozen solutions to extrapolate to 77 K.

A proposed mechanism can help us understand the data and may help us formulate a better buffer solution for the penicillin. Penicillin has a lactam structure—an amide ring. The strained four-member ring is easily hydrolyzed to give an inactive compound. The stoichiometric reaction is

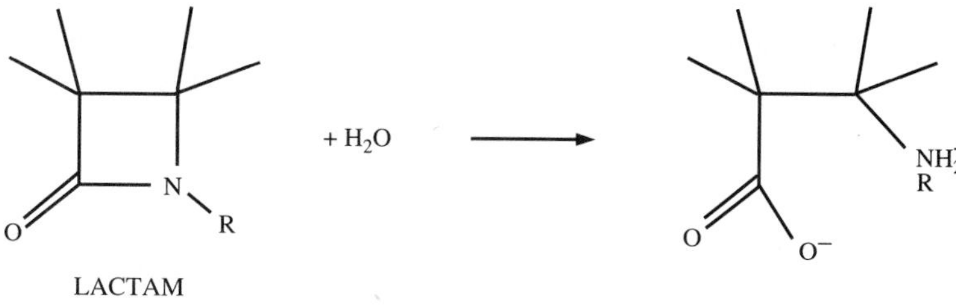

$+ H_2O \longrightarrow$

LACTAM

By analogy with other ester and amide hydrolysis mechanisms we propose

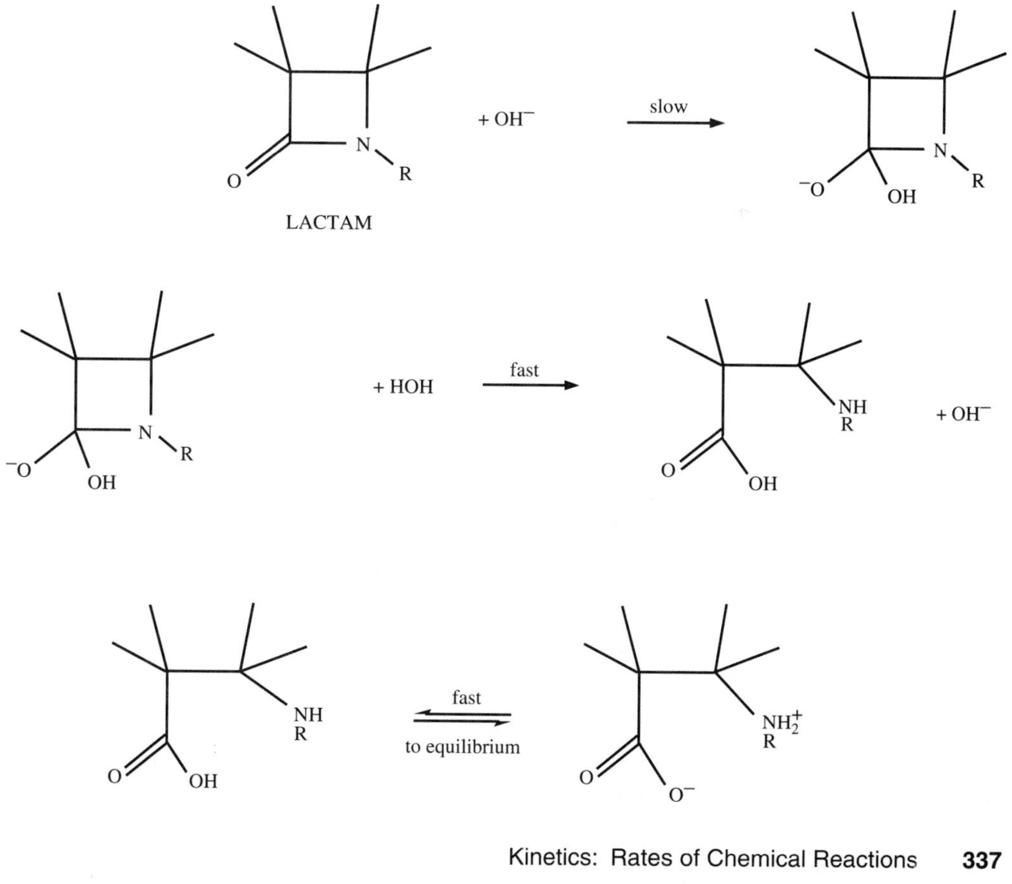

$+ OH^-$ $\xrightarrow{\text{slow}}$

LACTAM

$+ HOH$ $\xrightarrow{\text{fast}}$ $+ OH^-$

$\underset{\text{to equilibrium}}{\overset{\text{fast}}{\rightleftharpoons}}$

The reaction starts with a hydroxide ion adding to the carbonyl carbon of the lactam. This is proposed to be the slowest step in the reaction, therefore we expect that the rate of the reaction will be directly proportional to the concentration of OH⁻. The next two steps are much faster than the first step, so they do not affect the kinetics. We say that the first step is rate determining. If the buffer had a pH of 6 instead of 7 (and the proposed mechanism were correct), the rate of decomposition should be one-tenth as fast. The half-life would be 33.7 weeks instead of 3.37 weeks. Of course the rate of penicillin decomposition at pH 6 would have to be measured, and the effect of injection into patients of pH 6 solution would need to be determined, but clearly knowledge of kinetics has been helpful.

We will discuss reaction mechanisms in much more detail later in this chapter. Here we just want to give an idea of how kinetic measurements can be used.

Example 7.1 Carbon dioxide in the atmosphere contains a small but readily detectable amount of the radioactive isotope $^{14}_{6}C$. This isotope is produced by high-energy neutrons (in cosmic rays) that transform nuclei of nitrogen atoms by the process

$$^{14}_{7}N + ^{1}_{0}n \longrightarrow {}^{14}_{6}C + ^{1}_{1}H$$

The $^{14}_{6}C$ nucleus is unstable and decays by the first-order process

$$^{14}_{6}C \longrightarrow {}^{14}_{7}N + ^{0}_{-1}\beta$$

with a half-life of 5770 years. The production and decay of ^{14}C leads to a constant steady-state concentration. The amount of ^{14}C present in a carbon-containing sample can be determined by measuring its radioactivity (the rate of production of high-energy electrons, $^{0}_{-1}\beta$ particles).

In the atmosphere CO_2 contains a constant amount of ^{14}C. Once CO_2 is "fixed" by photosynthesis, however, it is taken out of the atmosphere and new ^{14}C is no longer added to it. The level of radioactivity then decreases by a first-order process with a 5770-year half-life.

A sample of wood from the core of an ancient bristlecone pine in the White Mountains of California shows a ^{14}C content that is 54.9% as great as that of atmospheric CO_2. What is the approximate age of the tree?

Solution Since the process is first order, we need know only the *ratio* of the present radioactivity to the original value. Assume that the level of ^{14}C in the atmosphere has not changed over the life of the tree. (This is not strictly true. Corrections can be made, but they are small.) Thus

$$\frac{[^{14}C]}{[^{14}C]_0} = 0.549$$

Using Eqs. (7.9) and (7.12), we obtain

$$\ln \frac{[^{14}C]}{[^{14}C]_0} = -\frac{0.693}{t_{1/2}} t$$

where t is the age of the wood. Therefore,

$$\ln(0.549) = -\frac{0.693}{5770 \text{ yr}} t$$

and

$$t = \frac{(0.600)(5770 \text{ yr})}{0.693} = 4990 \text{ yr}$$

We conclude that the carbon in the tree was taken from the atmosphere approximately 4990 years ago.

Second-Order Reactions

A *second-order reaction* corresponds to the rate law:

$$v = k_2 c^2 \quad \text{or} \quad v = k_2 c_A c_B$$

The units of k_2 are concentration^{-1} time^{-1}, such as M^{-1} s^{-1}.

It is useful to separate the treatment of second-order reactions into two major classifications depending on whether the rate law depends on (I) the second power of a single reactant species, or (II) the product of the concentrations of two different reagents.

Class I (A + A → P)

$$v = k_2[A]^2$$

Although the stoichiometric equation may involve either one or several components, the rate law for many reactions depends only on the second power of a single component. Some examples are:

$$2 \text{ proflavin} \longrightarrow [\text{proflavin}]_2 \; : \; v = k_2[\text{proflavin}]^2$$

$$\underset{\text{ammonium cyanate}}{\text{NH}_4\text{OCN}} \underset{\xrightarrow{\text{fast}}}{\overset{}{\rightleftharpoons}} \text{NH}_4^+ + \text{OCN}^- \xrightarrow{\text{slow}} \underset{\text{urea}}{\text{NH}_2\text{CONH}_2} \quad v = k_2[\text{NH}_4\text{OCN}]^2$$

$$2 \text{ A—A—G—C—U—U} \longrightarrow \begin{matrix} \text{A—A—G—C—U—U} \\ \vdots \; \vdots \; \vdots \; \vdots \; \vdots \; \vdots \\ \text{U—U—C—G—A—A} \end{matrix} \quad v = k_2[\text{A}_2\text{GCU}_2]^2$$

hexanucleotide*   hexanucleotide dimer

In each case the rate law is of the form

$$v = -\frac{d[A]}{dt} = k_2[A]^2 \tag{7.16}$$

The variables can be separated,

$$\frac{-d[A]}{[A]^2} = k_2 \, dt$$

* See the Appendix for the structure of the nucleotides.

and each side integrated, to give

$$\frac{1}{[A]} = k_2 t + C$$

Since $[A] = [A]_0$ when $t = 0$, the constant of integration is $1/[A]_0$, and we obtain an integrated form of the second-order rate equation:

$$\frac{1}{[A]} - \frac{1}{[A]_0} = k_2 t \qquad (7.17)$$

An alternative form is

$$\frac{1}{[A]_0 - x} - \frac{1}{[A]_0} = k_2 t \qquad (7.18)$$

where x is the concentration of A which has reacted. Note that for second-order kinetics one expects a linear relation between the reciprocal of the reactant concentration and time. This is in contrast to first-order kinetics where the logarithm of concentration is linear in time.

The half-life of any reaction is the time required for half of $[A]_0$ to react. For a second-order (class I) reaction we use Eq. (7.17)

$$\frac{1}{\frac{1}{2}[A]_0} - \frac{1}{[A]_0} = k_2 t_{1/2}$$

$$\frac{1}{[A]_0} = k_2 t_{1/2}$$

$$t_{1/2} = \frac{1}{k_2 [A]_0} \qquad (7.19)$$

For second-order reactions, the half-life depends on the initial concentration as well as on the rate constant. It is only first-order reactions where the half-life (or any other fractional life) is independent of initial concentration.

A graphical comparison of first- and second-order kinetics is instructive. Figure 7.5 shows concentration versus time plots to illustrate the differences. The time scales have

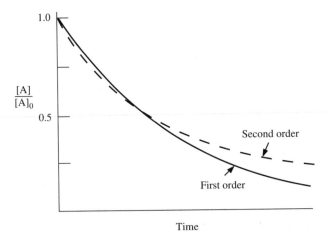

Fig. 7.5 Comparison of normalized first- and second-order kinetics with time scales adjusted to give the same $t_{1/2}$.

been adjusted so that the two half-lives are the same. Note that the second-order reaction proceeds more rapidly during the early stages of the reaction, but slows down and persists to longer times than does the first-order reaction. The difference in the two curves is subtle, however, and it is best seen using this superposition approach.

Class II (A + B → P)

$$v = k_2[A][B]$$

A reaction that is second order overall may be first order with respect to *each* of the two reactants. Some examples are

$$CH_3COOC_2H_5 + OH^- \longrightarrow CH_3COO^- + C_2H_5OH \qquad v = k_2[CH_3COOC_2H_5][OH^-]$$

$$NO(g) + O_3(g) \longrightarrow NO_2(g) + O_2(g) \qquad v = k_2[NO][O_3]$$

$$H_2O_2 + 2\,Fe^{2+} + 2\,H^+(excess) \longrightarrow 2\,H_2O + 2\,Fe^{3+} \qquad v = k_2[H_2O_2][Fe^{2+}]$$

Again we see that the overall stoichiometric equation is not a valid indicator of the rate law. We may expect that there are significant underlying differences in the mechanisms of these reactions.

$[a \neq b]$ In general, for reactions of the type

$$A + B \longrightarrow products$$

exhibiting class II kinetics, the initial concentrations of the reactants need not be the stoichiometric ratio. For this reason, we rewrite the rate law

$$v = k_2[A][B] \tag{7.20}$$

in terms of the reaction variable, where $x =$ the concentration of each species reacted

$$[A] = a - x; \qquad [A]_0 = a$$
$$[B] = b - x; \qquad [B]_0 = b$$

Following substitution into Eq. (7.20) and separation of variables, we obtain

$$\frac{dx}{(a - x)(b - x)} = k_2\,dt \tag{7.21}$$

A table of integrals gives the result

$$\frac{1}{a - b} \ln \frac{b(a - x)}{a(b - x)} = k_2 t \qquad [a \neq b] \tag{7.22}$$

or, alternatively,

$$\frac{1}{[A]_0 - [B]_0} \ln \frac{[B]_0[A]}{[A]_0[B]} = k_2 t \tag{7.23}$$

After separating the constant terms involving only initial concentrations, we see that class II second-order reactions exhibit a linear reaction between $\ln([A]/[B])$ and time.

[$a = b$]   Note that when $a = b$ (when the initial concentrations are stoichiometric), then Eqs. (7.22) and (7.23) do not apply. In this case, however, the method of class I is appropriate, because the values of [A] and [B] will be in a constant ratio throughout the entire course of the reaction. That is, if $a = b$, then [A] = [B] at all times and

$$v = k_2[A][B] = k_2[A]^2$$

which can be integrated exactly as in class I to give Eq. (7.17).

Example 7.2   Hydrogen peroxide reacts with ferrous ion in acidic aqueous solution according to the reaction

$$H_2O_2 + 2\,Fe^{2+} + 2\,H^+ \longrightarrow 2\,H_2O + 2\,Fe^{3+}$$

From the following data, obtained at 25°C, determine the order of the reaction and the rate constant.

$$[H_2O_2]_0 = 1 \times 10^{-5}\ M$$

$$[Fe^{2+}]_0 = 1 \times 10^{-5}\ M$$

$$[H^+]_0 = 1\ M$$

Time (min)	0	5.3	8.7	11.3	16.2	18.5	24.6	34.1
$10^5 \times [Fe^{3+}]\ M$	0	0.309	0.417	0.507	0.588	0.632	0.741	0.814

Solution   Because the H^+ concentration is in huge excess, it will not change significantly during the course of the reaction. As a consequence, we can write the velocity expression as

$$v = k[H_2O_2]^i[Fe^{2+}]^j$$

recognizing that this is the simplest general form. Note also that the initial concentration of H_2O_2 is twice the amount needed to react with all of the Fe^{2+} present.

1. Test for first order in $[Fe^{2+}]$, where $i = 0$ and $j = 1$. In this case, we expect from Eq. (7.9) that

$$\ln\frac{[Fe^{2+}]_0}{[Fe^{2+}]} = kt \quad \text{or} \quad \frac{1}{t}\ln\frac{[Fe^{2+}]_0}{[Fe^{2+}]} = k$$

2. Test for first order in $[H_2O_2]$, where $i = 1$ and $j = 0$. Then

$$\frac{1}{t}\ln\frac{[H_2O_2]_0}{[H_2O_2]} = k$$

The data for testing proposals 1 and 2 are as follows:

Time (min)	0	5.3	8.7	11.3	16.2	18.5	24.6	34.1
$10^5 \times [Fe^{2+}](M)$	1.00	0.691	0.583	0.493	0.412	0.368	0.259	0.186
$\ln \dfrac{[Fe^{2+}]_0}{[Fe^{2+}]}$	0	0.370	0.540	0.707	0.887	1.000	1.351	1.682
(1) $\dfrac{10^3}{t} \ln \dfrac{[Fe^{2+}]_0}{[Fe^{2+}]}$	—	69.8	62.0	62.6	54.7	54.0	54.9	49.3
$10^5 \times [H_2O_2](M)$	1.00	0.845	0.791	0.741	0.706	0.684	0.630	0.593
(2) $\dfrac{10^3}{t} \ln \dfrac{[H_2O_2]_0}{[H_2O_2]}$	—	31.8	26.9	26.5	21.5	20.5	18.8	15.3

The fact that the values in the rows designated (1) and (2) are not constant, but show a trend, indicates that neither of these predictions is valid for the reaction.

3. Test for second order overall, where $i = 1$ and $j = 1$. Because the stoichiometric coefficients are not unity, we need to modify the development of Eq. (7.22). For a reaction with stoichiometry $A + 2B \rightarrow$ product, we define

$$v = -\frac{d[A]}{dt} = k_2[A][B] = -\frac{1}{2}\frac{d[B]}{dt}$$

where
$$[A] = a - x \qquad [A]_0 = a$$
and
$$[B] = b - 2x \qquad [B]_0 = b$$

The equation analogous to Eq. (7.21) is now

$$\frac{dx}{(a - x)(b - 2x)} = k_2\, dt$$

or
$$\frac{dx}{(a - x)(b/2 - x)} = 2\, k_2\, dt$$

This is readily solved by noting that with respect to Eq. (7.21) we have just replaced b by $b/2$ and k_2 by $2\,k_2$. We simply make the same replacements in the solution, Eq. (7.22), to obtain the result

$$\frac{1}{a - \dfrac{b}{2}} \ln \frac{\dfrac{b}{2}(a - x)}{a\left(\dfrac{b}{2} - x\right)} = 2\,k_2 t$$

or
$$\frac{1}{2a - b} \ln \frac{b(a - x)}{a(b - 2x)} = k_2 t$$

or
$$\frac{1}{2[H_2O_2]_0 - [Fe^{2+}]_0} \ln \frac{[Fe^{2+}]_0[H_2O_2]}{[H_2O_2]_0[Fe^{2+}]} = k_2 t$$

Since $[Fe^{2+}]_0/[H_2O_2]_0 = 1.0$ from the initial conditions, we expect the relation

$$\frac{1}{t} \ln \frac{[H_2O_2]}{[Fe^{2+}]} = \text{constant}$$

to hold if the rate law is first order with respect to each reactant.
Proposal 3 is tested by the following tabulation:

Time (min)	0	5.3	8.7	11.3	16.2	18.5	24.6	34.1
$\dfrac{[H_2O_2]}{[Fe^{2+}]}$	1.00	1.223	1.357	1.503	1.714	1.859	2.432	3.188
(3) $\dfrac{1}{t} \ln \dfrac{[H_2O_2]}{[Fe^{2+}]}$ (min^{-1})	—	0.0380	0.0351	0.0361	0.0332	0.0335	0.0361	0.0340

The entries in the bottom row are approximately constant and show that the results do correspond to this rate law. Therefore,

$$v = k_2[H_2O_2][Fe^{2+}]$$

The value of k can be determined from

$$k_2 = \frac{1}{2[H_2O_2]_0 - [Fe^{2+}]_0} \left[\frac{1}{t} \ln \frac{[H_2O_2]}{[Fe^{2+}]} \right]$$

The average value of $(1/t) \ln ([H_2O_2]/[Fe^{2+}]_0)$ from the table is 0.0351 min^{-1}. Therefore,

$$k_2 = \frac{0.0351 \text{ min}^{-1}}{(2 \times 10^{-5} - 1 \times 10^{-5}) M} = 3.51 \times 10^3 \, M^{-1} \text{min}^{-1}$$

We could, of course, have solved this problem by graphical rather than numerical methods.

Renaturation of DNA as an Example of a Second-Order Reaction

A double-stranded DNA is made of two antiparallel chains with nucleotide sequences that are complementary to each other. If a linear DNA fragment is heated or subjected to high pH, the two chains separate and this disordered product is termed the denatured or coiled form. If the temperature or the pH is then lowered so that the double helix is again the stable form, pairing of the bases between chains of complementary sequences occurs and the chains will reassociate or *renature*. If we designate the strands with complementary sequences as A and A′, the renaturation reaction can be written as

$$A + A' \xrightarrow{k} AA'$$

which has been found experimentally to be second order. The most striking feature of renaturation kinetics is that when DNA from different sources is first broken down to about the same size (by sonication, for example) and then the rates of renaturation are measured at the same DNA concentrations in moles of nucleotides per liter, the rates are found to span a range of several orders of magnitude (Fig 7.6). The analysis below shows that the rate is expected to be inversely proportional to the *sequence complexity* of the DNA.

Figure 7.7 illustrates that the rate of renaturation depends on how repetitive the sequence of DNA is. When a long DNA molecule with no repetitive sequences is fragmented into pieces longer than 20 base pairs, each piece should be a unique sequence. For example, a DNA 10^6 base pairs long can be broken into essentially 10^6 (actually $10^6 - 19$) pieces of length 20. The number of possible DNA sequences which are 20 base pairs long is 4^{20} (any of the 4 bases can be in each position). As $4^{20} = 10^{12}$ which is much, much larger than 10^6, each piece is likely to be unique. When the pieces are denatured, the concentration of

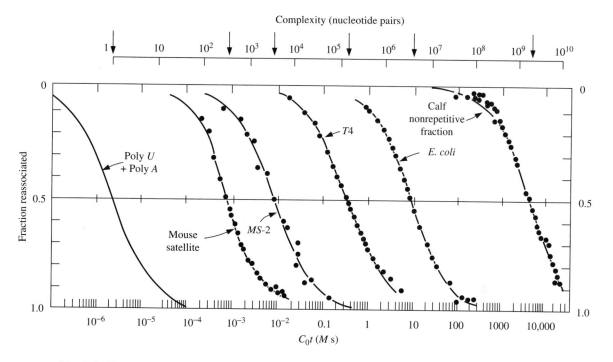

Fig. 7.6 Reassociation rates in solutions containing 0.18 M Na$^+$ for several double-stranded DNAs (mouse satellite—part of the mouse genome which has highly repetitive sequences, T4 bacteriophage, *E. coli* bacteria, and part of the calf genome). Data for two double-stranded RNAs (MS-2 viral RNA and a synthetic RNA, polyA·polyU) are also shown. The large nucleic acids were first broken into double-stranded fragments of about 400 base pairs by sonication. The fragments were then denatured into single strands of about 400 nucleotides by heating the solution to 90°C for a few minutes. The solution was cooled quickly and the fraction reassociated to double strands was measured vs. time, t. The rate of reassociation (renaturation) to double strands depends on the concentration of the single strands, C_0, and on the number of different sequences in the nucleic acids—the sequence complexity. C_0 is expressed in units of moles of nucleotides per liter. The complexity of the samples, indicated by arrows over the upper scale, range from 1 for polyA·polyU to over 10^9 for the nonrepetitive fraction of calf DNA. [From R. Britten and D. Kohne, *Science 161*, 529 (1968).]

DNA with nonrepetitive sequence of 10^6 base pairs

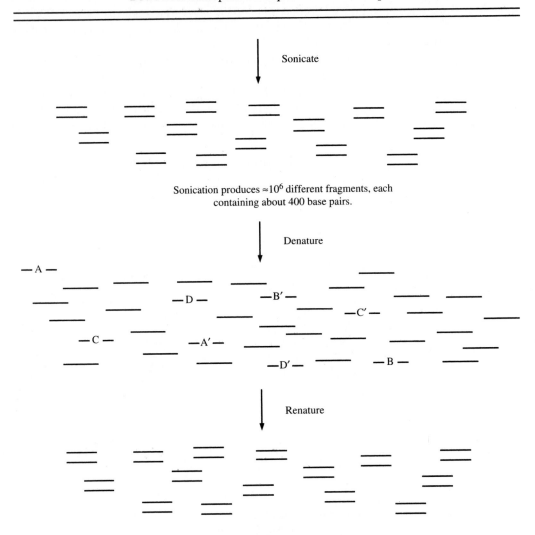

Sonicate

Sonication produces $\approx 10^6$ different fragments, each
containing about 400 base pairs.

Denature

—A—
—D—
—B'—
—C'—
—C—
—A'—
—D'—
—B—

Renature

Renaturation is very slow because strands A and A'
are in very low concentrations; many collisions are
needed to find complementary partners.

Fig. 7.7 Renaturation (or reassociation rates) depend on the sequence complexity of the DNA. In (a) a complex DNA is fragmented to give essentially all different fragments. The concentration of each fragment is very small compared to the total concentration of all fragments, therefore the rate of renaturation is slow. In (b) all the fragments are the same and the rate of renaturation is much faster.

(b)

DNA with repeating sequence of 10^6 base pairs—poly dA · poly dT

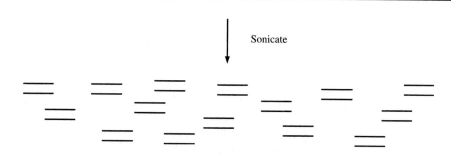

Sonicate

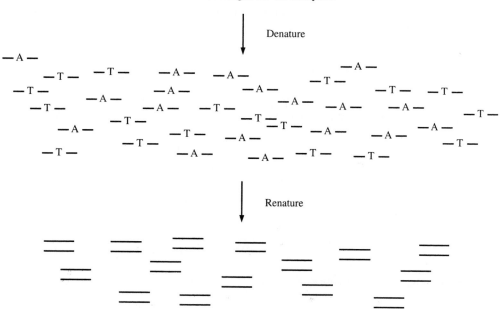

Sonication produces ≈10^6 identical fragments, each
containing about 400 base pairs.

Denature

Renature

Renaturation is very fast because strands
A and T are in high concentrations.

each unique strand is very small and the probability of finding its Watson-Crick complementary strand is small. The rate of renaturation will be very slow. (See Fig. 7.7a.) However, if there are repetitive sequences, the fragmented pieces can be identical and it will be easy for a denatured strand to find its complement (Fig. 7.7b). The analysis can be done quantitatively.

Let C_0 be the total concentration (in moles of nucleotide per liter) of all the single strands before any renaturation occurs. This total concentration can be measured easily from the ultraviolet absorbance of the solution. The rate of renaturation will depend on $[A]_0$, the initial concentration (in moles of nucleotide per liter) of fragment A which is complementary to fragment A'. We see from Fig. 7.7 that $[A]_0$ can vary from $C_0/2$ for poly dA·poly dT (Fig. 7.7b) to $C_0/2N$ for a DNA of N base pairs of a nonrepeating sequence. Therefore we define N as the sequence complexity—the total number of base pairs in the smallest repeating sequence in the DNA. The concentration $[A]_0$ is proportional to C_0/N. If the DNA is that of the bacterium *E. coli*, for example, there is very little repetition of the sequence and N is the same as the number of base pairs per genome, or about 3×10^6; if the DNA is polydeoxyadenylate (poly dA) in one strand and polydeoxythymidylate (poly dT) in the other, then N is 1, since the smallest repeating sequence is just a single base pair.

The rate of renaturation is

$$\frac{-d[A]}{dt} = \frac{-d[A']}{dt} = k[A][A'] = k[A]^2$$

The DNA is originally double stranded, so the concentrations of $[A]$ and $[A']$ must be equal. The equation is the same as Eq. (7.16), and the half-life of the reaction is, from Eq. (7.19)

$$t_{1/2} = \frac{1}{k[A]_0} \tag{7.19}$$

But $[A]_0$ is proportional to C_0/N

$$[A]_0 \propto \frac{C_0}{N}$$

substituting into the equation above gives

$$t_{1/2} \propto \frac{N}{kC_0}$$

or

$$C_0 t_{1/2} \propto N$$

Thus the half-life of renaturation times the total initial denatured strand concentration is proportional to the sequence complexity of the DNA. A long half-life means that each single strand will make many collisions before it finds its complement; the sequence is complex. The quantity $C_0 t_{1/2}$ can thus be used to measure how complex is the nucleotide sequence of the genome of an organism. The method is obviously applicable to double-stranded RNA as well. Figure 7.6 shows that the sequence complexity varies from 1 for the synthetic poly A · poly U to 10^9 for the nonrepetitive fraction of calf thymus DNA.

Reactions of Other Orders

It is easy to generalize the treatment of class I reactions of order n, where n is any positive or negative number except +1. The rate law is then

$$v = k[A]^n \tag{7.24}$$

which can be integrated to give

$$\frac{1}{n-1}\left[\frac{1}{[A]^{n-1}} - \frac{1}{[A]_0^{n-1}}\right] = kt \qquad n \neq 1 \tag{7.25}$$

The half-life will be

$$t_{1/2} = \frac{2^{n-1} - 1}{(n-1)k[A]_0^{n-1}} \tag{7.26}$$

For example, the thermal decomposition of acetaldehyde is $\frac{3}{2}$ order over most of the course of the reaction. A linear plot should be obtained, according to Eq. (7.25), if $1/\sqrt{[A]}$ is plotted versus time. The half-life will be given by

$$t_{1/2} = \frac{\sqrt{2} - 1}{\frac{1}{2}k[A]_0^{1/2}}$$

Determining the Order and Rate Constant of a Reaction

We have discussed various simple orders which can represent reaction kinetics, and we have given specific illustrations. Here we shall mention some general methods that can be used to determine the order and rate constant for a reaction. We need to know the stoichiometry of the reaction and we need to be able to measure the concentration of a product or reactant. To determine the order of the reaction with respect to each reactant, we actually need only to measure a property that is directly proportional to concentration. Except for first-order reactions, we do need to know concentrations to obtain rate constants. The characteristics of the different order reactions are given in Table 7.3.

Table 7.3 The characteristics of simple-order reactions

Reaction order	Linear plot	Rate proportional to	Units of k
Zero-Order	c vs. t	$dc/dt = k$	M time^{-1}
First-Order	$\ln c$ vs. t	$dc/dt = kc$	time^{-1}
Second-Order (I)	$1/c$ vs. t	$dc/dt = kc^2$	M^{-1} time^{-1}
Second-Order (II)	$\ln(c_A/c_B)$ vs. t	$dc_A/dt = kc_A c_B$	M^{-1} time^{-1}
nth-Order ($n \neq 1$)	$1/c^{n-1}$ vs. t	$dc/dt = kc^n$	M^{1-n} time^{-1}

Direct data plots

A plot of concentration versus time gives a clue to the order. If the plot is linear, a zero-order reaction is indicated. If a curved plot results, other plots, such as $\log c$ versus time or $1/c$ versus time, can be tried. One looks for a linear plot that indicates the order and whose slope is proportional to the rate constant.

Velocity versus concentration plots (differentiation of the data)

From measurements of the slopes of plots of concentration versus time, one can determine the velocity of the reaction at different times. These velocities can then be plotted versus the corresponding reactant concentration in an attempt to deduce the form of the rate law. This approach requires many experimental points or a continuous curve.

Method of initial rates

The velocity is measured during the earliest stages of the reaction. The concentrations do not change much with time, and they can be replaced in the rate law by the initial concentrations. This method is particularly useful when the rate law is not of the simplest form. For example, in many enzyme-catalyzed reactions the method of initial rates was used to find that

$$v = \frac{k[S]}{K + [S]}$$

where k and K are constants and $[S]$ is the substrate concentration. We shall discuss this Michaelis-Menten type of kinetics in Chapter 8.

Changes in initial concentrations

The influence of different initial concentrations on the initial rate, the half-life, or some other parameter can be determined. The method measures the order of the reaction with respect to a particular component directly. To find the order with respect to component A in the expression

$$v_1 = k[A]^a[B]^b[C]^c$$

we measure the initial rate v_1 with initial concentrations A, B, and C.

$$v_1 = k[A]^a[B]^b[C]^c$$

We then double the concentration of A, keeping the others constant, and measure the new initial rate v_2,

$$v_2 = k(2[A])^a[B]^b[C]^c$$

The ratio of these two rates leads to the value of a,

$$\frac{v_2}{v_1} = 2^a$$

because all other factors cancel.

Methods of reagents in excess

We can decrease the kinetic order of a reaction by choosing one or more of the reactant concentrations to be in large excess. Normally, the concentration of a reactant decreases steadily during the course of a reaction. This complicates the kinetic analysis, especially if the rate law is not a simple one. By using a large excess of one component, its concentration is maintained nearly constant and the corresponding term in the rate law now hardly changes during the course of the reaction. In the example above we could choose B and C to be 1 M and A to be 0.01 M. Concentrations of B and C would change only slightly (depending on the stoichiometry) during the complete reaction of A. We could essentially study the kinetics with respect to A independently of the B and C kinetics.

Once the order of the reaction is known with respect to each component, the value of the rate constant can be calculated. It is important to specify what rate and what species are being described. Uncertainty can occur when the stoichiometric coefficients for the reaction are not all the same. For example, the reaction

$$N_2O_5 \longrightarrow 2\,NO_2 + \tfrac{1}{2}O_2$$

exhibits first-order kinetics. If we express concentrations of these substances in mol L^{-1}, the rate of disappearance of reactant N_2O_5 is half as great as the rate of formation of the product NO_2 and twice as great as the rate of formation of O_2. Expressed mathematically, this is

$$-\frac{d[N_2O_5]}{dt} = \frac{1}{2}\frac{d[NO_2]}{dt} = 2\frac{d[O_2]}{dt}$$

The coefficients in this velocity equation are simply the reciprocals (apart from the signs) of the stoichiometric coefficients. There is no universal rule among kineticists as to which of these to use to express the velocity of the reaction quantitatively. Usually, the specific substance measured is used. In this book we will adopt the convention that, for a reaction whose stoichiometric equation is written as

$$mM + nN \longrightarrow pP + qQ$$

the velocity of the reaction will be taken as

$$v = -\frac{1}{m}\frac{d[M]}{dt} = -\frac{1}{n}\frac{d[N]}{dt} = \frac{1}{p}\frac{d[P]}{dt} = \frac{1}{q}\frac{d[Q]}{dt}$$

and the rate-law expression

$$v = k[M]^i[N]^j \cdots$$

will serve to define the relation of the rate constant to the velocity of the reaction. (Remember that the exponents i and j in the rate law are not related to the stoichiometric coefficients.)

REACTION MECHANISMS AND RATE LAWS

The form of the rate law determined from the kinetics of a reaction does not tell us the actual mechanism, although it gives important clues. A very complex process, such as the growth and multiplication of bacterial cells, can exhibit simple first-order kinetics. Growth, as measured by the increase in the number of cells, occurs in proportion to the number of cells present at any given time and with a rate constant determined by the turnaround time for the complex events in the cell cycle. We cannot hope for this single parameter, the rate constant, to give us detailed information about the individual steps of DNA replication, transcription into RNA, synthesis of protein, mitotic division, and so on, that are involved in the cell cycle. The rate law does, however, make a quantitative statement about how fast the cells grow.

Typically, a complex reaction sequence or mechanism is made up of a set of elementary reactions. Consider, for example, the process by which the ozone layer of the upper atmosphere is maintained. This ozone layer acts as an important shield, absorbing potentially harmful ultraviolet (uv) radiation from the sun and protecting organisms on the surface of the earth from dangerous biological consequences. Ozone is formed from atmospheric oxygen by a process that can be described by the stoichiometric equation

$$3\,O_2 \longrightarrow 2\,O_3$$

This reaction does not occur as such in the atmosphere. Instead, ozone is formed as a consequence of an elementary photochemical step resulting from the absorption of short uv radiation followed by a second elementary reaction in which the oxygen atoms produced in the first step go on to react with O_2 to form ozone.

$$O_2 \xrightarrow{\ h\nu\ \text{(below 242 nm)}\ } O + O$$

$$O + O_2 + M \longrightarrow O_3 + M$$

where M is a third body (any other gas molecule) whose function is to remove excess energy from the excited ozone initially formed. The protective role of ozone results from its absorption of somewhat longer-wavelength ultraviolet light not absorbed by O_2, and this radiation leads to the decomposition of ozone by the elementary reaction

$$O_3 \xrightarrow{\ h\nu\ \text{(190 to 300 nm)}\ } O_2 + O$$

The balance in the ozone concentration is achieved by these three elementary steps operating together, supplemented by lesser contributions of additional reactions involving other elements.

Johnston (1975) has pointed out that the flights of supersonic transports in the upper atmosphere pose a threat to the stability of the protective ozone layer. This comes about because of the role of oxides of nitrogen (NO_x) in the exhaust gases of the jets. The oxides catalyze the decomposition of ozone:

$$O + O_3 \xrightarrow{\ NO_x\ } 2\,O_2$$

This process can be understood in terms of the elementary reactions involved, which are

$$NO + O_3 \longrightarrow NO_2 + O_2$$

$$NO_2 + O \longrightarrow NO + O_2$$

Not only does this show in molecular detail how the oxides of nitrogen produce the ozone decomposition, but it also shows that the NO and NO_2 are not used up in the process. Together they serve a true catalytic function.

Another class of catalyst for the destruction of ozone are chlorofluorocarbons (Molina and Rowland, 1974; Anderson et al., 1991). Chlorine atoms are produced by the photochemical decomposition of, for example, dichlorodifluoromethane (Freon 12)

$$CCl_2F_2 \longrightarrow CClF_2 + Cl$$

The chlorine atoms catalyze the decomposition of ozone molecules by oxygen atoms.

$$Cl + O_3 \longrightarrow ClO + O_2$$

$$O + ClO \longrightarrow Cl + O_2$$

The net reaction is that ozone is converted to oxygen and the chlorine atoms are not used up. Because of the environmental dangers caused by the loss of ozone, the use of chlorofluorocarbons in aerosol cans has been banned in the United States.

It often happens that more than one mechanism can be written consistent with a given rate law. The best one can be chosen only if we learn more about the intermediates, the energetics of the process, and the rates of elementary steps determined from other sources (different reaction systems). The concept of elementary reactions is important to chemical kinetics for the same reason that the study of chemical bonds or of functional groups is important in simplifying and organizing the chemistry of the hundreds of thousands of different organic compounds. The kinetic parameters of an elementary step are the same regardless of what overall reaction it is participating in.

Elementary reactions are described in terms of their *molecularity:* the number of reactant particles involved in the elementary reaction. Reactions between two particles are called *bimolecular.* They may be thought of in terms of a collision between the two reactants to form a transition state. A transition state is an unstable arrangement of atoms intermediate between reactants and products. The transition state exists for no longer than the time it takes for a molecular vibration (of order 10 fs). For example, the reaction of ozone with nitric oxide in the gas phase occurs as a simple bimolecular process. The transition state is the transient species in which an O—O bond is being broken as an O—N bond forms.

Since the process is bimolecular, it follows that it must exhibit second-order kinetics. The converse is not true, however; many kinetically second-order reactions do not have simple bimolecular mechanisms but are more complex. *Hence one cannot use the form of the rate law alone to predict the mechanism; whereas knowing the mechanism does permit the rate law to be deduced.* This is an important distinction to learn.

From the example given above we can define molecularity as the number of particles that come together to form the transition state in an elementary reaction. Note that the term "molecularity" applies only to elementary reactions and not to the overall mechanism. *Unimolecular* reactions are those that involve only a single reactant particle, for example radioactive decay. True unimolecular processes are rare, because most reactions require some *activation energy* in order to proceed. This energy is commonly picked up by collisions with the surroundings (other molecules, walls of the vessel). *Termolecular* reactions involve collisions of three molecules and are more likely to occur at high pressures or in liquid solution.

The concept of molecularity was developed largely for gas-phase reactions. It is a difficult concept to carry over to liquid solutions, where the role of the solvent and other closely associated molecules in the medium makes it difficult to define or determine the reacting "particle."

The simplest examples of reaction mechanisms are those that involve only a single elementary reaction. In this case the rate law can be written by inspection. The velocity is proportional to the concentration of each of the species involved in forming the transition state, and the exponents are determined by the participating numbers of each kind of particle. For example, if the reaction

$$2\,A\, +\, B\, \longrightarrow\, \text{products}$$

is an *elementary* reaction, its velocity will be given by

$$v\, =\, k[A]^2[B]$$

The real problem occurs when the overall reaction is complex and consists of several connected elementary reactions. The overwhelming majority of reactions and processes in nature are complex, and it is a relatively rare one that turns out to be elementary.

To account for the kinetic observations, it is useful to formulate ("hypothesize") the mechanisms in terms of a set of elementary reactions. These often involve one or more intermediates, which may have been observed previously or may be postulated uniquely for this reaction. This proposed mechanism can then be used to predict a rate law that can be tested against experimental findings. Furthermore, one can look for any postulated intermediates by standard physical or chemical methods.

A few examples will illustrate some important methods of dealing with complex reactions. One general method is to see if one elementary reaction controls the overall kinetics, that is, to look for the *rate-determining step*. For certain mechanisms one elementary reaction is found to dominate the kinetics of the complex reaction. When this is so, the analysis of the mechanism is greatly simplified. We can write the rate for the rate-determining step and equate it to the overall rate. In the following examples we shall discuss the general case first and then indicate the simplification that occurs if there is a rate-determining step.

Parallel Reactions

Parallel reactions are of the following type:

$$A \overset{k_1}{\underset{k_2}{\rightleftarrows}} \begin{matrix} B \\ C \end{matrix}$$

The substance A can decompose by either of two paths, giving rise to different products, B and C. The rate expressions are

$$-\frac{d[A]}{dt} = k_1[A] + k_2[A] = (k_1 + k_2)[A] \tag{7.27}$$

$$\frac{d[B]}{dt} = k_1[A] \tag{7.28}$$

$$\frac{d[C]}{dt} = k_2[A] \tag{7.29}$$

The solution to the first equation, which has the *form* of the first-order rate law, Eq. (7.7), is

$$\ln \frac{[A]}{[A]_0} = -(k_1 + k_2)t$$

$$[A] = [A]_0 \exp[-(k_1 + k_2)t] \tag{7.30}$$

Thus [A] decays exponentially, as illustrated in Fig. 7.8. To find out how [B] and [C] increase with time, we need to solve Eqs. (7.28) and (7.29). This appears to present a problem, because each of these equations contains three variables; it cannot be solved as it stands. However, we can substitute for A using Eq. (7.30). Thus

$$\frac{d[B]}{dt} = k_1[A] = k_1[A]_0 \exp[-(k_1 + k_2)t]$$

and

$$\frac{d[C]}{dt} = k_2[A] = k_2[A]_0 \exp[-(k_1 + k_2)t]$$

We now separate variables and integrate. If $[B]_0 = [C]_0 = 0$, the solutions are

$$[B] = \frac{k_1[A]_0}{k_1 + k_2}\{1 - \exp[-(k_1 + k_2)t]\} \tag{7.31}$$

$$[C] = \frac{k_2[A]_0}{k_1 + k_2}\{1 - \exp[-(k_1 + k_2)t]\} \tag{7.32}$$

Note that B and C are formed in a constant ratio, $[B]/[C] = k_1/k_2$ throughout the course of the reaction. The extension of this method to three or more parallel reactions is straightforward.

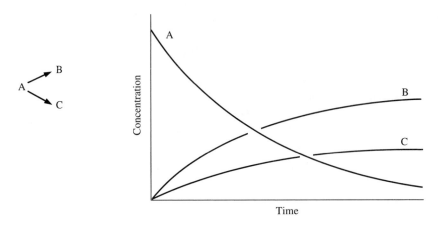

Fig. 7.8 Kinetic curves for reactant and products for two first-order reactions in parallel.

Example 7.3 Radioactive decay frequently occurs simultaneously by two separate first-order paths. For example $^{64}_{29}Cu$ is unstable against decay, emitting either an electron or a positron. The scheme can be written as follows:

$$^{64}_{29}Cu \quad \overset{k_-}{\nearrow} \quad ^{64}_{30}Zn + \beta^- \quad (62\%)$$
$$\underset{k_+}{\searrow} \quad ^{64}_{28}Ni + \beta^+ \quad (38\%)$$

The decay of $^{64}_{29}Cu$ occurs exponentially with a single half-life of 12.80 hr. Knowing the fraction of decays by each path then permits one to calculate k_+ and k_-. Do this for the example.

In parallel reactions, if one step is much faster than the others, this fast step dominates the reaction. This is very different from the case of series reactions, in which the slow step dominates, as we will see in the next section. In the example above, if 99% of the reaction goes through the β^- path, we can ignore the β^+ path in calculating the decay of $^{64}_{29}Cu$. We can see this quantitatively in Eq. (7.30). When $k_1 \gg k_2$, the decay of [A] with time depends only on k_1 and $[A] \cong [A]_0 \exp(-k_1 t)$.

Series Reactions (First Order)

Series reactions are of the type

$$A \overset{k_1}{\longrightarrow} B \overset{k_2}{\longrightarrow} C$$

Compound A reacts to form B, which then goes on to form C. The rate expressions are

$$v_1 = -\frac{d[A]}{dt} = k_1[A]$$

$$[A] = [A]_0 \exp(-k_1 t) \tag{7.33}$$

$$\frac{d[B]}{dt} = k_1[A] - k_2[B] = k_1[A]_0 \exp(-k_1 t) - k_2[B]$$

This differential equation is slightly more difficult to solve. We shall give the result. If $[B]_0 = 0$,

$$[B] = \frac{k_1[A]_0}{k_2 - k_1} [\exp(-k_1 t) - \exp(-k_2 t)] \tag{7.34}$$

Similarly,

$$v_2 = \frac{d[C]}{dt} = k_2[B]$$

$$= \frac{k_1 k_2 [A]_0}{k_2 - k_1} [\exp(-k_1 t) - \exp(-k_2 t)]$$

Variables $[C]$ and t can be separated here to give, for $[C]_0 = 0$,

$$[C] = [A]_0 \left\{ 1 - \frac{1}{k_2 - k_1} [k_2 \exp(-k_1 t) - k_1 \exp(-k_2 t)] \right\} \tag{7.35}$$

Where such complex formulas are involved, it is important to get some intuitive feeling for how these reactions behave. The graph in Fig. 7.9 illustrates the pattern for series first-order reactions. As the concentration of A decreases, it does so in a simple exponential fashion. If there is no B present initially, the rate of the second reaction will be negligible at the beginning, $[v_2]_0 = 0$. As B is formed by the first step, its concentration rises, reaches a maximum, and then declines to zero as the second step removes it from the reaction mixture. In accordance with the rate law, the curve for $[C]$ starts out with zero slope, where $[v_2]_0 = 0$. The rising portion of curve $[C]$ then undergoes an inflection (the point where $d^2[C]/dt^2 = 0$) when $[B]$ has its maximum value. Finally, $[C]$ approaches a limiting

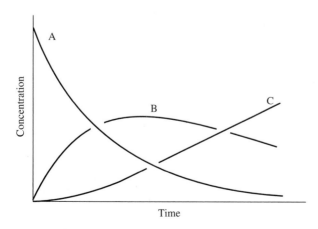

Fig. 7.9 Kinetic curves for reactant, A, intermediate, B, and product, C, for two first-order reactions in series. A → B → C.

value as the reaction nears completion. These properties are useful diagnostics for series reactions.

When two or more reactions occur in series, the slow reaction is the rate-determining step. It dominates in the kinetic control of the overall process. The analogy with a city water distribution system is appropriate. It does not matter how large the reservoir, water mains, and distribution pipes are. If the smallest cross section of the whole system is the $\frac{1}{2}$-in. pipe coming into your house, that will limit the maximum possible flow, and the only practical way to increase the flow is to replace the small pipe with one of larger diameter.

For the series reactions

$$A \xrightarrow{k_1} B \xrightarrow{k_2} C$$

we can consider the two cases $k_1 \gg k_2$ and $k_1 \ll k_2$. The first situation means that the first step is much faster than the second, and the second is the rate-determining step. Once the reaction is started, A will be converted to B rapidly in comparison with the subsequent reaction of B to C. During most of the course of the reaction, B undergoes a first-order conversion to C which is controlled by the rate constant k_2. If the formation of C is our measure of the velocity of the reaction, its appearance is nearly identical to that of a single-step (elementary) first-order reaction. This is illustrated schematically in Fig. 7.10. We arrive at the same conclusion from examination of Eq. (7.35) in the limit where $k_1 \gg k_2$. Under those conditions, Eq. (7.35) reduces to

$$[C] = [A]_0[1 - \exp(-k_2t)]$$

for $k_1 \gg k_2$, which has the same form as the simple first-order formation kinetics.

The second situation, where $k_1 \ll k_2$, starts out with a slow conversion of A to B and is followed by a very rapid reaction for B going on to C. In this case the concentration of B will remain low throughout the course of the reaction, and C will appear essentially as A disappears. This is illustrated in Fig. 7.11. It should be obvious that the velocity of the reaction can be measured either by the formation of C or by the disappearance of A, in this case. Thus Eq. (7.33) applies to the overall reaction.

The chief reason for emphasizing this point is because complex processes in biology frequently have rate-limiting steps. For our analysis, it does not matter how many steps are involved in the series mechanism; all that is required is that one step should be appreciably

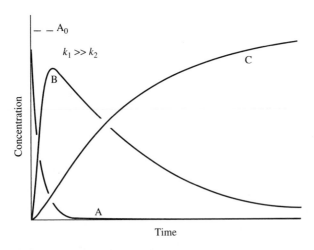

Fig. 7.10 Two first-order reactions in series, with the *second* as the rate-determining step.
A → B → C.

slower than any of the others. If the rate-limiting step is not first order, as in the example treated here, it must be treated using the kinetic equations for the appropriate order.

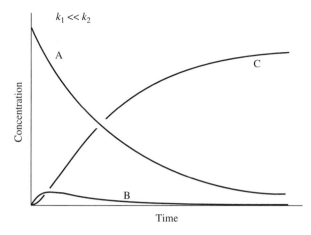

Fig. 7.11 Two first-order reactions in series, with the *first* as the rate-determining step. A → B → C.

Example 7.4   A complex but illustrative example of the kinetic behavior of series reactions occurs in the clotting of blood. The final step in the process is the conversion of fibrinogen to fibrin (the clotting material) catalyzed by the proteolytic (protein cutting) enzyme thrombin. Thrombin is not normally present in the blood, but is converted from a precursor, prothrombin, by the action of another activated proteolytic enzyme and calcium ion. The overall process involves a cascade of at least eight such steps, as shown in Fig. 7.12.

A consequence of this cascade of steps is the fact that the thrombin concentration, as measured by the decrease in clotting time for blood samples, increases dramatically following wounding and then decreases again once the clot is formed. The behavior, shown in Fig. 7.12, is very much like that of an intermediate in a series of reactions. In the mechanism outlined on the next page, the injury sets off a cascade of reactions involving a series of proteolytic enzymes, each of which activates a target protein. This cascade phenomenon serves to accelerate the mobilization of the clotting mechanism so that it can be switched from fully off to fully on in less than 1 min. The conversion of prothrombin to thrombin, catalyzed by active proaccelerin and Ca^{2+}, is the immediate cause of clotting. The level of thrombin in the blood must not remain high, however, or it will cause fibrin to form in the circulatory system and block the normal flow of blood. Antithrombin is an additional factor that inactivates thrombin, usually within a few minutes after it has reached its maximum level. In this way the action of thrombin is restricted to the period when it is critically needed.

While this overall process is much more elaborate than the simple example of series reactions considered on the next page, a similarity is seen when we inspect the formation and disappearance of thrombin as rate-limiting steps. For this purpose, prothrombin plays the role of A, thrombin is B, and inactive thrombin is C. The appearance of active proaccelerin turns on the first step (A → B), and the subsequent formation of antithrombin turns on the second step (B → C). The rate of clotting of fibrinogen serves (kinetically speaking) as a simple indicator of the level of thrombin present at any time in the course of this process.

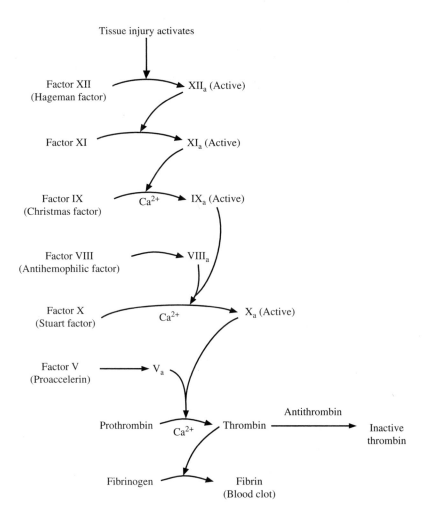

Tissue injury activates

Factor XII
(Hageman factor) $\longrightarrow$ XII$_a$ (Active)

Factor XI $\longrightarrow$ XI$_a$ (Active)

Factor IX
(Christmas factor) $\quad$ Ca^{2+} $\longrightarrow$ IX$_a$ (Active)

Factor VIII
(Antihemophilic factor) $\longrightarrow$ VIII$_a$

Factor X
(Stuart factor) $\quad$ Ca^{2+} $\longrightarrow$ X$_a$ (Active)

Factor V
(Proaccelerin) $\longrightarrow$ V$_a$

Prothrombin $\quad$ Ca^{2+} $\longrightarrow$ Thrombin $\xrightarrow{\text{Antithrombin}}$ Inactive thrombin

Fibrinogen $\longrightarrow$ Fibrin
(Blood clot)

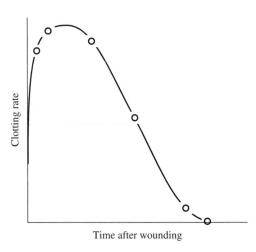

Clotting rate

Time after wounding

Fig. 7.12 Concentration of active thrombin after wounding. The clotting rate for the conversion of fibrinogen to fibrin is a direct measure of the thrombin concentration [From R. Biggs, *Analyst 78,* 84 (1953).]

Equilibrium and Kinetics

All reactions approach equilibrium. Therefore, in principle, for every forward reaction step, there is a backward step. In practice, we can sometimes ignore the backward step, because either the equilibrium constant is very large or the concentrations of products are kept very small. Nevertheless, it is important to know the general relation between kinetic rate constants (k) and thermodynamic equilibrium constants (K).

Let us consider an elementary first-order reversible reaction.

$$A \underset{k_{-1}}{\overset{k_1}{\rightleftharpoons}} B$$

The rate of disappearance of A is

$$-\frac{d[A]}{dt} = k_1[A] - k_{-1}[B]$$

At equilibrium, $-d[A]/dt = 0$; therefore,

$$\frac{[B]^{eq}}{[A]^{eq}} = \frac{k_1}{k_{-1}} = K$$

Exercise Show that for the elementary first-order reversible reaction

$$A \underset{k_{-1}}{\overset{k_1}{\rightleftharpoons}} B$$

$$\ln \frac{[A] - [A]^{eq}}{[A]_0 - [A]^{eq}} = \ln \frac{[B] - [B]^{eq}}{[B]_0 - [B]^{eq}} = -(k_1 + k_{-1})t$$

where $[A]_0$ and $[B]_0$ are the concentrations of A and B, respectively, at $t = 0$.
Hint: Let $x = [A] - [A]^{eq} = [B]^{eq} - [B]$, but because

$$-\frac{d[A]^{eq}}{dt} = k_1[A]^{eq} - k_{-1}[B]^{eq} = 0$$

therefore

$$-\frac{d[A]}{dt} = -\frac{dx}{dt} = (k_1 + k_{-1})t$$

Often there is more than one path for the reaction of A to form B. To be consistent with the principles of equilibrium thermodynamics, we must apply the principle of *microscopic reversibility.* The principle requires that, at equilibrium, the reactions between any pair of reactant, intermediate, or product species must occur with equal frequency in both directions. That is, the following mechanism is *not* possible:

Each elementary step in the reaction must be reversible to be consistent with the principle of microscopic reversibility. Thus we must formulate the mechanisms as follows:

$$
A \underset{k_{-1}}{\overset{k_1}{\rightleftharpoons}} B
$$

with

$$
A \underset{k_{-3}}{\overset{k_3}{\rightleftharpoons}} C \underset{k_{-2}}{\overset{k_2}{\rightleftharpoons}} B
$$

The relation to thermodynamics requires further that

$$
K = \frac{[B]^{eq}}{[A]^{eq}} = \frac{k_1}{k_{-1}} = \frac{k_2 k_3}{k_{-2}k_{-3}} \tag{7.36}
$$

The six rate constants are therefore not independent.

Complex Reactions

For many reactions, particularly those involving enzymes as catalysts, a series of reversible steps is involved. An example is the following set of coupled elementary reactions:

$$
A + B \underset{k_{-1}}{\overset{k_1}{\rightleftharpoons}} X
$$

$$
X \underset{k_{-2}}{\overset{k_2}{\rightleftharpoons}} P + Q
$$

In this case the exact solution to the rate equations is complex, because two of the elementary reactions are bimolecular and a total of five molecular species are involved. It is useful to learn some approximations that can be applied in cases like this.

Initial-velocity approximation

At the beginning of a reaction the product concentrations are usually very small or zero. For the set of opposed reactions above, the step designated -2 can be neglected in an analysis of *initial* velocities. This simplifies the kinetic expressions considerably. As the extent of reaction increases, the experimental results will begin to differ from the prediction of the approximate theory.

Prior equilibrium approximation

This approach is valuable when steps 1 and -1 are rapid in comparison with step 2.

$$
A + B \underset{k_{-1}}{\overset{k_1}{\rightleftharpoons}} X \qquad \text{(fast, equilibrium)}
$$

$$
X \overset{k_2}{\longrightarrow} P + Q \qquad \text{(slow)}
$$

A, B, and X rapidly attain a state of quasi-equilibrium, such that

$$
v_1 = v_{-1}
$$

and

$$k_1[A][B] = k_{-1}[X]$$

We can write this as an equilibrium expression:

$$K = \frac{k_1}{k_{-1}} = \frac{[X]}{[A][B]}$$

Step 2 is the rate-limiting step, and the rate of formation of product is given by

$$v = \frac{d[P]}{dt} = \frac{d[Q]}{dt} = k_2[X]$$

Substituting from above, we see that the velocity

$$v = \frac{k_2 k_1}{k_{-1}}[A][B] \tag{7.37}$$

can be expressed in terms of reactant concentrations only. This is particularly valuable if the relative concentration of X remains very small ($k_{-1} \gg k_1$), but is is just as valid when X is fairly large ($k_{-1} \sim k_1$). The important criterion for the prior equilibrium approximation to apply can be written

$$v \cong v_2 \ll v_1 \cong v_{-1}$$

This should be read: "The overall velocity of the reaction is limited by the slow step 2, and the velocity is much slower than the forward and reverse reactions of step 1, which are essentially in equilibrium."

Steady-state approximation

It frequently happens that an intermediate is formed that is very reactive. As a consequence, it never builds up to any significant concentration during the course of the reaction.

$$A + B \xrightarrow{k_1} X \qquad \text{(slow)}$$

$$X + D \xrightarrow{k_2} P \qquad \text{(fast)}$$

To a first approximation, X reacts as rapidly as it is formed:

$$v_1 = v_2$$
$$k_1[A][B] = k_2[X][D] \tag{7.38}$$

Hence,

$$v = \frac{d[P]}{dt} = k_1[A][B]$$

in terms of reactant concentrations only. A more general way of formulating the steady-state approximation is to consider all of the steps involving formation and disappearance of the reactive intermediate and to set the sum of their rates equal to zero. In this example X is formed in step 1 and it disappears in step 2. Thus

$$\frac{d[X]}{dt} = k_1[A][B] - k_2[X][D] \cong 0 \tag{7.39}$$

This gives the same result [Eq. (7.38)]. However, this second approach is more general. The significance of this equation is not that [X] is constant during the course of the reaction. Such is never the case. It is true, however, that the *slope*, $d[X]/dt$, of the curve of [X] versus time is much smaller than those typical of reactants or products. For example, this can be seen for component B for the scheme shown in Fig. 7.11. It is this nearly zero slope throughout the course of the reaction that is the characteristic of the steady-state condition.

The steady-state approximation can be applied in many chemical and biochemical reactions. Care must be taken that the intermediate satisfies the criterion that $d[X]/dt \cong 0$. Note that this is not necessarily the case where prior equilibrium is involved. It is also not true for the intermediate B illustrated in Fig. 7.9.

Deducing a Mechanism from Kinetic Data

There is no straightforward way to obtain a mechanism from kinetic data. There is always more than one mechanism which will be consistent with the kinetic data. Therefore, why propose a mechanism at all? The answer is that we like to think that we understand a reaction and that we can guess what the molecules are doing. A mechanism gives us some basis for predicting what should happen for other reactions.

The best way to illustrate how one hypothesizes a mechanism is to discuss specific examples. The only general advice that can be given is to think of a simple and plausible mechanism and then calculate the kinetics and see if they are consistent with the data. A unique rate equation can always be obtained from a proposed mechanism. If the rate law is simple first or second order, assume a unimolecular or bimolecular rate-determining step. Then, if necessary, add other steps to agree with the stoichiometry. For example, suppose that for the stoichiometric reaction

$$A + B \longrightarrow C$$

the rate law is found to be

$$-\frac{d[A]}{dt} = k[A][OH^-]$$

We can assume for a mechanism

$$A + OH^- \xrightarrow{k_1} M^-$$

$$M^- + B \xrightarrow{k_2} C + OH^-$$

The first elementary reaction postulates that $A + OH^-$ can react to form an intermediate, M^-; this gives the correct rate law. The second elementary reaction must be added to account for the stoichiometry. From the proposed mechanism we can now predict rate laws for [B] and [C] and experimentally check the predictions. These rate laws are complicated in the general case, so we shall not present them. Another mechanism consistent with the data is

$$A + OH^- \underset{}{\overset{K}{\rightleftharpoons}} N^- \qquad \text{(fast to equilibrium)}$$

$$N^- \xrightarrow{k_1} P^- \qquad \text{(slow, rate determining)}$$

$$P^- + B \xrightarrow{k_2} C + OH^- \qquad \text{(fast)}$$

The rate-determining step is

$$-\frac{d[N^-]}{dt} = k_1[N^-]$$

but

$$\frac{[N^-]}{[A][OH^-]} = K$$

Therefore,

$$-\frac{d[N^-]}{dt} = k_1K[A][OH^-]$$

Because all the other elementary reactions are fast, each time N^- reacts to form P^-, A and B react and C is formed. The rate laws are then

$$-\frac{d[A]}{dt} = -\frac{d[B]}{dt} = \frac{d[C]}{dt} = k_1K[A][OH^-]$$

To decide between the two mechanisms, other kinetic experiments can be done and attempts to detect the postulated intermediates can be made. Many other mechanisms could also be written. It should be clear that proposing reasonable mechanisms requires practice and experience.

TEMPERATURE DEPENDENCE

It seems almost intuitive that if you want a reaction to proceed faster, you should heat it. It came as a surprise when some chemical reactions were found to have negative temperature coefficients; that is, the reaction went *slower* at higher temperature. In biology this is a common occurrence. Few organisms or biochemical processes can withstand temperatures above 50°C for very long in the first place, but this results from irreversible structural changes that alter the mechanisms or destroy the organism. Processes like the coagulation of the protein egg albumin (egg white) occur during cooking, and they occur more rapidly at high temperatures. This seems normal. If, however, a protein is first carefully denatured, it can be renatured more rapidly at a lower temperature than at a higher one! There is a kinetic effect in addition to the shift in equilibrium constant for this process.

Observations by Arrhenius and others showed that for most chemical reactions there is a logarithmic relation between the rate constant and the reciprocal of the temperature in Kelvins. We can therefore write the Arrhenius result as

$$\ln k = -\left(\frac{E_a}{RT}\right) + \ln A \qquad (7.40)$$

or

$$k = A \exp\left(-\frac{E_a}{RT}\right) \qquad (7.41)$$

The *activation energy* for the reaction is E_a and A is called the pre-exponential factor. This is an empirical equation which represents the behavior of many chemical systems and even some rather complex biological processes (see Fig. 7.13). The usual way of obtaining

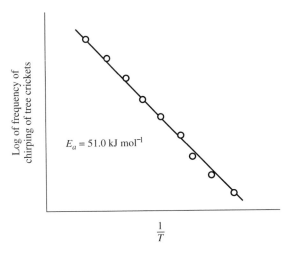

$E_a = 51.0 \text{ kJ mol}^{-1}$

$\frac{1}{T}$

Log of frequency of chirping of tree crickets

Fig. 7.13 Arrhenius plot of the chirping rate of tree crickets at different temperatures. The linearity of this plot suggests that the complex biological process is governed by a rate-limiting step having the activation energy shown. [From K. J. Laidler, *J. Chem. Educ. 49,* 343 (1972).]

the activation energy is to plot ln k versus $1/T$. The slope of the line is equal to $-E_a/R$, with the units of E_a dependent on the value of R. Once E_a is known, Eq. (7.41) can be solved for A. An alternative method of obtaining E_a and A is illustrated below.

Example 7.5   The decomposition of urea in 0.1 *M* HCl occurs according to the reaction

$$NH_2CONH_2 + 2\,H_2O \longrightarrow 2\,NH_4^+ + CO_3^{2-}$$

The first-order rate constant for this reaction was measured as a function of temperature, with the following results:

Expt.	Temperature (°C)	k (min^{-1})
1	61.0	0.713×10^{-5}
2	71.2	2.77×10^{-5}

Calculate the activation energy, E_a, and the A factor in Eq. (7.41) for this reaction.

Solution   This problem can be solved either numerically or graphically. The numerical solution is given here. Since ln $k = -E_a/RT + \ln A$, we will first calculate ln k and $1/T$ for each experiment.

Expt.	ln k	$\frac{1}{T}$ (K)$^{-1}$
1	-11.85	2.992×10^{-3}
2	-10.49	2.904×10^{-3}

We can write

$$\ln k_2 - \ln k_1 = -\frac{E_a}{R} \cdot \left(\frac{1}{T_2} - \frac{1}{T_1} \right)$$

Combining experiments 1 and 2 in this way, we obtain

$$E_a = -\frac{R \cdot (\ln k_2 - \ln k_1)}{1/T_2 - 1/T_1}$$

$$= -\frac{0.008314 \text{ kJ K}^{-1} \text{ mol}^{-1} \, (-10.49 + 11.85)}{(2.904 - 2.992) \times 10^{-3} \text{ K}^{-1}}$$

$$= 128.5 \text{ kJ mol}^{-1}$$

Using the data from either experiment, we can calculate a value for A from

$$\ln A = \ln k + \frac{E_a}{RT}$$

Using the data of experiment 2,

$$\ln A = -10.49 + \frac{128.5}{(0.008314)(344.4)}$$

$$= -10.49 + 44.84 = 34.35$$

Therefore,

$$A = 8.28 \times 10^{14} \text{ min}^{-1}$$

$$= 1.38 \times 10^{13} \text{ s}^{-1}$$

To understand the meaning of the activation energy, consider the elementary reaction $M + N \rightarrow P$ with an energy barrier between reactants and products as illustrated in Fig. 7.14. In this figure we plot the energy versus the *reaction coordinate*.

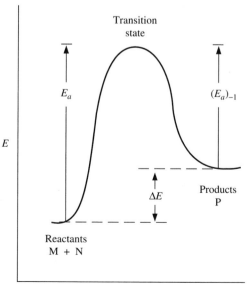

Fig. 7.14 Energy changes that occur throughout the course of a hypothetical reaction. The reaction coordinate is a fictitious variable that measures the progress from reactants to transition state to product. E_a is the activation energy for the forward reaction and $(E_a)_{-1}$ is the activation energy for the reverse reaction.

The reaction coordinate is a convenient way to represent the change of the reactants into products as a reaction takes place. It is not a simple coordinate, such as the x coordinate for a point in space; instead, it can represent the positions of all the significant atoms in the reacting molecules. For example, in the reaction

$$OH^- + CH_3Br \longrightarrow CH_3OH + Br^-$$

the reaction coordinate represents the concerted motions of the OH^- approaching the central carbon, while the Br^- is leaving and the three hydrogen atoms are also moving. We can plot any property versus the reaction coordinate, but energy or free energy is particularly interesting.

The molecules M and N must together have significant energy to react. Because the process must be reversible at the microscopic level, the same barrier will be present for the reverse reaction. The energy of the products of a reaction is usually different from that of the reactants

$$\Delta E = E_P - (E_M + E_N)$$

Therefore the activation energy for the reverse reaction is usually not the same as for the forward reaction. The relation

$$(E_a)_{\text{forward}} - (E_a)_{\text{reverse}} = \Delta E \tag{7.42}$$

does hold, however, at least for elementary reactions.

As the temperature is increased, the *fraction* of molecules having energy greater than E_a increases, and the rate of reaction correspondingly increases. In some cases, correlations can be made between the activation energy and the energy of a bond that must be broken in order for a reaction to proceed.

Very detailed theories of chemical kinetics exist, or are being developed, which can predict rate constants for simple, gas-phase reactions. We shall discuss more general theories which are useful mainly in helping us understand the factors that affect the kinetics.

In the *collision theory* of chemical kinetics we assume that, to react, molecules must collide (including collisions at the walls of the vessel in many cases). The rate of a reaction whose mechanism is

$$M + N \longrightarrow P$$

is given by

$$v = k[M][N] = A[M][N] \exp\left(-\frac{E_a}{RT}\right) \tag{7.43}$$

At high temperatures the exponential approaches unity, and $k \cong A$:

$$\lim_{T \to \infty} A \exp\left(-\frac{E_a}{RT}\right) = A$$

The maximum rate of this reaction at high temperature would be $A[M][N]$, which is therefore interpretable as a collision frequency for the molecules M and N. The exponential factor resembles a Boltzmann probability distribution (see Chapter 6). It can be considered to represent the fraction of molecules with proper orientation that have sufficient energy to react.

Examined in detail, the collision theory has serious shortcomings. Collision frequencies of molecules in the gas phase can be calculated quite well using kinetic theory of gases (see Chapter 6). From Eq. (7.43) theoretical values for the rate constant of a bimolecular gas-phase reaction can be calculated. These are on the order of 10^{11} mol^{-1} L s^{-1} for small molecules, somewhat dependent on the molecular radii. Experimentally, values of A range widely, from 10^2 to 10^{13} mol^{-1} L s^{-1} for reactions that are thought to be relatively simple. While there are few reactions that proceed much faster than the rates calculated from simple collision theory, there are many that are orders of magnitude slower. Attempts to correct this by introducing "steric factors" are simply assigning an adjustable parameter to our ignorance.

The situation is worse for liquid-phase reactions, where the nature of collisions is harder to describe or treat theoretically. Furthermore, it is impossible to interpret an inverse temperature dependence (negative activation energy) in a meaningful way by collision theory. The difficulty here arises because collision theory assumes that essentially only the molecular kinetic energy (translation, vibration, rotation) is involved in the activation process. There is no rational way of handling configurational or entropic contributions in the collision theory framework.

TRANSITION-STATE THEORY

An alternative model was introduced in 1935 by Eyring and others. Its significant new feature was that it considered the reactive intermediate or *transition state** (an unstable species at a free energy maximum) to be just like a stable molecule except for motion of atoms along the reaction coordinate between reactants and products. Thus the transition state is a molecule lasting during only a few molecular vibrations. When the transition-state theory was proposed, it was not possible to study a transition state because of its short lifetime. However, transition states in the gas phase now have been characterized spectroscopically (Zewail and Bernstein, 1988; Grubele and Zewail, 1990) using laser pulses of femtoseconds (1 fs = 10^{-15} s) to picoseconds (1 ps = 10^{-12} s) in length.

In the transition-state theory, every elementary reaction

$$M + N \xrightarrow{k} P$$

is rewritten as

$$M + N \xrightleftharpoons{K^{\ddagger}} MN^{\ddagger} \xrightarrow{k^{\ddagger}} P$$

It is assumed that reactants are in rapid equilibrium with the transition state, $MN^{\ddagger}$. Here $K^{\ddagger}$ is the equilibrium constant between the transition state and reactants, and $k^{\ddagger}$ is a universal rate constant.

$$K^{\ddagger} = \frac{[MN^{\ddagger}]}{[M][N]} \tag{7.44}$$

$$\frac{d[P]}{dt} = k^{\ddagger}[MN^{\ddagger}] \tag{7.45}$$

* The transition state was originally called an activated complex.

Neither $K^{\ddagger}$ nor $k^{\ddagger}$ are measurable parameters; they are only used to derive the results. Combining Eqs. (7.44) and (7.45) we obtain

$$\frac{d[P]}{dt} = k^{\ddagger}K^{\ddagger}[M][N] \tag{7.46}$$

The experimental rate expression is

$$\frac{d[P]}{dt} = k[M][N]$$

So we have related an experimental rate constant k to properties of the transition state.

$$k = k^{\ddagger}K^{\ddagger} \tag{7.47}$$

But every equilibrium constant is related to the standard free energy of the reaction.

$$\Delta G^{\ddagger} = -RT \ln K^{\ddagger}$$

and transition-state theory shows that the universal rate constant $k^{\ddagger}$ is equal to

$$k^{\ddagger} = \frac{k_B T}{h} \tag{7.48}$$

where k_B is the Boltzmann constant,* and h is Planck's constant. Combining Eqs. (7.46–7.48) we obtain

$$k = \frac{k_B T}{h} K^{\ddagger} = \frac{k_B T}{h} \exp\left(-\frac{\Delta G^{\ddagger}}{RT}\right) \tag{7.49}$$

This equation permits a calculation of the free energy of activation, $\Delta G^{\ddagger}$, from a measured rate constant. The rate constant must have units of seconds to be consistent with Boltzmann's and Planck's constants; the concentration units (usually M) specify the standard state for the free energy. The same derivation could be given for the reverse reaction. So for any reaction the picture shown in Fig. 7.15 holds. The free energy difference between products and reactants is equal to the difference in activation free energies

$$\Delta G_{react} = \Delta G^{\ddagger}_{forward} - \Delta G^{\ddagger}_{backward} \tag{7.50}$$

Additional insight comes if we consider the distinction between energy (enthalpy) and entropy contributions. At constant temperature,

$$\Delta G^{\ddagger} = \Delta H^{\ddagger} - T\,\Delta S^{\ddagger} \tag{7.51}$$

where $\Delta H^{\ddagger}$ and $\Delta S^{\ddagger}$ are the activation enthalpy and entropy, respectively. Substituting into Eq. (7.49),

$$k = \frac{k_B T}{h} \exp\left(\frac{\Delta S^{\ddagger}}{R}\right) \exp\left(-\frac{\Delta H^{\ddagger}}{RT}\right) \tag{7.52}$$

Transition-state theory predicts a slightly different temperature dependence from that of Arrhenius, although the main temperature dependence for both are in the exponential terms. According to Arrhenius (Eq. 7.40) a plot of $\ln k$ vs. $1/T$ gives a straight line with

* In this section only we use k_B for the Boltzmann constant to distinguish it from rate constants.

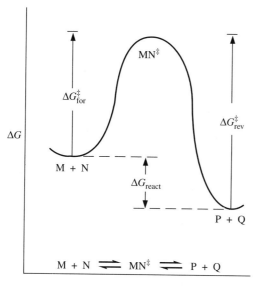

Fig. 7.15 Progress of a reaction according to the transition-state theory.

Reaction coordinate

slope equal to $-E_a/R$. In transition-state theory a plot of $\ln k$ vs. $1/T$ has a slope of $-(\Delta H^{\ddagger} + RT)/R$. This is seen by taking the derivative of $\ln k$ with respect to $1/T$. For most reactions $\Delta H^{\ddagger}$ is much larger than RT so the slope is nearly equal to $-\Delta H^{\ddagger}/R$, analogous to the Arrhenius theory. To obtain $\Delta H^{\ddagger}$ kineticists usually plot $\ln k$ vs. $1/T$ to obtain E_a, the Arrhenius activation energy, then calculate $\Delta H^{\ddagger}$ from

$$\Delta H^{\ddagger} = E_a - RT \approx E_a \tag{7.53}$$

Often the activation energy E_a is large enough so that subtracting RT does not change the value of $\Delta H^{\ddagger}$ with the experimental error of its measurement. The value of $\Delta S^{\ddagger}$ can be obtained from Eq. (7.52) and $\Delta H^{\ddagger}$; it can also be calculated from the Arrhenius pre-exponential term A. Substituting Eq. (7.53) into Eq.(7.52) and comparing the result with Eq. (7.41) we obtain

$$\Delta S^{\ddagger} = R\left(\ln \frac{Ah}{k_{\mathrm{B}}T} - 1\right) \approx R\left(\ln \frac{Ah}{k_{\mathrm{B}}T}\right) \tag{7.54}$$

In both Eqs. (7.54) and (7.53) we use an average T in the temperature range of interest.

The chief advantage of the transition-state theory is that it relates kinetic rates to thermodynamic properties of reactants and a transition intermediate. This can help us to understand some of the factors that influence the rates. The entropy of activation characterizes the changes in configurations of the reactants along the reaction path. Entropies of activation are typically negative for simple gas-phase reactions. They reflect the loss in randomness in the transition state relative to the reactants.

Consider the bimolecular dimerization of butadiene in the gas phase

$$2\,C_4H_6 \xrightarrow[-\Delta S^{\ddagger}\,=\,+68\,\text{J K}^{-1}\,\text{mol}^{-1}]{} (C_4H_6 \cdots C_4H_6)^{\ddagger} \longrightarrow C_8H_{12}$$

Kinetics: Rates of Chemical Reactions   **371**

The fairly large negative entropy of activation corresponds to the loss of three translational degrees of freedom in going from two independent reactant particles to the single one in the associated transition state.

Theoretical calculations for the energies along possible reaction paths have been done for reactions of diatomic or triatomic molecules in the gas phase. This leads to an energy surface for the reaction and predictions of the reactivity of each vibrational and rotational state of the molecules. The predictions can be tested by molecular beam experiments which provide kinetic data for molecules with different amounts of excitation and different orientations relative to each other (Lee, 1987). Values of the rate constant, its temperature dependence and thus, $\Delta H^{\ddagger}$ and $\Delta S^{\ddagger}$, have been calculated for reactions in the gas phase. These parameters represent combinations of all the detailed data that have been obtained. Theoretical calculations for solution reactions are much more difficult, but some success has been obtained for simple reactions such as an electron transfer from Fe^{2+} to Fe^{3+} in aqueous solution. Nevertheless, the transition state and its associated thermodynamic variables are mainly useful for comparisons and for qualitative conclusions.

Example 7.6 Calculate the entropy of activation at 71.2°C for the reaction in Example 7.5 involving the decomposition of urea.

Solution Using the approximation that $\Delta H^{\ddagger} = E_a$, we use Eq. (7.54), remembering that A must have units of s^{-1}. Therefore, for $A = 1.38 \times 10^{13}\ s^{-1}$

$$\Delta S^{\ddagger} = R \cdot \left(\ln A - \ln \frac{k_B T}{h} \right)$$

$$\Delta S^{\ddagger} = (8.314\ \text{J K}^{-1}\ \text{mol}^{-1}) \left[30.256 - \ln \frac{(1.381 \times 10^{-23})(344.4)}{6.626 \times 10^{-34}} \right]$$

$$= (8.314\ \text{J K}^{-1}\ \text{mol}^{-1})(30.256 - 29.602)$$

$$= 5.44\ \text{J K}^{-1}\ \text{mol}^{-1}$$

The entropy of activation for this reaction is small. For reactions in which the solvent water is able to interact less favorably with the transition state than with the reactant, a positive entropy of activation is often observed.

IONIC REACTIONS AND SALT EFFECTS

The effect of the nonideal behavior of reactants in solution can be treated by extending the transition-state theory to include activities and activity coefficients explicitly. Equation (7.45) is written as

$$v = k^{\ddagger}c^{\ddagger} = \frac{k_B T}{h} c^{\ddagger} \tag{7.55}$$

and Eq. (7.44) can be rewritten as

$$K^{\ddagger} = \frac{[a^{\ddagger}]}{[a_M][a_N]} = \frac{c^{\ddagger}\gamma^{\ddagger}}{c_M\gamma_M c_N\gamma_N} \tag{7.56}$$

where the c's are concentrations and the γ's are activity coefficients. Combining Eqs. (7.55) and (7.56) and substituting into Eq. (7.46), we obtain the result

$$v = \frac{k_B T}{h} K^{\ddagger} \frac{\gamma_M \gamma_N}{\gamma^{\ddagger}} c_M c_N \tag{7.57}$$

In this case the rate constant can be written as

$$k = \frac{k_B T}{h} K^{\ddagger} \frac{\gamma_M \gamma_N}{\gamma^{\ddagger}} = k_0 \frac{\gamma_M \gamma_N}{\gamma^{\ddagger}} \tag{7.55}$$

where k_0 is the rate constant extrapolated to infinite dilution of all species in solution; at infinite dilution all activity coefficients = 1.

For reactions involving ionic species, we can use the Debye-Hückel limiting law (Chapter 4) to express the dependence of the rate constant, k, on ionic strength. For each ionic species we use

$$\log \gamma_i = -0.51 Z_i^2 \sqrt{I}$$

$$I = \tfrac{1}{2} \sum_i^{\text{all species}} c_i Z_i^2 \tag{7.56}$$

where the constant 0.51 applies to water solutions at 25°C, Z_i is the charge on species i and I is the ionic strength. The charge relation for the reaction $M + N \rightarrow MN^{\ddagger}$ is clearly $Z_M + Z_N = Z^{\ddagger}$, and we can write the logarithm of Eq. (7.55) as

$$\log k = \log k_0 - 0.51[Z_M^2 + Z_N^2 - (Z_M + Z_N)^2] \sqrt{I}$$

$$= \log k_0 + 2(0.51)Z_M Z_N \sqrt{I} \tag{7.57}$$

Equation (7.57) predicts that a plot of $\log k$ versus $\sqrt{I}$ for dilute ionic solutions should give a straight line with a slope $2(0.51)Z_M Z_N \cong Z_M Z_N$. The slope then gives direct information about the charges of the reacting species. This has been verified for a variety of ionic reactions in dilute aqueous solutions. For example, when the rate of decomposition of H_2O_2 in the presence of HBr was studied as a function of ionic strength, it was found that the data fit Eq. (7.57) with a slope of -1. The stoichiometric reaction

$$H_2O_2 \longrightarrow H_2O + \tfrac{1}{2}O_2$$

does not involve ionic species, but the kinetics imply that H^+ and Br^- are involved in the formation of the transition-state species.

Example 7.7   The acid denaturation of CO-hemoglobin has been studied at pH 3.42 in a formic acid-sodium formate buffer as a function of sodium formate (Na^+ + $OOCH^-$) concentration (Zaiser and Steinhardt, 1951). The half-lives for the first-order denaturation are as follows:

NaOOCH (M)	0.007	0.010	0.015	0.020
$t_{1/2}$ (min)	20.2	13.6	8.1	5.9

Determine whether these results follow the Debye-Hückel limiting law in the dependence of the rate constant on ionic strength and, if so, what is the apparent charge on the CO-hemoglobin?

Solution Since NaOOCH is an uni-univalent electrolyte, the ionic strength, I, is equal to the molar concentration, c, of sodium formate (we ignore the small H^+ concentration). For a first-order reaction,

$$k = \frac{\ln 2}{t_{1/2}} = \frac{0.693}{t_{1/2}}$$

Therefore:

$\sqrt{I}$	0.084	0.100	0.122	0.141
k (min^{-1})	0.0343	0.0510	0.0856	0.117
log k	−1.465	−1.293	−1.068	−0.930

A plot of these data is shown in Fig. 7.16. Although the data show some curvature, there is a reasonably good fit to a straight line with a slope of +9. Since the denaturation occurs by reaction with monovalent H^+, this suggests a charge of approximately +9 for the CO-hemoglobin at pH 3.42.

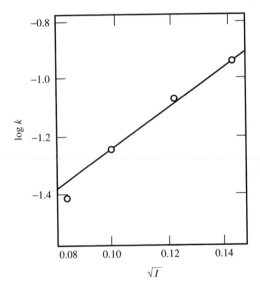

Fig. 7.16 Data for the rate of acid denaturation of CO-hemoglobin. [From E. M. Zaiser and J. S. Steinhardt, *J. Am. Chem. Soc. 73*, 5568 (1951).]

ISOTOPES AND STEREOCHEMICAL PROPERTIES

The rate law, the activation energy, the ionic effects, and so on, all provide important clues to the mechanism of a reaction. A number of other experimental and theoretical approaches have been devised; the use of isotopes and the examination of the stereochemical properties of reactants and products are two of the most powerful experimental methods to provide the structural details of a reaction. In the hydrolysis of an ester,

$$CH_3-\overset{\overset{\displaystyle O}{\|}}{C}-O-\overset{\overset{\displaystyle H}{|}}{\underset{\underset{\displaystyle R'}{\diagdown}}{C}}\overset{\diagup R}{} + H_2O \rightleftharpoons CH_3-\overset{\overset{\displaystyle O}{\|}}{C}-OH + \overset{R}{\underset{R'}{}}CHOH$$

the rate law cannot tell us whether the bond cleaved is indicated by the solid or dashed arrow. When the reaction is carried out in ^{18}O (a stable isotope of the more abundant ^{16}O) labeled H_2O, ^{18}O is found only in the acid and not in the alcohol produced. Thus the cleavage must be occurring at the position indicated by the solid arrow. In the reaction between acetaldehyde bound to an enzyme,

$$\overset{CH_3}{\underset{H}{\diagup}}C=O$$

and a nucleophilic (electron-donating) reagent X^-, the stereospecificity of the product is different depending on from which side of the plane containing the four atoms

$$\overset{C}{\underset{H}{\diagup}}C=O$$

the attack occurs. The two products are depicted below:

$$\overset{\overset{\displaystyle O^-}{|}}{\underset{H \quad X \quad CH_3}{C}} \qquad \overset{\overset{\displaystyle O^-}{|}}{\underset{H \quad CH_3}{C}}X$$

They are not superimposable but are mirror images (Chapter 10). Therefore, from the stereospecificity of the product we can deduce from which side of the plane the attacking group approaches. Kineticists have made extensive use of isotopic labeling and stereospecificity in deducing mechanisms; the two examples given provide a glimpse of how this might be achieved.

VERY FAST REACTIONS

The rates of many reactions can be measured by mixing two solutions and measuring the change of concentrations with time. The concentrations are measured continuously by absorbance or other spectroscopic method, or the reaction is quenched and an appropriate analytical method is used. However, some reactions are too fast for this technique. Consider the very simple and very important reactions

$$H^+ + OH^- \longrightarrow H_2O$$

$$CH_3COO^- + H^+ \longrightarrow CH_3COOH$$

The kinetics of neutralization in aqueous solution occur in the reaction mechanisms for many reactions. Yet the kinetics of these reactions cannot be studied by mixing sodium hydroxide and hydrochloric acid, or sodium acetate and hydrochloric acid. The reaction is over long before the solutions can be mixed thoroughly, therefore the apparent rate of reaction is actually a measure of the rate of mixing. Methods that do not require mixing two solutions are needed for these very fast reactions. *Relaxation methods* allow you to measure a rate by perturbing an equilibrium and following the return of the system to equilibrium.

Relaxation Methods

Consider a reversible system

$$A \underset{k_{-1}}{\overset{k_1}{\rightleftharpoons}} B$$

At equilibrium, the rates of the forward and back reactions are the same. The position of equilibrium is represented in terms of an equilibrium constant,

$$K = \frac{k_1}{k_{-1}} = \frac{[B]^{eq}}{[A]^{eq}}$$

which is the ratio of the rate constants k_1 and k_{-1}. The usual measurements of the "average" composition of the system at equilibrium can provide a value for only the ratio, not for the individual rate constants. Clearly, k_1 and k_{-1} can range from very large to very small and still give the same ratio (the equilibrium constant K). If we change the temperature or pressure of the system so as to attain a new equilibrium state, the new state will in general have a different value for the equilibrium constant. We are still completely ignorant of the individual rate constants, however.

Several approaches are possible to extract the desired rate constants and mechanistic information. One category involves the use of relaxation methods, in which the system at equilibrium is suddenly perturbed in such a way that it finds itself out of equilibrium under the new conditions. It then relaxes to a state of equilibrium under the new conditions, and the rate of relaxation is governed by the rate constants and mechanism of the process. Examples of relaxation processes include:

1. Temperature jump, which can be achieved by the discharge of an electric capacitor, the use of a rapid laser pulse, or by addition of hot water to produce a sudden increase in temperature of a system.

2. Pressure jump, where a restraining diaphragm is suddenly ruptured to release pressurization of the system.

3. Flash or laser pulse photolysis, where a pulse of light produces a sudden change in the population of electronically excited states of absorbing molecules. In cases where the excited state is more acidic than the ground state, this results in a jump in $[H^+]$.

In each of these cases the system suddenly finds itself out of equilibrium at the instant of the perturbation, and the experimenter uses a suitable experimental method to monitor the approach to a new equilibrium state under the altered conditions.

A second major category of methods for dealing with systems at equilibrium involves the study of fluctuations about the equilibrium state. As a consequence of both the dynamic and statistical nature of systems at equilibrium, the system and many of its properties undergo constant and rapid fluctuations about some average or median state (see Chapter 3). Any normal (long-term) measurements always give the same result if the system is truly at equilibrium, and these results serve to characterize the average equilibrium state. When we are able to study the system on a short-enough time scale, however, we observe that its properties exhibit a kind of Brownian motion, in the sense that they are constantly moving one way or the other with respect to their average equilibrium value. This can be true for such properties as

1. Refractive index, where fluctuations result in variations in scattering of light from laser beams.

2. Concentration, which can be detected as fluctuations in the absorption or fluorescence of particular species.

3. Position, involved in scattering of fine laser beams off single particles.

4. Pressure, where fluctuations can produce "noise" in a suitable acoustic detector.

In each of these cases there is no need to perturb the system artificially; one simply takes advantage of the fluctuations that occur naturally because of the particulate and statistical nature of matter.

Relaxation Kinetics

As an example of relaxation kinetics, consider an equilibrium of the type

$$A + B \underset{k_{-1}}{\overset{k_1}{\rightleftharpoons}} P$$

The rate expression is

$$\frac{d[P]}{dt} = k_1[A][B] - k_{-1}[P] \tag{7.58}$$

At equilibrium, $d[P]/dt$ is zero, and

$$K = \frac{k_1}{k_{-1}} = \frac{[\overline{P}]}{[\overline{A}][\overline{B}]} \tag{7.59}$$

where the bar over the concentration terms emphasizes the equilibrium values. Now generate a nearly discontinuous change (typically within 10^{-6} s, but it can be as short as 10^{-8} s) of temperature or pressure. Changing temperature or pressure will in general cause the equilibrium concentrations to change; we can use thermodynamics to tell us how large a change to expect. From Chapter 3 we remember that

$$\left(\frac{\partial \Delta G}{\partial T}\right)_P = -\Delta S \quad \text{and} \quad \left(\frac{\partial \Delta G}{\partial P}\right)_T = \Delta V \qquad (7.60)$$

Combining these equations with $\Delta G^0 = -RT \ln K$, we obtain

$$\left(\frac{\partial \ln K}{\partial T}\right)_P = \frac{\Delta H^0}{RT^2} \quad \text{and} \quad \left(\frac{\partial \ln K}{\partial P}\right)_T = -\frac{\Delta V^0}{RT} \qquad (7.61)$$

Thermodynamic standard states for solutions include the requirement that the pressure is 1 atm. Therefore, we note that the pressure dependence of K refers to change in equilibrium concentrations, not activities. So Eq. (7.61) tells us that the equilibrium constant, K, will be different at the new temperature or pressure immediately after the pulse. How much different will depend on the enthalpy change of the reaction for a T jump, or the volume change of the reaction for a P jump. The system finds itself suddenly out of equilibrium under the new conditions, and we can follow the concentration changes during the subsequent relaxation to the new equilibrium state.

Conditions are usually chosen so that the T jump or P jump produces relatively small displacements from equilibrium, that is, ΔT of 5 to 10 degrees or ΔP of 100 to 1000 atm. This simplifies not only the mathematics, but also the thermodynamic analysis of the results. Suppose that, following the sudden perturbation, the system relaxes so as to increase the concentration of A by a small amount $\Delta[A]$, B by $\Delta[B]$, and to decrease P by $\Delta[P]$, as shown in Fig. 7.17. The perturbation is small enough so that $(\Delta[A])^2$ is negligible compared to $\Delta[A]$, and so on. The values of $\Delta[P]$, $\Delta[A]$, and $\Delta[B]$ are all defined as positive, and because of the stoichiometry, they are all equal. It is then convenient to rewrite the instantaneous, time-dependent concentrations as

$$[P] = [\bar{P}] + \Delta[P]$$

$$[A] = [\bar{A}] - \Delta[A] = [\bar{A}] - \Delta[P]$$

$$[B] = [\bar{B}] - \Delta[B] = [\bar{B}] - \Delta[P]$$

The values of $[\bar{P}]$, $[\bar{A}]$, and $[\bar{B}]$ refer to the equilibrium concentrations at the final T or P. Therefore,

$$\frac{d[P]}{dt} = \frac{d([\bar{P}] + \Delta[P])}{dt} = \frac{d(\Delta[P])}{dt}$$

Because $d[\bar{P}]/dt = 0$,

$$\frac{d(\Delta[P])}{dt} = \frac{d[P]}{dt} = k_1[A][B] - k_{-1}[P]$$

$$= k_1([\bar{A}] - \Delta[P])([\bar{B}] - \Delta[P]) - k_{-1}([\bar{P}] + \Delta[P])$$

$$= k_1[\bar{A}][\bar{B}] - k_{-1}[\bar{P}] - k_1([\bar{A}]\Delta[P] + [\bar{B}]\Delta[P] - (\Delta[P])^2)$$
$$- k_{-1}\Delta[P] \qquad (7.62)$$

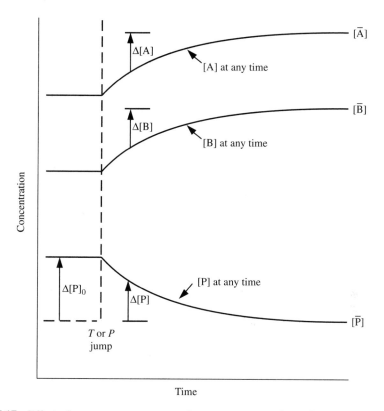

Fig. 7.17 Effect of temperature or pressure jump on concentrations of reactants and products for a system initially at equilibrium or in a steady state. The reaction is $A + B \rightleftarrows P$ and the perturbation causes the concentration of P to decrease slightly while A and B increase slightly. After a short time a new equilibrium is reached with concentrations $[\bar{A}]$, $[\bar{B}]$, $[\bar{P}]$ at the higher temperature or pressure.

$$-\frac{d(\Delta[P])}{dt} = \{k_1 \cdot ([\bar{A}] + [\bar{B}]) + k_{-1}\} \Delta[P] \tag{7.63}$$

To achieve the last equation, we have dropped the term involving $(\Delta[P])^2$, because it is small for small displacements from equilibrium, and we have used Eq. (7.58) to show that the first two terms of Eq. (7.62) sum to zero. The resulting Eq. (7.63) is a simple first-order equation, because $\{k_1([\bar{A}] + [\bar{B}]) + k_1\}$ is independent of time. If we rewrite Eq. (7.63) as

$$-\frac{d(\Delta[P])}{dt} = \frac{\Delta[P]}{\tau}$$

it can be integrated to give

$$\Delta[P] = \Delta[P]_0\, e^{-t/\tau} \tag{7.64}$$

where τ is the relaxation time for the process. For this example

$$\tau = \frac{1}{k_{-1} + k_1([\bar{A}] + [\bar{B}])} \tag{7.65}$$

Kinetics: Rates of Chemical Reactions   **379**

Results for other examples are presented in Table 7.4.

Table 7.4 Relaxation times for reactions involving single steps

Mechanism	Relaxation time*
$A \underset{k_{-1}}{\overset{k_1}{\rightleftarrows}} B$	$\tau = \dfrac{1}{k_1 + k_{-1}}$
$A + B \underset{k_{-1}}{\overset{k_1}{\rightleftarrows}} P$	$\tau = \dfrac{1}{k_{-1} + k_1([\bar{A}] + [\bar{B}])}$
$A + B + C \underset{k_{-1}}{\overset{k_1}{\rightleftarrows}} P$	$\tau = \dfrac{1}{k_{-1} + k_1([\bar{A}][\bar{B}] + [\bar{B}][\bar{C}] + [\bar{A}][\bar{C}])}$
$A + B \underset{k_{-1}}{\overset{k_1}{\rightleftarrows}} P + Q$	$\tau = \dfrac{1}{k_1([\bar{A}] + [\bar{B}]) + k_{-1}([\bar{P}] + [\bar{Q}])}$
$2\,A \underset{k_{-1}}{\overset{k_1}{\rightleftarrows}} A_2$	$\tau = \dfrac{1}{4k_1[\bar{A}] + k_{-1}}$

* $[\bar{A}], [\bar{B}]$, etc., represent the equilibrium concentrations after the perturbation.
SOURCE: M. Eigen and L. De Maeyer, in *Investigation of Rates and Mechanisms of Reactions,* 3rd ed., Vol. 6, Part II, G. G. Hammes (ed.), Wiley-Interscience, New York, 1974, Chapter 3.

A very important and at first surprising result of analyses of each of the examples listed in Table 7.4 is that the relaxation kinetics is always simple first order (exponential in time) regardless of the number of molecules involved as reactants or products. This is true because the perturbations produce only small changes in the equilibrium concentrations and we can ignore squared terms. The behavior is illustrated in Fig. 7.17 for the case where the perturbation shifts the equilibrium in favor of reactants. The reverse situation is also found, depending on the sign of ΔH^0 or ΔV^0.

An experimental trace is shown in Fig. 7.18 for the dimerization of proflavin, according to the reaction

$$2\,P \underset{k_{-1}}{\overset{k_1}{\rightleftarrows}} P_2$$

Because of the fast kinetics involved, it was necessary to use a pulsed laser to provide the temperature jump. For dimerizations it is possible to simplify the relaxation expression by writing it in terms of the total concentration $[P]_t$ of monomers and dimers. Turner et al. (1972) used the expression

$$\frac{1}{\tau^2} = k_{-1}^2 + 8k_1 k_{-1}[P]_t \tag{7.66}$$

The analysis according to Eq. (7.66) does not require a prior knowledge of the equilibrium constant to determine both rate constants. The data should fall on a straight line when plotted according to Eq. (7.66). The plot shown in Fig. 7.19 demonstrates not only that this is observed, but also that the rate constants are independent of pH between 4 and 7.8. From

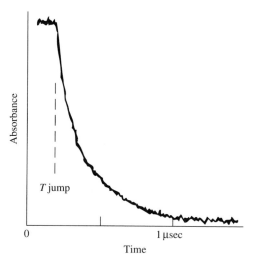

Fig. 7.18 Relaxation of proflavin dimerization. Total concentration $4.64 \times 10^{-3}\ M$, pH 7.8, final temperature 6°C. Absorbance monitored at 455 nm. [Curve from D. H. Turner, G. W. Flynn, S. K. Lundberg, L. D. Faller, and N. Sutin, *Nature 239*, 215 (1972).]

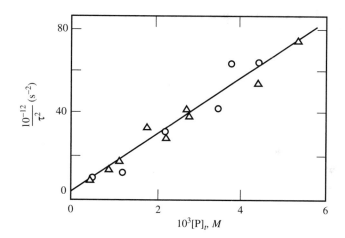

Fig. 7.19 Plot of relaxation data for proflavin dimerization as a function of total proflavin concentration, P_t; ○, pH 7.8; △, pH 4.0. Temperature 25°C. [From D. H. Turner, G. W. Flynn, S. K. Lundberg, L. D. Faller, and N. Sutin, *Nature 239*, 215 (1972).]

the slope and intercept of the line shown in Fig. 7.19, $k_1 = 8 \times 10^8\ M^{-1}\ s^{-1}$ and $k_{-1} = 2.0 \times 10^6\ s^{-1}$ at 25°C were obtained. The dimerization rate constant corresponds to a process that is almost at the limit placed by the diffusion of the monomeric species in aqueous solution, as discussed in the next section.

Exercise   Use the results of this kinetic analysis to determine the equilibrium constant for proflavin dimerization at 25°C.

Example 7.8 The dimerization of the decanucleotide A_4GCU_4 has been studied by Pörschke et al. (1973). The letters A, G, C, and U represent the bases adenine, guanine, cytosine, and uracil, and this oligonucleotide forms base-paired double-stranded helices, which are models for nucleic acids. The two strands are antiparallel; therefore, the oligonucleotide is self-complementary.

$$2\ A_4GCU_4 \underset{k_{-1}}{\overset{k_1}{\rightleftharpoons}} \begin{array}{cccccccccc} A{-}A{-}A{-}A{-}G{-}C{-}U{-}U{-}U{-}U \\[-2pt] \cdot\ \ \cdot\ \ \cdot\ \ \cdot\ \ \cdot\ \ \cdot\ \ \cdot\ \ \cdot\ \ \cdot\ \ \cdot \\[-2pt] U{-}U{-}U{-}U{-}C{-}G{-}A{-}A{-}A{-}A \end{array}$$

(a) A temperature jump was applied to a solution containing 7.45×10^{-6} M oligonucleotide at pH 7.0 and a final temperature of 32.4°C.

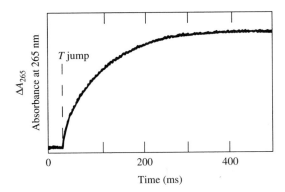

From the response in the uv absorption shown in the figure, the following data can be extracted:

Time (ms)	0	50	100	150	200	250	300	∞
$100 \times \Delta A_{265}$	0	2.0	3.2	3.9	4.28	4.47	4.57	4.70

The difference in absorbance at 265 nm (ΔA_{265}) is directly proportional to the concentration of product at time t minus the concentration at zero time. Use the results to calculate the relaxation time constant under these conditions.

(b) Similar experiments were carried out at 23.3°C and pH 7.0 for a series of concentrations of single-stranded oligomer. The observed relaxation times were as follows:

τ (ms)	455	370	323	244
$[\overline{M}]$ (μM)	1.63	2.45	3.45	5.90

Determine k_1 and k_{-1} for double-strand (dimer) formation and dissociation, respectively, under these conditions.

Solution

(a) First, test the data for first-order relaxation. The absorbance should approach its value at infinite time (its equilibrium value) exponentially.

$$\frac{\Delta A_\infty - \Delta A_t}{\Delta A_\infty} = e^{-t/\tau}$$

This is equivalent to Eq. (7.64).

Time (ms)	0	50	100	150	200	250	300
$100 \times (\Delta A_\infty - \Delta A_t)$	4.7	2.7	1.5	0.8	0.42	0.23	0.13

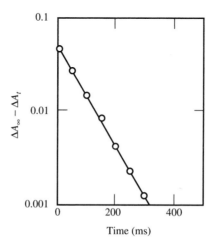

The data do fit a linear first-order plot (use semilog graph paper) with a slope of 5.19 s^{-1}. The relaxation time is

$$\tau = \frac{1}{(2.30)(5.19 \text{ s}^{-1})}$$

$$= 0.0838 \text{ s}$$

$$= 84 \text{ ms}$$

(b) We use the appropriate expression from Table 7.4,

$$\tau = \frac{1}{4k_1[\overline{M}] + k_{-1}}$$

where $[\overline{M}]$ is the concentration of oligonucleotide. Thus, a plot of $1/\tau$ versus $[\overline{M}]$ should give a straight line.

$[\overline{M}]$ (μM)	1.63	2.45	3.45	5.90
$1/\tau$ (s^{-1})	2.20	2.70	3.10	4.10

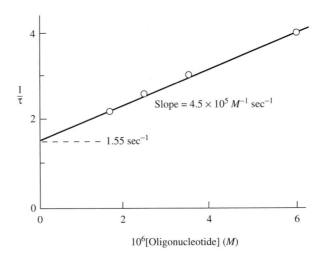

$$10^6[\text{Oligonucleotide}] \ (M)$$

The straight line corresponds to the equation

$$\frac{1}{\tau} = 4k_1[\overline{\text{M}}] + k_{-1}$$

Therefore, from the plot shown,

$$4k_1 = 4.5 \times 10^5 \ M^{-1} \ s^{-1}$$

$$k_1 = 1.12 \times 10^5 \ M^{-1} \ s^{-1}$$

$$k_{-1} = 1.55 \ s^{-1}$$

In this case, the dimerization step is several powers of 10 slower than the diffusion limit.

DIFFUSION-CONTROLLED REACTIONS

The concept of a diffusion-controlled reaction is useful for interpreting the kinetics of fast reactions. Imagine a simplified system where reactant species M and N initially move about in the system independently of one another and far enough apart so that they do not interact. In the gas phase the intervening space is mostly empty, and the molecules would come together, react, and the product(s) depart without interference from other collisions.

The situation in the liquid (or solid) phase is quite different. The reactant species are always in contact with the solvent or other solutes, and they are constantly being bumped and jostled by their neighbors (Fig. 7.20). As a consequence, the motion of any one reactant molecule is quite haphazard. This is known formally as a random-walk process. Because the spaces between molecules in the liquid phase are small (typically 10% or so of the molecular diameters), there results a kind of cage effect in which the molecule makes many collisions with its neighbors before the picture changes sufficiently to say that the molecule has moved away to a new site, where it encounters new neighbors. The motion is

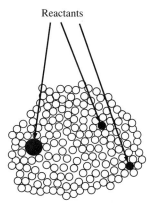

Reactants

Fig. 7.20 Schematic diagram of the environment of two reactant species in tight "cages" in liquid solution.

similar to that of people on a crowded dance floor. The time required to move across the dance floor is very great when the floor is crowded, whereas a nearly empty dance floor is easily traversed.

The consequences of random motions of molecules are described in terms of their diffusion coefficients (see Chapter 6). Since the diffusion coefficient, D, is defined as the proportionality between the flux or velocity of material flow, J_x in units of (concentration) $\times$ (cm s^{-1}), across a unit area in response to a concentration gradient dc/dx, we can write the equation for Fick's first law:

$$J_x = -D\frac{dc}{dx} \tag{6.37}$$

The diffusion coefficient is a measure of the mobility of molecules and thus characterizes how often they can collide. Diffusion coefficients for small molecules are typically on the order of 10^{-5} cm^2 s^{-1}. Large molecules have smaller diffusion coefficients compared with those of small species, but the numerical values depend on the nature of both the diffusing species and the medium (solvent) through which they are moving. Hydrogen ions in water diffuse with unusual ease.

This picture can be developed mathematically so as to calculate an encounter frequency for two solute species, M and N. Put into the units of the rate expression, we obtain a value for the A factor in the Arrhenius expression

$$A_{\text{diffusion}} = \frac{4\pi(r_{\text{MN}})(D_{\text{M}} + D_{\text{N}})N_0}{1000} \tag{7.67}$$

for reactions in liquid solution. Here r_{MN} is the encounter distance of the pair of reacting molecules M and N, and is typically several angstroms. N_0 is Avogadro's number. For $D_{\text{M}} = D_{\text{N}} = 1.5 \times 10^{-5}$ cm^2 s^{-1} and $r_{\text{MN}} = 4$ Å $= 4 \times 10^{-8}$ cm, the value of $A_{\text{diffusion}}$ is about 10^{10} L mol^{-1} s^{-1}. For a reaction with zero activation energy, this would be the value of the rate constant if every encounter led to reaction. It constitutes a kind of upper bound for reactions that are limited only by the frequency of encounters. These are known as *diffusion-controlled reactions*.

Rate constants for a variety of representative second-order reactions are listed in Table 7.5. Note that these are all rate constants at room temperature. Where activation energies are not zero, the Arrhenius A factors will be somewhat larger than the rate constants

Table 7.5   Rate constants for some second-order reactions in liquid solution

Reaction	k $(M^{-1}\,s^{-1})$
$H^+ + OH^-$	1.3×10^{11}
$H^+ + NH_3$	4.3×10^{10}
$H^+ + imidazole$	1.5×10^{10}
$OH^- + NH_4^+$	3.4×10^{10}
$OH^- + imidazole^+$	2.3×10^{10}
ribonuclease + cytidine 3′-phosphate	6×10^7
lactate dehydrogenase + NADH	1×10^9
creatine kinase + ADP	2.2×10^7
aspartate aminotransferase	
+ α-methylaspartate	1.2×10^4
+ β-erythrohydroxyaspartate	3.1×10^6
+ aspartate	$>1 \times 10^8$
catalase + H_2O_2	5×10^6
O_2 + hemoglobin	4×10^7

(The reactions ribonuclease through catalase are grouped under "Enzyme-substrate complex formation.")

given in Table 7.5. Despite this limitation it is clear that many reactions involving protonation or reaction with hydroxide operate near the diffusion limit. Because of electrostatic attractions, reactions involving oppositely charged ions or of ions with polar molecules are somewhat faster than calculations from simple diffusion theory would predict. While a few enzyme-substrate complex formation reactions appear to be diffusion-limited, many others are slower by several orders of magnitude. It should be obvious that this difference is at least partly associated with the appreciable negative entropy of activation [Eq. (7.54)] that is typically found for such reactions.

Example 7.8   Glucose binds to the enzyme hexokinase from yeast with a rate constant $k = 3.7 \times 10^6\ M^{-1}\,s^{-1}$. Obtain the necessary data to estimate the rate constant if the reaction were diffusion-limited, and compare this calculation with the experimental value.

Solution   To use Eq. (7.67) we need to know the diffusion coefficients of glucose and hexokinase, and to estimate the radius of a reactive encounter.

$$D(\text{glucose}) = 0.673 \times 10^{-5}\ cm^2\,s^{-1}$$

$$D(\text{hexokinase}) = 2.9 \times 10^{-7}\ cm^2\,s^{-1}$$

We assume that the encounter distance is of the order of 5×10^{-8} cm. Then, from Eq. (7.67),

$$A_{\text{diff}} = \frac{4\pi(5 \times 10^{-8}\,\text{cm})(0.673 + 0.029)(10^{-5}\,\text{cm}^2\,\text{s}^{-1})(6.0 \times 10^{23})}{1000}$$

$$= 2.6 \times 10^9\,M^{-1}\,\text{s}^{-1}$$

which is nearly 1000 times faster than the observed rate constant. Thus the reaction is not diffusion-limited under these conditions.

PHOTOCHEMISTRY AND PHOTOBIOLOGY

Light is a factor in our environment having both highly beneficial and potentially hazardous aspects. We survive on planet Earth because of the process of photosynthesis, which uses solar energy to form the chemical substances of the plants that are suitable as food. Our eyes employ rhodopsin to convert light stimuli into neural signals to the brain, enabling us to see. Solar irradiation of the upper atmosphere induces photochemistry that results in the formation of a complex distribution of oxygen, nitrogen, and hydrogen compounds. One consequence is the presence of the ozone shield, which screens out harmful ultraviolet radiation. The reason for concern about uv light is that its photons are of sufficiently high energy to induce mutations in both microorganisms and higher organisms. Such mutations are more likely to be harmful than beneficial. Furthermore, prolonged exposure to sunlight rich in uv components is known to lead to an increase in the incidence of skin cancer. Ultraviolet germicidal lamps are potent lethal weapons for microorganisms and are used for sterilizing food, drugs, serum solutions, and operating rooms. Ordinary fluorescent light has been used therapeutically for babies suffering from hyperbilirubinemia, a malfunctioning of the process of disposal of products of heme metabolism. Even this short list gives an indication of the wide variety of roles played by light in biology.

To understand the role of light in photoreactions we must realize that only radiation *absorbed* by the system is effective in producing photochemical change. The quantitative measure of light is its intensity, I, usually given in photons per square meter per second ($\text{m}^{-2}\,\text{s}^{-1}$), einsteins $\text{m}^{-2}\,\text{s}^{-1}$, or $\text{J}\,\text{m}^{-2}\,\text{s}^{-1}$ ($= \text{W}\,\text{m}^{-2}$). The relation among these quantities can be derived from the *Planck equation*,

$$\varepsilon = h\nu = \frac{hc}{\lambda} \tag{7.68}$$

which states that the energy of a single photon, ε, is proportional to its frequency ν (in s^{-1}) or inversely proportional to its wavelength λ (in m), h is Planck's constant ($6.626 \times 10^{-34}\,\text{J s}$), and c is the velocity of light ($3.0 \times 10^8\,\text{m}\,\text{s}^{-1}$ in vacuum). A mole of photons (6.022×10^{23} photons) is known as an *einstein*. Thus, for an einstein of radiation, the energy is

$$E = N_0 h\nu = \frac{N_0 hc}{\lambda} \tag{7.69}$$

Intensities of light are measured as the number of photons (or of einsteins, or of energy in J) crossing a 1-m^2 cross section in each second of time. Alternative measures of intensity are often used in different applications.

Light that is transmitted through the sample or is scattered by it does not become absorbed and, therefore, will not produce any photochemistry. For a photochemical reaction

$$B \xrightarrow{h\nu} P$$

the velocity or photochemical rate is given by

$$v = -\frac{d[B]}{dt} = \frac{\phi I_{abs}\,1000}{l} \tag{7.70}$$

where I_{abs} is the light absorbed by the sample in einsteins $cm^{-2}\ s^{-1}$, l is the path length in cm and 1000 converts cm^3 to liters, and ϕ is the quantum yield. The units of v are thus $M^{-1}\ s^{-1}$. The photochemical quantum yield is defined as the number of molecules reacted per photon absorbed; it is obviously unitless.

$$\phi = \frac{\text{number of molecules reacted}}{\text{number of photons absorbed}} \tag{7.71}$$

Some representative values are given in Table 7.6.

Table 7.6 Some representative quantum yields

Photochemical reaction	Quantum yield, ϕ^*
$NH_3(g) \longrightarrow \frac{1}{2} N_2(g) + \frac{3}{2} H_2(g)$	0.2
$S_2O_8^{2-}(aq) + H_2O \longrightarrow 2\,SO_4^{2-}(aq) + 2\,H^+(aq) + \frac{1}{2} O_2(g)$	1
$H_2(g) + Cl_2(g) \longrightarrow 2\,HCl(g)$	10^5
$H_2O + CO_2 \xrightarrow{\text{chloroplasts}} \frac{1}{x} [CH_2O]_x + O_2$ (in photosynthesis)	10^{-1}
Rhodopsin $\longrightarrow$ retinal + opsin (in mammalian eyes)	1
2 Thymine $\longrightarrow$ thymine dimer (in DNA)	10^{-2}
Hemoglobin $\cdot$ CO $\longrightarrow$ hemoglobin + CO(g)	1
Hemoglobin $\cdot$ O$_2$ $\longrightarrow$ hemoglobin + O$_2$(g)	10^{-2}

* Values depend on wavelength and other experimental conditions.

According to the *Einstein law of photochemistry,* in a primary photochemical process each molecule is excited by the absorption of one photon. However, the excited molecule can then undergo many different reactions. It can form product; it can catalyze a chain reaction; it can fluoresce; it can thermally de-excite, and so forth. Thus, the quantum yield, ϕ, can be much smaller or much greater than 1 as seen in Table 7.6. The value of ϕ must be determined experimentally for each photochemical reaction. Recently, exceptions to the Einstein law have been demonstrated at very high intensities using pulsed lasers. At super-high intensities two or three photons can be absorbed simultaneously by the same molecule.

The relation between the rate of a photochemical reaction and the absorbed light [Eq. (7.70)] can be written in a more useful form by introducing the Beer-Lambert law (see Chapter 10) for light absorption:

$$I = I_0 10^{-\epsilon lc} = I_0 10^{-A} \tag{7.72}$$

where I_0 = intensity of incident light

$\quad\ I$ = intensity of transmitted light

$\quad\ l$ = path length of light in sample

$\quad\ c$ = concentration of absorbing molecules, mol L^{-1}

$\quad\ \epsilon$ = molar absorptivity, or molar extinction coefficient, L mol^{-1} cm^{-1}

$\quad\ A$ = absorbance; it is defined by Eq. (7.72)

The intensity of light absorbed is $(I_0 - I)$; therefore,

$$-\frac{d[B]}{dt} = \phi(I_0 - I)\frac{1000}{l} = \phi I_0 (1 - 10^{-A})\frac{1000}{l} \tag{7.73}$$

The exponential term can be rewritten using a series expansion, as in

$$10^{-A} = 1 - 2.303A + \frac{(2.303A)^2}{2!} \cdots$$

For *dilute solutions,* where $A \ll 1$, the terms involving higher powers are negligible in comparison with the first-order term, and we can substitute:

$$1 - 10^{-A} \cong 2.303A$$

into Eq. (7.73) to obtain, for dilute solutions,

$$-\frac{d[B]}{dt} = \frac{(2.303)(1000)\phi I_0 A}{l} = (2.303)(1000)\phi I_0 \epsilon[B] \tag{7.74}$$

For constant incident light intensity, this has the form of the usual first-order rate expression. The concentration and time variables can be separated, and the equation can be integrated in the usual manner to give

$$\ln\frac{[B]}{[B]_0} = -2303\phi I_0 \epsilon t = -kt \tag{7.75}$$

Most, but not all, photochemical reactions are first order in reactant concentration at sufficiently low concentration. A common test that this is so consists of illuminating identical dilute samples at a set of different intensities and for different times such that $I_0 \cdot t$ is a single constant for all the experiments. If the reaction is of first order in the absorbing species, the extent of photochemical conversion will be the same for each of the experiments.

Example 7.10   In the presence of a reducible dye D, chloroplasts from spinach are able to photocatalyze the oxidation of water to O_2 by the reaction

$$2\ H_2O + 2\ D \xrightarrow[\text{chloroplasts}]{h\nu} O_2 + 2\ DH_2$$

The rate of this process at an incident intensity of 40×10^{-12} einstein cm^{-2} s^{-1} of 650-nm light was 6.5×10^{-12} equivalent cm^{-3} s^{-1} in a solution of 1-cm path length. The chloroplasts exhibited an absorbance of $A_{650}^{1\ cm} = 0.140$. Calculate the quantum efficiency for this reaction.

Solution Because the solution is optically dilute ($A_{650} \ll 1$), we can use Eq. (7.74). In a photocatalytic reaction the concentration of the absorbing chlorophyll in the chloroplasts does not change with time; that is, A is a constant during the process. Therefore,

$$v = \frac{2.303\phi I_0 A}{l}$$

and

$$\phi = \frac{vl}{2.303\phi I_0 A} = \frac{(6.5 \times 10^{-12} \text{ equivalent cm}^{-3} \text{ s}^{-1})(1 \text{ cm})}{2.303(40 \times 10^{-12} \text{ einstein cm}^{-2} \text{ s}^{-1})(0.140)}$$

$$= 0.50 \frac{\text{equivalent}}{\text{einstein}} = 0.50 \frac{\text{electron}}{\text{photon}}$$

[Note: We omitted the factor of 1000 in Eq. (7.74) because concentration units of mol cm^{-3} were given.]

We shall complete our discussion of photochemical kinetics by describing some biologically important processes.

Vision

Visible and near-ultraviolet light absorbed by the eye pigment rhodopsin produces a sequence of photoreactions that is still under active investigation. A recent proposal is shown in Fig. 7.21. The normal form of rhodopsin in dark-adapted mammalian rod outer segments

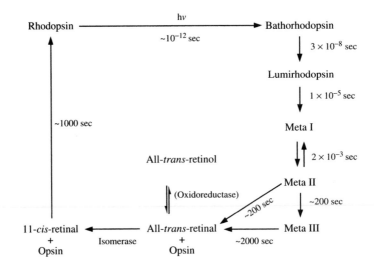

Fig. 7.21 Photobleaching and regeneration of rhodopsin. The lifetimes shown are only approximate to within a factor of about 3 in most cases. (For details, see H. Shichi, *Biochemistry of Vision,* Academic Press, New York, 1983.)

has retinal in the 11-*cis* form covalently attached as a chromophore ("color bearer") to the protein opsin. Following illumination, the characteristic absorption spectrum of rhodopsin undergoes a series of changes associated with a set of intermediates, as shown in Fig. 7.21. Some of these steps are very fast at room temperature and have been studied using either flash relaxation methods or by freezing them at low temperature. The primary step in the photo response of rhodopsin occurs with a quantum yield of about unity. In the eye, the light stimulus requires additional processing before it arrives as a signal at the brain. Nevertheless, an incidence of only a few photons per second is sufficient to give a significant visual sensation.

The visual process is a cyclic one, and the rhodopsin is regenerated in the dark by processes that are incompletely understood. It is likely that the reconstitution of active rhodopsin has a requirement for metabolic energy. There is no evidence that any of the energy of the photon is stored by the rhodopsin photochemistry; light simply serves to trigger or activate an otherwise exergonic ($\Delta G < 0$) reaction. The principal function of the photon is to overcome an activation barrier. This is the most common kind of photochemical process.

Photosynthesis

The process of photosynthesis in plants, algae, and bacteria involves the incorporation of carbon into the various compounds such as carbohydrates, proteins, lipids, nucleic acids, and so on, that make up the material substance and the essential functional components of the organism. The source of carbon for plants and most algae is carbon dioxide; in other organisms, small carbon-containing compounds (acetate, succinate, malate, etc.) are required. The representative reaction for higher plants is

$$CO_2 + H_2O \longrightarrow (CH_2O) + O_2$$

where (CH_2O) represents the fixed carbon of carbohydrate, the major metabolic product. This reaction is endothermic by about 485 kJ (mol CO_2)$^{-1}$, and the energy required to drive this and the other biosynthetic reactions comes ultimately from sunlight. By contrast with most photochemical reactions and with vision, a portion of the energy of the absorbed photons is retained in photosynthesis in the form of chemical potential of the metabolic products. [For purposes of estimating the efficiency of such processes as photosynthesis, we should compare ΔG for the carbon-fixation process with the free energy change associated with the absorption of radiation. The value of ΔG for carbon fixation is about 494 kJ (mol CO_2)$^{-1}$, but the free energy change for the absorption of radiation is more difficult to estimate. [A detailed analysis for solar irradiation is given by Ross and Calvin (1967).]

The overall process of photosynthesis can be divided into light-dependent and dark reactions. Light absorbed by chlorophyll pigments serves to split water molecules into molecular oxygen and hydrogen atom equivalents. The latter are transferred as electrons and hydrogen ions along a transport chain of cytochromes, quinones, and iron-, manganese-, and copper-containing proteins to nicotinamide adenine dinucleotide phosphate (NADP$^+$), which becomes reduced. During this process a portion of the energy is stored as the high-energy product adenosine triphosphate (ATP):

$$2\,H_2O + 2\,NADP^+ \xrightarrow[\substack{\text{chloroplast}\\\text{membranes}}]{hv} O_2 + 2\,NADPH + 2\,H^+$$

$$2(ADP + P_i) \overbrace{\qquad\qquad}\searrow 2\,ATP$$

The NADPH and ATP are then used in a series of dark enzymatic reactions to fix CO_2.

$$CO_2 + 2\,NADPH + 2\,ATP + 2\,H^+ \xrightarrow[\text{enzymes}]{\text{dark}}$$

$$(CH_2O) + 2\,NADP^+ + 2\,ADP + 2\,P_i + H_2O$$

The sum of these reactions is the overall process presented above.

Detailed and extensive studies of the kinetics of photosynthetic light reactions have been made, and the following list summarizes the current state of our knowledge of this subject:

1. The light reactions occur in lipoprotein membranes within chloroplasts in higher plants or within the cells of blue-green algae or photosynthetic bacteria.

2. Chlorophyll or bacteriochlorophyll are the essential pigments that absorb the light, although carotenoids and other accessory pigments can transfer absorbed photon excitation to the chlorophylls.

3. All wavelengths from the near-uv to the near-infrared can be effective. In particular, wavelengths as great as 700 nm ($170\ \text{kJ einstein}^{-1}$) are effective in higher plants and as great as 1000 nm ($120\ \text{kJ einstein}^{-1}$) in certain photosynthetic bacteria.

4. The quantum yield varies with growth conditions. Under optimal levels of CO_2 pressure, relative humidity, and soil nutrients, the yield at low light intensities corresponds to 1 mol of CO_2 fixed or O_2 evolved for each 8 to 9 einsteins absorbed. This represents a higher efficiency than it would seem, because each O_2 molecule is produced by the removal of four electrons from two water molecules and it is known that there are two photon acts or light reactions that operate in series for the transfer of each electron. A current view of the mechanism of photosynthetic electron transport is shown in Fig. 7.22.

5. The rates of photosynthesis reach a maximum at higher light intensities. In simple chloroplast photoreactions not connected to CO_2 fixation, this light saturation is hyperbolic in intensity and resembles the dependence of rates of enzymatic reactions on substrate concentration. When CO_2 fixation is included, the rates follow the same hyperbolic curve at low light intensities but then change abruptly to a constant, light-independent rate at higher intensities. This occurs when one of the steps in the carbon-fixation biochemistry becomes rate-limiting. The contrast between these two types of behavior is illustrated in Fig. 7.23.

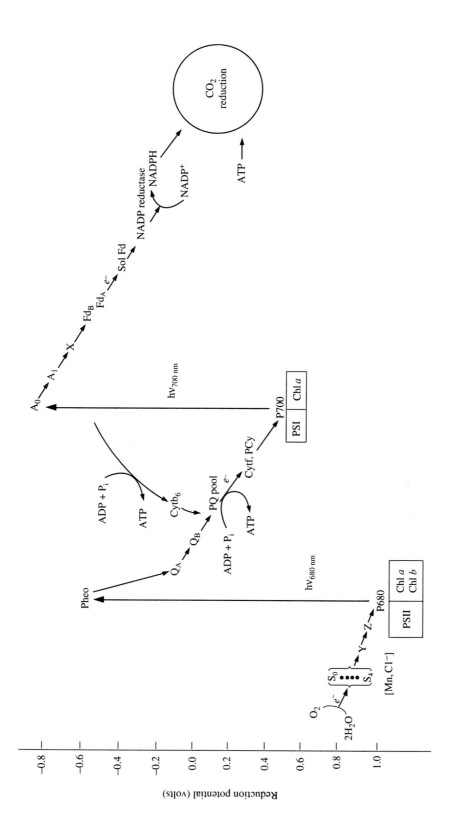

Fig. 7.22 Schematic diagram of the mechanism of photosynthetic electron transport involving two light reactions operating in series. P700 and P680 are the reaction-center chlorophylls (Chl) of photosystems I and II (PSI and PSII), respectively. The electron transport cofactors include cytochromes (Cyt f, Cyt b_6), plastocyanin (PCy), and so on, and a variety of components ($S_0, \ldots, S_4$, Z, Q, A, X) whose structures are still under investigation. Some identified components include pheophytin (Pheo) and iron-sulfur proteins [Fd].

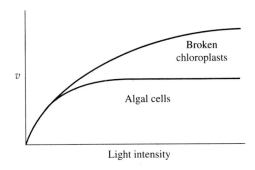

Broken
chloroplasts

Algal cells

v

Light intensity

Fig. 7.23 Velocity of oxygen evolution by broken chloroplasts using an added oxidant (ferricyanide) is compared with that for whole cells of algae, where the CO_2-fixing reactions are coupled.

SUMMARY

Note: We represent concentrations as c_A or [A], etc.

Zero-Order Reactions

Rate is constant for zero-order reactions:
For a product whose concentration is c

$$\frac{dc}{dt} = k \qquad (7.2)$$

$k \equiv$ rate constant; units are concentration time^{-1}, for example, $M\ s^{-1}$

Concentration is linear in time for zero-order reactions:

$$c_2 - c_1 = k(t_2 - t_1)$$

Plot of concentration versus time for zero-order reactions:

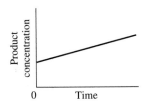

k = slope; concentration at zero time = y intercept

Possible mechanism for zero-order reaction:

$$A \longrightarrow B + C$$

but concentration of A is held constant:

$$\frac{dc_B}{dt} = k$$

$$c_B = kt + \text{constant}$$

First-Order Reactions

Rate is proportional to concentration for first-order reactions:
For a reactant whose concentration is c

$$-\frac{dc}{dt} = kc \tag{7.5}$$

$k \equiv$ rate constant; units are time^{-1}, for example, s^{-1}

Logarithm of reactant concentration linear in time for first-order reactions:

$$\ln c = -kt + \text{constant}$$

$$\ln \frac{c_2}{c_1} = -k(t_2 - t_2) \tag{7.10}$$

$\ln c \equiv$ natural logarithm of concentration

Plot logarithm of reactant concentration versus time for first-order reactions:

$k =$ slope; logarithm of reactant concentration at zero time is y intercept. Any concentration units can be used for the plot.

Reactant concentration is exponential in time:

$$\frac{c}{c_0} = e^{-kt} \quad \text{or} \quad \frac{c}{c_0} = 10^{-0.343kt} \tag{7.11}$$

$c =$ concentration at time t
$c_0 =$ concentration at zero time (note that only ratio of concentrations is important)

Half-life for first-order reactions:

$$\frac{c}{c_0} = 2^{-t/t_{1/2}} \tag{7.13}$$

$t_{1/2} \equiv$ half-life $=$ time necessary for concentration to become half of its original value, c_0
   $= 0.693/k$

Relaxation time for first-order reactions:

$$\frac{c}{c_0} = e^{-t/\tau} \tag{7.14}$$

$\tau \equiv$ relaxation time $=$ time necessary for concentration to become $1/e$ of its original value; $1/e = 0.368$

Possible unimolecular mechanism for first-order reactions:

$$A \longrightarrow B$$

$$-\frac{dc_A}{dt} = kc_A \quad \text{and} \quad \frac{dc_B}{dt} = kc_A$$

$$\frac{c_A}{(c_A)_0} = e^{-kt} \qquad \frac{c_B}{(c_A)_0} = 1 - e^{-kt}$$

c_A, c_B = concentrations of A, B at any time t
$(c_A)_0$ = concentration of A at zero time

Second-Order Reactions

Rate is proportional to concentration squared, or to the product of two different concentrations for second-order reactions:

$$\frac{-dc}{dt} = kc^2 \text{ (class I)} \quad \text{or} \quad \frac{-dc}{dt} = kc_A c_B \text{ (class II)}$$

$k \equiv$ rate constant; units are in concentration^{-1} time^{-1}, for example, M^{-1} s^{-1}

Reciprocal of reactant concentration linear in time for second-order reactions (class I):

$$\frac{1}{c} = kt + \text{constant}$$

Plot of reciprocal of reactant concentration versus time for second-order reactions (class I):

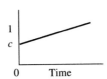

k = slope; reciprocal of concentration at zero time = y intercept

Possible bimolecular mechanism for second-order reactions (class I):

$$A + A \longrightarrow B$$

$$-\frac{dc_A}{dt} = kc_A^2 \quad \text{and} \quad \frac{dc_B}{dt} = \tfrac{1}{2} kc_A^2$$

$$\frac{1}{c_A} - \frac{1}{(c_A)_0} = kt \qquad c_B = \frac{(c_A)_0}{2} - \frac{c_A}{2}$$

c_A, c_B = concentrations of A, B at any time t
$(c_A)_0$ = concentration of A at zero time

Possible bimolecular mechanism for second-order reactions (class II):

$$A + B \longrightarrow C$$

$$-\frac{dc_A}{dt} = -\frac{dc_B}{dt} = kc_A c_B$$

$$\ln \frac{(c_A)(c_B)_0}{(c_A)_0(c_B)} = [(c_A)_0 - (c_B)_0]kt \qquad (7.23)$$

c_A, c_B = concentrations of A, B at any time t
$(c_A)_0$ = concentration of A at zero time
$(c_B)_0$ = concentration of B at zero time

If $(c_A)_0 = (c_B)_0$, then $c_A = c_B$ at all times, and

$$\frac{1}{c_A} - \frac{1}{(c_A)_0} = \frac{1}{c_B} - \frac{1}{(c_B)_0} = kt \qquad \text{as in Class I.}$$

Temperature Dependence

Arrhenius equation:

$$k = Ae^{-E_a/RT} \qquad (7.41)$$

k $\equiv$ rate constant
A $\equiv$ pre-exponential factor; units are same as rate constant
E_a = activation energy, units are J mol^{-1}
R = gas constant = 8.314 J K^{-1} mol^{-1}
T = absolute temperature

Plot of logarithm of k versus reciprocal of absolute temperature:

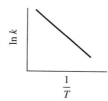

E_a = $-R$ slope
$\ln A$ = y intercept, but extrapolation is often impractical; therefore,
$\ln A$ = $\ln k + (1/T)(\text{slope})$ where $\ln k$ and $(1/T)$ are the coordinates of any point on the
    line

Activation energy:

$$E_a = \frac{RT_2 T_1}{T_2 - T_1} \ln \frac{k_2}{k_1}$$

E_a = activation energy, which is assumed independent of temperature
k_2, k_1 = rate constants at T_2, T_1
R = gas constant = 8.314 J K^{-1} mol^{-1}

Eyring equation:

$$k = \frac{k_B T}{h} e^{\Delta S^{\ddagger}/R} e^{-\Delta H^{\ddagger}/RT} \qquad (7.52)$$

k = rate constant; its units of time must be in seconds; its units of concentration specify the standard state for $\Delta H^{\ddagger}$, $\Delta S^{\ddagger}$

$\Delta H^{\ddagger}$ = enthalpy of activation = enthalpy difference per mole between activated complex and reactants, each in their standard states

$\Delta S^{\ddagger}$ = entropy of activation

T = absolute temperature

k_B = Boltzmann constant = R/N_0; its units are consistent with Planck's constant, h

$$\frac{k}{h} = \frac{R}{N_0 h} = \text{gas constant(Avogadro's number)}^{-1}(\text{Planck's constant})^{-1}$$

$$= \frac{8.3144 \text{ J K}^{-1} \text{ mol}^{-1}}{(6.0220 \times 10^{23} \text{ molecules mol}^{-1})(6.6262 \times 10^{-34} \text{ J s})}$$

$$= 2.0837 \times 10^{10} \text{ (s K)}^{-1}$$

Relation of $\Delta H^{\ddagger}$, $\Delta S^{\ddagger}$ to E_a

$$\Delta H^{\ddagger} = E_a - RT \approx E_a \qquad (7.53)$$

$$\Delta S^{\ddagger} \approx R\left(\ln \frac{Ah}{k_B T}\right) \qquad (7.54)$$

A = Arrhenius pre-exponential factor, its units of time must be in s^{-1}; its concentration units specify the standard states of $\Delta H^{\ddagger}$, $\Delta S^{\ddagger}$

Free energy of activation:

$$\Delta G^{\ddagger} = \Delta H^{\ddagger} - T\Delta S^{\ddagger} \qquad (7.51)$$

$$k = \frac{RT}{N_0 h} e^{-\Delta G^{\ddagger}/RT} \qquad (7.49)$$

$$\Delta G^{\ddagger} = -RT\left(\ln \frac{k}{T} - \ln \frac{R}{N_0 h}\right)$$

$$= -RT \ln K^{\ddagger}$$

$\Delta G^{\ddagger}$ = free energy of activation = free energy difference between transition state and reactants each in their standard states; the standard states are specified by the concentration units used in the rate constant k; if mol L^{-1} is used, the standard state is the infinitely dilute solution state on the molarity scale

$K^{\ddagger}$ = equilibrium constant for the formation of the transition state

Relaxation Kinetics

$$A + B \underset{k_{-1}}{\overset{k_1}{\rightleftharpoons}} P$$

$$\Delta[P] = \Delta[P]_0 \, e^{-(t/\tau)} \qquad (7.64)$$

$\Delta[P] = \Delta[A] = \Delta[B] =$ small displacement from equilibrium concentration
$\Delta[P]_0 =$ displacement at zero time
$1/\tau = k_{-1} + k_1(\overline{[A]} + \overline{[B]})$
$\overline{[A]}, \overline{[B]} =$ equilibrium concentrations of A, B after the perturbation

See Table 7.4 for other examples.

Diffusion-controlled Reactions

Diffusion-controlled reaction $M + N \rightarrow$ *products:*

$$A_{\text{diffusion}} = \frac{4\pi(r_{MN})(D_M + D_N)N_0}{1000} \qquad (7.67)$$

$A_{\text{diffusion}} =$ pre-exponential factor, in L mol^{-1} s^{-1}
$r_{MN} =$ encounter distance, cm
$D =$ diffusion constant, cm^2 s^{-1}
$N_0 =$ Avogadro's number $= 6.02 \times 10^{23}$ mol^{-1}

Absorption of Light

Beer-Lambert law:

$$A = \log \frac{I_0}{I} = \varepsilon c l \qquad (7.72)$$

$A =$ absorbance (or optical density)
$I_0 =$ incident intensity $\Big\}$ same units for both
$I =$ transmitted intensity
$\varepsilon =$ molar absorptivity, M^{-1} cm^{-1}
$c =$ concentration, M
$l =$ path length, cm

Photochemistry

Planck equation:

$$\varepsilon = h\nu = \frac{hc}{\lambda} \qquad (7.68)$$

$\varepsilon =$ photon energy, J photon^{-1}
$h =$ Planck's constant $= 6.6262 \times 10^{-34}$ J s
$\nu =$ frequency of radiation, s^{-1}
$c =$ velocity of light $= 2.9979 \times 10^8$ m s^{-1}
$\lambda =$ wavelength of radiation, m

$$E = N_0 h\nu = \frac{N_0 h c}{\lambda} \tag{7.69}$$

E = energy per einstein
N_0 = Avogadro's number = 6.022×10^{23} mol^{-1}
   = 6.022×10^{23} photons einstein^{-1}

Velocity of a photochemical reaction:

$$v = \frac{\phi I_{abs}\, 1000}{l} \tag{7.70}$$

v = velocity of reaction, M s^{-1}
ϕ = quantum yield, moles reacting per einstein absorbed
I_{abs} = flux of light absorbed, einstein cm^{-2} s^{-1}
l = path length in cm

Dilute solution photochemistry:

$$v = \frac{(2.303)(1000)\phi I_0 A}{l} \qquad (\text{where } A \ll 1) \tag{7.74}$$

v = velocity of reaction, M s^{-1}
ϕ = quantum yield, moles reacting per einstein absorbed
I_0 = incident intensity, einstein cm^{-2} s^{-1}
A = absorbance
l = path length in cm

MATHEMATICS NEEDED FOR CHAPTER 7

A differential equation is an equation containing derivatives. In this chapter, we apply only differential equations that can be solved by separating the variables and integrating each side of the equation separately. Consider a differential equation of the form

$$\frac{dx}{dt} = f(x) \cdot f(t)$$

where $f(x)$ is a function of x only and $f(t)$ is a function of t only. We separate the variables x and t,

$$\frac{dx}{f(x)} = f(t)\, dt$$

and integrate each side of the equation

$$\int \frac{dx}{f(x)} = \int f(t)\, dt$$

The integrals we use are

$$\int x^n \, dx = \frac{x^{n+1}}{n+1} + C \qquad \int_{x_1}^{x_2} x^n \, dx = \frac{x_2^{n+1} - x_1^{n+1}}{n+1} \qquad (n \neq -1)$$

$$\int \frac{dx}{x} = \ln x + C \qquad \int_{x_1}^{x_2} \frac{dx}{x} = \ln \frac{x_2}{x_1}$$

$$\int e^{ax} \, dx = \frac{e^{ax}}{a} + C \qquad \int_{x_1}^{x_2} e^{ax} \, dx = \frac{e^{ax_2} - e^{ax_1}}{a}$$

In addition,

$$e^a = 10^{0.4343a}$$

$$10^a = e^{2.303a}$$

$$\ln a = 2.303 \log a$$

REFERENCES

Kinetics Textbooks

CONNORS, K. A., 1990. *Chemical Kinetics,* VCH Publishers, New York.
ESPENSON, J. H., 1981. *Chemical Kinetics and Reaction Mechanisms,* McGraw-Hill, New York.
HAMMES, G. G., ed., 1974. *Investigation of Rates and Mechanisms of Reactions,* Part II: *Investigation of Elementary Reaction Steps in Solution and Very Fast Reactions,* John Wiley, New York.
LAIDLER, K. J., 1987. *Chemical Kinetics,* 3rd ed., Harper & Row, New York.
MOORE, J. W., and R. G. PEARSON, 1981. *Kinetics and Mechanism,* 3rd ed., John Wiley, New York.
STEINFELD, J. I., J. S. FRANCISCO, and W. L. HASE, 1989. *Chemical Kinetics and Dynamics,* Prentice-Hall, Englewood Cliffs, New Jersey.

SUGGESTED READINGS

ANDERSON, J. G., D. W. TOOHEY, and W. H. BRUNE, 1991. Free Radicals Within the Antarctic Vortex: The Role of CFCs in Antarctic Ozone Loss, *Science 251,* 39–46.
CLAYTON, R. K., 1980. *Photosynthesis: Physical Mechanisms and Chemical Patterns,* Cambridge University Press, New York.
DOOLITTLE, R. F., 1981. Fibrinogen and Fibrin, *Sci. Am.* (December), 126–135.
EIGEN, M., and L. DE MAEYER, 1974. Theoretical Basis of Relaxation Spectrometry, in *Investigation of Rates and Mechanisms of Reactions,* 3rd ed., Vol. 6, Part II, G. G. Hammes, ed., Wiley-Interscience, New York, pp. 63–146.
ELLIOTT, S., and F. S. ROWLAND, 1987. Chlorofluorocarbons and Stratospheric Ozone, *J. Chem. Ed. 64,* 387–391.
FASELLA, P., and G. G. HAMMES, 1967. A Temperature Jump Study of Aspartate Aminotransferase, *Biochemistry 6,* 1798–1904.
GRUEBELE, M., and A. ZEWAIL, 1990. Ultrafast Reaction Dynamics, *Physics Today 43,* 25–33.
HAMMES, G. G., and J. L. HASLAM, 1968. A Kinetic Investigation of the Interaction of α-Methylaspartic Acid with Aspartate Aminotransferase, *Biochemistry 7,* 1519–1525.

JACKSON, C. M., and Y. NEMERSON, 1980. Blood Coagulation, *Annu. Rev. Biochem. 49,* 765–811.

JOHNSTON, H. S., 1975. Ground-Level Effects of Supersonic Transports in the Stratosphere, *Acc. Chem. Res. 8,* 289–294.

LEE, Y. T., 1987. Molecular Beam Studies of Elementary Chemical Processes, *Science 236,* 793–799.

MOLINA, M., and F. S. ROWLAND, 1974. Stratospheric Sink for Chlorofluoromethanes: Chlorine Atom-Catalyzed Destruction of Ozone, *Nature 249,* 810–812.

NEWMAN, E. A., and P. H. HARTLINE, 1982. The Infrared "Vision" of Snakes, *Sci. Am.* (March), 116–127.

PÖRSCHKE, D., O. C. UHLENBECK, and F. H. MARTIN, 1973. Thermodynamics and Kinetics of the Helix-Coil Transition of Oligomers Containing GC Base Pairs, *Biopolymers 12,* 1313–1335.

ROSS, R. T., and M. CALVIN, 1967. Thermodynamics of Light Emission and Free Energy Storage in Photosynthesis, *Biophys. J. 7,* 595–614.

TURNER, D. H., G. W. FLYNN, S. K. LUNDBERG, L. D. FALLER, and N. SUTIN, 1972. Dimerization of Proflavin by the Laser Raman Temperature Jump Method, *Nature 239,* 215–217.

ZEWAIL, A. H., R. B. BERNSTEIN, 1988. Real-Time Laser Femtosecond Chemistry, *Chem. Engr. News 66* (45), 29–43.

PROBLEMS

1. Iodine reacts with a ketone in aqueous solution to give an iodoketone. The stoichiometric equation is

$$I_2 + \text{ketone} \longrightarrow \text{iodoketone} + H^+ + I^-$$

The rate of the reaction can be measured by measuring the disappearance of I_2 with time. Some data for initial rates and initial concentrations follow:

$-d[I_2]/dt$ (mol L^{-1} s^{-1})	$[I_2]$ (mol L^{-1})	[ketone] (mol L^{-1})	$[H^+]$ (mol L^{-1})
7×10^{-5}	5×10^{-4}	0.2	10^{-2}
7×10^{-5}	3×10^{-4}	0.2	10^{-2}
1.7×10^{-4}	5×10^{-4}	0.5	10^{-2}
5.4×10^{-4}	5×10^{-4}	0.5	3.2×10^{-2}

(a) Find the order of the reaction with respect to I_2, ketone, and H^+.

(b) Write a differential equation expressing your findings in part (a) and calculate the average rate constant.

(c) How long will it take to synthesize 10^{-4} mol L^{-1} of the iodoketone starting with 0.5 mol L^{-1} of ketone and 10^{-3} mol L^{-1} of I_2, if the H^+ concentration is held constant at 10^{-1} mol L^{-1}? Will the reaction go faster if we double the concentration of ketone? of iodine? of H^+? How long will it take to synthesize 10^{-1} mol L^{-1} of the iodoketone if all conditions are the same as above?

(d) Propose a mechanism consistent with the experimental results.

2. The saponification (hydrolysis) of ethyl acetate occurs according to the stoichiometric relation $CH_3COOC_2H_5 + OH^- \rightarrow CH_3COO^- + C_2H_5OH$. The reaction can be followed by monitoring the disappearance of OH^-. The following experimental results were obtained at 25°C:

Initial concentration of OH^- (M)	Initial concentration of $CH_3COOC_2H_5$ (M)	Half-life $t_{1/2}$ (s)
0.0050	0.0050	2000
0.0100	0.0100	1000

There is no significant dependence on the concentrations of products of the reaction.

(a) What is the overall kinetic order of the reaction?

(b) Calculate a value for the rate constant, including appropriate units.

(c) How long would it take for the concentration of OH^- to reach 0.0025 M for each experiment?

(d) Based on your answer to part (a), what are possible rate laws for this reaction?

(e) Carefully describe a single experiment that would enable you to decide which of the possibilities of part (d) is most nearly correct.

3. The kinetics of the reaction

$$I^- + OCl^- \rightleftarrows OI^- + Cl^-$$

was studied in basic aqueous media by Chia and Connick [*J. Phys. Chem. 63*, 1518 (1959)]. The initial rate of I^- disappearance is given below for mixtures of various initial compositions, at 25°C. (None of the solutions initially contained OI^- or Cl^-.)

Initial composition of I^- (M)	Initial composition of OCl^- (M)	Initial composition of OH^- (M)	Initial rate $\left(\dfrac{mol\ I^-}{L\ s}\right)$
2×10^{-3}	1.5×10^{-3}	1.00	1.8×10^{-4}
4×10^{-3}	1.5×10^{-3}	1.00	3.6×10^{-4}
2×10^{-3}	3×10^{-3}	2.00	1.8×10^{-4}
4×10^{-3}	3×10^{-3}	1.00	7.2×10^{-4}

(a) The rate law can be expressed in the form

$$-\frac{d[I\text{-}]}{dt} = k[I^-]^a[OCl^-]^b[OH^-]^c$$

Find the values of *a, b,* and *c.*

(b) Calculate the value of k including its units.

(c) Show whether this rate law is consistent with the mechanism

$$OCl^- + H_2O \xrightarrow{K_1} HOCl + OH^- \quad \text{(fast, equilibrium)}$$

$$I^- + HOCl \xrightarrow{k_2} HOI + Cl^- \quad \text{(slow)}$$

$$HOI + OH^- \xrightarrow{K_3} H_2O + OI^- \quad \text{(fast, equilibrium)}$$

4. The mechanism for a set of reactions is

$$A \xrightarrow{k_1} B$$

$$B + C \xrightarrow{k_2} D$$

(a) Write a differential equation for the disappearance of A.

(b) Write a differential equation for $d[B]/dt$.

(c) Write a differential equation for the appearance of D.

(d) If $[A]_0$ is the concentration of A at zero time, write an equation which gives $[A]$ at any later time.

5. The stoichiometric equation for a reaction is

$$A + B \longrightarrow C + D$$

The initial rate of formation of C is measured with the following results:

Initial concentration of A (M)	Initial concentration of B (M)	Initial rate (M s^{-1})
1.0	1.0	1.0×10^{-3}
2.0	1.0	4.0×10^{-3}
1.0	2.0	1.0×10^{-3}

(a) What is the order of the reaction with respect to A?

(b) What is the order of the reaction with respect to B?

(c) Use your conclusions in parts (a) and (b) to write a differential equation for the appearance of C.

(d) What is the rate constant k for the reaction? Do not omit the units of k.

(e) Give a possible mechanism for the reaction and discuss in words, or give equations, to show how the mechanism is consistent with the experiment.

6. A reaction is zero order in substance S. Starting with the differential rate law derive an expression for $t_{1/2}$ in terms of the starting concentration, $[S]_0$, and the zero-order rate constant, k.

7. Write an expression for the rate of appearance of D from the following mechanisms:

(a)
$$A \underset{k_2}{\overset{k_1}{\rightleftharpoons}} B$$

$$B + C \overset{k_3}{\longrightarrow} D \qquad \text{(assuming steady state of B)}$$

(b)
$$A + B \overset{K}{\rightleftharpoons} AB \qquad \text{(fast to equilibrium)}$$

$$AB + C \overset{k}{\longrightarrow} D$$

8. The following data were obtained for the concentration vs. time for a certain chemical reaction. Values were measured at 1.00 sec intervals, beginning at 0.00 and ending at 20.00 sec. Concentrations in mM are:

10.00, 6.91, 4.98, 4.32, 3.55, 3.21, 2.61, 2.50, 2.22, 1.91, 1.80, 1.65, 1.52, 1.36, 1.42, 1.23, 1.20, 1.13, 1.09, 1.00, 0.92.

(a) Plot concentration (c) vs. time, ln c vs. time, and $1/c$ vs. time.

(b) Decide whether the data best fit zero-order, first-order, or second-order kinetics. Calculate the rate constant (with units) for the reaction and write the simplest mechanism you can for the reaction.

(c) Describe an experimental method which you think might be used to measure the concentrations which are changing so rapidly. The reaction is nearly over in 20 seconds.

9. The kinetics of double strand formation for a DNA oligonucleotide containing a G·T base pair was measured by temperature-jump kinetics. The reaction is:

$$2\,CGTGAATTCGCG \underset{k_{-1}}{\overset{k_1}{\rightleftharpoons}} DUPLEX$$

The following data were obtained:

Temperature (°C)	$k_1\,(10^5\,M^{-1}\,s^{-1})$	$k_{-1}\,(s^{-1})$
31.8	0.8	1.00
36.8	2.3	3.20
41.8	3.5	15.4
46.7	6.0	87.0

(a) Determine E_a, $\Delta H^{\ddagger}$ and $\Delta S^{\ddagger}$ for the forward and reverse reactions; assume that the values are independent of temperature.

(b) Determine the standard enthalpy ΔH^0 and standard entropy ΔS^0 changes for the reaction (assumed independent of temperature), and calculate the equilibrium constant for duplex formation at 37°C.

(c) Discuss qualitatively what you would expect to happen to $\Delta H^{\ddagger}$ and the standard enthalpy change for the reaction if the duplex contained one more Watson-Crick base pair. Would each increase, decrease, or remain the same?

10. Equal volumes of two equimolar solutions of reactants A and B are mixed, and the reaction $A + B \rightarrow C$ occurs. At the end of 1 hr, A is 90% reacted. How much of A will be left *unreacted* at the end of 2 hr if the reaction is:
(a) First order in A and zero order in B?
(b) First order in A and first order in B?
(c) Zero order in both A and B?
(d) First order in A and one-half order in B?

11. In a second experiment the same reaction mixture of Problem 10 is diluted by a factor of 2 at the time of mixing. How much of A will be left unreacted after 1 hr for each of the assumptions (a) through (d) of Problem 10?

12. The reaction of a hydrogen halide, A, with an olefin, B, to give product P according to the stoichiometric relation $A + B \rightarrow P$ is proposed to occur by the following mechanism:

$$2\,A \overset{K_1}{\rightleftharpoons} A_2 \qquad \text{(fast to equilibrium)}$$

$$A + B \overset{K_2}{\rightleftharpoons} C \qquad \text{(fast to equilibrium)}$$

$$A_2 + C \overset{k_3}{\longrightarrow} P + 2\,A \quad \text{(slow)}$$

(a) Based on this mechanism derive an expression for the velocity of the reaction as a function of the concentrations of only reactants and stable product.

(b) In a particular experiment, equal concentrations of A and B are mixed together and both the initial rate, v_0, and the half-life, $t_{1/2}$, are determined. Based on your answer to part (a), predict for a subsequent experiment

(i) The effect on v_0 if the initial concentrations of both A and B are doubled.

(ii) The effect on $t_{1/2}$ if the initial concentrations of both A and B are doubled.

(iii) The effect on v_0 if the initial concentration of A is unchanged and the initial concentration of B is increased tenfold.

13. The age of water or wine may be determined by measuring its radioactive tritium ($^3_1$H) content. Tritium is present in a steady state in nature. It is formed primarily by cosmic irradiation of water vapor in the upper atmosphere, and it decays spontaneously by a first-order process with a half-life of 12.5 years. The formation reaction does not occur significantly inside a glass bottle at the surface of the earth. Calculate the age of a suspected vintage wine that is 20% as radioactive as a freshly bottled specimen. Would you recommend to a friend that he consider paying a premium price for the "vintage" wine?

14. A radioactive sample produced 1.00×10^5 disintegrations min^{-1}; 28 days later it produced only 0.25×10^5 disintegrations min^{-1}. (1 day = 1440 min.)

    (a) What is the half-life of the sample?

    (b) How many radioactive atoms were there in the sample that had 10^5 disintegrations min^{-1}?

15. Consider the following proposed mechanism for the reaction A → P:

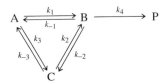

    (a) Write a differential equation for the rate of formation of B.

    (b) However, the formation of B from A and C is fast to equilibrium with equilibrium constants $K_1 = k_1/k_{-1}$, $K_2 = k_2/k_{-2}$, and $K_3 = k_3/k_{-3}$. k_4 is much smaller than all other k's. Write a differential equation for the formation of P in terms of concentration of A, equilibrium constants, and k_4.

    (c) Write an expression (containing no derivatives) for the concentration of P as a function of time. At zero time $[A] = [A]_0$, and $[B], [C], [P] = 0$.

16. There is evidence that a critical concentration of a trigger protein is needed for cell division (see *Molecular Biology of the Cell* by Alberts et al.). This unstable protein is continually being synthesized and degraded. The rate of protein synthesis controls how long it takes for the trigger protein to build up to the concentration necessary to start DNA synthesis and eventually to cause cell division.

    Let us choose a simple mechanism to consider quantitatively. The trigger protein U is being synthesized by a zero-order mechanism with rate constant k_0. It is being degraded by a first-order mechanism with rate constant k_1.

    (a) Write a differential equation consistent with the mechanism.

    (b) The solution to the correct differential equation is

    $$[U] = \frac{k_0}{k_1} (1 - e^{-k_1 t})$$

    if [U] is equal to zero at zero time. Show that your equation in (a) is consistent with this.

    (c) If U is being synthesized at a constant rate with $k_0 = 1.00$ nM s^{-1} and its half-life for degradation is 0.500 hr, calculate the maximum concentration that U will reach. How long will it take to reach this concentration? Make a plot of U vs. time.

(d) If a concentration of U of $1.00 \ \mu M$ is needed to trigger DNA synthesis and cell replication, how long will it take to reach this concentration?

(e) If the rate of synthesis is cut in half ($k_0 = 0.500 \ \text{n}M \ \text{s}^{-1}$), how long will it take for U to reach a concentration of $1.00 \ \mu M$?

(f) What is the smallest rate of U synthesis, k_0, that will allow (slow) cell replication. Assume k_1 remains constant.

17. A simple reaction for a DNA molecule is the exchange of a DNA proton with a water proton. In the DNA base pair shown below the imino proton (in bold face) exchanges with water at a rate measurable by NMR.

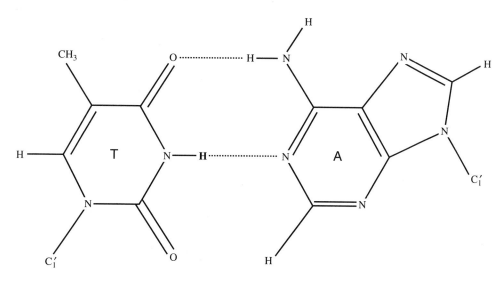

The stoichiometry of the reaction is

$$\mathbf{TH \cdot A} + \mathbf{H}^*OH^* \longrightarrow \mathbf{TH}^* \cdot A + \mathbf{HOH}^*$$

A mechanism for its rate of exchange is as follows.

$$(\mathbf{TH \cdot A})_{\text{closed}} \underset{k_{\text{cl}}}{\overset{k_{\text{op}}}{\rightleftarrows}} (\mathbf{TH \cdots A})_{\text{open}}$$

$$(\mathbf{TH \cdots A})_{\text{open}} + B + \mathbf{H}^*OH^* \overset{k_{\text{tr}}}{\rightleftarrows} \mathbf{TH}^* \cdot A + B + \mathbf{HOH}^*$$

$(\mathbf{TH \cdot A})_{\text{closed}}$ represents a closed base pair
$(\mathbf{TH \cdots A})_{\text{open}}$ is the open base pair ready to exchange
B is a base that catalyzes the transfer of $\mathbf{H}$ to $\mathbf{H}^*$

(a) Use a steady state approximation for $[(\mathbf{TH \cdots A})_{\text{open}}]$ to obtain an expression for the initial rate of exchange.

$$\frac{d[\mathbf{TH}^* \cdot A]}{dt} = k_{\text{ex}}[(\mathbf{TH \cdot A})_{\text{closed}}]$$

Write k_{ex} as a function of k_{op}, k_{cl}, k_{tr}, $[(\mathbf{TH \cdot A})_{\text{closed}}]$, $[B]$.

(b) Show how the rates of exchange vs. concentrations of $[(TH \cdot A)_{closed}]$ and [B] can be used to obtain k_{op} (the rate constant for base pair opening).

(c) The imino exchange rate for a synthetic polynucleotide, poly dA·dT, was measured as a function of pH at 27°C. Calculate a value for k_{op}, the rate constant for base pair opening, from the following data.

$[OH^-]$ (μM)	1.26	2.00	3.16	10.0	31.6
k_{ex} (s^{-1})	12.5	18.2	27.2	69.9	140.8

18. A reaction occurs with the following stoichiometry:

$$A + P \longrightarrow AP$$

The concentration of A was measured versus time after mixing; the data were:

Concentration (nM)	Time (s)
50.0	0
40.0	100
30.0	229
20.0	411
10.0	721

(a) What is the order of the reaction? Calculate the rate constant and give the units.

(b) Propose a mechanism that is consistent with the kinetics and stoichiometry of the reaction. Write differential equations for your mechanism and relate the measured k to the k's in your mechanism.

(c) The data given above were measured at 0°C. When the reaction was studied at 10°C, the rate constant doubled. Calculate the activation energy in kJ mol^{-1}.

19. The gas-phase decomposition of di-*tert*-butyl peroxide, $(CH_3)_3COOC(CH_3)_3$, is first order in the temperature range 110 to 280°C, with a first-order rate constant $k = 3.2 \times 10^{16}$ exp $[-(164 \text{ kJ mol}^{-1})/RT]$ s^{-1}.

(a) What are the values of $\Delta H^{\ddagger}$ and $\Delta S^{\ddagger}$ for this reaction at 110°C?

(b) Give a brief interpretation of the value of $\Delta S^{\ddagger}$.

(c) At what temperature will the reaction occur 10 times faster than at 110°C?

20. Calculate the activation energy which leads to a doubling of the rate of reaction with an increase in temperature from 25°C to 35°C.

21. The mechanism for a reaction is assumed to be

$$A + B \xrightarrow{k} P$$

$$k = 10^5 \text{ } M^{-1} \text{ s}^{-1} \text{ at } 27°C$$

(a) Calculate the initial rates of formation of P if 0.10 M A is mixed with 0.10 M B at 27°C. State the units.

(b) Calculate the initial rate of formation of P if 1.00×10^{-4} M A is mixed with 1.00×10^{-6} M B at 27°C. State the units.

(c) How long would it take (in seconds) to form 0.050 M product from 0.10 M A and 0.10 M B at 27°C?

(d) At 127°C the rate constant of the reaction increases by a factor of 10^3. Calculate E_a and $\Delta H^{\ddagger}$ at 27°C.

22. If A and B are mixed together in solution, it is found that the concentration of A decreases with time, but B remains constant. The stoichiometric equation is A $\rightarrow$ P.

(a) If the initial concentration of A is less than 0.01 M, it is found that the initial rate is $-d[A]/dt = k_1[A][B]$. Write the integrated form of this equation. What is the order of the reaction with respect to A? What is the order of the reaction with respect to B?

(b) If the initial concentration of A is greater than 1 M, it is found that the initial rate is nearly independent of A:

$$ -\frac{d[A]}{dt} = k_2[B] $$

Write the integrated form of this equation. What is the order of the reaction with respect to A? What is the order of the reaction with respect to B?

(c) Sketch a plot of the initial rate $-d[A]/dt$ versus initial concentration [A]. Write one rate equation which is consistent with experiment at both high and low concentrations of A.

23. (a) If we express the rate as the number of molecules reacting per unit time, for a unimolecular reaction the rate depends only on the *number* of reacting molecules present, but for a bimolecular reaction the rate depends on their *concentrations*. Explain.

(b) Consider the simple reaction

$$ A \underset{k_2}{\overset{k_1}{\rightleftharpoons}} B $$

It is proposed that the forward reaction (k_1) is first order and the reverse reaction (k_2) is zero order. Is this possible? Explain.

24. A reacts to form P. A plot of the reciprocal of the concentration of A versus time is a straight line. When the initial concentration of A is 1.0×10^{-2} M, its half-life is found to be 20 min.

(a) What is the order of the reaction?

(b) Write a one-line mechanism that is consistent with the kinetics.

(c) What is the value of the rate constant for your mechanism of part (b)?

(d) When the initial concentration of A is 3.0×10^{-3} M, what will be the half-life?

25. A and B react stoichiometrically to form P. If 0.01 M A and 10 M B are mixed, it is found that the log of the concentration of A versus time is a straight line.

(a) What is the order of the reaction with respect to A?

(b) Write a one-line mechanism that is consistent with the kinetics and stoichiometry.

(c) According to your mechanism, what is the order of the reaction with respect to B?

(d) The half-life for the disappearance of A is 100 s. What would be the predicted half-life if the concentration of B is changed from 10 M to 20 M?

(e) From the half-life given in part (d), calculate the rate constant for the reaction of A with B. Specify the units of the rate constant.

26. All living cells have approximately the same ^{14}C to ^{12}C ratio as found in CO_2 in the atmosphere. When an animal or plant or bacterial cell dies, it begins to lose ^{14}C by radioactive decay, thus over time the ^{14}C to ^{12}C ratio decreases. This ratio can be measured by a mass spectrometer. As a novice archaeologist you find an old piece of wood and have it analyzed for $^{14}C/^{12}C$. The ratio found is 28% of the present ratio. How long ago was the wood formed?

27. ^{131}I has a radioactive half-life of 7.80 days but iodine is also removed from the body by excretion at a first-order rate characterized by a "biological half-life" of 26 days.
   (a) What will be the "effective half-life" of ^{131}I in the body (the time required for half of the ^{131}I to be removed by the combination of excretion and radioactive decay)?
   (b) A laboratory animal injected with ^{131}I should not be reinjected with that isotope for 2 months. Using the result of part (a), explain why this should be so.

28. In a paper by Bada, Protsch, and Schroeder [*Nature 241,* 394 (1973)], the rate of isomerization of isoleucine in fossilized bone is used as an indication of the average temperature of the sample since it was deposited. The reaction

$$\text{L-isoleucine} \rightleftarrows \text{D-alloisoleucine}$$
$$\text{iso} \qquad\qquad \text{allo}$$

produces a nonbiological amino acid, D-alloisoleucine, that can be measured using an automatic amino acid analyzer. At 20°C, this first-order reaction has a half-life of 125,000 years and its activation energy is 139.7 kJ mol^{-1}. After a very long time, the ratio allo/iso reaches an equilibrium value of 1.38. You may assume that this equilibrium constant is temperature independent.
   (a) For a hippopotamus mandible found near a warm spring in South Africa, the allo/iso ratio was found to be 0.42. Assuming that no allo was present initially, calculate the ratio of the concentration of allo now present to the concentration of allo after a very long time (*Note:* The correct answer is between 0.40 and 0.60.)
   (b) Radiocarbon dating, which is temperature independent, indicated an age of 38,600 years for the hippo tooth. Using the result of part (a), estimate the half-life for the process.
   (c) Calculate the average temperature of this specimen during its residence in the ground. (The present mean temperature of the spring is 28°C.)

29. In aqueous solution the reaction of A to form B has the following rate expression:

$$-\frac{d[A]}{dt} = k[A]\{1 + k'[H^+]\}$$

   (a) Propose a mechanism that is consistent with this experimental rate. Relate the k's in your mechanism to k and k'.
   (b) From the pH dependence of the reaction k' was found to be $1.0 \times 10^5\ M^{-1}$. In a pH buffer it took 5 min for a 0.30 M solution of A to react to give 0.15 M B. Calculate the value of k and give its units.

30. Transfer RNA can exist in two forms which are in rapid equilibrium with each other; the equilibrium constant at 28°C, $K = [B]^{eq}/[A]^{eq} = 10$. A temperature-jump experiment is done to measure the rates of interconversion. A solution of the tRNA at a concentration of 10 μM is quickly (faster than 10 μs) raised in temperature from 25 to 28°C. An experiment signal is measured with a relaxation time, $\tau = 3$ ms. Assume that the mechanism is:

$$A \underset{k_{-1}}{\overset{k_1}{\rightleftarrows}} B$$

(a) What are the values of k_1 and k_{-1}, including units?

(b) At what temperature are these values measured?

(c) Would you expect doubling the concentration of tRNA to increase, decrease, or leave unchanged the values of τ, k_1, and k_{-1}?

31. The equilibrium

$$I_2 + I^- \underset{k_{-1}}{\overset{k_1}{\rightleftharpoons}} I_3^-$$

has been studied using a laser-induced temperature-jump technique. Relaxation times were measured at various equilibrium concentrations of the reactants at 25°C:

$[\bar{I}^-]$ (mM)	$[\bar{I}_2]$ (mM)	τ (ns)
0.57	0.36	71
1.58	0.24	50
2.39	0.39	39
2.68	0.16	38
3.45	0.14	32

SOURCE: Turner et al., *J. Amer. Chem. Soc. 94*, 1554 (1972).

(a) Calculate k_1 and k_{-1} for this system at 25°C.

(b) Compare your results with the value for the equilibrium constant, $K = 720\ M^{-1}$.

(c) Use simple diffusion theory to estimate a value for k_1. For I^- and I_2 the effective radii are 2.16 and 2.52×10^{-8} cm, respectively, and diffusion coefficients are 2.05 and 2.25×10^{-5} cm^2 s^{-1}, respectively. Comparing this result with the experimental value from part (a), what do you conclude about the effectiveness of collisions of I_2 and I^- in leading to reaction?

32. Consider the dimerization of proflavin

$$2P \underset{k_{-1}}{\overset{k_1}{\rightleftharpoons}} P_2$$

(a) Starting with the usual expression for the chemical relaxation time for a dimerization,

$$\tau = \frac{1}{4k_1[\bar{P}] + k_{-1}}$$

derive the following alternative form:

$$\frac{1}{\tau^2} = k_{-1}^2 + 8k_1 k_{-1}[P]_t$$

where $[P]_t = [P] + 2\,[P_2]$ is the total concentration of proflavin in the solution and $[\bar{P}]$, $[\bar{P}]_2$ are the concentrations of proflavin and its dimer at equilibrium.

(b) A bonus from this analysis is that both k_1 and k_{-1} can be obtained by measuring the relaxation rate as a function of $[P]_t$. Plot the following data and determine values for k_1 and k_{-1}. The temperature is 25°C.

τ (s)	$[P]_t$ (M)
3.2×10^{-7}	0.5×10^{-3}
1.8×10^{-7}	2.0×10^{-3}
1.4×10^{-7}	3.5×10^{-3}
1.2×10^{-7}	5.0×10^{-3}

(c) What is ΔG^0 at 25°C for the dimerization of proflavin?

33. The ionization constant for NH_4^+ is $K = 5.8 \times 10^{-10}$ at 25°C:

$$NH_4^+ \underset{}{\overset{K}{\rightleftarrows}} NH_3 + H^+$$

Table 7.5 gives

$$H^+ + NH_3 \xrightarrow{k_1} NH_4^+ \quad (k_1 = 4.3 \times 10^{10} \ M^{-1} \ s^{-1})$$

(a) A temperature jump from 20°C to 25°C is made on a 0.1 M NH_4Cl solution at pH 6. Calculate the relaxation time for the reaction assuming that the mechanism is given by the ionization equation shown.

(b) The relaxation time is a function of pH. At what pH will the relaxation time have its largest value?

(c) How would you measure the activation energy for k_1? Would you expect the activation energy to be less than 50 kJ? Greater than 100 kJ? Explain.

(d) According to Debye-Hückel theory, would you expect the relaxation time to increase, decrease, or remain unchanged if 0.1 M NaCl is added? Explain.

34. Gaseous ozone, O_3, undergoes decomposition according to the stoichiometric equation

$$2\ O_3(g) \longrightarrow 3\ O_2(g)$$

Two alternative mechanisms have been proposed to account for this reaction:

(I) $\qquad\qquad 2\ O_3 \xrightarrow{k} 3\ O_2 \qquad$ (bimolecular)

(II) $\qquad\qquad O_3 \overset{K_1}{\rightleftarrows} O_2 + O \qquad$ (fast, equilibrium)

$\qquad\qquad O + O_3 \xrightarrow{k_2} 2\ O_2 \qquad$ (slow)

(a) Derive rate laws for the formation of O_2 for each of these mechanisms.

(b) Thermodynamic measurements give standard enthalpies of formation for each of the following species at 298 K:

Substance	ΔH_{298}^0 (kJ mol^{-1})
$O_2(g)$	0.0
$O_3(g)$	142.3
O (g)	249.4

The observed activation enthalpy, $\Delta H^{\ddagger}$, for the overall reaction $2\ O_3 \rightarrow 3\ O_2$ is 125.5 kJ mol^{-1} of O_3. Sketch a curve of enthalpy (per mole of O_3) versus reaction coordinate for each of the two proposed mechanisms. Label the curves with numerical values for the $\Delta \bar{H}$ between reactants, products, intermediates, and transition states.

(c) Can you exclude either of these mechanisms on the basis of the thermodynamic and activation enthalpy values of part (b)? Explain your answer.

(d) Devise a kinetic procedure for distinguishing between the two mechanisms. State clearly the nature of the experiments you would perform and what results you would look for to make the distinction.

35. In the complete process of "simple" diffusion of a substance S across a membrane from side 1 to side 2, S first "jumps" into the membrane from the solution at side 1, then diffuses across the membrane interior, and finally "jumps" from the membrane at side 2 into the aqueous solution:

$$(S)_1 \underset{k_{-1}}{\overset{k_1}{\rightleftharpoons}} \Big| \ (S)_{m1} \underset{k_2}{\overset{k_2}{\rightleftharpoons}} (S)_{m2} \underset{k_1}{\overset{k_{-1}}{\rightleftharpoons}} \Big| \ (S)_2$$

$$\text{side 1} \qquad\qquad \text{membrane} \qquad\qquad \text{side 2}$$

For a symmetric membrane, the rate constants for "jumping" into and out of the membrane are the same at both surfaces, as indicated. At time zero, (S) is added to the solution on side 1. The area of the membrane is A.

(a) Give the rate law for the initial rate of appearance of (S) on side 2 taking k_1 as the rate-limiting step. Express the rate as

$$\frac{1}{A} \frac{dN_2}{dt}$$

where N_2 is the number of moles of S on side 2 and A is the area.

(b) Give the rate law for the initial rate of appearance of (S) on side 2 taking k_2 as the rate-limiting step. Again, express the rate as

$$\frac{1}{A} \frac{dN_2}{dt}$$

Your equation should contain the concentration of (S) on side 1 as the only concentration term.

(c) Experimentally, one can determine the initial rate of appearance of S on side 2 as a function of the initial concentration of S added to side 1. Using this type of kinetic data alone, can one deduce which step (k_1 or k_2) is rate-limiting? Explain briefly.

(d) Suppose that the membrane surfaces bear a net *negative* charge and that S is positively charged. What effect, if any, would decreasing the electrolyte concentration have on the initial rate of transport of S across the membrane? Briefly explain your answer.

36. Consider the reaction between substances $A^{\oplus}$ and $B^{\oplus}$, both positively charged. According to transition-state theory, $A^{\oplus}$ and $B^{\oplus}$ collide to form a transition-state complex which decomposes to product:

$$A^{\oplus} + B^{\oplus} \rightleftharpoons [A^{\oplus} \cdots B^{\oplus}] \longrightarrow \text{product}$$
$$\mapsto r^{\ddagger} \dashv$$

(a) Give an equation for the *electrostatic* contribution to the free energy of activation, $\Delta G^{\ddagger}$, in a solvent of dielectric constant ε and in extremely dilute solution so that the ionic strength is essentially zero. In the transition-state complex, the charge centers of $A^{\oplus}$ and $B^{\oplus}$ are $r^{\ddagger}$ apart, as indicated above. [*Hint:* Use Eq. (9.2.)]

(b) On the basis of electrostatic considerations, would you expect this reaction to be more rapid in a solvent of high or low dielectric constant? Explain briefly.

(c) Suggest an experimental approach that would allow determination of the electrostatic contribution to the free energy of activation.

37. Imidazole (Im) can react with H^+ or H_2O to form positively charged imidazole (ImH^+). The reaction mechanisms are

$$Im + H^+ \underset{k_{-1}}{\overset{k_1}{\rightleftharpoons}} ImH^+$$

$$Im + H_2O \underset{k_{-2}}{\overset{k_2}{\rightleftharpoons}} ImH^+ + OH^-$$

The rate constants in aqueous solution are: $k_1 = 1.5 \times 10^{10} \ M^{-1} \ s^{-1}$; $k_{-1} = 1.5 \times 10^3 \ s^{-1}$; $k_2 = 2.5 \times 10^3 \ s^{-1}$; $k_{-2} = 2.5 \times 10^{10} \ M^{-1} \ s^{-1}$.

(a) What is the value of the equilibrium constant for the ionization of imidazole ($ImH^+ \rightleftharpoons Im + H^+$)?

(b) Write the differential equation for the net rate of formation of ImH^+.

(c) If the pH is suddenly changed for a solution of $0.1 \ M$ imidazole in water from pH 7 to pH 4, what is the rate-determining step for the appearance of ImH^+ at pH 4?

(d) What is the value of the initial rate of increase of ImH^+ at pH 4?

(e) The rate constants k_1 and k_{-1} both depend on temperature. Would you expect them to decrease or increase with increasing temperature? Which would you expect to change most with temperature and why?

(f) Predict the sign of the heat of ionization for imidazole based on your answer to part (e). The experimental heat of ionization is positive.

38. In medical treatment drugs are often applied either continuously or in a series of small, closely spaced doses. Assume that oral ingestion and absorption of a particular drug are rapid compared with the rate of utilization and elimination, and that (after a single dose) the latter process occurs by a simple exponential decrease. We can represent this by a scheme

$$A \overset{k_1}{\longrightarrow} B \overset{k_2}{\longrightarrow} C$$

where [A] is the concentration of the drug at its site of introduction, [B] is the concentration of the drug in blood, and [C] represents the elimination of the drug. In this model the delivery, k_1, of the drug is zero order kinetically, and the elimination, k_2, is first order.

(a) Write the rate law, in terms of $d[B]/dt$, that is appropriate to this model.

(b) Assuming that the first introduction of the drug occurs at time zero, draw a curve showing how [B] will change with time during continuous, prolonged administration of the drug. [If this is not intuitively obvious to you, proceed first to part (f).]

(c) After some time the patient is cured and the use of the drug is stopped. Add to your drawing a sketch of how [B] changes subsequently.

(d) From an experimental record or monitor of [B] versus time, how would you determine k_2?

(e) How would you determine k_1?

(f) Derive a mathematical expression for [B] as a function of time during drug administration starting with the rate law from part (a). (A simple variable substitution should convert it to a form that was treated explicitly in this chapter.) Demonstrate that the result is consistent with your rise curve.

(g) To what level does [B] rise after a long time of administration?

39. In view of the current shortage of fossil fuels (chiefly petroleum and natural gas), alternative sources of energy are being sought. An imaginative proposal is to use rubber (latex), which is a plant hydrocarbon, polyisoprene. Latex represents about 50% of the carbon fixed in photosynthesis by rubber trees! Presumably this material could be cracked and processed in refineries in a fashion quite analogous to that used for petroleum. Under good conditions a rubber tree in Brazil can produce 10 kg of usable latex per year, growing at a density of about 100 trees per acre. The heat of combustion (fuel value) of hydrocarbons is about 644 kJ mol^{-1} of CH_2 or about 46,000 kJ kg^{-1}.

(a) What fraction of a year's solar energy is stored in rubber crop if the incident solar energy is 4 kJ min^{-1} ft^{-2} and the sun is shining 500 min day^{-1}, on the average? 1 acre = 43,560 ft^2. [Compare your answer with the value (energy stored)/(solar energy) = 0.3 that can be obtained from photosynthesis in the laboratory under optimal conditions.]

(b) In 1974 the total fossil fuel consumption in the United States was 3.8×10^{16} kJ yr^{-1}. What fraction of the area of the United States (3,000,000 mi^2) would have to be devoted to growing rubber trees (if they would grow in the climate of the United States, which they will not) to replace our present sources of fossil fuel by latex? (1 mi^2 = 640 acres.)

(c) In laboratory experiments it is found that photosynthesis occurs optimally with about 3% CO_2 in the (artificial) atmosphere. CO_2 occurs at only 0.03% in the natural atmosphere. Propose a scheme for increasing the energy yield of agricultural crops (not necessarily rubber trees) based on this observation. List some possible additional advantages and disadvantages of your scheme based on your background in kinetics and thermodynamics.

40. Consider the following reactions involved in the formation and disappearance of ozone in the upper atmosphere.

$$O_2 \xrightarrow{h\nu \text{ (below 242 nm)}} O + O \qquad \text{(A)}$$

$$O + O_2 + M \longrightarrow O_3 + M \qquad \text{(B)}$$

$$O_3 \xrightarrow{h\nu \text{ (190–300 nm)}} O_2 + O \qquad \text{(C)}$$

where the first and third reactions are photochemical and driven by sunlight, and the second reaction is termolecular. (The third body, M, may be N_2, O_2, or any other gas molecule in the atmosphere.) The potential danger of reducing the "ozone shield" by supersonic transports (SSTs) comes from the presence of nitrogen oxides (NO, NO_2, etc.) in the engine exhaust. Although the actual mechanism is more complex, two important reactions in the proposed ozone reaction are

$$NO + O_3 \xrightarrow{k_1} NO_2 + O_2 \qquad \text{(fast)} \qquad (1)$$

$$O + NO_2 \xrightarrow{k_2} NO + O_2 \qquad \text{(fast)} \qquad (2)$$

Note that NO is not used up in this pair of reactions, but it continues to decompose ozone in a pseudocatalytic fashion (until it eventually diffuses out of the ozone layer, which may be a very slow process taking many months).

(a) At a temperature of 217 K, characteristic of the 20-km altitude at which an SST might fly, the value of k_1 is 4.0×10^{-15} cm^3 $molecule^{-1}$ s^{-1}. From the data on exhaust composition, it is estimated that the steady-state concentration of NO would be about 10^{10} molecules cm^{-3}. If the NO were suddenly introduced at this level, what would be the half-time (in hours) for the reduction of ozone in the region of the SST flight paths? (*Note:* You do not need to know the ozone concentration to solve this problem.)

(b) The molecule NO_2 is unstable in the presence of sunlight, and decomposes photochemically according to the following reaction:

$$NO_2 \xrightarrow{h\nu \; (260-400 \text{ nm})} NO + O \tag{3}$$

The oxygen atoms are highly reactive and disappear by reactions (B) and (2) above. Consider carefully the contribution of reaction (3) to the overall rate of ozone disappearance and describe qualitatively the effect that you would expect if this reaction were added to the situation approximated in part (a).

41. The photosynthesis efficiency of a strain of algae was measured by irradiating it for 100 s with an absorbed intensity of 10 W and an average wavelength of 550 nm. The yield of O_2 was 5.75×10^{-4} mol. Calculate the quantum yield of O_2 formation. (The quantum yield per "equivalent" of photochemistry is four times this value, since four electrons must be removed from two water molecules to produce each O_2.)

42. The human eye, when completely adapted to darkness, is able to perceive a point source of light against a dark background when the rate of incidence of radiation on the retina is greater than 2×10^{-16} W.
    (a) Find the minimum rate of incidence of quanta of radiation on the retina necessary to produce vision, assuming a wavelength of 550 nm.
    (b) Assuming that all of this energy is converted ultimately to heat in visual cells having a total volume of 10^{-2} cm^3 and assuming that there are no losses of heat, by how much would the temperature of these cells rise in 1 s? Assume a heat capacity and density equal to that of water.
    (c) Estimate the safe range of visual intensities (those which you would not expect to cause permanent eye damage).

43. The protein rhodopsin is responsible for the absorption of light in photoreceptor cells. Following the absorption of light, the protein goes through the complex series of reactions shown in Fig. 7.21. Here we will use the symbols R, BR, LR, MI, and so on, to indicate the different forms of the protein, and we will assume that the lifetimes shown in Fig. 7.21 are the reciprocals of the corresponding rate constants $k_1, k_1, \ldots$ at 310 K.
    (a) Write the rate law for the formation of MI following a very short (10^{-12} s) flash of light that converts all of the R initially present to BR.
    (b) If R is initially present at 1×10^{-3} M, what is the initial rate of formation of MI? State any approximations or assumptions that you make.
    (c) Experimentally, it is found that the ratio [MII]/[MI] rapidly (within 5×10^{-3} s) reaches a constant value of 1.0 after light absorption, even though [MI] and [MII] individually decrease slowly with time. What is the value for k_{-4}, the rate constant for the reaction MII $\rightarrow$ MI?
    (d) It has been proposed that the transformation MI $\rightarrow$ MII involves large changes in protein structure. The enthalpy change for this step was determined to be ΔH^0_{310} (MI $\rightarrow$ MII) = 42 kJ mol^{-1}. Using the result of part (c), calculate ΔS^0_{310} for this transformation and interpret it in terms of a possible structural change.
    (e) Write the rate law for the disappearance of MII.

44. An electronically excited molecule A* can either emit fluorescence to return to its ground state, or it can lose its excitation by collision.

$$A* \xrightarrow{k_r} A + \text{photon}$$
$$A* \xrightarrow{k_T} A + \text{heat}$$

$$\left(\begin{array}{c} k_r \text{ and } k_T \text{ are rate} \\ \text{constants with units} \\ \text{of s}^{-1} \end{array} \right)$$

(a) The fluorescence intensity of A* can be measured as the number of photons emitted per second. Write an equation relating the fluorescence intensity to the concentration of A*.

(b) Derive an equation for the concentration of A* as a function of time, k_r, k_T, and $[A^*]_0$ (the concentration of A* at zero time).

45. Sunlight between 290 and 313 nm can produce sunburn (erythema) in 30 min. The intensity of radiation between these wavelengths in summer, at 45° latitude and at sea level, is about 50 μW cm^{-2}. Assuming that each incident photon is absorbed and produces a chemical change in 1 molecule, how many molecules per square centimeter of human skin must be photochemically affected to produce evidence of sunburn?

46. Psoralen is a three-ring heterocyclic molecule found in rotting celery. Psoralen reacts with DNA in the presence of light to form monoadducts which can react further to cross-link the DNA. This can lead to dermatitis and hyperpigmentation of the skin of celery workers.

A mechanism for the first step of the reaction with DNA is

$$\text{Psoralen} + \text{DNA} \underset{}{\overset{K}{\rightleftarrows}} \text{Intercalated complex}$$

$$\text{Intercalated complex} + \text{light} \longrightarrow \text{Monoadduct}$$

The equilibrium constant for intercalation into *E. coli* DNA is $K = 250\ M^{-1}$ at 25°C (the concentration of DNA is moles of base pairs per liter); the quantum yield for the photochemical reaction is $\phi = 0.030$.

(a) Calculate the concentration of complex at equilibrium in a solution of $1.00 \times 10^{-4}\ M$ of psoralen and $1.00 \times 10^{-3}\ M$ DNA at 25°C in the dark. Also calculate the fraction of psoralen bound to DNA and the fraction free.

(b) The solution in part (a) was irradiated with 365 nm light with an intensity of 1.00×10^{-11} einstein cm^{-2} s^{-1}. The absorbance of the solution in a 1 cm path length cell was 0.30 at 365 nm. At this wavelength only the free psoralen and the complex absorb. They both have the same molar extinction coefficient per psoralen. Calculate the initial rate of formation of monoadduct with units of $M\ s^{-1}$. Remember that only the light absorbed by the complex can lead to product.

(c) After half the psoralen has reacted to form monoadduct, the effective concentration of DNA has changed by only 5%. For these conditions calculate the rate of formation of adduct.

8

Enzyme Kinetics

CONCEPTS

Enzymes are biological catalysts. An enzyme can increase the rate of a specific reaction by many orders of magnitude, but it is left unchanged at the end of the reaction. Of course, enzymes can be inactivated by many substances present in the surroundings, such as oxygen, metal ion impurities, and so forth. Enzymes, like other catalysts, do not continue working indefinitely. An enzyme is a very specific catalyst; it binds a particular reactant (called a substrate) and facilitates its reaction, by stabilizing its transition state to a specific product. Most enzymes are proteins, but the active sites often contain groups other than amino acids. There are metalloenzymes in which zinc, iron, cobalt, manganese, or other transition metals are the catalytic centers. Many enzymes require cofactors—small molecules bound to the enzyme—to be effective catalysts. For example, the B-family of vitamins are all cofactors. Some enzymes are complexes of protein and RNA in which the RNA also participates in catalysis. One example is the ribosome, a complex assembly of many proteins and three RNAs. The ribosome has a key role in the translation of messenger RNA into proteins.

The discovery by Arthur Zaug and Thomas Cech (Zaug and Cech, 1986) that RNA by itself could be an enzyme has revolutionized the biochemists' view of biological catalysis. Now that we know that enzymes need not be proteins; it may be that all sorts of other biological molecules, such as carbohydrates, may also have catalytic activity. The catalytic activity of RNA has also provided new ideas about the evolution of enzymes; RNA enzymes—ribozymes—may have been primitive enzymes which have now been mainly replaced by the more efficient protein enzymes.

418

APPLICATIONS

Catalytic Antibodies and RNA Enzymes—Ribozymes

Antibodies are proteins synthesized by the immune system to protect an organism against foreign proteins or other foreign chemicals. Special cells synthesize antibodies which bind strongly to the foreign molecule. To produce antibodies to any desired chemical, the chemical is first covalently attached to a protein. The modified protein is used as an antigen to induce the cellular production of antibodies to the chemical. This means that proteins are produced which bind strongly and specifically to the chemical; this produces part of the requirements for an enzyme. Catalysts increase the rate of a reaction by decreasing the free energy of activation; the catalyst stabilizes the transition state. To produce an effective enzyme the chemical used to induce antibody production can be a transition-state analog of an enzyme substrate. For example, the transition state for hydrolysis of an ester is a tetrahedral species.

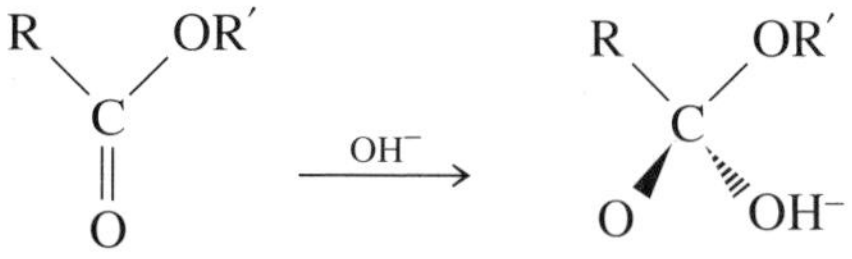

Substrate   Transition state

An excellent transition-state analog for ester hydrolysis is a phosphate ester.

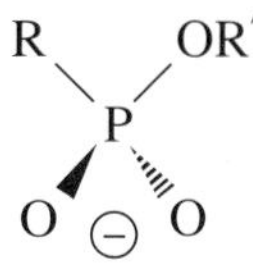

Transition-state analog

By attaching stable phosphonates (shown above) to proteins to form antigens, catalytic antibodies were produced which approach the catalytic efficiency of the protein enzyme chymotrypsin for ester substrates. Catalytic antibodies have been produced with enzymatic activity for many different types of substrates. One goal of the research is to produce enzymes which have catalytic activity not found in nature (Lerner et al., 1991; Schultz et al., 1990).

Proteins which have new catalytic activities do not overthrow long-held beliefs about enzymes, but RNA molecules that are catalytic clearly do. RNA molecules are transcribed from DNA molecules, but they are often modified—processed—before they are ready to function. A common occurrence is that the RNA is cut—to remove an internal piece—and spliced. Of course the cutting and splicing must be done very specifically, so the processing enzymes that do this are of interest. Thomas Cech was trying to isolate and purify the enzyme that processed a ribosomal RNA from an organism called *Tetrahymena thermophila*. He found that the enzymatic activity was not in any of the protein fractions; instead the activity was always part of the RNA itself. The easiest explanation for this was that the protein enzyme bound tightly to the RNA and could not be removed. Even though all standard methods of removing protein (such as using proteinases to hydrolyze the protein, or using hot detergent to denature the protein) did not work, one could still argue that the protein enzyme was either very stable or very active. Any explanation was better than to assume that the enzymatic activity was not a protein. The convincing evidence that the RNA itself was the enzyme came when Cech synthesized the RNA in vitro, and showed that it self-processed without any added protein. In 1989 he was awarded the Nobel Prize in Chemistry for his originality and perseverance; the award was shared with Sidney Altman, for his work on ribonuclease P, an RNA-containing protein which processes transfer RNA.

The catalytic unit of the *Tetrahymena* RNA is the part that is removed during processing; it normally excises itself from the pre-ribosomal RNA. Once free it can act as an enzyme—an RNA enzyme or ribozyme. It has the activity of a ribonuclease (it hydrolyzes RNA), and of an RNA ligase (it splices or ligates RNA). Its kinetics has been extensively studied, and detailed mechanisms have been proposed (Herschlag and Cech, 1990). Other ribozymes have since been found, and the expectation is that many more are still to be found.

ENZYME KINETICS

The role of enzymes as unusual catalysts is well illustrated by the decomposition of hydrogen peroxide:

$$2\ H_2O_2 \longrightarrow 2\ H_2O + O_2$$

This reaction occurs only very slowly in pure aqueous solution, but its rate is greatly increased by a large variety of catalysts. The increase in rate is typically first order in concentration of catalyst. It may be approximately first order in concentration of H_2O_2 as well. However, for catalysis by the enzyme catalase, the rate becomes zero order in H_2O_2 at higher initial concentrations of H_2O_2. This is a common behavior for enzyme-catalyzed reactions.

Table 8.1 gives approximate values for the kinetics of the rate expression

$$-\frac{d[H_2O_2]}{dt} = k[H_2O_2][\text{catalyst}]$$

In the table the rates for the extrapolated conditions of 1 M of H_2O_2 and 1 M of catalyst are given for comparison. For catalase, the maximum rate per mole of active sites is given. The rate for the uncatalyzed reaction, as shown in the table, corresponds to a decomposition of only 1% after 3 days at 25°C. Even then the reaction is probably "catalyzed" by dust particles that are difficult to remove completely from the solutions. Inorganic catalysts, such as iron salts or hydrogen halides, increase the rate of H_2O_2 decomposition by four to five orders of magnitude per mole of catalyst. The enzyme catalase, which occurs in blood and a variety of tissues (liver, kidney, spleen, etc.), increases the rate by more than 15 powers of 10 over the uncatalyzed rate! To appreciate the speed of this enzymatic reaction, consider that for the maximum rate each molecule of catalase is able to decompose more than 10 million molecules of H_2O_2 per second!

Table 8.1 Catalase and H_2O_2 decomposition: comparison of rates and activation energies at 25°C*

Catalyst	Rate, $-d[H_2O_2]/dt$ $(M\ s^{-1})$	E_a $(kJ\ mol^{-1})$
None	10^{-8}	71
HBr	10^{-4}	50
Fe^{2+}/Fe^{3+}	10^{-3}	42
Hematin or hemoglobin	10^{-1}	—
$Fe(OH)_2TETA^+$	10^3	29
Catalase	10^7	8

* The rate is calculated for 1 M of H_2O_2 and 1 M catalyst (except for catalase). For catalase, the maximum rate is given for 1 M concentration of active sites. This rate is numerically equal to the catalytic constant for catalase. TETA is triethylenetetramine. For a discussion of the mechanism of the reaction and reference to the earlier literature, see J. H. Wang, *J. Am. Chem. Soc. 77*, 4715 (1955).

We note that catalase is a hemoprotein with ferriprotoporphyrin (hematin) as a prosthetic group (the hematin group provides the catalytic function) at the active site. The hematin can be separated from the protein. Alone in solution, hematin exhibits catalytic activity toward H_2O_2 decomposition that is two orders of magnitude higher than that of the inorganic catalysts, but still more than a millionfold smaller than that of catalase. Not just any protein will enhance the catalytic activity of the iron protoporphyrin either. Heme (like hematin, but with the iron in the ferrous state) is the prosthetic group present in hemoglobin, but there is no catalytic enhancement of H_2O_2 decomposition in comparison with free hematin.

Chemists have risen to the challenge of trying to simulate the catalytic activity of enzymes, short of reproducing the entire biological molecule. One such attempt by Wang (1955) produced an iron complex $Fe(OH)_2TETA^+$ (TETA = triethylenetetramine), which exhibited a rate of $1.2 \times 10^3\ M\ s^{-1}$ for the decomposition of 1 M H_2O_2. That is pretty good compared with the other nonenzymatic catalysts listed in Table 8.1, but still less than 10^{-3} of the value for catalase.

The thermodynamics of the decomposition of hydrogen peroxide is also relevant to the function of catalysts. The reaction

$$H_2O_2(aq) \longrightarrow H_2O(l) + \tfrac{1}{2}O_2(g)$$

is exergonic, with $\Delta G^0_{298} = -103.10$ kJ mol^{-1}. The major contribution comes from the enthalpy change, which is $\Delta H^0_{298} = -94.64$ kJ mol^{-1}. Thus the reaction should go strongly to the right, if there is a suitable reaction path. The slowness of the uncatalyzed reaction must be associated with a large activation barrier. The experimental activation energy is 71 kJ mol^{-1} (Table 8.1). Using the observed rate of the uncatalyzed decomposition, $v < 4 \times 10^{-8}\ M\ s^{-1}$, we can calculate an upper limit for the Arrhenius pre-exponential factor of $A < 1 \times 10^5\ s^{-1}$, assuming first-order kinetics for the "uncatalyzed" reaction. While the A

factor is significantly less than that observed for other first-order reactions, nevertheless it is clear that the activation energy provides the larger contribution to the reaction barrier. This is illustrated in Fig. 8.1.

By contrast with the uncatalyzed reaction, the activation energy for the H_2O_2 reaction catalyzed by Fe^{2+}/Fe^{3+} or HBr is only two-thirds as large. In the presence of catalase, the activation energy is only 8 kJ mol^{-1}, and the Arrhenius pre-exponential factor is 1.6×10^8 M^{-1} s^{-1}. It appears that the enzyme (and to a lesser extent, the other catalysts) succeeds in speeding the reaction by providing a path with a substantially smaller energy of activation. This is also shown in Fig. 8.1. The entropy barrier is also reduced (hence the large A factor), but its effect on the reaction velocity is less significant at room temperature. Note that the decrease in barrier height for the forward reaction implies a decrease in activation energy for the reverse reaction as well. For this example, the reverse reaction is strongly endergonic and even when the barrier height for the forward reaction is reduced the reverse reaction is still very slow. However, for most other biochemical reactions that we will be considering, the overall reaction has a standard free energy change closer to zero. In those cases both the forward and reverse reactions occur at appreciable rates in the presence of the enzyme.

A typical behavior for enzyme-catalyzed reactions is that the reaction is first order in substrate at low substrate concentrations and then becomes zero order in substrate at high substrate concentrations. This means the reaction reaches a maximum velocity with increasing substrate concentration for a constant enzyme concentration. The *catalytic constant,* or *turnover number,* is defined as the maximum rate (M^{-1} s^{-1}) divided by the concentration of enzyme active sites (M^{-1}). Its units are obviously s^{-1}. Concentration of active sites is used rather than concentration of enzyme so that a fair comparison can be made with enzymes that have more than one active site. Catalase has four active sites per molecule.

It would be misleading to give the impression that catalase activity is entirely typical of enzyme-catalyzed reactions (Table 8.2). Catalase is the fastest by several orders of magnitude. A more typical value for the catalytic constant of enzymes is 10^3 s^{-1}, and most

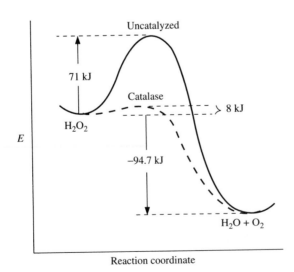

Fig. 8.1 Role of catalase in lowering the activation energy for the decomposition of H_2O_2.

Table 8.2 Catalytic constants for several enzymes

Enzyme	Substrate	Catalytic constant, k_{cat} (s^{-1})	Reference*
Catalase	H_2O_2	9×10^6	a
Acetylcholinesterase	Acetylcholine	1.2×10^4	b
Lactate dehydrogenase (chicken)	Pyruvate	6×10^3	c
Chymotrypsin	Acetyl-L-tyrosine ethyl ester	4.3×10^2	d
Myosin	ATP	3	e
Fumarase	L-Malate Fumarate	1.1×10^3 2.5×10^3	f
Carbonic anhydrase (bovine)	CO_2 HCO_3^-	8×10^4 3×10^4	g

* (a) Bonnichsen, R. K., B. Chance, and H. Theorell, *Acta Chem. Scand. 1,* 685, (1947); (b) Froede, H. C., and I. B. Wilson, *The Enzymes 5,* 87 (1971); (c) Everse, J., and N. O. Kaplan, *Advan. Enzymol. 37,* 61 (1973); (d) Kaufman, S., H. Neurath, and G. W. Schwert, *J. Biol. Chem. 177,* 793 (1949); (e) Brahms, J., and C. M. Kay, *J. Biol. Chem. 238,* 198 (1963); (f) Brant, D. A., L. B. Barnett, and R. A. Alberty, *J. Am. Chem. Soc. 85,* 2204 (1963); (g) DeVoe, H., and G. B. Kistiakowsky, *J. Amer. Chem. Soc. 83,* 274 (1961).

values fall within a factor of 10 of that value. This is not to say that these other enzymes are poor catalysts. The enzyme fumarase catalyzes the hydration of fumarate to L-malate with a catalytic constant of 2.5×10^3 s^{-1} at 25°C:

The rate constants for 1 *M* fumarate hydrolysis catalyzed by acid (1 *M*) or base (1 *M*) are about 10^{-8} s^{-1}. This "ordinary" enzyme still wins by 11 powers of 10.

MICHAELIS-MENTEN KINETICS

Because enzymes are such enormously effective catalysts, they are able to exert their influence at exceedingly low concentrations, typically 10^{-10} to 10^{-8} *M*. At this level, especially if the enzyme is present in a complex cellular soup, it is difficult to make direct measurements of what the enzyme itself is doing. Enzymology began, therefore, with a study of the kinetics of the disappearance of substrate and formation of products, because their concentrations are typically 10^{-6} to 10^{-3} M. Early observations resulted in the following generalizations about enzyme reactions:

1. The rate of substrate conversion increases linearly with increasing enzyme concentration (Fig. 8.2).

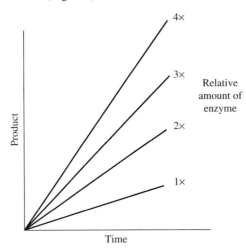

Fig. 8.2 Increase of product concentration with time for four different amounts of enzyme.

2. For a fixed enzyme concentration, the rate is linear in substrate concentration [S] at low values of [S] (Fig. 8.3).
3. The rates of enzyme reactions approach a maximum or saturation rate at high substrate concentration (Fig. 8.3).

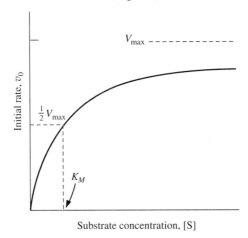

Fig. 8.3 Initial rate, v_0, of an enzyme-catalyzed reaction as a function of substrate concentration [S] for a fixed amount of enzyme.

At low concentrations,

$$v_0 = k[S] \tag{8.1}$$

where v_0 is the initial rate. At high substrate concentrations,

$$v_0 = V_{max} \quad \text{(maximum rate)} \tag{8.2}$$

This behavior is expected if the enzyme forms a complex with the substrate. At high substrate concentrations essentially all of the enzyme is tied up in the enzyme-substrate complex. Under these conditions the enzyme is working at full capacity, as measured by its

$$\text{catalytic constant} = k_{\text{cat}} = \frac{V_{\text{max}}}{[E]_0} \tag{8.3}$$

where $[E]_0$ is the total enzyme site concentration. At low substrate concentration the enzyme is not saturated, and turnover is limited partly by the availability of substrate.

A kinetic formulation of these ideas was presented by Michaelis and Menten, with significant improvements by Briggs and Haldane (1925). In the simplest picture, the enzyme and substrate reversibly form a complex, followed by dissociation of the complex to form the product and regenerate the free enzyme.

$$E + S \underset{k_{-1}}{\overset{k_1}{\rightleftharpoons}} ES \tag{8.4}$$

$$ES \xrightarrow{k_2} E + P \tag{8.5}$$

Because the second step must also be reversible, this mechanism applies strictly only to the initial stages of the reaction before the product concentration has become significant. Under these conditions

$$v_0 = \left(\frac{d[P]}{dt}\right)_0 = k_2[ES] \tag{8.6}$$

We apply the steady-state approximation:

$$\frac{d[ES]}{dt} = k_1[E][S] - k_{-1}[ES] - k_2[ES] \cong 0$$

Therefore,

$$[ES] = \frac{k_1[E][S]}{k_{-1} + k_2} \tag{8.7}$$

[E] and [S] refer to the concentrations of *free* enzyme and substrate; however, these are difficult to measure, so we write the equation in terms of the measurable *total* enzyme, $[E]_0$, and substrate, $[S]_0$, concentrations:

$$[E]_0 = [E] + [ES]$$
$$[S]_0 = [S] + [ES] \cong [S] \tag{8.8}$$

We can equate the total substrate concentration to the free substrate concentration, because the enzyme-substrate concentration is typically small compared to the substrate concentration. By substitution into Eq. (8.7) and rearranging terms, we obtain

$$[ES] = \frac{[E]_0}{1 + \dfrac{k_{-1} + k_2}{k_1[S]}} \tag{8.9}$$

Substitution into Eq. (8.6) gives

$$v_0 = \frac{k_2[E]_0}{1 + \dfrac{k_{-1} + k_2}{k_1[S]}} = \frac{V_{max}}{1 + \dfrac{K_M}{[S]}} \tag{8.10}$$

where $K_M = (k_{-1} + k_2)/k_1$ is the *Michaelis constant* for the enzyme and substrate combination, and V_{max} is $k_2[E]_0$. Equation (8.10) is termed the Michaelis-Menten equation. The limiting conditions given by Eqs. (8.1) and (8.2) follow readily from Eq. (8.10). At low substrate concentrations, such that $(K_M/[S]) \gg 1$, Eq. (8.10) gives

$$v_0 = \frac{V_{max}[S]}{K_M} = k[S] \tag{8.1}$$

where $k = V_{max}/K_M$ is a constant at a fixed total enzyme concentration $[E]_0$. At high substrate concentrations such that $(K_M/[S]) \ll 1$, Eq. (8.10) is reduced to

$$v_0 = V_{max} \quad \text{(maximum rate)} \tag{8.2}$$

$$k_{cat} = \frac{V_{max}}{[E]_0} = k_2 \tag{8.3}$$

When the substrate concentration [S] is equal to K_M, Eq. (8.10) gives

$$v_0 = \tfrac{1}{2} V_{max}$$

Thus the Michaelis constant is equal to the substrate concentration that is sufficient to give half the maximum rate for the enzyme. This is shown graphically in Fig. 8.3. One should get an intuitive feeling for the magnitude of K_M. A small value of K_M means that the enzyme binds the substrate tightly and small concentrations of substrate are sufficient to saturate the enzyme and to reach the maximum catalytic efficiency of the enzyme.

In general, enzyme-catalyzed reactions are reversible, and one needs to consider contributions from the reverse reaction once a significant amount of product is present. The use of initial rates avoids this complication by focusing on early times when the product concentration is still negligible. If the reverse reaction is not important (for example, if the position of equilibrium lies strongly toward products), it is possible to use an integrated form of the Michaelis-Menten equation that gives the substrate concentration as a function of time. Rewriting Eq. (8.10), we obtain

$$-\frac{d[S]}{dt} = \frac{V_{max}[S]}{K_M + [S]}$$

By separating the variables in this equation, we obtain

$$\left(\frac{K_M}{[S]} + 1 \right) d[S] = -V_{max}\, dt$$

which can be integrated, assuming that $[S] = [S]_0$ when $t = 0$,

$$K_M \ln \frac{[S]}{[S]_0} + [S] - [S]_0 = -V_{max}t$$

Without knowing K_M, there is no simple way to plot data so as to obtain a straight-line relation between a function of [S] and time. Using a computer it is straightforward to obtain the best values of K_M and V_{max} for a given data set, however.

It has become increasingly clear that very few enzymes act by a simple mechanism with a single enzyme-substrate complex. Even in the more complex situations, however, the kinetic equation often obeys the form given by Eq. (8.10). The exercise below illustrates one such case. Additional examples are provided by Eqs. (8.16) and (8.20) in a later section.

Exercise 8.1   The enzyme chymotrypsin (represented by EOH below, where OH signifies the hydroxyl group of the serine residue at position 195 of the protein) catalyzes the hydrolysis of peptides. It is one of a group of enzymes called serine proteases. A noncovalent complex (EOH·RCONHR') is formed rapidly between the enzyme (EOH) and the peptide substrate (RCONHR'). This is followed by the acylation of Ser-195 by the bound substrate to give a covalent acyl-enzyme intermediate (EOCOR), and the amine NH_2R' is released. The acyl-enzyme then hydrolyzes to give the enzyme-product complex EOH·RCO_2H, which dissociates rapidly to give the enzyme and RCO_2H. These steps are as follows:

$$EOH + RCONHR' \xrightleftharpoons{K_s} EOH \cdot RCONHR' \xrightarrow{k_2} \underset{\substack{+ \\ NH_2R'}}{EOCOR} \xrightarrow[k_3]{+H_2O}$$

$$EOH \cdot RCO_2H \xrightarrow{fast} EOH + RCO_2H$$

By applying the pre-equilibrium assumption for the K_s step and the steady-state assumption to [EOCOR], show that the overall kinetics of the scheme described above yields the Michaelis-Menten equation with

$$K_M = K_s \frac{k_3}{k_2 + k_3}$$

and

$$V_{max} = \frac{k_2 k_3}{k_2 + k_3} [E]_0$$

Kinetic Data Analysis

For quantitative purposes it is useful to rewrite Eq. (8.10) in a form that suggests a straight-line plot of the data. Several such approaches are used; the most popular was proposed by Lineweaver and Burk (1934).

$1/v_0$ versus $1/[S]$: Lineweaver-Burk

Taking the reciprocal of both sides of Eq. (8.10) and rearranging, we obtain

$$\frac{1}{v_0} = \frac{1}{V_{max}} + \frac{K_M}{V_{max}} \cdot \frac{1}{[S]}$$

(8.11)

Thus a plot of the reciprocal of initial rate versus the reciprocal of initial substrate concentration for experiments at fixed enzyme concentration should give a straight line. Furthermore, the intercept with the ordinate gives $1/V_{max}$, and the slope is K_M/V_{max} from which these two constants can be determined. Alternatively, K_M can be determined by extrapolation to the abscissa intercept, to give $-1/K_M$, as shown in Fig. 8.4.

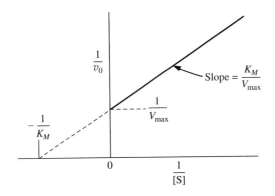

Fig. 8.4 Lineweaver-Burk plot of the reciprocal of initial reaction rate versus reciprocal of initial substrate concentration for a series of experiments at fixed enzyme concentration.

Exercise 8.2 Show on a graph such as Fig. 8.4 that $1/v_0 = 2/V_{max}$ when $[S] = K_M$ and explain how this could be used to determine V_{max} and K_M by a simple procedure.

Other procedures for manipulating the data are also used.

$[S]/v_0$ versus $[S]$: Dixon

Starting with Eq. (8.11) multiply both sides by $[S]$ to obtain

$$\frac{[S]}{v_0} = \frac{[S]}{V_{max}} + \frac{K_M}{V_{max}}$$

(8.12)

v_0 versus $v_0/[S]$: Eadie-Hofstee

Multiply both sides of Eq. (8.11) by $V_{max} \cdot v_0$ and rearrange to

$$v_0 = -K_M \cdot \frac{v_0}{[S]} + V_{max}$$

(8.13)

Ideally, with no experimental error each of these equations gives exactly the same desired information from a straight-line plot. In practice, one or the other may be preferable because of the nature of the data involved. Qualitatively, the Eadie-Hofstee plot spreads the values at high substrate concentration (where $v_0 \rightarrow V_{max}$), whereas the Lineweaver-Burk plot compresses the points in this region. Analysis using different plots may provide different values for K_M and V_{max} for the same set of data.

The usual method to obtain values of K_M and V_{max} is to obtain a least-squares-fit to the data that produce the linear plots. The values of K_M and V_{max} obtained by least-squares-fits to the different plots described by Eqs. (8.11), (8.12), and (8.13) will be different because different sums are minimized to obtain these values. [The equations are given in the section on Mathematics Needed in this Chapter.] The slope (m) and intercept (b) of a line are chosen so as to minimize the sum of the squares of the differences between the experimental y-values and the values calculated from the equation $y = mx + b$. Clearly our choice of which y and x to plot will affect these values. In the usual least-squares-fit programs (linear regressions) each point is weighted equally; the fact that different experimental points have different precision is ignored. Because of the statistical bias introduced by the different plotting methods, a statistically unbiased method has been developed (Cornish-Bowden, 1988). Only two experimental points of initial rate, v_0, and substrate concentration, [S], are needed to calculate K_M and V_{max}. If all possible combinations of data points are used to calculate these parameters, many different estimates of the parameters are obtained. The equations are

$$V_{max} = \frac{v_{01} v_{02}([S]_2 - [S]_1)}{[S]_2 v_{01} - [S]_1 v_{02}}$$

(8.14)

$$K_M = \frac{[S]_1 [S]_2 (v_{02} - v_{01})}{[S]_2 v_{01} - [S]_1 v_{02}}$$

(8.15)

where v_{01} and v_{02} are any two rates corresponding to two substrate concentrations $[S]_1$ and $[S]_2$. The many calculated values of K_M and V_{max} are ordered and the median values found.

The *median values* of K_M and V_{max} are the best statistically unbiased representation of the kinetic data. The median value is the middle of an ordered set of numbers; there are as many values above it as below it. Comparison of different plotting methods and the statistically unbiased method is illustrated in Example 8.1.

Example 8.1 The hydrolysis of carbobenzoxyglycyl-L-tryptophan catalyzed by pancreatic carboxypeptidase occurs according to the reaction

Carbobenzoxyglycyl-L-tryptophan + H_2O $\longrightarrow$

carbobenzoxyglycine + L-tryptophan

The following data on the rate of formation of L-tryptophan at 25°C, pH 7.5, were obtained by R. Lumry, E. L. Smith, and R. R. Glantz, *J. Am. Chem. Soc. 73*, 4330 (1951):

Substrate concentration (mM)	2.5	5.0	10.0	15.0	20.0
Rate (mM s^{-1})	0.024	0.036	0.053	0.060	0.064

(a) Plot these data according to the Lineweaver-Burk method and determine values for K_M and V_{max}.
(b) Repeat the analysis using the Eadie-Hofstee method.
(c) Find the median of all 10 possible values of K_M and V_{max}.

Solution For the purpose of making these plots, the data are reformulated in the following table:

[S] (mM)	2.5	5.0	10.0	15.0	20.0
$\dfrac{1}{v_0}$ (mM^{-1} s)	41.7	27.8	18.9	16.7	15.6
$\dfrac{1}{[S]}$ (M^{-1})	400	200	100	66.7	50
$10^3 \times \dfrac{v_0}{[S]}$ (s^{-1})	9.6	7.2	5.3	4.0	3.2

(a) The Lineweaver-Burk plot is as shown.

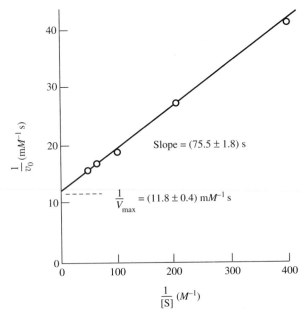

Lineweaver-Burk plot

From the slope and intercept, we obtain

$$V_{max} = (0.0847 \pm 0.0027) \text{ mM s}^{-1}$$

$$\frac{K_M}{V_{max}} = (75.5 \pm 1.8) \text{ s} \quad \text{and} \quad K_M = (6.39 \pm 0.36) \text{ mM}$$

(b) The Eadie-Hofstee plot gives

$$V_{max} = (0.0856 \pm 0.0022) \text{ mM s}^{-1}$$

$$K_M = (6.52 \pm 0.35) \text{ mM}$$

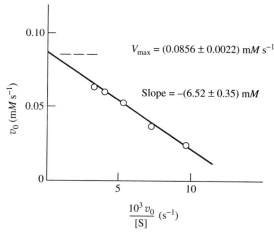

Eadie-Hofstee plot

(c) Applying Eqs. (8.14) and (8.15) to the data gives the following 10 values for K_M and for V_{max}; each set of numbers has been ordered independently. For five sets of points there are $4 + 3 + 2 + 1 = 10$ possible pairs; for N sets there are $N(N-1)/2$ possibilities.

V_{max} (mM s^{-1})	K_M (mM)
0.0720	5.0000
0.0800	5.0000
0.0808	5.2381
0.0815	5.3846
0.0840	**6.2500**
0.0857	**6.4286**
0.0864	6.7442
0.0887	7.0000
0.0900	7.5000
0.1004	8.9474

The median of each column is the average of the two numbers in the middle (in bold-face); for an odd number of values the median would be simply the center value of the ordered set. The unbiased values are

$$V_{\text{max}} = 0.0848 \text{ m}M \text{ s}^{-1}$$

$$K_M = 6.34 \text{ m}M$$

We note that the values obtained by all three methods agree well. The values are within the standard deviations calculated from the least-squares-fits to the plots, and the ranges estimated from the columns of numbers above. What is the best method to use? There is no absolute answer. Graphical methods are the most commonly used, and they make it easy to see data points which are obviously in error. However, as the use of computers continues to increase, the statistically unbiased method may become more popular.

Two Intermediate Complexes

Because most enzymatic reactions are readily reversible, it should not be surprising that there is also evidence for an enzyme-product complex. This is often kinetically distinct from the enzyme-substrate complex. The action of the enzyme fumarase illustrates this nicely.

fumarate   L-malate

The enzyme fumarase catalyzes the hydration of fumarate to form malate, and of course it must also catalyze the reverse reaction. At equilibrium, the mixture consists of about 20% fumarate and 80% malate at room temperature.

Alberty and collaborators (see Alberty and Peirce, 1957) considered the following mechanism:

$$E + F \underset{k_{-1}}{\overset{k_1}{\rightleftarrows}} EF \underset{k_{-2}}{\overset{k_2}{\rightleftarrows}} EM \underset{k_{-3}}{\overset{k_3}{\rightleftarrows}} E + M$$

There are six kinetic constants in this formalism and, as in the case of the simpler mechanism [Eqs. (8.4) and (8.5)], we cannot determine all of them by steady-state experiments. By fairly straightforward treatment similar to that of the earlier section, we can develop the following relations:

Forward reaction, $[M]_0 = 0$

$$v_F = \left(\frac{d[M]}{dt}\right)_0 = k_3[EM]$$

$$= \frac{k_2 k_3 [E]_0 [F]}{(k_2 + k_{-2} + k_3)\left\{\dfrac{k_{-1}k_{-2} + k_{-1}k_3 + k_2 k_3}{k_1(k_2 + k_{-2} + k_3)} + [F]\right\}}$$

which can be written

$$v_F = \frac{V_F[F]}{K_M^F + [F]} \tag{8.16}$$

where

$$V_F = \frac{k_2 k_3}{k_2 + k_{-2} + k_3}[E]_0 \tag{8.17}$$

and

$$K_M^F = \frac{k_{-1}k_{-2} + k_{-1}k_3 + k_2 k_3}{k_1(k_2 + k_{-2} + k_3)} \tag{8.18}$$

Despite the seeming differences, Eq. (8.16) has precisely the same form as Eq. (8.10), and there is no way to detect the presence of the additional intermediate by simple rate measurements *in steady-state experiments.* Methods like those of relaxation kinetics are needed to resolve this question.

Reverse reaction, $[F]_0 = 0$

$$v_R = \left(\frac{d[F]}{dt}\right)_0 = k_{-1}[EF] \tag{8.19}$$

which can be written

$$v_R = \frac{V_R[M]}{K_M^R + [M]} \tag{8.20}$$

where

$$V_R = \frac{k_{-1}k_{-2}}{k_{-1} + k_2 + k_{-2}}[E]_0 \tag{8.21}$$

and

$$K_M^R = \frac{k_{-1}k_{-2} + k_{-1}k_3 + k_2 k_3}{k_{-3}(k_{-1} + k_2 + k_{-2})} \tag{8.22}$$

It is also possible to formulate an expression for the net rate as

$$v = \frac{V_F K_M^R[F] - V_R K_M^F[M]}{K_M^F K_M^R + K_M^R[F] + K_M^F[M]} \tag{8.23}$$

This reduces to the previous equations in the appropriate limits. Note that when equilibrium is reached , $v = 0$ and therefore

$$V_F K_M^R[F]^{eq} - V_R K_M^F[M]^{eq} = 0$$

The equilibrium constant K for the reaction is

$$K = \frac{[M]^{eq}}{[F]^{eq}} = \frac{V_F K_M^R}{V_R K_M^F} \tag{8.24}$$

In a process involving a complex sequence of steps, such as occurs in enzymatic reactions, it is difficult to determine the detailed mechanism from steady-state kinetic studies. The reason is that only one, or a very few, of the steps in the sequence is rate determining. Changes in concentration of substrate or enzyme will produce consequences only in terms of the effect on this rate-determining step.

It is difficult, but sometimes possible, to design steady-state studies so as to extract further information about other steps. For example, some of the individual rate constants in the mechanism of the fumarase reaction were estimated from a study of the temperature and pH dependence of the steady-state rates (Brant et al., 1963). In other cases, direct measurement of an enzyme complex is possible, and this provides additional information that can be used to establish relations among the rate constants.

MECHANISMS OF ENZYMATIC REACTIONS

For any number of enzyme-substrate complexes that exist between substrates and products, the Michaelis-Menten equation still applies.

$$E + S \underset{k_{-1}}{\overset{k_1}{\rightleftarrows}} ES \overset{k_2}{\longrightarrow} ES' \overset{k_3}{\longrightarrow} ES'' \overset{k_4}{\longrightarrow} E + P$$

$$v_0 = \frac{k_{cat}[E]_0[S]}{K_M + [S]}$$

The parameters k_{cat} and K_M are functions of all the rate constants in the proposed mechanism. However, there are some generalizations that can be made for any mechanism. If there is one slow step (a rate-limiting step) in the reaction path from ES to product, the first-order rate constant for that step is equal to k_{cat}. In the simplest mechanism with only one enzyme-substrate complex, $k_{cat} = k_2$. The Michaelis constant, K_M, can always be treated as an apparent dissociation constant for all enzyme-bound species. It is equal to the true dissociation constant for dissociation of ES only if the rate constant for its dissociation, k_{-1}, is much greater than the rate constants for the forward reaction to products; then $K_M = k_{-1}/k_1$, the equilibrium constant for dissociation of the enzyme-substrate complex.

The ratio k_{cat}/K_M is called the *specificity constant*. It is important because it relates the rate of the reaction to the concentration of *free* enzyme [E] rather than total enzyme $[E]_0$.

$$v_0 = \frac{k_{cat}[E][S]}{K_M}$$

This equation allows us to write the relative rates of reaction for two competing substrates, A and B.

$$\frac{v_A}{v_B} = \frac{(k_{cat}/K_M)_A[A]}{(k_{cat}/K_M)_B[B]} \tag{8.25}$$

Equation (8.25) tells us that the ratio of rates, v_A/v_B, for two substrates is equal to the ratio of their specificity constants times the ratio of concentrations of the substrates. We see that the ratio of k_{cat} to K_M determines the selectivity (or specificity) for competing substrates. There are nearly always competing substrates for each enzyme in a biological environment. A very important example is DNA polymerase whose biological function is to incorporate the correct nucleotide at each position in the DNA strand being synthesized. The correct nucleotide is the one that is complementary to the template strand. A misincorporation, if not corrected, will lead to a mutation with possible serious results. Enzymatic studies have shown that the selectivity of a DNA polymerase (from the fruit fly *Drosophila*) for equal concentrations of substrates is about 1000 to 1 for the correct substrate (Boosalis et al., 1987). Although this is a high selectivity, it is not enough to keep an organism alive; other mechanisms—called proofreading—are necessary to produce new DNA with the required fidelity.

The great specificity and selectivity of enzymatic reactions have been well documented. A consequence of this closeness of fit between enzyme and substrate occurs in the action of inhibitors. Several classes of inhibitors are known, and studies of their behavior made a major early contribution to our knowledge of enzyme mechanisms.

Competitive Inhibition

A molecule that resembles the substrate may be able to occupy the catalytic site because of its similarity in structure, but may be nearly or completely unreactive. By occupying the active site, this molecule acts as a *competitive inhibitor* in preventing normal substrates from being examined and catalyzed. Operationally, competitive inhibitors are those that bind reversibly to the active site. The inhibition can be reversed by (1) diluting the inhibitor, or (2) adding a large excess of substrate.

Mechanistically, we can write a step

$$E + I \rightleftharpoons EI \tag{8.26}$$

in addition to the usual Michaelis-Menten formulation. Thus

$$
\begin{array}{c}
E + S \rightleftharpoons ES \rightleftharpoons E + P \\
+ \\
I \\
\updownarrow \\
EI
\end{array}
\tag{8.27}
$$

where EI is an inactive form of the enzyme. The kinetic behavior in the presence of the competitive inhibitor is seen in Figs. 8.5 and 8.6. Because the equilibria are reversible, the same maximum velocity is reached at sufficiently high substrate concentrations whether the inhibitor is present or not. But the Michaelis constant is different in the presence of the

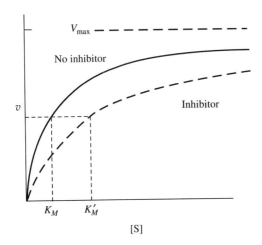

Fig. 8.5 Effect of competitive inhibitor on rate of substrate reaction: increase of Michaelis constant, but no change in maximum velocity.

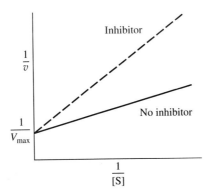

Fig. 8.6 Competitive inhibition illustrated by a Lineweaver-Burk plot. The extrapolated maximum rate, V_{max}, is not changed by a competitive inhibitor.

inhibitor; that is, the concentration of substrate required to reach one-half V_{max} is greater in the presence of competitive inhibitors than in their absence. It is not difficult to show that the new apparent Michaelis constant, K'_M, is given by

$$K'_M = K_M \left(1 + \frac{[I]}{K_I}\right) \tag{8.28}$$

where [I] is the inhibitor concentration and K_I is the dissociation constant for the enzyme-inhibitor complex. We obtained Eq. (8.28) by writing the total enzyme concentration as

$$[E]_0 = [E] + [ES] + [EI]$$

and using the definition of K_I,

$$K_I = \frac{[E][I]}{[EI]}$$

A classic example of competitive inhibition occurs in the reaction catalyzed by the enzyme succinic dehydrogenase. The normal substrate of this enzyme is succinate, the ionized form of succinic acid, and the enzyme catalyzes its oxidation to fumarate.

$$
\begin{array}{ccc}
\mathrm{COO^-} & & \mathrm{COO^-} \\
| & & | \\
\mathrm{CH_2} & \xrightarrow[\text{dehydrogenase}]{\text{succinic}} & \mathrm{CH} \\
| & & \| \\
\mathrm{CH_2} & & \mathrm{HC} \\
| & & | \\
\mathrm{COO^-} & & \mathrm{COO^-} \\
\text{succinate} & & \text{fumarate}
\end{array}
$$

Malonate (from malonic acid), which has the structure

$$
\begin{array}{c}
\mathrm{COO^-} \\
| \\
\mathrm{CH_2} \\
| \\
\mathrm{COO^-} \\
\text{malonate}
\end{array}
$$

acts as a competitive inhibitor for succinic dehydrogenase. There is a difference of only one methylene group between the structures of succinate and malonate; hence the two ions resemble one another. In this case it is clear that malonate cannot act as an alternative substrate, because there is no oxidized form of the molecule analogous to fumarate. The value of K_I for inhibition of yeast succinic dehydrogenase by malonate is 1×10^{-5} M, which means that a concentration of 1×10^{-5} M malonate decreases the apparent affinity of the enzyme for succinate by a factor of 2.

Inhibition by the product of the reaction is another common example of competitive inhibition. Again, a similarity in structure is implicit in the relation between a substrate and its product. Furthermore, inhibition by the product can serve the useful purpose of turning off (or down) an enzyme's action when it has made sufficient product for the biochemical needs of the cell. It is one of the most immediate forms of regulation or metabolic control.

Noncompetitive Inhibition

Several types of inhibition occur that cannot by overcome by large amounts of substrate. These may occur as a consequence of (1) some permanent (irreversible) modification of the active site; (2) reversible binding of the inhibitor to the enzyme, but not at the active site itself; (3) reversible binding to the enzyme-substrate complex.

The simplest form of noncompetitive inhibition occurs when only the value of $V_{\max}$ is affected, but the affinity of the enzyme for substrate, as measured by $1/K_M'$, is not affected. The kinetic behavior in this case is illustrated in Figs. 8.7 and 8.8. As an example,

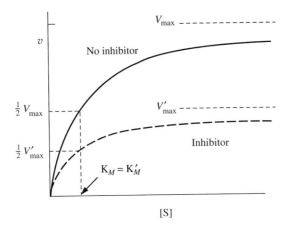

Fig. 8.7 Effect of noncompetitive inhibitor on rate of substrate reaction: decrease of maximum rate, but no change in Michaelis constant.

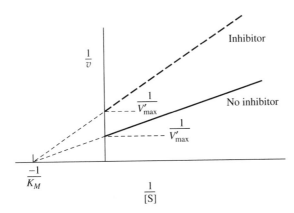

Fig. 8.8 Noncompetitive inhibition illustrated by Lineweaver-Burk plot.

consider the action of an irreversible modifier, such as the chemical alkylating agent iodoacetamide. This compound reacts with exposed sulfhydryl groups, in particular with cysteine residues of the enzyme protein, to form a covalently modified

$$\text{enzyme} - \text{S} - \text{CH}_2\text{CNH}_2$$
$$\underset{\text{O}}{\overset{\|}{}}$$

In triose phosphate dehydrogenase and in many other enzymes, the chemically modified enzyme is inactive in its catalytic role. In this case it is clear why the V_{max} decreases, because the inhibitor effectively inactivates a portion of the enzyme molecules irreversibly. At the same time, the unreacted enzyme molecules are perfectly normal, in the sense that the K_M value is unaffected by the prior addition of the inhibitor.

Other forms of noncompetitive inhibition occur in which both V_{max} and K_M are affected by the inhibitor. A type designated *uncompetitive* results in Lineweaver-Burk plots that are parallel but displaced upward in the presence of inhibitors. In addition, where the enzyme has more than one substrate, the action of an inhibitor must be evaluated with respect to each of the substrates and to the order of addition of components.

An interesting form of inhibition is known as *substrate inhibition.* In this case the rate of the reaction reaches a maximum and then declines at high substrate concentrations. The consequence in terms of the Lineweaver-Burk plot is shown in Fig. 8.9. The origin of this behavior can be that the substrate binds at a second site on the enzyme, and this site indirectly modifies the main catalytic site so as to render it less effective.

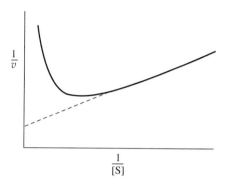

Fig. 8.9 Substrate inhibition illustrated by a Lineweaver-Burk plot.

We can interpret the term "inhibitor" in a more general sense. There are many substances that serve to promote a reaction. These may be essential cofactors, such as pyridine nucleotides or ATP, that are really second substrates but operate in cyclic fashion in the overall biochemistry of the cytoplasm. Various ions, such as H^+, OH^-, alkali metals, alkaline earths, and so on, can bind to the enzyme and cause specific and profound changes in the activity of the catalytic site. Transition metals or organic functional prosthetic groups may be involved directly in reactions catalyzed by certain enzymes. The subject of biochemistry possesses a rich and continuing history of examples of such behavior, and we can only begin to appreciate the variety that nature has incorporated into enzyme function and control.

SUMMARY

Typical Enzyme Kinetics

Rate of substrate conversion is linear in enzyme concentration:

$$v = k'[E]$$

Rate is linear in substrate concentration at low [S]; first-order reaction:

$$v = k''[S] \tag{8.1}$$

Rate approaches saturation at high [S]; zero-order reaction:

$$v = V_{max} \quad \text{(maximum rate)} \tag{8.2}$$

$$\text{catalytic constant} = k_{\text{cat}} = \frac{V_{\text{max}}}{n[\text{E}]_0} \quad (\text{in } s^{-1}) \tag{8.3}$$

$[\text{E}]_0$ = total enzyme concentration, M^{-1}
 n = number of active centers per molecule

Michaelis-Menten Mechanism

Initial rate of forward reaction:

$$\text{E} + \text{S} \underset{k_{-1}}{\overset{k_1}{\rightleftharpoons}} \text{ES}$$

$$\text{ES} \xrightarrow{k_2} \text{E} + \text{P}$$

$$v_0 = \frac{V_{\text{max}}}{1 + K_M/[\text{S}]} \tag{8.10}$$

$V_{\text{max}} = k_2[\text{E}]_0$ = maximum rate, concentration time^{-1}
$[\text{E}]_0 = [\text{E}] + [\text{ES}]$ = total enzyme concentration
$K_M = \dfrac{k_{-1} + k_2}{k_1}$ = Michaelis constant, M

Lineweaver-Burk:

$$\frac{1}{v_0} = \frac{1}{V_{\text{max}}} + \frac{K_M}{V_{\text{max}}} \cdot \frac{1}{[\text{S}]} \tag{8.11}$$

Eadie-Hofstee:

$$v_0 = -K_M \cdot \frac{v_0}{[\text{S}]} + V_{\text{max}} \tag{8.13}$$

Michaelis-Menten mechanism, including reverse reaction:

$$\text{E} + \text{S} \underset{k_{-1}}{\overset{k_1}{\rightleftharpoons}} \text{ES} \underset{k_{-2}}{\overset{k_2}{\rightleftharpoons}} \text{E} + \text{P}$$

Rate in either direction:

$$v = \frac{V_S K_M^P [\text{S}] - V_P K_M^S [\text{P}]}{K_M^S K_M^P + K_M^P [\text{S}] + K_M^S [\text{P}]} \tag{8.23}$$

V_S = maximum velocity for utilization of S
V_P = maximum velocity for utilization of P
$K_M^S = \dfrac{k_{-1} + k_2}{k_1}$
$K_M^P = \dfrac{k_{-1} + k_2}{k_{-2}}$

The relative rates of reaction for two competing substrates, A and B.

$$\frac{v_A}{v_B} = \frac{(k_{cat}/K_M)_A[A]}{(k_{cat}/K_M)_B[B]} \tag{8.25}$$

Reversible inhibition:

$$E + S \underset{k_{-1}}{\overset{k_1}{\rightleftharpoons}} ES \underset{k_{-2}}{\overset{k_2}{\rightleftharpoons}} E + P$$

$$+$$

$$I$$

$$k_{-3} \updownarrow k_3$$

$$EI$$

$$v_0 = \frac{V_S}{1 + K'_M/[S]}$$

$$K'_M = K_M \cdot (1 + [I]/K_I) \tag{8.28}$$

$$K_I = \frac{k_{-3}}{k_3} = \frac{[E][I]}{[EI]}$$

MATHEMATICS NEEDED FOR CHAPTER 8

Least-squares programs for fitting the best line to a set of points are available in most programmable calculators. To fit a line to a set of points labeled (x_i, y_i) where each x_i is an abscissa and each y_i is an ordinate, calculate

$$\text{slope} = m = \frac{n \sum_i y_i x_i - \sum_i y_i \sum_i x_i}{D}$$

$$\text{intercept} = b = \frac{\sum_i y_i \sum_i x_i^2 - \sum_i y_i x_i \sum_i x_i}{D}$$

where $D = n \sum_i x_i^2 - \left(\sum_i x_i\right)^2$

The sums are over the total number, n, of points in the set. The slope and intercept determine the line that minimizes the sum of the squares of the vertical (y_i) distances of each point from the line; this is the best least-squares line. It is always a good idea to plot this line $(y = mx + b)$ on the same graph as the points to check that the fit is reasonable.

REFERENCES

DRESSLER, D., and D. H. POTTER, 1991. *Discovering Enzymes,* Scientific American Library, W. H. Freeman, New York, is a beautiful book that not only describes the discovery of enzymes and their importance in medicine, physiology and industry, but also relates enzyme mechanism to structure.

BOYER, P. D., ed., *The Enzymes,* 3rd ed., Academic Press, New York, is a useful many-volume series.

Short papers on many different aspects of enzymology are presented in the continuing two series entitled *Advances in Enzymology* (John Wiley and Sons) and *Methods in Enzymology* (Academic Press). Volumes 63, 64 and 87 of *Methods in Enzymology* are devoted to kinetics and mechanism. They are entitled:

PURICH, D. L., ed., *Enzyme Kinetics and Mechanism*, Academic Press, New York
  Part A (1979) Initial Rate and Inhibitor Methods
  Part B (1980) Isotopic Probes and Complex Enzyme Systems
  Part C (1982) Intermediates, Stereochemistry and Rate Studies
The following books treat enzymes and enzyme kinetics in great detail.

CORNISH-BOWDEN, A., 1979. *Fundamentals of Enzyme Kinetics,* Butterworths, Boston.

CORNISH-BOWDEN, A., and C. W. WHARTON, 1988. *Enzyme Kinetics,* IRL Press, Washington, DC.

FERSHT, A., 1985. *Enzyme Structure and Mechanism,* 2nd ed., W. H. Freeman, New York.

See also the biochemistry texts listed in the References for Chapter 1.

SUGGESTED READINGS

ALBERTY, R. A., and W. H. PEIRCE, 1957. Studies of the Enzyme Fumarase: V. Calculation of Minimum and Maximum Values of Constants for the General Fumarase Mechanism, *J. Am. Chem. Soc. 79,* 1526–1530.

BOOSALIS, M. S., J. PETRUSKA, and M. F. GOODMAN, 1987. DNA Polymerase Insertion Fidelity: Gel Assay for Site-Specific Kinetics, *J. Biol. Chem. 262,* 14,689–14,696.

BRANT, D. A., L. B. BARNETT, and R. A. ALBERTY, 1963. The Temperature Dependence of the Steady State Kinetic Parameters of the Fumarase Reaction, *J. Am. Chem. Soc. 85,* 2204–2209.

BRIGGS, G. E., and J. B. S. HALDANE, 1925. A Note on the Kinetics of Enzyme Action, *Biochem. J. 19,* 338-339.

HERSCHLAG, D., and T. R. CECH, 1990. Catalysis of RNA Cleavage by the *Tetrahymena thermophila* Ribozyme. 1. Kinetic Description of the Reaction of an RNA Substrate Complementary to the Active Site, *Biochemistry 29,* 10,159-10,171; 2. Kinetic Description of the Reaction of an RNA Substrate That Forms a Mismatch at the Active Site, *Biochemistry 29,* 10,172–10,180.

LERNER, R. A., S. J. BENKOVIC, and P. G. SCHULTZ, 1991. At the Crossroads of Chemistry and Immunology: Catalytic Antibodies, *Science 252,* 659–667.

LINEWEAVER, H., and D. BURK, 1934. The Determination of Enzyme Dissociation Constants, *J. Am. Chem. Soc. 56,* 658–666.

SCHULTZ, P. G., R. A. LERNER, and S. J. BENKOVIC, 1990. Catalytic Antibodies, *Chem. Engr. News,* May 28, 26–40.

WANG, J. H., 1955. On the Detailed Mechanism of a New Type of Catalase-like Action, *J. Am. Chem. Soc. 77,* 4715–4719.

WANG, J. H., 1970. Synthetic Biochemical Models, *Acc. Chem. Res. 3,* 90–97.

ZAUG, A. J., and T. R. Cech, 1986. The Intervening Sequence RNA of *Tetrahymena* Is an Enzyme, *Science 231,* 470–475.

PROBLEMS

1. The decarboxylation of a β-keto acid catalyzed by a decarboxylation enzyme can be measured by the rate of CO_2 formation. From the initial rates given in the table, determine the Michaelis-Menten constant for the enzyme and the maximum velocity by a graphical method.

Keto acid concentration [M]	Initial velocity (μmol CO_2/2 min)
2.500	0.588
1.000	0.500
0.714	0.417
0.526	0.370
0.250	0.256

2. The hydration of CO_2,

$$CO_2 + H_2O \rightleftharpoons HCO_3^- + H^+$$

is catalyzed by the enzyme carbonic anhydrase. The steady-state kinetics of the forward (hydration) and reverse (dehydration) reactions at pH 7.1, 0.5°C, and 2×10^{-3} M phosphate buffer were studied using bovine carbonic anhydrase [H. DeVoe and G. B. Kistiakowsky, *J. Am. Chem. Soc. 83*, 274 (1961)]. Some typical results for an enzyme concentration of 2.3×10^{-9} M are:

Hydration		Dehydration	
$\dfrac{1}{v}$	$[CO_2]$	$\dfrac{1}{v}$	$[HCO_3^-]$
$(M^{-1}\ s)$	(mM)	$(M^{-1}\ s)$	(mM)
36×10^3	1.25	95×10^3	2
20×10^3	2.5	45×10^3	5
12×10^3	5	29×10^3	10
6×10^3	20	24×10^3	15

Using graph paper make suitable plots of data and determine the Michaelis constant K_M, and the rate constant, k_2, for the decomposition of the enzyme-substrate complex to form product for:

(a) The hydration reaction.
(b) The dehydration reaction.
(c) From your results, calculate the equilibrium constant for the reaction

$$CO_2 + H_2O \rightleftharpoons HCO_3^- + H^+$$

(*Hint:* Start by writing an equation that defines the equilibrium condition in terms of the Michaelis-Menten formulation of this reaction. Note that the kinetics were measured at pH 7.1.)

3. At pH 7 the measured Michaelis constant and maximum velocity for the enzymatic conversion of fumarate to L-malate,

$$\text{fumarate} + H_2O \longrightarrow \text{L-malate}$$

are 4.0×10^{-6} M and 1.3×10^3 $[E]_0$ s^{-1}, respectively, where $[E]_0$ is the total molar concentration of the enzyme. The Michaelis constant and maximum velocity for the reverse reaction are

1.0×10^{-5} M and $800[E]_0$ s^{-1}, respectively. What is the equilibrium constant for the hydration reaction? (The activity of water is set equal to unity; see Chapter 4.)

4. The enzyme pyrophosphatase catalyzes the hydrolysis of inorganic pyrophosphate $[PP_i]$ to orthophosphate $[P_i]$ according to a reaction that can be written

$$PP_i + H_2O \longrightarrow 2\,P_i$$

The standard free energy change for this reaction is $\Delta G^{0\prime}_{310} = -33.5$ kJ-(mol $PP_i)^{-1}$ at pH 7. The pyrophosphatase from *E. coli* bacteria has a molecular weight of 120,000 daltons and consists of six identical subunits. Purified enzyme has a maximum velocity, V_{max}, of 2700 units per milligram of enzyme, where a unit of activity is defined as the amount of enzyme that hydrolyzes 10 µmol of PP_i in 15 min at 37°C under standard assay conditions. As the concentration of the substrate, PP_i, is decreased, the initial velocity decreases to 1350 units per milligram of enzyme when $[PP_i] = 5.0 \times 10^{-6}$ M.
   (a) How many moles of substrate are hydrolyzed per second per milligram of enzyme at maximum velocity?
   (b) How many moles of active site are there in 1 mg of enzyme? (Assume that each subunit has one active site.)
   (c) What is the catalytic constant of each active site of the enzyme?
   (d) Estimate the Michaelis constant, K_M, for this enzyme.

5. Consider the simple Michaelis-Menten mechanism for an enzyme-catalyzed reaction,

$$E + S \underset{k_{-1}}{\overset{k_1}{\rightleftharpoons}} ES \overset{k_2}{\longrightarrow} E + P$$

   The following data were obtained:

   k_1, k_{-1} very fast
   $k_2 = 100$ s^{-1}, $K_M = 1.0 \times 10^{-4}$ M at 280 K
   $k_2 = 200$ s^{-1}, $K_M = 1.5 \times 10^{-4}$ M at 300 K

   (a) For $[S] = 0.10$ M and $[E]_0 = 1.0 \times 10^{-5}$ M, calculate the rate of formation of product at 280 K.
   (b) Calculate the activation energy for k_2.
   (c) What is the value of the equilibrium constant at 280 K for the formation of the enzyme-substrate complex ES from E and S?
   (d) What is the sign and magnitude of the standard thermodynamic enthalpy ΔH^0 for the formation of ES from E and S?

6. Assuming that a catalyst is not itself affected by temperature, do you expect the rate of the catalyzed or uncatalyzed reaction to be more sensitive to temperature changes? Explain.

7. The enzyme urease catalyzes the hydrolysis of urea to ammonia and carbon dioxide:

$$H_2O + H_2NCNH_2 \longrightarrow 2\,NH_3 + CO_2$$
$$\overset{\|}{O}$$

   The enzyme has a molecular weight of 473,000 and it contains eight independent active sites per molecule.
   (a) Temperature-dependence studies show that the value of the maximum velocity at 25°C is 2.2 times greater than it is at 10°C. The value of K_M is not significantly affected by temperature. Calculate the enthalpy of activation, $\Delta H^{\ddagger}$, for this reaction.

(b) The catalytic constant per active site is about 560 s^{-1} at 25°C. Calculate the value of $\Delta S^{\ddagger}$ for the reaction using the result of part (a). Interpret your answers to parts (a) and (b).

(c) The hydrolysis of urea also occurs in acidic solutions in the absence of any enzyme, but the rate constant is estimated to be 10^{14} times slower at 25°C. The activation energy for the acid-catalyzed reaction is 105 kJ mol^{-1}. What temperature would be required for the acid-catalyzed reaction to occur as rapidly as the enzyme-catalyzed reaction does at room temperature?

8. Consider the following mechanism for the role of an inhibitor, I, of an enzyme,

$$\text{E} + \text{S} \underset{k_{-1}}{\overset{k_1}{\rightleftharpoons}} \text{ES} \xrightarrow{k_2} \text{E} + \text{P}$$

$$\text{E} + \text{I} \underset{k_{-3}}{\overset{k_3}{\rightleftharpoons}} \text{EI}$$

The concentrations of [S] and [I] are much larger than the total enzyme concentration $[\text{E}]_0$. Derive an expression for the rate of appearance of products.

9.

$$\text{E} + \text{F} \underset{k_{-1}}{\overset{k_1}{\rightleftharpoons}} \text{EF} \underset{k_{-2}}{\overset{k_2}{\rightleftharpoons}} \text{EM} \underset{k_{-3}}{\overset{k_3}{\rightleftharpoons}} \text{E} + \text{M}$$

(a) What rate constant would you need to measure as a function of temperature to determine the $\Delta H^{\ddagger}$ for formation of EF from E + F? Formation of EM from EF? Formation of EM from E + M?

(b) What rate constants would you need to measure as a function of temperature to determine ΔH^0 for formation of EF from E + F? Formation of EM from E + M?

(c) How could you determine ΔH^0 for formation of M from F using only kinetic experiments?

10. Penicillinase is an enzyme secreted by certain bacteria to inactivate the antibiotic penicillin. A typical form of the enzyme with a molecular weight of 30,000 has a single active site, a catalytic constant of 2000 s^{-1} and a Michaelis constant, $K_M = 5.0 \times 10^{-5}$ M. In response to treatment with 5 μmol of penicillin, a 1.00-mL suspension of certain bacterial cells release 0.04 μg of the enzyme.

(a) Assuming that the enzyme equilibrates quickly with its substrate, what fraction of the enzyme will be complexed with penicillin in the early stages of the reaction?

(b) How long will it take for half of the penicillin present initially to become inactivated?

(c) What concentration of penicillin would cause the enzyme to react at half its maximum velocity, V_{max}?

(d) A second suspension of bacterial cells releases 0.06 μg/mL of the enzyme. Will this affect the answer to parts (b) or (c)? If so, by how much?

(e) An antibiotically inactive analog of penicillin serves as a competitive inhibitor to penicillinase. If the enzyme has the same affinity for inhibitor as for substrate, what concentration of inhibitor is required to decrease the rate of disappearance of penicillin fivefold at low (subsaturating) penicillin concentration?

11. The hydrolysis of sucrose by the enzyme invertase was followed by measuring the initial rate of change in polarimeter (optical rotation) readings, α, at various initial concentrations of sucrose. The reaction is inhibited reversibly by the addition of urea.

[Sucrose]$_0$ (mol L^{-1})	0.0292	0.0584	0.0876	0.117	0.175	0.234
Initial rate, $\left(\dfrac{d\alpha}{dt}\right)_0 = v_0$	0.182	0.265	0.311	0.330	0.372	0.371
Initial rate (2 M urea), v_0'	0.083	0.119	0.154	0.167	0.192	0.188

(a) Make a plot of the data in the absence of urea and determine the Michaelis constant K_M for this reaction.

(b) Carry out an analysis of the data in the presence of urea and determine whether urea is a competitive or a noncompetitive inhibitor of the enzyme for this reaction. Justify your answer.

12. The reaction of an enzyme E with a substrate S typically passes through at least two intermediates, ES and EP, and three transition states before reaching the dissociated product P:

$$E + S \longrightarrow [ES]^{\ddagger} \longrightarrow ES \longrightarrow [EZ]^{\ddagger} \longrightarrow EP \longrightarrow [EP]^{\ddagger} \longrightarrow E + P$$

From the following data, prepare a diagram of the energy of reaction of fumarate to malate catalyzed by the enzyme fumarase as a function of the reaction coordinate. (See Fig. 8.1 for an example of a simpler reaction.) Carefully label all enthalpy or energy differences for which you have information.

	ΔH (kJ)	E_a (kJ)
Fumarate + H$_2$O $\rightarrow$ malate	−15.1	
Fumarate + fumarase $\rightarrow$ fumarate-fumarase complex, ES	+17.6	
Fumarate-fumarase complex $\rightarrow$ transition state, [EZ]‡		+25.5
Malate + fumarase $\rightarrow$ malate-fumarase complex, EP	−5.0	
Malate-fumarase complex $\rightarrow$ transition state, [EZ]‡		+63.0

13. A certain enzyme has an ionizable group at the active site such that the enzyme is active only when this ionizable group is protonated:

$$E + S \rightleftharpoons ES$$
$$\Updownarrow H^+ \qquad \Updownarrow H^+$$
$$H^+E + S \rightleftharpoons H^+ES \xrightarrow{k} products + H^+E$$

Assume that the conversion of H$^+$ES to products and H$^+$E is slow, so that equilibria are established for the four reversible steps shown.

(a) What is the relation among the equilibrium constants of the four steps forming a cyclic path?

(b) Show that the enzyme obeys Michaelis-Menten kinetics, and express K_M and V_{max} in terms of the equilibrium constants of part (a) in addition to k, [H$^+$], and [E]$_0$.

(c) How does the rate depend on pH at high substrate concentrations?

14. The initial rate of ATP dephosphorylation by the enzyme myosin can be estimated from the amount of phosphate produced in 100 s, [P$_i$]$_{100}$. Use the following data and graphical methods to evaluate K_M and V_{max} for this reaction at 25°C. [Myosin]$_0$ = 0.040 g L^{-1}.

$$\text{ATP} \xrightarrow{\text{myosin}} \text{ADP} + \text{P}_i$$

[ATP]$_0$ (μM)	7.1	11	23	40	77	100
[P$_i$]$_{100}$ (μM)	2.4	3.5	5.3	6.2	6.7	7.1

Explain any curvature that you observe in the plot of the data. Justify the slope that you choose to characterize the reaction.

15. Since it was first discovered in 1968, the enzyme superoxide dismutase, SOD, has been found to be one of the most widely dispersed in biological organisms. In one form or another, its presence has been demonstrated in mammalian tissues (heart, liver, brain, etc.), blood, in invertebrates, plants, algae, and aerobic bacteria. It catalyzes the reaction

$$\text{O}_2^- + \text{O}_2^- + 2\,\text{H}^+ \xrightarrow{\text{SOD}} \text{O}_2 + \text{H}_2\text{O}_2$$

where O_2^- represents an oxygen molecule with an extra (unpaired) electron. SOD is, therefore, of great importance in detoxifying tissues from the potentially harmful O_2^-. (Several of the proteins had been known since 1933, but their enzymatic activity went unrecognized for 35 years!)

The enzyme has some unusual properties, as illustrated by the following data (for the velocity of the reaction in terms of O_2 formed) taken from a paper by Bannister et al. [*FEBS Lett. 32*, 303 (1973)]. The enzyme kinetics is independent of pH in the range 5 to 10.

$$[\text{SOD}]_0 = 0.4\ \mu M. \qquad \text{Buffer pH} = 9.1.$$

$\dfrac{1}{v_0}$ (s M^{-1})	$\dfrac{1}{[\text{O}_2^-]}$ (M^{-1})
260	13 $\times 10^4$
175	8.5 $\times 10^4$
122	6.0 $\times 10^4$
60	3.0 $\times 10^4$
10	0.50 $\times 10^4$

(a) Use graph paper to plot these data according to the Lineweaver-Burk method.
(b) What values do you obtain for V_{max} and K_M for this reaction? (Do not be surprised if they are somewhat unusual.)
(c) How can you interpret the results of part (b) in terms of a Michaelis-Menten type of mechanism?
(d) What is the kinetic order of the reaction with respect to O_2^-? Explain your answer.
(e) The reaction is independent of pH (in the region investigated) and first order in enzyme concentration. Write the simplest rate law consistent with the experimental observations. What is the value of the rate constant, including appropriate units?

Klug-Roth, Fridovich, and Rabani [*J. Am. Chem. Soc. 95*, 2768 (1973)] have proposed a ping-pong-like mechanism for this reaction, involving the steps

$$E + O_2^- \xrightarrow{k_1} E^- + O_2$$

$$E^- + O_2^- \xrightarrow[k_2]{2\,H^+} E + H_2O_2$$

where $k_2 = 2\,k_1$. (The enzyme contains copper that is oxidized in E and reduced in E^-.)

(f) Show that this mechanism is consistent with the rate data of Bannister et al.

(g) What is the ratio $[E]/[E^-]$ under steady-state conditions?

(h) Given the answer in part (e) for the "observed" rate constant, calculate k_1 and k_2.

16. RNA polymerase is thought to bind rapidly to DNA to form an initial complex. This complex then rearranges slowly to form an "open" complex, with the DNA strands now available to act as templates for the synthesis of RNA. A proposed mechanism is:

$$A + B \underset{k_{-1}}{\overset{k_1}{\rightleftharpoons}} C \qquad \left\{ \begin{array}{l} \text{(very fast, equilibrium)} \\ K_1 = k_1/k_{-1} \end{array} \right.$$

$$C \underset{k_{-2}}{\overset{k_2}{\rightleftharpoons}} D \qquad \left\{ \begin{array}{l} \text{(slow to equilibrium)} \\ K_2 = k_2/k_{-2} \end{array} \right.$$

At the beginning of the reaction 1.00 nM A and 1.00 nM B are present. At final equilibrium, the concentrations are $[\overline{A}] = [\overline{B}] = 0.60$ nM, $[\overline{C}] = 0.010$ nM, and $[\overline{D}] = 0.39$ nM. The bar over a concentration is to remind you that it is an equilibrium value.

(a) Calculate K_1 and K_2 from the equilibrium concentrations.

(b) Write the differential equations for

$$\frac{d[C]}{dt} \quad \text{and} \quad \frac{d[D]}{dt}$$

in terms of k's and concentrations.

(c) Sketch curves for [A], [B], [C], and [D] versus time on the same plot. You have been given their initial concentrations and their final (equilibrium) concentrations. The main questions to consider are which concentrations have minima or maxima, roughly where they occur in time, and what are the minimum or maximum concentrations.

(d) The value of k_2 was found to be 0.050 s^{-1}. Calculate the value of k_{-2} in s^{-1}.

17. Sodium succinate, S, is oxidized in the presence of dissolved oxygen to form sodium fumarate, F, in the presence of the enzyme succinoxidase, E. The following table gives the initial velocity, v_0, of this reaction at several succinate concentrations.

[S] (mM)	10	2.0	1.0	0.50	0.33
v_0 (μM s^{-1})	1.17	0.99	0.79	0.62	0.50

(a) Make a suitable plot of the data to test whether they follow the Michaelis-Menten mechanism. Draw the best line or curve through the data.

(b) Calculate values for the maximum velocity, V_{max}, and the Michaelis constant, K_M, for this reaction. Show how you obtain them from the plot.

(c) Malonate is a competitive inhibitor of this reaction. In an experiment with initially 1.0 mM succinate, the initial velocity is decreased by 50% in the presence of 30 mM malonate. Use this result to construct a second curve or line on the plot showing the expected kinetic behavior of this reaction in the presence of 30 mM malonate.

18. Alcoholism in humans is a disorder that has serious sociological as well as biochemical consequences. Much research is currently under way to discover the nature of the biochemical consequences to understand better how this disorder can be treated. You can analyze this problem using a variety of kinetic approaches.

Alcohol taken orally is transferred from the gastrointestinal tract (stomach, etc.) to the bloodstream by a first-order process with $t_{1/2} = 4$ min. (These and subsequent figures are subject to individual variations of ±25%.) The transport by the bloodstream to various aqueous body fluids is very rapid; thus, the ethanol becomes rapidly distributed throughout the approximately 40 L of aqueous fluids of an adult human. These fluids behave roughly like a sponge from which the alcohol must be removed.

The removal occurs in the liver, where the alcohol is oxidized in a process that follows zero-order kinetics. A typical value for this rate of removal is about 10 mL ethanol hr^{-1}, or 4×10^{-3} mol (liter of body fluid)$^{-1}$ hr^{-1}. Consumption of about 1 mol (46 g, or 60 mL) of ethanol produces a state defined as legally intoxicated. (This amount of alcohol is contained in about 4.5 oz of 80-proof liquor.) At this level, alcohol is rapidly taken up by the body fluids and only slowly removed by the liver.

In the liver, ethanol is oxidized to acetaldehyde in the presence of the enzyme liver alcohol dehydrogenase (LADH). The overall process may be represented by the equation

$$C_2H_5OH \text{ (stomach)} \xrightarrow[\text{transfer}]{k_1} C_2H_5OH \text{ (body fluids)} \xrightarrow[\substack{\text{LADH}\\\text{liver}}]{k_0} CH_3CHO \downarrow \text{acetate, etc.}$$

(a) On the basis of the data above, calculate at least three points that will enable you to construct a semiquantitative (±10%) sketch of the concentration of ethanol in body fluids as a function of time following the consumption of 60 mL of C_2H_5OH at time zero. Pay particular attention to the rising portion, the maximum level reached, and the decaying portion of the curve.

(b) What is the maximum concentration of ethanol attained in the body fluids after consuming 60 mL of C_2H_5OH? Take this to be the threshold level defining legal intoxication and draw a horizontal line at this level. Approximately how many hours are required to reduce the ethanol concentration essentially to zero? This is the recuperation time or hangover period.

(c) On the same graph add a curve showing the time dependence following initial consumption of 120 mL of ethanol. (Label the two curves "60 mL" and "120 mL," respectively.) The concentration in body fluids remains above the "legally intoxicated" level for about how many hours? The hangover period is how long?

(d) Without constructing further plots, estimate the length of time that a person would remain "legally intoxicated" if initially 180 mL of ethanol is consumed.

(e) If 60 mL of ethanol is consumed initially, what is the maximum amount that this person may drink at subsequent hourly intervals and still avoid exceeding the level of legal intoxication?

(f) It is a popular opinion that a "cocktail party" drinker is less susceptible to intoxication because the consumption of alcohol is stretched over a longer period interspersed with conversation. Using dashes, add a curve to the graph that shows the body fluid concentration for the social drinker who extends the consumption of 120 mL of ethanol over a period of 1 hr in 15-min intervals. Justify your plot and comment on the effect of the social drinker's tactic on the intoxication period.

19. The removal of ethanol in the liver involves its oxidation to acetaldehyde by nicotinamide adenine dinucleotide (NAD$^+$) catalyzed by the enzyme liver alcohol dehydrogenase (LADH). The overall reaction is

$$C_2H_5OH + NAD^+ \underset{}{\overset{LADH}{\rightleftharpoons}} CH_3CHO + NADH + H^+$$

The reaction follows a sequential or ordered mechanism wherein NAD$^+$ must bind to the enzyme before C_2H_5OH binds to form a ternary complex, and CH_3CHO and H$^+$ dissociate from the ternary complex before the NADH is released.

(a) Write a detailed set of reactions involving intermediate complexes so as to represent the mechanism of this reaction. Indicate each step as being reversible and designate the rate constants (in order) by k_1, k_2, etc., for forward reaction steps and k_{-1}, k_{-2}, etc., for reverse steps.

(b) Give a plausible explanation for the observation that the removal of ethanol from the body obeys zero-order kinetics.

(c) The rate-limiting step for the overall reaction under physiological conditions and in the presence of an intoxicating level of ethanol is the final dissociation of the LADH·NADH complex. The rate constant for this step has been measured to be 3.1 s^{-1}. Using appropriate data given in Problem 18, calculate the amount (in μmol) of the LADH enzyme present in the liver.

(d) Relaxation measurements using the temperature-jump technique were applied to the binding of NAD$^+$, at two different concentrations, to LADH in the absence of ethanol. The relaxation times measured were

$$\tau = 1.65 \times 10^{-3} \text{ s} \qquad \text{at } [NAD^+] = 1.0 \text{ m}M$$

$$\tau = 7.9 \ \times 10^{-3} \text{ s} \qquad \text{at } [NAD^+] = 0.1 \text{ m}M$$

where [LADH] $\ll$ [NAD$^+$]. Use these relaxation times to calculate the constants k_1 and k_{-1} for the reactions

$$LADH + NAD^+ \underset{k_{-1}}{\overset{k_1}{\rightleftharpoons}} [LADH \cdot NAD^+]$$

(e) By contrast with ethanol, other simple alcohols, such as methanol (CH_3OH) and ethylene glycol ($HOCH_2CH_2OH$), are highly toxic to humans. Methanol is oxidized to formaldehyde, which reacts irreversibly with proteins (it is a commonly used cross-linking agent, or fixative). Ethylene glycol is not toxic but its oxidation product oxalic acid is. Each of these alcohols is about as good as ethanol as a substrate for LADH. A common antidote for methanol or ethylene glycol poisoning is to administer ethanol in large quantities. Propose a rationale for this therapy.

20. Pyrazole has been proposed as a possible nontoxic inhibitor of LADH-catalyzed ethanol oxidation. Its kinetics was studied by Li and Theorell [*Acta Chem. Scand. 23*, 892 (1969)]. In separate series of experiments, the velocity of the reaction was measured as a function of [C_2H_5OH] or of [NAD$^+$]. Portions of their results are tabulated on the next page.

Ethanol as Variable Substrate
[LADH] = 4 μg mL^{-1},
[NAD$^+$] = 350 μM, pH 7.4, 23.5°C

$\dfrac{1}{[C_2H_5OH]}$ (M^{-1})	1/v (Relative units)	
	Control (no pyrazole)	1 × 10^{-5} M pyrazole
0.125 × 10^3	0.47	0.59
0.5 × 10^3	0.55	0.97
1.0 × 10^3	0.66	1.45
1.5 × 10^3	0.74	1.91

NAD$^+$ as Variable Cofactor
[LADH] = 4 μg mL^{-1},
[C$_2$H$_5$OH] = 5 mM, pH 7.4, 23.5°C

$\dfrac{1}{[NAD^+]}$ (M^{-1})	1/v (Relative units)	
	Control (no pyrazole)	4 × 10^{-5} M pyrazole
0.9 × 10^4	0.59	1.17
1.8 × 10^4	0.70	1.41
3.0 × 10^4	0.77	1.59
6.0 × 10^4	1.05	2.15

(a) Plot these data as Lineweaver-Burk plots on two separate graphs.

(b) What are the Michaelis constants K_M(C$_2$H$_5$OH) and K_M(NAD$^+$), in the absence of inhibitor?

(c) What type of inhibition is exhibited by pyrazole against C$_2$H$_5$OH as a substrate?

(d) What type of inhibition is exhibited against NAD$^+$ as a cofactor?

(e) With reference to your answer to Problem 19(a) suggest a mechanism for the action of this inhibitor.

(f) It has been proposed that the damaging effects of ethanol on the liver result from the pronounced decrease in the ratio [NAD$^+$]/[NADH] caused by the oxidation of ethanol. (NAD$^+$ is required for several key dehydrogenation steps in the pathway of glucose synthesis. NADH promotes the accumulation of fat in the liver cells.) Assuming that pyrazole proves to be nontoxic and that it is transported to the liver, evaluate its use as an antidote to liver damage in treating chronic alcoholism.

21. A carboxypeptidase was found to have a Michaelis constant, K$_M$, of 2.00 μM and a k_{cat} of 150 s^{-1} for its substrate, A.

(a) What is the initial rate of reaction for a substrate concentration of 5.00 μM and an enzyme concentration of 0.01 μM? Give units.

(b) The presence of 5.00 mM of a competitive inhibitor decreased the initial rate above by a factor of 2. What is the dissociation constant for the enzyme-inhibitor complex, K_I, where $K_I = [E][I]/[EI]$?

(c) A competing substrate, B, is added to the solution in part (a). Its K_M is 10.00 μM and its k_{cat} is 100 s^{-1}. Calculate the relative rates of substrate reaction for equal concentrations of substrates; that is, calculate v_B/v_A.

22. For the enzyme succinoxidase, which adds oxygen to succinate to give fumarate, the following rate data were determined for a solution in which the enzyme concentration is 10.0 μM.

[S] (mM)	0.33	0.50	1.00	2.00	10.00
v (μM s^{-1}) at 20°C	0.50	0.62	0.79	0.99	1.17
v (μM s^{-1}) at 40°C	2.41	2.99	3.81	4.78	5.65

(a) What are the values of the Michaelis constants, K_M, for this enzyme at the two temperatures?

(b) What are the values of the rate constants, k_{cat}, for this enzyme at the two temperatures?

(c) What is the activation energy for the reaction?

(d) Would you expect the activation energy to be higher or lower for the same reaction in the absence of enzyme? Explain.

23. A simple noncompetitive inhibitor of acetylcholinesterase binds to the enzyme to affect V_{max} only; it does not affect K_M. For an inhibition constant of $K_I = 2.9 \times 10^{-4}\ M$, what concentration of inhibitor is needed to give a 90% inhibition of the enzyme?

24. The catalytic activity of a ribozyme (an RNA enzyme) from *Tetrahymena thermophila* is discussed in detail by Herschlag et al. [*Biochemistry 29*, 10159 (1990)]. The ribozyme uses a guanosine cofactor (G) to cleave a ribo-oligonucleotide substrate. A simplified mechanism for the reaction is:

$$
\begin{array}{ccc}
\text{S + E} & \underset{k_{-1}}{\overset{k_1}{\rightleftharpoons}} & \text{ES} \\
+ & & + \\
\text{G} & & \text{G} \\
K_G \updownarrow & & \updownarrow K_G \\
\text{S + GE} & \underset{k_{-1}}{\overset{k_1}{\rightleftharpoons}} \text{GES} \xrightarrow{k_c} \text{EP} \xrightarrow{k_p} \text{E + P}
\end{array}
$$

At 50°C, neutral pH, and 10 mM Mg^{2+}, the following data were obtained.
Kinetics of binding of substrate S, E + S → ES, and GE + S → GES:

$$k_1 = 9 \times 10^7\ M^{-1}\ \text{min}^{-1};\ k_{-1} = 0.2\ \text{min}^{-1}$$

Binding of G is fast to equilibrium with dissociation constant, $K_G = 0.5$ mM.

$$K_G = \frac{[E][G]}{[GE]} = \frac{[ES][G]}{[GES]}$$

We note that substrate and cofactor bind independently.
Chemical step for cleavage of substrate: $k_c = 350$ min^{-1}
Release of product: $k_p = 0.06$ min^{-1}

(a) Derive an expression for the steady state concentration of GES in terms of the total concentration of enzyme, $[E_0]$, $[S]$, $[G]$, k_1, k_{-1}, k_c, and K_G. Remember that $[E_0] = [E] + [ES] + [GE] + [GES]$.

(b) In the steady state the rate of formation of product can be written as

$$\frac{d[P]}{dt} = k_P [GES]$$

Use your expression from part (a) and the values of rate constants and the equilibrium constant to calculate $d[P]/dt$ as a function of $[S]$ for a constant value of $[E_0] = 100\ \mu M$ at three different concentrations of G: $[G] = 0.05$ mM; $[G] = 0.5$ mM; $[G] = 5$ mM. Make an approximate plot of the rate vs. $[S]$ for each value of $[G]$; use enough values of $[S]$ to be able to see the trends.

(c) Describe the experiments you would need to do to measure k_1 and k_{-1} under different conditions of temperature and solvent.

25. In a study of the force-generating mechanism in muscle, Goldman, Hibberd, McCray, and Trentham [*Nature 300*, 701 (1982)] first soaked rabbit skeletal muscle fibers in a solution containing an analog of ATP, P^3-1-(2-nitro)phenylethyladenosine 5′-triphosphate, which has no effect on muscle contraction or relaxation. When the muscle fibers equilibrated with this analog are irradiated with a pulse of intense 347-nm uv light from a frequency-doubled ruby laser, photolysis of the analog occurs and ATP is suddenly generated:

It has been observed that in the presence of excess Mg^{2+}, this generation of ATP leads to relaxation of the muscle fiber according to first-order kinetics. By varying the intensity of the laser beam, different concentrations of ATP can be generated and the first-order relaxation time τ has been measured as a function of ATP concentration:

ATP (μM)	65	107	119	125	168	205	277
τ (ms)	230	59	40	56	42	23	22

(a) From the data given, obtain the best second-order rate constant for the relaxation of the muscle.

(b) Describe briefly why it is important to use a short pulse of light in this experiment.

26. Consider the enzyme-catalyzed hydrolysis of an ester which is competitively inhibited by Mg^{2+}. The following data were obtained: $K_M = 10^{-3}\ M$, $K_I = 10^{-4}\ M$, $V_{max} = 1\ M\ s^{-1}$.

(a) Estimate how long it will take for $10^{-5}\ M$ substrate to become $10^{-6}\ M$ in the absence of inhibitor.

(b) Estimate how long it will take 1 M substrate to become $10^{-1}\ M$ in the absence of inhibitor.

(c) Repeat calculations for parts (a) and (b) in the presence of $10^{-4}\ M\ Mg^{2+}$.

27. Consider the stoichiometric reaction (S → P), which is enzyme-catalyzed and is found to follow Michaelis-Menten kinetics. Its maximum velocity, V_{max}, is equal to 100 M s^{-1}. In the presence of a *competitive* inhibitor (I) the following data are obtained:

Concentration of S (M)	Concentration of I (M)	Velocity (M s^{-1})
3×10^{-3}	0	50
6×10^{-3}	1.0×10^{-4}	50

(a) What is the value of the Michaelis constant, K_M, in the absence of an inhibitor?
(b) What is the value of the dissociation constant for the enzyme-inhibitor complex, K_I?
(c) The proposed mechanism for the reaction is

$$E + S \underset{k_{-1}}{\overset{k_1}{\rightleftarrows}} ES \underset{k_{-2}}{\overset{k_2}{\rightleftarrows}} EP \underset{k_{-3}}{\overset{k_3}{\rightleftarrows}} E + P$$

Assume that there is a transition state between each reactant, intermediate, and product. They are, in order, (ES)‡, (EZ)‡, and (EP)‡.

From the following thermodynamic and kinetic data, plot energy versus reaction coordinate. Label the position of the substrate, each intermediate, and each transition state with its energy value relative to product, P, taken as the zero of energy.

Reaction	ΔH^0 (kJ)	$\Delta H^{\ddagger}$ (kJ)
S → P	−30	—
S → ES	—	+12
ES → EP	+4	+46
EP → P	−25	+75

(d) Which kinetic rate constant would you measure as a function of temperature to obtain $\Delta H^{\ddagger}$ for EP → P? What is the value of $\Delta H^{\ddagger}$ corresponding to k_{-1}?
(e) Which step in the reaction do you think is rate determining? Why?

28. You have measured the following data on the destruction of thiamine by an enzyme, both with and without 0.20×10^{-4} M inhibitor (*o*-aminobenzyl-4-methylthiazolium chloride).

[Thiamine] (mM)	Velocity (M min^{-1})	
	Uninhibited	Inhibited
0.10	0.746×10^{-6}	0.136×10^{-6}
0.25	0.992×10^{-6}	0.285×10^{-6}
0.50	1.12×10^{-6}	0.618×10^{-6}
1.00	1.26×10^{-6}	0.758×10^{-6}
2.00	1.36×10^{-6}	1.29×10^{-6}

(a) Make suitable plots and use the V_{max} and K_M values to determine the type of inhibition.
(b) Based on these results, determine the binding constant for the inhibitor.

29. Methylglyoxal, as well as many other 2-oxoaldehydes, reacts readily with gluthathione (γ-L-glutamyl-L-cysteinylglycine, denoted by HSG below) to give a hemimercaptal adduct:

$$CH_3\overset{\overset{\text{O}}{\|}}{-C}\overset{\overset{\text{O}}{\|}}{-C}-H + HSG \underset{\text{equilibrium}}{\overset{\text{fast}}{\rightleftharpoons}} CH_3\overset{\overset{\text{O}}{\|}}{-C}\overset{\overset{\text{OH}}{|}}{\underset{\underset{\text{H}}{|}}{-C}}-SG \qquad (1)$$

The enzyme glyoxalase I catalyzes the isomerization of the adduct to give S-D-lactoyl-glutathione:

$$CH_3\overset{\overset{\text{O}}{\|}}{-C}\overset{\overset{\text{OH}}{|}}{\underset{\underset{\text{H}}{|}}{-C}}-SG \underset{}{\overset{\text{glyoxalase I}}{\rightleftharpoons}} CH_3\overset{\overset{\text{OH}}{|}}{\underset{\underset{\text{H}}{|}}{-C}}\overset{\overset{\text{O}}{\|}}{-C}-SG$$

Hemimercaptal S-D-Lactoylglutathione
adduct

$$(2)$$

The forward reaction of (2) catalyzed by glyoxalase I purified from human erythrocytes is found to obey Michaelis-Menten kinetics with $K_M^F = 0.057$ mM and $(V_{max}^F/[E]_0) = 1.17 \; 10^3 \; s^{-1}$ [Sellin et al., *J. Biol. Chem.* **258**, 2091 (1983)]. The reverse reaction of (2) can be demonstrated if free glutathione (HSG) is removed quantitatively from the solution with a thiol reagent such as 5,5′-dithiobis(2-nitrobenzoate), denoted Nbs$_2$, which in turn removes the hemimercaptal adduct according to (1).

The rate v of disappearance of S-D-lactoylglutathione (S) has been measured as a function of [S] upon addition of Nbs$_2$. The double reciprocal plot shown below for data obtained at several Nbs$_2$ concentrations indicates that Nbs$_2$ is also a competitive inhibitor of the enzyme.

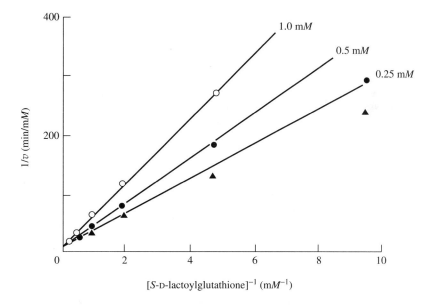

(a) Evaluate K_M^R, V_{max}^R, and K_I from the data.
(b) Calculate the equilibrium constant for reaction (2).

9

Molecular Structures and Interactions: Theory

CONCEPTS

Molecules are made of electrons and nuclei. The nuclei attract the electrons but, of course, the electrons repel each other and the nuclei repel each other. The electrons move relative to the nuclei. The balance of the forces determines the structure and the chemical reactivity of each molecule. The nuclei and electrons arrange themselves to obtain the lowest energy possible. This produces the bonding and the electron distribution of the molecule in its most stable form. Higher energies correspond to excited states of the molecule caused by molecular collisions or by photoexcitation; these states are important in the reactions of molecules. The energies of molecules (and of everything else) are quantized. Each molecule is found in a unique set of stable states with definite energy levels; a molecule does not have a continuum of energies. It is as if the molecule were on a set of steps with each step corresponding to a different energy level. The molecule is between steps only for a short time. Because the differences between energy levels depend on the type of energy (translational, rotational, vibrational, electronic), sometimes the steps are so close together that the stairway looks like a continuous ramp. This means that for many problems we can ignore quantization; classical physics applies accurately.

Quantum mechanics is necessary to describe quantized energy levels and therefore to understand bonding and electronic orbitals of atoms and molecules. However, classical physics is sufficient to understand many of the interactions between molecules. The forces between molecules in solution can be treated in terms of electrostatic and van der Waals potential energies. The effects of these forces can be used to understand noncovalent interactions such as hydrogen bonds and hydrophobic effects. Simple potential energy equations, which describe bonds as atoms connected by springs, can be added to the electrostatic and van der Waals energies to understand the conformations and flexibilities of molecules.

456

APPLICATIONS

Thermodynamic measurements can give us values for the free energy of hydrolysis of a peptide bond, or for the enthalpy of binding of a mutagen to a nucleic acid. Thermodynamic measurements describe which reactions occur and which do not. Kinetic measurements can tell us how fast the reactions occur. But in order to make sense of the data we need a basic understanding of how molecules—any molecules—interact. As all molecules are collections of electrons and nuclei, we expect that there are general rules for their interaction. The rules are deduced from the theories of quantum mechanics, classical mechanics, and electrostatics.

Proteins and nucleic acids are polypeptides and polynucleotides which have specific conformations (shapes) determined by their sequences. The precise folding of an antibody to bind specifically and tightly to any antigen, as well as the compact and similar forms of all transfer RNAs, are governed by their sequences. The interactions among the amino acids of the proteins, or the bases of the nucleic acids, control their conformations. There are further effects caused by interactions with the surrounding water molecules, with ions, and with any other molecules present in the environment. We are not yet able to predict the conformations of proteins or nucleic acids simply from their sequences, but this is the goal. At present we can make educated guesses about what will happen if we mutate an amino acid or a base. We have a logical basis for designing new proteins with improved properties as enzymes. Or we can engineer a messenger RNA with a greatly increased stability. Understanding the different contributions to intermolecular forces makes it easier to explain and to remember the consequences of any molecular reactions.

A metal ion is at the catalytic center of many enzyme-catalyzed reactions. The reactants are arranged for optimal interaction to lead to products by the arrangement of the electron orbitals of the metal. Knowing the orientations and energies of electronic orbitals for all the elements allows us to design better enzymes or better inhibitors of enzymes. We can state which metal ions should be able to replace naturally occurring ones, and which should resist oxidation better.

THE PROCESS OF VISION

One area in which quantum mechanics has played an important role involves attempts to understand the process of visual stimulation. In the eyes of mammals and other animals the visual pigment rhodopsin absorbs visible light and undergoes photochemical changes that result in electrical signals being transmitted by the optic nerves to the brain. The rhodopsin molecule is a protein, opsin, that is associated with the chromophore retinal, a polyene closely related to Vitamin A (see Fig. 9.1). In the dark-adapted retina, the pigment is present in rhodopsin as 11-*cis*-retinal; it is connected at one end by a Schiff's base link to a lysine amino acid side chain of the opsin protein. Upon illumination, one of the earliest events of the visual process is the conversion of the 11-*cis*-retinal to all-*trans*-retinal. Each photon absorbed produces a photoexcited rhodopsin which causes a phosphodiesterase to be activated. This enzyme hydrolyzes hundreds of guanosine 3′–5′ cyclic phosphate molecules (cyclic GMP). Cyclic GMP keeps sodium channels open across membranes, but

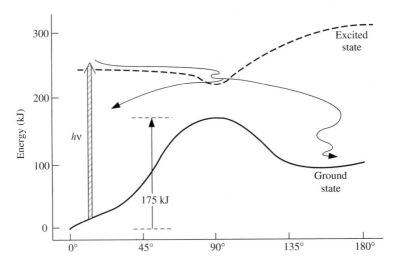

cis

H

1 7 11
2 6 H
9 10 12
3 5 8
4 14 13

H
N⁺
|
R

11-*cis*-retinal
(attached to opsin, R)

hν →

trans

H
11

12

H

N⁺ R
|
H

All-*trans*-retinal
(attached to opsin, R)

Fig. 9.1 Light-driven isomerization of 11-*cis*-retinal attached to opsin (R) by a Schiff's base linkage. Theoretical calculations of the energy of both the ground state and an excited electronic state suggest a possible path whereby absorption of a photon will lead to conversion from the *cis* to the *trans* configuration about 60% of the time. The angle of rotation about the 11,12-double bond is plotted as the horizontal axis.

Following the absorption of a photon of visible light (heavy vertical arrow at left of diagram) by the chromophore in the 11-*cis* configuration, the calculations indicate that the molecule follows a trajectory such as shown by the paths (solid curves in upper part of diagram) leading first to the minimum of the excited-state energy surface at a torsion angle of about 90°. After a few oscillations in this geometry, it continues to lose energy by returning to the ground-state energy surface. The branch in the trajectory at this point indicates that it may continue on to the all-*trans* configuration (180° torsion angle) or it may return to the original 11-*cis* form (0° torsion angle). The branching ratio is determined by quantum mechanical considerations. There is experimental evidence to support the conclusion that the entire process depicted occurs within a few picoseconds (see Fig. 7.21). (Reproduced, with permission, from R. B. Birge in Annual Review of *Biophysics and Bioengineering,* Volume 10, pp. 315–354, ©1981 by Annual Reviews.)

when the GMP is hydrolyzed the sodium channels close and the passage of millions of sodium ions is blocked. Thus the absorption of each photon is amplified many times to produce a nerve impulse (Stryer 1987, 1988).

The *cis* → *trans* isomerization caused by photoexcitation involves a significant change in the geometry of the chromophore at its attachment site to the protein; nevertheless it occurs exceedingly rapidly (less than 6 ps) and without dissociation of the link between retinal and opsin. At the same time, events are triggered that result in an electrical signal, associated with the translocation of protons across the membrane in which the rhodopsin molecule is intrinsically embedded. Quantum mechanics can contribute in several ways to understanding how this process may occur. Consider, for example, the light absorption process itself. Visual pigments occur in a variety of forms: rod cells contain rhodopsin that absorbs maximally at about 500 nm, whereas cone pigments may have absorption maxima at longer wavelengths, to 580 nm. All of these contain 11-*cis*-retinal, which absorbs light maximally only at 380 nm in the ultraviolet region as the isolated pigment. Even as the Schiff's base, it absorbs only at 450 nm in ordinary solution. Clearly, interaction with the protein opsin produces not only the red shift in absorption that makes rhodopsin exquisitely sensitive to visible light, but also the variety of absorptions in different cone cells, which is the origin of color vision. Theoretical analysis by Honig and coworkers (Nakanishi et al., 1980) indicates that the origin of these spectroscopic red shifts is electrostatic in nature. It is known that the Schiff's base nitrogen atom is protonated in rhodopsin, which places a positive charge on that region of the molecule. The compensating negative charge, which usually comes from a counterion in solution, appears to be associated with a particular negatively charged amino acid (for example, aspartate or glutamate) in the immediate vicinity of the retinal chromophore. Calculations show that the placement of a charge near the 11,12-double bond should produce the kind of red shift that is observed.

Calorimetric studies of rhodopsin photochemistry show that the photoisomerized product, bathorhodopsin (all *trans*), has 145 kJ mol^{-1} more energy than the starting rhodopsin (11-*cis*). By comparison, there is only a few kJ mol^{-1} difference for *cis/trans* isomers of analogous polyenes in solution. Where does this extra "stored" energy reside in bathorhodopsin? One proposal (Birge, 1981) suggests that it resides in increased separation of charge between the protonated Schiff's base nitrogen and negatively charged amino acid side chains of the opsin. The *cis* → *trans* isomerization produces a change in location of the Schiff's base relative to the opsin matrix to which it is bound. If this region of rhodopsin does not incorporate solvent water, then an increased charge separation of only a few ångstroms in the low-dielectric medium could result in a substantial "storage" of energy. In visual transduction the process does not stop there, of course. Other processes follow in which this stored energy is converted, ultimately to the translocation of a proton across the 40- to 50-Å thickness of the embedding membrane.

Studies of temperature effects on the early events in rhodopsin photochemistry show that there is actually an energy barrier that lies at least 175 kJ mol^{-1} above the dark-adapted (11-*cis*) state. It is most important that this barrier is there; otherwise, our eyes would have to deal with an enormous "background" signal generated by thermal energy in the environment. You would see this background even with your eyes closed. The high barrier effectively prevents appreciable triggering by thermal excitation at room temperature. Photons of visible light have energies corresponding to 200 kJ mol^{-1} or more. Thus the photon energy is more than sufficient to allow the molecule to surmount the barrier between the

11-*cis* and the all-*trans* forms (see Fig. 9.1). The "noise" level is sufficiently low that light fluxes of a few photons per second per visual cell can be discriminated from the background. Furthermore, the quantum yield of this photoisomerization is quite high. Over one-third of the photons absorbed by rhodopsin lead to bathorhodopsin. A sketch of potential energy surfaces for the ground electronic state and the first-excited electronic state of rhodopsin as a function of the angle of rotation of the chromophore about the 11,12-double bond is shown in Fig. 9.1. Also shown are some calculated "trajectories" which attempt to describe typical routes that might be taken during the photoisomerization (or the return to the original 11-*cis* form following those events that failed to lead to isomerization). Studies such as these are examples of the kinds of contributions that quantum mechanics can make to understanding important biological phenomena.

INTERMOLECULAR AND INTRAMOLECULAR FORCES

Electrostatic Energy and Coulomb's Law

The forces between all molecules or parts of molecules are ultimately described in terms of quantum mechanics, but it is often easier to use more intuitive classical methods. We start by remembering from elementary physics that a force can be related to a potential energy of interaction. A potential energy well, a decrease in potential energy, causes an attractive force; a potential energy hill is the equivalent of a repulsive force. We will therefore consider intermolecular (between different molecules) interaction energies and intramolecular (among parts of a molecule) interaction energies.

  Like charges repel and unlike charges attract; the electrostatic potential energy of interaction depends on the reciprocal of the distance between charges. This is Coulomb's law. For two charges separated by a distance r, the potential energy is

$$U(r) = \frac{q_1 q_2}{4\pi\varepsilon_0 r} \tag{9.1}$$

The charges q_1 and q_2 are in coulombs; r is in meters; ε_0 (the permittivity constant) $= 8.854 \times 10^{-12}\,C^2\,N^{-1}\,m^{-2}$; the potential energy is in joules. Because we most often deal with interactions between electrons, it is convenient to write Coulomb's law in terms of electronic charges and in terms of distances in ångstroms, Å (1 Å $= 0.1$ nm). Ångstroms are often used for molecular interactions because bond distances are 1–2 Å. Furthermore, for comparison with thermodynamic energies in kilojoules mol^{-1}, it is convenient to have the potential energy $U(r)$ in kJ mol^{-1} or kcal mol^{-1}.

$$U(r) = 1389\,\frac{q_1 q_2}{r} \qquad \text{(Energy in kJ mol}^{-1}\text{)}$$

$$\tag{9.2}$$

$$U(r) = 331.8\,\frac{q_1 q_2}{r} \qquad \text{(Energy in kcal mol}^{-1}\text{)}$$

  Here the charges q_1 and q_2 are in multiples of an electronic charge and the distance is in Å. For example, for two electrons separated by 1 Å ($q_1 = q_2 = r = 1$) the electrostatic interaction energy is $+1389$ kJ mol^{-1}. The positive sign means that the electrons repel each other.

If the two charges are separated by a medium, such as a solvent, the potential energy of interaction is decreased. Equations (9.2) must be divided by the dielectric constant (relative permittivity) ε of the medium.

$$U(r) = 1389 \frac{q_1 q_2}{\varepsilon r} \qquad \text{(Energy in kJ mol}^{-1})$$

$$U(r) = 331.8 \frac{q_1 q_2}{\varepsilon r} \qquad \text{(Energy in kcal mol}^{-1})$$

The dielectric constant of a vacuum is defined as 1.000. Dielectric constants of solvents vary from about 2 for nonpolar liquid hydrocarbons to about 80 for liquid water. Clearly, the interaction of two charges in a nonpolar membrane can be much larger than when they are surrounded by water.

For inter- and intramolecular interactions for most molecules of biological interest (such as proteins and nucleic acids) simplifying approximations must be made. Instead of trying to calculate the interactions of all electrons and all nuclei, the net charge on each atom is considered. An isolated atom is neutral, but electrons are not shared equally between the atoms in a molecule. Some atoms such as oxygen or nitrogen are usually negative; some are positive, such as hydrogen; and some are near neutral but can be positive or negative, such as carbon. The net charges on the atoms can be calculated for amide groups or phenyl rings, for example, and then used to help calculate the folding of a protein. The simplification of using net charges on atoms compared with trying to calculate the interactions of all electrons and nuclei is tremendous. There are 156 electrons plus 39 nuclei in the dipeptide sweetener Aspartame (aspartyl phenylalanine methyl ester), but only 39 atoms. As the number of interactions depends on the square of the number of particles, the saving in computer effort is large. Once there are further constraints placed on the atoms (equilibrium bond lengths and bond angles), a conformational calculation on a dipeptide becomes practical.

Example 9.1 Calculate the electrostatic potential energy of interaction for two water molecules arranged (a) favorably for hydrogen bonding (Fig. 9.2a) and (b) with the same distance between oxygen atoms, but rotated so that a hydrogen is not between the O's (Fig. 9.2b).

Fig. 9.2 Two water molecules separated by 2.500 Å between oxygen atoms. In (a) an attractive hydrogen bond is present; in (b) a repulsive orientation is shown.

Solution The O—H bond distance in H—O—H is 0.957 Å and the bond angle is 104.52°. A charge of −0.834 (in units of electronic charges) on the oxygen and

+0.417 on each hydrogen atom has been found to give reasonable calculated properties for liquid water. See Jorgensen et al., *J. Chem. Phys. 79*, 926 (1983) for further details. We calculate the electrostatic energy for two planar orientations of the water molecules at a typical $O \cdots H$—O hydrogen bond distance of 2.500 Å between oxygen atoms. In Fig. 9.2a the acceptor water molecule is symmetrical with respect to the $O \cdots H$—O line. The first step is to find x, y coordinates for the atoms so that interatomic distances can be calculated. We place the acceptor oxygen (O1) at the origin of the coordinate system. The x coordinates of its two hydrogen atoms (H1a, H1b) are at $-0.957 \cos(104.52°/2)$; the y coordinates are at $\pm 0.957 \sin(104.52°/2)$. The donor oxygen (O2) is placed at $x = 2.500$ and $y = 0.000$ and the bonding hydrogen (H2a) at $x = (2.500 - 0.957)$. The other hydrogen (H2b) is at $x = [2.500 + 0.957 \cos(180° - 104.52°)]$ and $y = -0.957 \sin(180° - 104.52°)$. The values are

	x	y
H1a	−0.586	0.757
H1b	−0.586	−0.757
O1	0.000	0.000
H2a	1.543	0.000
H2b	2.740	−0.926
O2	2.500	0.000

The distances between each atom of the donor water molecule and each atom of the acceptor water molecule are calculated by

$$r_{ij} = \sqrt{(x_j - x_i)^2 + (y_j - y_i)^2}$$

The nine distances (in Å) are

	H1a	H1b	O1
H2a	2.259	2.259	1.543
H2b	3.727	3.330	2.892
O2	3.177	3.177	2.500

The electrostatic energies are calculated using Eq. (9.2).

$$U(\text{H—H terms}) = 1389(0.417)^2 \left[\frac{1}{r(1a-2a)} + \frac{1}{r(1a-2b)} + \frac{1}{r(1b-2a)} + \frac{1}{r(1b-2b)} \right]$$

$$= +351.1 \text{ kJ mol}^{-1}$$

$$U(\text{O—H terms}) = 1389(0.417)(-0.834) \left[\frac{1}{r(O1-2a)} + \frac{1}{r(O1-2b)} + \frac{1}{r(O2-1a)} + \frac{1}{r(O2-1b)} \right]$$

$$= -784.2 \text{ kJ mol}^{-1}$$

$$U(\text{O—O terms}) = 1389(-0.834)^2 \left[\left[\frac{1}{r(O1-O2)} \right] \right]$$

$$= +386.5 \text{ kJ mol}^{-1}$$

$$U(\text{total}) = -46.6 \text{ kJ mol}^{-1}$$

There is an attractive force between the water molecules in this orientation. Bringing the two separated water molecules to this distance and this orientation decreases the electrostatic energy by 46.6 kJ mol^{-1}. This is not the hydrogen bond energy of forming water dimers because there are additional energy contributions. If only electrostatics were acting, the H atom and O atom would prefer to have zero distance between them; there must also be a repulsive contribution to their interaction. There are also other attractive forces. We will describe these shortly.

A similar calculation for the orientation of two water molecules shown in Fig. 9.2b with the oxygen atoms at the same distance, but with no hydrogen in between leads to

$$U(\text{H—H terms}) = +253.2 \text{ kJ mol}^{-1}$$

$$U(\text{O—H terms}) = -608.2 \text{ kJ mol}^{-1}$$

$$U(\text{O—O terms}) = +386.5 \text{ kJ mol}^{-1}$$

$$U(\text{total}) = +31.5 \text{ kJ mol}^{-1}$$

This arrangement produces a repulsive force on the water molecules with an unfavorable energy of 31.5 kJ mol^{-1}. Unless other forces are stronger the water molecules will not dimerize in this orientation.

This calculation illustrates that intermolecular electrostatic energy is important in determining how two molecules will approach each other. The possibility of forming a stable complex and its structure can be investigated by the simple application of Coulomb's law.

To summarize: We can calculate the electrostatic potential energy of interaction between two molecules, or between two parts of a single molecule, using Coulomb's law between net charges on each atom. We sum over all the atoms and their corresponding distances.

Net Atomic Charges and Dipole Moments

A molecule is an arrangement of nuclei surrounded by electrons; the electron distribution is characterized by the quantum mechanical wavefunction which we will define shortly. The electron distribution determines the partial charge on each atom. An atom with more electrons than the charge on its nucleus has a negative net charge; an atom with a deficiency of electrons has a positive net charge. Usually each net charge is only a fraction of an electron. The net charges are useful in providing a qualitative picture of how molecules interact. Obviously the negative region of one molecule will be attracted to the positive region of another. We gave an example of this in the calculation of the electrostatic contribution to hydrogen bonding (Example 9.1). Water is such a good solvent for ionic salts because the negative oxygen atom is strongly attracted to positive ions and the positive hydrogens are strongly attracted to negative ions. The ion-water attractions in solution are stronger than the ion-ion attractions in the crystal.

Net charges are useful in other ways. Figure 9.3 shows calculated net charges on the nucleic acid base adenine. This gives us some ideas about relative reactivity of various positions on adenine. We cannot make firm predictions because (1) the electron distribution is very approximate, (2) the solvent is important, and (3) calculations on the possible product species would need to be made. However, the charge densities are consistent with our chemical intuition. The nitrogens are all negative, so positive substituents would be attracted there. Because the three unsubstituted ring nitrogens are more negative than the amino nitrogen, we could predict that protonation occurs first on the ring. This is found experimentally. Calculation of ionization constants for each nitrogen could be made from the partial charges. One consequence of the charge distribution is easily measured and can be used as a check on the calculated electron distribution. The *dipole moment*, μ, is a measure of the polarity of a molecule. If the center of positive charge from all the nuclei is at the same position as the center of negative charge from all the electrons, the molecule has zero dipole moment; it is nonpolar. Otherwise, it is polar and it has a dipole moment. The simplest example is where two charges, $+q$ and $-q$, are separated by a distance r. In this case the dipole moment is given by $\mu = qr$. The dipole moment is a vector, because it has a magnitude (as calculated above) and a direction (from the center of positive charge to the center of negative charge). Symmetry is very helpful in calculating dipole moments. Any molecule with a point, or center, of symmetry has zero dipole moment. This includes H_2, N_2,

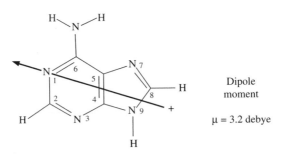

Fig. 9.3 Approximate quantum mechanical electronic wavefunctions were used to calculate the charge distribution and dipole moment of adenine. 1 debye is 10^{-18} esu cm. The charge distribution is given as the fraction of an electronic charge at each atom relative to the charge (equal to zero) on the isolated atom. (The charge on an electron is -4.803×10^{-10} esu.) [From B. Pullman and A. Pullman, *Prog. Nucleic Acid Res. 9*, 327 (1969).]

O_2, CH_4, benzene, naphthalene, and so on. For a molecule with a plane of symmetry (such as adenine), the dipole moment, if it exists, must lie in the plane. To calculate the magnitude and direction of the dipole moment for a molecule such as adenine or retinal, we need to know the locations of all the negative and positive charges in the molecule. The components of the dipole moment vector are

$$\mu_x = \sum_i q_i x_i$$

$$\mu_y = \sum_i q_i y_i \qquad (9.3)$$

$$\mu_z = \sum_i q_i z_i$$

Each q_i is the charge at position x_i, y_i, z_i. For each nucleus we know its charge and its position, so the sum over nuclei is easy to do. For the electrons we use the electronic wavefunction.

The electronic wavefunction for all of the electrons in a molecule tells us the electron density at every point in space. All we really need to know is whether a given atom in a molecule has an excess of electronic charge or a deficiency relative to the corresponding neutral atom. Then, by using the coordinates of the nuclei of the atoms, we can estimate the dipole moment. This method is illustrated in Fig. 9.3.

The magnitude of the dipole moment is

$$|\mu| = \sqrt{\mu_x^2 + \mu_y^2 + \mu_z^2}$$

The magnitude of the dipole moment can be determined from dielectric measurements on solutions containing the molecule. As we mentioned before, it provides one check on the electron distribution. The calculated dipole moment for adenine is shown in Fig. 9.3. The dimensions of dipole moment are charge times distance; chemists have traditionally used electrostatic units (esu) times centimeters. The charge on the electron is -4.803×10^{-10} esu. The magnitudes of dipole moments are thus about 10^{-18} esu cm, so a special unit has been defined in honor of Peter Debye, who first explained the difference between polar and nonpolar molecules. One debye (D) equals 10^{-18} esu cm. In SI units the electronic charge is expressed in coulombs (C) and the distance in meters. Thus 1 D $= 3.33564 \times 10^{-30}$ C m.

Example 9.2 Estimate the magnitude of the dipole moment of adenine using the calculated charge distribution shown in Fig. 9.3. The molecule is assumed to be planar, and the atom position coordinates given in the table below are taken from analysis of x-ray diffraction studies of crystals of an adenine derivative.

Solution We compute the magnitude of μ_x, μ_y, and $|\mu|$. The position of each atom of adenine is designated in a coordinate system with its origin at the H atom attached to N9 and with the H—N bond along the y axis. (For neutral molecules the calculation of dipole moments is independent of where the origin is placed.) For the net charge densities in Fig. 9.3 the contributions of both the positive and negative charges at each atom position have already been included in the values given.

Atom	q_i (electron charge)	x_i (Å)	y_i (Å)	$q_i x_i$	$q_i y_i$
N_1	-0.52	-2.791	3.937	1.451	-2.047
C_2	0.31	-3.201	2.664	-0.992	0.826
N_3	-0.51	-2.391	1.608	1.219	-0.820
C_4	0.33	-1.079	1.828	-0.356	0.603
C_5	0.12	-0.604	3.113	-0.073	0.374
C_6	0.34	-1.500	4.163	-0.510	1.415
N_7	-0.56	0.763	3.128	-0.427	-1.752
C_8	0.32	1.055	1.810	0.338	0.579
N_9	-0.19	0.000	1.000	0.000	-0.190
H_2	0.07	-4.284	2.473	-0.300	0.173
$N(H_2)$	-0.44	-1.041	5.422	0.458	-2.386
$(N)H$	0.23	-1.773	6.104	-0.408	1.404
$(N)H'$	0.23	-0.067	5.647	-0.015	1.299
H_8	0.07	2.089	1.434	0.146	0.100
H_9	0.20	0.000	0.000	0.000	0.000
	$\sum q_i = 0.00$			$\sum q_i x_i = 0.531$	$\sum q_i x_i = -0.422$

Thus

$$\mu_x = (0.531)(4.803) = 2.55 \text{ D}$$

$$\mu_y = (-0.422)(4.803) = -2.03 \text{ D}$$

$$|\mu| = \sqrt{(2.55)^2 + (-2.03)^2} = 3.26 \text{ D}$$

The calculated adenine dipole moment is 3.26 D, and the measured value in CCl_4 solvent is 3.0 D; this agreement is considered satisfactory for such calculations.

Dipole-Dipole Interactions

A further simplification in calculating the electrostatic potential energy of interaction between two molecules can be made by approximating the effect of all the charges on the two molecules by two dipoles. The charge distribution on a molecule is replaced in this approximation by a dipole—two equal and opposite charges which are very close together. The dipole has an orientation and a magnitude; it is a vector. We can thus replace a sum over all atoms (all charges) of each molecule by one term that includes the distance between the centers of the two molecules (the positions of the vectors) and their orientations. The dipole-dipole approximation is most useful when the distance between centers of the molecules is larger than the distance between any two atoms in each molecule. The approximation improves as the distance between the molecules increases.

The electrostatic potential energy of interaction between two dipoles depends on the reciprocal of the cube of the distance between them, on the angle between the two dipoles, and on the angle each dipole makes with the line joining them. As shown in Fig. 9.4, at a fixed distance two dipoles can attract each other, can repel each other, or have zero interaction energy.

If each of two interacting charge distributions (molecules or groups within a molecule) are replaced with point dipoles, then we can represent these by vectors, $\boldsymbol{\mu}_1$ and $\boldsymbol{\mu}_2$ respectively, having both magnitude and direction. Additionally, we can define a position vector $\mathbf{R}_{12}$ which is drawn from dipole 1 to dipole 2, as illustrated in Fig. 9.4. The coulombic interaction energy between these point dipoles is given in vector notation by

$$U(R) = \frac{1}{R_{12}^3} [\boldsymbol{\mu}_1 \cdot \boldsymbol{\mu}_2 - 3(\hat{\mathbf{R}}_{12} \cdot \boldsymbol{\mu}_1)(\hat{\mathbf{R}}_{12} \cdot \boldsymbol{\mu}_2)]$$

where $\hat{\mathbf{R}}_{12}$ is a unit vector (a vector with magnitude equal to 1) in the direction of $\mathbf{R}_{12}$

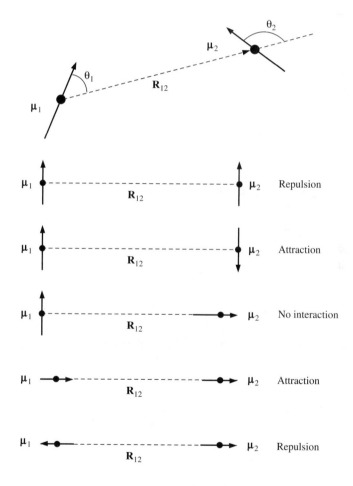

Fig. 9.4 Schematic arrangement of two point dipole moments, $\boldsymbol{\mu}_1$ and $\boldsymbol{\mu}_2$, separated by a distance $\mathbf{R}_{12}$. The interaction energy depends on the inverse cube of the distance between the dipoles, and on their relative orientation.

$$\hat{\mathbf{R}}_{12} \equiv \frac{\mathbf{R}_{12}}{|R_{12}|} \qquad \text{where } |R_{12}| \text{ is the absolute magnitude of } \mathbf{R}_{12}$$

The dot product of two vectors $\mathbf{a} \cdot \mathbf{b}$, is a scalar (a number) that is equal to the product of the magnitudes of the vectors times the cosine of the angle between them:

$$\mathbf{a} \cdot \mathbf{b} \equiv |a| \, |b| \, \cos \theta$$

Therefore, the dipole-dipole interaction energy in units of kJ mol^{-1} can be written as

$$U(R) = \frac{1389 \, \mu_1 \mu_2}{R_{12}^3} \, (\cos \theta_{12} - 3 \cos \theta_1 \cos \theta_2) \tag{9.4}$$

Here μ_1, μ_2, (eÅ) and R_{12} (Å) are the absolute magnitudes of the corresponding vectors; θ_1 and θ_2 are shown in Fig. 9.4; θ_{12} is the angle between μ_1 and μ_2. Because μ_1 and μ_2 will not, in general, lie in the same plane, one way to visualize the angle between them is to draw a vector parallel to μ_2 but placed at the position of μ_1. This new vector does lie in the same plane as μ_1, and θ_{12} is the angle between them.

We note that the dipole-dipole interaction decreases with the cube of the distance of separation between the dipoles. The interaction energy depends on the magnitudes of the two dipoles and on their orientation. Figure 9.4 illustrates some orientations that are attractive (negative interaction energy) and some that are repulsive (positive interaction energy).

Bond Stretching and Bond Angle Bending

Bond lengths and bond angles are determined by the interactions of electrons and nuclei as described quantum mechanically, but the forces which change them can be treated by simple physical models. Bond stretching and bond angle bending can be treated as if the atoms were connected by springs. The energy of moving the atoms so as to stretch or compress a bond, or to change a bond angle, depends on the square of the change in bond length, or the square of the change in bond angle.

$$U = k_r \, (r - r_{eq})^2 \tag{9.5}$$

$$U = k_b \, (\theta - \theta_{eq})^2 \tag{9.6}$$

Here $r - r_{eq}$ is the difference between the perturbed bond length and the equilibrium one, and $\theta - \theta_{eq}$ is the difference between the perturbed bond angle and the equilibrium bond angle. The constants k_r and k_b are positive which means the energy increases when the bonds are perturbed from their equilibrium positions. It takes more energy to change the bond length slightly by a chosen percent than it does to change the bond angle by the same percent. The reason for including the word slightly in the previous sentence is that Eqs. (9.5) and (9.6) only apply to small changes in the bond lengths and bond angles. Changes that would lead to bond breaking have more complicated potential energy curves.

Figures 9.5 and 9.6 plot the bond stretching and bond bending potential energies vs. the changes in bond length or bond angle for typical bonds. Stretching or compressing a bond requires a large amount of energy; note that changing the bond length of a single bond by 0.1 Å requires about 10 kJ mol^{-1}; a double bond requires about twice as much energy.

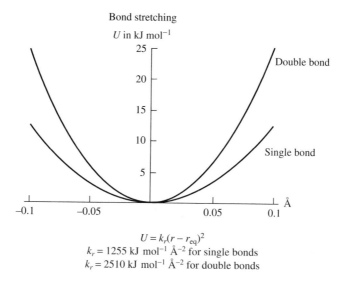

Bond stretching

U in kJ mol^{-1}

Double bond

Single bond

$$U = k_r(r - r_{eq})^2$$
$k_r = 1255$ kJ mol^{-1} Å^{-2} for single bonds
$k_r = 2510$ kJ mol^{-1} Å^{-2} for double bonds

Fig. 9.5 Potential energy curves are shown for stretching bonds away from their equilibrium bond lengths. It takes more energy to stretch a double bond than to stretch a single bond. Typical values for bond stretching force constants are shown.

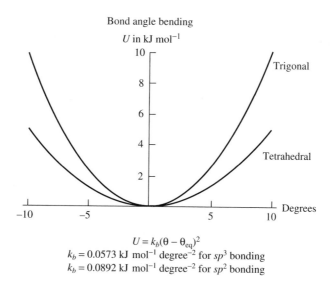

Bond angle bending

U in kJ mol^{-1}

Trigonal

Tetrahedral

$$U = k_b(\theta - \theta_{eq})^2$$
$k_b = 0.0573$ kJ mol^{-1} degree^{-2} for sp^3 bonding
$k_b = 0.0892$ kJ mol^{-1} degree^{-2} for sp^2 bonding

Fig. 9.6 Potential energy curves are shown for bending bond angles away from their equilibrium positions. It takes more energy to bend a trigonal bond with sp^2 bonding than to bend a tetrahedral bond with sp^3 bonding. Typical values for bond bending force constants are given.

Bending a bond requires much less energy than stretching or compressing it. Changing the bond angle 10° requires about 5 kJ mol^{-1} for a typical tetrahedral bond; a trigonal bond, such as the O—C—N bond of an amide requires more energy to bend. A bond angle is defined by two bonds as shown below.

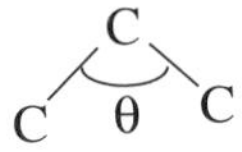

Figure 9.6 shows the energy vs. bond angle for tetrahedral and trigonal bonds.

Rotation Around Bonds

Rotation around single bonds can cause large changes in the conformation of a molecule, and the rotation does not require much energy. For example, at room temperature the benzene ring in phenylalanine rotates rapidly around the bond joining it to the rest of the molecule. This means that the energy barriers to rotation are no larger than the thermal kinetic energy, an RT value of about 2.5 kJ mol^{-1}. The energy necessary to rotate around a C—C single bond will have maxima and minima corresponding to different orientations of the substituents on the carbon atoms. Figure 9.7 illustrates the shape of the energy (called the torsion energy) curve as the angle of rotation around a C—C bond changes. For two tetrahedral carbons there are three minima corresponding to substituents being *trans* ($\phi = 180°$), *gauche* plus ($\phi = +60°$) or *gauche* minus ($\phi = -60°$). The equation corresponding to this shape is

$$U = (V_3/2)\,[1 + \cos(3\phi)] \tag{9.7}$$

The value of V_3 characterizes the barrier to internal rotation; note that at $\phi = 0°$ the energy $U = V_3$.

A torsion angle is shown below; the angle phi, ϕ, can vary from 0° to 360° (or from 0° to plus 180° and to minus 180°). Angle ϕ is defined as 0° when atoms A, B, C, and D are in the same plane and A and D are on the same side of bond B—C; bonds A—B and C—D are eclipsed.

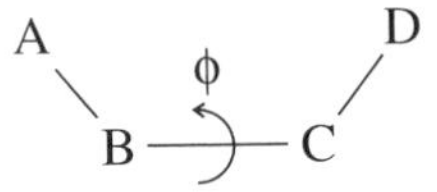

For a tetrahedral C—C bond in an aliphatic chain . . .C—CH$_2$—CH$_2$—C. . . the end carbon atoms are most stable in *gauche* or *trans* conformations as shown in Fig. 9.8.

It takes a large amount of energy to rotate around a double bond; you have to break the pi bond. For a bond with partial double bond character such as the C—N bond in formamide, the energy required to rotate around it is in between that of a double bond and a single bond. There is another difference; both the C and the N have bonding with trigonal symmetry. Thus the NH$_2$ plane will tend to be parallel to the HCO plane.

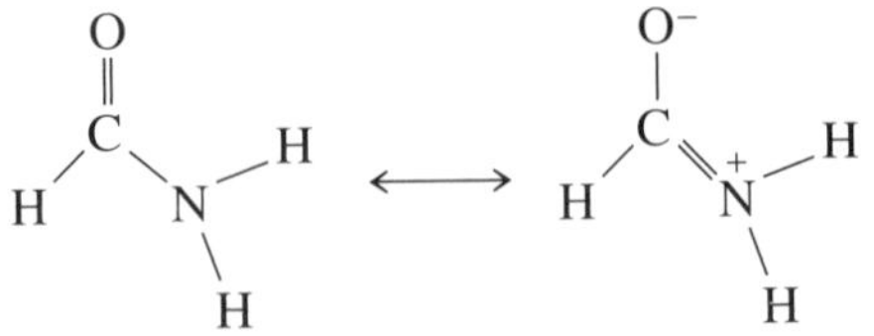

This is characterized by the equation

$$U = (V_2/2) [1 + \cos(2\phi - 180)] \tag{9.8}$$

The potential energy curve is given in Fig. 9.9. Now there are only two minima—at 0° and 180°—both corresponding to the planar molecule. For a peptide bond with one carbon atom attached to the N and another to the carbonyl carbon, steric effects make the two minima have different energies. The conformation with the attached carbons on opposite sides of the C—N bond—the *trans* conformation—is lower energy than the *cis* conformation with the carbon atoms on the same side of the C—N bond.

The values for V_3 and V_2 are representative; they depend on the substituents on the bond. The magnitudes—the barriers to internal rotation—correspond to the maxima in Figs. 9.7 and 9.9. As a torsion angle is varied the energy will change because of the barrier to internal rotation as seen in Figs. 9.7 and 9.9, but interactions from other atoms in the molecule will also contribute to the energy change. These are called nonbonded interactions.

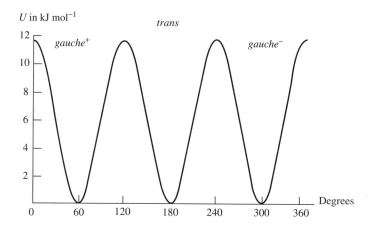

Fig. 9.7 The potential energy for rotation around a tetrahedral C—C single bond. Equation (9.7) is used with the value of $V_3 = 11.6$ kJ mol^{-1}. Note the minima in energy corresponding to *gauche$^+$*, *trans,* and *gauche$^-$* conformations as shown in Fig. 9.8.

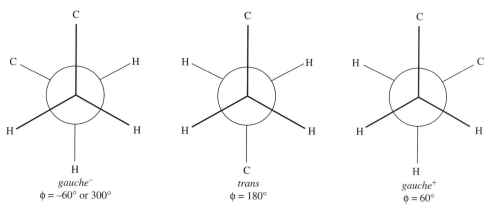

Fig. 9.8 The most stable conformations in an aliphatic C—C chain. In the picture above we are looking along the central carbon-carbon bond; these are the atoms labeled B and C in the diagram showing the torsion angle ϕ. The two end carbons are most likely to be found near $\phi = 180°$, or $\pm 60°$.

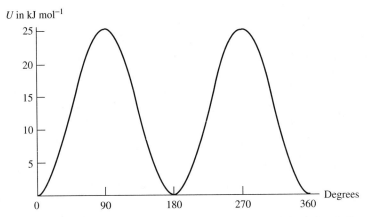

Fig. 9.9 The potential energy for rotation around a trigonal single bond. Equation (9.8) is used with $V_2 = 25$ kJ mol^{-1}.

Nonbonded Interactions

There are general interactions that occur between all molecules, and all parts of molecules. *Intra*molecular interactions occur between parts of the same molecule; *inter*molecular interactions occur between two different molecules. The forces can be attractive or repulsive depending on the properties of the groups and the distances between the interacting groups.

van der Waals Repulsion

All atoms or molecules repel at short distances. The energy of two atoms increases very rapidly if they are squeezed into the same space. The repulsion can be calculated using quantum mechanics; it follows from the fact that you cannot put more than two electrons in one electronic orbital. The repulsion increases nearly exponentially at short distances. For mathematical convenience simple models of the electron-overlap repulsion have been used. The simplest model of molecular structure is known as the "hard sphere" model for atoms. Two atoms are assumed to have zero interaction energy until they come in contact at the "hard sphere" distance; the interaction energy then goes suddenly to infinity. This is the picture that serves as a basis for the space-filling models that are designed to give a feeling for the size and shape of a molecule.

A better approximation is a repulsion that depends on the inverse twelfth power of the distance; it falls off very rapidly but not as rapidly as a hard sphere. The repulsion between any two atoms—the steric repulsion—is represented as

$$U(r) = \frac{B_{ij}}{r_{ij}^{12}} \qquad B_{ij} \text{ is positive} \tag{9.9}$$

The distance between atoms i and j is r_{ij}. The constant B_{ij} is positive and depends on both atoms; its units must be energy times (distance)^{12}.

Figure 9.10 shows the repulsion potential energy between two oxygen atoms in two water molecules. The interaction energy is essentially zero until the distance between the atoms is 3 Å; then the energy increases rapidly. Remembering from elementary physics that the force is the negative of the slope of an energy vs. distance plot, one sees from Fig. 9.10 that the repulsive force increases rapidly for distances less than 3 Å. The repulsion can be between two atoms in one molecule (*intra*molecular repulsion) or between two atoms in two different molecules (*inter*molecular repulsion). This universal repulsion of atoms and molecules at short distances is called van der Waals repulsion after the Dutch scientist who studied it experimentally. The van der Waals radius of an atom is determined from the distance between two atoms where the repulsion energy starts rising rapidly. For two oxygen atoms the repulsion energy starts rising rapidly at about 3 Å, therefore the van der Waals radius of an oxygen atom is about 1.5 Å. One way of detecting a hydrogen bond between two atoms is to find them closer than the sum of their van der Waals radii. If two O's, or two N's, or an N and an O are within 2.8 Å of each other, they are nearly always hydrogen bonded.

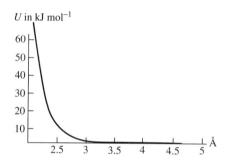

Fig. 9.10 Repulsive potential energy of interaction for two atoms. Equation (9.9) is used with $B_{ij} = 412{,}800$ kJ mol^{-1} Å^{12}.

London Attraction

All atoms attract each other if they are not too close. The attraction is caused by fluctuations in the electronic motions. An instantaneous dipole occurs in one atom which induces an oppositely directed dipole in a neighboring atom. This fluctuation dipole-induced dipole interaction energy always becomes more negative as the distance between the atoms decreases; the force is always attractive. The energy depends on the inverse sixth power of the distance between the two atoms. This universal attraction is named after Fritz London, who derived the equation explaining the effect. The theory is purely quantum mechanical; there is no classical analog. The magnitude of the interaction is proportional to the polarizability of each of the groups involved. The polarizability can be thought of as a measure of the mobility or ease of delocalization of the electrons in a molecule. The London attraction energy between two oxygen atoms in two water molecules is shown in Fig 9.11. The equation is

$$U(r) = -\frac{A_{ij}}{r_{ij}^6} \qquad A_{ij} \text{ is positive} \tag{9.10}$$

The constant A_{ij} depends on both atoms i and j; it has units of energy times (distance)^6. Values of A_{ij} and B_{ij} are given in Table 9.1.

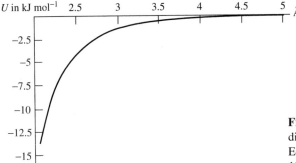

Fig. 9.11 The London fluctuation dipole-induced dipole attraction. Equation. 9.10 is used with $A_{ij} =$ 1046 kJ mol^{-1} Å^6.

Table 9.1 Parameters A_{ij} and B_{ij} for the London-van der Waals interaction energy between atoms in molecules. The values will depend on the bonding around each atom; these values are representative (only three figures are significant). The interaction energy is calculated from Eqs. (9.9) and (9.10); it is called a 6-12 potential. The last row gives the sum of the van der Waals radii for the atoms.

	H$\cdots$H	O$\cdots$O	C$\cdots$C	O$\cdots$H	C$\cdots$H	C$\cdots$O
A_{ij} (kJ mol^{-1} Å^6)	222	1048	1796	440	613	1348
B_{ij} (kJ mol^{-1} Å^{12})	21,300	413,500	1,387,000	89,200	182,300	751,400
sum of vdW radii (Å)	2.40	3.04	3.40	2.72	2.90	3.22

London-van der Waals Interaction

The attractive and repulsive energies of interaction can be summed to learn the net interaction, and the net force on the interacting parts. This is illustrated in Fig. 9.12 where the r^{-6} attraction of Fig. 9.11 is added to the r^{-12} repulsion of Fig. 9.10. The curve in Fig. 9.12 is called a 6-12 potential. Note the minimum in energy at 3 Å. If there are no other forces

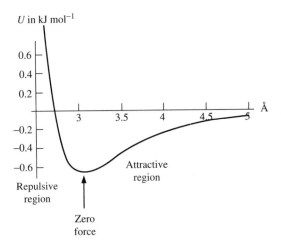

Fig. 9.12 The London-van der Waals interaction energy of two oxygen atoms in two water molecules. The curve is $U = B/r^{12} - A/r^6$; it is a 6-12 potential.

acting, the equilibrium distance between the two interacting atoms is 3 Å. If the separation distance starts to increase, an attractive force is felt. If the distance starts to decrease, a repulsion is felt. Clearly the minimum in energy is the preferred position.

The minimum in the energy curve is characterized by the sum of the *van der Waals radii* of the atoms. This is the closest the atoms approach unless there are other forces acting on them. Covalently bonded oxygens are separated by only about 1.3 Å. Hydrogen-bonded oxygen atoms can be 2.8 Å apart, but unbonded oxygens remain 3.0 Å apart. Two carbon atoms which are not bonded to each other do not approach much closer than 3.4 Å; this is the sum of their van der Waals radii.

The Lowest Energy Conformation

We have learned about the different contributions to molecular interactions. They can be divided into two types. Bonded interactions include bond stretching, bond angle bending, and torsion angle rotation. Nonbonded interactions include London-van der Waals and Coulomb interactions. Sometimes all of the nonbonded interactions are simply called van der Waals interactions. The total energy is simply the sum of all of the bonded and nonbonded interactions.

$$
\begin{aligned}
U = \; & + k_r (r - r_{eq})^2 && \text{bond stretching} \\
& + k_b (\theta - \theta_{eq})^2 && \text{bond angle bending} \\
& + V_n/2[1 + \cos(n\phi - \phi_0)] && \text{torsion angle rotation } (n = 2 \text{ or } 3) \\
& + \frac{B_{ij}}{r_{ij}^{12}} \quad B_{ij} \text{ is positive} && \text{van der Waals repulsion} \\
& - \frac{A_{ij}}{r_{ij}^{6}} \quad A_{ij} \text{ is positive} && \text{London attraction} \\
& + C \frac{q_i q_j}{r_{ij}} && \text{Coulomb interaction (C depends on the units)}
\end{aligned}
$$

$$(9.11)$$

These interactions can be used to calculate the low energy conformations of a molecule. For proteins or nucleic acids you may know an approximate conformation from experimental data such as nuclear magnetic resonance spectra or low resolution x-ray diffraction data. You vary the torsion angles and calculate the energy, searching for a minimum. You include nonbonded interactions only between atoms that are separated by more than two bonds. The interactions of atoms that are separated by only one or two bonds are already taken into account in the bond stretching and bond angle bending terms. It is easy to find a minimum energy for a conformation not very different from the starting structure; however, there may be a very different structure with a lower energy. We say that it is easy to find a local minimum, but to find the global minimum for a large molecule is very difficult. Thus it is not possible to predict the three-dimensional structure of a polypeptide with 100 residues. It is possible to calculate how its structure will change when a mutation changes one amino acid to another, or when a substrate is bound to it. Energy minimization is routinely used to improve approximate structures obtained from x-ray diffraction of crystals or nuclear magnetic resonance data of solutions.

Energy calculations can predict how a protein or nucleic acid will bind a drug, or in general how two molecules will interact. The procedure is called docking. The interaction energies are calculated as the molecules are brought near each other. The molecules are rotated and moved relative to each other to find a minimum energy orientation. To calculate the interaction energy between two molecules, you first calculate the energy of each separately. Then you calculate the energy when the two molecules are docked together. The difference is the interaction energy. A likely conformation for the complex of two molecules will minimize the interaction energy.

We have not explicitly discussed two important interactions; they are hydrogen bonding and hydrophobic interactions. A hydrogen bond is a noncovalent bond mediated by a hydrogen atom between two electronegative atoms such as O and N. Its contribution to the energy of a structure can be calculated as a sum of nonbonded interactions, but with parameters different from non-hydrogen-bonded atoms. Experimental values of hydrogen-bond energies are discussed below.

The interactions of biological molecules occur in aqueous solution, so the effect of water must be taken into account. Nonpolar molecules tend to be insoluble in water and thus lower the free energy of a mixture by avoiding the water. This means that the transfer of a nonpolar molecule from a nonpolar environment to an aqueous environment raises the free energy. This hydrophobic (literally water-fearing) effect helps determine how proteins fold, how they fit into membranes, and how lipids interact in water. The interactions of water molecules with themselves and with the nonpolar molecules involve the nonbonded interactions discussed previously; they are not new types of interactions. However, they are interactions that involve many water molecules; more macroscopic descriptions of the interaction are useful.

At this point the reader will begin to understand the complexity of trying to calculate the overall interaction energy of a large molecule, or even its minimum-energy configuration. Nevertheless, efforts have been made to do just this for nucleic acids and proteins with the aid of much calculation time on fast computers. Descriptions of these procedures can be found in Weiner et al. (1984) and Némethy and Scheraga (1990). An even more difficult calculation is involved in trying to understand the stability of condensed-phase structures that are not covalently linked, such as phospholipid bilayers, micelles, and even the liquid phase itself.

Hydrogen Bonds

A hydrogen atom placed so as to interact simultaneously with two other atoms is said to form a hydrogen bond. Hydrogen bonds are extremely important in influencing the structure and chemistry of most biological molecules. The hydrogen bonds of water give rise to unique physical properties that make it ideal as a medium for life processes.

To be capable of forming hydrogen bonds, the H atom in a molecule, HA, must be somewhat acidic. The acidic hydrogen will interact with an electron donor species, B, to form the hydrogen bond

$$B + HA \longrightarrow B \cdots HA$$

Hydrogen-bond donors include strong acids (HC1), weak acids (H_2O), and even molecules containing acidic C—H bonds ($HCCl_3$). Hydrogens attached to atoms of low electronegativity (CH_4) do not form hydrogen bonds. The best hydrogen-bond acceptors are atoms of the most electronegative elements (F, O, N). The bond energies of some typical hydrogen bonds are given in Table 9.2. Others are listed in Table 3.1.

Because solvents such as water can serve as both hydrogen-bond donors and acceptors, the enthalpies listed in Table 9.2 are strongly solvent dependent. The values given for amide or for the nucleoside base pairs are measured in relatively nonpolar solvents ($HCCl_3$ or CCl_4). Measured in water, the ΔH^0 for the amide hydrogen bond is zero, which means

Table 9.2 Bond enthalpies for some hydrogen bonds*

	ΔH^0 (kJ mol^{-1})		ΔH^0 (kJ mol^{-1})
Water	21	Alcohol	10
Acids	58(=2 × 29)	Amides	15
Nucleic acid base pairs			
G·C	40–48	U·A	26

*The hydrogen-bond enthalpies are given for water, carboxylic acids, alcohols, and amides in the gas phase. The nucleic acid base pairs were measured in deuteriochloroform at 25°C. In aqueous solution the net hydrogen-bond strengths would be much weaker, because of competition with water.

that there is essentially no stabilization of amide intermolecular hydrogen bonding relative to the solvation of individual amides by water. This is not true in the case of nucleoside base pairs, where intramolecular interactions may involve two (U·A) or three (G·C) hydrogen bonds per base pair. In polynucleotides as well as in proteins there are cooperative interactions involving the formation of many hydrogen bonds per molecule. These are supported, in the case of the polynucleotides, by significant stacking interactions (van der Waals) of the aromatic bases in helical structures.

Even in liquid water there are extensive networks of clustered molecules that interact cooperatively by hydrogen bonding. As a consequence, water has physical properties that are anomalous considering the related molecules H_2S, H_2Se, and H_2Te formed from atoms in the same column of the periodic table. These anomalous properties of water include a high boiling point, high viscosity, and a high entropy of vaporization. This latter, in particular, arises from the necessity of breaking up the ordered clusters of molecules in liquid water to form the vapor.

Hydrogen bonds are directional. The maximum stability (lowest energy) occurs when the three atoms A—H···B all lie on a straight line. Bent hydrogen bonds occur, but they usually have decreased stability. The directionality of hydrogen bonds is important for determining the crystal structure of ice. Each water molecule is capable of donating two H atoms to form hydrogen bonds and each O atom is capable of accepting two hydrogen bonds. To maximize the number of hydrogen bonds in ice, each O atom is surrounded by four other O atoms placed tetrahedrally with respect to the central atom. Between each pair of O atoms lies an H atom involved in hydrogen bonding. This is quite compatible with the molecular structure of water, whose H—O—H bond angle of 105° is close to the tetrahedral angle, 109.5°. Thus each H atom will lie nearly on the line joining the adjacent O atoms. This results in a relatively open, cagelike structure that accounts for the unusual property that the molar volume of ice is about 10% larger than that of liquid water at the freezing point. The stability of the hydrogen-bonded structure prevents the water molecules in ice from forming a more compact structure; when pressure in excess of 2000 atm is applied to ice, however, the ice does undergo a phase transition to a structure with a lower molecular volume.

Molecules that are apolar will not interact strongly with water. Nevertheless, some apolar molecules can form stable crystalline hydrates. These are found to consist of a cagelike shell of ordered water molecules that surrounds the apolar molecule. The stability of such water clusters can be appreciated by noting that the enthalpies of hydration of molecules such as Ar, Kr, CH_4, C_2H_6, etc., are all about 60 kJ mol^{-1}. When other, more complex molecules expose an apolar group, such as an aromatic amino acid side chain of a protein, to solvent water, a complete cage of solvent water cannot form. Nevertheless, there may be appreciable stabilization energy that results from smaller, transient water clusters stabilized by their mutual hydrogen bonding at the interface with the apolar residue. This is thought to make an important contribution to hydrophobic interactions, discussed in the next section.

The experimental criterion which establishes that a hydrogen bond exists between A and B (A, B = N or O) is that the A····B distance in A—H···B is about 2.8 to 3.0 Å. Without a hydrogen bond, that distance is greater than 3.0 Å. A quantum mechanical description of a hydrogen bond involves the electrostatic attraction of the positively charged hydrogen nucleus of the donor to a negatively charged nonbonding orbital on the acceptor.

This attraction of one proton to two nuclei can provide an efficient path for moving charges; the proton can move from one nucleus to another.

$$O—H\cdots O \longrightarrow O\cdots H—O$$

The very high mobility of H^+ in water is explained by this mechanism.

In biological systems the most important hydrogen bonds occur with the lone-pair electrons of O and N, in whatever molecules these are available. The strongest hydrogen bonds occur with F, but fluorine is not commonly encountered biologically. Carbon and elements in the next row in the periodic table (e.g., S or Cl) form very weak hydrogen bonds, if at all.

Hydrophobic and Hydrophilic Environments

The terms *hydrophobic* (water fearing) and *hydrophilic* (water loving) are often encountered in the biochemical literature. The concepts involved are obviously very important, but their description and origins are complicated. Excellent detailed analyses are given by Tanford (1980) and Muller (1990).

Because water is the medium in which biological organisms have developed, the interactions of biomolecules with water are very important. To understand the distinctions between terms like hydrophobic and hydrophilic, we need to recognize that there is always a competition between the interactions of molecules of two substances, for instance hexane and water, with one another and their interactions with molecules of the same type. Water molecules are *hydrophilic,* in the sense that there are strong interactions among the water molecules in either pure water or aqueous solutions. Hexane and other nonpolar substances also contain molecules that interact with one another by nonbonding interactions of the *van der Waals* type. It is a mistake to describe such interactions as resulting from "hydrophobic forces," because they occur completely and naturally in the absence of any water whatsoever.

When samples of the two liquids, water and hexane, are brought together, they do not mix. In fact, a small amount of each component is dissolved in the bulk phase of the other at equilibrium, although the two liquids form separate phases with an interface between them. The reason they do not mix is that the interactions of water molecules with other water molecules are more favorable (lower ΔG, incorporating contributions both from ΔH and $-T\Delta S$) than are the interactions of water with hexane molecules. A similar situation occurs for the interactions of the hexane molecules. Another way of looking at this is that the transfer of additional water from the liquid water phase to the liquid hexane phase results in an increase in the chemical potential, μ. Because this is characteristic of an unfavorable process, it will not happen spontaneously.

Many biomolecules have both hydrophilic and hydrophobic character in the same molecule. Such molecules are said to be amphiphilic; examples include phospholipids, many amino acids, nucleotide bases, and even certain proteins and mucopolysaccharides. Detergents are common synthetic examples. In some amphiphiles one end of the molecule is hydrophilic and the other end is hydrophobic. Such molecules are good surfactants, and we have seen in Chapter 5 some examples of their ability to concentrate at the interface between a lipid or hydrocarbon phase and an aqueous phase. We have also seen that such molecules play important roles in forming biological membranes.

An important characteristic of proteins is the way in which the hydrophilic or hydrophobic amino acid side chains are distributed along the polypeptide backbone. The specific sequence of amino acids, or course, is what distinguishes one protein from another. For water-soluble, globular proteins of known structure it is seen that hydrophobic amino acids are typically buried inside the protein, whereas hydrophilic amino acids tend to be at the surface of the protein, where they interact with solvent water. A quantitative description of this phenomenon has been developed in the form of hydropathy plots (Kyte and Doolittle, 1982), where a numerical index of the hydrophilic/hydrophobic character of short protein segments is plotted along the chain length and compared with the location of those segments in the native protein structure. A hydropathy plot for lactate dehydrogenase from dogfish is shown in Fig. 9.13. In other applications it can be seen that intrinsic membrane proteins contain extensive stretches of 20 or more hydrophobic amino acids in regions where the proteins pass through the interior part of the membrane lipid bilayer. An isolated polar or charged amino acid side chain in such a hydrophobic stretch may be paired with an oppositely charged residue on a neighboring chain, or it may serve as a specific ligand to an associated prosthetic group such as heme or chlorophyll.

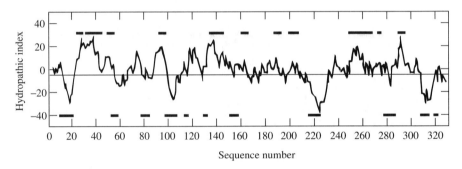

Fig. 9.13 Hydropathy profile of the enzyme dogfish lactate dehydrogenase. The hydropathy index, which is positive for hydrophobic amino acid side chains and negative for hydrophilic, is averaged over nine adjacent amino acids and plotted at the sequence number of the central amino acid. Bars at the top and bottom of the figure indicate those regions of the protein known to be on the inside and outside, respectively, based on the known x-ray crystallographic structure of the protein. A reasonably good correlation between the hydropathy index and peptide location is found. (Adapted from Kyte and Doolittle, 1982, *J. Mol. Biol. 170*, 723.)

It is essential to maintain a balanced view of the interplay of hydrophilic and hydrophobic concepts. One should not picture water as being entirely excluded from a hydrophobic environment, for example. The solubility of water in liquid hydrocarbons is sufficient that there is a rapid exchange across the interface between the small amount of dissolved water and that in the bulk phase. In fact, most biological membranes are quite freely permeable to water, despite the fact that the membrane has a hydrocarbon-like interior. The permeability to charged ions and to large hydrophilic solute molecules is very much less. The rapid movement of water across membranes is an important aspect of *homeostasis,* whereby individual cells and biological organisms respond to biochemical stress in a way that does not lead to a serious and potentially disruptive imbalance of forces.

MOLECULAR DYNAMICS SIMULATION

We can improve our understanding of the structures and interactions of large molecules by simulating their motions on a computer. The free energy of the system characterizes which reactions occur and which species or conformations are present at equilibrium. To calculate a Gibbs free energy it is necessary to know both the energy (actually enthalpy) and the entropy. We have given formulas for the energy terms [Eq. 9.11)]. The entropy requires knowledge of how many ways the system can change without affecting the energy; the more rearrangements that are possible for the system, the higher the entropy and the lower the free energy. There are two methods for calculating entropy and free energy for a system by simulating the motion of the molecules: *Monte Carlo* and *molecular dynamics*. Both methods calculate the energies for different possible arrangements of the system. The free energy depends on the *average* of $e^{-U/RT}$ averaged over the different arrangements of the system.

$$G = -RT \ln \langle e^{-U/RT} \rangle$$

where $\langle \rangle$ represents the average over the possible states of the system. The important concept is that the averaging includes the contribution of the entropy to the free energy. We can understand this by considering two molecules, both of which have the same minimum energy, U. For one of them any change in conformation raises the energy greatly; the molecule is rigid. For the other, changes in conformation do not raise the energy much; it is a flexible molecule. Clearly the rigid molecule will have a higher average value of $\langle e^{-U/kT} \rangle$ and thus a higher free energy than the flexible molecule. We describe the flexible molecule with the lower free energy as having a higher entropy.

Monte Carlo

The Monte Carlo method is named after the famous gambling city in Europe; an equivalent name could be the Las Vegas method. In this method you start with an arbitrary arrangement of the system and perturb it by slightly moving one molecule, or one part of a large molecule. After each move you calculate the change in energy. If the energy has decreased you allow the move; if the energy has increased you compare the increase with the thermal energy, kT, to decide whether to accept the move or not. You compare the value of $e^{-(\Delta U/kT)}$ with a random number between 0 and 1; here ΔU is the increase in energy per particle. If $e^{-(\Delta U/kT)}$ is larger than the random number, you allow the move; otherwise you try another move. The randomness is where the gambling comes in. It is obvious that this procedure will make the system approach a minimum energy arrangement. Perturbations that lower the energy are always allowed, but perturbations that raise the energy are sometimes allowed. Changes in the system that raise the energy are necessary to prevent it from getting trapped in a local energy minimum. By repeating this procedure many, many times you will eventually find the lowest *free energy* for the system rather than the lowest energy. The number of perturbations that do not increase the system energy near the minimum is a measure of the entropy. This allows calculation of a free energy.

Molecular Dynamics

The molecular dynamics method directly simulates the motions of all the molecules in the system. Newton's law is used for each molecule. The force on each atom of a molecule is calculated from the potential energy using Eq. (9.11); the force depends on the derivatives of the potential energy along different directions in space. The motion of each atom is then calculated from Newton's law:

$$\textbf{Force} = (\text{mass}) \cdot \textbf{(acceleration)}$$

Both **force** and **acceleration** are vectors with a magnitude and a direction. Therefore, they will determine how fast and in what direction each atom in a molecule will move. The velocities and thus average kinetic energy of the molecules in the system determine the temperature, so the motion can be simulated at any temperature. The system is started in an arbitrary arrangement at a temperature near absolute zero; the atoms are nearly stationary. The velocities of the atoms are then allowed to increase so that the average kinetic energy of the system is increased to correspond to the temperature of interest, such as 300 K. Now the motion of all the atoms in the system is simulated at that temperature. The system could be a protein or a nucleic acid fragment surrounded by water molecules. A great deal of computer power is needed to simulate the motion even for a few nanoseconds, because Newton's law must be applied to each atom every one or two femtoseconds to give realistic behavior. However, the simulation allows you to learn how the shape of the molecule fluctuates, and how the molecule and the solvent rearrange to facilitate binding of a substrate (Karplus and Petsko, 1990). The molecular motions as well as the free energy depend on the potential energy, so it is not surprising that analysis of the motions of all the molecules can provide the free energy of the system.

Only differences in free energies can be measured experimentally, so it is important to be able to calculate these differences. For example, the free energy change of unfolding a protein could be calculated; this would involve changes only in conformation—rotation about single bonds. The binding of a substrate to an enzyme can be calculated, but it is easier to calculate the difference in free energies of binding two substrates, S_1 and S_2. The following thermodynamic cycle is used.

$$
\begin{array}{ccc}
E + S_1 & \xrightarrow{\Delta G_1} & ES_1 \\
\Big\downarrow \Delta G_3 & & \Big\downarrow \Delta G_4 \\
E + S_2 & \xrightarrow{\Delta G_2} & ES_2
\end{array}
$$

The difference in standard binding free energies, $\Delta\Delta G^\circ$, can be obtained experimentally by measuring the equilibrium constants.

$$\Delta\Delta G^0 \text{(experimental)} = \Delta G_2^0 - \Delta G_1^0 = -RT \ln K_2/K_1$$

where K_2 and K_1 are the equilibrium binding constants. The difference can be calculated theoretically by simulating the conversion of free substrate S_1 to S_2 (ΔG_3), and the conversion of S_1 to S_2 bound to the enzyme (ΔG_4).

$$\Delta\Delta G \text{(calculated)} = \Delta G_4 - \Delta G_3$$

The free energy of conversion is simulated by continuously varying, in the calculation, one atom into another. For example, to calculate the difference in free energy of binding of formate ion vs. acetate ion, requires the conversion of a methyl group to a hydrogen atom. In the calculation the parameters that specify the interactions of the methyl with its surroundings are changed slightly towards that of a hydrogen atom. After each slight change enough molecular dynamics is done so that the slight change in free energy can be calculated. Thus the total free energy change, ΔG, for the conversion of a methyl to a hydrogen is calculated as the sum of the small changes in free energy caused by each perturbation in parameters. The change in free energy is independent of path; this is true even if a path is chosen that is not experimentally possible. Thus the free energy difference of binding formate vs. acetate in an enzyme site is calculated by: (1) changing formate into acetate surrounded by solvent—this is ΔG_3; (2) changing formate into acetate surrounded by the enzyme binding site—this is ΔG_4; (3) subtracting (1) from (2). Free energies of chemical reactions can also be calculated by this method. This is done by simulating the conversion of one molecule into another (described by Bash et al., 1987).

These types of calculations have already provided some quantitative predictive power (McCammon, 1987; Jorgensen, 1989; Kollman and Merz, 1990). However their most important contribution is to provide a qualitative understanding of how reactions and conformations depend on molecular interactions. We can learn to predict qualitatively how a change in a substrate, or a protein or nucleic acid, will affect the reaction.

QUANTUM MECHANICAL CALCULATIONS

We have described how to calculate the conformations of molecules—the shapes of molecules that depend on relatively weak interactions. By rotating torsion angles and by small perturbations of bond angles and bond lengths the most stable conformation is attained. But how do we understand the forces that determine the bond lengths and bond angles? We must solve the Schrödinger equation for the interactions of nuclei and electrons that make up the molecule. By solving the Schrödinger equation for the three nuclei and ten electrons that constitute H_2O we can calculate the lowest energy H—O—H bond angle and O—H bond length. This provides the equilibrium geometry of a water molecule. We can also calculate the energy for any bond length and bond angle to learn how distortion of a water molecule increases its energy. In fact we can calculate any property of a water molecule, often to greater accuracy than can be measured experimentally. The electron distribution in the lowest energy state (the ground state) of water can be obtained, as well as electron distributions in excited states. Excited states are important for understanding the chemical reactivity of water, and its spectroscopic properties like absorption of light.

In principle we can calculate any property of any molecule using quantum mechanics. In practice the ability to obtain useful results from these *a priori* calculations decreases rapidly as the number of electrons increases. Calculations on a nucleic acid base pair of 30 nuclei and 136 electrons are possible but extremely difficult to do accurately. That is why we use the more empirical methods described earlier to investigate the conformations of proteins or nucleic acids. However, we can directly apply quantum mechanical methods to parts of a large molecule, such as the retinal in rhodopsin, the metal-containing catalytic site of an enzyme, or the special chlorophyll in the reaction center of the photosynthetic apparatus.

Biologists should know at least a minimum of quantum mechanics so they can feel confident in using the many spectroscopic methods. Spectra do not make sense without quantum mechanics. It is also important to know the vocabulary of bonding, orbitals, electron distribution, and charge densities to understand reaction mechanisms and molecular interactions. These concepts are based on relatively simple quantum mechanical descriptions of molecules. The goal of molecular biologists is to explain biological functions in terms of molecular interactions. The goal of quantum biologists, or submolecular biologists, is to explain molecular interactions in terms of electron distribution and motion. In this chapter we present the fundamental aspects of quantum mechanics that provide a basis for this powerful method.

Electrons as Waves

In the 1920s it became clear that for many purposes an electron (and any other particle, for that matter) is better understood as a wave phenomenon. One of the most direct experimental verifications of this wavelike character was the observation of electron diffraction by Davisson and Germer in 1927. They projected a fine beam of electrons onto a single crystal of nickel metal and observed a diffraction pattern of concentric rings around the transmitted beam. Such diffraction patterns were well known from studies of light, sound, x rays, ripples on the surface of a liquid, and other phenomena. From their measurements, Davisson and Germer were able to calculate a wavelength for the electron, and the wavelength λ turned out to be inversely proportional to the momentum (mv) as predicted two years earlier by de Broglie. The *de Broglie relation* is stated as

$$\lambda = \frac{h}{mv}$$

where h is Planck's constant, (6.626×10^{-34} J s), m is the rest mass of the electron (9.11×10^{-31} kg), and v is its velocity (m s^{-1}). The wavelength thus has units of meters. Since the velocity is determined by the electric potential used to accelerate the electron, it is easy to change v by changing the electric potential and to observe the related change in the spacing of the diffraction pattern. Subsequent experiments showed that other particles, such as protons and neutrons, also exhibit wavelike behavior and fit the de Broglie relation when the appropriate rest masses are used. Because these are much heavier than the electron, their characteristic wavelengths are much shorter for the same velocity. The nuclei of atoms are thus particles with wavelengths that are typically very much shorter than those of the lighter and faster electrons.

Wave Mechanics and Wavefunctions

Once the wave nature of matter was recognized, it became possible for Heisenberg and Schrödinger, independently, to formulate mathematical descriptions of the electron wave motion. The Schrödinger formulation, which is the more familiar of the two, assigns an amplitude ψ to the electron wave; ψ is known as the *wavefunction* of the system. For classical wave motion, the wavefunction is simply the displacement of the system from its equilibrium position; for example, the height of waves on the surface of a lake. For electron wavefunctions there is no simple interpretation of the wavefunction itself, but ψ^2 characterizes the distribution of the electron in space.

In general the wavefunction of a system is a function both of the space coordinates and of time. However, we shall consider only time-independent wavefunctions in this section. Once we know the wavefunction for a molecule, we can calculate all the properties of the molecule. To illustrate this, consider a possible molecule, HeH^{2+} (Fig. 9.14). This molecule has one electron, so its electronic wavefunction is a function of the x, y, z coordinates of the electron.

$$\psi = \psi(x, y, z)$$

The method of obtaining wavefunctions will be illustrated in several of the following sections of this chapter. For the moment, we consider the He and H nuclei fixed, and we are interested in the electron distribution around the nuclei. We will see that for each position of the two nuclei there is a set of time-independent states of the electron characterized by a definite energy, E_n, and wavefunction, ψ_n. The lowest energy and the electron distribution corresponding to it is the ground state. There are excited states corresponding to higher energies and to different electron distributions.

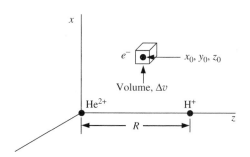

Fig. 9.14 Diatomic molecular HeH^{2+} with a given internuclear distance, R. The probability of finding the electron in a volume, Δv, around point x_0, y_0, z_0 is just the square of the electronic wavefunction multiplied by the volume. HeH^{2+} is a thermodynamically stable molecular ion with an equilibrium internuclear distance of 0.774 Å and bond dissociation energy of 155 kJ.

Because there is only one electron in this molecule, it is fairly easy to obtain $\psi_n(x, y, z)$ for many internuclear distances. That is, we obtain ψ_n as a function of possible bond lengths. What properties can we now calculate?

The *electron density or distribution* is given by $[\psi_n(x, y, z)]^2$. For a chosen internuclear distance and a particular ground or excited state, the probability of finding the electron in a small volume of size Δv around point x_0, y_0, z_0 is

$$[\psi_n(x_0, y_0, z_0)]^2 \, \Delta v$$

The probability of finding the electron in a finite volume of space is obtained by integration over the volume:

$$\int [\psi_n(x, y, z)]^2 \, dv \tag{9.12}$$

We use one integral sign to represent integration over all variables and dv to represent all the coordinates of integration. The probability of finding the electron in all of space is equal to 1. This is called the *normalization condition*. It states that the electron must be found somewhere; the probability of finding it somewhere is unity.

The electron distribution ψ_n^2 directly determines the scattering of x rays by the molecule. The electron distribution, of course, also determines the chemical reactivity of the molecule, but this is harder to quantify.

The *average position* of the electron can be calculated from the wavefunction. As a diatomic molecule is symmetric around the bond, the average position of the electron will be on the z axis, the internuclear axis. The average position for the positive charges of the nuclei is $R/3$, where R is the internuclear distance measured from He. The average position of negative charge (the electron) may change markedly upon excitation from the ground to the excited state. This was first predicted and then found to be the case for retinal. Evidence about the excited state of retinal is obtained from spectroscopic measurements, such as the effect of an intense electric field on the absorption of light (Mathies and Stryer, 1976). A large shift in charge density was observed for the Schiff base of retinal; the positive charge is moved from the N^+ toward the six-membered ring. In this way the photon energy has been changed to an electrical signal. The hope is that further experimental and theoretical investigations based on quantum mechanics will eventually help us to understand the transmission of the signal across the retinal rod membrane and along the optic nerve to the brain.

The *energy* of the electron in each electronic state can be calculated from the corresponding wavefunction. The energy of the electron is of vital importance because it tells us if the molecule is stable and what its stable bond distance is. We can calculate the electronic energy as a function of internuclear distance. This energy will decrease (become more negative) as the internuclear distance decreases. The repulsion between the positively charged nuclei, which is easily calculated from Coulomb's law for two point charges, will increase as the internuclear distance decreases. If there is a minimum in the total (electron plus nuclear) energy at some internuclear distance, a stable molecule can exist (Fig. 9.15). The energy versus distance can also be calculated for excited electronic states. This tells us when excitation (by light, for example) will cause dissociation of the molecule. If dissociation does not occur, how does the bond distance change on electronic excitation? The position of the minimum in energy for the excited electronic state tells us the stable bond distance in the excited state. We can see how electronic excitation in a molecule such as retinal leads to a change in conformation.

Calculation and measurement of energies of molecules in their ground and excited states involves all of spectroscopy; we will save detailed discussions of this for Chapter 10.

We have treated a simple, one-electron, diatomic molecule to make the notation simpler, but the same principles apply to more complicated molecules. For a two-electron molecule, the electronic wavefunction depends on the positions of both electrons.

$$\psi = \psi(x_1, y_1, z_1, x_2, y_2, z_2)$$

The subscripts refer to electrons 1 and 2. The equations for calculating electron distribution, position, energy, and so on, are the same, but we now must integrate over six coordinates instead of three. It is clear why many-electron molecules quickly begin to tax present-day computers. Adenine (Fig. 9.3) has 70 electrons, so we have electronic wavefunctions

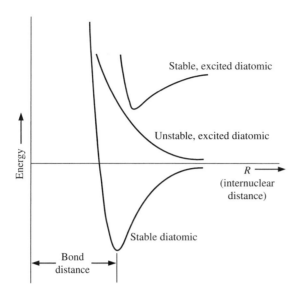

Energy

Stable, excited diatomic

Unstable, excited diatomic

R ⟶
(internuclear
distance)

Stable diatomic

Bond
distance

Fig. 9.15 Plot of calculated energy against distance, R, between nuclei in a diatomic molecule.

that depend on 210 coordinates. Therefore, at present we need to make approximations to calculate useful energies, bond distances, electron densities, and so on. A useful approximation for many applications is to consider only the π electrons. This approximation is adequate for treating the ultraviolet spectrum of adenine, for example.

For the larger molecules encountered by the biochemist or biologist, it is necessary to make approximations to tackle the problems at all. Among the most useful approximations are those that correspond to our chemical intuition.

1. Molecules can be thought of as collections of atoms or groups (such as sulfate, ammonium) whose internal structure (nuclei, electrons) need not be known in detail.
2. Electrons may be localized primarily on individual atoms, they may be shared between pairs of atoms, or they may be delocalized over a large portion of the molecule.
3. The motion of electrons is very rapid in comparison with that of the much heavier nuclei. As a consequence, one may assume that changes in electronic structure or distribution (caused, for example, by photon absorption) will occur with the nuclear arrangement essentially frozen.

Schrödinger's Equation

Schrödinger's equation is the differential equation whose solutions give the wavefunctions. From the wavefunctions, measurable properties can be calculated. These properties are functions of the positions and momenta of the electrons and nuclei in the molecule. Schrödinger (and Heisenberg) deduced the form of *operators,* which operate on a wavefunction to give a measurable property. We use operators constantly in mathematics. For example, $\ln x$, $\sqrt{x}$, dx, and $3x$ are all examples of operators operating on x. The *position operator* in quantum mechanics is simply the coordinate itself. For example, to find the average position of the electron in the molecule HeH^{2+} (Fig. 9.14), we use*

$$\bar{z} = \int \psi_n z \psi_n \, dv$$

Here z operates on ψ_n simply by multiplying ψ_n. The quantity $\bar{z}$ is known as the *expectation value* or average value of z. If ψ_n is the wavefunction of an electron, for example, then $\bar{z}$ gives us the *average* position of the electron along the z axis. Expectation values can be calculated similarly for the momentum, the energy, the dipole moment, and many other useful properties that can be compared directly with experimental physical measurements. In each case the evaluation involves integrating the product of operator and wavefunction over all space—at least, everywhere that the product has a significant value.

The momentum operator in quantum mechanics is slightly more complicated. In classical mechanics the momentum is mass times velocity; in the x direction it is $p_x = mv_x$. In quantum mechanics the *momentum operator* in the x direction is

$$p_x = \frac{\hbar}{i} \frac{d}{dx}$$

$$\hbar \equiv \frac{h}{2\pi} = 1.055 \times 10^{-34} \text{ J s}$$

$$i \equiv \sqrt{-1}, \text{ imaginary} \qquad i^2 = -1$$

The quantum mechanical momentum operator operates on ψ and involves taking the derivative of ψ with respect to a space coordinate and multiplying by $\hbar/i$.

The Schrödinger equation is simply an equation describing the conservation of energy for a system written in terms of its wavefunction, ψ. In the operator notation that we have developed, and for a single particle of mass m moving in one dimension, this is

$$\text{kinetic energy } + \text{ potential energy } = \text{ total energy}$$

$$T_x \psi \qquad + \qquad U(x)\psi \qquad = \qquad E\psi$$

$$\frac{p_x^2}{2m}\psi \qquad + \qquad U(x)\psi \qquad = \qquad E\psi$$

$$-\frac{\hbar^2}{2m}\frac{d^2\psi}{dx^2} \qquad + \qquad U(x)\psi \qquad = \qquad E\psi$$

$$\left[-\frac{\hbar^2}{2m}\frac{d^2}{dx^2} \qquad + \qquad U(x) \right]\psi \qquad = \qquad E\psi$$

$$\mathcal{H}\psi \qquad = \qquad E\psi$$

$$(9.13)$$

---

* We shall consider only real wavefunctions here. In general, the equation must be written as $\int \psi_n^* z \psi_n \, dv$, where ψ_n^* is the complex conjugate of ψ_n. This means that wherever the imaginary, i, appears in ψ_n it is replaced by $(-i)$ in ψ_n^*.

The operator $\mathcal{H}$, which operates on the wavefunction to give the energy, is called the *Hamiltonian*. T_x is the kinetic energy operator; $U(x)$ is the potential energy operator; E is the total energy operator. Note that

$$T_x = \tfrac{1}{2} m v_x^2 = \frac{p_x^2}{2m}$$

is the expression for the kinetic energy in classical mechanics for a particle of mass m moving with a velocity v_x.

When E is constant, Eq. (9.13) is the time-independent Schrödinger equation in one dimension for a particle of mass m. It is a differential equation whose solutions give the wavefunctions, ψ_n, and energies, E_n, for the particle in its ground and excited states. To emphasize this, we can write the Schrödinger equation as

$$\mathcal{H}\psi_n = E_n\psi_n$$

$$-\frac{\hbar^2}{2m}\frac{d^2\psi_n}{dx^2} + U(x)\psi_n = E_n\psi_n \tag{9.14}$$

The subscript n is the *quantum number* for the wavefunction or *eigenfunction*, ψ_n, and the *energy* or *eigenvalue*, E_n. It is a central postulate of wave mechanics that all of the measurable information about a system is contained in its wavefunction. Before considering applications of the Schrödinger equation to simple problems, let us first note that we are not restricted to working in one dimension or to using an x, y, z-coordinate system. For example, the Schrödinger equation written for a particle in three coordinates in the x, y, z system is

$$-\frac{\hbar^2}{2m}\left(\frac{\partial^2\psi}{\partial x^2} + \frac{\partial^2\psi}{\partial y^2} + \frac{\partial^2\psi}{\partial z^2}\right) + U(x, y, z)\psi = E_{n_x,n_y,n_z}\psi \tag{9.15}$$

where $\psi = \psi(x, y, z)$. There are three quantum numbers, n_x, n_y, n_z, associated with the system. Many problems involving central symmetry, such as a hydrogen atom, are best solved using spherical coordinates, r, θ, ϕ.

Other coordinate systems are also useful. To write the Schrödinger equation in a form independent of the coordinate system, the Laplace operator is introduced. The *Laplace operator*, ∇^2, indicates the operation of taking second derivatives with respect to all coordinates. In Cartesian coordinates, it has the form

$$\nabla^2 = \frac{\partial^2}{\partial x^2} + \frac{\partial^2}{\partial y^2} + \frac{\partial^2}{\partial z^2}$$

Using this operator we can write the time-independent Schrödinger equation as

$$-\frac{\hbar^2}{2m}\nabla^2\psi + U\psi = E\psi \tag{9.16}$$

or, even more compactly, as

$$\mathcal{H}\psi = E\psi \tag{9.13}$$

where $\mathcal{H}$ is the Hamiltonian operator

$$\mathcal{H} = -\frac{\hbar^2}{2m} \nabla^2 + U \qquad (9.17)$$

representing the sum of kinetic and potential energy. For more than one particle in the system, we sum over all particles:

$$\mathcal{H} = -\frac{\hbar^2}{2} \sum_i \frac{\nabla_i^2}{m_i} + U$$

Solving Wave Mechanical Problems

An understanding of the basic elements and procedures of wave mechanics can be of great value in appreciating the contribution of theoretical chemistry to problems of biological interest. It also serves as a basis for understanding the origins of molecular spectroscopy and the kinds of information that can be obtained from spectroscopic studies. We shall choose a few simple examples to illustrate the approach. The examples include a particle in a potential well, an electron in a central force field, a harmonic oscillator, and molecular orbitals constructed as linear combinations of atomic orbitals. We shall then point out what approximations are needed to approach the real molecular situation. The simple examples are more than mere exercises; insights derived from them will be used to describe and interpret important aspects of biomolecular structure and spectra.

First, let us examine the general strategy used in solving real problems in wave mechanics. We shall consider only stationary states of molecules, so that time will not enter explicitly as a variable. If the system does change with time (by the absorption or emission of radiation), we can still describe the initial and final states in this stationary approximation. The dynamics (time course) of the changes can be treated wave mechanically also, but the calculations are mathematically more involved than we need to consider here.

Outline of Wave Mechanical Procedures

I. Definition of the problem

A. *Write the appropriate potential for the particular problem.* Although the kinetic-energy operator has the same form for all problems, the potential-energy operator distinguishes each problem.

1. Particle in a box: We consider a particle that is contained in a box but is otherwise free. The potential energy is zero inside the box but infinite at the walls and outside the box. For a one-dimensional box,

$$U(x) = 0 \qquad \text{for } 0 < x < a$$

$$U(x) = \infty \qquad \text{for } x \leq 0 \quad \text{and} \quad x \geq a$$

Think of a bead on a wire capped at each end. If there is no friction between the bead and wire, the bead is free to move along the wire between the caps.

2. Harmonic oscillator: A harmonic oscillator represents a particle attached to a spring. The potential energy increases whenever the spring is stretched or compressed. This problem is easy to solve and it approximately represents the vibrations of nuclei of a molecule:

$$U(x) = \tfrac{1}{2} kx^2 \quad k = \text{force constant}$$

3. Coulomb's law: This is the dominant potential involving the interactions between nuclei and electrons in molecules; it is the only one we will consider. Other much weaker interactions depend on the spin of the electrons and nuclei. For two charges separated by a distance r, the potential energy is

$$U(r) = \frac{q_1 q_2}{r}$$

For charges q_1 and q_2 in electrostatic units (esu) and r in centimeters, the potential energy is in ergs. This is a common form used by chemists and biologists. It is consistent with using m in grams and $\hbar = 1.055 \times 10^{-27}$ erg s in the kinetic-energy operator. With SI units the potential energy is

$$U(r) = \frac{q_1 q_2}{4\pi\varepsilon_0 r} \tag{9.1}$$

Here the charges are in coulombs, r is in meters, ε_0 (the permittivity constant) = 8.854×10^{-12} C^2 N^{-1} m^{-2}, and the energy is in joules. It is consistent with using m in kilograms and $\hbar = 1.055 \times 10^{-34}$ J s in the kinetic-energy operator.

We can now write the potential-energy operator for any atom or molecule. For a hydrogen atom, in cgs units,

$$U(r) = -\frac{e^2}{r}$$

where e = electronic charge in esu and r is the distance between the proton and the electron in cm. For a helium ion, He$^+$,

$$U(r) = -\frac{2e^2}{r}$$

For a helium atom,

$$U(r_1, r_2) = -\frac{2e^2}{r_1} - \frac{2e^2}{r_2} + \frac{e^2}{r_{12}}$$

The coordinates r_1 and r_2 refer to electrons 1 and 2 in the helium atom; r_{12} is the absolute (positive) distance between the electrons. The three terms represent, respectively, the electron-nuclear attraction for each electron and the electron-electron repulsion. For a molecule we can easily write the potential energy just by adding terms corresponding to each electron-nucleus attraction, each electron-electron repulsion, and each nucleus-nucleus repulsion.

B. *Establish the boundary conditions for the wavefunction.* The wavefunctions are solutions to the Schrödinger equation, but in addition they must satisfy some auxiliary conditions. Examples of these are:

1. The wavefunction should be single-valued and finite everywhere.
2. The wavefunction of a bound electron, molecule, and so on, should vanish at infinite distance and in any region where the potential is plus infinity.
3. The wavefunction should be continuous and have a continuous first derivative, except where the potential energy becomes plus or minus infinity, such as at the walls of the box or at the nucleus of an atom, respectively.

II. Writing the Schrödinger equation for the problem

Write down the Schrödinger equation in a suitable coordinate system. Some examples are:

1. Particle in a box (one dimension):

$$-\frac{\hbar^2}{2m}\frac{d^2\psi_n(x)}{dx^2} = E_n\psi_n(x) \qquad \text{for } 0 < x < a$$

2. Harmonic oscillator (one dimension) of reduced mass μ:

$$-\frac{\hbar^2}{2\mu}\frac{d^2\psi_v(x)}{dx^2} + \tfrac{1}{2}kx^2\psi_v(x) = E_v\psi_v(x)$$

3. Hydrogen atom:

$$-\frac{\hbar^2}{2\mu}\nabla^2\psi_{n,\,l,\,m}(r,\,\theta,\,\phi) - \frac{e^2}{r}\psi_{n,\,l,\,m}(r,\,\theta,\,\phi) = E_n\psi_{n,\,l,\,m}(r,\,\theta,\,\phi)$$

In spherical polar coordinates, where r is the radius, θ is a colatitude angle, and ϕ is a longitude, the Laplace operator is given by

$$\nabla^2 = \frac{1}{r^2}\frac{\partial}{\partial r}\left(r^2\frac{\partial}{\partial r}\right) + \frac{1}{r^2\sin^2\theta}\frac{\partial^2}{\partial\phi^2} + \frac{1}{r^2\sin\theta}\frac{\partial}{\partial\theta}\left(\sin\theta\frac{\partial}{\partial\theta}\right)$$

We have introduced a new symbol, μ, to represent the reduced mass. In the Schrödinger equations for the harmonic oscillator and hydrogen atom we are interested in relative motions of particles. For example, in the hydrogen atom we are solving for the wavefunction of the electron relative to the nucleus; we ignore the motion of the hydrogen atom through space. However, both the electron and the nucleus move relative to the center of mass of the hydrogen atom. Because of this, we must use the reduced mass of the two particles in the Schrödinger equation. The reduced mass, μ, for two particles is

$$\frac{1}{\mu} = \frac{1}{m_1} + \frac{1}{m_2} \tag{9.18}$$

Because the proton is nearly 2000 times more massive than the electron, for a hydrogen atom the reduced mass is nearly equal to the mass of an electron.

III. Solving for the eigenfunctions and for the eigenvalues of energy

Solve the Schrödinger equation using standard methods of solving differential equations. Obtain the set of wavefunctions, ψ_n, and eigenvalues, E_n, that satisfy the Schrödinger equation and the boundary conditions. We will simply present some of the earlier solutions that were obtained. It is relatively simple to convince yourself that they are indeed valid; simply substitute the solutions into the Schrödinger equation and show that they satisfy it exactly.

IV. Interpretation of the wavefunctions

The interpretation of the solutions to the Schrödinger equation involves comparison with available experimental data on the system. A great variety of properties can be tested. Some of the most important of these are:

1. Energies (eigenvalues)—comparison with values obtained from spectroscopy, ionization potentials, electron affinities, and so on.
2. Electron distribution (eigenfunctions)—comparison with atomic and molecular dimensions, bond dissociation energies, dipole moments, probabilities of transition from one state to another, directional character of bonding, intermolecular forces.
3. Systematic changes for different, but related systems—comparison of spectra, ionization potentials, bond lengths, bond energies, force constants, dipole moments, and so on, for atoms or molecules that differ from one another in a simple manner.

PARTICLE IN A BOX

The particle-in-a-box problem is perhaps the simplest of any that can be solved using the wave equation. In this problem the *particle,* which may be an electron, a nucleus, or even a baseball, is considered to move freely ($U = 0$) within a defined region of space, but is prohibited ($U = \infty$) from appearing outside that region. The problem can be solved readily in one, two, or three dimensions and for any shape of rectangular box.

Some biologically relevant applications of this problem include: the behavior of π electrons that are delocalized over large portions of molecules, as in linear polyenes (carotenoids, retinal), planar porphyrins (heme, chlorophyll), and large aromatic hydrocarbons; conduction electrons that may move over extensive regions of biopolymers; and the exchange of protons involved in hydrogen bonding between two nucleophilic atoms, as between the oxygens of adjacent water molecules. Even in cases where the real potential well is not infinite and square, the particle-in-a-box calculation is a useful approximation for obtaining the order of magnitude of the energies involved.

The potential energy for a particle in a one-dimensional box,

$$U = 0 \qquad 0 < x < a$$
$$U = +\infty \quad \begin{cases} x \leq 0 \\ x \geq a \end{cases}$$

is shown diagrammatically in Fig. 9.16. The solution of the Schrödinger equation outside the box is trivial; because the particle cannot exist there, its wavefunction must be zero

everywhere outside the box. Within the box, it must satisfy the one-dimensional Schrödinger equation for $U \equiv 0$:

$$-\frac{\hbar^2}{2m} \frac{d^2\psi_n(x)}{dx^2} = E_n\psi_n(x) \qquad (9.19)$$

Rather than present the rigorous solution to this second-order differential equation by conventional methods, which may be unfamiliar to the reader, let us look at the requirements
for a wavefunction that solves Eq. (9.19). It must be a function of x such that its second derivative, $d^2\psi/dx^2$, is equal, apart from constant factors, to the original function, ψ. Two such
functions are $\sin bx$ and $\cos bx$, since

$$\frac{d^2}{dx^2} (\sin bx) = -b^2 \sin bx$$

and

$$\frac{d^2}{dx^2} (\cos bx) = -b^2 \cos bx$$

In fact, a general solution can be written in the form

$$\psi = A \sin bx + B \cos bx$$

We can immediately simplify the solution by applying the continuity condition (I.B.3 in the
previous section). We know that the wavefunction must be zero at the boundary of the box
(since it is zero everywhere outside); therefore, $\psi(0) = \psi(a) = 0$. When $x = 0$, $\sin bx = 0$, but
$\cos bx = 1$. Therefore, we must set $B = 0$ in the equation above to satisfy the boundary condition at $x = 0$. Thus

$$\psi = A \sin bx$$

To satisfy the boundary condition at $x = a$, $\sin bx$ must $= 0$ when $x = a$. We know that
$\sin n\pi = 0$ for any integer, n; therefore it is easy to see that the function

$$\psi_n = A \sin \frac{n\pi}{a} x \qquad n = 1, 2, 3, 4, \dots$$

is the only one that satisfies the second part of the boundary conditions as well. Here a is
the length of the box and n can be any positive integer; it is the quantum number for the
problem. Note that the solution for $n = 0$ requires that $\psi_0 = 0$, which is not meaningful because it is associated with zero probability for finding the particle in the box. We obtain A
by using the normalization condition:

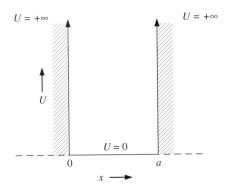

Fig. 9.16 Square potential well of
length a.

$$A^2 \int_0^a \sin^2 \frac{n\pi x}{a}\, dx \ = \ 1$$

Using a table of integrals, we obtain $A = \sqrt{2/a}$. We now have the wavefunction for a particle in a box in its ground ($n = 1$) and excited ($n = 2, 3, 4, \ldots$) states:

$$\psi_n \ = \ \sqrt{\frac{2}{a}}\, \sin \frac{n\pi x}{a} \qquad\qquad (9.20)$$

The solutions to any Schrödinger equation—the eigenfunctions—have the following properties: they are *orthogonal* and it is convenient to *normalize* them. Two functions are orthogonal when the integral of their product is zero. For a particle in a box it is easy to show that any two states are orthogonal,

$$\int_0^a \psi_n \psi_m\, dx \ = \ 0 \qquad (n \neq m)$$

and that all states are normalized.

$$\int_0^a \psi_n^2\, dx \ = \ 1 \qquad (\text{any } n)$$

The correct wavefunctions for any two different states of a molecule are orthogonal: the orthogonality condition is often used as a criterion to test approximate wavefunctions.

Figure 9.17 shows the behavior of ψ versus x for several values of n. (In each case, the negative of the ψ shown is also a valid solution.) The analogy of a standing wave for the classical vibrating string is worth noting here. The violin string or a rope fixed at both ends can be set into motion as a standing wave with nodes (points where the string has zero displacement) only at the ends, or with additional nodes between the ends. These are called the *fundamental mode* ($n = 1$), the *first harmonic* or *overtone* ($n = 2$), the *second overtone* ($n = 3$), and so on. The requirements that the ends are fixed and undergo no displacement

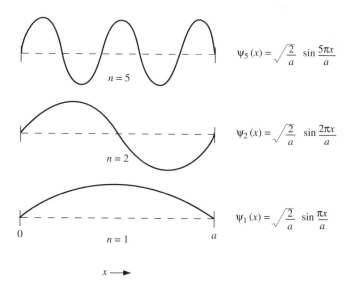

Fig. 9.17 Variation of $\psi_n(x)$ with x for $n = 1$, 2, and 5. The magnitude of $\psi_n(x)$ is plotted versus x. The extreme values of $\psi_n(x)$ are $\pm \sqrt{2/a}$.

Molecular Structures and Interactions: Theory   **495**

are the boundary conditions that limit standing waves to these particular (integral) modes. The properties of the harmonic progression arise naturally from the constraints on the system, just as the quantum numbers arise naturally in wave mechanics from the boundary conditions characteristic of the particular problem.

Let us turn now to the energies associated with the particle-in-a-box problem. Having determined the eigenfunctions [Eq. (9.20)] that constitute the solutions to the Schrödinger equation for this problem, we proceed to substitute them into Eq. (9.19). Thus

$$-\frac{\hbar^2}{2m}\frac{d^2}{dx^2}\left(A \sin \frac{n\pi x}{a}\right) = E_n\left(A \sin \frac{n\pi x}{a}\right)$$

$$\frac{\hbar^2}{2m}\frac{n^2\pi^2}{a^2}\left(A \sin \frac{n\pi x}{a}\right) = E_n\left(A \sin \frac{n\pi x}{a}\right)$$

from which we can readily extract the eigenvalue solutions

$$E_n = \frac{\hbar^2\pi^2 n^2}{2ma^2} = \frac{h^2 n^2}{8ma^2} \qquad n = 1, 2, 3, \ldots \qquad (9.21)$$

Each eigenfunction has an associated energy or eigenvalue, determined in part by the quantum number, n. In fact, the energy values increase with the square of the quantum number, as illustrated in Fig. 9.18. Apart from the fundamental constants that enter into Eq. (9.21), notice the dependence on m and a. The spacing depends inversely on the mass of the particle and inversely on the square of the length of the box. A light particle such as an electron will have widely spaced energy levels in comparison with a heavy particle like a nucleus or a baseball. A given particle in a small box will have widely spaced eigenvalues compared with the same particle in a larger box. It is this property that makes quantization important for light particles (electrons) in small regions of space (atoms or molecules). For a baseball

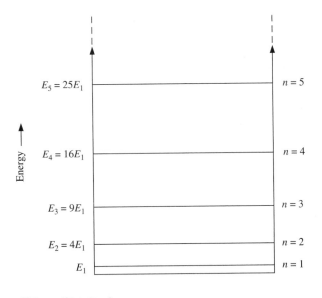

$E_5 = 25E_1$ $n = 5$

$E_4 = 16E_1$ $n = 4$

$E_3 = 9E_1$ $n = 3$

$E_2 = 4E_1$ $n = 2$

E_1 $n = 1$

Energy

Fig. 9.18 Eigenvalues of energy for a particle in a one-dimensional potential well.

in a room or an electron traversing a vacuum tube, the eigenvalues are so closely spaced that they may be considered to provide a continuum of energy. This is just the approach used by classical physics, and it is entirely appropriate for problems of large dimensions. The wave mechanical analysis allows us to determine just when the classical picture breaks down and gives us a method of dealing with truly submicroscopic problems.

Example 9.3 (a) Calculate the energies of the two states of lowest energy for an electron in a one-dimensional box of length 2.0 Å. (b) What is the probability that the electron is within 0.5 Å of the center of the box in the lowest energy state? (c) Repeat the calculation of part (b) for the second-lowest energy state.

Solution

(a) Using Eq. (9.21) we obtain

$$E_1 = \frac{(6.626 \times 10^{-34} \text{ J s})^2 (1)^2}{8(9.11 \times 10^{-31} \text{ kg})(2.0 \times 10^{-10} \text{ m})^2} = 1.506 \times 10^{-18} \text{ J}$$

$$E_2 = (1.506 \times 10^{-18})(2)^2 = 6.024 \times 10^{-18} \text{ J}$$

(b) The probability of finding the electron between $x = \frac{1}{4}a$ and $x = \frac{3}{4}a$ is obtained by evaluating Eq. (9.12) for the appropriate wavefunction (Fig. 9.17).

$$\int_{a/4}^{3a/4} \psi_1^2(x) \, dx = \frac{2}{a} \int_{a/4}^{3a/4} \sin^2 \frac{\pi x}{a} \, dx$$

$$= \frac{2}{a} \left(\frac{x}{2} - \frac{a}{4\pi} \sin \frac{2\pi x}{a} \right) \Big|_{a/4}^{3a/4} \qquad \text{(see Summary at end of chapter)}$$

$$= \frac{2}{a} \left(\frac{3a}{8} - \frac{a}{4\pi} \sin \frac{3\pi}{2} - \frac{a}{8} + \frac{a}{4\pi} \sin \frac{\pi}{2} \right)$$

$$= 2 \left(\frac{1}{4} + \frac{1}{2\pi} \right) = \frac{1}{2} + \frac{1}{\pi} = 0.818$$

Note that the probability of finding a particle in the central half (or any other fraction) of the box is independent of the length of the box.

(c) For $n = 2$

$$\int_{a/4}^{3a/4} \psi_2^2(x) \, dx = \frac{2}{a} \int_{a/4}^{3a/4} \sin^2 \frac{2\pi x}{a} \, dx$$

$$= \frac{2}{a} \left(\frac{x}{2} - \frac{a}{8\pi} \sin \frac{4\pi x}{a} \right) \Big|_{a/4}^{3a/4}$$

$$= \frac{2}{a} \left(\frac{3a}{8} - \frac{a}{8\pi} \sin 3\pi - \frac{a}{8} + \frac{a}{8\pi} \sin \pi \right)$$

$$= 0.500$$

This result is exactly what you expect if you inspect the behavior of $\psi_2(x)$ in Fig. 9.17. The area under the central half of a plot of $\psi_2^2(x)$ will be exactly equal to the sum of the area under the two outer quarters. It is also easy to see that the probability that the electron will be in the central portion of the box is greater in the state $n = 1$ than for the state $n = 2$, again confirmed by our calculation.

The eigenfunctions and eigenvalues for the particle-in-a-box problem are more than mathematical abstractions. To appreciate their physical significance, we choose to look at the family of carotenoids, which are examples of linear polyenes. A widely distributed example is β-carotene, whose molecular structure is as follows:

It is a symmetrical molecule containing 40 carbon atoms. If the molecule is oxidized, it breaks at the center (dashed line) and forms two molecules of retinal, or vitamin A_1, which is closely related to the pigment of rhodopsin. Many isomers and chemically modified derivatives of β-carotene occur in nature. They play important roles in vision, as sensitizers in photosynthesis, and as protective agents against harmful biological oxidations.

β-Carotene and the other carotenoids possess two classes of bonding electrons. For the present, consider one kind, the sigma (σ) electrons, to be responsible for the single bonds and for half of each double bond. These σ electrons are localized in the small regions of space between adjacent atoms. The other type, the pi (π) electrons, contribute the other halves of the double bonds. In systems like β-carotene or benzene, where double and single bonds alternate in the classical molecular structure, the π electrons are not strongly confined to regions between adjacent atoms. In fact, they are nearly free to roam the full length of the conjugated system. Thus they are confined approximately in a one-dimensional potential well which is about as long as the conjugated system and within which the potential energy is virtually constant. The particle-in-a-box model provides a reasonable approach to the eigenfunctions and eigenvalues for these π electrons. Having adopted this model, we can now proceed to examine its predictions and compare them with physically measurable properties.

β-Carotene has 11 conjugated double bonds, and thus has 22 π electrons. We shall assume that the energies and wavefunctions for these 22 electrons can be *very* approximately described by the energies and wavefunctions of a particle in a box. The energy-level pattern shown in Fig. 9.18 can be extended to any value of n. The electron configuration of the lowest-energy state for the 22 π electrons of β-carotene is obtained by placing the electrons in the levels or orbitals of lowest available energy, such that each orbital has two electrons with opposite spins. This causes 11 orbitals ($n = 1$ to 11) to be completely filled and just uses up the 22 π electrons. If we could somehow measure the energy of the electrons in the $n = 11$ level, we could use Eq. (9.21) to calculate a, the length of the potential well, and compare it with the sum of bond lengths for the conjugated system. There is, in fact, no good way to measure directly the energy of the bound electrons in a molecule such as β-carotene. We can, however, relate this model to an *energy difference* that can be measured from the absorption spectrum of the molecule.

When light of the appropriate wavelength is incident on a molecule, it may absorb some of the radiation, resulting in the promotion or excitation of electrons to higher-energy orbitals. The process occurs only if the energy of the incident photons corresponds to the difference between two possible states (eigenstates) of the molecule. Because there are many filled and many empty states available, the absorption spectrum is potentially a rich source of information about the location of the energy levels. In practice, the lowest-energy transition, corresponding to the longest-wavelength electronic absorption band, is the easiest to identify and observe. The lowest-energy absorption occurs when an electron in the highest filled energy level is excited to the lowest unfilled energy level. This is shown in Fig. 9.19 for that segment of the energy-level diagram. Thus the energy associated with the photons absorbed in the long-wavelength absorption band is just the difference between energy level $n = 11$ and level $n = 12$. As we shall see, this is just as useful as knowing E_{11} or E_{12} for the purposes of making an estimate of the length a.

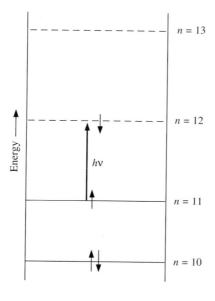

Fig. 9.19 Longest-wavelength electronic transition for the particle-in-a-box approximation to β-carotene.

Using Eq. (9.21), we can easily derive a general expression for the energy difference between adjacent levels for the particle-in-a-box problem. Let N represent the quantum number of the highest filled level; then $N + 1$ is the quantum number for the lowest unfilled level. The energy difference between these two levels is

$$\Delta E = E_{N+1} - E_N = \frac{h^2}{8ma^2}(N + 1)^2 - \frac{h^2}{8ma^2}N^2$$

$$= \frac{h^2}{8ma^2}(N^2 + 2N + 1 - N^2) = \frac{h^2}{8ma^2}(2N + 1)$$

Since

$$\Delta E = \frac{hc}{\lambda}$$

we obtain the result that

$$\lambda\,(N \rightarrow N + 1) = \frac{8mca^2}{h(2N + 1)} \tag{9.22}$$

For linear polyenes with different extents of conjugation, both the length a and the number of π electrons ($2N$) will be different. The wavelength increases in proportion to the square of the length a and inversely with the number of π electrons. Since the length of the π system increases *linearly* with the number of π bonds, the long-wavelength absorption band increases in wavelength (shifts toward the red) with increasing extent of conjugation.

The long-wavelength absorption of β-carotene is shown in Fig. 9.20. The structure of the band actually arises from additional vibrational effects and need not concern us here. The long-wavelength maximum is readily identified and occurs at about 480 nm. By substitution into Eq. (9.22) and solving for a, we obtain for $N = 11$,

$$a^2 = \frac{h\lambda\,(2N + 1)}{8mc} = \frac{(6.626 \times 10^{-34}\ \text{J s})(4.8 \times 10^{-7}\ \text{m})(23)}{8(9.11 \times 10^{-31}\ \text{kg})(3.0 \times 10^{8}\ \text{m s}^{-1})}$$

$$a = 1.83 \times 10^{-9}\ \text{m} = 18.3\ \text{Å}$$

This is an effective length for the one-dimensional free-electron model of carotene.

The actual length of the zigzag chain can also be calculated from the sum of the lengths of the single and double bonds. For such a conjugated system, the length of a single bond is 1.46 Å and the length of a double bond is 1.35 Å. There are 11 double bonds and 10 single bonds for the conjugated chain in β-carotene; therefore, the calculated length is 10×1.46 Å $+ 11 \times 1.35$ Å or 29 Å. The actual length and effective length are different

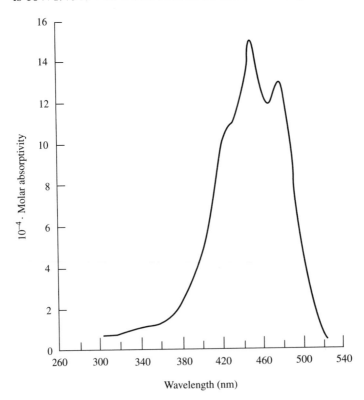

Fig. 9.20 Absorption spectrum of all-*trans* β-carotene.

for several reasons. The actual molecular potential does not have a flat bottom and infinitely steep sides. There are also important electron-electron repulsions. Nevertheless, if we compare the spectra of carotenoids with different lengths of conjugation, there is a very good correlation with the form of Eq. (9.22).

Particles in two- and three-dimensional boxes can be solved just like those in the one-dimensional box. An electron in a two-dimensional box provides a reasonable model for discussing the spectrum of such molecules as benzene, naphthalene, and anthracene. A particle in a three-dimensional box accurately describes a dilute gas in a real container. For a box of sides $a, b, c,$ the wavefunctions are

$$\psi(x, y, z) = \sqrt{\frac{8}{abc}} \sin \frac{n_x \pi x}{a} \sin \frac{n_y \pi y}{b} \sin \frac{n_z \pi z}{c} \tag{9.23}$$

The energies are

$$E = \frac{h^2}{8m} \left(\frac{n_x^2}{a^2} + \frac{n_y^2}{b^2} + \frac{n_z^2}{c^2} \right) \tag{9.24}$$

These energies correspond to the translational energies of an ideal gas. They can be used to derive the ideal gas equation and to predict when quantum effects will cause deviations from the classical ideal gas behavior. These quantum effects can become important for He gas at temperatures near absolute zero.

Particles in two- and three-dimensional boxes introduce the concept of *degeneracy*. Whenever several wavefunctions have the same energy, the wavefunctions are said to be degenerate. For a particle in a three-dimensional box, different values of n_x, n_y, n_z—and therefore different wavefunctions—can have the same energy. For example, the three different wavefunctions with quantum numbers, n_x, n_y, $n_z = 2, 1, 1,$ or $= 1, 2, 1$ or $= 1, 1, 2$ all have the same energy. The energy level is threefold degenerate. We will find that for the hydrogen atom the lowest energy state (1s) is nondegenerate, but the 2s, $2p_x$, $2p_y$, and $2p_z$ orbitals all have the same energy; there is a fourfold degeneracy.

Using a modified coordinate system, it is possible to describe the motion of an electron on a circular ring as a one-dimensional particle-in-a-box problem. Here the box has no ends with infinite potential; the potential is zero everywhere on the circumference of the ring. The continuity condition requires that the wavefunctions are continuous with themselves for each 2π rotation about the ring. Some additional states (degeneracies) arise because the particle can move either clockwise or counterclockwise around the ring with wavefunctions having the same energy. The electron-on-a-ring model has been used as a model for the behavior of the π electrons in large aromatic molecules such as porphyrins, heme, and chlorophyll. For more compact molecules such as CH_4 or CCl_4 it is possible to treat the electrons as particles in a spherical (three-dimensional) potential well, but this becomes less useful where the distribution of nuclear charge is highly nonuniform.

The infinite potential well is a convenient abstraction, but in real systems the heights of the barriers are always finite. This leads to the important phenomenon of "tunneling" of the particle into the barrier and even out the other side. Tunneling of electrons and even protons is currently of interest in many biological and biochemical systems. It has been studied with respect to electron transport associated with respiration and photosynthesis and with respect to proton translocation across membranes driven by molecular pumps, such as bacteriorhodopsin.

TUNNELING

We discussed a particle-in-a-box in detail because it is the simplest quantum mechanics problem to solve. From now on we will only give the results of solving Schrödinger's equation, and we will discuss the applications of the solutions. The box we considered has walls which are infinite in height; thus, the particle cannot escape. Its wavefunction is zero outside the box. What happens for a box with walls of finite height? Classically, the particle can only escape if its energy is equal to or greater than the energy of the walls. However, quantum mechanics shows that the particle can escape even if its energy is less than the repulsion of the walls. The wavefunction has a nonzero value outside the walls. The particle is said to tunnel through the barrier. The probability of tunneling depends on the energy and mass of the particle, and on the height and width of the barrier. The mass of the particle is important because, as we stated earlier, de Broglie found that the wavelength of a particle is inversely proportional to its momentum, mv. Because the kinetic energy, E, is equal to $(1/2)mv^2$, we can combine these two equations to obtain

$$\lambda = h/(2mE)^{1/2} \tag{9.25}$$

We see that for a constant energy, the wavelength of the wavefunction of the particle increases as the mass of the particle decreases. The wavelength of a 20 kJ mol^{-1} electron is 27 Å; the wavelength of a proton of the same energy is only 0.63 Å.

Electron tunneling occurs in many oxidation-reduction reactions catalyzed by enzymes. Rapid electron transfer between donor and acceptor sites separated by 10–20 Å in a protein has been established (Axup et al., 1988). Proton tunneling can also occur. Consider the energy barrier which separates reactants from products for a proton transfer from one carbon to another as shown in Fig. 9.21. An example is the oxidation of ethanol by the enzyme liver alcohol dehydrogenase to form acetaldehyde. The proton is transferred from the alcohol to the enzyme cofactor nicotinamide adenine dinucleotide, NAD$^+$ (the structure is given in Appendix A.9).

$$CH_3CH_2OH + NAD^+ \longrightarrow CH_3HC = O + NADH + H^+$$

Classically, the reactants need to have an energy equal to or larger than the activation energy to react, but by tunneling, the proton can sneak through the activation barrier instead of going over it. This will speed up the reaction. It is not possible to calculate absolute rates of reaction for the classical model or the quantum mechanical model, but it is possible to calculate relative rates for a triton (T = ^{3}H), vs. a deuteron (D = ^{2}H), and vs. a proton (H = ^{1}H). The only difference in the reactions is the mass of the tunneling particle; all the electronic distributions and the intermolecular interactions are the same. By experimentally measuring the kinetic isotope effects, k_H/k_D, k_H/k_T and k_D/k_T, Klinman and her coworkers found that proton tunneling occurs in several different alcohol dehydrogenases (Klinman, 1989; Cha et al., 1989). The effects are large. For the oxidation of benzyl alcohol to benzaldehyde by yeast alcohol dehydrogenase a $k_H/k_T = 7.13 \pm 0.07$ was measured; without tunneling the ratio would be only $k_H/k_T = 5.91 \pm 0.20$. Clearly quantum mechanical tunneling can have a major effect on the rates of reactions involving protons (and by extrapolation an even larger effect on electron transfer). Tunneling becomes negligible for larger particles. A qualitative understanding of the phenomenon is sufficient for our purposes.

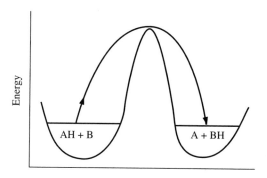

Reaction coordinate
Classical crossing of transition state

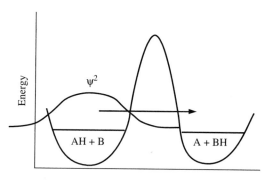

Reaction coordinate
Quantum mechanical tunneling

Fig. 9.21 The transfer of a proton in a reaction can occur by tunneling through the barrier, instead of by going over the barrier. The square of the wavefunction of the proton on the reactant illustrates that there is a small probability that the proton is found on the product instead of the reactant. This speeds the reaction for protons (H) more than for the heavier deuterons (D) or tritons (T).

SIMPLE HARMONIC OSCILLATOR

We used a simple harmonic oscillator [Eq. 9.5] to describe the stretching or compressing of a bond when we discussed molecular interactions. We can gain a clearer understanding of the vibration of molecules by considering the quantum mechanics of a simple harmonic oscillator. Each oscillator has a fundamental vibration frequency, ν_0, which depends on the strength of the bond and the masses attached to the bond. The energies and vibration frequencies of the C—H, C—D, and C—T bonds determine the kinetic isotope effects in the absence of tunneling. A harmonic potential energy can be written in the form

$$U = \tfrac{1}{2} kx^2 \tag{9.26}$$

where k is the force constant for the bond and the origin of the coordinate system is chosen at the equilibrium bond length.

The harmonic oscillator Schrödinger equation is thus

$$-\frac{\hbar^2}{2\mu}\frac{d^2\psi_v}{dx^2} + \tfrac{1}{2}kx^2\psi_v = E_v\psi_v \qquad (9.27)$$

We use the subscript v for the vibrational quantum number. This differential equation can be solved by standard methods. Some of the wavefunctions and their corresponding energies are tabulated in Table 9.3. In Fig. 9.22, the wavefunctions, $\psi_v(x)$ and their squares, $\psi_v^2(x)$, which give the probability of finding the particle at x, are plotted for several quantum numbers.

Table 9.3 Some eigenfunctions and eigenvalues for a simple harmonic oscillator*

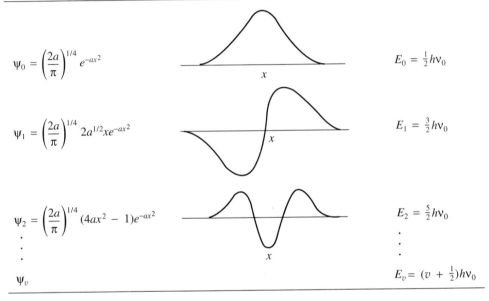

$$\psi_0 = \left(\frac{2a}{\pi}\right)^{1/4} e^{-ax^2} \qquad\qquad E_0 = \tfrac{1}{2}h\nu_0$$

$$\psi_1 = \left(\frac{2a}{\pi}\right)^{1/4} 2a^{1/2}xe^{-ax^2} \qquad\qquad E_1 = \tfrac{3}{2}h\nu_0$$

$$\psi_2 = \left(\frac{2a}{\pi}\right)^{1/4}(4ax^2 - 1)e^{-ax^2} \qquad\qquad E_2 = \tfrac{5}{2}h\nu_0$$

$$\psi_v \qquad\qquad E_v = (v + \tfrac{1}{2})h\nu_0$$

$$* \; \nu_0 = \frac{1}{2\pi}\left(\frac{k}{\mu}\right)^{1/2}, \; a = \frac{\pi}{h}(k\mu)^{1/2}.$$

It is interesting to compare the quantum mechanical results with the classical mechanical results. First, the energy is quantized into energy levels, as is always true in quantum mechanics. The energy of the harmonic oscillator can be only

$$E_v = (v + \tfrac{1}{2})h\nu_0 \qquad (9.28)$$

where the quantum number, v, is a positive integer or zero. According to classical mechanics, the energy can be any value depending on the amplitude. Second, classical mechanics allows the oscillator to be at rest and therefore have an energy equal to zero, but according to quantum mechanics, the lowest energy level is $\tfrac{1}{2}h\nu_0$, called the *zero-point energy*. Both quantum mechanics and classical mechanics give the same fundamental vibration frequency, ν_0, of the oscillator, however.

$$\nu_0 = \frac{1}{2\pi}\left(\frac{k}{\mu}\right)^{1/2} \qquad (9.29)$$

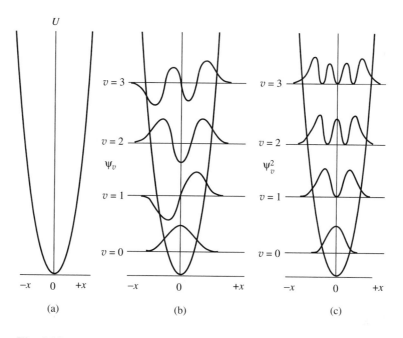

Fig. 9.22 (a) Potential well for a classical harmonic oscillator. (b) Allowed energy levels and wavefunctions for a quantum mechanical harmonic oscillator. $\psi_v(x)$ is plotted versus x; each different wavefunction is arbitrarily displaced vertically. (c) Probability functions for a quantum mechanical harmonic oscillator. The square of each wavefunction $[\psi_v(x)]^2$ is plotted versus x.

Here μ is the reduced mass of the oscillator [Eq. (9.18)] and k is the force constant for the bond. It is also true that when v is large or when (k/μ) is small, the quantum mechanics and classical mechanics pictures become very similar.

The quantum mechanical harmonic oscillator is used in interpreting infrared spectra which correspond to molecular vibrations. The difference in energy corresponds to a change in vibrational quantum number $\Delta v = 1$:

$$\Delta E = h\nu_0$$

Therefore, light will be absorbed if its frequency ν is equal to a vibrational frequency of a harmonic oscillator. A harmonic oscillator is a good approximation for the vibration of diatomic molecules and the vibration of bonds in polyatomic molecules. Vibrational frequencies are characteristic of a particular type of bond such as C—H, C—C, C=C, and C—N, but they can also be used to determine force constants, k, for the particular vibration. For example, when the force constant for a double-bond vibration decreases (the frequency decreases), it means that the bond has become weaker; the double-bond character has decreased. These differences in bond strength for the same type of bond in different molecules are a useful clue to the distribution of electrons in the molecules. An important application of infrared vibrational spectra is the study of hydrogen bonding. The spectra of O—H and N—H vibrations are very different for the hydrogen-bonded and non-hydrogen-bonded species.

Example 9.4 Calculate the fundamental vibrational frequency of carbon monoxide, CO, considering it to behave like a harmonic oscillator with a force constant 1902.5 N m^{-1}. By how much will this frequency change if the molecule contains the isotope ^{13}C instead of the more abundant ^{12}C?

Solution For the principal isotopic species ^{12}C^{16}O the expression for the reduced mass is

$$\mu = \frac{m_C m_O}{m_C + m_O} = \left[\frac{(12.0)(16.0)}{28.0} \times \frac{1}{6.023 \times 10^{26}}\right] \frac{kg}{molecule}$$

$$= 1.1385 \times 10^{-26} \text{ kg molecule}^{-1}$$

According to Eq. (9.29),

$$\nu_0 = \frac{1}{2\pi}\left(\frac{1902.5 \text{ N m}^{-1}}{1.1385 \times 10^{-26} \text{ kg}}\right)^{1/2} = 6.506 \times 10^{13} \text{ s}^{-1}$$

Infrared spectroscopists usually express this in wavenumbers $\bar{\nu}_0$:

$$\bar{\nu}_0 = \frac{\nu_0}{c} \qquad \text{where } c = \text{velocity of light}$$

$$= \frac{6.506 \times 10^{13} \text{ s}^{-1}}{2.9979 \times 10^{10} \text{ cm s}^{-1}} = 2170 \text{ cm}^{-1}$$

If the molecule contains ^{13}C instead of ^{12}C, the reduced mass will be altered, but not the force constant. (The force constant is the measure of the stiffness or strength of the electronic bond holding the nuclei together.) Thus

$$\frac{\nu_0(^{13}C^{16}O)}{\nu_0(^{12}C^{16}O)} = \left[\frac{\mu(^{12}C^{16}O)}{\mu(^{13}C^{16}O)}\right]^{1/2}$$

$$\nu_0(^{13}C^{16}O) = \left[\frac{(12.0)(16.0)}{28.0} \times \frac{29.0}{(13.0)(16.0)}\right]^{1/2} (2170 \text{ cm}^{-1})$$

$$= (0.9778)(2170 \text{ cm}^{-1}) = 2122 \text{ cm}^{-1}$$

This difference is readily detected in the infrared spectrum.

RIGID ROTATOR

Another mode of motion for which quantum mechanics provides important insights is rotation. For gas-phase molecules at low pressure, rotation is relatively free, at least during the time between collisions with other molecules or with the walls. A simple model that reasonably describes such rotational motion is that of the free rigid rotator, a collection of masses (atoms) at fixed distances and relative orientation. This problem can readily be solved using the methods just illustrated for the harmonic oscillator, but using a potential function ($U = 0$) and kinetic energy appropriate to a freely rotating system. The solution is given in standard physical chemistry texts. Again, a set of quantized energy levels is obtained, with $E_{rot} = J(J + 1)\hbar^2/2I$, where I is the moment of inertia of the molecule and J is a rotational quantum number that takes on integer values $J = 0, 1, 2, 3, \ldots$. The moment of inertia of a diatomic molecule of reduced mass μ is $I = \mu r_0^2$, where r_0 is the equilibrium internuclear distance. You can readily determine that the energy levels are not uniformly spaced, by contrast with the harmonic oscillator energy levels. Furthermore, for all but the lightest molecules, rotation-energy-level spacings are much smaller than those for typical vibrations. In fact, the rotational spacings are small compared with thermal energy, kT, at room temperature, which means that many rotational levels are well populated in gases under these conditions. This means that gas-phase molecules will be distributed over many rotational levels at room temperature. Because of the wide spacing of vibrational levels, most molecules will be in one, or only a few, vibrational levels at room temperature. The spacing of energy levels has important effects on the statistical interpretation of energy and entropy, as described in Chapter 11.

The situation for liquids or solids is very different. Here the model of a freely rotating molecule is not appropriate, because as soon as an individual molecule in a condensed phase begins to rotate, it bumps into its neighbors or experiences a restoring force that changes its direction. Such interrupted or irregular motion does not produce the sharply defined quantized energy levels characteristic of gas-phase molecules. In liquids and to some extent in solids the rocking motions or librations that do occur result in a broadening out or smearing of the sharp electronic and vibrational energies. As a consequence, the spectra of molecules in the condensed phase are usually much broader and lack fine resolution compared with gas-phase spectra. There are some exceptions in highly ordered crystals where molecules in a uniform environment may exhibit quite sharp spectra.

Librational motion in pure liquids and solutions is still an important energy reservoir. It reflects the nature of intermolecular forces and local molecular ordering, and it is an important contributor to the motions that lead to chemical reactions.

HYDROGEN ATOM

A hydrogen atom consists of two particles, a nucleus of mass m_1 and charge $+e$, and an electron of mass m_2 and charge $-e$. The force acting between the particles is coulombic, and the potential energy U is equal to $-e^2/r$, where r is the distance between the particles. For SI units, $U = -e^2/4\pi\varepsilon_0 r$. The location of the electron relative to the nucleus (actually relative to the center of mass of the hydrogen atom) can be specified by a wavefunction of three coordinates. Because the potential energy depends only on r (it is spherically symmetrical), the wavefunction can be written as a product of two parts; one depends only on r, the other on two angular coordinates. Solving the Schrödinger equation for a hydrogen atom (see the references at the end of the chapter), one obtains the wavefunctions and the energy levels.

Three quantum numbers, n, l, and m, characterize the wavefunction; l, m characterize the angular part of the wavefunction, while n characterizes the distance dependence.

These quantum numbers are called the *principal quantum number, n;* the *angular momentum quantum number, l;* and the *magnetic quantum number, m.* The permissible values of n are positive integers 1, 2, 3, . . .; the permissible values of l for a given n are 0 and positive integers up to $n - 1$; and the permissible values of m for a given l are integers from $-l$ to $+l$. The symbols s, p, d, and f are assigned to functions with $l = 0$, 1, 2, and 3, respectively.

The energy E of the hydrogen atom is quantized, as is the energy of the particle in a box, the simple harmonic oscillator, and all other systems. The eigenvalues for E for the H atom are

$$E_n = -\frac{\mu e^4}{2\hbar^2 n^2} \tag{9.30}$$

Note that the energy of a hydrogen atom depends only on the principal quantum number n. All three quantum numbers, n, l, and m, characterize the electron distribution, but the energy depends only on n. With μ in g, $\hbar$ in erg s, and e in esu, energy is obtained in ergs. In SI units, we have for the H atom,

$$E_n = -\frac{\mu e^4}{32\pi^2 \varepsilon_0^2 \hbar^2 n^2} \tag{9.31}$$

Now μ is in kilograms, e in coulombs, and $\hbar$ in J s; E_n is in joules. As chemists, spectroscopists, and others each seem to use different energy units in discussing the hydrogen atom, it is convenient to write

$$E_n = -\frac{R}{n^2} \tag{9.32}$$

$$R = 2.179 \times 10^{-11} \text{ erg molecule}^{-1}$$

$$R = 2.179 \times 10^{-18} \text{ J molecule}^{-1}$$

$$R = 1312 \text{ kJ mol}^{-1}$$

$$R = 13.60 \text{ eV molecule}^{-1}$$

The energy is given relative to the separated proton and electron as the zero of energy; the negative value means that the hydrogen atom is stable. The values of the Rydberg constant, R, given are the ones most used. The first two values correspond to Eqs. (9.30) and (9.31). The value of R in kJ mol^{-1} is useful for comparison with bond energies and heats of chemical reactions. The ionization energy of an atom or molecule is the energy required to remove an electron; it is traditional to quote ionization energies in electron volts. The value of R in eV tells us that the ionization energy of the hydrogen atom is $+13.6$ eV.

The Schrödinger equations for other single-electron species (He^+, Li^{2+}, Be^{3+}, etc.) can be solved with the potential energy equal to $-Ze^2/r$, where Z is the atomic number. The hydrogen atom itself corresponds to $Z = 1$. Several of the hydrogen-like wavefunctions are tabulated in Table 9.4. The energy of a hydrogen-like orbital is dependent on the quantum number n only, and is

$$E = -\frac{Z^2 \mu e^4}{2\hbar^2 n^2} = -\frac{Z^2 R}{n^2} \tag{9.33}$$

The value of R depends slightly on the mass of the nucleus, because of the reduced mass μ; however, to an accuracy of 0.1%, μ equals the mass of the electron.

Table 9.4 Hydrogenic wavefunctions*

n	l	m	
1	0	0	$\psi(1s) = \dfrac{1}{\sqrt{\pi}}\left(\dfrac{Z}{a_0}\right)^{3/2} e^{-Zr/a_0}$
2	0	0	$\psi(2s) = \dfrac{1}{4\sqrt{2\pi}}\left(\dfrac{Z}{a_0}\right)^{3/2}\left(2 - \dfrac{Zr}{a_0}\right) e^{-Zr/2a_0}$
2	1	0	$\psi(2p_z) = \dfrac{1}{4\sqrt{2\pi}}\left(\dfrac{Z}{a_0}\right)^{5/2} (z)\, e^{-Zr/2a_0}$
2	1	±1	$\psi(2p_x) = \dfrac{1}{4\sqrt{2\pi}}\left(\dfrac{Z}{a_0}\right)^{5/2} (x)\, e^{-Zr/2a_0}$ $\psi(2p_y) = \dfrac{1}{4\sqrt{2\pi}}\left(\dfrac{Z}{a_0}\right)^{5/2} (y)\, e^{-Zr/2a_0}$

$* z = r\cos\theta$

$x = r\sin\theta\cos\phi$

$y = r\sin\theta\sin\phi$

$r^2 = x^2 + y^2 + z^2$

$a_0 = $ Bohr radius $= \dfrac{\hbar^2}{me^2}$

$= 0.529\ \text{Å} = 5.29 \times 10^{-2}\ \text{nm}$

$Z = $ charge on nucleus

ELECTRON DISTRIBUTION

The electron distribution in molecules is responsible for the enormous range of their properties. The chemical bonds that hold atoms together, the distribution of charge that produces a dipole moment, the forces between molecules that lead to association or to solvation, the interaction with light to produce color or to induce photochemistry—all of these depend critically on how the electrons are distributed in the molecule. Inevitably this is a matter for quantum mechanical description; however, even the simplest molecules cannot yet be solved exactly by quantum mechanics. An important part of our task now is to develop reasonable models or approximations that will allow us to calculate molecular properties to the desired accuracy.

We will first look at how electrons in atoms can be described using quantum mechanics. Then we will consider molecules as collections of atoms having somewhat modified characteristics. We will treat some electrons as localized in chemical bonds, others as delocalized over large regions or, perhaps, all of the molecule. Ultimately, the electron distribution determines what is the most stable configuration or geometry of the molecule. It also determines bond lengths, force constants, and bond angles.

Electron Distribution in a Hydrogen Atom

The square of the wavefunction $\psi(r, \theta, \phi)$ is proportional to the probability of finding the electron at a position r, θ, ϕ, where the origin of the spherical coordinates is at the nucleus (Fig. 9.23). We want to be able to visualize the hydrogen wavefunctions and their squares, because we use hydrogen-like orbitals to represent electron distributions in molecules. Figure 9.24 plots $\psi(1s)$, $\psi(2s)$, and $\psi(3s)$ for a hydrogen atom and $\psi(1s)$ for a helium ion. Each ψ is a maximum at $r = 0$ (the position of the nucleus) and decreases rapidly as the distance from the nucleus increases. The ground state of the hydrogen atom ($n = 1$) has the electron closest to the nucleus. For successive excited states ($n = 2, 3, \ldots$), the electron is spread out over more space away from the nucleus. Note how the $\psi(1s)$ for He$^+$ with a nuclear charge of +2 is pulled toward the nucleus relative to $\psi(1s)$ for H with a nuclear charge of +1.

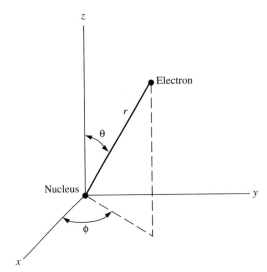

Fig. 9.23 Spherical coordinate system. For an atom the origin should be at the center of mass, but negligible error is made by having it at the nucleus.

A better feeling for the shape of the wavefunctions can be obtained from computer plots of the functions as illustrated in Fig. 9.25. The pictures for $\psi(1s)$ and $\psi(2s)$ are just the first two plots in Fig. 9.24 rotated about the origin. The $\psi(2p_x)$ picture shows a wavefunction that is not symmetrical about the origin. $\psi(2p_y)$ and $\psi(2p_z)$ are identical to $\psi(2p_x)$ but are oriented along y and z. The wavefunctions $\psi(2s)$, $\psi(2p_x)$, $\psi(2p_y)$, and $\psi(2p_z)$ all correspond to the same energy for hydrogen atoms; $E_2 = -R/4$.

The probability of finding the electron at a distance r from the nucleus is proportional to the volume of a spherical shell of radius r around the nucleus. The volume of a spherical shell is directly proportional to r^2; therefore, the probability of finding the electron at distance r is proportional to $r^2\psi^2$. This is plotted versus r/a_0 in Fig. 9.26 for $\psi(1s)$, $\psi(2s)$, and $\psi(3s)$. The probability of finding the electron is a maximum at $r = a_0$ for the ground state, $\psi(1s)$, of the hydrogen atom. Bohr had deduced a circular orbit at this distance for the electron in a ground-state hydrogen atom; therefore, this distance is called the *Bohr radius*. Quantum mechanics states that the electron can be anywhere relative to the nucleus, but its most probable value occurs at a_0. For the excited state $\psi(2s)$ the most probable value of r for the electron is at $5.24a_0$.

Still another way of visualizing an electron distribution in space is shown in Fig. 9.27. Here the enclosed, shaded areas represent the space where there is 90% probability of finding the electron.

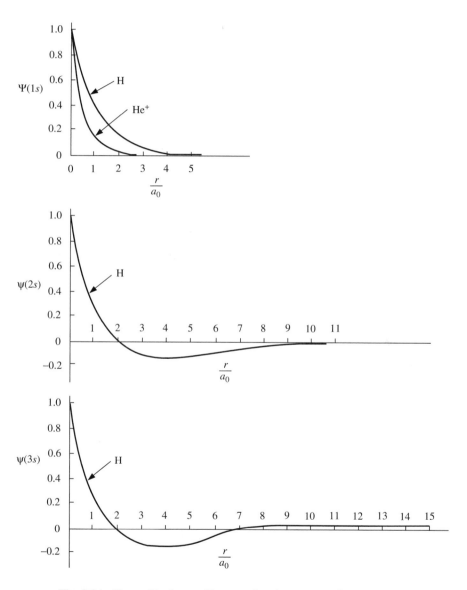

Fig. 9.24 Plots of hydrogen-like wavefunctions versus distance for spherically symmetrical wavefunctions, $\psi(ns)$. The distance scale has units of $a_0 = 0.529$ Å. ψ is set equal to 1 at $r = 0$. ψ can be positive or negative; the number of radial nodes ($\psi = 0$) is equal to the principal quantum number, n.

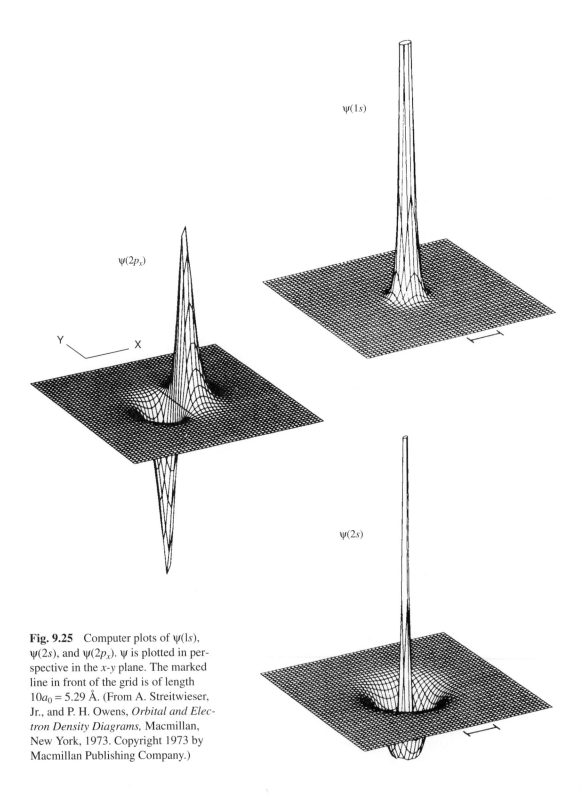

$\psi(1s)$

$\psi(2p_x)$

Y   X

$\psi(2s)$

Fig. 9.25 Computer plots of $\psi(1s)$, $\psi(2s)$, and $\psi(2p_x)$. ψ is plotted in perspective in the x-y plane. The marked line in front of the grid is of length $10a_0 = 5.29$ Å. (From A. Streitwieser, Jr., and P. H. Owens, *Orbital and Electron Density Diagrams,* Macmillan, New York, 1973. Copyright 1973 by Macmillan Publishing Company.)

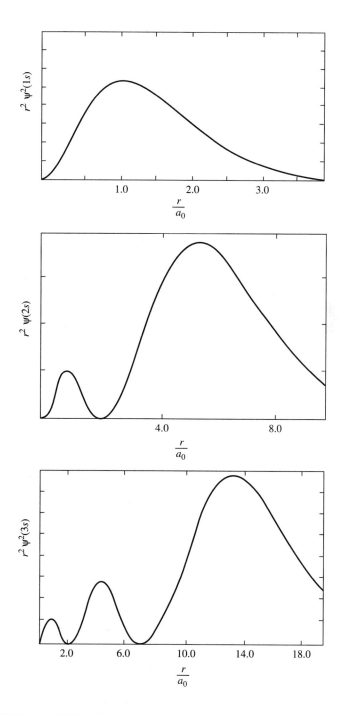

Fig. 9.26 Probability of finding the electron in a hydrogen atom in a spherical shell of radius r around the nucleus. $r^2\psi^2$, which is proportional to this probability, is plotted versus distance in units of $a_0 = 0.529$ Å. Note that the nodes occur at the same locations as in Fig. 9.24.

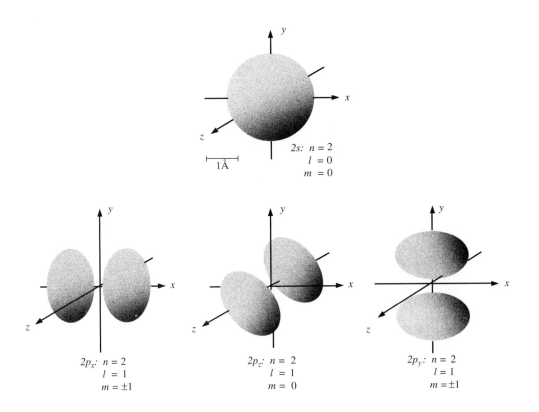

2s: n = 2
l = 0
m = 0

1Å

2p$_x$: n = 2
l = 1
m = ±1

2p$_z$: n = 2
l = 1
m = 0

2p$_y$: n = 2
l = 1
m = ±1

Fig. 9.27 Pictures that show where there is 90% probability of finding the electron in a hydrogen atom in (2s), (2p$_x$), (2p$_y$) orbital. (From G. C. Pimentel and R. D. Spratley, *Understanding Chemistry,* Holden-Day, Oakland, 1971, p. 486.)

Example 9.5   For the 1s orbital of the hydrogen atom, calculate the probability of finding the electron within a distance a_0 (the Bohr radius) from the nucleus.

Solution   Taking $\psi(1s)$ from Table 9.4 and using Eq. (9.12) adapted to a spherical coordinate system, we obtain

$$\int [\psi(1s)]^2 \, dv = 4\pi \int_0^{a_0} \left[\frac{1}{(\pi a_0^3)^{1/2}} e^{-r/a_0} \right]^2 r^2 \, dr$$

$$= 4 \int_0^1 e^{-2\rho} \rho^2 \, d\rho \qquad \text{where } \rho = \frac{r}{a_0}$$

$$= -\frac{e^{-2\rho}}{2} (4\rho^2 + 4\rho + 2) \bigg|_0^1 = 1 - 5e^{-2}$$

$$= 0.323$$

Thus the probability of finding the electron within one Bohr radius of the nucleus of the hydrogen atom in the $1s$ state is 32.3%.

So far we have not mentioned the spin of the electron. Experimental results as well as theoretical considerations indicate that an electron has an intrinsic magnetic moment and behaves as if it were a small magnet. In addition to the quantum numbers n, l, and m, a fourth quantum number, m_s, called the *spin quantum number,* is used to specify the intrinsic magnetic moment. The spin quantum number can be either $+\frac{1}{2}$ or $-\frac{1}{2}$. We have ignored the spin of the electron up to now, because it has an extremely small effect on the energy levels of the hydrogen atom. However, as soon as we consider more than one electron in an atom, the effect of spin becomes important.

Many-Electron Atoms

For atoms with more than one electron we can easily write the Schrödinger equation, but we cannot solve it exactly. Consider, for example, the helium atom that has two electrons. The Hamiltonian operator for the two electrons is

$$\mathscr{H} = -\frac{\hbar^2}{2\mu} \cdot (\nabla_1^2 + \nabla_2^2) - \left(\frac{2e^2}{r_1} + \frac{2e^2}{r_2} - \frac{e^2}{r_{12}}\right)$$

where μ, the reduced mass, is essentially the mass of an electron, the subscripts 1 and 2 refer to quantities for electrons 1 and 2, respectively. The distance from the nucleus to electron 1 is r_1, the distance to electron 2 is r_2, and r_{12} is the distance between the electrons.

Because of the term containing r_{12}, it is not possible to separate the variables in the Schrödinger equation as we can do for the hydrogen atom. Exact solutions are therefore unavailable. This means we cannot write an exact ψ for He explicitly as we can for a hydrogen atom. However, we can calculate energy levels, electron distributions, and other electronic properties for helium or any atom to a high degree of accuracy. The calculated values agree with experiments within experimental error. Therefore, we are confident in the predicted values for properties that have not yet been measured.

We will discuss many-electron atoms qualitatively; we will not go into details about the approximations used in obtaining wavefunctions and energy levels. We consider a many-electron atom to have hydrogen-like wavefunctions. These orbitals have the shapes shown in Figs. 9.24 through 9.27, but their energies are very different from those of a hydrogen atom. The ordering of the energy levels is $1s$, $2s$, $2p_x = 2p_y = 2p_z$, $3s$, $3p$ (three orbitals), $4s$, $3d$ (five orbitals), and so on. The *Pauli exclusion principle* states that one, or at most two, electrons can be in each orbital; two electrons in the same orbital must have opposite spins. Pauli deduced that no two electrons in an atom can have the same four quantum numbers: n, l, m, and m_s. Two electrons in the same orbital have identical values of n, l, and m; therefore, they must have antiparallel spins ($m_s = +\frac{1}{2}$ for one and $-\frac{1}{2}$ for the other). These simple facts are enough to understand the arrangement of elements in the periodic table. The student is encouraged to review an elementary chemistry book to obtain a more extensive discussion of atomic orbitals and the periodic table. Here we will just introduce the notation for specifying the electron distribution in a many-electron atom. Figure 9.28 illustrates this notation.

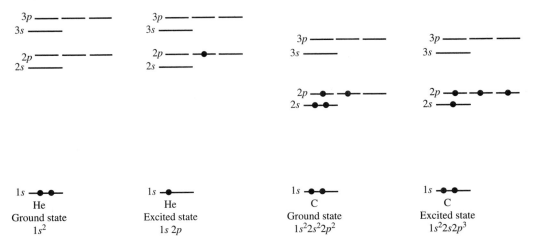

Fig. 9.28 Energy ordering and electron orbital notation for many-electron atoms. At most, two electrons are placed in each orbital in accord with the Pauli exclusion principle. The electronic state of the atom is specified by the number of electrons in each orbital. For example, the fluorine ground state is $1s^2 2s^2 2p^5$.

Molecular Orbitals

The electronic wavefunctions of a molecule can be approximated by a set of molecular orbitals (MOs) in which one, or at most two, electrons can be placed. The electronic wavefunctions depend on the internuclear distances, but we can consider the nuclei fixed in determining the electronic wavefunctions and orbitals. Each molecular orbital can be represented as a *linear combination of atomic orbitals* (LCAOs). This just means that we can write the molecular orbital as a sum of atomic orbitals on each nucleus. We know what each of the atomic orbitals looks like (Figs. 9.24 to 9.27); therefore we know the shapes of the molecular orbitals. From a chosen number of atomic orbitals we can write the same number of molecular orbitals.

The first step in using and understanding molecular orbitals is to learn the vocabulary. Consider the H_2 molecule with two nuclei labeled A and B. The simplest LCAO we can write is

$$\psi_{H_2} = \frac{1}{\sqrt{2}} [\psi_A(1s) + \psi_B(1s)] \tag{9.34}$$

This says that an MO of H_2 can be approximated by a $1s$ hydrogen atomic orbital (AO) on nucleus A plus a $1s$ hydrogen AO on nucleus B. The $1/\sqrt{2}$ is necessary to normalize the H_2 MO. Because we have chosen to use two AOs, we can write two MOs. The other one is

$$\psi_{H_2} = \frac{1}{\sqrt{2}} [\psi_A(1s) - \psi_B(1s)] \tag{9.35}$$

We can calculate the electron distributions corresponding to these MOs by squaring them, and we can calculate their energy levels by using the Hamiltonian operator for an H_2 molecule. These two MOs are examples of sigma, σ, and sigma star, σ^*, MOs, respectively.

Equations (9.34) and (9.35) give approximations to the wavefunctions of H_2. To calculate reasonable values for the electronic energies of H_2, much more complicated wavefunctions would be needed. Nevertheless, MOs that are LCAOs are useful concepts to understand electron distributions in molecules qualitatively.

The names of MOs depend on their symmetry properties. A *sigma-bonding orbital* (σ-bonding MO) has cylindrical symmetry around the line joining the nuclei and has no node between the nuclei. A *sigma-star antibonding orbital* (σ^* antibonding MO) has cylindrical symmetry around the line joining the nuclei but has a node between the nuclei. Pictures of a σ and a σ^* MO are shown in Fig. 9.29.

The electron density of a σ bond has cylindrical symmetry around the bond and there is a high probability of finding the electrons between the nuclei. For a σ^* orbital there is still cylindrical symmetry around the bond, but the probability of finding the electrons between the nuclei becomes zero between the nuclei (at the node). The energy calculated for a σ MO is less (more negative) than the sum of the energies for the two AOs, while the energy of a σ^* MO is greater than the sum of the two AOs.

A π (*pi*)-*bonding molecular orbital* has a node along the line joining the two nuclei. The electron density for a π-bonding MO has a plane of symmetry passing through the bond. A π^* (*pi star*)-*antibonding molecular orbital* has a node along the line joining the nuclei and also has a node between the two nuclei. Pictures of π and π^* MOs are shown in Fig. 9.30.

A molecule can be thought of as a collection of molecular orbitals. We have discussed σ, σ^*, π, and π^* MOs; there are other MOs with different symmetry which are sometimes important in metal coordination complexes. The distribution of electrons among these orbitals determines the bonding and stability of the molecule. In general, the proper atomic orbitals to be used for the formation of the molecular orbitals are those with comparable energies and the same symmetry with respect to the bond axis. Atomic orbitals of greatly different energies are poor choices for constructing molecular orbitals.

As an example, let us consider 10 MOs which can be formed from combinations of 5 AOs ($1s$, $2s$, $2p_x$, $2p_y$, $2p_z$) on each of two identical nuclei A and B. These 10 MOs can hold up to 20 electrons, so they allow us to discuss the bonding and stability of possible homonuclear diatomic molecules from H_2 to Ne_2. We start adding electrons to the MOs and see what qualitative conclusions we can make about the resulting molecules. Figure 9.31 shows the usual energy-level pattern obtained for the 10 MOs. The actual spacing will depend on which nuclei are involved, and for some molecules $\sigma(2p_x)$ may be above the $\pi(2p_y)$, $\pi(2p_z)$ levels. However, this pattern of levels is a useful one to remember for most molecules, not just diatomics. Figure 9.31 is consistent with the known data on the diatomic molecules of the first row of the periodic table. He_2, Be_2, and Ne_2 are unstable; H_2, Li_2, B_2, and F_2 form single bonds; C_2 and O_2 form double bonds; and N_2 forms a triple bond. The number of bonds, sometimes called the *bond order,* is one-half the number of electrons in bonding MOs minus one-half the number of electrons in antibonding orbitals. Each pair of filled bonding and antibonding MOs does not contribute to any bonds. For molecules containing C, N, O, F, and higher atomic-number elements, the $\sigma(1s)$ and $\sigma^*(1s)$ MOs involving these atoms are filled and are not involved in bonding. The $1s$ electrons are then called *core electrons* and are not explicitly considered in the MOs.

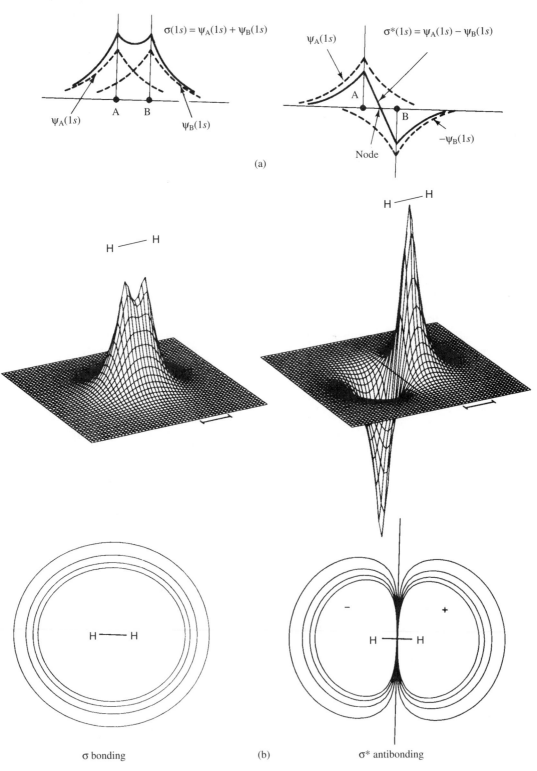

$\sigma(1s) = \psi_A(1s) + \psi_B(1s)$

$\psi_A(1s)$

$\psi_A(1s)$

$\psi_B(1s)$

A   B

$\sigma^*(1s) = \psi_A(1s) - \psi_B(1s)$

$\psi_A(1s)$

A

B

$-\psi_B(1s)$

Node

(a)

H —— H

H —— H

σ bonding

(b)

σ* antibonding

H — H

H — H

−

+

518   *Chapter 9*

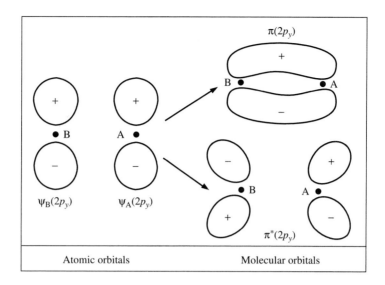

Atomic orbitals   Molecular orbitals

Fig. 9.30 Combination of two $2p_y$ atomic orbitals into a bonding π and antibonding π^* molecular orbital. The drawings represent the angular distributions of the electron densities; the signs + and − specify the signs of the wavefunction.

Figure 9.31 indicates what kind of electronic transitions to expect on excitation by light. For H_2 the longest-wavelength absorption corresponds to a $\sigma \rightarrow \sigma^*$ transition. The excited state with one electron in a σ MO and one in a σ^* MO is unstable, as expected. For O_2, a $\pi \rightarrow \pi^*$ transition occurs, and the double-bond character decreases on excitation. This results in a weaker bond in the excited molecule.

We must emphasize again that this MO picture is an approximation. However, it can help us remember and systematize the experimental results. There are other reasonable explanations of the bonding in O_2 that we could give. One is based on hybridization of AOs before forming MOs. We will define the various hybrid AOs in the next section, but here we will mention that an alternative description of O_2 involves first forming an sp hybrid on each O atom. The $\sigma(2p_x)$ MO in Fig. 9.31 is replaced by a $\sigma(sp)$, and the $\sigma(2s)$ plus $\sigma^*(2s)$ become two *nonbonding orbitals, n(sp)*, on each O atom. A *nonbonding MO* is an orbital that is located on a single atom only and is thus not involved in bonding.

Fig. 9.29 (a) Sum [$\sigma(1s)$] and difference [$\sigma^*(1s)$] of two $1s$ orbitals. These are the simplest examples of a σ and a σ^* molecular orbital. (b) Computer plots of LCAO approximations to the ground state (σ bonding) and first excited state (σ^* antibonding) for the H_2 molecule. In the middle part of the figure ψ is plotted in perspective in the x-y plane. The marked line in front of the grid is $2a_0 = 1.058$ Å long. The bottom part of the figure shows contour lines on which the value of the wavefunction is constant. The + and − on the σ^* orbital indicate the signs of the wavefunction. (Reprinted with permission from A. Streitwieser, Jr., and P. H. Owens, *Orbital and Electron Density Diagrams*, Macmillan Publishing Company, New York, 1973. Copyright 1973 by Macmillan Publishing Company.)

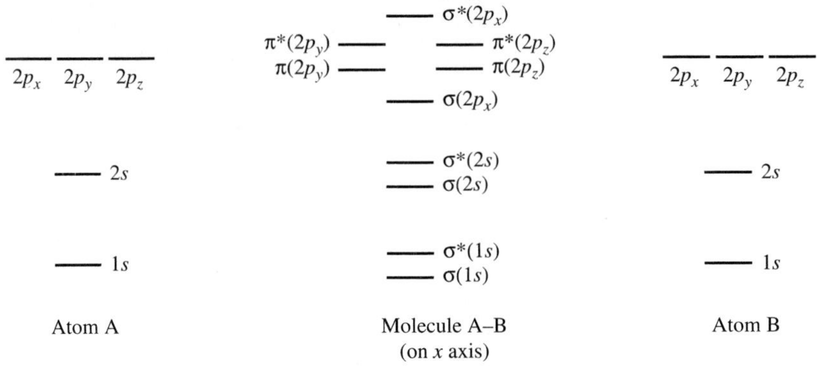

$\sigma^*(2p_x)$
$\pi^*(2p_y)$
$\pi^*(2p_z)$
$\pi(2p_y)$
$\pi(2p_z)$
$\sigma(2p_x)$
$\sigma^*(2s)$
$\sigma(2s)$
$\sigma^*(1s)$
$\sigma(1s)$
$2p_x$ $2p_y$ $2p_z$
$2s$
$1s$
$2p_x$ $2p_y$ $2p_z$
$2s$
$1s$
Atom A
Molecule A–B (on x axis)
Atom B

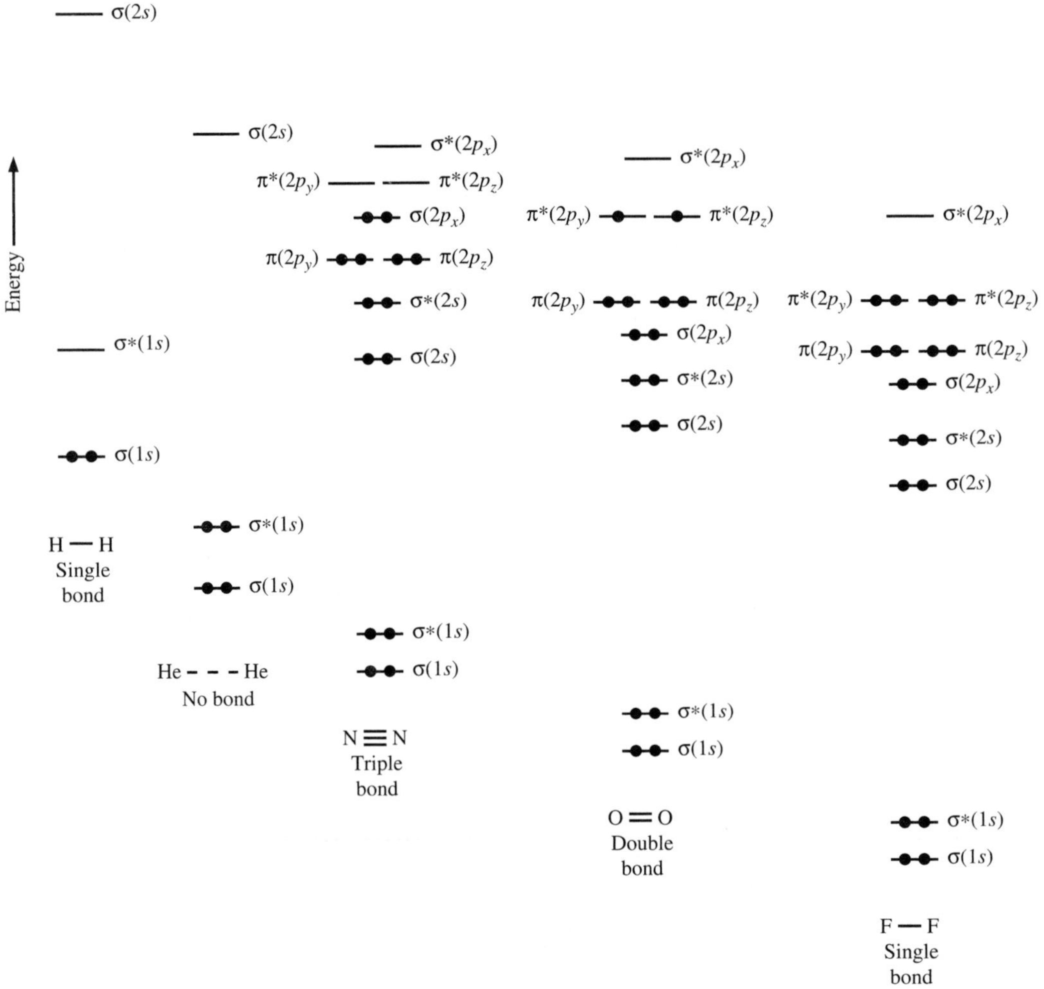

Energy
$\sigma(2s)$
$\sigma^*(1s)$
$\sigma(1s)$
H — H
Single bond
$\sigma(2s)$
$\sigma^*(1s)$
$\sigma(1s)$
He - - - He
No bond
$\sigma^*(2p_x)$
$\pi^*(2p_y)$ $\pi^*(2p_z)$
$\sigma(2p_x)$
$\pi(2p_y)$ $\pi(2p_z)$
$\sigma^*(2s)$
$\sigma(2s)$
$\sigma^*(1s)$
$\sigma(1s)$
N ≡ N
Triple bond
$\sigma^*(2p_x)$
$\pi^*(2p_y)$ $\pi^*(2p_z)$
$\pi(2p_y)$ $\pi(2p_z)$
$\sigma(2p_x)$
$\sigma^*(2s)$
$\sigma(2s)$
$\sigma^*(1s)$
$\sigma(1s)$
O = O
Double bond
$\sigma^*(2p_x)$
$\pi^*(2p_y)$ $\pi^*(2p_z)$
$\pi(2p_y)$ $\pi(2p_z)$
$\sigma(2p_x)$
$\sigma^*(2s)$
$\sigma(2s)$
$\sigma^*(1s)$
$\sigma(1s)$
F — F
Single bond

Example 9.6   The molecule C_2 is known in the gas phase. Assuming that the ordering of the MO energy levels is the same as that shown for N_2 in Fig. 9.31, show the electron occupancy of the ground state of C_2. Do you expect C_2 to be paramagnetic or diamagnetic? If C_2^+ is formed by removing an electron from the highest energy MO of C_2, will the bond energy be smaller or larger than that of C_2?

Solution   The molecule C_2 has two fewer electrons than does N_2. Thus the electron occupation will be the same as for N_2 except that the $\sigma(2p_x)$ MO will be empty for C_2. All of the electrons are paired, and the molecule should be diamagnetic, which it is. The bond dissociation energy of C_2 is 600 kJ mol^{-1}, which is reasonable for a double bond between the carbon atoms.

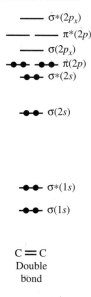

$$C = C$$
Double
bond

If one electron is removed to form the ion C_2^+, the easiest to remove is one from a $\pi(2p)$ orbital. This is a bonding orbital. Removing an electron from it decreases the bond order from 2 in C_2 to $\frac{3}{2}$ in C_2^+. Thus the bond should be weaker in C_2^+. A value of 510 kJ mol^{-1} has been determined, which is indeed somewhat weaker than for C_2.

Fig. 9.31 Energy-level patterns and electron occupation for homonuclear diatomic molecules. (Top) Typical energy levels of molecular orbitals formulated in linear combinations of orbitals of the separated atoms. (Bottom) The type of bonding and the relative bond energies can be understood for the first 10 elements in the periodic table. The energy of a given type of MO decreases from left to right because increasing nuclear charge results in greater electrostatic interaction with the electrons in the MOs. The molecules are all shown in their ground (lowest energy) electronic state with no more than two electrons per MO, according to the Pauli principle. Note that O_2 contains two "unpaired" electrons, which are located in two π^* orbitals that have the same energy but different spatial orientations. This is the basis of our understanding of the fact that O_2 exhibits *paramagnetism* in its interactions with a magnetic field. The other molecules shown are *diamagnetic* in their ground states; all electrons are paired.

Hybridization

In principle, the molecular orbitals of a multinuclear molecule can be obtained from the appropriate combinations of the atomic orbitals of the proper symmetry types. It is easier to visualize a complex molecule if the atomic orbitals of a given atom are combined into a set of *hybrid* orbitals first. These hybrid orbitals are then used to form molecular orbitals with neighboring atoms. A hybrid AO is just a linear combination of AOs on the same atom. The purpose of forming hybrid orbitals is to rationalize the observed geometry of the molecule. Hybridization can make more orbitals available for bonding (as in CH_4) or it can produce a more favorable bond angle (as in H_2O) than would be the case using unhybridized orbitals.

The three types of hybrid AOs that we will consider are sp, sp^2, and sp^3. There are two sp hybrids made from possible combinations of a $2s$ and a $2p$ AO. There are three sp^2 hybrids from possible combinations of a $2s$ and two $2p$ AOs. There are four sp^3 hybrids from possible combinations of a $2s$ and three $2p$ AOs. All four sp^3 hybrids have the same shape, but they are oriented in different directions in space. The three sp^2 hybrids are also identical to each other in shape, as are the two sp hybrids. They are similar in shape to the sp^3 hybrid (Fig. 9.32), but they are oriented differently in space. Equations for one of each set of hybrid orbitals are

$$\psi(sp) = \frac{1}{\sqrt{2}} \psi_A(2s) + \frac{1}{\sqrt{2}} \psi_A(2p)$$

$$\psi(sp^2) = \frac{1}{\sqrt{3}} \psi_A(2s) + \sqrt{\frac{2}{3}} \psi_A(2p) \qquad (9.36)$$

$$\psi(sp^3) = \frac{1}{\sqrt{4}} \psi_A(2s) + \sqrt{\frac{3}{4}} \psi_A(2p)$$

Figure 9.32 can represent any one of these hybrid AOs.

The important difference among the different hybrids is their orientation in space. The two sp hybrids point in opposite directions. The bond angle between two sp hybrids is 180°. The three sp^2 hybrids point toward the corners of an equilateral triangle centered on the nucleus. The bond angle between two sp^2 hybrids is 120° and all three orbitals lie in the same plane. The four sp^3 hybrids point toward the corners of a tetrahedron centered on the nucleus. The bond angle between two sp^3 hybrids is 109°28'. Examples of molecules that can be described in terms of hybrid orbitals are shown in Fig. 9.33.

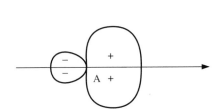

Fig. 9.32 Angular dependence of an sp^3 hybrid orbital; sp and sp^2 orbitals are similar. The arrow indicates the orientation of the hybrid. The plus and minus indicate the sign of the wavefunctions. A bond can be formed by combining this atomic orbital with an atomic orbital on another nucleus in the direction of the arrow.

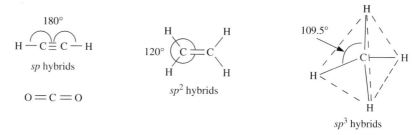

Fig. 9.33 Examples of molecules whose σ bonds can be represented with hybrid atomic orbitals. In acetylene each C—H bond is a linear combination of an *sp* hybrid on C plus a 1*s* orbital on H. In CO_2 the *sp* hybrids on C interact with valence shell orbitals on each O to form σ bonds. In ethylene the six σ bonds involve sp^2 hybrids on C. In methane the four σ bonds are combinations of the four sp^3 hybrids on C plus a 1*s* orbital on each H. The π bonds in acetylene, CO_2, and ethylene are sums of 2*p* orbitals.

Hybrid orbitals of a more complex nature can be written for transition metals such as Fe, Cu, Mo, and others that play important biochemical roles. In many such cases *d* orbitals will participate in forming hybrids, such as the d^2sp^3 hybrids that result in octahedral coordination of six ligands, along three mutually perpendicular axes. The use of such atomic orbitals to account for bonding is known as the *valence-bond* (V-B) approach. It works best where electrons are localized in the region between two participating atoms. For molecules where electrons are delocalized over several nuclei, the *molecular orbital* (MO) approach (below) is more satisfactory.

Real molecules, even those with localized electrons, are not always easy to describe using simple atomic or hybrid orbitals. For example, consider the important molecule H_2O, which is a bent molecule with two equal O—H bonds and two pairs of nonbonded electrons associated with the oxygen. Without hybridization the 2*p* orbitals of O would lead to a 90° bond angle for H_2O. This is because any two 2*p* orbitals chosen to form the two bonds to the two H atoms would be perpendicular to one another. If, on the other hand, sp^3 hybrids are used to form the bonds, then a 109.5° angle is expected. The lone-pair electrons would then occupy the two sp^3 hybrids that are not involved in bonding to H. The measured bond angle for H_2O is 104.5°, indicating that the true valence-bond picture lies somewhere in between. The fact that the accurate description of real molecules can become quite elaborate should not obscure the fact that the simple V-B and MO pictures provide clear insights into the general aspects of bonding and structure.

Delocalized Orbitals

We have previously thought of MOs as two-centered, with each orbital localized around two neighboring atoms. Although this approximation is often suitable for σ bonds, it may not be useful for conjugated π bonds. We have already seen, in discussing the free-electron model, that the π electrons in a conjugated system behave as if they are free to move throughout the conjugated system. Let us therefore consider the molecular orbitals of a simple conjugated system, 1,3-butadiene (CH_2=CH—CH=CH_2). Butadiene is a planar molecule, with angles between adjacent σ bonds close to 120°. This immediately suggests that sp^2-type hybrid orbitals of the carbon atoms are involved: each carbon atom forms three σ bonds with its three neighboring atoms. If the neighboring atom is a hydrogen, the bonding σ orbital is formed from the $1s$ orbital of the hydrogen and the sp^2 orbital of the carbon. If the neighboring atom is a carbon, the bonding σ orbital is formed from two sp^2 orbitals, one from each carbon. These bonds are localized; each includes two nuclei of the molecule. They are illustrated in Fig. 9.34a.

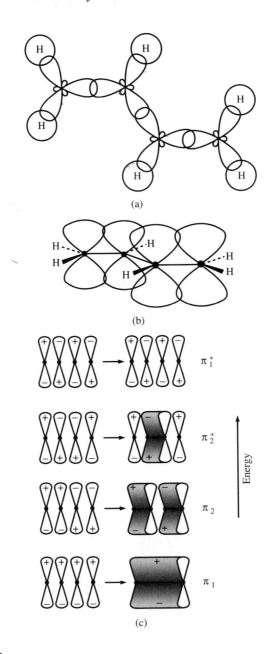

(a)

(b)

(c)

For each carbon atom, the formation of three sp^2 orbitals from a $2s$ and two $2p$ orbitals leaves it with one $2p$ orbital. If we choose the molecular plane as the xy plane, the $2p_z$ orbitals of the carbon atoms are not involved in the formation of the sp^2 orbitals and the σ bonds. The $2p_z$ orbitals form a set of multicentered π orbitals, as illustrated in Fig. 9.34b. The four $2p_z$ orbitals from the four carbons can form four MOs with π or π^* symmetry. Two turn out to be bonding, π, and two antibonding, π^*. The two bonding ones are filled. The delocalized orbitals mean that writing the double bonds as fixed is not very accurate. It is better to think of the double-bond character also being partly on the central C—C bond. Because the four π electrons occupy the two lowest-energy orbitals, π_1 and π_2, it should be clear that the central bond has less double-bond character than either of the two end C—C bonds. This is consistent with the experimental bond lengths of 1.476 Å for the central bond and 1.337 Å for the end bonds. The experimental C—C—C bond angle is 122.9°, which is close to that required for sp^2 hybrids to form the σ bond structure.

On the whole, the MO picture is a much better way of describing the properties of a molecule like 1,3-butadiene than is the use of classical resonance structures, such as

$$CH_2\!=\!\!CH\!-\!CH\!=\!\!CH_2 \;\longleftrightarrow\; \cdot CH_2\!-\!CH\!=\!\!CH\!-\!CH_2 \;\longleftrightarrow$$

$$\ominus\!:\!CH_2\!-\!CH\!=\!\!CH\!-\!CH_2\!\oplus$$

There is no experimental evidence for contributions of unpaired electrons or significant charge separation in 1,3-butadiene; as we have seen, the MO description does not require any.

We can label the carbon atoms, a, b, c, and d, starting at the left, and assign each an atomic $2p_z$ orbital, ψ_a, ψ_b, etc. Four normalized π molecular orbitals for butadiene, obtained using methods that are beyond the scope of this book, are

$$\psi(\pi_1) = 0.372\psi_a + 0.602\psi_b + 0.602\psi_c + 0.372\psi_d$$

$$\psi(\pi_2) = 0.602\psi_a + 0.372\psi_b - 0.372\psi_c - 0.602\psi_d$$

$$\psi(\pi_2^*) = 0.602\psi_a - 0.372\psi_b - 0.372\psi_c + 0.602\psi_d$$

$$\psi(\pi_1^*) = 0.372\psi_a - 0.602\psi_b + 0.602\psi_c - 0.372\psi_d$$

MOLECULAR STRUCTURE AND MOLECULAR ORBITALS

The language of molecular orbitals is used to help us systematize and remember facts of chemical structure. It presents us with a framework that simplifies the logical ordering of the data. The two main ideas we want to emphasize are:

Fig. 9.34 (a) The σ bonds in butadiene $CH_2\!=\!\!CH\!-\!CH\!=\!\!CH_2$ viewed face on. (b) The molecule viewed nearly edge on, showing the arrangement of the p_z orbitals relative to the σ-bond skeleton. (c) The π molecular orbitals in butadiene. The four π electrons occupy the two bonding orbitals, π_1 and π_2. The dots ($\cdot$) represent the positions of the carbon atom nuclei.

1. The electron distribution in a molecule can be described in terms of a set of molecular orbitals and corresponding energy levels. In the *ground state* of the molecule these MOs are filled by adding electrons to the lowest energy levels with at most two electrons in each level. In an *excited electronic state* one or more of the electrons is placed in an orbital of higher energy. The properties of the molecule depend on which MOs are occupied and which excitations of electrons to unfilled MOs can occur.

2. Each molecular orbital can be thought of as a sum of hydrogen-like atomic orbitals on different nuclei (except for nonbonding MOs which are localized on a single nucleus).

Geometry and Stereochemistry

It is straightforward to correlate bond angles, bond lengths, and orbitals. We have mentioned the correlation of sp^3, sp^2, and sp hybrids with 109.5°, 120°, and 180° bond angles. We also know that two unhybridized p orbitals on the same atom will form bonds at 90°. We can now use measured bond angles to help us decide on a consistent set of MOs, and we can use assumed MOs to predict bond angles. Similarly, bond types and bond distances can be correlated with bonding orbitals. A single bond (a sigma bond) is longer than a double bond (a sigma bond plus a pi bond), which is longer than a triple bond (a sigma bond plus two pi bonds). Rotation about a single bond is easy, whereas rotation about a double bond requires enough energy to break the π bond. Figure 9.35 shows some measured bond angles and bond lengths. The expected trends are seen. The C≡C bond is 20% shorter than the C—C bond; the benzene C—C bond is intermediate in length between ethane and ethylene. The H—C—H bonds in ethane are tetrahedral. In ethylene the H—C—H is less than the 120° trigonal angle expected, but in benzene the C—C—H angles are precisely 120°.

The structure of formamide is informative. The H—N—H angle of 119° implies that the bonding around the N is trigonal, not tetrahedral pyramidal as in NH_3. The O—C—N angle is also trigonal planar, as expected. The C—N bond distance is 1.343 Å, which is shorter than a C—N single bond (1.47 Å) such as is found in methyl amine. These data indicate that formamide is planar and that the C—N bond is partially a double bond. In resonance theory this is described in terms of migration of the lone pair of electrons on the N atom.

In the MO picture the π orbitals assumed for formamide would be delocalized over three nuclei (C, N, O). The peptide bond is similar to formamide in that it is planar with delocalized π MOs.

In butadiene the delocalized π MOs are consistent with the fact that the two terminal double bonds in CH_2=CH—CH=CH_2 are slightly longer than a C=C double bond in a

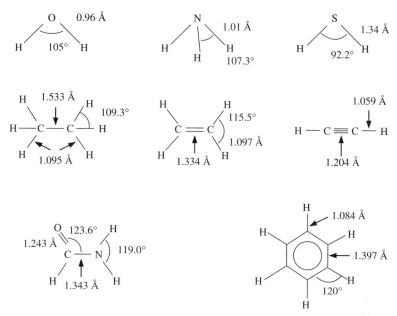

Fig. 9.35 Molecular geometry. Bond angles are given in degrees, and bond distances are in angstroms (1 Å = 0.1 nm).

nonconjugated system: 1.34 Å in the former versus 1.33 Å in the latter. On the other hand, the central C—C bond of the molecule is considerably shorter than the typical C—C single bond: 1.48 Å versus 1.54 Å. Also, while rotation around a regular single bond has a low energy barrier, the π-bond character of the central C—C link prohibits rotation around this bond.

A very important application of molecular orbital methods is in the understanding of chemical reactions and reaction mechanisms. The Woodward-Hoffman rules (Hoffman and Woodward, 1968) use the symmetry characteristics of molecular orbitals to rationalize and predict the stereochemical course of concerted organic reactions. An example is the conversion of cyclobutene to its isomer, butadiene. An isotopically labeled reactant can form either of two products:

The important MOs are the π and π^* of the $C_2{=}C_3$ double bond, the σ and σ^* of the C_1—C_4 single bond of cyclobutene, and the four π, π^* MOs of butadiene. If we can find a path that converts occupied bonding MOs in the reactant to occupied bonding MOs in the product, the reaction should occur easily. Hoffman and Woodward conclude that ground-state cyclobutene should form product (A), while excited-state cyclobutene should form product (B). They thus predict, and find, that thermal reaction gives (A), but photochemical reaction gives (B).

The catalytic efficiency and stereochemical selectivity of enzymes can be related to the ability of the enzyme to constrain reactants and products to particular orientations. This specific orientation will greatly enhance certain reactions and reaction paths over the more random collisions produced by thermal motion. The literature in this field is large and it often contains molecular orbital language [see for example, Lolis and Petsko (1990), Knowles (1989), and Stubbe (1989)].

Transition Metal Ligation

The molecular orbital picture is very important for describing the interactions between ligands and transition metals, such as Fe in myoglobin, hemoglobin, or cytochromes, Cu in plastocyanin or cytochrome oxidase, Co in vitamin B_{12}, and many others. These metal atoms contain incompletely filled d orbitals in their valence shells: they can accept electrons into these and other low-lying unfilled orbitals from electron-rich ligands, such as Cl^-, OH^-, pyridine, CO, O_2, and the central nitrogens of the heme porphyrin ring. The metal atoms can, at the same time, form favorable interactions by donating some of their valence electrons into favorably oriented, unoccupied orbitals of the ligands. The *ligand field strength* is a measure of the ability of a given ligand to coordinate well with a metal atom. This in turn influences the energy levels of the metal atom and the geometry of the resulting complex.

An illustration of how the presence and nature of ligands influences the d-orbital energies of metal atoms is shown in Fig. 9.36. In the unliganded atom or ion, the five equivalent d orbitals, designated d_{xy}, d_{xz}, d_{yz}, d_{z^2}, and $d_{x^2-y^2}$, are all degenerate in energy. The presence of ligand interactions breaks this degeneracy in one of several characteristic patterns shown. Which of these is formed by a particular metal-ligand interaction depends on several factors, but ultimately the structure giving the lowest energy will occur.

1. Nature of the ligand. Ligand field strength goes, from "weak field" to "strong field," in the order

$$I^- < Br^- < Cl^- < F^- < OH^- < H_2O < NCS^- < NH_3$$
$$< \text{ethylene diamine} < NO_2^- < CN^- \sim CO$$

Strong-field ligands produce larger energy splittings. Ligand field strength is partly a measure of the electrostatic field created by the ligand. Small ions, such as F^-, concentrate the field better than do equivalent large ions, such as I^-. Ligands such as CN^- or CO that have an unfilled antibonding π^* orbital exhibit enhanced ligand field strength because of interactions with d electrons in the metal orbitals that are favorably oriented.

2. Oxidation state of the metal. A larger charge on the metal produces a greater electrostatic interaction with the electrons of the ligand.
3. Number of electrons available to occupy the MOs.
4. Geometric placement of the ligands relative to the regions of highest electron density
   of the orbital. Overlap occurs best for metal orbitals oriented along axes where ligands are located.
5. Both σ contributions and π contributions (especially in ligands such as NO_2^-, CN^-,
   and CO) are important in maximizing the interactions.

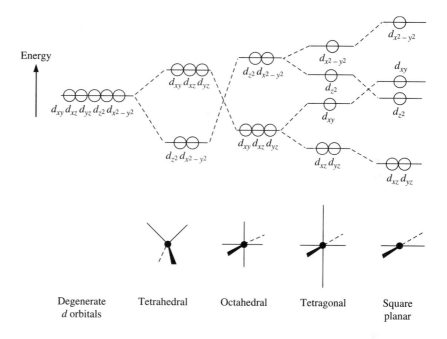

Fig. 9.36 Ligand-induced splitting of d orbitals of transition metal complexes
and the relation to geometry. Electrons in orbitals that point in the directions of
ligands are increased in energy relative to those that are localized primarily away
from the ligand direction. The d electrons of the transition metal tend to occupy
orbitals of the lowest available energy. There is, however, an additional tendency
for electrons to unpair (Hund's rule), even if it costs a small amount of energy to
promote one of them to a higher orbital to achieve this unpairing. The interplay
of these factors is the basis of the rich assortment of spectroscopic, magnetic, and
chemical properties exhibited by transition metal complexes.

    In addition to the geometry of the ligand-metal coordination complex, other proper-
ties affected by these factors include the color of the species, the spin state or paramagnetism, the standard reduction potential, and the chemical reactivity. The familiar differences
between the green color of solutions containing $CuCl_4^{2-}$ and the deep blue of $Cu(NH_3)_4^{2+}$ or
the pink color of $Co(H_2O)_6^{2+}$ and the deep blue of $CoCl_4^{2-}$ arise from differences in the
energy-level splittings caused by the different ligands. The origins of such color differences
will be explored further in the next chapter. The spin state or paramagnetism of transition

metals is controlled by how the electrons in the valence shell are distributed among the available d orbitals. As always, the configuration of lowest overall energy will occur, but several factors contribute. In general, electrons will go into the available orbitals of lowest energy, but Hund's rule states that it costs some energy to pair electrons in the same orbital rather than have them remain unpaired in different orbitals of about the same energy. Thus the five d electrons of Fe(III) in $Fe(H_2O)_6^{3+}$ or in the heme-containing enzyme horseradish peroxidase are all unpaired and distributed one in each of the available d orbitals (Fig. 9.37). This gives the "high-spin" state of maximum paramagnetism. This state is designated as the $S = \frac{5}{2}$ state, where S, the spin quantum number for the ion or molecule as a whole, is the sum of the spin quantum numbers $m_s = \frac{1}{2}$ for the individual electrons. Thus, for $Fe(H_2O)_6^{3+}$, the value $S = 5(\frac{1}{2}) = \frac{5}{2}$ is obtained. By contrast, for $Fe(CN)_6^{3-}$ or the heme in ferricytochrome c only one of the five d electrons remains unpaired, giving a low-spin state with $S = \frac{1}{2}$. The difference occurs because CN^-, for example, is a strong-field ligand that splits the d orbitals by such a large amount that it costs too much energy to unpair the electrons and promote them from the lower orbital to the upper one. With a weak-field ligand like H_2O, however, the orbital splitting is sufficiently small that a state of maximum unpairing has the lowest overall energy. Many examples of this have been documented, and measurements of the paramagnetic state provide very important information about the structure and chemistry of biochemical substances that contain transition metals.

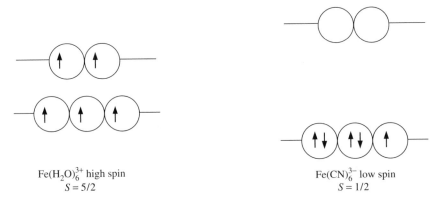

$Fe(H_2O)_6^{3+}$ high spin
$S = 5/2$

$Fe(CN)_6^{3-}$ low spin
$S = 1/2$

Fig. 9.37 Electron distribution among d orbitals in octahedral Fe(III) complexes with a weak-field ligand (H_2O) and with a strong-field ligand (CN^-).

Charge Distributions and Dipole Moments

An LCAO-MO description of a molecule characterizes the electron distribution in the molecule. When we write, for example, a π MO for formamide (Fig. 9.35), it will have the form

$$\psi(\pi) = C_N \phi_N(2p_z) + C_C \phi_C(2p_z) + C_O \phi_O(2p_z)$$

where N, C, and O refer to the nitrogen, carbon, and oxygen nuclei in formamide; $\phi_N(2p_z)$, etc., are corresponding $2p_z$ AOs on each nucleus; and C_N, C_C, and C_O are numerical coefficients. Each electron that occupies that MO is distributed according to the square of the wavefunction. Therefore, a π electron in that MO will be distributed as follows: C_N^2

probability on N, C_C^2 probability on C, and C_O^2 probability on O. The sum of probabilities is, of course, equal to 1. The total electronic charge on each nucleus is just the sum of charges contributed by each occupied MO. In this approximation the net charge on each nucleus is the nuclear charge minus the electronic charge calculated from the MOs. The dipole moment in this approximation is given by Eq. (9.3), with each charge being the net charge on each nucleus. An example of this kind of calculation is illustrated for adenine in Fig. 9.3.

A more general discussion of applications of charge distributions is given by Wiberg et al. (1991).

SUMMARY

Coulomb's Law

The energy of interaction between charges

$$U(r) = \frac{q_1 q_2}{4\pi\varepsilon_0 r} \qquad \text{(Energy in joules)} \qquad (9.1)$$

q_1 and q_2 are in coulombs; r is in meters
ε_0 (the permittivity constant) = $8.854 \times 10^{-12}\,C^2\,N^{-1}\,m^{-2}$

$$U(r) = 1389\,\frac{q_1 q_2}{r} \qquad \text{(Energy in kJ mol}^{-1}\text{)}$$

$$(9.2)$$

$$U(r) = 331.8\,\frac{q_1 q_2}{r} \qquad \text{(Energy in kcal mol}^{-1}\text{)}$$

q_1 and q_2 are in multiples of an electronic charge; r is in Å.

For charges interacting in a medium of dielectric constant ε, the right side of Eq. (9.1) or (9.2) is divided by ε.

Dipoles and Their Interaction Energy

The components of the dipole moment vector are

$$\boldsymbol{\mu}_x = \sum_i q_i x_i$$

$$\boldsymbol{\mu}_y = \sum_i q_i y_i \qquad (9.3)$$

$$\boldsymbol{\mu}_z = \sum_i q_i z_i$$

$$U(R) = \frac{1389\,\mu_1\mu_2}{R_{12}^3}\,(\cos\theta_{12} - 3\cos\theta_1\cos\theta_2) \qquad \text{(Energy in kJ mol}^{-1}\text{)} \qquad (9.4)$$

μ_1 and μ_2 are in electron Å (1 electron Å = 4.803 debye); R_{12} is in Å
θ_1 and θ_2 are the angles each dipole makes with the line joining the two point dipoles; θ_{12} is the angle between the two dipoles. See Fig. 9.4.

Intramolecular (Within) and Intermolecular (Between) Interactions

$$U = + k_r(r - r_{eq})^2 \qquad \text{bond stretching}$$

$$+ k_b(\theta - \theta_{eq})^2 \qquad \text{bond angle bending}$$

$$+ V_n/2[1 + \cos(n\phi - \phi_0)] \qquad \text{torsion angle rotation } (n = 2 \text{ or } 3)$$

$$+ \frac{B_{ij}}{r_{ij}^{12}} \qquad B_{ij} \text{ is positive} \qquad \text{van der Waals repulsion}$$

$$- \frac{A_{ij}}{r_{ij}^{6}} \qquad A_{ij} \text{ is positive} \qquad \text{London attraction}$$

$$+ C\frac{q_i q_j}{r_{ij}} \qquad \text{Coulomb interaction (C depends on the units)}$$

$$(9.11)$$

All terms can contribute to intramolecular interactions; only the last three can contribute to intermolecular interactions.

Schrödinger's Equation

Schrödinger's time-independent equation:

$$\mathscr{H}\psi = E\psi \qquad (9.13)$$

$$\mathscr{H} = \text{Hamiltonian operator}$$

$$\equiv -\frac{h^2}{8\pi^2} \sum_i \frac{1}{m_i} \cdot \left(\frac{\partial^2}{\partial x_i^2} + \frac{\partial^2}{\partial y_i^2} + \frac{\partial^2}{\partial z_i^2} \right) + U(x_i, y_i, z_i)$$

h = Planck's constant

Σ_i = sum over all particles in system

m_i = mass of each particle i in system

x_i, y_i, z_i = coordinates of each particle i in system

$U(x_i, y_i, z_i)$ = potential-energy operator for system

ψ = eigenfunction or wavefunction, a function of coordinates of all particles in system; there will be an infinite number of wavefunctions corresponding to different states of the system and usually each wavefunction will correspond to a different energy of the system

E = energy of the system

The value of any measurable property for a system characterized by wavefunction ψ can be calculated from the appropriate operator:

$$\text{value} = \int \psi^* \text{ operator } \psi \, dv$$

ψ = normalized wavefunction of system; ψ^* = its complex conjugate; $\int \psi^* \psi \, dv = 1$; for real wavefunctions, $\psi^* = \psi$.

The probability of finding an electron in a small volume of size Δv around point x_0, y_0, z_0 is $[\psi(x_0, y_0, z_0)]^2 \Delta v$, where $\psi(x_0, y_0, z_0)$ is the wavefunction of the electron evaluated at point x_0, y_0, z_0. To calculate the probability of finding an electron in a finite volume of space, the square of the wavefunction is integrated over the volume.

Some Useful Operators

Energy E:
   Operator for Cartesian coordinates for one particle:

$$\mathcal{H} = -\frac{h^2}{8\pi^2 m} \left(\frac{\partial^2}{\partial x^2} + \frac{\partial^2}{\partial y^2} + \frac{\partial^2}{\partial z^2} \right) + U$$

Position x, y, z:
Momentum: p_x, p_y, p_z:
   The linear momentum components along x, y, z are p_x, p_y, p_z. The classical definition of momentum is mass $\cdot$ velocity. Units are, for example, g cm s^{-1} or kg m s^{-1}.
   Operators in Cartesian coordinates:

$$p_x = \frac{h}{2\pi i} \frac{\partial}{\partial x} \qquad p_y = \frac{h}{2\pi i} \frac{\partial}{\partial y} \qquad p_z = \frac{h}{2\pi i} \frac{\partial}{\partial z}$$

Systems Whose Schrödinger Equation Can Be Solved Exactly

Particle in a box: potential energy for a one-dimensional box is $U(x) = 0$ in box and $U(x) = \infty$ outside box:

$$-\frac{h^2}{8\pi^2 m} \frac{d^2\psi}{dx^2} = E\psi \tag{9.19}$$

$$\frac{h^2}{8\pi^2} = \frac{\hbar^2}{2} = 5.561 \times 10^{-69} \text{ J}^2 \text{ s}^2$$

m = mass of particle
ψ = wavefunction of particle in box
E = energy of particle in box

Normalized wavefunctions for a particle in a one-dimensional box of length a:

$$\psi_n = \sqrt{\frac{2}{a}} \sin \frac{n\pi x}{a} \tag{9.20}$$

ψ_n = normalized wavefunction of particle in box in quantum state n
n = quantum number, a positive integer
a = length of box

Energies for a particle in a one-dimensional box of length a are proportional to n^2:

$$E_n = \frac{n^2 h^2}{8ma^2} \tag{9.21}$$

E_n = energy of particle in quantum state n; each energy corresponds to a unique
  quantum state; that is, no energy value is degenerate
n = quantum number, a positive integer
$h^2/8$ = 5.4883×10^{-68} J^2 s^2
m = mass of particle
a = length of box

Normalized wavefunctions for a particle in a two-dimensional box:

$$\psi_{n_x n_y} = \frac{2}{\sqrt{ab}} \sin \frac{n_x \pi x}{a} \sin \frac{n_y \pi y}{b}$$

$\psi_{n_x n_y}$ = normalized wavefunction of particle in box of quantum state specified by
  quantum numbers n_x and n_y
n_x, n_y = quantum numbers, positive integers
a = length of box in x direction
b = length of box in y direction

Energies for a particle in a three-dimensional box:

$$E_{n_x n_y n_z} = \left(\frac{n_x^2}{a^2} + \frac{n_y^2}{b^2} + \frac{n_z^2}{c^2} \right) \frac{h^2}{8m} \tag{9.24}$$

Schrödinger equation for a one-dimensional harmonic oscillator:

$$-\frac{h^2}{8\pi^2\mu} \frac{d^2\psi}{dx^2} + \frac{k}{2} x^2\psi = E\psi \tag{9.27}$$

ψ = wavefunction for a linear oscillator whose potential energy is directly proportional
  to the square of its displacement
E = energy of harmonic oscillator
k = force constant of harmonic oscillator, mass time^{-2}
h = Planck's constant
μ = reduced mass of oscillator (molecule)

Energies for a harmonic oscillator are equally spaced:

$$E_v = (v + \tfrac{1}{2})h\nu_0$$

$$\nu_0 = \frac{1}{2\pi}\sqrt{\frac{k}{\mu}}$$

(9.28)

$E_0 = h\nu_0/2 =$ zero-point energy
$v =$ quantum number for harmonic oscillator $= 0$, or 1, or 2, etc.
$h =$ Planck's constant
$\nu_0 =$ fundamental vibration frequency of oscillator
$k =$ force constant of harmonic oscillator, mass time^{-2}
$\mu =$ reduced mass of oscillator

Energies for an electron in a hydrogen atom are inversely proportional to n^2:

$$E_n = -\frac{2\pi^2\mu e^4}{(4\pi\varepsilon_0)^2 h^2 n^2} \qquad \text{SI units}$$

(9.31)

$$E_n = -\frac{2\pi^2\mu e^4}{h^2 n^2} \qquad \text{cgs-esu units}$$

(9.30)

$\mu =$ reduced mass of electron and proton $= \dfrac{m_p m_e}{m_p + m_e}$
$m_p =$ mass of proton
$m_e =$ mass of electron
$e =$ elementary charge
$h =$ Planck's constant
$\varepsilon_0 =$ permittivity of vacuum
$n =$ principal quantum number

For SI units, $2\pi^2\mu e^4/(4\pi\varepsilon_0)^2 h^2 = 2.179 \times 10^{-18}$ J; $m_e = 9.110 \times 10^{-31}$ kg; $e = 1.602 \times 10^{-19}$ C; $h = 6.626 \times 10^{-34}$ J s; $\varepsilon_0 = 8.8542 \times 10^{-12}$ C^2 N^{-1} m^{-2}.

For cgs-esu units, $2\pi^2\mu e^4/h^2 = 2.179 \times 10^{-11}$ erg; $m_e = 9.110 \times 10^{-28}$ g; $e = 4.803 \times 10^{-10}$ esu; $h = 6.626 \times 10^{-27}$ erg s.

$$E_n = \frac{-2.179 \times 10^{-18}}{n^2} \qquad \text{J atom}^{-1}$$

$$E_n = \frac{-2.179 \times 10^{-11}}{n^2} \qquad \text{erg atom}^{-1}$$

$$E_n = \frac{-313.6}{n^2} \qquad \text{kcal mol}^{-1}$$

$$E_n = \frac{-1312}{n^2} \qquad \text{kJ mol}^{-1}$$

$$E_n = \frac{-13.60}{n^2} \qquad \text{eV atom}^{-1}$$

MATHEMATICS NEEDED FOR CHAPTER 9

In this chapter for the first time we use a few trigonometric identities.

$$\cos n\pi = 1 \qquad \text{for } n = 0 \text{ or an even integer}$$
$$= -1 \qquad \text{for } n = \text{an odd integer}$$
$$\sin n\pi = 0 \qquad \text{for } n = \text{any integer}$$
$$\sin (\theta + 2\pi) = \sin \theta$$
$$\cos (\theta + 2\pi) = \cos \theta$$
$$\sin^2 \theta + \cos^2 \theta = 1 \qquad \text{for any value of } \theta$$

We discuss differential equations, but the reader is not expected to solve any. The reader should be able to use calculus to verify the solutions, however. Remember that

$$\frac{d}{d\theta} (\sin a\theta) = a \cos a\theta$$

$$\frac{d}{d\theta} (\cos a\theta) = -a \sin a\theta$$

$$\frac{d}{dx} (e^{ax}) = ae^{ax}$$

$$\frac{d}{dx} (e^{ax^2}) = 2axe^{ax^2}$$

Integrals of functions other than powers of x are used in solving some of the problems.

$$\int_0^\infty e^{-a^2x^2} dx = \frac{1}{2a} \sqrt{\pi}$$

$$\int \sin^2 ax \, dx = \frac{x}{2} - \frac{\sin 2ax}{4a} + C$$

$$C = \text{a constant}$$

The complex conjugate of a function is obtained by changing i (the imaginary) to $-i$ in the function.

We use vectors to represent dipoles. A vector has a magnitude and a direction; it is represented by a boldface letter such as $\boldsymbol{\mu}$. We can think of a vector as an arrow. The length of the arrow is the absolute magnitude; it points along the direction. The direction of a vector can be characterized quantitatively by specifying its components along the x, y, z axes of a coordinate system. The absolute magnitude is

$$|\mu| = \sqrt{\mu_x^2 + \mu_y^2 + \mu_z^2}$$

where μ_x, μ_y, μ_z are the components along x, y, z. The dot product (scalar product) of two vectors, $\mathbf{a} \cdot \mathbf{b}$, produces a scalar (a number). Its numerical value is obtained by either of two equivalent formulas. The components can be used:

$$\mathbf{a} \cdot \mathbf{b} = a_x b_x + a_y b_y + a_z b_z$$

or the angle between the two vectors can be used:

$$\mathbf{a} \cdot \mathbf{b} = |a| \, |b| \cos \theta$$

REFERENCES

The fundamentals of forces and energies are presented in

HALLIDAY, D., R. RESNICK, and J. WALKER, 1993. *Fundamentals of Physics,* 4th ed., John Wiley, New York.

Good discussions of atomic and molecular orbitals can be found in the following introductory chemistry books and organic chemistry books.

DICKERSON, R. E., H. B. GRAY, M. Y. DARENSBOURG, and D. J. DARENSBOURG, 1984. *Chemical Principles,* 4th ed., Benjamin/Cummings, Menlo Park, California.

MAHAN, B. H., and R. J. MYERS, 1987. *University Chemistry,* 4th ed., Benjamin/Cummings, Menlo Park, California.

OXTOBY, D. W., and N. H. NACHTRIEB, 1990. *Principles of Modern Chemistry,* 2nd ed., Saunders College Publishing, Philadelphia.

STREITWIESER, A., JR., C. H. HEATHCOCK, and E. M. KOSOWER, 1992. *Introduction to Organic Chemistry,* 4th ed., Macmillan, New York.

VOLLHARDT, K. P. C., and N. E. SCHORE, 1994. *Organic Chemistry,* 2nd ed., W. H. Freeman, New York.

Books which provide more depth to the topics discussed here include

DEKOCK, R. L., and H. B. GRAY, 1989. *Chemical Structure and Bonding,* 2nd ed., University Science Books, Mill Valley, California.

MCQUARRIE, D. A., 1983. *Quantum Chemistry,* University Science Books, Mill Valley, California.

TANFORD, C., 1980. *The Hydrophobic Effect,* 2nd ed., John Wiley, New York.

SUGGESTED READINGS

AXUP, A. W., M. ALBIN, S. L. MAYO, R. J. CRUTCHLEY, and H. B. GRAY, 1988. Distance Dependence of Photoinduced Long-Range Electron Transfer in Zinc/Ruthenium-Modified Myoglobins, *J. Am. Chem. Soc. 110,* 435–439.

BASH, P. A., U. C. SINGH, R. LANGRIDGE, and P. A. KOLLMAN, 1987. Free Energy Calculations by Computer Simulation, *Science 236,* 564–568.

BIRGE, R. R., 1981. Photophysics of Light Transduction in Rhodopsin and Bacteriorhodopsin, *Annu. Rev. Biophys. Bioeng. 10,* 315–354.

CHA, Y., C. J. MURRAY, and J. P. KLINMAN, 1989. Hydrogen Tunneling in Enzyme Reactions, *Science 243,* 1325–1330.

HOFFMANN, R., and R. B. WOODWARD, 1968. The Conservation of Orbital Symmetry, *Acc. Chem. Res. 1,* 17–22.

JORGENSEN, W. L., 1989. Free Energy Calculations: A Breakthrough for Modeling Organic Chemistry in Solution, *Acc. Chem. Res. 22,* 184–189.

KARPLUS, M., and G. A. PETSKO, 1990. Molecular Dynamics Simulation in Biology, *Nature 347,* 631–638.

KLINMAN, J. P., 1989. Quantum Mechanical Effects in Enzyme-Catalyzed Hydrogen Transfer Reactions, *Trends Biochem. Sci. 14,* 368–373.

KNOWLES, J. R., 1989. The Mechanism of Biotin-Dependent Enzymes, *Annu. Rev. Biochem. 58,* 195–221.

KOLLMAN, P. A., and K. M. MERZ, JR., 1990. Computer Modeling of the Interactions of Complex Molecules, *Acc. Chem. Res. 23,* 246–252.

KYTE, J., and R. F. DOOLITTLE, A Simple Method for Displaying the Hydropathic Character of a Protein, *J. Mol. Biol. 170,* 723–764.

LOLIS, E., and G. A. PETSKO, 1990. Transition-State Analogues in Protein Crystallography: Probes of the Structural Source of Enzyme Catalysis, *Annu. Rev. Biochem. 59,* 597–630.

MATHIES, R., and L. STRYER, 1976. Retinal Has a Highly Dipolar Vertically Excited Singlet State, *Proc. Natl. Acad. Sci. USA 73,* 2169–2173.

McCAMMON, J. A., 1987. Computer-Aided Molecular Design, *Science 238,* 486–491.

MULLER, N., 1990. Search for a Realistic View of Hydrophobic Effects, *Acc. Chem. Res. 23,* 23–28.

NAKANISHI, K., V. BALOGH-NAIR, M. ARNABOLDI, K. TSUJIMOTO, and B. HONIG, 1980. An External Point-Charge Model for Bacteriorhodopsin to Account for Its Purple Color, *J. Am. Chem. Soc. 102,* 7945–7947.

NÉMETHY, G., and H. A. SCHERAGA, 1990. Theoretical Studies of Protein Conformation by Means of Energy Computations, *FASEB J. 4,* 3189–3197.

RICHARDS, W. G., ed., 1989. *Computer-Aided Molecular Design,* VCH Publishers, New York.

STRYER, L., 1987. The Molecules of Visual Excitation, *Sci. Am.,* July, 42–50.

STRYER, L., 1988. Molecular Basis of Visual Excitation, *Cold Spring Harbor Symp. Quant. Biol. 52,* 283–294.

STUBBE, J. A., 1989. Protein Radical Involvement in Biological Catalysis, *Annu. Rev. Biochem. 58,* 257–285.

WEINER, S. J., P. J. KOLLMAN, D. A. CASE, U. C. SINGH, C. GHIO, G. ALAGONA, S. PROFETA, JR., and P. WEINER, 1984. A New Force Field for Simulations of Proteins and Nucleic Acids, *J. Am. Chem. Soc. 106,* 765–784.

WIBERG, K. B., C. M. HADAD, C. M. BRENEMAN, K. E. LAIDIG, M. A. MURCKO, and T. J. LePAGE, 1991. The Response of Electrons to Structural Changes, *Science 252,* 1266–1272.

PROBLEMS

1. (a) Coulomb's law is an important element in determining molecular structure. To understand this better, calculate the interaction energy in kJ mol^{-1} for an electron and a proton at a distance of the Bohr radius of 0.529 Å. Calculate the Coulombic interaction energy between atoms in kJ mol^{-1} for an O$\cdots$H hydrogen bond distance of 1.5 Å. The partial charge on O is –0.834; the partial charge on H is +0.417 in units of electronic charges.

   (b) For a molecule to be in a conformation which maximizes attractive forces and minimizes repulsive forces, which bonding interaction is most likely to change: bond stretching, bond angle bending, or torsion angle rotation? Which of these is the most difficult to change? How do you know?

2. (a) A carbon-carbon single bond has a length of 1.540 Å. For a quadratic potential with a force constant of 1200 kJ mol^{-1} Å^{-2}, how much energy (kJ mol^{-1}) would it take to stretch the bond length to 1.750 Å?

   (b) A C—C—C bond angle is nearly tetrahedral with a bond angle of 109.5°. For a quadratic potential with a force constant of 0.0612 kJ mol^{-1} degree^{-2}, how much energy (kJ mol^{-1}) would it take to distort the bond angle to a right angle (90°)?

(c) The potential energy, U, for rotation around a single bond can be represented by an equa-tion of the form $U = k[1 + \cos(3\phi)]$ where ϕ is the torsion angle. Use this equation to cal-culate the values of ϕ which give minima and maxima in the potential energy curve vs. ϕ.

(d) All atoms have an attractive interaction with one another that was first explained by Fritz London; it depends on r^{-6}, where r is the distance between the two atoms. Explain what causes this universal attraction.

3. When spacing between translational energy levels is small compared to random thermal ki-netic energies, classical mechanics is a good approximation for quantum mechanics. Thermal kinetic energy is of order of magnitude of kT, where k is Boltzmann's constant. Consider which particles in which boxes can be treated classically in their translational motion. For each system listed calculate the energy of transition from the ground state to the first excited state, and calculate the ratio of this transition energy to kT. State whether the system can be treated classically or not.

(a) A helium atom in a 1 μm cube at 25°C, and at 1 K.

(b) A protein with a molecular weight of 25,000 in a 100 Å cube at 25°C, and at 1 K.

(c) A helium atom in a 1 Å cube at 25°C, and at 1 K.

4. Consider a one-dimensional box of length l. For parts (a), (b), and (c), calculate the energy of the system in the ground state and the wavelength (in Å) of the first transition (longest wave-length).

(a) $l = 10$ Å, and the box contains one electron.

(b) $l = 20$ Å, and the box contains one electron.

(c) $l = 10$ Å, and the box contains two electrons (assume no interaction between the electrons).

(d) Write the complete Hamiltonian for the system described in part (c).

(e) What will happen to the answers in part (c) if the potential of interaction between the elec-trons is included in this Hamiltonian?

5. Consider a particle in a one-dimensional box.

(a) For a box of length 1 nm, what is the probability of finding the particle within 0.01 nm of the center of the box for the lowest energy level?

(b) Answer part (a) for the first excited state.

(c) The longest-wavelength transition for a particle in a box [not the box in part (a)] is 200 nm. What is the wavelength if the mass of the particle is doubled? What is the wave-length if the charge of the particle is doubled? What is the wavelength if the length of the box is doubled?

6. The exact ground-state energy for a particle in a box is $0.125h^2/ma^2$; the exact wavefunction is $(\sqrt{2/a}) \sin(n\pi x/a)$. Consider an approximate wavefunction $= x \cdot (a - x)$.

(a) Show that the approximate wavefunction fits the boundary conditions for a particle in a box.

(b) Given this approximate wavefunction, $\psi(x) = x \cdot (a - x)$, the approximate energy E' of the system can be calculated from

$$E' = \int_0^a \psi(x)\, \mathcal{H}\psi(x)\, dx \bigg/ \int_0^a \psi^2(x)\, dx$$

This is called the *variational method*. Calculate the approximate energy E'.

(c) What is the percent error for the approximate energy?

(d) According to the variational theorem, the more closely the wavefunction used approxi-mates the correct wavefunction, the lower will be the energy calculated according to part (b). The converse is also true. Propose another approximate wavefunction which could be used to calculate an approximate energy. How can you find out whether it is a better approximation?

7. The π electrons of metal porphyrins, such as the iron-heme of hemoglobin or the magnesium-porphyrin of chlorophyll, can be visualized using a simple model of free electrons in a two-dimensional box.

    (a) Obtain the energy levels of a free electron in a two-dimensional square box of length a.

    (b) Sketch an energy-level diagram for this problem. Set $E_0 = h^2/8ma^2$ and label the energy of each level in units of E_0.

    (c) For a porphyrin-like hemin that contains 26 π electrons, indicate the electron population of the filled or partly filled π orbitals of the ground state of the molecule. (Note that orbitals that are degenerate in energy, $E_{12} = E_{21}$, can still hold two electrons each.)

    (d) The porphyrin structure measures about 1 nm on a side ($a = 1$ nm). Calculate the longest-wavelength absorption band position for this molecule. (Experimentally these bands occur at about 600 nm.)

8. Nitrogenase is an enzyme that converts N_2 to NH_3. Extraction procedures have shown that nitrogenase contains an iron-sulfur cluster, Fe_4S_4, where the atoms of each type are arranged alternately at the corners of a cube. We wish to attempt to model the bonding in this iron-sulfur cube with the wavefunctions derived for an electron in a cubical (three-dimensional) box whose edges have length a.

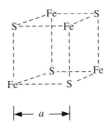

    (a) Write the Hamiltonian for the electron in the box.

    (b) Write general expressions for the energies (eigenvalues) and wavefunctions (eigenfunctions) for the electron in a three-dimensional cubical box as a function of the quantum numbers.

    (c) Construct an energy-level diagram showing the relative ordering of the lowest 11 "molecular orbitals" of this cube. Label the levels in terms of their quantum numbers. Be alert for degeneracies.

    (d) Assuming that the total number of valence electrons available is 20, show the electron configuration on the plot in part (c).

    (e) If $a = 3$ Å, calculate the transition energy for excitation of an electron from the highest filled orbital to the lowest unfilled orbital. At what wavelength in nm should this transition occur?

9. If G is a measurable parameter and g is the quantum mechanical operator corresponding to this parameter, the expected value of G, $\langle G \rangle$, can be calculated from

$$\langle G \rangle = \int \psi^* g \psi \, dv \bigg/ \int \psi^* \psi \, dv$$

where the integration extends over all space important to the problem and ψ^* is the complex conjugate of ψ (if ψ has no imaginary part, $\psi^* = \psi$). Calculate the expected position $\langle x \rangle$ and momentum $\langle p \rangle$ for a harmonic oscillator in its ground state.

10. For a diatomic species A—B, the vibration of the molecule can be treated approximately by considering the molecule as a harmonic oscillator.

(a) Calculate the reduced mass for HF.

(b) Take $k = 970$ m^{-1}; plot the potential energy versus internuclear distance near the equilibrium distance of 0.92 Å.

(c) Sketch in the same diagram horizontal lines representing the allowed vibrational energy levels.

11. The C=C stretching vibration of ethylene can be treated as a harmonic oscillator.

(a) Calculate the ratio of the fundamental frequency for ethylene to that for completely deuterated ethylene.

(b) Putting different substituents on the ethylene can make the C=C bond longer or shorter. For a shorter C=C bond, will the vibration frequency increase or decrease relative to ethylene? State why you think so.

(c) If the fundamental vibration frequency for the ethylene double bond is 2000 cm^{-1}, what is the wavelength in nm for the first harmonic vibration frequency?

12. The fundamental vibration frequency of gaseous ^{14}N^{16}O is 1904 cm^{-1}.

(a) Calculate the force constant, using the simple harmonic oscillator model.

(b) Calculate the fundamental vibration frequency of gaseous ^{15}N^{16}O.

(c) When ^{14}N^{16}O is bound to hemoglobin A, an absorption band at 1615 cm^{-1} is observed and is believed to correspond to the bound NO species. Assuming that when an NO binds to hemoglobin, the oxygen is so anchored that its effective mass becomes infinite, estimate the vibration frequency of bound ^{14}N^{16}O from the vibration frequency of gaseous NO.

(d) Using another model, that binding of NO to hemoglobin does not change the reduced mass of the NO vibrator, calculate the force constant of bound NO.

(e) Estimate the vibration of ^{15}N^{16}O bound to hemoglobin using, respectively, the assumptions made in parts (c) and (d).

13. The ionization potential of an electron in an atom or molecule is the energy required to remove the electron. Calculate the ionization potential for a $1s$ electron in He$^+$. Give your answer in units of eV and kJ mol^{-1}.

14. A physicist wishes to determine whether the covalently bonded species (HeH)$^{2+}$ could be expected to exist in an ionized gas.

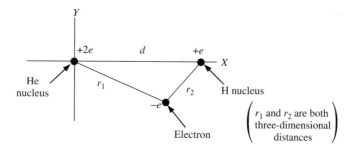

(a) Assuming that the nuclei remain separated by a fixed distance d (the nuclei do not move) and that the electron is free to move in three dimensions, write the Hamiltonian operator for this molecule.

(b) Write out fully the equation that must be solved exactly to determine the various wavefunctions and allowed energies of (HeH)$^{2+}$.

(c) Suppose that you have obtained all the exact solutions to the equation in part (b) How can you tell whether the covalently bonded species (HeH)$^{2+}$ is energetically stable?

15. Consider the diatomic molecules NO, NO^-, and NO^+.
    (a) Give MO diagrams for each of these molecules similar to those in Fig. 9.31.
    (b) Rank these species in increasing order of bond energies and in increasing order of bond lengths.

16. The Schrödinger equation can give you the electronic energy levels and wavefunctions for the hydrogen atom.
    (a) The electron distribution is characterized by the wavefunctions. Write the equation whose solution would allow you to calculate the radius of the sphere around the hydrogen nucleus within which the electron will be found 90% of the time. What is the radius of the sphere around the hydrogen atom within which the electron will be found 100% of the time?
    (b) Calculate the longest wavelength absorption (in nm) for an electronic transition in a hydrogen atom. In what region of the electromagnetic spectrum does this occur?
    (c) Write the Schrödinger equation whose solution will give you the translational energy levels and wavefunctions for a hydrogen atom in a cubic box with sides of length of 1 cm. Define all symbols in your equation.
    (d) Calculate the longest wavelength absorption (in m) for a translational energy transition of a hydrogen atom in a cubic box of length 1 cm. In what region of the electromagnetic spectrum does this occur?

17. Sketch the spatial distribution of the electron density for all of the valence orbitals of one of the carbon atoms in each of the following molecules. Identify each of the orbitals.
    (a) Acetylene.
    (b) Ethylene.
    (c) Ethane.

18. In a protein, a basic structural unit is the amide residue:

    where C_α represents a carbon at the α position to the carbonyl group. A key clue to the solution of the three-dimensional structures of proteins was the realization that the six atoms shown lie in a plane. By constructing the molecular orbitals around the atoms O, C, and N, explain why the planar structure is expected. What are the expected bond angles between adjacent bonds?

19. The molecule crotonaldehyde is an unsaturated aldehyde with the structural formula

    (a) Draw a picture showing the σ-bond structure for crotonaldehyde.
    (b) Crotonaldehyde has four π electrons associated with the two double bonds. Draw diagrams that illustrate the four π-type molecular orbitals for crotonaldehyde, showing the atomic orbitals from which they were constructed, their phase (sign) relations, and with which atoms they are associated. Rank the MOs in energy in terms of bonding versus antibonding character.
    (c) Describe the geometrical configuration that you expect for crotonaldehyde. Which groups, if any, will undergo free rotation with respect to the rest of the molecule?

(d) Aldehydes such as crotonaldehyde react with primary aliphatic amines such as *n*-butyl amine to form substituted imines, or Schiff bases.

$$R - C \overset{H}{\underset{O}{\diagdown}} + H_2NR' \longrightarrow RCH = NR' + H_2O$$

Schiff base

What geometry do you expect for the atoms associated with the imine function in a Schiff base?

(e) Schiff bases readily undergo protonation by H^+ in acidic solution. What atom is protonated, and what geometrical structure do you expect for the product? Is the resulting bond σ- or π-type? What type of orbital (p, sp, sp^2, sp^3, . . .) of the protonated atom is involved?

20. The nucleic acid base adenine is a planar aromatic molecule containing 70 electrons, of which 20 are core electrons. The σ bonding around each aromatic C and N can be represented by sp^2 hybrid orbitals.

(a) How many π electrons are there in adenine? State which atoms contribute one electron to the π MOs and which atoms contribute two electrons.
(b) How many π or π* MOs are there in adenine if we consider linear combinations of 2*s* and 2*p* atomic orbitals only?
(c) How many MOs of other types (σ, σ*, *n*) derived from 2*s* and 2*p* atomic orbitals are there in adenine?
(d) Which atoms, if any, have nonbonding orbitals localized on them?

21. An approximate MO calculation gives the following π-electrons charges on an amide group:

	Ground-state charges	Excited-state charges
O	−0.33*e*	+0.08*e*
C	+0.18*e*	−0.16*e*
N	+0.15*e*	+0.08*e*

$e = 4.8 \times 10^{-10}$ esu. Use the bond lengths given in Fig. 9.35 and 120° for the O—C—N bond angle.

(a) Calculate the components of the π dipole moment parallel and perpendicular to the C—N bond for the ground and excited state. Use debyes for units.
(b) Calculate the magnitudes of the π dipole moment for the ground state, for the excited state, and for the change in dipole moment.
(c) Plot the magnitudes and the directions to scale.

22. Consider the molecular orbitals of the planar peptide bond. The hypothesis is that the O and the central C and N atoms are sp^2 hybridized.

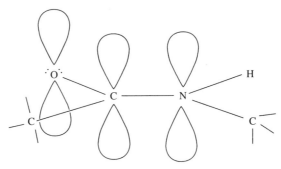

(a) Based on intuition derived from the particle-in-a-box wavefunctions, write the three π-type molecular orbitals for the "peptide bond" (ψ_1, ψ_2, ψ_3) in terms of the three available p orbitals (ϕ_O, ϕ_C, ϕ_N).

(b) Normalize ψ_1, ψ_2, and ψ_3. [To be properly normalized, the LCAO molecular orbitals

$$\psi_i = C_{i1}\phi_1 + C_{i2}\phi_2 + \ldots + C_{ij}\phi_j + \ldots$$

must satisfy three conditions:

(1) Normalization $\sum_j C_{ij}^2 = 1$

(2) Orthogonality $\int \phi_r \phi_s d\tau = 0$, for $r \neq s$

(3) Unit atomic orbital contribution $\sum_i C_{ij}^2 = 1$

Considerable advantage can be taken of symmetry inherent in the problem; e.g., in this case $C_{13} = \pm C_{11} = \pm C_{31} = \pm C_{33}$ and $C_{12} = \pm C_{32}$ and $C_{21} = \pm C_{23}$. You may wish to practice by testing the normalization conditions above for the MO expressions presented in connection with Fig. 9.34.]

(c) Rank these molecular orbitals in order of their energies. Determine how many electrons each atom (O, C, N) contributes to the π system. Give the electron configuration in the π MOs of the peptide bond.

(d) What is the π-electron density for each of the *filled* MOs as a function of ϕ_O, ϕ_C, ϕ_N? State and justify any approximations that you make.

(e) Verify that the integrals of these densities over all space give the total number of π electrons in each orbital.

(f) Integration of the total π electron density in a region around each atom yields the π-electron density of the atom. Calculate separately the π-electron charge on the O, C, and N atoms.

(g) What is the net charge on these atoms? (Remember that the molecule is neutral.) Calculate the peptide dipole moment given the coordinates $(x:y)$ for the three atoms (in Å): C(0:O), N(1.34:O), and O(−0.62, −1.07).

10

Molecular Structures and Interactions: Spectroscopy

CONCEPTS

Spectroscopy is the study of the interaction of electromagnetic radiation with matter. Figure 10.1 shows the wavelength regions of the electromagnetic spectrum that are useful in spectroscopy. Note that the visible light region is a very small part of this wide range. *Absorption* occurs when a molecule undergoes a transition to a higher energy level. The molecule is excited and absorbs energy from the radiation. *Nuclear magnetic resonance* is the absorption of radiofrequency electromagnetic radiation by nuclei placed in a static magnetic field. *Circular dichroism* is the differential absorption of right- and left-circularly polarized light. *Fluorescence* or *phosphorescence* occurs when an excited molecule undergoes a transition to a lower energy level and emits light. Excited molecules can also lose energy as heat and by chemical reactions. *Photochemistry* is the study of reactions caused by light.

All matter scatters electromagnetic radiation. In *Rayleigh scattering* only the direction of incident light is changed; in *Raman scattering* there is also a change in wavelength. For absorption to occur there must be an energy level spacing corresponding to the wavelength of the incident light, but scattering occurs at all wavelengths. The *refractive index,* the ratio of the speed of light in vacuum to the speed of light in a medium, is a measure of the light scattered in the direction of the incident beam. *Linear birefringence* is the difference in refractive index for light polarized in planes perpendicular to each other. *Circular birefringence* is the difference in refractive index for right- and left-circularly polarized light. *Optical rotation,* which is directly proportional to the circular birefringence, is the rotation of linearly polarized light by matter.

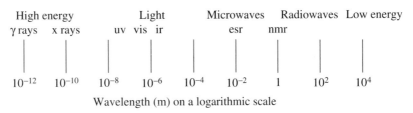

ELECTROMAGNETIC SPECTRUM

| High energy | | | Light | | Microwaves | Radiowaves | Low energy |
| γ rays | x rays | uv vis ir | | | esr | nmr | |

10⁻¹² 10⁻¹⁰ 10⁻⁸ 10⁻⁶ 10⁻⁴ 10⁻² 1 10² 10⁴

Wavelength (m) on a logarithmic scale

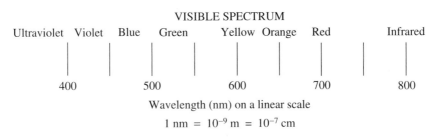

VISIBLE SPECTRUM

Ultraviolet Violet Blue Green Yellow Orange Red Infrared

400 500 600 700 800

Wavelength (nm) on a linear scale

$1 \text{ nm} = 10^{-9} \text{ m} = 10^{-7} \text{ cm}$

Fig. 10.1 Wavelength regions of the electromagnetic spectrum.

APPLICATIONS

Essentially all the spectroscopic methods can be used to detect and quantitate molecular species. The identities and concentrations of molecules in any environment are usually determined spectroscopically. For example, we can detect molecules in biological cells with a fluorescence microscope, or we can detect CO_2 in the atmosphere by infrared absorption. Changes in conformations of proteins and nucleic acids are easily followed by absorption, fluorescence, or circular dichroism. Nuclear magnetic resonance studies of macromolecules can provide their structures in solutions at atomic resolution.

ELECTROMAGNETIC SPECTRUM

We can characterize the electromagnetic spectrum in terms of wavelength, frequency, or energy. Different types of spectroscopists prefer different units. In the visible and near-ultraviolet region, wavelengths in nanometers are used: 1 nm, 10^{-9} m ≡ 1 mμm (millimicron) ≡ 10 Å ≡ 10^{-7} cm. In the microwave and radiowave region, frequencies are usually used.

$$\nu = \frac{c}{\lambda} \tag{10.1}$$

where ν = frequency, Hz (s^{-1})
   c = speed of light in vacuum = 3×10^8 m s^{-1}
   λ = wavelength in vacuum, m

Infrared spectroscopists plot their spectra versus wavenumbers, $\bar{\nu}$:

$$\bar{\nu} = \frac{1}{\lambda} \qquad (10.2)$$

where $\bar{\nu}$ = "frequency," cm^{-1} (wavenumbers)
λ = wavelength in vacuum, cm

Wavenumber is the number of wave amplitude maxima per 1-cm length along the wave. *Wavelength* is the distance between successive wave amplitude maxima. Units of energy are popular for spectra in the far-ultraviolet, x-ray, and γ-ray regions. Planck's equation, $\Delta\varepsilon = h\nu$, is used to convert frequencies to energies. For example, 200-nm ultraviolet light corresponds to about 6 eV or 598 kJ mol^{-1}. Interaction of radiation with matter may result in the absorption of incident radiation, emission of fluorescence or phosphorescence, scattering into new directions, rotation of the plane of polarization, or other changes. Each of these interactions can give useful information about the nature of the sample in which it occurs. Before discussing spectroscopy in detail, let us look at a few familiar examples.

COLOR AND REFRACTIVE INDEX

One of the most obvious properties of a substance is its color. The green color of leaves is due to chlorophyll absorption, the orange of a carrot or tomato arises from carotenes, and the red of blood results from hemoglobin. The color is characteristic of the spectrum (in the visible region) of light transmitted by the substance when white light (or sunlight) shines through it, or when light is reflected from it. The quantitative measure of color that we use is the absorption spectrum or reflectance spectrum. This is a plot of the variation with wavelength or frequency of the absorption of radiation by the substance. The absorption spectrum is complementary to the transmission of light. Chlorophyll is green because it absorbs strongly in the blue (435 nm) and red (660 nm) regions of the spectrum. Figure 10.2 compares the visible spectra of green chlorophyll, orange β-carotene, and red oxyhemoglobin. Note that the spectra of the three different molecules have both similarities and differences; the spectra are much richer in information than the colors would indicate.

The bent appearance of a pencil partly immersed in a cup of water results from the difference in refractive index of light in air and in water. The refractive index, *n,* is simply a measure of the ratio of the velocity of light in vacuum, *c,* to the velocity in the medium, *v.* The velocity of light in vacuum is approximately 3×10^8 m s^{-1} (3×10^{10} cm s^{-1}); this corresponds to about 1 ft ns^{-1}:

$$n = \frac{c}{v} \qquad (10.3)$$

The refractive index of air for visible light is about 1.00028 at 15°C and 1 atm pressure, so the velocity of light in air is very close to that in vacuum. The refractive index at a wavelength of 589 nm for some common substances is given in Table 10.1. The reason that we are able to see objects that are not colored is because the changes of refractive index at the boundaries result in bending of the light rays. Techniques such as phase contrast have been used in microscopy to increase the contrast between different regions of biological

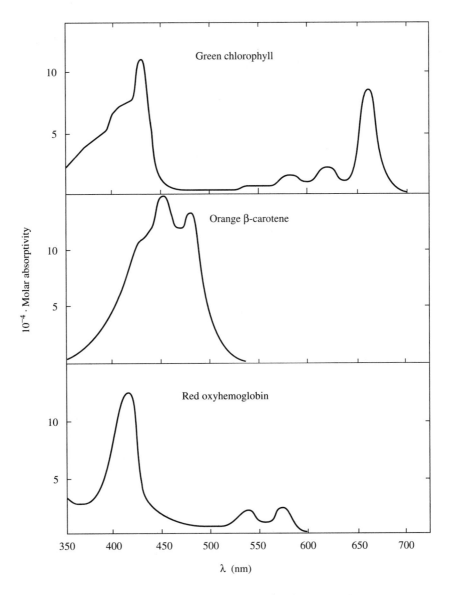

Fig. 10.2 Comparison of the visible spectra of three molecules.

organelles.* The appearance of detailed internal structure in such organelles occurs because their refractive index differs in the different regions. Because refractive index is an intrinsic property of a material that is related to its composition, a measurement in a refractometer can be used to characterize the substance.

---

* An organelle is an organized structure in a cell with a specialized biological function and surrounded by a limiting membrane. Examples are mitochondria, chloroplasts, and nuclei.

Table 10.1 Refractive indexes for some common substances at 589 nm

Substance	Conditions	Refractive index, n_{589}
Air	1 atm, 15°C	1.0002765
Water	Liquid, 15°C	1.33341
Ethanol	Liquid, 20°C	1.3611
Carbon tetrachloride	Liquid, 20°C	1.4601
Hexane	Liquid, 20°C	1.37506
Quartz	Fused	1.45843
Protein		1.51–1.54
Sucrose	Crystal, 20°C	1.5376
Lipids		1.40–1.44

Even more information can be gained by examination of the refractive index using polarized light. Plane-polarized light is light in which the electric field of the light oscillates in only one direction; natural light becomes plane polarized upon passage through Polaroid or other polarizing materials. Objects such as single crystals or biological cells exhibit different refractive indices depending on the direction of polarization of the light relative to the object. This anisotropy of refractive index is known as *birefringence* (literally, double refraction). It is an indication of ordering of molecules within the sample and can give useful information about intermolecular associations in cell membranes, nerve fibers, muscle, and the like. Crystals of some substances, such as quartz, fluorite (CaF_2), and gypsum ($CaSO_4 \cdot 2\,H_2O$), exhibit birefringence that is related to the crystal axes.

ABSORPTION AND EMISSION OF RADIATION

Light that is incident on a colored sample is partly absorbed by it. This means that at certain wavelengths there is less intensity transmitted. Careful examination shows that there is no breakdown in the law of conservation of energy, however. The result of the absorption may appear as *heat* producing a temperature rise in the sample, *luminescence* in which a photon of the same or lower energy is emitted, *chemistry* that incorporates energy into altered bonding structures, or a combination of these. The total energy change of the processes is always exactly equal to the energy of the photon that was absorbed.

The interactions leading to absorption of light are essentially electromagnetic in origin. The oscillating electromagnetic field associated with the incoming photon generates a force on the charged particles in a molecule. If the electromagnetic force results in a change in the arrangement of the electrons in a molecule, we say that a transition to a new electronic state has occurred. The new state will be of higher energy if radiation is absorbed, and the process can be illustrated using an energy-level diagram, as shown in Fig. 10.3. The absorbed photon results in the excitation of the molecule from its normal or ground state, G, to a higher energy or excited electronic state, E. The excited electronic state has a rearranged electron distribution. It can be described approximately as a promotion of one of the ground-state electrons to an orbital of higher energy. This completes the absorption process, which usually occurs very rapidly (within 10^{-15} s), but the excited state is usually

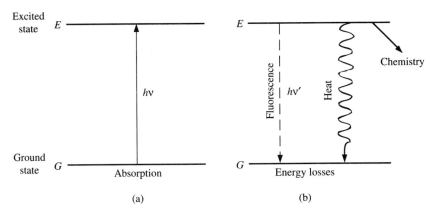

Fig. 10.3 Energy-level diagram: (a) the absorption of a photon of energy $h\nu$, and (b) three important processes by which the excitation energy is subsequently released or converted.

not stable for very long. Typically, within 10^{-8} s it has disappeared by fluorescence emission, the dissipation of heat, or the initiation of photochemistry. Fluorescence and heat conversion result in the return of the molecule to the ground state, and the only net effect is the conversion of energy of the absorbed photon into heat and (perhaps) a photon of lower energy. Figure 10.4 shows the energy of a molecule (versus time), which has absorbed a photon and emitted some of the energy as fluorescence.

At ordinary intensities the relaxation of the system back to the ground state is so rapid that the light produces very little change in the population of molecules in the ground state. At the very high intensities encountered in flash-lamp or laser experiments, it is possible to excite nearly all of the molecules out of the ground state simultaneously, but this occurs only under unusual conditions.

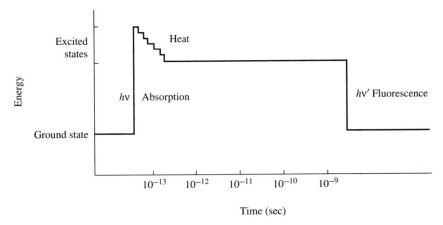

Fig. 10.4 Energy versus time for a single molecule in the presence of light. The molecule absorbs a photon and is excited to a vibrational level of an excited electronic state. The excess vibrational energy is rapidly lost as heat, and the molecule reaches the relatively stable lowest vibrational level of the excited electronic state. After a time the molecule emits a photon and returns to its ground state. Each molecule will have a different excited-state lifetime, but the collection of molecules will have a characteristic relaxation time or half-life.

Radiation-Induced Transitions

Einstein considered the dynamics of an atom or molecule placed in a radiation field using a rather simple model. Consider any two states of a molecule, such as the excited (E) state and ground (G) state of Fig. 10.3. The states will differ in energy by an amount $E_E - E_G = h\nu$. If the energy separation is large (compared with kT), virtually all of the molecules will be in the ground state in the dark at ordinary temperatures. In the presence of a radiation field, this population distribution will be slightly disturbed. We can characterize the radiation field by a radiation energy density $\rho(\nu)$, which is the amount of energy near frequency ν and per unit volume of the sample. The dimensions of $\rho(\nu)$ are energy volume^{-1}. It is related to the more familiar intensity of illumination of a surface $I(\nu)$, in energy area^{-1} time^{-1}, by the equation

$$I(\nu) = \frac{c}{n}\,\rho(\nu) \qquad (10.4)$$

where c is the velocity of light and n is the refractive index of the medium. Some representative values of approximate illumination intensities and radiation-energy densities are given in Table 10.2. Values are also given in terms of quanta or photons using the Planck equation.

Table 10.2 Radiation densities and illumination intensities values for polychromatic sources are given per nanometer of spectral bandwidth

Source of phenomenon	λ (nm)	Illumination intensity, $I(\lambda)$*		Radiation density, $\rho(\lambda)$*	
		μW cm^{-2}	quanta s^{-1} cm^{-2}	J cm^{-3}	quanta cm^{-3}
Threshold of completely dark-adapted human eye	507	4×10^{-9}	1×10^4	1.3×10^{-25}	3×10^{-7}
Firefly luminescence	562	5×10^{-2}	1.5×10^{11}	1.7×10^{-18}	5
Full moon at surface of earth	500	0.3	1×10^{12}	1×10^{-17}	30
Bright summer sunlight	500	1×10^5	2×10^{17}	3×10^{-12}	1×10^7
Ruby laser, 1 kJ, 1 ms duration	694.3	1×10^{12}	4×10^{24}	3×10^{-5}	1.2×10^{14}

* $I(\lambda)$ and $\rho(\lambda)$ are given for a wavelength range of 1 nm around the λ quoted.

As a consequence of the introduction of the radiation field to an absorbing sample, three processes are induced:

1. *Absorption.* The radiation induces transitions from the ground state G to the excited state E. The rate of this absorption process will depend on the number of molecules in state G, on the radiation density, and on the absorption coefficient of the molecule.
2. *Stimulated emission.* Once molecules are present in the excited state E, the radiation field can stimulate transitions down to the lower state, giving emission of radiation.
3. *Spontaneous emission.* Molecules put into the excited state E by the radiation field are substantially out of equilibrium. Spontaneous relaxation processes restore the system to equilibrium. These are spontaneous in the sense that they do not depend on the radiation density.

Classical Oscillators

The classical theory of light absorption considers matter to consist of an array of charges that can be set into motion by the oscillating electromagnetic field of the light. The electric-dipole oscillators set in motion by the field have certain characteristic, or natural, frequencies, v_i, that depend on the substance. When the frequency of the radiation is near the oscillation frequency, absorption occurs, and the intensity of the radiation decreases on passing through the substance. The refractive index of the substance also undergoes large changes, called *dispersion anomalies,* in the same spectral region. The intensity of the interaction is known as the *oscillator strength, f_i,* and it can be thought of as characterizing the number of electrons per molecule that oscillate with the characteristic frequency, v_i.

Although the classical picture has been replaced by the much more informative quantum mechanical description of the absorption of radiation, there are classical holdovers in the current literature. For example, a transition that is fully allowed quantum mechanically is said to have an *oscillator strength* of 1.0. Operationally, the oscillator strength, f, is related to the intensity of the absorption, that is, to the area under an absorption band plotted versus frequency:

$$ f = \frac{2303cm}{\pi N_0 e^2 n} \int \varepsilon(v) \, dv \qquad (10.5) $$

where ε is the molar absorptivity, c is the velocity (cm s^{-1}) of light, m the mass (g) of the electron and e its charge (esu), n the refractive index of the medium, and N_0 is Avogadro's number. The integration is carried out over the frequency range associated with the absorption band. Oscillator strengths can be observed from magnitudes of unity down to very small values ($<10^{-4}$). Measured values greater than unity, as for the blue bands of hemoglobin or chlorophyll, usually indicate the overlap of two or more electronic transitions.

Quantum Mechanical Description

The most satisfactory and complete description of the absorption of radiation by matter is based on time-dependent wave mechanics. It analyzes what happens to the wavefunctions of molecules in the presence of an external oscillating electromagnetic field. Because it requires knowledge of differential equations to follow the development of time-dependent wave mechanics, we will simply cite some of the useful results obtained from the detailed treatment.

Because the interaction is electromagnetic, the operator that conveniently describes it is the electric dipole-moment operator, $\mu = er$. This operator is just the position operator, $r = x + y + z$, multiplied by the electronic charge. For example, the *permanent dipole moment* of a molecule in its ground state is obtained by using the ground-state wavefunction of the molecule to calculate the average position of negative charge. When the centers of positive and negative charge coincide, the permanent dipole moment is zero.

A transition from one state to another occurs when the radiation field connects the two states. In wave mechanics, the means for making this connection is described by the *transition dipole moment*

$$\mu_{0A} = \int \psi_0 \mu \psi_A \, dv \tag{10.6}$$

where ψ_0, ψ_A = wavefunctions for the ground and excited state, respectively
dv = volume element; integration is over all space

The transition dipole moment will be nonzero whenever the symmetry of the ground and excited state differ. A hydrogen atom will always have zero permanent dipole moment, but if ψ_0 is a $1s$ orbital and ψ_A is a $2p$ orbital, the transition dipole does not equal zero. Similarly, ethylene has zero permanent dipole moment, but if ψ_0 is a π MO and ψ_A is a π^* MO, then $\mu_{\pi\pi^*}$ is not zero. The transition moment has both magnitude and direction; the direction is characterized by the vector components: $\langle\mu_x\rangle_{0A}$, $\langle\mu_y\rangle_{0A}$, and $\langle\mu_z\rangle_{0A}$. Most electronic transitions are polarized, which means that the three vector components are not all equal. In the examples above, the hydrogen-atom transition from s to p_x is polarized along x. The ethylene $\pi \rightarrow \pi^*$ transition is polarized along the C=C double bond. The magnitude of the transition is characterized by its absolute-value squared, which is called the *dipole strength*, D_{0A}:

$$D_{0A} = |\mu_{0A}|^2 \tag{10.7}$$

This is a scalar quantity rather than a vector. It has the units of a dipole moment squared. Dipole moments are frequently expressed in *debyes*, D, where $1\,\text{D} = 10^{-18}$ esu cm $= 3.336 \times 10^{-30}$ C m, and the units of the dipole strength are then debye².

The relations among dipole strength, oscillator strength, and spectra are given in Table 10.3.

Table 10.3 Spectroscopic relations*

1. Dipole strength $\leftrightarrow$ transition dipole moment
$$D_{0A} = |\mu_{0A}|^2$$

2. Dipole strength $\leftrightarrow$ absorption spectrum
$$D_{0A} = \frac{3\hbar \cdot 2303c}{4\pi^2 N_0 n} \int \frac{\varepsilon(\nu)}{\nu}\, d\nu$$
$$= \frac{9.185 \times 10^{-39}}{n} \int \frac{\varepsilon(\nu)}{\nu}\, d\nu \qquad (\text{esu cm})^2$$

3. Oscillator strength $\leftrightarrow$ absorption spectrum
$$f_{0A} = \frac{2303cm}{\pi N_0 e^2 n} \int \varepsilon(\nu)\, d\nu$$
$$= \frac{1.441 \times 10^{-19}}{n} \int \varepsilon(\nu)\, d\nu \qquad \text{unitless}$$

* Frequency of light, ν, in s⁻¹; N_0 = Avogadro's number; n = refractive index; m, e = mass (g), charge (esu) of electron; $\hbar = h/2\pi$, $c=$ speed of light in cm s⁻¹; $\varepsilon(\nu)$ is the molar absorptivity in M^{-1} cm⁻¹.

Beer-Lambert Law

Consider a sample of an absorbing substance (liquid solution, solid, or gas) placed between two parallel windows that will transmit the light, as shown in Fig. 10.5. (Although the discussion involves "light," these principles apply equally well to radiation in the uv, ir, or other spectral regions.) Suppose that light of intensity I_0 is incident from the left, propagates along the x direction, and exits to the right, with decreased intensity, I_t. At any point x within the sample, there is intensity I, which decreases smoothly from left to right. (We ignore for the moment the discontinuous intensity decrease, usually about 10%, which results from reflections at the windows.)

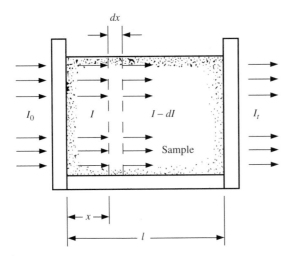

Fig. 10.5 Transmission of light by an absorbing sample.

If the sample is homogeneous, the *fractional* decrease in the light intensity is the same across a small interval dx, regardless of the value of x. The fractional decrease for a solution depends linearly on the concentration of the absorbing substance. This can be written

$$-\frac{dI}{I} = \alpha c \, dx \qquad (10.8)$$

where dI/I is the fractional change in light intensity, c is the concentration of absorber, and α is a constant of proportionality. Equation (10.8) is the differential form of the Beer-Lambert law. It is straightforward to integrate it between limits I_0 at $x = 0$ and I_t at $x = l$, where l is the optical path length. Since neither α nor c depends on x, we can write

$$-\int_{I_0}^{I_t} \frac{dI}{I} = \alpha c \int_0^l dx$$

$$(10.9)$$

$$\ln \frac{I_0}{I_t} = \alpha c l \qquad\qquad I_t = I_0 e^{-\alpha c l}$$

For measurements made with cuvets (optical sample cells) of different path lengths, the transmitted intensity, I_t, decreases exponentially with increasing path length (Fig. 10.6).

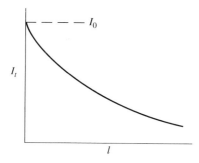

Fig. 10.6 Dependence of transmitted light intensity on path length.

Alternatively, for a single-cuvet path length, the transmitted intensity decreases exponentially with increasing concentration of an absorbing solute. The *absorbance* is defined as base 10 rather than natural logarithms,

$$A = \log \frac{I_0}{I_t} = \varepsilon c l \tag{10.10}$$

where A is the *absorbance* or *optical density*, and $\varepsilon = \alpha/2.303$ is the molar absorptivity (or molar extinction coefficient), with units M^{-1} cm^{-1}, when the concentration, c, is in molarity, M. Equation (10.10) is the Beer-Lambert law and shows that the absorbance is linearly related to concentration (or path length). The relation between absorbance and transmission ($T = I_t/I_0$) is exactly analogous to that between pH and (H$^+$):

$$A = -\log T$$

Because absorption depends strongly on wavelength for nearly all compounds, we must specify the wavelength at which measurement is made. This is usually done using a subscript λ, indicating the particular wavelength, as A_λ or ε_λ. For a single substance at a specified wavelength, ε_λ is a constant, characteristic of the absorbing sample, and is *independent* of both c and l. The wavelength dependence of ε_λ (or of A_λ) is known as the *absorption spectrum* of the compound.

It was mentioned earlier that a small but significant portion of light is lost by reflection at the cuvet windows. Corrections for this effect, as well as for absorption by the solvent, are usually made by substituting solvent for the solution in the same cuvet and making a second measurement. The transmitted intensity of this second measurement is then used as I_0 in the Beer-Lambert expressions. Alternatively, a double-beam method may be used, where the solution (sample) and solvent (reference) are placed in matched cuvets and their transmissions are measured simultaneously.

Deviations from the Beer-Lambert law can occur for a variety of reasons. Problems arising from insufficiently monochromatic radiation, divergent light beams, or imprecise sample geometry are minimized by the design of most spectrophotometers. Individual samples should be checked to determine whether corrections need to be applied. Beer-Lambert law deviations can arise from inhomogeneous samples, light scattering by the sample, dimerization or other aggregation at high concentrations, or changes in equilibria involving dissociable absorbing solutes. The most common consequence is that the measured absorbance does not increase linearly with increasing concentration or path length. This behavior can be used to obtain useful information about the sample. It is essential to understand the origin of such effects in order for spectrophotometry to be a useful tool for experimental investigations.

Quantitative determinations using Beer's law

One of the most widely used applications of spectroscopy is for the quantitative determination of the concentration of substances in solution. Noting that both the absorbance, A, and the molar absorptivity, ε, depend on wavelength, we now write

$$A_\lambda = \varepsilon_\lambda c l$$

A plot of A_λ versus λ (or A_ν versus ν) is the absorption spectrum of the solution. For a single substance, the absorbance A_λ at any wavelength is directly proportional to concentration c for a fixed path length, l. Experimentally, the absorbance is determined by taking the log of the ratio of the incident to the transmitted light intensities.

The absorbance of a solution of more than one independent species is additive. For two components, M and N,

$$A_\lambda = A_\lambda^M + A_\lambda^N = \varepsilon_\lambda^M l[M] + \varepsilon_\lambda^N l[N]$$
$$= (\varepsilon_\lambda^M[M] + \varepsilon_\lambda^N[N])l \tag{10.11}$$

If measurements are made at two (or more) wavelengths where the ratios of extinction coefficients differ, the resulting equations,

$$A_1 = (\varepsilon_1^M[M] + \varepsilon_1^N[N])l$$
$$A_2 = (\varepsilon_2^M[M] + \varepsilon_2^N[N])l$$

can be solved for the concentrations of the absorbing solutes,

$$[M] = \frac{1}{l} \frac{\varepsilon_2^N A_1 - \varepsilon_1^N A_2}{\varepsilon_1^M \varepsilon_2^N - \varepsilon_2^M \varepsilon_1^N}$$
$$[N] = \frac{1}{l} \frac{\varepsilon_1^M A_2 - \varepsilon_2^M A_1}{\varepsilon_1^M \varepsilon_2^N - \varepsilon_2^M \varepsilon_1^N} \tag{10.12}$$

These relations are widely used in the spectrophotometric analysis of mixtures of absorbing species.

A wavelength at which two or more components have the same extinction coefficient is known as an *isosbestic wavelength*. $[\varepsilon_\lambda^M = \varepsilon_\lambda^N = \varepsilon_{iso}]$. At this wavelength, the absorbance can be used to determine the total concentration of the two components.

$$\lambda = \text{isosbestic:} \quad A_{iso} = \varepsilon_{iso} l[M] + \varepsilon_{iso} l[N] = \varepsilon_{iso} l \cdot ([M] + [N]) \tag{10.13}$$

Measurements at an isobestic wavelength plus one other wavelength where the extinction coefficients differ for the two components provide a particularly simple solution to the Beer-Lambert law equations:

$$[M] + [N] = \frac{A_{iso}}{\varepsilon_{iso} l}$$

$$\frac{[M]}{[N]} = \frac{\varepsilon_1^N A_{iso} - \varepsilon_{iso} A_1}{\varepsilon_{iso} A_1 - \varepsilon_1^M A_{iso}}$$

Example 10.1 Solutions containing the amino acids tryptophan and tyrosine can be analyzed under alkaline conditions (0.1 M KOH) from their different uv spectra. The extinction coefficients under these conditions at 240 and 280 nm are

$$\varepsilon_{240}^{Tyr} = 11,300 \ M^{-1} \ cm^{-1} \qquad \varepsilon_{240}^{Trp} = 1960 \ M^{-1} \ cm^{-1}$$

$$\varepsilon_{280}^{Tyr} = \ \ 1500 \ M^{-1} \ cm^{-1} \qquad \varepsilon_{280}^{Trp} = 5380 \ M^{-1} \ cm^{-1}$$

A 10-mg sample of the protein glucagon is hydrolyzed to its constituent amino acids and diluted to 100 mL in 0.1 M KOH. The absorbance of this solution (1-cm path) was 0.717 at 240 nm and 0.239 at 280 nm. Estimate the content of tryptophan and tyrosine in μmol (g protein)$^{-1}$.

Solution Neither of the wavelengths is an isosbestic for these amino acids, so we use Eq. (10.12):

$$[Tyr] = \frac{(5380)(0.717) - (1960)(0.239)}{(11,300)(5380) - (1500)(1960)}$$

$$= \frac{3.39 \times 10^3}{57.9 \times 10^6}$$

$$= 5.85 \times 10^{-5} \ M$$

$$[Trp] = \frac{(11,300)(0.239) - (1500)(0.717)}{57.9 \times 10^6}$$

$$= 2.81 \times 10^{-5} \ M$$

Since 10 mg of protein was hydrolyzed and diluted to 100 mL of solution, the estimated contents are:

$$585 \ \text{μmol of Tyr (g protein)}^{-1}$$

$$281 \ \text{μmol of Trp (g protein)}^{-1}$$

When only two absorbing compounds are present in a solution, one or more isosbestics are frequently encountered if we examine the entire wavelength range of the uv, visible, and ir regions. The isosbestics do not *necessarily* occur, however, for the molar absorptivity of one compound may be less than that of the other in every accessible wavelength region. An easy way to spot isosbestics is to superimpose plots of ε versus λ for the two compounds. Wherever the curves cross, there is an isosbestic. An example is shown in Fig. 10.7 for cytochrome c.

The additivity of absorbance can be generalized to any number of components. Thus, for n components,

$$A = A_1 + A_2 + A_3 + \cdots + A_n = (\varepsilon_1 c_1 + \varepsilon_2 c_2 + \varepsilon_3 c_3 + \cdots + \varepsilon_n c_n)l$$

To determine the n concentrations by absorption methods, we need to make measurements at a minimum of n wavelengths, each characterized by a unique set of ε's. This may be impractical when n is large. Even so, a single component can sometimes be measured accurately in a complex mixture if there is a wavelength where its absorbance is much greater

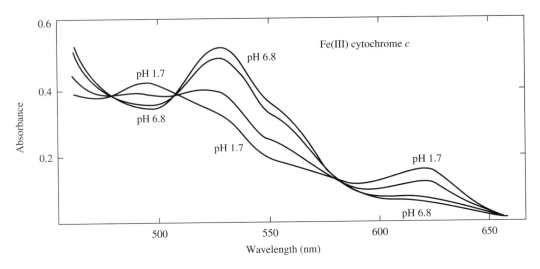

Fig. 10.7 Absorption spectra of Fe(III) cytochrome c (5.3×10^{-5} M) at various pH values from 6.8 to 1.7. Isosbestic points occur at 477, 509, and 584 nm. (From E. Yang, unpublished spectra.)

than that of any of the other components. For example, hemoglobin in red blood cells can be measured spectrophotometrically at 541 or 577 nm (Fig. 10.2), because the other cell constituents (proteins, lipids, carbohydrates, salts, water, and so on) do not absorb significantly at those wavelengths.

Programmable calculators may have a program for solving a set of simultaneous linear equations. Then n values of the measured absorbances and n^2 values of the ε's can be entered to obtain n concentrations. Digitizing spectrophotometers which can measure absorbances at many wavelengths simultaneously can calculate concentrations of a mixture directly from the absorbances.

In multicomponent solutions isosbestic points almost never occur. The probability that three or more compounds have identical molar absorbances at any wavelength is exceedingly small. The probability of two or more isosbestic points for the spectra of the set of compounds is so small as to be completely negligible. This property is useful for diagnostic purposes and results in the following generalization: *The occurrence of two or more isosbestics in the spectra of a series of solutions of the same total concentration demonstrates the presence of two and only two components absorbing in that spectral region.*

It is possible, in principle, for this generalization to be violated, but the chances are vanishingly small. Note that the rule does not necessarily apply to components that do not absorb at all in the wavelength region investigated. It also does not apply if two chemically distinct components have identical absorption spectra (such as ADP and ATP). In this case, the entire spectrum is a set of isosbestics for these two components alone.

Two common examples of the usefulness of isosbestics are the study of equilibria involving absorbing compounds and investigations of reactions involving absorbing reactants and products. The presence of isosbestics is used as evidence that there are no intermediate species of significant concentration between the reactants and products.

Exercise Consider the reaction of an absorbing reactant M to give an absorbing product P by the reaction

$$M \xrightarrow{k_1} N \xrightarrow{k_2} P \qquad\qquad k_1 \sim k_2$$

Show that no isosbestic points would be expected for this system, even if the intermediate N has no measurable absorbance in the same spectral region.

PROTEINS AND NUCLEIC ACIDS: ULTRAVIOLET ABSORPTION SPECTRA

Most proteins and all nucleic acids are colorless in the visible region of the spectrum; however, they do exhibit absorption in the near-ultraviolet region. These spectroscopic absorptions contain information relevant to the composition of these complex molecules, as well as about the conformation or three-dimensional structure of the molecules in solution. Figure 10.8 shows the uv absorption of serum albumin, a representative protein. There is a distinctive absorption maximum at around 280 nm and a much stronger maximum at 190 to 210 nm. The 280-nm band is assigned to $\pi \rightarrow \pi^*$ transitions in the aromatic amino acids, the 200-nm band to $\pi \rightarrow \pi^*$ transitions in the amide group. These features are sufficiently characteristic that they are often used in the preliminary identification of proteins in biological materials. The spectra of nucleic acids in the same region (Fig. 10.9) show somewhat different characteristics. An absorption maximum occurs at 260 nm, and a shoulder is seen near 200 nm on a background rising to shorter wavelengths. Again these features are assigned to $\pi \rightarrow \pi^*$ transitions and are characteristic of all nucleic acids. The observations

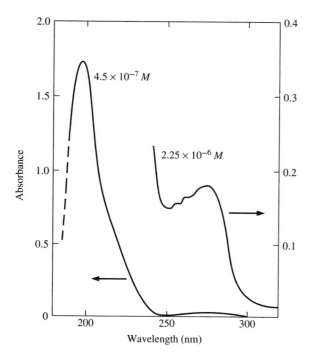

Fig. 10.8 Ultraviolet absorption spectrum of bovine serum albumin. Solution in 10^{-3} M phosphate buffer pH 7.0, 1.0-cm path. The wavelength region above 240 nm was measured at a higher concentration and on an expanded absorbance scale (right) so that the weaker absorption bands in that region would be visible. (From E. Yang, unpublished results.)

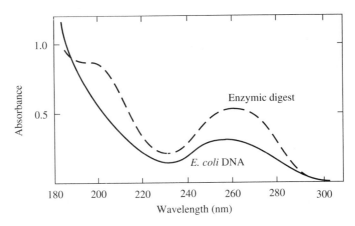

Fig. 10.9 Ultraviolet absorption spectrum of DNA from *E. coli* in the native form at 25°C (solid curve) and as an enzymic digest of nucleotides (dashed curve). [From D. Voet, W. B. Gratzer, R. A. Cox, and P. Doty, *Biopolymers 1,* 193 (1963).] Reprinted with permission of John Wiley and Sons, Inc.

are sufficiently general that the ratio of absorbances A_{260}/A_{280} has been used to determine quantitatively the ratio of nucleic acid to protein in a mixture of the two. This requires careful calibration, however, because proteins differ significantly in their content of aromatic amino acids.

Proteins are natural polyamino acids. That is, they are polymers in which an assortment of the 20 natural amino acids are connected by amide linkages in a long chain. Each protein has a characteristic sequence, but there are thousands of different proteins in nature and accordingly many different compositions and sequences. Nucleic acids are also linear polymers, but the monomer unit is the nucleotide, consisting of an aromatic heterocyclic base attached to a sugar phosphate. Only four different bases commonly occur, but there are many different ways in which they can be combined and arranged in the linear sequence. The chain involves covalent links between the sugars (ribose for RNA, deoxyribose for DNA) and the phosphate linking groups. The structures of amino acids and nucleotides are given in the Appendix. To understand the uv absorption of proteins or nucleic acids, we need to examine the various contributions to the spectra. We shall consider the following important factors:

  1. Absorption spectra of individual monomers.
  2. Contribution of the polymer backbone.
  3. Secondary and tertiary structure, including helix formation.
  4. Hydrogen bonding and solvent effects.

Amino Acid Spectra

The uv spectra of various amino acids are summarized in Figs. 10.10 and 10.11. Several distinctive categories of spectra appear. The aromatic amino acids tryptophan, tyrosine, and phenylalanine are the only ones absorbing significantly at wavelengths longer than 230 nm.

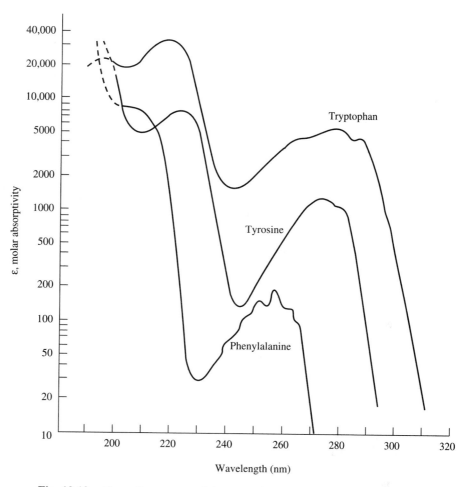

Fig. 10.10 Absorption spectra of the aromatic amino acids (tryptophan, tyrosine, and phenylalanine) at pH 6. [From S. Malik, cited by D. B. Wetlaufer in *Adv. Protein Chem. 17,* 303 (1962).]

In particular, at the "characteristic" protein wavelength of 280 nm, absorbance reflects the presence only of tryptophan and tyrosine in a protein and cannot be used for quantitative purposes without supplementary analysis. For example, the absorbance in a 1-cm cell at 280 nm for 1% solutions of proteins in water varies from 3.1 for NAD nucleosidase from pig to 27 for egg white lysozyme. Between 200 and 230 nm, the aromatic amino acids (especially tryptophan) have absorption and so do histidine, cysteine, and methionine. Between 185 and 200 nm, only phenylalanine and tyrosine have distinct maxima, but the curves of all the other amino acids rise sharply to shorter wavelengths as illustrated by alanine in Fig. 10.11. Extensive tables of amino acid absorption data are given in the *Handbook of Biochemistry and Molecular Biology,* 3rd ed. (CRC Press, Cleveland, Ohio, 1976).

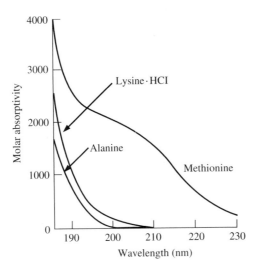

Fig. 10.11 Ultraviolet absorption spectra of three α-amino acids in aqueous solution at pH 5.

Polypeptide Spectra

The contribution of the amide linkages to the absorption spectra can be seen by comparing the spectrum of lysine hydrochloride in Fig. 10.11 with that of poly-L-lysine hydrochloride in the random-coil form (Fig. 10.12). The broad absorption centered at 192 nm ($\varepsilon_{192} = 7100\ M^{-1}\ cm^{-1}$) is characteristic of the amide linkage in poly-L-lysine and increases the absorbance in this region by eightfold over that of the free amino acid. All proteins have contributions to the absorption spectra in the region around 190 nm (180 to 200 nm) from the polypeptide backbone; however, they are accompanied by absorption contributions from certain of the side chains, especially the aromatic ones.

Secondary Structure

The secondary structure of a protein describes which residues are in helices or other ordered conformations. The conformation of a protein is sensitively detected by uv spectroscopy. Figure 10.12 shows the effect on the poly-L-lysine spectrum resulting from raising the pH to induce helix formation (by reducing the net positive charge on the lysine side chains) and by raising the temperature, which converts the polypeptide to the β-sheet structure. Proteins are more complicated in their native secondary structure and they usually contain simultaneously several elements of secondary structure, at different locations along the peptide chain. The process of denaturation destroys much of this secondary structure, and changes in the absorption spectra occur as a consequence. The absorption features associated with helix, β sheet, or random coil are sufficiently distinctive to be used for diagnostic purposes for the native protein (Rosenheck and Doty, 1961); circular dichroism spectra (see later in this chapter) can be more informative, however.

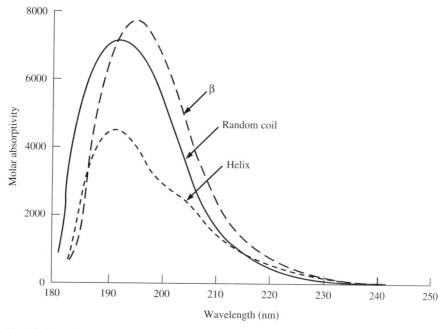

Fig. 10.12   Ultraviolet absorption spectra of poly-L-lysine hydrochloride in aqueous solution; random coil, pH 6.0, 25°C; helix, pH 10.8, 25°C; β form, pH 10.8, 52°C. [From K. Rosenheck and P. Doty, *Proc. Natl. Acad. Sci. USA 47*, 1775 (1961).]

Origin of Spectroscopic Changes

It is meaningful to analyze the various factors that contribute to the spectroscopic changes that occur when a protein undergoes a change in secondary structure. Ultimately, most of the effects are electrical in origin, because the spectra are sensitive to charge effects on the ground and excited electronic states. A list of such influences includes (1) changes in local charge distribution; (2) changes in dielectric constant; (3) changes in bonding interactions, such as hydrogen bonds; and (4) changes in dynamic (or resonance) coupling between different parts of the molecule.

The uv or visible spectrum of a substance depends markedly upon such factors as its state (gas, liquid, solid), the solvent, temperature, extent of aggregation or dissociation, and specific molecular complexes. Figure 10.13 illustrates this behavior for the molecule anisole

$$CH_3O-\langle\!\!\!\!\!\!\!\bigcirc\!\!\!\!\!\!\!\rangle$$

and is representative of other aromatic molecules. The spectrum of the vapor (bottom) is typically the "best resolved," in the sense that the absorption-band envelope contains a number of sharp spikes or peaks. These are characteristic of particular vibrations associated with the molecular structure and are superimposed on the underlying electronic transition. The energy of the different vibrations is added to the energy of the electronic transition when both the electronic and vibrational quantum numbers change as a result of absorption of a photon. This is called a *vibronic transition*. The energy of the electronic

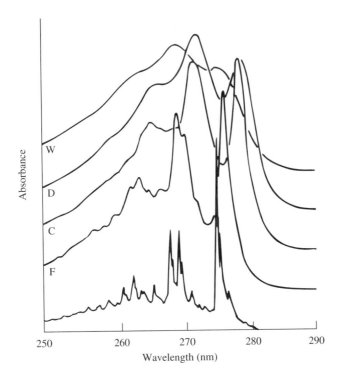

Absorbance

250   260   270   280   290

Wavelength (nm)

W

D

C

F

Fig. 10.13 Absorption spectra of anisole vapor and of anisole dissolved in perfluorooctane, F; in cyclohexane, C; in dioxane, D; and in water, W. All at 30°C. The spectra of anisole solutions are displaced vertically to decrease overlap. (From G. L. Tilley, Ph.D. thesis, Purdue University, 1967; cited by M. Laskowski, Jr., in *Spectroscopic Approaches to Biomolecular Conformation,* D. W. Urry, ed., American Medical Association, Chicago, 1970.)

transition itself (without added vibrational energy) is determined from the low-frequency or long-wavelength edge of the vibronic spectrum. A detailed vibrational analysis may be needed to locate this energy precisely. Because there are many ways in which large molecules can vibrate (many *normal modes* of vibration), the envelope of an electronic transition of a molecule in the gas phase is often very complex. The reason that the vibronic structure is so sharp is that most of the molecules in the gas phase are relatively isolated from one another; only a small fraction is in the process of undergoing a collision during the very brief time (about 10^{-15} s) that is required for the photon to be absorbed. Thus, nearly all of the gas-phase molecules are in essentially the same environment and not interacting with one another.

In the liquid state, either pure liquid or in solution, the environment of the molecules is quite different. Close contacts exist between molecules in liquids, and essentially all of the molecules are in the process of undergoing collisions. The forces between molecules are strong enough to perturb the molecular energy levels significantly and, together with the much greater variety of environments present at the instants of photon absorption, this leads to a broadening and shifting of the spectra. The effects depend on the polarity and hydrogen-bonding ability of the solvent, as can be seen in the upper spectra in Fig. 10.13.

The local environment of the amino acids in a native protein depends sensitively on the electric properties of the nearby peptide chain and associated solvent. Careful measurements of the absorption spectrum of a protein near 280 nm show small but distinct differences from the spectrum of the constituent aromatic amino acids. These differences reflect the sum of the effects of the local environments on the aromatic amino acids. Changes of the same magnitude are encountered upon denaturation of the protein.

Nucleic Acids

By contrast with the amino acid units of proteins, the nucleotides that make up the nucleic acid polymers have rather similar absorption spectra. The aromatic bases that are attached to the ribose- or deoxyribose-phosphates all have absorption maxima near 260 nm (Table 10.4). The free base, the nucleoside (the base attached to the sugar), the nucleotide (the base attached to the sugar phosphate), and the denatured polynucleotide all have very similar absorption spectra in this region. For example, the wavelength of maximum absorption is at 260.5 nm for adenine ($\varepsilon = 13.4 \times 10^3$), at 260 nm for adenosine ($\varepsilon = 14.9 \times 10^3$), at 259 nm for adenosine-5'-phosphate ($\varepsilon = 15.4 \times 10^3$), and at 257 nm for the tetranucleotide pApApApA. The latter, however, exhibits a lower absorbance per nucleotide in aqueous solution ($\varepsilon_{259} = 11.3 \times 10^3$ M^{-1} cm^{-1}). In general, polynucleotides and nucleic acids absorb less per nucleotide than their constituent nucleotides. Also, native double-stranded DNA absorbs less per nucleotide than denatured ("melted") DNA strands. A decrease in absorptivity is called *hypochromicity* or *hypochromism;* an increase in absorptivity is *hyperchromicity* or *hyperchromism.* For example, the % hyperchromicity for the melting of a double-stranded DNA is

$$\% \text{ hyperchromicity} = \frac{A \text{ (melted DNA)} - A \text{ (native DNA)}}{A \text{ (native DNA)}} \times 100$$

Here each absorbance, A, is per nucleotide. The % hypochromicity of the tetranucleotide relative to the mononucleotide is

$$\% \text{ hypochromicity} = \frac{A \text{ (mononucleotide)} - A \text{ (tetranucleotide)}}{A \text{ (mononucleotide)}} \times 100$$

$$= \frac{15,400 - 11,300}{15,400} \times 100 = 27\%$$

There is a 27% decrease in absorbance of the tetranucleotide relative to the mononucleotide.

Table 10.4 Absorption properties of nucleotides*

	λ_{max} (nm)	$10^{-3} \varepsilon_{max}$ (M^{-1} cm^{-1})
Ribonucleotides		
Adenosine-5'-phosphate	259	15.4
Cytidine-5'-phosphate	271	9.1
Guanosine-5'-phosphate	252	13.7
Uridine-5'-phosphate	262	10.0
Deoxyribonucleotides		
Deoxyadenosine-5'-phosphate	258	15.3
Deoxycytidine-5'-phosphate	271	9.3
Deoxyguanosine-5'-phosphate	—	—
Thymidine-5'-phosphate	267	10.2

* Wavelengths of maxima and molar absorptivities of nucleotides at pH 7.

The hypochromicity of the polynucleotides or nucleic acids relative to the nucleotides results primarily from interactions between adjacent bases in the stacked arrangement of the helical polymer. The change upon denaturation to single strands or upon hydrolysis to the mononucleotides is easily measured (typically 30 to 40% change) and is often used to follow the kinetics or thermodynamics of the denaturation process. The origin of the hypochromism is electromagnetic in nature. It involves interactions between the electric-dipole transition moments of the individual bases with those of their neighbors. Thus, it depends not only on the intrinsic transition moments of each base, which differ both in magnitude and direction for the chemically distinct bases, but also on the relative orientations of the interacting bases. Stacked bases absorb less per nucleotide than unstacked bases; a stack of two bases has an absorbance slightly less than twice the absorbance of one base, etc. Thus stacked base pairs in a double helix absorb less than partially stacked bases in single strands, which absorb less than the mononucleotides.

Rhodopsin: A Chromophoric Protein

Many proteins include groups other than the common amino acids. Often, but not always, these groups are covalently linked to a polypeptide chain. Some examples and the nature of the attached group are: glycoproteins (sugars), hemeproteins (iron porphyrins), flavoproteins (flavin), and rhodopsin (retinal; vitamin A). In the last three cases the group is a *chromophore,* and contributes to the absorption spectrum in the visible or near-uv regions. We shall examine some features of the visual pigment protein rhodopsin, which not only absorbs radiation in the visible region of the spectrum but also undergoes a photochemical transformation that triggers the visual stimulus.

Rhodopsins in nature occur widely in both vertebrates and invertebrates. A form of rhodopsin, called "bacteriorhodopsin," has been found in the outer membranes of certain halophilic (salt-loving) bacteria. Rhodopsins consist of a colorless protein (or glycoprotein), opsin, to which the chromophore retinal is covalently attached. In mammals the isomer 11-*cis*-retinal is attached to the ε-amino group of a lysine side chain of the opsin by a protonated Schiff's base linkage:

11-*cis*-Retinal   Protonated Schiff's base

In other organisms the chromophore appears to be the 9-*cis* or 13-*cis* isomer or the molecule 3-dehydroretinal. Rhodopsins have intense broad absorption bands with maxima lying between 440 and 565 nm, depending on the specific chromophore and its environment. Cow rhodopsin absorbs at 500 nm (Fig. 10.14) and is a bright red-orange color; animals that are deficient in pigments (many albinos) show this red-orange in the pupil.

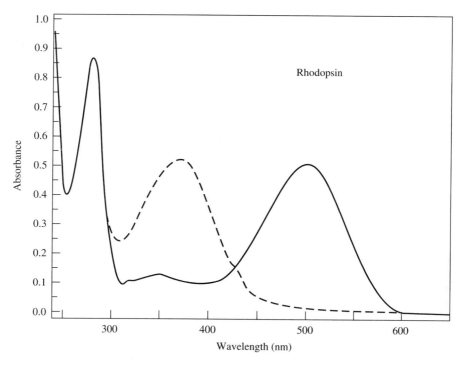

Fig. 10.14 Absorption spectrum of cow rhodopsin in the uv and visible. Purified rhodopsin was measured as isolated from dark-adapted cow retinas (solid curve) and again following its conversion by illumination (dashed curve). [From M. L. Applebury, D. M. Zuckerman, A. A. Lamola, and T. M. Jovin, *Biochemistry 13,* 3448 (1974). Copyright by the American Chemical Society.]

Retinal is an example of the linear polyenes that we examined in Chapter 9. The six conjugated double bonds of the aldehyde contain π-electrons that can be treated approximately by the particle-in-a-box model. The long-wavelength absorption band (a $\pi \rightarrow \pi^*$ transition) of isolated retinal in a variety of organic solvents occurs between 366 and 377 nm, which is the expected region for polyenes of this length. In fact, the spectrum of rhodopsin after it has been photoconverted (Fig. 10.14, dashed curves) is very similar to that of retinal in an organic solvent. The origin of the shift of this band to 500-nm or even longer wavelengths in unilluminated rhodopsin has been connected to the presence of a negative charge in the opsin near the C-12, C-13, and C-14 atoms in retinal (see Fig. 9.1). Derivatives of retinal, with different double bonds reduced, were bound to opsin and the absorption spectrum was measured. This indicated which part of the π system of electrons was needed to produce the large shift on binding (the opsin shift). Then approximate quantum mechanical calculations were done to show that a single negative charge about 0.3 nm distant from both C-12 and C-14 could account for the observed shift. The same method was applied to bacteriorhodopsin to show that its purple color can be explained by the presence of a negative charge near the β-ionone ring (Nakanishi et al., 1980; Honig et al., 1979).

FLUORESCENCE

Many biological substances emit characteristic fluorescence. Chlorophyll from leaves emits red fluorescence; proteins fluoresce in the ultraviolet primarily from tryptophan residues; reduced pyridine nucleotides, flavoproteins, and the Y base of transfer RNA also exhibit characteristic fluorescence emission. Interesting and useful information can be obtained from fluorescent species that are introduced to biochemical molecules or systems. Fluorescent "labels" have been prepared by equilibrium binding or covalent attachment of fluorescent chromophores to enzymes, antibodies, membranes, and so on. Modified fluorescent substrate analogs, such as ε(etheno)-ATP, are useful as probes of enzyme active sites. In some cases natural chromophores that are not themselves particularly fluorescent, such as retinal in rhodopsin, can serve as quenchers of the fluorescence of added probe molecules.

Bioluminescence occurs in a wide variety of biological organisms. Although fireflies are among the most spectacular of these species, by far the greatest number of bioluminescent species occur in the ocean. The luminescence is produced biochemically and usually does not require prior illumination. It serves a variety of purposes, including visual assistance in minimum-light environments, communication for social or sexual purposes, as camouflage, and for repelling predators. The full role of light emission is only beginning to be appreciated.

In many respects fluorescence or luminescence spectroscopy is even richer than absorption in the variety of information and the range of sensitivity that it can provide. As with absorption, changes in the shape or position of the fluorescence spectrum reflect the environment in which the fluorescing molecule (*fluorophore*) exists. In addition, the intensity of fluorescence or the fluorescence yield (per photon absorbed) can vary over a large range. In some cases changes of 100-fold or more can be produced. These changes in fluorescence yield are accompanied by changes in the fluorescence lifetime that can be measured directly. Another property that can be measured is the *excitation spectrum*. The fluorescence intensity is measured as the wavelength of the excitation light is varied. Fluorescence cannot occur unless light is first absorbed, so the excitation spectrum is related to the absorption spectrum. As the wavelength of excitation scans the absorption band, the fluorescence intensity will change proportionally to the amount of absorption.

The polarization or depolarization of the fluorescent light reflects the relative immobilization of the fluorophore, the extent to which the excitation is transferred to other molecules, and relative conformational changes among the different parts of a complex fluorophore. Quenching of fluorescence by added molecules provides evidence of molecular interactions, of the accessibility of the surface of organelles, and of the presence of intermediates such as triplet states or free radicals. Fluorescent properties can be used as an accurate meter stick to measure distances at the molecular level.

Excited-State Properties

The origin of fluorescence can be seen in Fig. 10.3. One of the principal paths by which the energy of electronic excited states may be released is by the emission of radiation. If a sample is illuminated to produce excited electronic states which fluoresce and then the light is extinguished, the fluorescence will decay. Kinetically, the process is usually first order and is characterized by a rate law of the form

$$F \propto -\frac{d[M^*]}{dt} = k_d[M^*] \tag{10.14}$$

where F = intensity of fluorescence
$[M^*]$ = concentration of the excited electronic state that undergoes fluorescence

In accordance with this rate law, the decay is commonly exponential with time, described by a *fluorescence decay time*, τ, given by

$$\tau = \frac{1}{k_d} \tag{10.15}$$

Measurements of the decay of fluorescence of simple molecules in dilute solution after excitation by a flash of light usually give log F versus t plots that are linear. An example is shown for anthracene in Fig. 10.15.

If fluorescence is the only decay path for the excited state, then the fluorescence decay rate constant (designated k_f) is the reciprocal of the *natural fluorescence lifetime*, τ_0. Thus

$$k_f = \frac{1}{\tau_0} \tag{10.16}$$

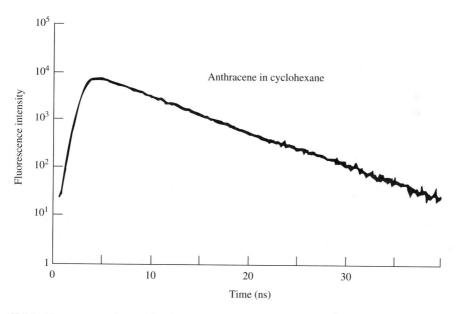

Fig. 10.15 Fluorescence decay of anthracene in cyclohexane (1.7×10^{-3} M), excited by a light flash of 1.4-ns duration (half-width) and at approximately 360 nm. Emission measured at 450 nm. Logarithmic decay curve, obtained using single-photon counting method. [From P. R. Hartig, K. Sauer, C. C. Lo, and B. Leskovar, *Rev. Sci. Instrum. 47,* 1122 (1976).]

In most cases, however, there are significant nonradiative processes competing for the decay of the excited state. These include thermal deactivation, photochemistry, and the quenching by other molecules, Q. The overall rate of decay of the excited state is, therefore, the sum of the rates of all of these processes:

$$-\frac{d[M^*]}{dt} = k_f[M^*] + k_t[M^*] + k_p[M^*] + k_Q[M^*][Q]$$

$$= k_d[M^*]$$

where k_f, k_t, k_p, and k_Q = rate constants for fluorescence, thermal deactivation, photochemistry, and quenching, respectively, and

$$k_d = k_f + k_t + k_p + k_Q[Q]$$

The observed lifetime of fluorescence is then

$$\tau = \frac{1}{k_f + k_t + k_p + k_Q[Q]} \tag{10.17}$$

The *quantum yield* of fluorescence, ϕ_f, is the fraction of absorbed photons that lead to fluorescence; it is the number of photons fluoresced divided by the number of photons absorbed. Obviously, the quantum yield is less than or equal to 1.

$$\phi_f = \frac{\text{number of photons fluoresced}}{\text{number of photons absorbed}} \tag{10.18}$$

The quantum yield can also be considered as the ratio of the rate of fluorescence to the rate of absorbance. But the rate of absorbance must equal (in the steady state) the rate of decay of the excited state. Therefore,

$$\phi_f = \frac{k_f[M^*]}{k_d[M^*]} = \frac{k_f}{k_d} \tag{10.19}$$

But, using Eqs. (10.15) and (10.16),

$$\phi_f = \frac{\tau}{\tau_0} \tag{10.20}$$

Equation (10.20) provides a relation between the quantum yield of fluorescence and the fluorescence lifetime. Table 10.5 gives a summary of fluorescence quantum yields for a variety of fluorophores commonly encountered in biological studies.

Fluorescence almost always occurs from the lowest excited (singlet) state of the molecule. We might otherwise expect to see fluorescence from each of the excited states reached by progressively greater frequency (photon energy) of the exciting radiation absorbed by the fluorophore, but this is never the case. (The molecule azulene is an example of one of the rare exceptions; azulene fluorescence comes predominantly from the second excited singlet state.) The reason that only the lowest state normally emits radiation is that the processes of internal conversion of the higher states (thermal deactivation from higher electronic states to the lowest excited state) are exceedingly rapid. This is illustrated for bacteriochlorophyll in Fig. 10.16, where the absorption and fluorescence spectra are plotted in the vertical direction (turned 90° from the usual orientation) to correspond to the energy-level diagram. Internal conversion from the lowest excited state to the ground state also occurs. It is one of the important sources of thermal deactivation, $k_t[M^*]$, that compete

Table 10.5 Fluorescence quantum yields and radiative lifetimes

Compound	Medium	τ, ns	ϕ	Reference*
Fluorescein	0.1 M NaOH	4.62	0.93	a
Quinine sulfate	0.5 M H_2SO_4	19.4	0.54	a
9-Aminoacridine	Ethanol	15.15	0.99	a
Phenylalanine	H_2O	6.4	0.004	b
Tyrosine	H_2O	3.2	0.14	b
Tryptophan	H_2O	3.0	0.13	b
Cytidine	H_2O, pH 7	—	0.03	c
Adenylic acid (AMP)	H_2O, pH 1	—	0.004	c
Etheno-AMP	H_2O, pH 6.8	23.8	1.00	d
Chlorophyll a	Diethyl ether	5.0	0.32	e
Chlorophyll b	Diethyl ether	—	0.12	e
Chloroplasts	H_2O	0.35–1.9	0.03–0.08	f
Riboflavin	H_2O, pH 7	4.2	0.26	g
DANSYL sulfonamide†	H_2O	3.9	0.55	h
DANSYL sulfonamide + carbonic anhydrase	H_2O	22.1	0.84	h
DANSYL sulfonamide + bovine serum albumin	H_2O	22.0	0.64	h

* (a) W. R. Ware and B. A. Baldwin, *J. Chem. Phys. 40,* 1703 (1964); (b) R. F. Chen, *Anal. Letters 1,* 35 (1967); (c) S. Udenfriend, *Fluorescence Assay in Biology and Medicine,* Vol. II, Academic Press, New York, 1969; (d) R. D. Spencer et al., *Eur. J. Biochem. 45,* 425 (1974); (e) G. Weber and F. W. J. Teale, *Trans. Faraday Soc. 53,* 646 (1957); (f) A. Müller, R. Lumry, and M. S. Walker, *Photochem. Photobiol. 9,* 113 (1969); (g) R. F. Chen, G. G. Vurek, and N. Alexander, *Science 156,* 949 (1967); (h) R. F. Chen and J. C. Karnohan, *J. Biol. Chem. 242,* 5813 (1967).

†DANSYL sulfonamide is 1-dimethylaminonaphthalene-5-sulfonamide.

with fluorescence. The rate is often slower for this step, however, partly because of the greater energy separation between the ground state and the first excited electronic state compared with the energy differences among the excited states.

Fluorescence Quenching

A decrease in fluorescence intensity or quantum yield occurs by a variety of mechanisms. These include collisional processes with specific quenching molecules, excitation transfer to nonfluorescent species, complex formation or aggregation that forms nonfluorescent species (concentration quenching), and radiative migration leading to self-absorption. There are important biological applications or consequences of each of these quenching mechanisms.

Quenching of fluorescence by added substances or by impurities can occur by a collisional process. Because it is the excited state of the fluorescent molecule that must undergo collisional quenching, the encounters must occur frequently and the quenching process must be efficient. Molecular oxygen is one of the most widely encountered quenchers. This is because O_2 is a triplet species (two unpaired electrons) in its ground

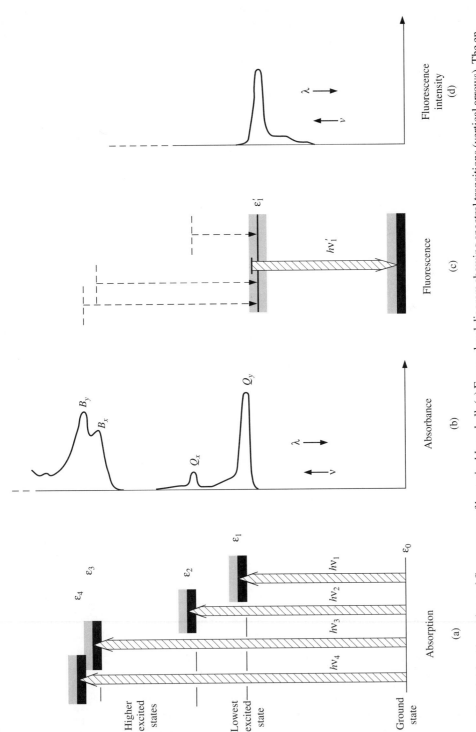

Fig. 10.16 Absorption and fluorescence of bacteriochlorophyll. (a) Energy-level diagram showing spectral transitions (vertical arrows). The energy levels are broadened (shading) by vibrational sublevels that are not usually resolved in solution spectra. (b) Absorption spectrum corresponding to energy levels of part (a). This spectrum is turned 90° from the usual orientation to show the relation to the energy levels. (c) Radiationless relaxation (dashed arrows) and fluorescence (shaded arrow). (d) Fluorescence emission spectrum corresponding to part (c). Note the red shift of the fluorescence compared with the corresponding Q_y absorption illustrated in parts (a) and (b). (From K. Sauer, in *Bioenergetics of Photosynthesis*, Govindjee, ed., Academic Press, New York, 1975, pp. 115–181.)

electronic state. Most excited fluorophores emit from a singlet state (no unpaired electrons), and O_2 quenches the fluorescence by means of the reaction

$$M^* \text{ (singlet)} + O_2 \text{ (triplet)} \longrightarrow M^* \text{ (triplet)} + O_2 \text{ (singlet)}$$

Normally, M^* (singlet) is fluorescent and M^* (triplet) is not.

Collisional quenching is a bimolecular process kinetically; however, the excited oxygen molecules quickly return to the ground triplet state upon subsequent collisions or interactions with the solvent. As a consequence, they are not consumed in the process, and collisional quenching obeys pseudo-first-order kinetics. For a generalized quencher molecule, Q, the relevant equations are

$$M + h\nu \longrightarrow M^* \qquad\qquad \text{excitation}$$

$$M^* \xrightarrow{k_f} M + h\nu' \qquad \text{fluorescence}$$

$$M^* + Q \xrightarrow{k_Q} M + Q^* \qquad \text{quenching}$$

In the absence of quenchers,

$$\phi_f^0 = \frac{k_f}{k_f + k_t + k_p} \tag{10.21}$$

In the presence of a quencher at concentration [Q],

$$\phi_f = \frac{k_f}{k_f + k_t + k_p + k_Q[Q]} \tag{10.22}$$

Therefore, we obtain a result known as the Stern-Volmer relation

$$\frac{\phi_f^0}{\phi_f} = 1 + \frac{k_Q[Q]}{k_f + k_t + k_p} = 1 + K[Q] \tag{10.23}$$

We can also write the equation in terms of the lifetime, τ', in the absence of quencher.

$$\frac{\phi_f^0}{\phi_f} - 1 = k_Q \tau'[Q]$$

where

$$\tau' = \frac{1}{k_f + k_t + k_p}$$

Because the intensity of fluorescence, F, is proportional to the quantum yield, ϕ_f, a plot of F^0/F versus [Q] will give a straight line with slope $k_Q\tau'$. Thus, the longer the lifetime of the excited state, the greater is the probability of quenching.

The consequences of concentration quenching can be quite dramatic. The fluorescence of benzene in oxygen-free solutions occurs with a lifetime of 29 ns; in a solution in equilibrium with O_2 at 1 atm pressure, this lifetime is decreased fivefold, to 5.7 ns. Chlorophyll a is strongly fluorescent in dilute solutions ($\phi_f \cong 0.3$), but the fluorescence intensity is quenched essentially to zero with added quinones or carotenoids. (In this case, the mechanism may involve complex formation between chlorophyll and the quencher molecule.) It is clear from these examples that great care must be taken to remove all extraneous quenching species in the determination of the intrinsic properties of fluorescing molecules.

Concentration quenching may occur as a consequence of aggregation, dissociation, or other changes in the fluorophore itself. *Excimers* (excited dimers) may form because of greater interactions of the excited-state species. In each case there will result a concentration-dependent quantum yield of fluorescence. For example, if nonfluorescent excimers form according to

$$M^* + M \xrightarrow{k_e} [M \cdot M]^*$$

then

$$\phi_f = \frac{k_f}{k_f + k_t + k_p + k_e[M]}$$

Chlorophyll a at concentrations of about 10^{-2} M exhibits concentration quenching in most solvents. Depending on the particular medium, the aggregated chlorophylls may be completely nonfluorescent or they may have a weak but distinctive fluorescence of their own. In photosynthetic membranes the chlorophyll concentrations locally are typically 0.05 to 0.1 M, and fluorescence yields are only about one-tenth of the monomer value.

Excitation transfer processes, which will be discussed next, provide additional paths for fluorescence quenching.

Excitation Transfer

Some of the most valuable applications of fluorescence to biochemical systems involve the transfer of excitation from one chromophore to another. Because such transfer processes depend strongly on the distance between the chromophores and on their relative orientations, experiments can be designed to obtain useful information concerning macromolecular geometry. Excitation transfer has been treated theoretically by a number of authors, and several of the most important relations have been verified quantitatively in carefully designed model experiments (Stryer, 1978).

We shall consider the transfer of excitation from an excited donor molecule, D^*, to an acceptor, A, which then fluoresces.

$$D \xrightarrow{h\nu} D^* \qquad \text{(absorption by donor)}$$
$$D^* \xrightarrow{k_f} D + h\nu' \qquad \text{(fluorescence of donor)}$$
$$D^* + A \xrightarrow{k_T} D + A^* \qquad \text{(excitation transfer)}$$
$$D^* \longrightarrow D \qquad \text{(all other de-excitation)}$$
$$A^* \longrightarrow A + h\nu'' \qquad \text{(fluorescence of acceptor)}$$

The transfer of excitation can be measured in three different ways: (1) the decrease in fluorescence quantum yield of the donor due to the presence of the acceptor, (2) the decrease in lifetime of the donor due to the presence of the acceptor, and (3) the increase in fluorescence of the acceptor due to the presence of the donor. The efficiency of excitation transfer is defined as the fraction of $D*$ that is de-excited by transfer. It is simply related to the rate constants for excitation transfer, k_T, and for all other processes of de-excitation, including fluorescence, k_d.

$$\text{efficiency } (Eff) = \frac{k_T}{k_T + k_d} \qquad (10.24)$$

From Eq. (10.19) the quantum yield for fluorescence of the donor alone is

$$\phi_D = \frac{k_f}{k_d} \qquad (10.19)$$

and for the donor in the presence of acceptor is

$$\phi_{D+A} = \frac{k_f}{k_d + k_T}$$

Therefore, the efficiency of excitation transfer is directly related to the ratio of quantum yields for the donor in the presence (ϕ_{D+A}) and absence of the acceptor (ϕ_D).

$$Eff = 1 - \frac{\phi_{D+A}}{\phi_D} \qquad (10.25)$$

As quantum yields and lifetimes are proportional to each other [Eq. (10.20)], the efficiency is similarly related to the fluorescence lifetimes.

$$Eff = 1 - \frac{\tau_{D+A}}{\tau_D} \qquad (10.26)$$

where τ_{D+A} is the fluorescence lifetime in the presence of acceptor, and τ_D is the lifetime in the absence of acceptor. The efficiency of excitation transfer depends on the distance between donor and acceptor. In the range from about 1 to 10 nm, *resonance energy transfer* (*Förster transfer*) occurs. For each donor-acceptor pair, the efficiency of transfer depends on r^{-6}, where r is the distance between them. The energy transfer efficiency is

$$Eff = \frac{r_0^6}{r_0^6 + r^6} \qquad (10.27)$$

where Eff = efficiency of transfer ($0 \leq Eff \leq 1$)
$\quad\quad r_0$ = characteristic distance for the donor-acceptor pair; it is the distance for which $Eff = 0.5$
$\quad\quad r$ = distance between donor and acceptor

The value of r_0 depends on the amount of overlap between the fluorescence spectrum of the donor and the absorption spectrum of the acceptor. It also depends on the angular orientation between donor and acceptor.

$$r_0(\text{nm}) = 8.79 \times 10^{-6}(J\kappa^2 n^{-4}\phi_D)^{1/6} \qquad (10.28)$$

where $\quad J = \int \varepsilon_A(\lambda) \, F_D(\lambda)\lambda^4 \, d\lambda$
$\quad\quad \varepsilon_A(\lambda)$ = absorption spectrum of acceptor
$\quad\quad F_D(\lambda)$ = fluorescence spectrum of donor

n = refractive index of medium

κ^2 = an orientation factor which depends on the angle between transition dipoles [see Eq. (10.6)] of the donor and acceptor; it is zero, if the transition dipoles are perpendicular to each other, and four, if they are parallel; for random orientations $\langle \kappa^2 \rangle = \frac{2}{3}$

ϕ_D = quantum yield of donor

Molecular Rulers

Extensive experimental tests of resonance or Förster transfer have been carried out. Förster (1949) verified the expected concentration dependence in a study of the quenching of fluorescence of trypaflavin in methanol by the dye rhodamine B. In this case a value $r_0 = 5.8$ nm was obtained, which shows that transfer can occur over distances several times larger than the actual molecular dimensions. The dependence on the inverse sixth power of distance has been tested in an elegant series of experiments by Latt et al. (1965) and by Stryer and Haugland (1967). In these studies two chromophores are attached covalently to a rigid molecular framework. Then excitation transfer is measured for a series of synthetic species where the distance between the donor and acceptor molecule is different. Stryer and Haugland examined transfer from a dansyl group at one end to a naphthyl group at the other end of oligoprolines with 1 to 12 monomer units in the rigid chain. The results showed excellent agreement with the r^{-6} dependence from 1.2 to 4.6 nm separation, as shown in Fig. 10.17.

An important determinant of excitation transfer is the spectral overlap between the donor emission and the acceptor absorption spectrum. This overlap was varied 40-fold by solvent effects for a modified steroid by Haugland et al. (1969), and the transfer rate varied almost in parallel. These extensive tests have confirmed the validity of the Förster resonance transfer mechanism, at least over distances in the range 1 to 10 nm.

Genetic recombination occurs when two DNA double strands come together and by a process of cutting and splicing exchange parts of their sequence; genes are exchanged. The structure of the intermediate is important for understanding the genetic outcome of the

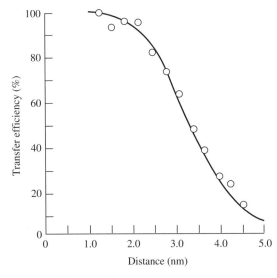

Fig. 10.17 Efficiency of energy transfer as a function of distance in dansyl-(L-prolyl)$_n$-α-naphthyl, $n = 1$ to 12. Energy transfer is 50% efficient at 3.46 nm. Solid line corresponds to r^{-6} dependence. [From L. Stryer and R. P. Haugland, *Proc. Natl. Acad. Sci. USA* **58**, 719 (1967).]

process. Fluorescence energy transfer was used to determine that the sequences of genes on the two double strands are arranged in antiparallel directions, and that the junction forms a right-handed cross (Murchie et al., 1989). A model recombination junction of two DNA fragments of about 30 base pairs each was synthesized. Fluorescein (the donor) was attached to the 5′ end of one strand of one DNA, and rhodamine (the acceptor) was attached to the 5′ end of a strand of the other DNA. The parallel, or antiparallel, arrangement could thus easily be detected. The amount of fluorescence energy transfer provides the distance between the labeled ends. To decide between a right-handed or left-handed cross, the length of one of the DNA fragments was systematically increased. As DNA is a right-handed helix, lengthening the helix also means that the end rotates around the helix axis. This rotation will move the acceptor closer to, or farther from, the donor on the other DNA depending on whether the DNAs form a right- or left-handed cross.

Another application of excitation transfer has been to the determination of the arrangement of proteins in the 30S *E. coli* ribosome (Huang et al., 1975). The ribosome contains 21 proteins bound to a 16S RNA molecule; the ribosome can be dissociated and reconstituted to form a biologically active particle. Ribosomes were reconstituted using 20 different pairs selected from a set of proteins which had been covalently linked to either donors or acceptors. From the measured excitation transfer efficiencies, which ranged from 0 to 75%, some protein pairs were determined to be in direct contact, some were separated by an RNA strand (but not by a protein molecule), and some were far apart. This method is generally applicable to any biochemical structure that can be specifically labeled. For a general review, see Stryer (1978).

Phosphorescence

An excited singlet state with all electrons paired can become an excited triplet state with two unpaired electrons. The paired electrons (with opposite spins) can occupy two different orbitals, or be in the same orbital. The unpaired electrons (with identical spins) must be in separate orbitals as required by the Pauli exclusion principle. The triplet state is thus usually of lower energy than the singlet state. The unpaired electrons in different orbitals stay farther apart from each other, reducing the electron-electron coulomb repulsion and lowering the energy. Although formation of the triplet is favored by the decrease in energy, the probability of a singlet-triplet transition is usually small. It is a forbidden transition. The singlet-triplet conversion is catalyzed by certain molecules, such as O_2, which is a triplet in the ground state, and by species with a high atomic number, such as I^-. Once a molecule is in an excited triplet state, it can emit light and return to its ground state. The emission of light from an excited triplet state is *phosphorescence*. Experimentally, phosphorescence is usually distinguished from fluorescence by its lifetime. Phosphorescence lifetimes typically occur in the millisecond and longer range, while fluorescence lifetimes are in the microsecond and shorter range. Phosphorescence occurs at a longer wavelength than do the fluorescence or the absorption, as is illustrated in Fig. 10.18. The proof of whether luminescence is phosphorescence or fluorescence is to measure the magnetic properties of the excited state. A molecule in a triplet state is paramagnetic and will be attracted by a magnet; its energy is lowered by the magnetic field.

At room temperature DNA does not phosphoresce; the triplet state is quickly quenched. However, in frozen solution DNA shows a characteristic phosphorescence due

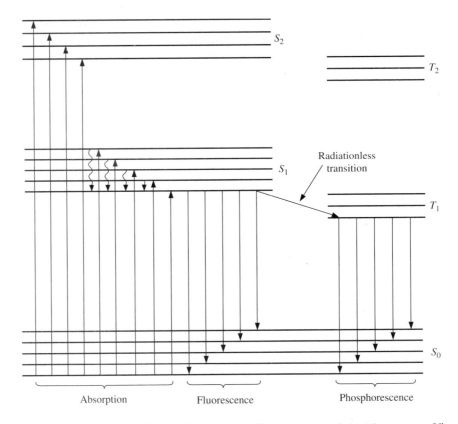

| Absorption | Fluorescence | Phosphorescence |

Fig. 10.18 Energy-level diagram for absorption, fluorescence, and phosphorescence. Vibronic transitions from a ground-state singlet level, S_0, with all electrons paired, to an excited-state singlet level, S_1 or S_2, with all electrons paired, lead to absorption of light. The transition from an excited singlet to a ground-state singlet leads to fluorescence. A radiationless transition can occur from the excited singlet, S_1, to the excited triplet state, T_1, with two unpaired electrons. Vibronic transitions from an excited triplet to the ground-state singlet lead to phosphorescence. Other transitions are possible. Direct transitions from the ground-state singlet, S_0, to an excited-state triplet, T_1 or T_2, lead to weak (forbidden) singlet-triplet absorption bands. Transitions between excited triplet states are allowed (as are transitions between excited singlet states). Measurement depends on the ability to produce sufficient population of the excited state. The longer lifetime of the triplet makes $T_1 \rightarrow T_2$ absorption easier to measure than that of $S_1 \rightarrow S_2$.

to thymidylic acid. The triplet states of the other nucleotides are either quenched or they transfer excitation to the thymidylic nucleotides, which then emit the phosphorescence.

OPTICAL ROTATORY DISPERSION AND CIRCULAR DICHROISM

A property of most biological molecules is molecular asymmetry or *chirality;* such molecules are not identical to their mirror images. The simplest examples result from the occurrence of asymmetric carbon atoms in these molecules. A carbon that is tetrahedrally bonded to four different atoms or groups can exist in two different structures that are mirror images of one another. They are frequently referred to as left- and right-handed forms. The rules for characterizing such molecules are described in organic chemistry texts.

Chiral molecules have distinctive properties that bear emphasis. First, consider some common examples of such "handedness" that are more familiar. Screws or nuts and bolts can be cut with a right-hand or a left-hand thread. A tumbled assortment of left- and right-handed bolts could be sorted by using a test nut. Once the bolts are separated they will have different interactions with nuts of a given handedness. In much the same way, chiral molecules in solution and with random, constantly changing orientations interact differently with light that is polarized in a chiral way (circularly polarized light).

These examples emphasize also that chirality of materials does not necessarily depend on the presence of asymmetrically substituted carbon atoms, although that is commonly the origin for small molecules. In the case of nucleic acids, an important source of chirality is whether the helical polynucleotide winds in a left- or a right-handed sense. Nucleic acids are mainly right-handed, but certain base sequences can form left-handed double helices. This switch from a right- to a left-handed structure may have an important control function in the genetic expression of nearby base sequences. Proteins in the α-helical conformation wind in a right-handed sense. The handedness of the helices depends on the stereochemistry of the monomer units: the L (levo or left) amino acid isomer for proteins and the D (dextro) sugar for nucleic acids. The overall structure of biologically active proteins (enzymes, antibodies) depends on additional determinants of tertiary structure that are not so easily characterized. These are very important, however, for they serve to make the detailed surface structure of an enzyme or antibody nonsuperimposable with its mirror image. As a consequence, the active sites of these molecules are able to distinguish between mirror-image substrate molecules, much as left-handed nuts will interact only with left-handed bolts. Enzymes that are involved with the synthesis of chiral molecules may make a single chiral isomer from a substrate that is not itself asymmetric. It is clear from these few examples that the property of chirality is of fundamental and wide-reaching importance in biology.

Polarized Light

Chiral structures can be distinguished and characterized by polarized light. The optical properties usually measured are *optical rotation,* the rotation of linearly polarized light by the sample, and *circular dichroism* (CD), the difference in absorption of left- and right-circularly polarized light. The spectrum of optical rotation is known as *optical rotatory dispersion* (ORD).

Electromagnetic radiation can be described by an electric field vector that oscillates with a characteristic frequency in time and space. For *unpolarized* light, the electric vector may oscillate in any direction perpendicular to the direction of propagation. For a large number of photons in an unpolarized beam, all directions are equally represented. The electric vectors can be pictured as radiating spokes on a many-spoked wheel, as shown in Fig. 10.19 (top). For *plane-polarized* light, the electric vector oscillates in a single plane that includes the propagation direction. A pictorial description of vertically plane-polarized light is shown in Fig. 10.19 (top and middle). For an observer moving at the photon's velocity,

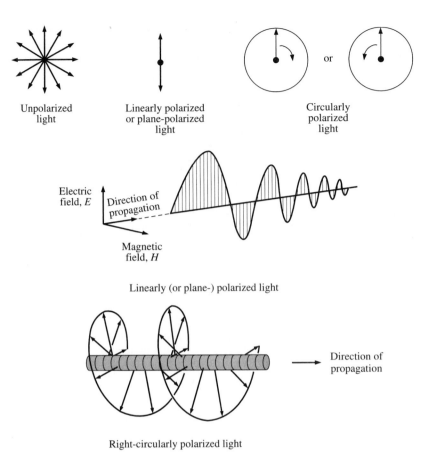

Unpolarized light

Linearly polarized or plane-polarized light

Circularly polarized light

or

Electric field, E

Direction of propagation

Magnetic field, H

Linearly (or plane-) polarized light

Direction of propagation

Right-circularly polarized light

Fig. 10.19 Different types of polarized light. At the top the arrows represent the electric vector of the light as seen by an observer moving with the light. The light is moving forward, out of the page. At the middle and bottom, the light is seen by a stationary observer.

the electric vector appears to be oscillating back and forth along a line. For this reason, plane-polarized light is also referred to as being linearly polarized. *Circularly polarized* light propagates so that the tip of its electric vector sweeps out a helix (Fig. 10.19, bottom); right-circularly polarized light produces a right-handed helix. To the observer moving with the photons, the electric vector appears to be moving in a circle, like the hands of a clock (Fig. 10.19, top). The convention is that for left-circularly polarized light, the electric vector moves counterclockwise as the light moves away from the observer; for right-circularly polarized light, it moves clockwise. Elliptically polarized light, which we shall not consider in detail here, propagates such that its electric vector sweeps out an ellipse, as seen by the observer moving with the light. Elliptically polarized light is either right- or left-handed, and it may be thought of as intermediate between circularly and plane-polarized. In fact, optical retardation plates will progressively produce plane, elliptical, circular, elliptical, plane, elliptical, and so on, polarizations as the thickness of the retarder is increased.

We can represent the electric vector of light in terms of its components. For light propagating in the z-direction and linearly polarized in the x-direction, the x- and y-components of the electric vector for linearly polarized light are:

Linearly polarized light:

$$E_x(z,\,t) \,=\, E_0 \sin 2\pi v \left(\frac{z}{c} - t \right) = E_0 \sin \frac{2\pi}{\lambda} (z - ct)$$

$$E_y(z,\,t) \,=\, 0$$
(10.29)

c is the speed of light in vacuum, and v is the frequency and λ the wavelength of light in vacuum. The x- and y-components of the oscillating electric vector are $E_x(z,\,t)$ and $E_y(z,\,t)$; E_0 is the maximum amplitude of the vector. Because the oscillating magnetic field is perpendicular to the electric field, its components are the same as the electric components except that x and y are interchanged. We do not have to consider the magnetic field explicitly if we specify the electric field, because Maxwell's equations characterize the magnetic field. The x- and y-components of the electric vector for circularly polarized light are:

Left-circularly polarized light:

$$E_x(z,\,t) \,=\, E_0 \sin 2\pi \left(\frac{z}{\lambda} - \frac{ct}{\lambda} \right)$$

$$E_y(z,\,t) \,=\, E_0 \sin 2\pi \left(\frac{z}{\lambda} - \frac{ct}{\lambda} + \frac{1}{4} \right)$$
(10.30a)

Right-circularly polarized light:

$$E_x(z,\,t) \,=\, E_0 \sin 2\pi \left(\frac{z}{\lambda} - \frac{ct}{\lambda} \right)$$

$$E_y(z,\,t) \,=\, -E_0 \sin 2\pi \left(\frac{z}{\lambda} - \frac{ct}{\lambda} + \frac{1}{4} \right)$$
(10.30b)

The y-component of the light is one-fourth of a cycle ahead of (or behind) the x-component. The resultant of the components is an electric field whose magnitude is constant, but which rotates counterclockwise [Eq. (10.30a)] or clockwise [Eq. (10.30b)] as it moves forward in space (along z) and in time. Addition of Eqs. (10.30a) and (10.30b) makes it clear that linearly polarized light is equivalent to the sum of left- and right-circularly polarized light beams propagating in the same direction with the same wavelength. Linearly and circularly polarized light can be produced easily from unpolarized light by transmission through appropriate films (such as Polaroid) or crystals (Nicol prism, quarter-wave plate).

We have described previously materials that are (linearly) birefringent toward plane-polarized light. These materials must be geometrically anisotropic; they have different properties along different directions. *Linear birefringence* occurs because the refractive index, and hence the light propagation velocity, is different for polarization planes oriented along different directions of the material. In an absorption band of an anisotropic material *linear dichroism* occurs; the absorbance is different for different orientations of the plane of polarization of the light. Most crystals are anisotropic and therefore linearly birefringent and linearly dichroic. Polaroid sheets contain oriented polymers which make them linearly dichroic (and linearly birefringent) in the visible. They absorb light polarized parallel to the oriented polymer and transmit light polarized perpendicular to the polymer; thus a Polaroid sheet is a linear polarizer for incident unpolarized light.

Substances that are optically active (chiral) exhibit *circular birefringence,* where circular birefringence is $n_L - n_R$, the different between the refractive index for left-circularly polarized and the refractive index for right-circularly polarized light. In other words, the velocities of propagation for left- and right-circularly polarized light are different. There is an important difference between circular birefringence and linear birefringence. A homogeneous chiral sample, such as a solution of an optically active solute, is geometrically isotropic; its optical properties are identical viewed from any direction. The circular birefringence results from an intrinsic property of the material that persists even though the orientations of the molecules are random. Chiral molecules also show *circular dichroism* $(A_L - A_R)$, a difference in absorbance for left- and right-circularly polarized light.

A quantum mechanical derivation shows how circular birefringence or dichroism is related to electronic wavefunctions of the ground and excited states of the molecule. An optically active transition must have an electric-dipole transition moment and a magnetic-dipole transition moment that are not perpendicular to each other. This can occur only for molecules that are not superimposable with their mirror images. A qualitative understanding of the electronic motion in these molecules, during electronic excitation by incident light, is that the electrons do not move in a line or in a circle; instead, they move in a helical path.

Optical Rotation

Optical rotation by chiral samples results from, and is a measure of, their circular birefringence. The term "optical rotation" comes from the usual experimental measurement procedure. If plane-polarized light is propagated through a transparent chiral sample, the emerging beam will also be plane-polarized, but its plane of polarization (still including the direction of propagation) will be rotated by an angle ϕ with respect to the direction of the polarization of the incident light. If the rotation is clockwise as seen by the observer, ϕ is positive; counterclockwise rotation is assigned a negative value of ϕ (Fig. 10.20). The origin of optical rotation can be rationalized by considering the incident plane-polarized light to be made up of two opposite circularly polarized, but in-phase, components, as represented by Eq. (10.30). Because the chiral sample is circularly birefringent ($n_L \neq n_R$), the two circular-polarized component electric vectors will propagate through the sample with different velocities (c/n) and they will have different wavelengths, λ_m, in the medium ($\lambda_m = \lambda/n$). As a consequence, one of them becomes advanced in phase with respect to the other. The resultant light, which remains plane-polarized, becomes rotated progressively upon passage through the sample. When it emerges from the other side, it has undergone a rotation. The angle of rotation can be obtained by replacing z in Eq. (10.30b) by $n_R z$ and z in Eq. (10.30a) by $n_L z$, and adding the x-components and the y-components. The resultant light is linearly polarized, but rotated by an angle ϕ given by

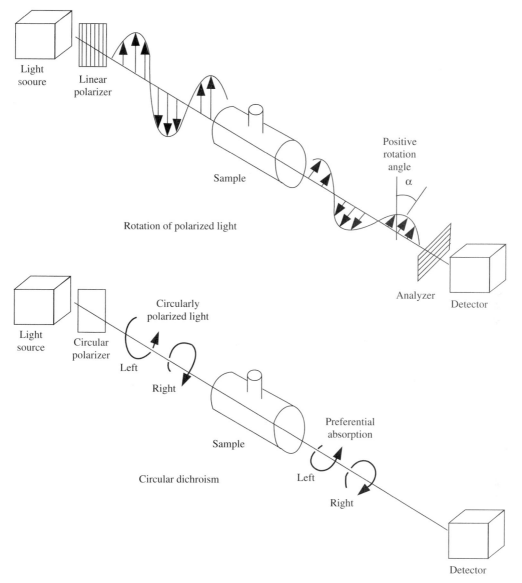

Fig. 10.20 (Top) Measurement of the rotation of linearly polarized light. If the sample in the top cell has significant absorbance, the transmitted light will be elliptically polarized. (See Fig. 10.21.) (Bottom) Measurement of circular dichroism, the preferential absorption of circularly polarized light. The detector measures the difference in absorbance of the right- and left-circularly polarized light.

$$\text{rotation (rad cm}^{-1}) = \phi = \frac{\pi}{\lambda}(n_L - n_R) \tag{10.31}$$

where λ is the wavelength of the light in vacuum. Note that ϕ is given per centimeter of pathlength in the sample. The actual angle of rotation clearly increases linearly with the path length through the sample. The rotation is measured by placing a polarizing element (suitable crystal or Polaroid sheet) called an *analyzer* in the emergent beam. By turning the analyzer so that it is crossed (perpendicular) to the direction of polarization of the emergent beam, the intensity is extinguished and the angle of rotation can be measured to better than 1 millidegree. To appreciate how sensitively circular birefringence can be measured, consider Eq. (10.31) for a measurement made at $\lambda = 314$ nm ($= \pi \times 10^{-5}$ cm). A rotation of 1 rad cm^{-1} (1 rad = 57.3 deg) corresponds to $n_L - n_R = 10^{-5}$.

Circular Dichroism

Materials that exhibit linear dichroism toward plane-polarized light do so because they have different absorbances as a function of the orientation of the sample with respect to the polarization direction of the incident light. Polaroid sheets are examples of dichroic materials. As with birefringence, linear dichroism results from geometric anisotropy in the sample.

Circular dichroism, analogously, results from a differential absorption of left- and right-circularly polarized light (Fig. 10.20) by a sample that exhibits molecular asymmetry. A simple expression of the circular dichroism is given by

$$\Delta A = A_L - A_R \qquad (10.32)$$

where A_L and A_R are the absorbances of the sample for pure left- and right-circularly polarized light, respectively. As with circular birefringence, circular dichroism may be either positive or negative. Circular dichroism occurs only in a region of the spectrum where the sample absorbs, whereas circular birefringence occurs in all wavelength regions of an optically active substance. This latter property is of obvious advantage for transparent substances such as sugars, in which the lowest energy electronic transitions occur in the far ultraviolet.

Referring again to Fig 10.20, the passage of plane-polarized light, E_0, through a circularly dichroic sample produces not only a phase shift due to the circular birefringence (which occurs in the absorption region as well) but also a differential decrease of the amplitudes (the electric vector magnitudes) of the left- and right-circularly polarized components. The emerging beam, E, is found to be elliptically polarized (Fig. 10.21) as a consequence. A detailed analysis shows that the ellipticity is given by

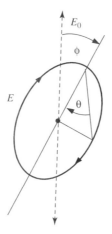

Fig. 10.21 Elliptically polarized light emerging toward the observer from a circularly dichroic sample. The sign convention is that ϕ is positive for clockwise rotation of the major axis and θ is positive for right-elliptically polarized light, as shown. The rotation angle is also called α; the ellipticity angle is also called ψ.

$$\theta \text{ (rad cm}^{-1}) = \frac{2.303(A_L - A_R)}{4l} \tag{10.33}$$

where l is the path length through the sample.

Experimental measurements of optical rotation and ellipticity are burdened with a history of measurements in cells of length 10 cm (path length d, in decimeters) and the use of different symbols for the same quantity. The equations summarized in Table 10.6 represent a consensus of current usage.

Table 10.6 Summary of experimental parameters for optical rotation and circular dichroism*

Optical rotation

Specific rotation $= [\alpha] = \dfrac{\alpha}{dc}$

Molar rotation $= [\phi] = \dfrac{100\alpha}{l\text{M}}$

Circular dichroism

Molar ellipticity $[\theta] = \dfrac{100\psi}{l\text{M}}$

Circular dichroism $= \Delta\varepsilon = \varepsilon_L - \varepsilon_R = \dfrac{A_L - A_R}{l\text{M}}$

$[\theta] = 3298(\varepsilon_L - \varepsilon_R)$

$*$
α = rotation angle, degrees	l = path length, cm
ψ = ellipticity, degrees	c = concentration, g cm^{-3}
ε = molar absorptivity	M = concentration, mol L^{-1}
d = path length, dm	

Circular Dichroism of Nucleic Acids and Proteins

The optical activity of a nucleic acid or protein is the sum of the individual contributions from the monomeric units and the contribution from their interactions in the polymeric arrangement. By comparing the circular dichroism (CD) of the native polymer to that of its monomeric units, their separate contributions can be determined experimentally and subtracted from the optical activity of the intact polymer. The difference then gives the contribution from the interactions in the native polymer conformation. This is illustrated for DNA and RNA in Fig. 10.22. The dramatic differences between the solid and dashed curves demonstrate that the principal contribution to the CD of the nucleic acids arises from interactions in the polymer.

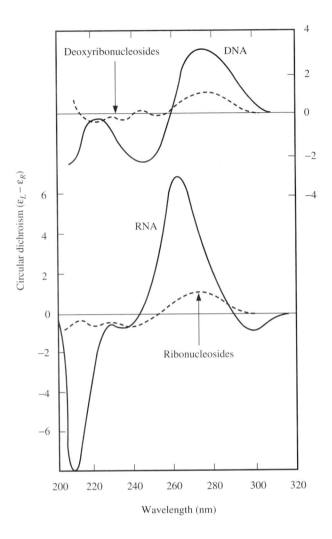

Fig. 10.22 Circular dichroism of double-stranded DNA and RNA compared with their component mononucleosides. *M. lysodeikticus* DNA data from F. Allen et al., *Biopolymers 11,* 853 (1972). Rice dwarf virus RNA data from T. Samejima et al., *J. Mol. Biol. 34,* 39 (1968). Nucleoside spectra calculated from the base composition (72% G + C for the DNA; 44% G + C for the RNA) and CD data of C. R. Cantor et al., *Biopolymers 9,* 1059, 1079 (1970). [From V. A. Bloomfield, D. M. Crothers, and I. Tinoco, Jr., *Physical Chemistry of Nucleic Acids,* Harper & Row, New York, 1974, p. 134.]

Figure 10.23 shows an important contribution of CD to molecular biology. The synthetic polynucleotide poly(dG-dC) · poly(dG-dC) is a double-stranded helix with a sequence of alternating guanine and cytosine bases on each strand. At high salt concentration its unusual CD (curve Z in Fig. 10.23) suggested that the helix could be left-handed. Because this was the first experimental indication of a left-handed DNA helix, there was much

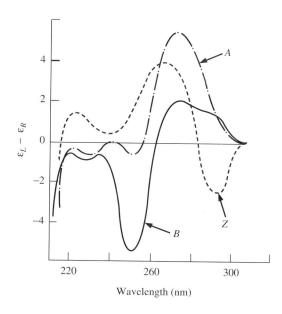

Fig. 10.23 Circular dichroism of the synthetic polynucleotide poly(dG-dC) · poly(dG-dC) in different conformations. The polynucleotide is a double-stranded helix; each strand has a sequence of alternating deoxyguanylic acid (dG) and deoxycytidilic acid (dC). Different conformations are obtained by changing the solvent. The B form is obtained in 0 to 40% ethanol or 10^{-3} M to 2 M NaCl; it is a right-handed helix with about 10 base pairs per turn of the double helix. The Z form is obtained in 56% ethanol or 3.9 M NaCl; it is a left-handed helix. The A form is obtained in 80% ethanol; it is a right-handed helix with about 11 base pairs per turn. [From F. M. Pohl, *Nature 260,* 365 (1976). © 1976 Macmillan Journals Limited.]

skepticism. However, the crystal structure of a double helix of a six-base-pair fragment, $(dC\text{-}dG)_3 \cdot (dC\text{-}dG)_3$, confirmed the presence of a left-handed helix and provided a detailed structure (Wang et al., 1979). Subsequently, left-handed DNA has been found *in vivo,* and there is speculation that a right- to left-handed transformation is involved in regulation of which regions of DNA are transcribed to RNA (Rich, 1983). RNA can also exist in a left-handed form (Hall et al., 1984). Regardless of whether left-handed DNA (or RNA) turns out to be biologically important, the knowledge that DNA conformation is so dependent on the environment is valuable. The existence of the different conformations (A, B, Z) represented in Fig. 10.23 is critically dependent on the composition of the solvent and the temperature. Small amounts of molecules that bind strongly to DNA can cause a change from one conformation to another.

Several forms of secondary structure are present in native proteins. They include α-helix, parallel and antiparallel β-pleated sheets, and random coil. By measuring homopolypeptides under conditions where the conformation is uniform throughout the polymer, Gratzer and Cowburn (1969) prepared plots of ORD and CD spectra in the peptide region that show the range of values encountered for polypeptides of different amino acids (Fig.

10.24). The ranges are relatively small compared with the differences among the CD spectra of the different conformations. This finding implies that the measured CD of a protein can be used to obtain the amount of each type of secondary structure. Hennessey and Johnson (1981) showed that after calibration with proteins of known structure, the measured CD from 178 nm to 250 nm yielded correct secondary structures.

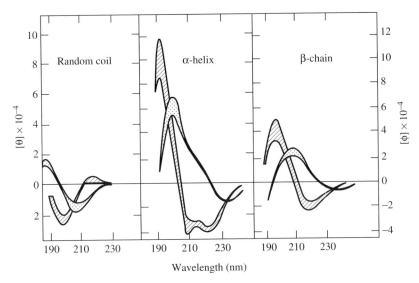

Fig. 10.24 CD (crosshatched) and ORD (dotted) of homopolypeptides in the random coil, α-helical, and β conformations. The shaded areas indicate the range of values. [From W. B. Gratzer and D. A. Cowburn, *Nature 222*, 426 (1969) © Macmillan Journals Limited.]

Induced Circular Dichroism of Chromophores

The CD spectra of chromophores in biochemical systems can indicate interactions with molecules in the immediate vicinity. An illustration of this characteristic is the phenomenon of induced CD. If a symmetric dye molecule or other absorbing solute is dissolved in a transparent chiral solvent, such as amyl alcohol or menthol, then circular dichroism is observed in the region of the absorption band of the solute. This induced CD is not present if the same chromophore is dissolved in nonchiral solvents. The induced CD results from an interaction of the transition moment of the chromophore with the asymmetric molecules in its environment (Hayward and Totty, 1971).

Certain dye molecules, such as acridine, proflavin, and ethidium, exhibit strong binding to biopolymers, such as proteins, nucleic acids, or polysaccharides in aqueous solutions. Such interactions can lead to induced CD in the absorption region of the bound dye. Many biological molecules of interest contain native chromophores that exhibit induced CD. Examples in which the separated chromophore itself is nonchiral include hemoglobin, myoglobin, cytochromes, hemocyanin, rhodopsin, ferredoxins, transferrin, copperproteins, vitamin B_{12}, and flavoproteins. The absorption and CD of frog rhodopsin are shown in Fig. 10.25. This example is particularly illustrative because, upon bleaching of

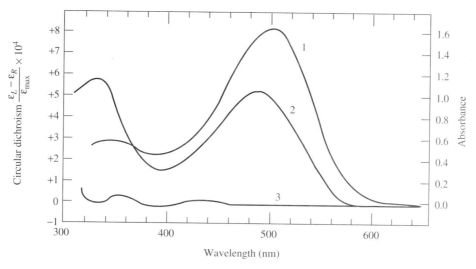

Fig. 10.25 Circular dichroism of rhodopsin from *Rana pipiens*. Curve 1: absorbance curve of rhodopsin in 2% digitonin, pH = 6.7. Curve 2: circular dichroism of same solution expressed in terms of a rhodopsin with unit absorbance at the maximum; 10-nm path length. Curve 3: Same solution after exposure to tungsten light of 60 W until all purple had faded. [From F. Crescitelli, W. F. H. M. Mommaerts, and T. I. Shaw, *Proc. Natl. Acad. Sci. USA 56*, 1729 (1966).]

the rhodopsin by light, both the absorption and the CD (bottom curve in Fig. 10.25) in the visible region disappear. The isolated chromophore, retinal, does not exhibit any measurable circular dichroism, as expected for this achiral molecule.

NUCLEAR MAGNETIC RESONANCE

We have been discussing electronic spectroscopy in which the energy levels correspond to different arrangements of electrons in the molecule. We shall now describe a spectroscopic method (nuclear magnetic resonance, NMR) in which the energy levels correspond to different orientations of the magnetic moment of a nucleus in a magnetic field.

Some nuclei have magnetic dipole moments that will orient in a magnetic field. The possible energies of the nuclear magnetic moment in the magnetic field are of course quantized. Therefore, characteristic frequencies of electromagnetic radiation are absorbed. The frequencies depend on the nucleus, the strength of the magnetic field, and, most important to us, the chemical and magnetic environment of the nucleus. Nuclei with an even number of protons and an even number of neutrons have zero nuclear magnetic moments; no nuclear magnetic spectra are possible. These nuclei include many isotopes that commonly occur in nature: $^{12}_{6}C$, $^{16}_{8}O$, and $^{32}_{16}S$. (The atomic number is the subscript; the mass number is the superscript.) Other nuclei have magnetic moments and their nuclear magnetic resonance (NMR) can be studied. The simplest nuclei to consider have a spin of $\frac{1}{2}$; these include $^{1}_{1}H$, $^{13}_{6}C$, $^{15}_{7}N$, $^{16}_{9}F$, and $^{31}_{15}P$. (The electron also has a spin of $\frac{1}{2}$ and a magnetic moment that allows study of electron spin resonance in a magnetic field.) Nuclei with spins greater than $\frac{1}{2}$, such as $^{14}_{7}N$, have more complicated spectra, which we will not discuss. The most extensively studied nucleus has been the proton. The ^{31}P isotope is also 100% naturally abundant; it is increasingly used in studying biochemical metabolism. The spin $\frac{1}{2}$ isotopes of carbon and nitrogen are present in slight amounts at natural abundance—1% ^{13}C and 0.4% ^{15}N—so NMR experiments are much easier with isotopically enriched samples. Moreover, there are tremendous advantages in structure determination of macromolecules that are uniformly labeled with 100% ^{13}C and ^{15}N. A whole series of new NMR experiments becomes possible which involves correlations between protons, carbons, nitrogens, and phosphorus. These experiments are described in a following section on two-dimensional and three-dimensional NMR.

The reason that the magnetic spectrum of nuclei of spin $\frac{1}{2}$ are easy to interpret is that the nucleus can have only two orientations in the magnetic field: with the field or against the field. The magnetic quantum number, which specifies the orientation, can have only the values of $\pm \frac{1}{2}$. Therefore, an isolated nucleus of spin $\frac{1}{2}$ in a magnetic field will have a single absorption frequency, as shown in Fig. 10.26a. The absorption frequency, ν_0, is

$$h\nu_0 = 2\mu_m H_0 \qquad (10.34)$$

where μ_m is the component of the nuclear magnetic moment along the direction of the magnetic field H_0. It is characteristic of each different nucleus.

In an equivalent classical description of NMR, each magnetic nucleus is considered to be a spinning bar magnet. In a magnetic field H_0, the magnet will precess around the direction of the magnetic field. Its motion is analogous to a spinning gyroscope in the earth's gravitational field. Its precession frequency is identical to that defined quantum mechanically by Eq. (10.34). This precession frequency, usually given as the angular frequency $(\omega_0 = 2\pi\nu_0)$, is called the Larmor frequency:

$$\omega_0 = \gamma H_0 \qquad (10.35)$$

where γ is the gyromagnetic ratio of the nucleus; it is related to the magnetic moment of a spin $\frac{1}{2}$ nucleus by $\mu_m = \frac{1}{2}\gamma\hbar$ with $\hbar = h/2\pi$.

An NMR spectrum consists of a set of peaks corresponding to the resonance frequencies of, for example, the protons in the sample; the area of each peak is proportional to the number of nuclei in the sample with each frequency. In a constant magnetic field of 11.74 Tesla all protons absorb near 500 MHz, but the exact position of each resonance peak depends on the structures and environments of the molecules. An NMR spectrum can be measured by varying the frequency of the applied radiofrequency radiation in a constant magnetic field; this is analogous to the method used in visible absorption spectroscopy. A much more efficient method uses *Fourier transform spectroscopy* to measure the spectrum. A short (microsecond) pulse of radiofrequency radiation is applied to the sample in the magnetic field; this pulse is equivalent to a range of frequencies. The response of the sample to the pulse depends on the absorption of the sample over the range of frequencies applied—for example, 500 MHz ± 2500 Hz. The information obtainable by varying the frequency point by point is obtained instead by using only one pulse; this gives an increase in efficiency of about 1000. The pulse is actually applied many times and the results averaged to improve the signal-to-noise in the spectrum. The Fourier transform method provides a great increase in sensitivity over the old method of frequency variation.

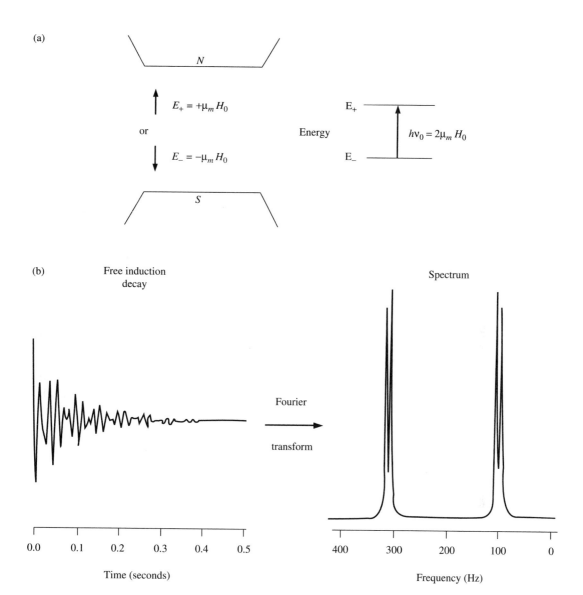

(a)

$$E_+ = +\mu_m H_0$$

or

$$E_- = -\mu_m H_0$$

$$E_+$$

Energy

$$h\nu_0 = 2\mu_m H_0$$

$$E_-$$

N

S

(b) Free induction decay

Spectrum

Fourier

transform

Time (seconds)

0.0   0.1   0.2   0.3   0.4   0.5

Frequency (Hz)

400   300   200   100   0

Fig. 10.26 (a) Orientation of a nuclear magnetic dipole (represented by an arrow) of spin $\frac{1}{2}$ in a magnetic field. The resonance condition for absorption of radiation ($h\nu = 2\mu_m H_0$) shows that the NMR spectrum will depend both on the applied magnetic field and the properties of the nucleus. (b) An NMR spectrum is obtained by applying a short pulse (~1 μs) of radiofrequency radiation (for example 500 MHz), then measuring the free induction decay as a function of time. The Fourier transform [Eq. (10.36)] of the free induction decay is the NMR spectrum as a function of frequency. The figure was kindly supplied by Professor David Wemmer, U. of California, Berkeley.

Fourier transform spectroscopy depends on the mathematical fact that any time-dependent function is equivalent to a frequency-dependent function. The Fourier transform of $f(t)$ is

$$F(\omega) = \int_{-\infty}^{+\infty} f(t)e^{i\omega t}\, dt \tag{10.36}$$

i = the imaginary = $\sqrt{-1}$
ω = frequency = $2\pi\nu$

The applied pulse, a function of time, $f(t)$, is equivalent to a range of frequencies given by its Fourier transform, $F(\omega)$, from Eq. (10.36). The response of the sample—called the *free induction decay*—is converted to a frequency spectrum by a Fourier transform: an amplitude vs. time signal is transformed into an amplitude vs. frequency signal. The Fourier transform is a complex function of the form $(a + ib)$ which has a real part a and an imaginary part b because of the i in Eq. (10.36). The real part gives the absorption spectrum of the sample as a function of frequency. A free induction decay and the NMR spectrum obtained from its Fourier transform is shown in Fig. 10.26b.

Chemical Shifts

The fact that makes NMR useful is that the nuclei in molecules are not isolated. They interact with the surrounding electrons and with the other nuclei in the molecule. The applied magnetic field induces a magnetic moment in the electrons which is in the opposite direction to the applied field. The electrons thus shield the nucleus from the magnetic field; the shielding is small but measurable. The amount of shielding is characterized by a unitless shielding constant, σ. The resonance condition is now

$$h\nu = 2\mu_m H_0(1 - \sigma) \tag{10.37}$$

Each magnetically distinct nucleus in a molecule will have a different amount of shielding and thus a different value of σ. It is impractical to measure σ directly, so a relative shielding is measured. A standard molecule is introduced into the NMR spectrometer, and resonance frequencies are measured relative to it. The chemical shift can be measured as the difference in frequency for the nucleus relative to the standard. The shift is measured in hertz (Hz) as upfield (toward a higher magnetic field) or downfield (toward a lower magnetic field) relative to the standard. The chemical shift in hertz is directly proportional to the applied magnetic field; therefore, the spectrometer frequency must also be given. To avoid an instrument-dependent parameter, a unitless chemical shift, δ, has been defined as

$$\delta \text{ (ppm)} = \frac{\nu \text{ (sample, Hz)} - \nu \text{ (reference, Hz)}}{\nu \text{ (spectrometer, MHz)}} \tag{10.38}$$

$$\delta \text{ (ppm)} = [\sigma \text{ (reference)} - \sigma \text{ (sample)}] \times 10^6$$

The spectrometer frequency is an instrument characteristic which depends on the magnitude of the magnetic field and the nucleus of interest. The chemical shift, δ, is a unitless number given in parts per million (ppm). It is defined to be characteristic of the molecule and independent of applied magnetic field.

The chemical shift of a nucleus depends on many factors, but the electron density at the nucleus is often dominant. A high electron density causes a large shielding and means that the applied magnetic field must be increased to obtain resonance. This upfield shift means a decrease in the magnitude of δ. Conversely, a low electron density at a nucleus means a downfield shift and an increase in δ. For example, the chemical shifts of the methyl protons (underlined below) in CH_3X relative to tetramethylsilane become larger as X becomes a better electron-withdrawing group: $\delta(C\underline{H}_3—CH_3) = 1$, $\delta(C\underline{H}_3—C_6H_5) \approx 2$, $\delta(C\underline{H}_3—OH) \approx 4$. A proton attached directly to an electronegative atom such as in a carboxyl group has a very low electron density (which is why it is acidic); it can have a $\delta(COO\underline{H}) \approx 10$. The range of 0 to 15 ppm nearly covers the range of δ for protons. The chemical shift in Hz for protons ranges from 0 to 900 Hz for a 60-MHz spectrometer and up to 7500 Hz for a 500 MHz spectrometer. Figure 10.27 shows the approximate values of chemical shifts for various types of protons.

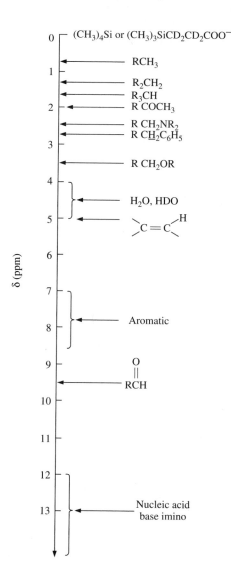

Fig. 10.27 Approximate chemical shifts in ppm relative to a reference for different types of protons. The smaller the value of δ, the greater the chemical shielding and the more upfield the proton signals occur. Increasing δ means decreasing electron density at the proton. The reference signal is tetramethylsilane (TMS) for nonaqueous solutions or trimethylsilylpropionate-d_4 (TSP) for aqueous samples.

For biochemical macromolecules we are often interested in the effect of conformation on chemical shifts. For example, the chemical shift of the methyl protons of alanine in a protein will be affected differently by a nearby aromatic ring from phenylalanine, or by a carboxylate from glutamic acid. The NMR spectrum can reveal conformational changes in a protein due to binding of a substrate, for example. Any influence that affects the magnetic field sensed by the nucleus will change the chemical shift. The formation of a hydrogen bond decreases the electron density at the proton and causes a downfield shift in the resonance (an increase in δ). A nearby paramagnetic ion (such as Fe^{3+} or Ni^{2+}) produces a large magnetic field, which causes very large changes in chemical shifts. One of the most useful conformation-dependent chemical shifts is caused by aromatic rings. When a planar aromatic molecule with delocalized π molecular orbitals is placed in a magnetic field, a magnetic field is induced around the aromatic system as shown in Fig. 10.28. It is called the ring current effect. The induced field is in the opposite direction to the applied field above and below the plane of the molecule, and in the same direction as the applied field on the sides of the molecule. Therefore, a proton near an aromatic group will be shielded (decrease in δ) if it is above or below the center of the planar ring, and its δ will be increased if it is at the outside of the planar ring. The effect decreases approximately as the inverse cube of the distance between the proton and the center of the aromatic ring. Ring-current chemical shifts are most useful for interpreting conformations of proteins near phenylalanine or tyrosine residues, and for conformations of nucleic acids, where the purines, adenine and guanine, have the largest ring-current effects.

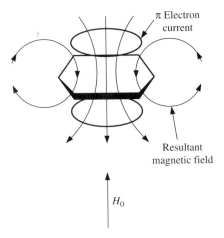

Fig. 10.28 The ring-current effect in NMR. When a planar aromatic molecule is placed in a magnetic field H_0, the mobile π electrons can be considered to flow in a current around the periphery of the planar molecule. Classically a current loop generates a magnetic field as shown in the figure. The magnitude of the field decreases approximately as the inverse cube from the center of the loop. This model is approximate, but it provides a good qualitative understanding of the effect of neighboring aromatic molecules on chemical shifts.

Spin-Spin Splitting

The chemical shift is a measure of nuclear-electron interactions. The *spin-spin splitting, J,* is a measure of the interaction of two or more nuclei, where the interaction is transmitted by the intervening electrons. The two possible orientations of a spin $\frac{1}{2}$ nucleus in a magnetic field can split the energy levels of neighboring nuclei. This means that the absorption line of a set of equivalent nuclei is split into a multiplet. The frequency separation between the lines of the multiplet is the spin-spin splitting, J, in hertz. If the spin-spin splitting is less than one-tenth the frequency difference due to the chemical shifts, simple first-order theory, as illustrated in Fig. 10.29, can be applied. This means that the effects of the chemical shifts and the spin-spin splittings are additive. If the spin-spin splitting between protons A and B (J_{AB}) is comparable to or larger than the difference in their chemical shifts ($v_A - v_B$), then the spectrum depends on the ratio of J_{AB} to ($v_A - v_B$) as shown in Fig. 10.30. The first-order theory used in Fig. 10.29 is no longer adequate if the spin-spin splitting J_{AB} is comparable or larger than ($v_A - v_B$). We can see in Fig. 10.30 that magnetically equivalent nuclei do not split each other; because their chemical shifts are identical, they have only one line as shown at the bottom of Fig. 10.30. This means, for example, that methane, ethane, and benzene in the gas phase each show only one peak in their NMR spectrum. Similarly, the three protons of a methyl group do not split each other, just as the two protons of a methylene do not split each other (Fig. 10.29). However, chemically equivalent protons are not always magnetically equivalent. If there is not free rotation around C—C bonds, the two protons on a methylene can be in different environments and have different chemical shifts.

The spin-spin splitting in hertz (unlike the chemical shift in hertz) is independent of the applied magnetic field. The values of J for protons range from 0 to about 20 Hz. If we are measuring proton NMR in a hydrocarbon or carbohydrate, there are no effects from the carbon or oxygen nuclei, because they have no magnetic moment (except for negligible amounts of ^{13}C and ^{17}O.) Naturally occurring nitrogen (^{14}N) has a spin of 1 and tends to broaden neighboring proton lines rather than split them. Consequently, we often need to consider only proton-proton splittings. Figure 10.29 gives several examples of spin-spin splittings. Each set of n magnetically equivalent protons, such as the three on a methyl group, split neighboring protons into a multiplet of ($n + 1$) lines. The number of lines and their relative intensities can be easily derived by counting the number of ways or arranging the nuclear spins in a magnetic field. One proton creates a doublet of intensity 1 : 1. Two equivalent protons create a triplet of intensity 1 : 2 : 1. Three equivalent protons create a quadruplet of intensity 1 : 3 : 3 : 1. The student should easily be able to obtain the result that n protons create ($n + 1$) lines with relative intensities corresponding to the coefficients of a binomial expansion $(1 + x)^n$.

For protons on adjacent atoms, the spin-spin coupling constant, J_{AB}, is of order of magnitude 10 Hz, but its exact value is very characteristic of the configuration and conformation. For example, the protons on a carbon-carbon double bond have a $J(cis) \cong 12$ Hz and a $J(trans) \cong 19$ Hz. For protons on a carbon-carbon single bond, the coupling constant is related to the torsion angle between the protons. The torsion angle is defined as follows:

$$\phi = 0° \qquad\qquad \phi = 180°$$

When this angle is 90°, the spin-spin coupling constant is near zero; when the torsion angle is 0° or 180°, the coupling constant is near 10 Hz. If there is free rotation about the bond an

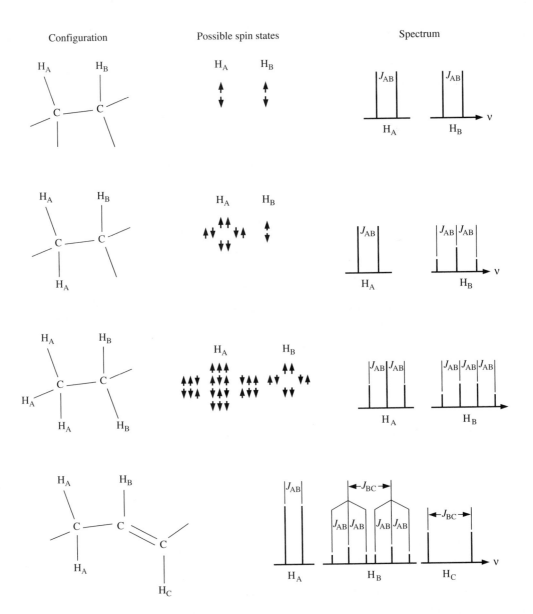

Fig. 10.29 Spin-spin splittings. A group of n equivalent protons will have $(n + 1)$ possible spin states in a magnetic field. This will give rise to a multiplet of $(n + 1)$ lines in adjacent protons. Note that the magnitude of the splitting of A by B equals that of B by A. The central position of the multiplet depends on the chemical shift. The total intensity of each multiplet is a measure of the number of equivalent protons that are split. In the example at the bottom of the figure, proton B is split by two equivalent protons and by a different, single proton. This gives rise to a pair of triplets.

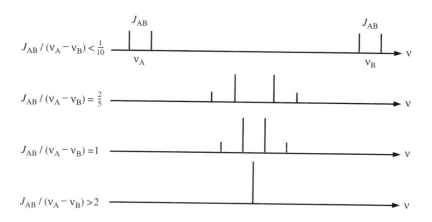

Fig. 10.30 NMR spectrum of two protons as a function of their ratio of spin-spin splitting (J_{AB}) in hertz to the difference in their chemical shifts ($\nu_A - \nu_B$) in hertz. When the splitting is small compared to the separation of the peaks we have the simple first-order theory illustrated in Fig. 10.29. As ($\nu_A - \nu_B$) decreases the inner lines increase in magnitude and the outer lines decrease. For magnetically equivalent protons ($\nu_A - \nu_B$) only one line (with twice the intensity) is seen. Note that ν_A and ν_B are directly proportional to the magnetic field, but J_{AB} is independent of field. Therefore, a 500-MHz spectrum may show the simple first-order pattern, while a 100-MHz spectrum will not. Remember that ($\nu_A - \nu_B$) is five times larger at 500 MHz than at 100 MHz.

average coupling constant is measured. The form of the dependence of coupling constant on conformation is given by the Karplus equation:

$$J = A + B \cos \phi + C \cos 2\phi \qquad (10.39)$$

where A, B, and C are parameters that can be calculated from quantum mechanics, or can be established from measurements on molecules of known conformation. Their values depend on the other substituents on the carbon atoms. Protons separated by four bonds (H_A—C—C—C—H_B) often produce negligible spin-spin splitting. In aromatic or conjugated systems however, a splitting of up to 1 or 2 Hz may be seen.

Spin-spin decoupling is a useful method for assigning protons in a complicated spectrum. If one proton (A) has been assigned, a proton (B) which is coupled to it can be determined. Intense radiation is applied at a frequency which proton A absorbs. This saturates A and decouples it from the other protons; the splitting of B by A disappears in the spectrum. This proves that proton B was indeed coupled to A.

Nuclear magnetic resonance is a very powerful method for determining the structure and conformation of molecules in solution. The chemical shifts are characteristic, the spin-spin splittings indicate neighboring nuclei, and the area under each peak gives the number of equivalent nuclei.

Example 10.2 Once we have measured an NMR spectrum the first thing we have to do is assign the NMR peaks to the protons of the molecule. Figure 10.31 shows measured spectra for two amino acids in D_2O solution. The reference is the methyl protons of the water-soluble trimethylsilylpropionate-d_4. The large peak at 4.8 ppm

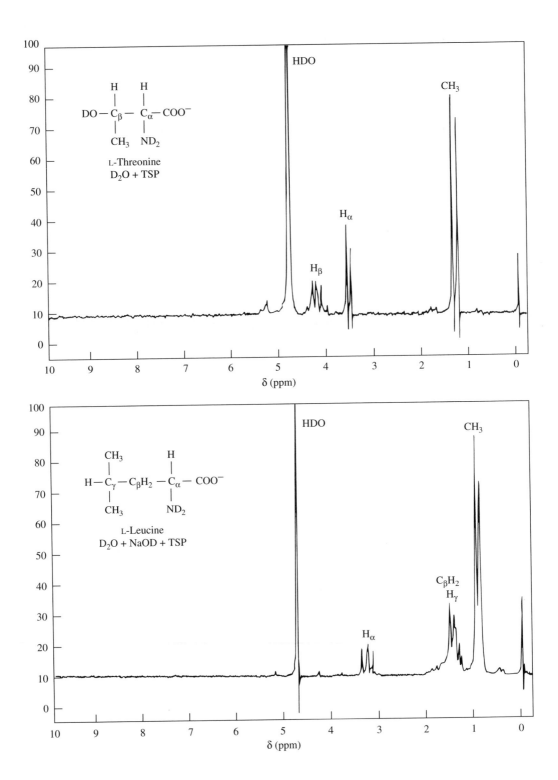

is HDO from residual protons in D_2O; we measured the pure solvent to establish this. We can assign the spectrum mainly on the basis of the number of protons and the spin-spin splittings.

In threonine there are three types of nonexchangeable protons (the protons on the hydroxyl and amino groups exchange quickly with the deuterons from the solvent), so we expect three peaks with the following characteristics:

Threonine

Protons	Relative areas	Splittings
C_α—H	1	Doublet (by C_β—H)
C_β—H	1	Doublet of quartets (by C_α—H and CH_3)
CH_3	3	Doublet (by C_β—H)

The large doublet at 1.3 ppm is obviously CH_3; the small doublet at 3.6 ppm is C_α—H and the multiplet at 4.3 ppm must be the C_β—H. These chemical shifts are in general agreement with the rules shown in Fig. 10.27. Methyls are near 1 ppm and protons on carbon bonded to oxygen or nitrogen are around 3 ppm or higher.

The same logic applied to leucine yields:

Leucine

Protons	Relative areas	Splittings
C_α—H	1	Triplet (by CH_2)
CH_2	2	Doublet of doublets (by C_α—H and C_γ—H)
CH_3	6	Doublet (by C_γ—H)
C_γ—H	1	Triplet of heptuplets (by CH_2 and two CH_3's)

The assignment is methyls at 0.9 ppm; C_α—H is at 3.3 ppm and the multiplet near 1.5 ppm is a combination CH_2 and C_γ—H. A doublet of doublets will often look like a triplet, so the assignment of the triplet at 3.3 ppm is made on the basis of relative areas. The chemical shifts are consistent with the general values in Fig. 10.27.

Figure 10.32 shows the measured spectrum of a mutagenic molecule, ethidium. The spectrum is complicated, but all of the peaks can be assigned to protons. The H4 and H7 protons are the only ones without protons on neighboring carbons, so they show only 1- or 2-Hz splittings. The H7 proton is assigned to the highest upfield peak ($\delta = 6.544$ ppm) because of its magnetic shielding by the phenyl ring. This leaves $\delta = 7.407$ as H4. The phenyl

Fig. 10.31 Proton NMR spectrum of two amino acids in aqueous solution. Measurements were made at 60 MHz with TSP $\equiv (CH_3)_3SiCD_2CD_2COO^-$ as reference at 0 ppm. The very small peaks in the spectrum on either side of the large peaks are an artifact in the data caused by rapid spinning of the sample tube. The sample tube is spun to average over small inhomogeneities in the magnetic field at the sample. (From the *Aldrich Library of NMR Spectra*, Edition II, Vol. 1, 1983.)

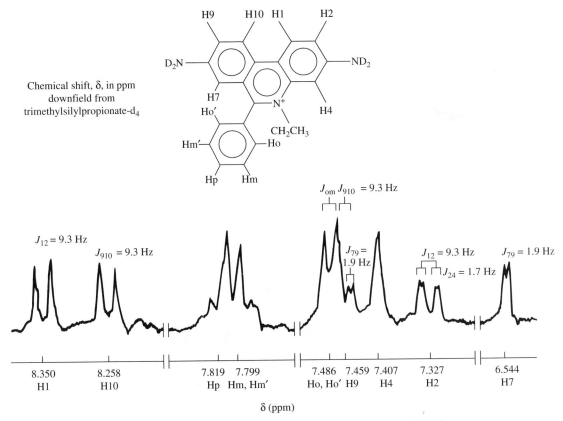

Fig. 10.32 Proton magnetic resonance spectrum of ethidium in D_2O solution at 4°C. The spectrum was measured in an 8.45 T magnet at 360 MHz. The original spectrum was plotted with a scale of 11.654 Hz cm^{-1}. As 1 Hz is $\frac{1}{360}$ ppm, the chemical shifts can easily be measured to ± 0.003 ppm. The ethyl protons are further upfield (smaller δ) and are not shown on this spectrum. (Data provided by Che-Hung Lee.)

ring can rotate relative to the phenanthridene group; therefore, the two ortho (Ho, Ho′) and the two meta (Hm, Hm′) are equivalent. The three similar protons on the phenyl (Hp, Hm, Hm′) are not resolved in this spectrum and are assigned to the $\delta = 7.81$ region (three protons in area) by analogy with other substituted benzene rings. The two ortho protons are assigned to the doublet with area corresponding to two protons at $\delta = 7.486$ ppm. The assignments of H1, H2, H9, and H10 are also done by analogy and previous experience. Proof of the assignments can be obtained by study of selectively deuterated derivatives. Spin-spin splittings and spin-spin decoupling are also helpful. The 9-10 and 1-2 spin-spin splittings are about 9 Hz, characteristic of a three-bond separation in an aromatic system. The aromatic four-bond spin-spin splitting of about 2 Hz is clearly seen for the 2-4 and 7-9 interactions.

This molecule binds strongly to DNA, and various spectroscopic and hydrodynamic methods have shown that it intercalates into the DNA. That is, the flat aromatic rings fit between two adjacent base pairs in the DNA. NMR studies on ethidium bound to DNA fragments are consistent with this interpretation. Furthermore, the relative shifts of the different protons give very detailed information about the structure of the complex. The H2 and

H9 protons have the largest shifts on binding to DNA. The H4, H7, and H1, H10 show similar, but smaller, shifts. When ethidium is bound to DNA, Ho and Ho′ are no longer equivalent; the binding immobilizes the phenyl group. There is no appreciable effect on the Hm, Hm′, and Hp. Similar NMR studies of the nucleic acid part of the complex give further useful information.

Molecules that intercalate into DNA can cause frame-shift mutations. These are mutations in which base pairs are added to, or deleted from, the replicated DNA at this site. NMR (and other spectroscopic) studies of mutagen-DNA complexes should help in understanding how frame-shift mutations occur.

A valuable application of NMR is to the study of metabolic processes in living organisms (Shulman, 1983). Here protons are less useful because of their presence in all molecules, including H_2O, but ^{13}C and ^{31}P are very valuable. The metabolic fate of ^{13}C-labeled glucose, for example, can be followed with time in different bacterial cells. Figure 10.33 shows the ^{31}P NMR spectrum of a frog muscle. The main phosphorus-containing species observed in muscle are creatine phosphate, adenosine triphosphate (ATP), adenosine diphosphate (ADP), and inorganic phosphate, which is a mixture of $H_2PO_4^-$ and HPO_4^{2-} in rapid equilibrium. The positions and magnitudes of the peaks are very informative.

Two species in rapid equilibrium produce one NMR peak with a chemical shift weighted by the fraction of each species present in the sample. Rapid equilibrium in NMR

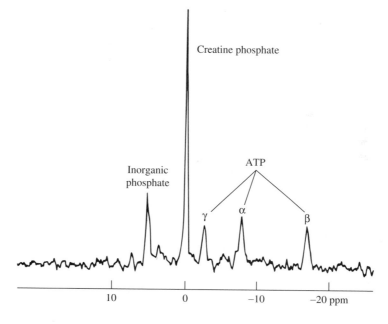

Fig. 10.33 ^{31}P NMR spectrum of intact frog muscles. The phosphorus peaks are from inorganic phosphate (a rapid equilibrium between $H_2PO_4^-$ and HPO_4^{2-}), creatine phosphate, and adenosine triphosphate (including some adenosine diphosphate). The resolution is not high enough to show that the α and γ peaks of ATP are doublets split by the two neighboring ^{31}P, and that the β peak is a triplet split by the two neighboring ^{31}P. (Figure from D. G. Gadian, *Nuclear Magnetic Resonance and Its Applications to Living Systems,* Clarendon Press, Oxford, 1982.)

means that the species are interchanging rapidly compared to the frequency difference of their NMR resonance peaks. For example, two species with chemical shifts which differ by 1 ppm at 300 MHz have a 300-s^{-1} difference in resonance peaks. If they interchange much more rapidly than 300 times per second, only one peak will be observed. If they interchange much more slowly than 300 times per second, two peaks are seen. For intermediate kinetics broadened spectra are seen. The kinetics of proton transfer for phosphate ions is very rapid compared to the NMR time scale. Therefore, the chemical shift of the inorganic phosphate peak gives a direct measure of the ratio of ($H_2PO_4^-/HPO_4^{2-}$) and thus of the pH of the muscle. The α and β phosphorus NMR signals of ADP have nearly the same chemical shifts as the α and γ phosphorus NMR signals of ATP. However, the relative amounts of creatine phosphate, ATP + ADP, and inorganic phosphate (as measured from the areas of the NMR peaks) reveals the metabolic state of the muscle. Studies on a human forearm during exercise under reduced blood flow showed a rapid drop in pH and creatine phosphate, and an increase in inorganic phosphate. Various muscular and circulatory disorders can be diagnosed by NMR; nothing need enter the body except the magnetic field (Ross et al., 1981).

Relaxation Mechanisms

In the presence of a magnetic field a nucleus of spin $\frac{1}{2}$ has two energy levels (Fig. 10.26a). Radiation of the resonance frequency causes transitions to occur between the two energy levels. As there are more nuclei in the lower energy level than in the upper level, the transitions lead to a net absorption of energy from the incident radio frequency radiation. This is the signal we measure. However, if the radiation is intense enough, the number of nuclei can become equal in the upper and lower energy levels. Then the transitions do not produce any net absorption (as many photons are emitted as are absorbed). We say that the signal is saturated. There are always spontaneous transitions from the upper to the lower energy level. These spontaneous transitions allow the spins of the saturated signal (no absorption) to return to their equilibrium population. The return of a spin population in a magnetic field to its equilibrium population follows first-order kinetics; its relaxation time is defined as the *spin-lattice relaxation time*, T_1. One way to measure T_1 is, as implied above, to saturate a signal (NMR peak area is zero), then remove the saturating radiation and measure the peak area with time. The first-order kinetics is exponential.

$$M = M_{eq}(1 - e^{-t/T_1}) \tag{10.40}$$

The magnetization (measured by peak area) at any time t is M, and M_{eq} is the magnetization at equilibrium.

We can measure the spin-lattice relaxation time, T_1, for a nucleus, but what does it mean? The magnetic moment of a nucleus relaxes by interacting with all the nuclei around it. The nuclei can be on the same molecules or on different molecules, such as solvent molecules. The important quantity is the magnitude of the magnetic field (caused by surrounding nuclei) at the transition frequency of the nucleus of interest. So, for a nucleus to contribute effectively to spin-lattice relaxation it should be nearby and it should be in a molecule that rotates near the transition frequency—of order 100 MHz. The magnitude of the magnetic field created by a nucleus depends on the inverse sixth power of the distance from the nucleus. The frequencies of the magnetic field created by a nucleus will depend on how fast the nucleus reorients—how fast the molecule or group containing the nucleus rotates. The values of T_1's for the nuclei in a molecule will thus depend on the size of the molecule (how fast it rotates), its structure (how close neighboring nuclei are), and on the solvent. For example, measurement of T_1 can provide rotational diffusion coefficients for rigid molecules. Measurements of T_1 values for a molecule of known size can be used to determine the microscopic viscosity of the solvent in a complex mixture such as blood.

There is another contribution to T_1. If protons whose spin population is saturated exchange with protons with an equilibrium population, this provides a chemical exchange contribution to spin-lattice relaxation, T_1. For exchangeable protons, measurement of T_1 by NMR can provide the kinetics of the rate of chemical exchange. This has been used to measure the exchange lifetimes for imino protons on guanine and thymine in nucleic acids in H_2O. Because a G·C base pair must open for the guanine imino proton to exchange with H_2O and an A·T base pair must open for the thymine imino proton to exchange, the exchange rate is in turn a measure of the opening rate of the base pairs. Molecules which bind to DNA, such as ethidium or antibiotics, can stabilize or destabilize the base pairs (Pardi et al., 1983).

There is another characteristic relaxation time which is measurable in NMR. It is the *spin-spin relaxation time, T_2.* It characterizes interactions between spins on equivalent nuclei; it does not involve exchange of energy with the environment (the lattice). The spin-spin relaxation time can be measured from the width of the NMR peak

$$\frac{1}{T_2} = \pi \, \Delta\nu_{1/2} \tag{10.41}$$

where $\Delta\nu_{1/2}$ is the peak width at half-height. Chemical exchange processes can also contribute to NMR line widths and therefore to T_2. In fact, all processes that affect T_1 also affect T_2; T_2 is always smaller than T_1. These two relaxation times can be used to study chemical kinetics and rotational and conformational motion of molecules.

Nuclear Overhauser Effect

We have discussed chemical shifts, spin-spin splittings, and T_1 and T_2; these are all NMR parameters which are used to study conformations and reactions in macromolecules. The nuclear Overhauser effect is a very specific NMR method used to determine which nuclei are near each other. The change of intensity of one NMR peak when another is irradiated is called the nuclear Overhauser effect (NOE). Intense radiation corresponding to the transition frequency of one type of proton is applied to the sample. This saturates the NMR peak; it changes the spin population of those protons so that there is an equal number of nuclei in the upper and lower energy levels. These nuclei interact with neighboring nuclei and change the spin population of the neighboring nuclei from their equilibrium distribution. The intensity of the NMR peak of each nearby nucleus will change. Because the interaction between magnetic nuclei (the square of the magnetic dipole-dipole interaction) depends on r^{-6} where r is the distance between the nuclei, the nuclear Overhauser effect is appreciable only for very close neighbors. Normally, the effect is seen only for nuclei closer than 5 Å (0.5 nm). The short range of the effect limits its applications, but it provides great specificity. Although quantitative distance information can be obtained, NOE studies can be used simply to identify protons which are closer than 5 Å. For example, once an amide proton peak is identified in a protein, NOE measurements can be used to assign NMR peaks of protons which are nearby.

Example 10.3   A frame-shift mutation is caused when there is an error in replication of DNA in which one of the two strands has an extra nucleotide. The conformation of this extra nucleotide is of interest. Is the extra base stacked in the double helix, or is it outside the double helix? An NOE study of the imino proton, which is hydrogen bonded in each base pair, can answer this question. Figure 10.34 shows the NMR spectra of the eight imino protons in a synthetic base-paired DNA double helix with an extra C on one strand.

$$
\begin{array}{ccccccccc}
 & & & & \text{C} & & & & \\
 & & & & \diagup \quad \diagdown & & & & \\
\text{C} & \!-\!\text{A}\!-\! & \text{A}\!-\! & \text{A} & \quad \text{A} & \!-\!\text{A}\!-\! & \text{A}\!-\! & \text{G} \\
\cdot & \cdot & \cdot & \cdot & \cdot & \cdot & \cdot & | \\
\text{H} & \text{H} & \text{H} & \text{H} & \text{H} & \text{H} & \text{H} & \text{H} \\
| & | & | & | & | & | & | & \cdot \\
\text{G} & \!-\!\text{T}\!-\! & \text{T}\!-\! & \text{T} & \!-\!\text{T}\!-\! & \text{T}\!-\! & \text{T}\!-\! & \text{C} \\
1 & 2 & 3 & 4 & 5 & 6 & 7 & 8
\end{array}
$$

NOE can determine whether the extra cytosine base is stacked between base pairs 4 and 5, or is outside the helix. The NMR spectrum of the molecule in H_2O is shown at the top of Fig. 10.34. There are eight peaks in the imino region. The NOE experiment is done by adding power at the transition frequency of each peak (to saturate its signal) and looking for changes in intensity in other peaks. NOE spectra are shown as different spectra—the spectrum before irradiation minus the spectrum after saturating one peak. The largest peak in the difference spectrum will be the peak saturated (its signal decreases to zero at saturation); any other peaks that appear correspond to protons very near the proton saturated. Spectrum (c) in Fig. 10.34 shows an NOE from imino proton 6 to imino protons 5 and 7, as expected for neighboring stacked base pairs. Spectrum (b) shows an NOE from imino proton 4 to imino protons 3 and 5. The NOE from base pair 4 to base pair 5 shows that they are stacked on each other, and thus that the extra cytosine is outside the helix. If the extra cytosine were stacked in the helix, the imino proton on base pair 4 would be at least 7 Å away from the imino proton on base pair 5. No NOE would result. The extra-helical cytosine is corroborated by NOEs from imino protons to nonexchangeable aromatic protons, and from ring-current shielding effects on the extra cytosine.

Fig. 10.34   Nuclear Overhauser effect study of the hydrogen-bonded imino protons of guanine and thymine in a DNA double helix containing an extra cytosine on one strand. See Fig. 3.7 for the structure of DNA base pairs. The spectra were measured at 500 MHz in aqueous (H_2O) buffer at 0°C, pH = 7; the reference was $(CH_3)_3SiCD_2CD_2COO^-$. (a) The spectrum of the molecule in the imino region; the base aromatic protons and the deoxyribose protons appear below 9 ppm and are not shown. (b) A plot of the normal spectrum minus the spectrum after imino proton 4 is saturated. The peaks at 3 and 5 indicate an intensity change and thus spin interaction with proton 4. Base pairs 4 and 5 are within 5 Å of each other. (c) A similar NOE spectrum when proton 6 is saturated; intensity changes occur at 5 and 7, as expected. [Data taken from Morden, Chu, Martin, and Tinoco, *Biochemistry* **22**, 5557 (1983).]

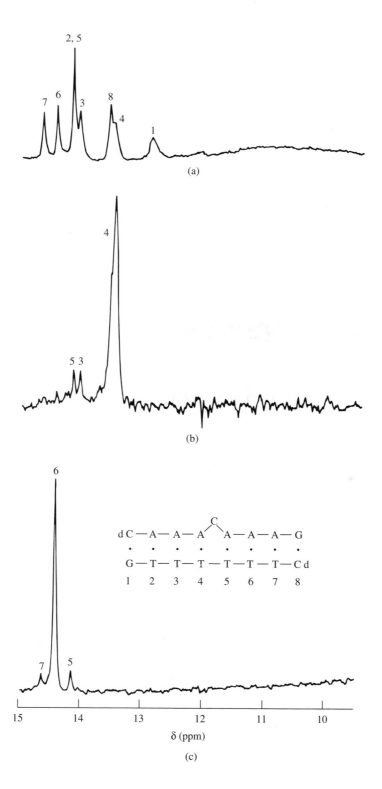

(a)

(b)

dC — A — A — A $\overset{\text{C}}{\diagup}$ A — A — A — G
 •   •   •   •   •   •   •   •
 G — T— T— T — T— T— T—C d
 1   2   3   4   5   6   7   8

(c)

15   14   13   12   11   10

δ (ppm)

Two- and Three-Dimensional NMR

We have been describing one-dimensional NMR (1D-NMR) spectra until now; these are plots of absorption vs. frequency (the frequency axis is usually labeled as chemical shifts). In a multi-dimensional spectrum a series of pulses is used to learn about the interactions of nuclei through bonds (spin-spin splitting or J-coupling) or through space (nuclear Overhauser effect). The through-bond correlations can be homonuclear, ^1H—^1H for example, or heteronuclear, ^1H—^{13}C for example. Spreading a spectrum into two or three dimensions produces a large increase in resolution. Two different protons in a 1D spectrum can have the same chemical shift, although the carbons to which they are bonded have different chemical shifts. A two-dimensional ^1H—^{13}C NMR spectrum uses the carbon frequency dimension to resolve the overlapping proton resonances in the proton frequency dimension. Bax and Lerner (1986) and the book by Derome (1987) give very clear descriptions of 2D-NMR.

To obtain a two-dimensional spectrum (2D-NMR) two or more pulses are used; this is illustrated in Fig. 10.35. A 2D homonuclear correlated spectrum—called a COSY spectrum—is obtained by applying two short pulses to the sample before the free induction decay (FID) is measured as a function of time, t_2. The FID depends on the time t_1 between the pulses; the range of t_1 values used covers the same time scale as the FID. Thus, as t_1 is varied the FID changes, so that the FID becomes a function of two time variables, t_1 and t_2. A two-dimensional Fourier transform [similar to the 1D transform in Eq. (10.36)] is done to obtain a 2D spectrum in terms of two frequencies. A 2D nuclear Overhauser effect spectrum—called a NOESY—is obtained by adding a third pulse which is separated from the last pulse by a constant time, the mixing time τ_{mix}. The mixing time determines how long the nuclei are allowed to interact through the nuclear Overhauser effect; the optimum value used (of order 100 ms) depends on the size of the molecule and the main purpose of the experiment—assigning resonances or measuring interproton distances.

Figure 10.36 shows a two-dimensional nuclear Overhauser effect spectrum (2D-NOESY) below the one-dimensional spectrum of a DNA oligonucleotide of 17 base pairs. There are about 300 protons in the molecule, and one sees that the proton spectrum is hopelessly overlapped in one dimension. The 2D spectrum is shown as a contour plot looking down on the spectrum; each dot or set of closed contours is a peak in the 2D spectrum. The diagonal in the 2D spectrum is the contour view of the 1D spectrum; the 2D spectrum is symmetric about this diagonal. Most of the 300 protons can be assigned from the 2D spectrum. Each off-diagonal cross-peak is labeled by two frequencies (in ppm) corresponding to two protons interacting through the nuclear Overhauser effect. Because the effect depends on the inverse sixth power of the distance (r^{-6} where r is the distance between two protons), the protons must be within 5 Å to interact significantly. The volume of each cross-peak (this corresponds to the area of a 1D peak) is proportional to the inverse sixth power of the distance between the two protons. Thus the 2D-NOESY provides a large number of distances between pairs of protons. Mainly local conformational information is obtained. However, when two protons are found far apart in the sequence but close in space, this places a very strong constraint on possible structures.

Each peak in a 2D spectrum corresponds to the interaction between two protons; the type of interaction detected depends on the pulses used (Fig. 10.35). A 2D-COSY has cross-peaks for nuclei that interact through spin-spin splitting or J-coupling. Measurable J-coupling occurs for nuclei separated by one bond, two bonds, three bonds, or even four

1D-NMR

2D-NMR   COSY

NOESY

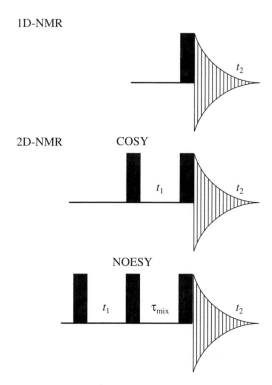

Fig. 10.35   The pulse sequences which produce one- and two-dimensional NMR spectra. A single short pulse produces a 1D spectrum after Fourier transformation of the free induction decay (FID) as shown in Fig. 10.26b. A COSY (correlated spectroscopy) 2D spectrum identifies nuclei that are correlated by J-coupling (through-bond coupling). It requires two short pulses separated by variable time t_1 before the FID is measured. The FID depends on two times (t_1, t_2), so a 2D Fourier transform is used to produce a 2D frequency spectrum. A NOESY (nuclear Overhauser effect spectroscopy) 2D spectrum identifies nuclei that are dipole coupled through space; this corresponds to protons less than 5 Å apart. Three pulses are needed; the third one adds a constant mixing time, τ_{mix}, before acquisition of the FID and Fourier transformation to produce a 2D frequency spectrum.

bonds, with generally decreasing coupling constants. A COSY spectrum is the most rigorous method for assigning protons because there must be two or three bonds between them to produce a cross-peak. The pattern of bond connectivities is characteristic of the molecule or group. Consider the COSY connectivity of the amino acids threonine and leucine whose low resolution 1D spectra are given in Fig. 10.31. Idealized 2D-COSY spectra have the patterns shown in Fig. 10.37. Once one peak is assigned—such as the methyl resonance with a characteristic chemical shift near 1 ppm—the others follow from the COSY connectivity. In threonine (Fig. 10.37a) the methyl protons are connected only to H_β, and H_β is connected to both H_α and the methyl protons. In leucine (Fig. 10.37b) the COSY connectivity is methyl to H_γ to H_β to H_α. We have assumed in the diagram that the two methyl groups of leucine are magnetically equivalent and that the two methylene protons on the β carbon are equivalent. Restricted torsional rotation around the C_α—C_β and C_β—C_γ bonds can make these nonequivalent.

The COSY spectrum provides assignments, but it also provides the coupling constants for the correlated protons. Each cross-peak in the 2D spectrum (shown simply as a circle in Fig. 10.37) actually shows the characteristic splittings which provide the J-coupling constants. We remember that the coupling constants are related to torsion angles for rotations around bonds by the Karplus equation [Eq. 10.39]. These means we can determine torsion angles, or at least ranges of torsion angles, from a COSY spectrum.

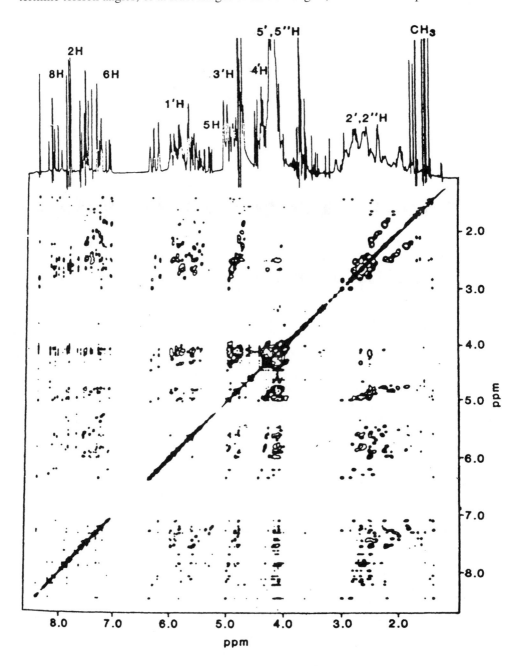

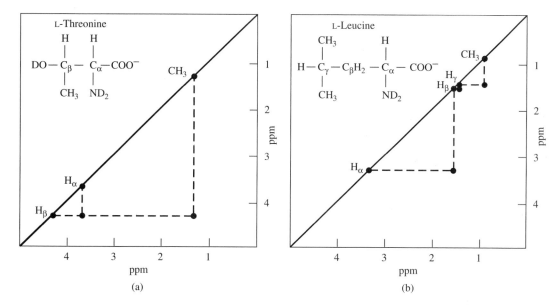

Fig. 10.37 (a) The proton connectivity seen in a COSY spectrum of L-threonine is illustrated. The spots along the diagonal represent the 1D spectrum shown in Fig. 10.31. The off-diagonal cross-peaks show which protons are correlated by three-bond J-coupling (α-β, β-methyl). (b) The proton COSY connectivity for L-leucine. The off-diagonal cross-peaks show three-bond correlations between protons α-β, β-γ and γ-methyl. The cross-peaks on only one side of the diagonal are shown.

We mentioned that a 2D heteronuclear experiment could correlate each proton with the ^{13}C atom to which it was bonded. The 2D heteronuclear spectrum has one axis labeled with proton chemical shifts, the other with carbon chemical shifts. Each cross-peak relates a proton to a carbon; thus, assignment of either a proton or a carbon chemical shift immediately identifies the bonded partner. A 2D proton-proton COSY or NOESY can be edited by correlating the protons to ^{13}C atoms. A molecule is synthesized that is specifically labeled with ^{13}C; for example, one peptide chain in a protein could be uniformly labeled, or one type of nucleotide in a nucleic acid could be uniformly labeled. The proton-proton

Fig. 10.36 A two-dimensional nuclear Overhauser effect spectrum (2D-NOESY) of a 17-base-pair fragment of DNA; the one-dimensional spectrum is at the top of the figure. The 2D spectrum is a contour plot with the 1D spectrum along the diagonal; we see the tops of the peaks in the 2D spectrum. The 1D spectrum at the top shows distinct classes of protons: 1–2 ppm (thymine methyls); 2–3 ppm (deoxyribose H2′, H2″ protons); 3–5 ppm (deoxyribose H3′, H4′, H5′, H5″ protons); 5-6 ppm (deoxyribose H1′, cytosine H5 protons); 7–8 ppm (adenine and guanine H8, adenine H2, and thymine and cytosine H6 protons). Each peak in the 2D-spectrum reveals a nuclear Overhauser effect interaction between two protons. For example, the peaks in the upper left (or lower right) are the interactions between H2, H6, and H8 protons (7–8 ppm) with methyls and H2′, H2″ (1–3 ppm). The numbering for the bases and sugars is given in the Appendix. [From A. Bax and L. Lerner, *Science 232*, 960–967 (1986).] The figure was supplied by Professor D. Wemmer.

COSY or NOESY experiments can be done so that only protons that are attached to ^{13}C appear in the spectra, or only those bound to ^{12}C appear in the spectra. The 2D spectra are thus ^{13}C-edited into simpler 2D spectra.

A three-dimensional NMR (3D NMR) spectrum adds another variable time to the pulse sequence which leads by a Fourier transform to a 3D frequency spectrum. There are many combinations of nuclei that can be used and different correlations that can be determined. An extension of the ^{13}C-edited 2D spectrum is a 3D spectrum where the first dimension is carbon chemical shifts and the next two dimensions are proton-proton COSY or NOESY interactions. A 2D slice through the 3D spectrum contains only the protons attached to carbons with the chosen carbon chemical shift. Each 2D proton-proton slice through the 3D spectrum thus tells us about protons attached to one particular type of carbon. We can think of a 3D spectrum as a series of 2D spectra stacked on top of each other; each 2D spectrum corresponds to a different carbon chemical shift. Many other combinations of nuclei—including ^1H, ^{13}C, ^{15}N and ^{31}P—are used for studying biological macromolecules. Four-dimensional spectra can be obtained by adding more pulses and a fourth time variable.

Determination of Macromolecular Structure by NMR

NMR can be used to obtain high resolution structures of macromolecules in solution (Bax, 1989; Wüthrich, 1989). The accessible size of the molecules keeps increasing as the resolution of the instruments has increased for protons from 500 MHz to 600 MHz to 750 MHz. Isotope labeling plus multi-dimensional spectra allow assignment and correlation of an increasing number of nuclei. Proteins containing 100 to 200 amino acids, or structural elements in nucleic acids, such as double strands, or loops of 25 to 50 nucleotides can be determined. Complexes of proteins and nucleic acids, or drugs bound to proteins or nucleic acids can be studied by isotope labeling only one of the reactants and using isotope-edited spectra. The use of NMR to determine structures in solution, and x-ray diffraction to determine structures in crystals (Chapter 12), is providing the knowledge necessary for molecular biologists to be able to understand biological and biochemical reactions.

The steps in determining a structure by NMR follow a straightforward logic. First the NMR spectra must be assigned. This means that each resonance must be assigned to a particular nucleus in the molecule; the proton assignments are the most important and most useful in determination of structure. The assignments are best done through bond connectivities as determined by COSY-type experiments, but NOESY interactions are also very useful. Once the spectra are assigned, torsion angles are estimated from the J-coupling constants (spin-spin splittings), and interproton distances are estimated from the NOESY spectra. The distances and torsion angles characterize the structure. We assume that we know the primary structure—the sequence of the protein or nucleic acid; we want to know the conformation—the secondary and tertiary structure. This is specified by the torsion angles, but the NMR spectra do not provide all the torsion angles. There is no information from J-coupling constants available for some torsion angles—such as the bonds on either side of the phosphorus atom in a nucleic acid.

The final step in the structure determination is to find the conformation, or conformations, that are consistent with the NMR-derived distances and torsion angles. One way to do this is to add NMR-derived restraints to the potential energy functions to calculate molecular dynamics and minimum-energy structures as described in Chapter 9. For example, if a NOESY cross-peak places two protons within 3 ± 1 Å, a potential energy term is added to the usual bond stretching terms in Eq. (9.11) that greatly increases the calculated energy of a conformation that has the protons outside this range. Similarly, if a COSY cross-peak specifies a torsion angle to be $60° \pm 30°$, a potential energy term is added that favors conformations with this angle. Molecular dynamics and energy minimization calculations are done which find structures consistent the NMR data and, of course, with the usual bond lengths and bond angles. Structures with the fewest violations of the NMR restraints are found; these are the structures with the lowest calculated energies. With enough NMR distance and angle restraints only one structure may be obtained, but with fewer restraints a number of structures can be found, all of which are equally consistent with the data. It may be that the molecule is flexible, and the NMR data indicates this flexibility. Or it may be that not enough NMR data were obtained, and that more extensive assignments and correlations will lead to a unique structure.

Example 10.4   An important element of RNA structure is the hairpin: a loop of unpaired bases closed by a base-paired, double-stranded stem. In 16S ribosomal RNAs nearly all the hairpin loops have four bases, and the sequence UUCG occurs very often in the loops. In fact the UUCG hairpin loop has been found to be involved in many different biological roles; its structure is clearly of interest. An RNA oligonucleotide with sequence 5′GGAC(UUCG)GUCC was synthesized from a DNA template by RNA polymerase to produce a hairpin with a stem of four base pairs. One- and two-dimensional NMR measurements were done using proton, phosphorus, and ^{13}C spectra. Over 100 spin-spin splittings between protons and between protons and phosphorus were measured. These splittings were related to torsion angles around bonds by empirical Karplus equations. Over 300 proton-proton NOEs were measured; these provided constraints on the distances between protons. The significant variables in the structure are the 86 torsion angles for rotation about backbone and ribose bonds. The NMR constraints produced reasonable conformations which were then energy-minimized (see Chapter 9) to obtain an optimum structure consistent with all the NMR data (Varani et al., 1991). The resulting structure for the loop and one base pair of the stem is shown in Fig. 10.38. The most interesting aspects of the loop structure are: (1) The occurrence of a non-Watson-Crick $G \cdot U$ base pair. (2) The presence of a specific base-backbone interaction in the loop via a cytosine amino to phosphate oxygen hydrogen bond. (3) The fact that the second base of the loop has no specific interactions with the rest of the loop. This explains why in ribosomal RNAs any base can occur here and why this is the most chemically reactive base in the loop. A hairpin loop with a different sequence (GAAA), which is also very common in ribosomal RNA, has been found to have a very similar structure (Heus and Pardi, 1991). This implies that structural similarity rather than sequence similarity is important in the function of these loops.

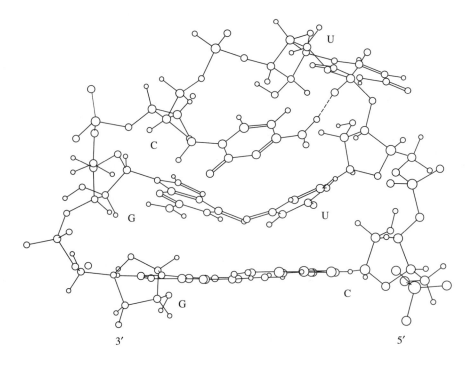

Fig. 10.38 The structure of an RNA hairpin loop determined by 2D-NMR. A Watson-Crick C·G base pair (bottom) closes the loop of four nucleotides 5'-C(UUCG)G-3'. Three additional base pairs complete the stem of the hairpin loop. Torsion angles determined from spin-spin splittings plus interproton distances determined from nuclear Overhauser effects provided the coordinates. [From G. Varani, C. Cheong, and I. Tinoco, Jr., *Biochemistry 30*, 3280–3289 (1991).]

Magnetic Resonance Imaging

Nuclear magnetic resonance is being used to obtain images of the soft tissues of the human body. (See the book by Morris, 1986, or the articles by Moonenen et al., 1990, or Radda, 1986.) X rays provide good images of bones, but are not as good for imaging slight differences in water content of different tissues. Magnetic resonance imaging (MRI) is used most often to image protons (mainly water), but it can also be used to image ^{13}C or phosphorus. The key to MRI is the use of a magnetic field gradient. The resonance frequency of a nucleus is directly proportional to the external magnetic field [Eq. (10.34)]. Therefore if an external magnetic field is not uniform in space, the resonance frequency of a nucleus will depend on its position in the field. This is illustrated in Fig. 10.39.

The NMR spectrum of a flask of water placed in a uniform field is compared with the spectrum of a flask in a linear field gradient. In a uniform field one peak is seen; all the protons have the same resonance frequency. In the field gradient an image of the water in the flask is produced by the spectrum of intensity vs. frequency. The protons on the left side of the flask are in a smaller magnetic field and have a lower resonance frequency than those on the right side. The intensity of absorption at each frequency is proportional to the number of protons at each value of the magnetic field. There are more protons in the center of the flask because of the water in the neck, so the center frequencies of the spectrum have

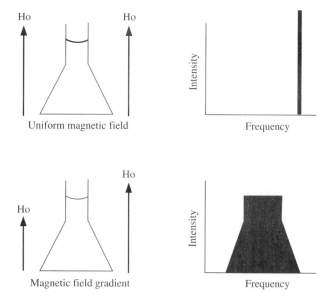

Uniform magnetic field

Magnetic field gradient

Fig. 10.39 Magnetic resonance imaging of a flask of water placed in a magnetic field that increases linearly from left to right. A magnetic field gradient means that the resonance frequency of a proton depends on its position in the field. In a magnetic field gradient a spectrum of intensity of absorption vs. frequency represents the number of protons vs. position in the field. The spectrum is thus a projection of the image of the water in the flask. The apparent distortion shown in the figure is caused by summing all the water molecules at each position in the gradient.

the highest intensity. The image obtained is a projection through the flask. To get a complete three-dimensional image, the flask or the field gradient must be rotated to produce several cross sections along different directions. In imaging, the spectrum is produced by the different magnitudes of the magnetic field; effects of chemical shifts of parts per million are completely negligible. An image of a human head is shown in Fig. 10.40.

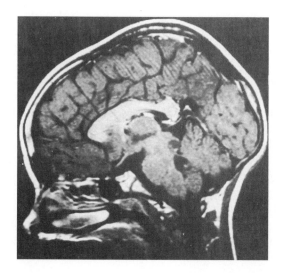

Fig. 10.40 Magnetic resonance image of a human head. The head is facing left and a cross section of the brain is obtained without the need to inject dyes, or to use any other invasive techniques.

VIBRATIONAL SPECTRA, INFRARED ABSORPTION, AND RAMAN SCATTERING

Vibrational spectra arise from transitions between vibrational energy levels. These energy levels include bond stretching, bond bending, and other internal motions of the molecules. Some vibrational frequencies identify particular groups in the molecule, such as a C—H bond, a C=C double bond, or a phenyl group. Other frequencies characterize conformations, such as amide frequencies in proteins or phosphate ester frequencies in nucleic acids. Vibrational spectra occur in the infrared region, which may be considered to include the region from approximately 10^3 nm to 10^5 nm. The traditional units used by infrared spectroscopists are either microns (μm) or wavenumbers (cm^{-1}). In these units the infrared range is 1 to 100 μm, or 10,000 to 100 cm^{-1}.

We learned in Chapter 9 that a harmonic oscillator has a vibration frequency equal to

$$\nu_0 = \frac{1}{2\pi}\left(\frac{k}{\mu}\right)^{1/2} \tag{9.29}$$

where k is the force constant for the oscillator and μ is its reduced mass. This equation applies approximately to vibrations in molecules. It states that stronger bonds with larger force constants have higher vibrational frequencies, and that if the vibration involves a proton attached to a heavy atom, the frequency will decrease by a factor of $\sqrt{2}$ on substitution of hydrogen (mass m_H) by deuterium (m_D).

$$\frac{\nu_0(H)}{\nu_0(D)} = \sqrt{\frac{m_D}{m_H}} = \sqrt{2}$$

This is useful in assigning spectra to vibrations in a molecule.

There are two principal ways to measure vibrational spectra: infrared absorption and Raman scattering.

Infrared Absorption

Figure 10.41 illustrates at the left the transitions directly measured in infrared absorption. There are selection rules that describe the allowed transitions. A vibration must cause a change of electric dipole moment of a molecule for there to be absorption of light. Thus N_2 and O_2 do not absorb in the infrared, but H_2O and CO_2 do. N_2 and O_2 have no electric dipole moment and a vibration leaves the dipole moment zero. H_2O does have an electric dipole moment and vibrations change its magnitude. CO_2 has no electric dipole moment, but an asymmetric stretch of its C=O bonds, or an O=C=O bend will break its symmetry and lead to infrared absorption. The infrared spectrum is measured in a spectrophotometer just like those used in the visible and ultraviolet region. The Beer-Lambert law is generally valid. Infrared is used routinely, as is NMR, in the identification and characterization of organic molecules. One disadvantage for studying biological macromolecules is the strong and broad absorption of liquid H_2O in the infrared region. Some very useful data on proteins and nucleic acids have been obtained in H_2O or D_2O solution, but only in a limited wavelength region. The entire infrared region is not accessible in aqueous solution.

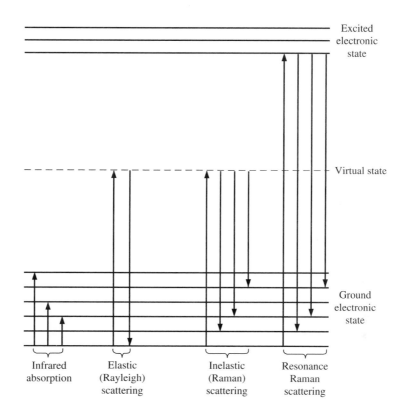

Fig. 10.41 Energy-level diagram for infrared absorption, elastic (Rayleigh) scattering, and inelastic (Raman) scattering. Transitions between vibrational energy levels of the ground electronic state absorb infrared radiation; not all transitions are allowed. For elastic scattering the frequency of the incident light is equal to the frequency of the scattered light. For inelastic scattering the frequency of the scattered light is different from that of the incident light. The virtual state shown can have any energy; the scattering of light occurs in any region of the spectrum.

Raman Scattering

All molecules scatter light. Elastic scattering occurs when a photon interacts with a molecule, but no absorption occurs. The electric field of the light perturbs the electron distribution of the molecule momentarily, but no transition occurs. Because the molecule does not go from one stationary state (ground-state energy level) to another (excited-state energy level), there are no selection rules. Figure 10.41 indicates scattering occurring by a transition to a virtual state (nonstationary state); the molecule immediately returns to the ground electronic state and the photon is scattered. If there is no change in energy of the scattered photon, the scattering is elastic; it is called *Rayleigh scattering*. The intensity of the scattered light is proportional to the square of the polarizability of the molecule, and if the wavelength is not too close to an absorption band, the scattering is inversely proportional to the fourth power of the wavelength. This wavelength dependence explains blue skies and red sunsets, although for the best red sunsets the scattering particles are dust, not molecules, and the wavelength dependence is more complicated.

The polarizability, α, is a measure of how easy it is to induce a dipole in a molecule by applying an electric field. The polarizability tells us how easy it is to distort the distribution of electrons of the molecule. The quantitative definition is

$$\mu_{ind} = \alpha E$$

where μ_{ind} is the dipole induced and E is the electric field. The usual units for α are cm^3. The electric field is then in esu-volt cm^{-1} and μ_{ind} is esu cm. In SI units α is in m^3 and

$$\mu_{ind} = 4\pi\varepsilon_0\alpha E$$

with μ_{ind} in coulomb m and E in volt m^{-1}.

In Raman scattering the molecule returns to a different energy level after interaction with the light. In Fig. 10.41 the final state is an excited vibrational level of the ground electronic state. This is the most common occurrence in Raman scattering. The scattered photon has a longer wavelength (lower energy) than does the incident photon. The scattered photon can have a shorter wavelength if the molecule was originally in an excited vibrational state and returned to the ground vibrational state. In either case the scattering is inelastic; there has been energy transfer between the molecule and the light. The selection rule for Raman scattering is that the polarizability of the molecule must change with the vibration in order to have a transition to a different energy level. Thus N_2 and O_2 will show vibrational Raman spectra, because their vibrations cause a change in polarizability. For CO_2 its asymmetric stretch will cause no change in polarizability, but its symmetric stretch and its bend will. We see that some vibrations can be measured using Raman, but not infrared; some can be measured using infrared, but not Raman; and some can be measured both ways.

In addition to the fundamental difference between Raman and infrared based on selection rules, there is a practical difference. Raman scattering is detected by shining light of one wavelength on the sample and measuring the wavelengths of the scattered light. Most of the light scattered will have the wavelength of the incident light; this is the Rayleigh scattering. There will also be light scattered at slightly different wavelengths; this is the Raman scattering. The difference in frequencies between the incident and scattered light is the frequency of the vibrational transition being measured. Note that using Raman scattering we learn about vibrational transitions (and infrared frequencies) using visible or ultraviolet light. It does not matter where the solvent absorbs; we can use a wavelength of light away from that absorption band for the Raman studies. This allows great flexibility in the Raman studies, which is particularly useful for biological samples.

Figure 10.42 compares the infrared (bottom) and Raman (top) spectrum of all-*trans*-retinal. Note the large peak at 1008 cm^{-1} in the Raman spectrum that does not occur in the infrared. The strong absorption at 966 cm^{-1} in the infrared has a corresponding peak in the Raman; it is assigned to out-of-plane vibrations of *trans*-ethylenic protons at $HC_7{=}C_8H$ and $HC_{11}{=}C_{12}H$ (see Fig. 9.1). Deuterium substitution and synthesis of demethyl analogs have led to a complete assignment of the vibrational spectra of retinal. Because the Raman spectra of visual receptors and bacteria which use retinal as a key chromophore are being studied in single cells, the changes of conformation of the retinal can be deduced during its biological function (Barry and Mathies, 1982).

Resonance Raman spectra occur when the exciting light is in an electronic absorption band of the molecule. There can be gains in sensitivity of a factor of 1000, or more, for vibrations within the absorbing chromophore. This allows one type of molecule to be studied preferentially in a very complicated mixture. Resonance Raman has been applied to heme proteins and to rhodopsin. For a review, see Spiro (1974).

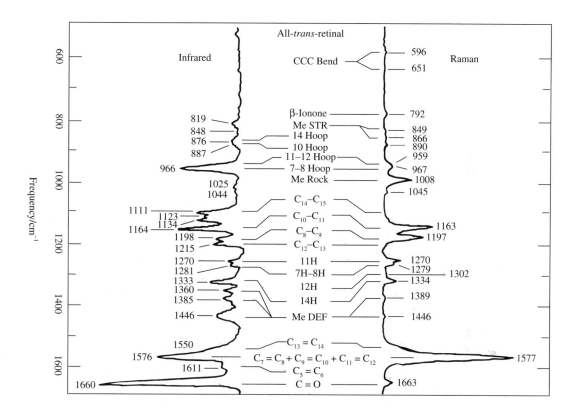

Fig. 10.42 Raman and infrared spectra of all-*trans*-retinal in the region from 600 cm^{-1} to 1600 cm^{-1}. The infrared absorbance is inverted to facilitate comparison with the Raman. The Raman spectrum was measured in CCl$_4$ using 676.4-nm excitation from a krypton ion laser. Solvent peaks were subtracted. The infrared spectrum was measured with a thin film deposited by evaporation from a pentane solution on a KBr window. All peaks have been assigned to particular vibrations of the molecule. Me STR, Me Rock, and Me DEF refer to the stretching, rocking and the deformation of methyl groups; Hoop means out-of-plane motion of H atoms. Note that double bond vibrations occur at higher frequencies than single bond vibrations. (From Curry, Palings, Broek, Pardoen, Mulder, Lugtenburg, and Mathies, *J. Phys. Chem 88,* 688–702 (1984). The figure was kindly supplied by Prof. Richard Mathies, University of California, Berkeley.)

ADDITIONAL SPECTROSCOPIC METHODS

The variety of spectroscopic tools that have been applied to biochemical and biological problems is enormous. We cannot hope to cover them all in sufficient detail to introduce them properly here. Instead, we shall list a few representative books, reviews, and surveys and suggest that the reader read them for further enlightenment.

Magnetic circular dichroism (MCD) can be induced in the absorption regions of any chromophoric molecule, whether it possesses intrinsic chirality or not. This property, and the related magnetic optical rotation first discovered by Michael Faraday in 1845, results from an asymmetric coupling between the applied external magnetic field and the light-driven transition moments. It measures very different properties from the normal CD and has been of considerable value in interpreting the spectra and electronic properties of heme proteins. For a review, see Stephens (1974).

Electron paramagnetic resonance (EPR) can be applied to free radicals or other molecular species that possess unpaired electrons (transition metal complexes, triplet states). Although its range of applicability is necessarily narrower than that of NMR, it can exhibit great sensitivity and is particularly valuable in detecting intermediate or transitory species where electron rearrangements are occurring. It has been applied to heme proteins, ferredoxins, copper-protein complexes, and so on (Swartz et al., 1972). Furthermore, it has been developed as a probe technique. Paramagnetic ions, such as Mn^{2+} or Eu^{3+}, can be used to characterize the nature of ion binding and its role in the catalytic sites of many enzymes. Stable free radicals, such as nitroxide derivatives, have been incorporated into the fatty acid chains of lipids (Hubbell and McConnell, 1969). When these are incorporated into artificial or natural membranes, their EPR spectra give detailed information about the flexibility of the lipid chains and the rotational and translational mobility within the lipid phase.

A survey of these and other spectroscopic techniques and their applications is provided in Biochemical Spectroscopy (1995), K. Sauer, ed. These include studies of the absorption spectra of x rays, x-ray fluorescence, x-ray absorption fine structure, photoelectron spectroscopy (stimulated by x rays, uv photons, or electron bombardment), picosecond spectroscopy stimulated by mode-locked laser pulses, electron-nuclear double resonance, and more.

SUMMARY

Absorption and Emission

Oscillator strength:

$$f = \frac{2303cm}{\pi N_0 e^2 n} \int \varepsilon(\nu)\, d\nu \tag{10.5}$$

c = velocity of light
m = mass of electron
N_0 = Avogadro's number
e = electronic charge
n = refractive index
ε = molar absorptivity
ν = frequency, Hz

Transition dipole moment:

$$\boldsymbol{\mu}_{0A} = \int \psi_0 \boldsymbol{\mu} \psi_A \, dv \tag{10.6}$$

$$\boldsymbol{\mu} = e \sum_i \mathbf{r}_i$$

Dipole strength:

$$D_{0A} = |\boldsymbol{\mu}_{0A}|^2 \tag{10.7}$$

Transformation relations:

See Table 10.3.

Beer-Lambert law:

$$A = \varepsilon c l \tag{10.10}$$

$A = \log \dfrac{I_0}{I_t}$ = absorbance
ε = molar absorptivity or molar extinction coefficient, $M^{-1}\,cm^{-1}$
c = concentration, M
l = path length, cm

Experimentally measured fluorescence decay constant:

$$k_d = \frac{1}{\tau} \tag{10.15}$$

k_d = first-order rate constant for decay of fluorescence. De-excitation can occur by fluorescence, thermal de-excitation, photochemistry, quenching, etc.
τ = experimentally measured lifetime of decay

Fluorescence rate constant:

$$k_f = \frac{1}{\tau_0} \tag{10.16}$$

k_f = first-order rate constant for fluorescence; de-excitation occurs only by fluorescence
τ_0 = natural lifetime of fluorescence

Fluorescence quantum yield:

$$\phi_f = \frac{\text{number of photons fluoresced}}{\text{number of photons absorbed}} = \frac{\tau}{\tau_0} \qquad (10.18),\ (10.20)$$

Excitation Transfer

Energy-transfer efficiency:

$$Eff = 1 - \frac{\tau_{D+A}}{\tau_D} \qquad (10.26)$$

τ_D = fluorescence lifetime of donor
τ_{D+A} = fluorescence lifetime of donor in the presence of acceptor

$$Eff = \frac{r_0^6}{r_0^6 + r^6} \qquad (10.27)$$

Eff = fractional efficiency of transfer
r_0 = characteristic distance for the donor-acceptor pair
r = distance between donor and acceptor

Optical Rotatory Dispersion and Circular Dichroism

$$\text{Circular birefringence} = n_L - n_R$$
$$\text{Optical rotation} = \phi = \frac{\pi}{\lambda}(n_L - n_R),\ \text{rad cm}^{-1} \qquad (10.31)$$

n_L and n_R are refractive indices for left- and right-circularly polarized light, respectively, at wavelength λ.

$$\text{Circular dichroism} = \Delta A = A_L - A_R$$
$$\text{Ellipticity} = \theta = \frac{2.303(A_L - A_R)}{4l},\ \text{rad cm}^{-1} \qquad (10.33)$$

A_L and A_R are absorbances for left- and right-circularly polarized light, respectively; l is the path length in cm.

Experimental parameters:

See Table 10.6.

Nuclear Magnetic Resonance

Fourier transform:

$$F(\omega) = \int_{-\infty}^{+\infty} f(t)e^{i\omega t}\ dt \qquad (10.36)$$

Chemical shift:

$$\delta \text{ (ppm)} = \frac{\nu \text{ (sample, Hz)} - \nu \text{ (reference, Hz)}}{\nu \text{ (spectrometer, MHz)}} \tag{10.38}$$

Spin-lattice relaxation time T_1:

$$M = M_{eq}(1 - e^{-t/T_1}) \tag{10.40}$$

M = magnetization at time t
M_{eq} = equilibrium magnetization

Spin-spin relaxation time T_2:

$$T_2 = 1/(\pi \Delta \nu_{1/2}) \tag{10.41}$$

$\Delta \nu_{1/2}$ = NMR peak width at half-height

REFERENCES

BOVEY, F. A., 1988. *Nuclear Magnetic Resonance Spectroscopy,* 2nd ed., Academic Press, San Diego.

CAMPBELL, I. D., and R. A. DWEK, 1984. *Biological Spectroscopy,* Benjamin/Cummings, Menlo Park, California.

CANTOR, C. R., and P. R. SCHIMMEL, 1980. *Biophysical Chemistry,* Part II. *Techniques for the Study of Biological Structure and Function,* W. H. Freeman, San Francisco.

CAREY, P. R., 1982. *Biochemical Applications of Raman and Resonance Raman Spectroscopies,* Academic Press, New York.

DEROME, A. E., 1987. *Modern NMR Techniques for Chemistry Research,* Pergamon Press, Oxford.

GADIAN, D. G., 1982. *Nuclear Magnetic Resonance and Its Applications to Living Systems,* Clarendon Press, Oxford.

LAKOWICZ, J. R., 1983. *Principles of Fluorescence Spectroscopy,* Plenum Press, New York.

MORRIS, P. G., 1986. *Nuclear Magnetic Resonance Imaging in Medicine and Biology,* Oxford University Press, Oxford.

ROBERTS, G. C. K., ed., 1993. *NMR of Macromolecules, A Practical Approach,* IRL Press at Oxford University Press, Oxford.

SAUER, K., ed., 1995. Biochemical Spectroscopy. *Methods in Enzymology,* Vol. 246, Academic Press, Orlando, Florida.

WÜTHRICH, K., 1986. *NMR of Proteins and Nucleic Acids,* John Wiley, New York.

SUGGESTED READINGS

Absorption

HONIG, B., R. DANIEL, K. NAKANISHI, V. BALOGH-NAIR, M. A. GAWINOWICZ, M. ARNABOLDI, and M. G. MOTTO, 1979. An External Point-Charge Model for Wavelength Regulation in Visual Pigments, *J. Am. Chem. Soc. 101,* 7084–7086.

LEVINE, J. S., and E. F. MacNICHOL, JR., 1982. Color Vision in Fishes, *Sci. Am.,* February, 140–149.

NAKANISHI, K., V. BALOGH-NAIR, M. ARNABOLDI, K. TSUJIMOTO, and B. HONIG, 1980. An External Point Charge Model for Bacteriorhodopsin to Account for Its Purple Color, *J. Am. Chem. Soc. 102,* 7947–7949.

ROSENHECK, K., and P. DOTY, 1961. The Far Ultraviolet Absorption Spectra of Polypeptide and Protein Solutions and Their Dependence on Conformation, *Proc. Natl. Acad. Sci. USA 47,* 1775–1785.

SHICHI, H., 1983. *Biochemistry of Vision,* Academic Press, New York.

Fluorescence

MURCHIE, A. I. H., R. M. CLEGG, E. VON KITZING, D. R. DUCKETT, S. DIEKMANN, and D. M. J. LILLEY, 1989. Fluorescence Energy Transfer Shows that the Four-Way DNA Junction Is a Right-Handed Cross of Antiparallel Molecules, *Nature 341,* 763–766.

HAUGLAND, R. P., J. YGUERABIDE, and L. STRYER, 1969. Dependence of the Kinetics of Singlet-Singlet Energy Transfer on Spectral Overlap, *Proc. Natl. Acad. Sci. USA 63,* 23–30.

HUANG, K., R. H. FAIRCLOUGH, and C. R. CANTOR, 1975. Singlet Energy Transfer Studies of the Arrangements of Proteins in the 30S *Escherichia coli* Ribosome, *J. Mol. Biol. 97,* 443–470.

LATT, S. A., H. T. CHEUNG, and E. R. BLOUT, 1965. Energy Transfer: A System with Relatively Fixed Donor-Acceptor Separation, *J. Am. Chem. Soc. 87,* 995–1003.

STRYER, L., 1978. Fluorescence Energy Transfer as a Spectroscopic Ruler, *Annu. Rev. Biochem. 47,* 819–846.

STRYER, L., and R. P. HAUGLAND, 1967. Energy Transfer: A Spectroscopic Ruler, *Proc. Natl. Acad. Sci. USA 58,* 719–726.

Optical Rotatory Dispersion and Circular Dichroism

GRATZER, W. B., and D. A. COWBURN, 1969. Optical Activity of Biopolymers, *Nature 222,* 426–431.

HALL, K., P. CRUZ, I. TINOCO, JR., T. M. JOVIN, and J. H. VAN DE SANDE, 1984. Z-RNA: Evidence for a Left-Handed RNA Double Helix, *Nature 311,* 584–586.

HAYWARD, L. D., and R. N. TOTTY, 1971. Optical Activity of Symmetric Compounds in Chiral Media: I. Induced Circular Dichroism of Unbound Substrates, *Can. J. Chem. 49,* 624–631.

HENNESSEY, J. P., JR., and W. C. JOHNSON, JR., 1981. Information Content in the Circular Dichroism of Proteins, *Biochemistry 20,* 1085–1094.

WANG, A. H.-J., G. J. QUIGLEY, F. J. KOLPAK, J. L. CRAWFORD, J. H. VAN BOOM, G. VAN DER MAREL, and A. RICH, 1979. Molecular Structure of a Left-Handed Double Helical DNA Fragment at Atomic Resolution, *Nature 282,* 680–686.

NMR

BAX, A., 1989. Two-Dimensional NMR and Protein Structure, *Annu. Rev. Biochem. 58,* 223–256.

BAX, A., and L. LERNER, 1986. Two-Dimensional Nuclear Magnetic Resonance Spectroscopy, *Science 232,* 960–967.

HEUS, H. A., and A. PARDI, 1991. Structural Features That Give Rise to the Unusual Stability of RNA Hairpins Containing GNRA Loops, *Science 253,* 191–194.

MOONENEN, C. T. W., P. C. M. VAN ZIJL, J. A. FRANK, D. LE BIHAN, and E. D. BECKER, 1990. Functional Magnetic Resonance Imaging in Medicine and Physiology, *Science 250,* 43–61.

PARDI, A., K. M. MORDEN, D. J. PATEL, and I. TINOCO, JR., 1983. The Kinetics for Exchange of the Imino Protons of the d(C-G-C-G-A-A-T-T-C-G-C-G) Double Helix in Complexes with the Antibiotics Netropsin and/or Actinomycin, *Biochemistry 22,* 1107–1113.

RADDA, G. K., 1986. The Use of NMR Spectroscopy for the Understanding of Disease, *Science 233,* 640–645.

ROSS, B. D., G. K. RADDA, D. G. GADIAN, G. ROCKER, M. ESIRI, and J. FALCONER-SMITH, 1981. Examination of a Case of Suspected McArdle's Syndrome by ^{31}P Nuclear Magnetic Resonance, *N. Engl. J. Med. 304,* 1338–1342.

SCHULMAN, R.G., 1983. NMR Spectroscopy of Living Cells, *Sci. Am. 248* (January), 86–93.

SCHULMAN, R.G., A. M. BLAMIRE, D. L. ROTHMAN, and G. MCCARTHY, 1993. Nuclear Magnetic Resonance Imaging and Spectroscopy of Human Brain Function. *Proc. Natl. Acad. Sci. USA 90,* 3127–3133.

VARANI, G., C. CHEONG, and I. TINOCO, JR., 1991. Structure of an Unusually Stable RNA Hairpin, *Biochemistry 30,* 3280–3289.

WÜTHRICH, K., 1989. The Development of Nuclear Magnetic Resonance Spectroscopy as a Technique for Protein Structure Determination, *Acc. Chem. Res. 22,* 36–44.

Raman and Infrared

BARRY, B., and R. MATHIES, 1982. Resonance Raman Microscopy of Rod and Cone Photoreceptors, *J. Cell Biol. 94,* 479–482.

CURRY, B., A. BROCK, J. LUGTENBURG, and R. MATHIES, 1982. Vibrational Analysis of All-*Trans*-Retinal, *J. Am. Chem. Soc. 104,* 5274–5286.

SPIRO, T. G., 1974. Raman Spectra of Biological Materials, in *Chemical and Biochemical Applications of Lasers,* Vol. 1, C. B. Moore, ed., Academic Press, New York, pp. 29–70.

Other Spectroscopic Methods

HUBBELL, W. L., and H. M. MCCONNELL, 1969. Orientation and Motion of Amphiphilic Spin Labels in Membranes, *Proc. Natl. Acad. Sci. USA 64,* 20–27.

STEPHENS, P. J., 1974. Magnetic Circular Dichroism, *Annu. Rev. Phys. Chem. 25,* 201–232.

SWARTZ, H. M., J. R. BOLTON, and D. C. BORG, eds., 1972. *Biological Applications of Electron Spin Resonance,* Academic Press, pp. 213–264.

PROBLEMS

1. The longest-wavelength absorption band of chlorophyll *a* peaks *in vivo* at a wavelength of about 680 nm.

   (a) For photons with a wavelength of 680 nm, calculate the energy in J photon^{-1}, in electron volts, and in kJ einstein^{-1}.

(b) CO_2 fixation in photosynthesis can be represented by the reaction

$$CO_2 + H_2O \longrightarrow (CH_2O) + O_2$$

where (CH_2O) represents $\frac{1}{6}$ of a carbohydrate molecule such as glucose. The enthalpy change for this endothermic reaction is $\Delta H = +485$ kJ mol^{-1} of CO_2 fixed. What is the minimum number of einsteins of radiation that need to be absorbed to provide the energy needed to fix 1 mol of CO_2 via photosynthesis?

(c) Experimentally, many values have been determined for the number of photons required to fix one CO_2 molecule. The presently accepted "best values" lie between 8 and 9 photons per CO_2 molecule. What is the photochemical quantum yield, based on your answer to part (b), for this process?

2. Using the data given in Table 10.2 for a 1-kJ ruby laser pulse of 1-ms duration, estimate the absorbance of safety goggles needed to protect the operator's eyes against damage from a potential accident. Base your estimate on the information in Problem 42(b) of Chapter 7.

3. The mechanism of sun tanning is initiated by the absorption of the 0.2% of solar radiation energy lying in the range 2900 to 3132 Å. The earth's atmosphere cuts off radiation <2900 Å and the burning efficiency drops from 100% at 2967 Å to 2% at 3132 Å. The *average* efficiency across the 2900 to 3132 Å band is 50%. The total incident radiation from sunlight at the surface of the earth corresponds to 2100 J cm^{-2} day^{-1}.

(a) Calculate the energy equivalent (in J cm^{-2} s^{-1}) of the radiation involved in sunburning. (Assume that a typical day length is 12 hr.)

(b) Many commercial ointments used to protect against sunburn contain *o*- or *p*-amino benzoates:

or titanium dioxide, TiO_2. The latter is a solid powder often used as a "superwhite" pigment for paints. Suggest mechanisms by which these substances might work.

4. The molar absorptivity (molar extinction coefficient) of benzene equals 100 at 260 nm. Assume that this number is independent of solvent.

(a) What concentration would give an absorbance of 1.0 in a 1-cm cell at 260 nm?

(b) What concentration would allow 1% of 260-nm light to be transmitted through a 1-cm cell?

(c) If the density of liquid benzene is 0.8 g cm^{-3}, what thickness of benzene would give an absorbance of 1.0 at 260 nm?

5. Indicator dyes are often used to measure the hydrogen-ion concentration. From the measured spectra of an indicator in solution, the ratio of acid to base species and the pH can be calculated:

$$HIn \rightleftharpoons In^- + H^+$$

The molar extinction coefficients are:

	ε (300 nm)	ε (400 nm)
HIn	10,000	2000
In$^-$	500	4000

(a) Calculate the ratio of concentrations [In$^-$]/[HIn] for a solution of the indicator which has A (300 nm) = 1.2 and A (400 nm) = 7 in a 1-cm cell.

(b) What is the pH if the pK of the indicator is 5.0?

6. The content of chlorophylls a and b in leaves or green algae is determined by extracting the pigments into acetone:water (80:20 v/v) solution. The specific absorptivities of the pigments in this solvent system reported by Mackinney [*J. Biol. Chem. 140*, 315 (1941)] at 645 and 663 nm are as follows:

Wavelength (nm)	Specific absorptivity (L g^{-1} cm^{-1})	
	Chl a	Chl b
645	16.75	45.60
663	82.04	9.27

Derive expressions, with numerical coefficients evaluated, that give the concentrations in μg mL^{-1} of chlorophyll a, of chlorophyll b, and of total chlorophylls ($a + b$) for specified values of A_{645} and A_{663} measured using a 1-cm path length.

7. The enzyme chymotrypsin is a protease which is competitively inhibited by a red dye. A solution containing 1.00×10^{-4} M total protein and 2.00×10^{-3} M total dye has an absorbance in a 1-cm cell at 280 nm of $A = 1.4$.

$$\text{Protein} + \text{Dye} \rightleftharpoons \text{Complex}$$

The molar absorptivities at 280 nm are (units of M^{-1} cm^{-1}): ε (Protein) = 12,000; ε (Dye) = 0; ε (Complex) = 15,000.

(a) Calculate the concentration of the complex in solution.

(b) Calculate the equilibrium constant (K with units of M^{-1}).

(c) When the dye binds to the enzyme the α-helix content of the protein changes. Describe how you would use measurements of circular dichroism to determine the change in α-helix content. What would you measure? How would you relate it to α-helix content?

8. An indicator is a dye whose spectrum changes with pH. Consider the following data for the absorption spectrum of an indicator (pK = 4) in its ionized and nonionized form.

$$\text{InH}^+ \rightleftharpoons \text{In} + \text{H}^+$$

λ (nm)	Molar absorptivity ε (M^{-1} cm^{-1})	
	InH$^+$	In
400	10,000	0
420	15,000	2000
440	8000	8000
460	0	12,000
480	0	3000

The absorbance of the indicator solution is measured in a 1-cm cell and found to be

λ	400	420	440	460	480
A	0.250	0.425	0.400	0.300	0.075

(a) Calculate the pH of the solution.

(b) Calculate the absorbance of the solution at 440 nm and pH 6.37 for the same total indicator concentration.

(c) If you want to measure the quantum yield of fluorescence, of In and InH$^+$ independently, what wavelength would you choose for exciting InH$^+$? For exciting In? Why?

9. Radioactivity is a very sensitive method of detection; let us compare it with fluorescence detection.

(a) ^{32}P has a half-life of 14.3 days. How many disintegrations per second will you obtain from 1000 atoms of ^{32}P? What is the maximum number of counts you can get?

(b) Fluorescein has a molar absorption coefficient of 70,000 M^{-1} cm^{-1} at 485 nm and a quantum yield for fluorescence of 0.93. What is its molecular absorption coefficient in cm^2 per molecule?

(c) 1000 molecules of fluorescein are irradiated with an argon ion laser which has an intensity of 2 mW cm^{-2} at 485 nm. How many fluorescent photons per second will be emitted?

(d) The quantum yield for photodegradation to nonfluorescent products for fluorescein is 10^{-6}. What is the maximum number of fluorescent photons you can get?

(e) Discuss the relative merits of detecting small numbers of molecules by radioactivity and by fluorescence. What other factors besides counting rate may be important?

10. In a fluorescence lifetime experiment a very short pulse of light is used to excite a sample. Then the intensity of the fluorescence is recorded as a function of time. Suppose we study a dansyl fluorophore. The intensity of the fluorescent light versus time is given below.

Fluorescence intensity	Time ns
1.1×10^4	0
4.9×10^3	5
2.3×10^3	10
1.3×10^3	15
5.6×10^2	20

(a) What is the observed fluorescence lifetime?

(b) The fluorescence quantum yield under the same conditions was determined to be 0.7. What is the intrinsic rate constant for radiative fluorescence decay?

(c) It has been determined separately that when the dansyl chromophore is 20 Å away from the 11-cis-retinal chromophore in rhodopsin, the efficiency of fluorescence energy transfer is 50%. In a second experiment rhodopsin is covalently labeled with a dansyl chromophore and the observed fluorescence lifetime for the dansyl-rhodopsin complex is 6 ns. What is the distance between the dansyl label and the retinal chromophore in rhodopsin?

11. Tyrosine has a pK_a of 10.0 for the ionization of the phenolic hydrogen. The approximate absorption spectra for the protonated and deprotonated forms of tyrosine follow.

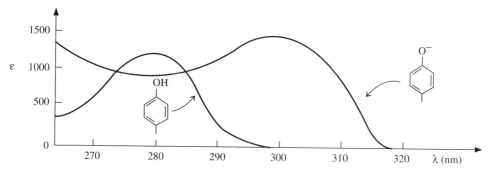

Suppose we obtain an absorption spectrum of a peptide that contains three tyrosine residues and no tryptophan or phenylalanine. The absorption spectrum of this sample in a 1-cm cell is given below.

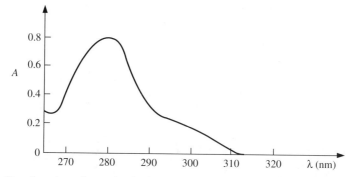

(a) Based on these data, what is the concentration of the peptide? Explain your method clearly.
(b) What is the pH of the solution? Explain your method clearly.

12. Quinine has a fluorescence quantum yield approximately equal to 1. Light at 10^{15} photons s^{-1} is incident on a solution of quinine in a 1-cm cell, and the concentration of quinine is 1.0×10^{-2} M. The wavelength is 300 nm and the extinction coefficient is 1.0×10^4 at this wavelength.
(a) How many photons are fluoresced per second?
(b) If you place an efficient detector at right angles to the incident light to measure the fluorescence, only a very small fraction of these photons is detected. Why?
(c) If any phosphorescence occurs, is the wavelength of phosphorescence greater or less than that of fluorescence? Why?
(d) A search is being made for mutants of photosynthetic bacteria which lack reaction centers where the first steps in photosynthesis occur. In these mutants the proteins that convert light energy into chemical energy are missing. The search involves looking for clones which have increased fluorescence. Can you explain why this method can be used?

13. Consider a protein which dimerizes.

$$\alpha + \alpha \underset{k_{-1}}{\overset{k_1}{\rightleftarrows}} \alpha_2$$

Describe how you could measure the equilibrium constant for the dimerization using a spectroscopic method. What would you measure? Write an equation relating the measured properties to the equilibrium constant.

14. The following equilibrium occurs in aqueous solution:

$$A + B \overset{K}{\rightleftharpoons} C$$

The molar extinction coefficients of A and B are ε_A and ε_B. Describe how you could use measurements of absorbance to obtain the equilibrium constant for the reaction. Tell what you would measure and how K would be calculated from the measurements.

15. A mixture of *M. lysodeikticus* DNA and rice dwarf virus RNA gave the following circular dichroism data in a 1-cm cell:

λ (nm)	300	280	260	240
$A_L - A_R$	-0.250×10^{-4}	1.16×10^{-4}	3.12×10^{-4}	-0.24×10^{-4}

The pure components gave the following results for solutions containing $1.00 \times 10^{-4}\ M$ concentrations of nucleotides per liter.

λ (nm)	300	280	260	240
$A_L - A_R$ (DNA)	0.10×10^{-4}	3.00×10^{-4}	0.00×10^{-4}	-2.20×10^{-4}
$A_L - A_R$ (RNA)	-0.50×10^{-4}	1.60×10^{-4}	6.00×10^{-4}	0.00×10^{-4}

(a) What is $\varepsilon_L - \varepsilon_R$ for the DNA at 300 nm?
(b) What are the concentrations of DNA and RNA in the mixture?
(c) If you hydrolyze the mixture to its component nucleotides, will the circular dichroism at 260 nm increase or decrease?
(d) Will the absorbance at 260 nm increase or decrease?
(e) Will the optical rotatory dispersion increase or decrease?

16. For the following questions, a single sentence or phrase should be sufficient as an answer.

Glycine   Phenylalanine   Adenine

(a) Which, if any, of these molecules would you expect to rotate the plane of linearly polarized light? Why?
(b) Which, if any, of these molecules would you expect to be circularly dichroic? Why?
(c) Which, if any, of these molecules could in principle emit fluorescence at wavelengths greater than 250 nm when appropriately excited? Why?
(d) If a molecule emitted light on excitation, what experiment would you do to tell if it was fluorescence or phosphorescence?

17. A fluorescent dansyl group has a lifetime of 21.0 ns when attached to a protein. When a naphthyl group is attached to the amino group of the terminal amino acid in the protein, the

dansyl lifetime becomes 15.0 ns. Use the data in Fig. 10.17 to calculate the distance between the dansyl and the naphthyl group.

18. Cytidylic acid (molecular weight = 323.2) in water at pH 7 has the following optical properties at 280 nm: $\varepsilon = 8000\ M^{-1}\ cm^{-1}$; $\varepsilon_L - \varepsilon_R = 3\ M^{-1}\ cm^{-1}$; $[\phi] = 7500\ deg\ M^{-1}\ cm^{-1}$.
    (a) Calculate the specific rotation, $[\alpha]$, and the molar ellipticity, $[\theta]$.
    (b) For $10^{-4}\ M$ cytidylic acid in a 1-cm cell, calculate the absorbance, A, the circular dichroism $(A_L - A_R)$, and the rotation angle in degrees at 280 nm.
    (c) Calculate the rotation in rad cm^{-1}, the circular birefringence $(n_L - n_R)$, and the ellipticity in rad cm^{-1} for the solution in part (b).

19. The fluorescence lifetime of benzene in cyclohexane is 29 ns when air is completely removed. In the presence of 0.0072 M dissolved O_2, the measured fluorescence lifetime is 5.7 ns because of quenching. Calculate the rate constant k_Q for the quenching reaction. If the relative fluorescence intensity is 100 for pure benzene, what is the relative fluorescence intensity of benzene containing 0.0072 M dissolved O_2?

20. When a double-stranded oligonucleotide (duplex D) in solution is heated, it "melts" to two single strands, S and R. The absorbance at 260 nm increases as the equilibrium between duplex and single strands is shifted toward the single strands.

$$D \rightleftharpoons S + R$$

$$K = [S][R]/[D]$$

A duplex whose concentration is 100 μM is melted in 1 M NaCl, pH 7. Below 10°C the solution contains only duplex; above 60°C the solution contains only single strands. Assume that the molar extinction coefficients of the duplex and its single strands are independent of temperature.
    (a) Use the data below to calculate the equilibrium constant as a function of temperature for the duplex-to-single-strands equilibrium.
    (b) Calculate the melting temperature of the duplex. The melting temperature is defined as the temperature at which the duplex is half-melted.
    (c) From the temperature dependence of the equilibrium constant, calculate ΔG^0, ΔH^0, and ΔS^0, all at 25°C. Assume that ΔH^0 and ΔS^0 are independent of temperature.

Temperature (°C)	Absorbance
10	0.790
15	0.800
20	0.813
25	0.822
30	0.852
35	0.890
40	0.940
45	0.971
50	0.990
55	1.003
60	1.114

21. When DNA or RNA is treated with chloracetaldehyde, a fluorescent derivative of adenine (εA) is formed. The following data were measured for the dinucleotide phosphate (ApεA) shown on the next page.

A

εA

	$T = 2°C$	$T = 80°C$
Molar extinction coefficients, $\varepsilon(260\text{ nm})$	$3.0 \times 10^4\ M^{-1}\ cm^{-1}$	$3.3 \times 10^4\ M^{-1}\ cm^{-1}$
Molar circular dichroism, $\varepsilon_L - \varepsilon_R(240\text{ nm})$	$32\ M^{-1}\ cm^{-1}$	$2.5\ M^{-1}\ cm^{-1}$
Fluorescence quantum yield	0.10	0.06
Fluorescence lifetime	8.0 ns	—

(a) Calculate the absorbance (A) of a 0.10 M solution of ApεA at 260 nm and 2°C in a 1-cm cell. Would it be experimentally possible to measure the temperature dependence of the absorbance of this concentration in a 1-cm cell? Explain.

(b) Calculate the circular dichroism ($A_L - A_R$) of a 0.10 M solution of ApεA at 240 nm and 80°C in a 1-cm cell. Would it be experimentally possible to measure the temperature dependence of the circular dichroism of this concentration in a 1-cm cell? Explain.

(c) Would it be experimentally possible to measure the temperature dependence of the fluorescence at 390 nm for this molecule in a 1-cm cell? Explain.

(d) The change in optical properties of this molecule with temperature is due to a stacking-unstacking equilibrium.

$$S \underset{}{\overset{K}{\rightleftharpoons}} U$$

Under appropriate conditions any optical property could be used to measure K. If you had instruments which could measure ε, $\varepsilon_L - \varepsilon_R$ or fluorescence with equal accuracy, which would you use to measure K? Why?

If P_s = property of stacked molecule, independent of T,

if P_u = property of unstacked molecule, independent of T, and

if $P(T)$ = property of mixture of stacked and unstacked species as a function of T,

(e) Write an equation for K, the equilibrium constant, in terms of the fraction of molecules unstacked.

(f) Write an equation for $K(T)$, the equilibrium constant at any T, in terms of P_s, P_u, $P(T)$.

22. The proton magnetic resonance of selectively deuterated methyl ethyl ether was measured in a 100-MHz spectrometer to give the following spectrum:

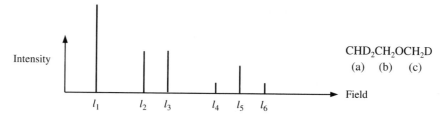

CHD$_2$CH$_2$OCH$_2$D
(a)   (b)   (c)

Intensity

Field

l_1 l_2 l_3 l_4 l_5 l_6

(a) Assign each of the six lines to protons (a), (b), or (c).
(b) If the separation between lines l_1 and l_5 is 200 Hz and is due to a chemical shift, what will the separation be in a 220-MHz spectrometer?
(c) If the spin-spin splitting of proton (a) is 5 Hz, what is the spin-spin splitting of proton (b)? What will the spin-spin splitting be in a 220-MHz spectrometer?
(d) Sketch the proton magnetic resonance spectrum for CD$_3$CH$_2$OCD$_3$.

23. Deduce the structure of the following compounds from their NMR spectra. Designate which protons go with which NMR lines. Chemical shifts are given in ppm relative to water.

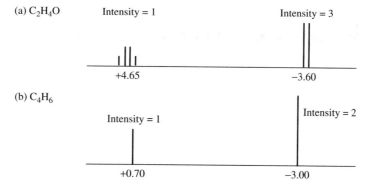

(a) C$_2$H$_4$O   Intensity = 1   Intensity = 3

                        +4.65   −3.60

(b) C$_4$H$_6$

            Intensity = 1   Intensity = 2

            +0.70   −3.00

24. Two possible isomers of C$_4$H$_9$Cl are 1-chlorobutane and 2-chlorobutane. Their proton NMR spectra are given in ppm relative to tetramethylsilane. The relative intensities are in parentheses above the peaks.

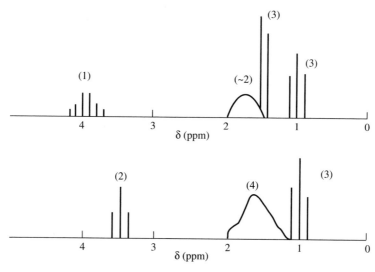

(1)   (3)

                                    (~2)   (3)

4   3   2   1   0
                δ (ppm)

(2)   (3)

                                    (4)

4   3   2   1   0
                δ (ppm)

(a) Decide which spectrum goes with which molecule, then assign the protons to their NMR peaks.

(b) Sketch the proton NMR spectrum for another isomer of C_4H_9Cl: trimethylchloromethane.

25. Sketch the proton NMR spectrum of the sugar portion of a deoxynucleoside in D_2O solution with the following parameters. Assume all two-bond spin-spin splittings are 10 Hz; all three-bond splittings are 4 Hz, and four or more bond splittings are zero.

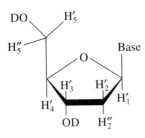

Proton	δ (ppm)
H1′	6.0
H2′	2.0
H2″	2.5
H3′	4.5
H4′	4.0
H5′	3.8
H5″	3.8

(a) A 500-MHz spectrometer is used. Plot a line spectrum versus frequency in Hz showing the position of each multiplet. For each multiplet show the splitting pattern on an expanded frequency scale with correct first-order intensities.

(b) What would change in the spectrum in a 600-MHz spectrometer? Give values of any parameters that would change.

26. Double-stranded nucleic acids can adopt different double helical structures, usually referred to as A-form and B-form. Among the differences are the conformations of the sugars; one conformation is called 2′-endo (the 2′-carbon is above the approximate plane of the four other atoms of the ribose ring); the other is 3′-endo (the 3′-carbon is above the plane). A ribose in an RNA chain is shown below.

DO, D′₅

D″₅   Base

O

H′₃   H′₂

H′₄   H′₁

OD   OD

Values of chemical shifts	
Proton	δ (ppm)
H1′	6.0
H2′	5.0
H3′	4.5
H4′	4.0

Proton ^1H—^1H splittings (in Hz)		
	A-form	B-form
$J_{1'2'}$	0	8
$J_{2'3'}$	5	5
$J_{3'4'}$	9	0

(a) Sketch the proton NMR spectrum of an A-form RNA; label all splittings. For each proton state whether it is a singlet, doublet, triplet, etc. in the spectrum.

(b) Sketch the proton NMR spectrum of a B-form RNA; label all splittings. For each proton state whether it is a singlet, doublet, triplet, etc. in the spectrum. B-form RNA has not been observed yet.

(c) A Karplus-type equation relates the ^1H—^1H splittings to a torsion angle, ϕ, which is a quantitative measure of the sugar conformation:

$$J_{1'2'} = 0.9 + 7.7 \cos^2 \phi$$

Calculate the value of ϕ predicted for B-RNA.

27. Consider the side chain of the amino acid arginine which can be modeled in D_2O as:

(a) Which type of proton would resonate farthest downfield (large ppm) in the proton NMR spectrum? Which would resonate farthest upfield? Explain, or provide evidence for, your answer.

(b) Draw a line NMR spectrum with the lengths of the lines to scale. Label the peaks and indicate upfield (low ppm) and downfield (high ppm).

(c) Draw the first-order splitting pattern for each peak assuming $J_{ab} = 10$ Hz, $J_{bc} = 3$ Hz, $J_{ac} = 1$ Hz. Show the relative magnitudes and splittings for each line in the pattern for peak a, peak b, and peak c.

28. Consider the two molecules shown below. The α- and β-carbons are tetrahedral and there is free rotation around the bonds connected to them. Sketch the proton NMR spectrum of each of the molecules dissolved in D_2O. The important elements of your spectrum are the number of peaks, their relative magnitudes, and the splittings. Consider only proton-proton splittings for protons separated by three or fewer bonds. Label the splittings as J_{ab}, etc. The chemical shifts are less important, but they can be estimated from data in the text.

Glycine   Phenylalanine

29. (a) NMR is very useful in determining molecular structure, but it can also be used in imaging. Describe how an NMR spectrum of intensity vs. frequency can be used to obtain an image. What is the main difference between NMR and MRI (magnetic resonance imaging)?

(b) A sphere of paraffin (C_nH_{2n+2}) is placed in a magnetic imaging instrument. Plot the image of intensity vs. frequency you would obtain.

30. The nuclear Overhauser effect (NOE) is a very powerful technique to investigate protein structure because different structures (α-helices, β-sheets) have characteristically different NOE patterns.

(a) Describe how an NOE is measured experimentally.

(b) NOEs are used to determine molecular structure. Describe how the measured NOE depends on molecular structure. Be quantitative.

(c) Typical proton-proton distances (in Å) for the proton on the α carbon (αH) and the amide proton (NH) in α-helices and β-sheets are summarized below:

	α-helix	β-sheet (antiparallel)	β-sheet (parallel)
NH—NH	2.8	4.3	4.2
αH—NH	3.5	2.2	2.2
αH—NH$_{i+3}$	3.4	>5	>5

(the subscript $i+3$ refers to an amide proton three residues away from the α proton)

In studying the small model oligopeptide: KAALAKAALAKAALAK (K is lysine, A is alanine, L is leucine), the following NOEs are observed:

I. Each amide proton has a very intense NOE to the next amide proton; for instance, each lysine amide (NH) has a strong NOE to an alanine amide (NH) proton.

II. Each αH proton has an NOE of moderate intensity to the next amide proton; for instance, each lysine α proton has an NOE to an alanine amide proton.

III. Each αH proton has an NOE of moderate intensity to an amide proton three residues away; for instance, each lysine αH proton has an NOE to the leucine.

What is the structure of the peptide? Is it α-helix, parallel β-sheet or antiparallel β-sheet? Explain your answer in terms of the measured NOEs.

31. Leucine and isoleucine are amino acids which are isomers of each other; their structures are given in the Appendix. The side chain of leucine has two H$_\beta$ protons, one H$_\gamma$ proton, and two δ methyls. The side chain of isoleucine has one H$_\beta$ proton, two H$_\gamma$ protons, a γ methyl, and a δ methyl. Of course each amino acid also has one H$_\alpha$ proton. Incorporated into a protein they give characteristic patterns of cross-peaks in a COSY experiment.

(a) Sketch the COSY spectrum for leucine assuming that each proton has a measurably different chemical shift (except the three protons on a methyl group). The order of chemical shifts can be taken as H$_\alpha$, H$_\beta$, H$_\beta'$, H$_\gamma$, δ methyl, δ methyl' in order of decreasing chemical shift. Figure 10.37b shows the COSY pattern for leucine assuming that the two β protons are not resolved, and that the two δ methyls are not resolved.

(b) Sketch the COSY spectrum for isoleucine assuming that each proton has a measurably different chemical shift (except the three protons on a methyl group). The order of chemical shifts can be taken as H$_\alpha$, H$_\beta$, H$_\gamma$, H$_\gamma'$, γ methyl, δ methyl in order of decreasing chemical shift.

(c) Describe how the different patterns of cross-peaks allow you to distinguish between leucine and isoleucine. If the two β protons of leucine, the two δ methyls of leucine and the two γ protons of isoleucine are not resolved, can the two amino acids still be distinguished? Explain.

32. The NMR spectrum from a single methyl group (on **T**) in a DNA oligomer

$$d(CGCGTA\textbf{T}ACGCG)$$

is shown as a function of temperature. In general terms explain what may be occurring to give rise to these spectra.

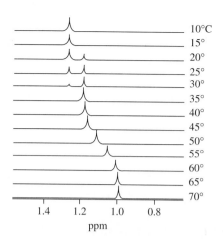

33. The NH resonances in the NMR spectrum at 25°C of a 5 amino acid (containing 3 NH groups) peptide YPIDV = Tyr Pro Ile Asp Val are shown below. P = Pro = Proline is unusual among amino acids in that the *cis* and *trans* forms of the peptide bond are more equal in energy than for other amino acids.

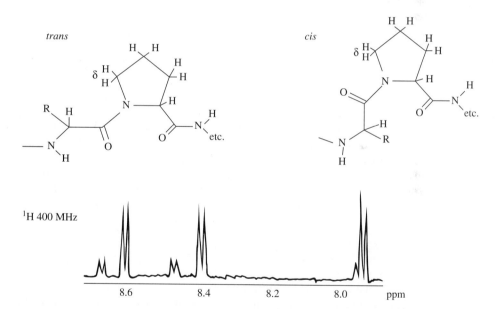

(a) Use the data to estimate the difference in standard free energy between the *cis* and *trans* forms of the peptide at 25°C.

(b) Explain what NMR experiment you would use to distinguish the resonances of the *cis* and *trans* forms. (Explain the experiment and what you would expect to see.)

(c) From these data what can you say about the rate of interconversion between the *cis* and *trans* isomers? (Explain.)

(d) From the structure of proline shown, predict what the splitting pattern for the δH resonances would be. Explain; state any assumptions.

11

Molecular Distributions and Statistical Thermodynamics

CONCEPTS

All molecules have a range of conformations produced by the vibrations of all the bonds and the torsional rotations about the single bonds. Macromolecules can have a very wide range of conformations because they contain so many bonds. DNA from the bacterial virus T4 is a double helix with a molecular weight of about 10^8; in solution this threadlike molecule can range from a tight coil to an extended form. Some properties of the molecule, such as how fast it can diffuse or sediment in solution, depend on the average dimension of the molecule. Others depend on the range of possible conformations of the molecule. For example, the probability that the molecule will be broken if the solution is stirred or poured depends on the fraction of molecules in an extended conformation, as well as the average dimension. Statistical methods can provide the distribution of molecular conformations and thus allow calculation of measurable properties. The concept of a random walk is very useful in understanding the possible conformations of a flexible macromolecule.

The binding of small molecules to many sites on a macromolecule, and the disruption of hydrogen bonds which lead to helix-coil transitions of polypeptides and polynucleotides, can be understood in terms of statistical distributions of bonds formed or broken.

Thermodynamic properties, such as energy, entropy, and free energy, are directly related to the average energy of all the molecules in a system and to the distribution of the molecules among the possible energies. The heat and work done when there is a change in the system can be calculated from the changes in the energy levels of the system and the changes in distribution of the molecules. These concepts connect molecular interactions and molecular properties to the thermodynamic laws we explored at the beginning of this book.

636

APPLICATIONS

If DNA in solution is heated, at a certain temperature the double helix begins to "melt." Base pairing is disrupted, regions of single-stranded loops form, and, if the temperature is high enough, the two strands of the double helix dissociate completely. A schematic illustration of the melting phenomenon is shown in Fig. 11.1. The transition from the helix to the dissociated strands occurs in a temperature range of only a few degrees; hence the term "melting" is used to describe this transition. Statistical analysis provides an explanation for the sharp transition. Furthermore, some of the thermodynamic parameters for the transition can be evaluated by a statistical analysis. These parameters are helpful in understanding the replication of the DNA, because the replication of DNA molecules involves the dissociation of the two parent strands and the formation of new double strands for each of the two daughter DNA molecules.

Either inside or outside the cell, the DNA interacts with a variety of small and not so small molecules. Among the small molecules the antibiotic actinomycin binds to many sites of the DNA and can block transcription of the DNA into RNA. Many heterocyclic dye molecules and polycyclic aromatic hydrocarbons, such as 9-amino-acridine and benzo[a]pyrene, can bind to the DNA and cause mutations. Such molecules are often carcinogenic. The binding of small molecules to DNA can be understood by statistical analy-

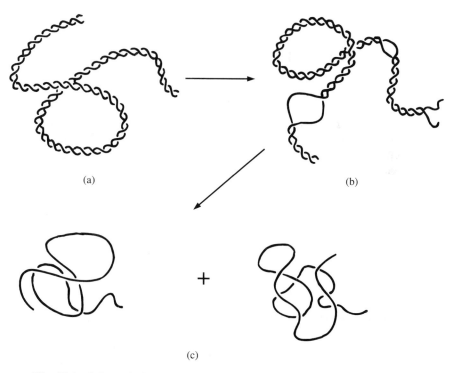

(a)   (b)

(c)

Fig. 11.1 Schematic illustration of the thermal "melting" of a double-stranded DNA (a) to its complementary single strands (c) At an intermediate temperature, single-stranded loops and/or ends can form, as illustrated in (b). See the text for a more detailed discussion.

ses of the binding experiments. Among the larger molecules which interact with the DNA, some bind to only a few specific sites; others bind with much less discrimination. The RNA polymerase of the host bacterium of T4, for example, initiates RNA synthesis (transcription) from only a few sites on the DNA. The product of the T4 gene 32 (the gene 32 protein), which is involved in DNA replication and recombination, is much less site-specific. It does, however, bind preferentially to single-stranded DNA in a highly "cooperative" fashion. If gene 32 protein is mixed with an excess of single-stranded DNA, some DNA molecules will be completely covered with the protein while other DNA molecules are bare. Statistical analysis again provides insight into such different patterns of interactions.

BINDING OF SMALL MOLECULES BY A POLYMER

For the interaction between two simple species, A and B, we found in Chapter 4 that the system at equilibrium can be described by an equilibrium constant K, and that the standard free energy change ΔG^0 is related to the equilibrium constant K by the equation $\Delta G^0 = -RT \ln K$. As an example, for the combination of A and B to form AB,

$$A + B \rightleftharpoons AB$$

the equilibrium constant and the standard free energy change are simply

$$K = \frac{[AB]}{[A][B]} \tag{11.1}$$

and

$$\Delta G^0 = -RT \ln K \tag{11.2}$$

In many cases of biological interest, one of the two interacting species is a complex macromolecule and the other a relatively small molecule. Usually, the macromolecule has more than one site where it can bind the small molecule. As a result, many different molecular complexes can form, depending upon which sites of the macromolecule are occupied by the small molecules. For example, a hemoglobin molecule has four sites for oxygen. A high-molecular-weight DNA has numerous sites for the binding of actinomycin. In such cases an expression such as Eq. (11.1) is not very descriptive in comparison with the richness of detail of the different species at the molecular level. One could write an equilibrium constant for each possible species, but the numbers often are very large and such cases can be treated more easily by the use of statistical methods. In the sections below we will treat a number of examples.

Identical-and-Independent-Sites Model

Suppose that a polymer molecule P has a number of identical and independent sites for binding a smaller molecule, A. If the sites are *identical* (each has the same affinity for A) and *independent* (the occupation of one site has no effect on the binding to another site), a very simple description of this system can be obtained. Examples for which the identical-and-independent-sites model is applicable are the binding of DNA polymerase molecules to a single-stranded DNA, the binding of the drug ethidium to a DNA double helix, and the binding of substrates to certain enzymes or proteins containing several identical subunits per molecule.

As a simple example, let us consider a polymer P with four identical and independent sites. This polymer molecule might be a tetrameric protein, with each of the four subunits having one site for the binding of a substrate A. Suppose that the four sites are distinguishable and can be labeled 1, 2, 3, and 4. Since the sites are identical, the equilibrium constants of the following reactions are identical:

$$P + A \rightleftharpoons PA^{(1)} \qquad K = \frac{[PA^{(1)}]}{[P][A]}$$

$$P + A \rightleftharpoons PA^{(2)} \qquad K = \frac{[PA^{(2)}]}{[P][A]}$$

$$P + A \rightleftharpoons PA^{(3)} \qquad K = \frac{[PA^{(3)}]}{[P][A]}$$

$$P + A \rightleftharpoons PA^{(4)} \qquad K = \frac{[PA^{(4)}]}{[P][A]}$$

The superscript (1, 2, 3, or 4) in each reaction specifies which site of the polymer is involved.

The total concentration of species with one A molecule per P molecule, [PA], is

$$[PA] = [PA^{(1)}] + [PA^{(2)}] + [PA^{(3)}] + [PA^{(4)}]$$

$$= 4K[P][A]$$

In other words, for the reaction

$$P + A \overset{K_1}{\rightleftharpoons} PA \qquad \text{(any site)}$$

$$K_1 = \frac{[PA]}{[P][A]} = 4K \tag{11.3}$$

The factor 4 in Eq. (11.3) can also be thought of as a result of the fact that there are four ways for the formation of a PA (by binding of an A to any one of the four sites of P), but only one way for any PA to dissociate into P and A.

For the reaction

$$PA \text{ (any site)} + A \overset{K_2}{\rightleftharpoons} PA_2 \qquad \text{(any two sites)}$$

$$K_2 = \frac{[PA_2]}{[PA][A]} = \left(\frac{4-1}{2}\right)K = \frac{3}{2}K \tag{11.4}$$

There are $(4-1)$, or three, ways of picking one of the three remaining sites in PA for the formation of a PA_2, and there are two ways for the dissociation of a PA_2 into PA and A. Similarly, the equilibrium constants K_3 and K_4 are

$$K_3 = \left(\frac{4-2}{3}\right)K = \frac{2}{3}K \tag{11.5}$$

and

$$K_4 = \left(\frac{4-3}{4}\right) K = \frac{1}{4} K \qquad (11.6)$$

The concentrations of the various species are:

$$[PA] = 4K[P][A] = 4[P](K[A])$$

$$[PA_2] = \frac{3}{2} K[PA][A] = \left(\frac{4 \cdot 3}{2}\right) K^2[P][A]^2 = 6[P](K[A])^2$$

$$[PA_3] = \frac{2}{3} K[PA_2][A] = \left(\frac{4 \cdot 3 \cdot 2}{3 \cdot 2}\right) K^3[P][A]^3 = 4[P](K[A])^3 \qquad (11.7)$$

$$[PA_4] = \frac{1}{4} K[PA_3][A] = K^4[P][A]^4 = [P](K[A])^4$$

Let $K[A] = S$. The total concentration of the polymer is

$$\begin{aligned}[P]_T &= [P] + [PA] + [PA_2] + [PA_3] + [PA_4] \\ &= [P](1 + 4S + 6S^2 + 4S^3 + S^4) \qquad (11.8) \\ &= [P](1 + S)^4 \end{aligned}$$

The total concentration of A molecules that are bound to the polymer molecules is

$$\begin{aligned}[A]_{\text{bound}} &= [PA] + 2[PA_2] + 3[PA_3] + 4[PA_4] \\ &= [P](4S + 12S^2 + 12S^3 + 4S^4) \\ &= 4[P]S(1 + 3S + 3S^2 + S^3) \qquad (11.9) \\ &= 4[P]S(1 + S)^3 \end{aligned}$$

The number of bound A molecules per polymer molecule is v, as defined in Eq. (5.10):

$$v \equiv \frac{[A]_{\text{bound}}}{[P]_{\text{total}}} = \frac{4[P]S(1 + S)^3}{[P](1 + S)^4} = \frac{4S}{1 + S} \qquad (11.10)$$

One can see that, in general, if there are N identical and independent sites,

$$\begin{aligned}v &= \frac{NS}{1 + S} \\ &= \frac{NK[A]}{1 + K[A]} \end{aligned} \qquad (11.11)$$

From Eq. (11.11), it follows that

$$v(1 + K[A]) = NK[A]$$

$$v = (N - v)K[A]$$

and

$$\frac{v}{[A]} = K(N - v) \qquad (11.12a)$$

or

$$\frac{v}{[A]} = KN - Kv \qquad (11.12b)$$

This is the *Scatchard equation*. When N is greater than 2, the parameter v (the number of bound A molecules per polymer molecule) is usually used to express the extent of binding. If $v/[A]$ is plotted versus v, according to Eq. (11.12a) or (11.12b), a straight line results. The slope of this line is $-K$, where K is the equilibrium constant for binding to one site. The intercept of the line with the v axis is N, where N is the total number of sites per polymer molecule. Such a plot is called a *Scatchard plot*. The Scatchard equation and Scatchard plot were first discussed in Chapter 5; see, for example, Figs. 5.4–5.6.

Langmuir Adsorption Isotherm

A binding curve (amount of material bound versus concentration) at a constant temperature is called an *isotherm*. In the independent-sites model, since the occupation of one site has no effect on binding to other sites, the spatial distribution of the sites has no effect on the final binding equation (11.12).

Equation (11.12) can also be changed into the following form:

$$\frac{v/N}{1 - v/N} = K[A]$$

Since v is the number of sites occupied per polymer molecule and N is the total number of sites per polymer molecule, v/N is simply the fraction of sites occupied, which is designated as f. We obtain, then,

$$\frac{f}{1 - f} = K[A] \qquad (11.13)$$

Equation (11.13) was first derived by Langmuir in 1916 for the adsorption of a gas on a solid surface.

Nearest-Neighbor Interactions and Statistical Weights

In the previous sections, the many binding sites on a macromolecule are assumed to be independent or noninteracting. Frequently, however, the binding of a small molecule to one site affects the binding of small molecules to other sites. The interactions between the sites greatly increase the complexity of the problem, and analytical solutions of the binding isotherms can be obtained only for special cases. However, a general method for simplifying the problem involves the use of statistical weights. The statistical weight, ω_i, of a species is defined as its concentration relative to a reference concentration.

$$\omega_i \equiv \frac{c_i}{c_{reference}}$$

Statistical weights can simplify the algebra in problems that contain many species. In equations involving ratios of concentrations, statistical weights can replace concentrations. For example, v can be written in terms of statistical weights instead of concentrations. Because the statistical weight of each species is the relative concentration of that species, the mole fraction of any species X_i is the statistical weight of the species, ω_i, divided by the sum of the statistical weights of all species:

$$X_i = \frac{\omega_i}{\sum_i \omega_i} \tag{11.14}$$

Since $\sum_i \omega_i$ appears in many calculations, we define a function called the sum over states:

$$Z \equiv \sum_i \omega_i \tag{11.15}$$

Now, we will consider a special case with N identical but interacting sites arranged in a linear array. In other words, the identical sites form a one-dimensional lattice. This model was first used by Ising and is called the *Ising model*. To simplify matters we will first consider only nearest-neighbor interactions. Longer-range interactions will be discussed in a later section.

The N identical sites in a linear array can be expressed by an N-digit number of 0's and 1's such as 00111000 . . . 1011100011, in which the symbol 0 represents a free site and the symbol 1 represents an occupied site. The number written represents a linear polymer with N identical sites, such that the first two sites are unoccupied, the next three sites are occupied, and so on. Let K be the equilibrium constant for the binding of an A molecule to a site with no occupied nearest neighbors, and τK be the equilibrium constant for the binding of an A molecule to a site with one adjacent site occupied. The parameter τ accounts for the interaction between the two adjacent occupied sites. If τ is less than 1, the binding is anticooperative (it is more difficult to bind next to an occupied site); if τ is greater than 1, the binding is cooperative (it is easier to bind next to an occupied site); if τ is equal to 1, there is no interaction between sites.

With these definitions the equilibrium constant for any reaction can be written quickly. A few examples follow for a simple case with $N = 3$:

Reaction	Equilibrium constant
000 + A $\rightleftarrows$ 001	K
001 + A $\rightleftarrows$ 101	K
001 + A $\rightleftarrows$ 011	τK
101 $\rightleftarrows$ 110	τ
101 + A $\rightleftarrows$ 111	$\tau^2 K$

The ratio of concentrations of [001] to [000], the statistical weight of [001], is

$$\omega_i = \frac{[001]}{[000]} = K[A]$$

Similarly, the statistical weights of [101], [011], and [111] are $(K[A])^2$, $\tau(K[A])^2$, and $\tau^2(K[A])^3$, respectively. The free species [000] has a statistical weight of one by convention. Defining the product $K[A]$ as S, the statistical weight of each state for the case $N = 3$ is given in Table 11.1.

Table 11.1 Statististical weights of species resulting from binding to a trimer

Number of sites occupied	Species	Statistical weight
0	000	1
1	001	S
	010	S
	100	S
2	101	S^2
	011	τS^2
	110	τS^2
3	111	$\tau^2 S^3$

Note that the statistical weight of each state can be written as $\tau^i S^j$, where j is the number of 1's (occupied sites) in the state and i is the number of 1's following 1 (number of nearest-neighbor interactions).

For our present case, summing the statistical weights listed in Table 11.1 gives

$$Z = 1 + 3S + S^2 + 2\tau S^2 + \tau^2 S^3 \tag{11.16}$$

The average number of bound A per P is

$$\begin{aligned}
\nu &= \frac{1[PA] + 2[PA_2] + 3[PA_3]}{[P] + [PA] + [PA_2] + [PA_3]} \\
&= \frac{(3S) + 2(S^2 + 2\tau S^2) + 3(\tau^2 S^3)}{1 + 3S + S^2 + 2\tau S^2 + \tau^2 S^3} \\
&= \frac{3S + 2(S^2 + 2\tau S^2) + 3\tau^2 S^3}{Z}
\end{aligned} \tag{11.17}$$

Equation (11.17) can be written in a simple form. Because

$$Z = 1 + 3S + S^2 + 2\tau S^2 + \tau^2 S^3$$

$$\left(\frac{\partial Z}{\partial S}\right)_\tau = 3 + 2S + 4\tau S + 3\tau^2 S^2$$

and

$$S\left(\frac{\partial Z}{\partial S}\right)_\tau = 3S + 2(S^2 + 2\tau S^2) + 3\tau^2 S^3 \tag{11.18}$$

The right side of Eq. (11.18) is identical to the numerator in Eq. (11.17). Thus

$$v = S \frac{(\partial Z/\partial S)_\tau}{Z}$$

$$= \left[\frac{\partial Z/Z}{\partial S/S} \right]_\tau \qquad\qquad (11.19)$$

$$= \left(\frac{\partial \ln Z}{\partial \ln S} \right)_\tau$$

Although Eq. (11.19) is derived for a special case, it applies to any system in which there is binding to N identical sites with interaction only between adjacent sites.

Exercise   Show that for the case of four identical and independent sites discussed previously, Eq. (11.19) gives the same result as Eq. (11.10).

Cooperative Binding, Anticooperative Binding, and Excluded-Site Binding

It is informative to examine the shape of a plot of the fraction of binding sites occupied, f ($f \equiv v/N$) as a function of $S = K[A]$. With $N = 3$, f is plotted against S for the case $\tau = 1$ and $\tau = 10^3$ in Fig. 11.2. The case $\tau = 1$ is, of course, equivalent to the case when the sites are independent. For such a case, f increases gradually with S and approaches 1 asymptotically. The case $\tau = 10^3$, or in general for $\tau > 1$, is termed *cooperative binding,* because the binding of A to one site makes it easier to bind another A to an adjacent site. Note that the shape of the f versus S plot is sigmoidal (S-shaped). As S increases, f increases slowly at first, showing an upward curvature, and then increases more rapidly. At still higher values of f, a decrease occurs in the slope of the f versus S plot with increasing S, showing a downward curvature. Such a sigmoidal plot of f versus S is indicative of cooperative binding.

For the independent-sites model, the shape of the f versus S curve is not affected by the number of sites N per polymer molecule, because N is absent in the binding equation (11.13). For cooperative binding the shape of the f versus S curve is dependent on N. As N increases, the number of terms in Eqs. (11.15) to (11.18) increases. For the same value of the interaction parameter τ, the difference in S for a given change in f is less for larger N. Thus the abruptness of the cooperative binding is even more extreme. If N is very large, it can be shown that f changes from essentially zero to essentially 1 in the range $(1 - 2/\sqrt{\tau}) < \tau S < (1 + 2/\sqrt{\tau})$. Thus a sharp transition occurs at $\tau S = 1$ for large values of τ. At low concentrations of [A], $\tau S \equiv \tau K[A]$ is less than 1 and very little binding occurs, but as the concentration of [A] is increased so that τS becomes slightly greater than 1, suddenly the cooperative binding is triggered and nearly all sites are filled.

Another important difference between independent-sites binding (non-cooperative binding) and cooperative binding is the relative amounts of the various species. For $N = 3$, let us examine the situation when $v = \frac{3}{2}$ (when half of the sites are occupied). The mole fractions of the species with 0, 1, 2, and 3 sites occupied are $1/Z$, $3S/Z$, $(S^2 + 2\tau S^2)/Z$, and

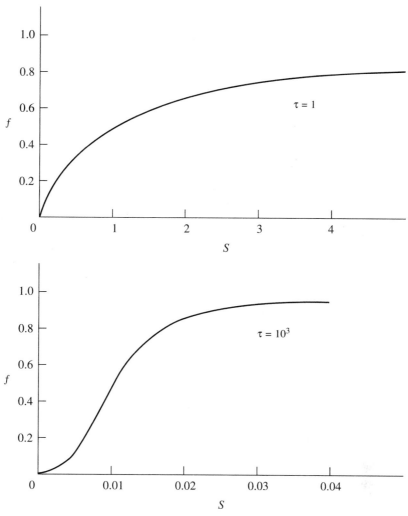

Fig. 11.2 Fraction of sites occupied, f, as a function of the substrate concentration times the binding constant for a polymer with three binding sites ($N = 3$). The curves shown are for the cases of no interaction between occupied sites ($\tau = 1$), and for positive interaction between adjacent sites with a value of $\tau = 10^3$.

$\tau^2 S^3 / Z$, respectively. These quantities are calculated for the case $\tau = 1$ ($v = \frac{3}{2}$ at $S = 1$) and $\tau = 10^3$ ($v = \frac{3}{2}$ at $S = 9.75 \times 10^{-3}$), and are plotted in Fig. 11.3.

The difference between the two distributions is evident. In the case $\tau = 1$, the most abundant species center around $v = \frac{3}{2}$. In the case $\tau = 10^3$, however, the predominant species are the one with all sites free (47 mol %) and the one with all sites occupied (43.6 mol %). Such an uneven distribution is a consequence of the cooperative nature of binding. Because the occupation of one site makes the binding to adjacent sites more favorable, the bound A molecules tend to cluster. When τ is very large, only the species with all sites occupied and the species with all sites free are of importance. This is the *all-or-none limit*.

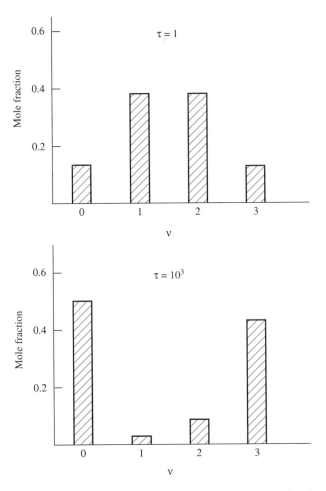

Fig. 11.3 Distribution of species at half saturation for $N = 3$. Independent sites ($\tau = 1$); cooperative binding ($\tau = 10^3$). Note that the cooperative binding leads to an all-or-none type binding.

In the all-or-none limit, for our case of $N = 3$, we can simply consider the reaction as

$$P + 3A \longrightarrow PA_3$$

because neither of the other species, PA and PA_2, is present in significant amounts. The equilibrium constant for the reaction is

$$K' = \frac{[PA_3]}{[P][A]^3}$$

and

$$\nu = \frac{3[PA_3]}{[P] + [PA_3]} = \frac{3K'[A]^3}{1 + K'[A]^3} \tag{11.20}$$

We use K' in these equations to remind us that this is not the equilibrium constant for binding at a single site.

Because $\nu/3 = f$, the fraction of sites occupied, simple algebra gives

$$\frac{f}{1 - f} = K'[A]^3 \tag{11.21}$$

and

$$\frac{d \log [f/(1 - f)]}{d \log [A]} = 3$$

In general, if there are N sites per polymer molecule, the all-or-none model (maximum cooperativity) yields

$$\frac{d \log [f/(1 - f)]}{d \log [A]} = N \tag{11.22}$$

For no cooperativity ($\tau = 1$), the left-hand side of Eq. (11.22) is equal to 1. An experimental measure of cooperativity in binding is obtained by a plot of $\log [f/(1 - f)]$ versus $\log [A]$. This plot is a *Hill plot* [see Fig. (5.9)] and the slope of the plot is n, the *Hill constant*. If the slope, n, is 1, the sites are independent. If the slope is greater than 1, there is some cooperativity. If the slope is equal to N, the binding can be approximated by the all-or-none model.

The case $\tau < 1$ is also of interest. This occurs when binding to one site reduces the affinity of adjacent sites (*anticooperative* or *interfering binding*). In the limit $\tau = 0$, the occupation of one site excludes binding to adjacent sites, and the case is referred to as the *excluded-site model*.

N Identical Sites in a Linear Array with Nearest-Neighbor Interactions

In the preceding section, we discussed quantitatively the binding of small molecules to a polymer molecule with three identical sites in a linear array. A number of important concepts came out of the discussion. These include cooperative and anticooperative binding, the all-or-none limit, and the excluded-site model. The transition from a polymer with $N = 3$ to one with a large number of monomer units requires no *new* physical concepts, but the mathematics is more involved. This is described in detail in the references given at the end of the chapter. Instead of repeating the lengthy derivations here, we shall examine some illustrative examples.

Example 11.1 Adenosine molecules can form hydrogen-bonded complexes with the bases of polyuridylic acid. In Fig. 11.4, the fraction of sites occupied in polyuridylic acid is shown as a function of the concentration of free adenosine. The shape of the curve clearly indicates cooperative binding. The interaction parameter τ is calculated to be $\sim 10^4$. The cooperativity is interpreted as a result of the favorable "stacking" interactions between adjacent adenosine molecules.

Example 11.2 The bacteriophage T4 gene 32 protein is involved in both DNA recombination and DNA replication. If the amount of the protein is insufficient to saturate the amount of the single-stranded DNA present, the distribution of the protein molecules among the DNA molecules is far from random. Some DNA molecules are nearly saturated with the protein, while others are essentially free of bound protein molecules, as shown in the electron micrograph in Fig. 11.5.

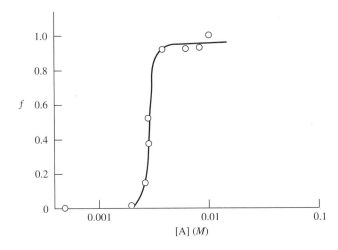

Fig. 11.4 Binding of adenosine, A, to polyuridylic acid. The fraction of sites occupied, *f,* is plotted against the concentration of free adenosine, [A]. [Data from W. M. Huang and P. O. P. Ts'o, *J. Mol. Biol. 16,* 523 (1966).]

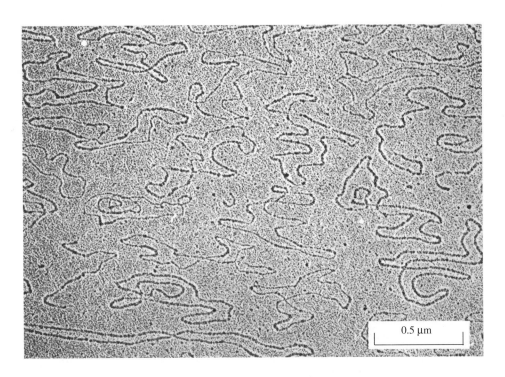

0.5 μm

Fig. 11.5 Electron micrograph, showing loops of single-stranded circular DNA molecules. Look carefully and note that some segments of the DNA appear dark and thick, while others are light and thin. The thick segments show where the DNA gene 32 protein forms complexes with the DNA. The protein distribution is far from uniform along the DNA. You can find some DNA molecules that are completely thick and others that consist mostly of thin segments. [From H. Delius, N. J. Mantell, and B. Alberts, *J. Mol. Biol. 67,* (1972).]

THE RANDOM WALK

In the preceding sections we have discussed the binding of small molecules to a polymer with many sites. We will now discuss an important problem in statistics: the random-walk problem or the problem of random flight. The simplest example of such a problem involves someone walking along a sidewalk but unable to decide at each step whether to go forward or back. The direction of each step is random. This is a *one-dimensional random walk*. The *two-dimensional random walk* is analogous to a dazed football player who may take steps randomly in any direction on the field. The problem is also easily generalized to three (or more) dimensions.

While the formulation of the random-walk problem may seem rather abstract and to bear little relation to physical problems, we will soon see that this is not so. A number of phenomena of interest, such as molecular diffusion and the average dimension of a long polymer molecule, can be treated as "random walk problems" in three dimensions.

To introduce an analysis of the random-walk problem, it is best to start with the one-dimensional version. If we use the symbol 1 for a step forward and the symbol 0 for a step back, the record of a walk of N steps is represented by an N-digit number, such as 11000101011 . . . 0011. Let p be the probability of a step forward and q be the probability of a step back. When we said that the direction of each step was "random," we implied that $p = q$. In general, though, p does not have to be the same as q.

Before going further, let us elaborate on what we mean by "the probability of a step forward being p." Imagine that N similar fleas (N is a very large number) are hopping randomly forward or backward. These N fleas form an ensemble. After each has taken one hop, m are found to have moved forward and the rest, $N - m$, to have moved backward. The fraction of fleas that moved forward, m/N, is the probability p. Similarly, $(N - m)/N = q$. Obviously, $p + q = 1$, a consequence of the fact that we allow only two possibilities; either a hop forward or a hop back. This imaginary experiment can be performed in a different way. Instead of having N fleas take one hop each, we can allow one flea to take N hops. We expect that, if there are no time-dependent changes, the fraction of steps forward is p. *It is a basic postulate of statistical mechanics that the time average of a property of a system is the same as the ensemble average at any instant.* This means that watching one molecule over a long period of time and averaging its property will give the same result as an average at one time over many molecules.

Now we come back to the one-dimensional random-walk problem. We need to be able to calculate the probability of taking any number of steps forward and any number backward. Then we will be able to find how far we an expect to move from the starting point after N random steps. The answer is that after N random steps we will just as likely have moved forward as backward. However, this does not mean that we end up at the starting point after N random steps. The square root of the mean-square displacement (a measure of how far we are from the start) is equal to the square root of the number of steps times the length of each step. We will derive these results in detail.

For a given sequence of N steps represented by 11000101011 . . . 0011, the probability for this sequence to occur is $ppqqqpqpqpp \ldots qqpp = p^m q^{N-m}$, where m is the number of 1's. There are many ways that we can take N steps with m of them forward. For example, if $N = 4$ and $m = 2$, the possible ways are 0011, 0110, 1100, 1010, 0101, and 1001. Several additional examples are presented in Table 11.2.

Table 11.2 Some random walk examples

Total number of steps	Record of random walk	Probability	Sum of probabilities of all possibilities
1	0	q	$q + p$
	1	p	
2	00	q^2	$q^2 + 2qp + p^2$
	01, 10	$2qp$	$= (q + p)^2$
	11	p^2	
3	000	q^3	$q^3 + 3q^2p + 3qp^2 + p^3$
	001, 010, 100	$3q^2p$	$= (q + p)^3$
	011, 110, 101	$3qp^2$	
	111	p^3	
N	0000 ...	q^N	$(q + p)^N$
	...	...	
	...	...	
	1111 ...	p^N	

In Table 11.2 we see that the sum of the probabilities of all possible outcomes of an N-step walk can be represented in the form $(q + p)^N$. Because $q + p = 1$, this sum is equal to 1, but we want to know how the components of the sum depend on q, p, and N. The sum can be expanded by the binomial theorem:

$$(q + p)^N = q^N + Nq^{N-1}p + \frac{N \cdot (N - 1)}{2!} q^{N-2}p^2 + \cdots$$

$$+ \frac{N!}{m!(N - m)!} q^{N-m}p^m + \cdots + p^N \qquad (11.23)$$

Remember that

$$N! = N \cdot (N - 1) \cdot (N - 2) \cdots (3) \cdot (2) \cdot (1)$$

Each term in the expansion gives the probability of the corresponding outcome. The term q^N, for example, is the probability that all N steps are backward. The term

$$W(m) = \frac{N!}{m!(N - m)!} q^{N-m}p^m \qquad (11.24a)$$

gives the probability that m steps are forward and $N - m$ steps are backward. The coefficient

$$\frac{N!}{m!(N - m)!} \qquad (11.24b)$$

is the number of ways one can take N steps with m of them forward. Because $q + p = 1$, the numerical value of $(q + p)^N$ is of course always 1.

Calculation of Some Mean Values for the Random-Walk Problem

The important equation is Eq. (11.24a). Knowing the probability of a random walk of N steps with m steps forward, we can now calculate averages of interest, such as the average number of steps forward:

$$\langle m \rangle = \sum_{m=0}^{N} mW(m) \qquad (11.25a)$$

and the mean of the square of the number of steps forward:

$$\langle m^2 \rangle = \sum_{m=0}^{N} m^2 W(m) \qquad (11.25b)$$

Mean displacement

First, let us calculate the average of the displacement. If the pace of each step is l, the net displacement for m steps forward is $[m - (N - m)]l$ or $(2m - N)l$. So if we determine $\langle m \rangle$, the average value of m, the *mean displacement* forward is $(2\langle m \rangle - N)l$.

Substituting Eq. (11.24a) into Eq. (11.25a), we obtain

$$\langle m \rangle = 0 \cdot q^N + 1 \cdot Nq^{N-1}p + 2 \cdot \frac{N \cdot (N - 1)}{2!} q^{N-2}p^2 + \cdots$$
$$+ m \cdot \frac{N!}{m!(N - m)!} q^{N-m}p^m + \cdots + Np^N \qquad (11.26)$$

We can use this equation to calculate the average number of steps forward, but we note that its terms look like those in Eq. (11.23), except that each term is multiplied by the exponent of p in the term. This suggests that the expression for $\langle m \rangle$ is related to the derivative of Eq. (11.23) and can be greatly simplified. We denote Eq. (11.23) by Z, because it is a sum of statistical weights of "species" with m forward steps.

$$Z = (q + p)^N = q^N + Nq^{N-1}p + \frac{N \cdot (N - 1)}{2!} q^{N-2}p^2 + \cdots$$
$$+ \frac{N!}{m!(N - m)!} q^{N-m}p^m + \cdots + p^N \qquad (11.23)$$

$$\frac{\partial Z}{\partial p} = Nq^{N-1} + 2 \cdot \frac{N \cdot (N - 1)}{2!} q^{N-2}p + \cdots$$
$$+ m \cdot \frac{N!}{m!(N - m)!} q^{N-m}p^{m-1} + \cdots + Np^{N-1}$$

$$p\left(\frac{\partial Z}{\partial p}\right) = Nq^{N-1}p + 2 \cdot \frac{N \cdot (N - 1)}{2!} q^{N-2}p^2 + \cdots$$
$$+ m \cdot \frac{N!}{m!(N - m)!} q^{N-m}p^m + \cdots + Np^N \qquad (11.27)$$

Comparing Eqs. (11.27) and (11.26), we see that

$$\langle m \rangle = p\left(\frac{\partial Z}{\partial p}\right) \qquad (11.28)$$

But

$$Z = (q + p)^N$$

Therefore,

$$\left(\frac{\partial Z}{\partial p}\right) = N(q + p)^{N-1}$$

and

$$\langle m \rangle = p\left(\frac{\partial Z}{\partial p}\right) = Np(q + p)^{N-1} = Np \tag{11.29}$$

The last equality comes from the fact that $(q + p) = 1$; this leads to the simple and intuitively reasonable result. The average number of steps forward is the probability of a forward step times the total number of steps. For the random (unbiased) walk, $p = \frac{1}{2}$, and

$$\langle m \rangle = \frac{N}{2} \tag{11.30}$$

and the net displacement forward is

$$(2\langle m \rangle - N)l = 0 \tag{11.31}$$

This seems completely reasonable. If the probability of going forward is the same as that of going backward, on the average half of the total number of steps taken will be forward. After taking N steps, sometimes one might be a few paces in front of the origin, sometimes a few paces behind the origin. The average or mean displacement comes out to be zero.

Mean-square displacement

After taking a larger and larger number of steps, N, at random, it becomes increasingly unlikely that we shall end up exactly where we started. The *mean-square displacement* is a measure of how far from the origin we can expect to be, on the average. The importance of the difference between the average distance from the origin and the average net displacement forward is that in calculating the average distance we do not distinguish whether we finally end up in front of or behind the origin. For example, if after N steps person A ends up three steps forward of the origin and person B ends up three steps behind the origin, the average of their net displacements is zero; but the average of their distances from the origin is clearly not zero; it is three steps.

One way to characterize the average distance from the origin is to calculate the *mean-square displacement*. In a single journey of N steps, if m steps are forward, the net displacement forward is

$$d = (2m - N)l \tag{11.32}$$

and the square of the displacement is

$$d^2 = (2m - N)^2 l^2 \tag{11.33}$$

Whether $m > N/2$ (a net displacement forward) or $m < N/2$ (a net displacement backward), d^2 is positive.

The mean value of $(2m - N)^2$ can be calculated as follows, where $\langle \rangle$ designates the average of that quantity:

$$\langle (2m - N)^2 \rangle = 4\langle m^2 \rangle - 4\langle m \rangle N + N^2 \tag{11.34}$$

But $\langle m \rangle = N/2$ for a random walk; thus

$$\langle (2m - N)^2 \rangle = 4\langle m^2 \rangle - 2N^2 + N^2 = 4\langle m^2 \rangle - N^2 \tag{11.35}$$

To calculate $\langle m^2 \rangle$, we start with its definition given in Eq. (11.25b),

$$\langle m^2 \rangle = \sum_{m=0}^{N} m^2 W(m)$$

$$= 0 \cdot q^N + 1 \cdot Nq^{N-1}p + 2^2 \cdot \frac{N \cdot (N-1)}{2!} q^{N-2}p^2 + \cdots \tag{11.36}$$

$$+ m^2 \cdot \frac{N!}{m!(N-m)!} q^{N-m}p^m + \cdots + N^2 p^N$$

If we differentiate Eq. (11.27) with respect to p, we obtain

$$\frac{\partial}{\partial p}\left[p\left(\frac{\partial Z}{\partial p} \right) \right] = Nq^{N-1} + 2^2 \cdot \frac{N \cdot (N-1)}{2!} q^{N-2}p + \cdots$$

$$+ m^2 \cdot \frac{N!}{m!(N-m)!} q^{N-m}p^{m-1} + \cdots \tag{11.37}$$

$$+ N^2 p^{N-1}$$

Comparing Eqs. (11.36) and (11.37), we note that multiplying the right side of (11.37) by p gives the right side of (11.36). Therefore,

$$\langle m^2 \rangle = p\left\{ \frac{\partial}{\partial p}\left[p\left(\frac{\partial Z}{\partial p} \right) \right] \right\} \tag{11.38}$$

From Eq. (11.29),

$$p\left(\frac{\partial Z}{\partial p} \right) = Np(q + p)^{N-1}$$

Therefore,

$$\frac{\partial}{\partial p}\left[p\left(\frac{\partial Z}{\partial p} \right) \right] = N\left[p\frac{\partial (q + p)^{N-1}}{\partial p} + (q + p)^{N-1} \right]$$

$$= N[p(N - 1)(q + p)^{N-2} + (q + p)^{N-1}]$$

$$= N[p(N - 1) + 1]$$

and

$$p \frac{\partial}{\partial p} \left[p \left(\frac{\partial Z}{\partial p} \right) \right] = Np[p(N - 1) + 1] \qquad (11.39)$$

For $p = \frac{1}{2}$,

$$\langle m \rangle = p \frac{\partial}{\partial p} \left[p \left(\frac{\partial Z}{\partial p} \right) \right] = \frac{N}{2} \left(\frac{N - 1}{2} + 1 \right) = \frac{N(N + 1)}{4} \qquad (11.40)$$

but

$$\langle (2m - N)^2 \rangle = 4 \langle m^2 \rangle - N^2 \qquad (11.35)$$

and, after all these complicated manipulations, we obtain the strikingly simple result that

$$\langle (2m - N)^2 \rangle = N(N + 1) - N^2$$
$$= N \qquad (11.41)$$

Therefore, the mean-square displacement $\langle d^2 \rangle$ is

$$\langle d^2 \rangle = Nl^2 \qquad (11.42)$$

The root-mean-square displacement $\langle d^2 \rangle^{1/2}$ is

$$\langle d^2 \rangle^{1/2} = N^{1/2} l \qquad (11.43)$$

Equation (11.43) says that after N random steps, on the average, one is $\sqrt{N}$ paces from where one started. Although we have derived this equation for a one-dimensional random walk, it can be shown that it holds for a random walk in two- or three-dimensional space as well. We presented the derivations for the random-walk displacements in detail to illustrate how average properties are obtained from statistical weights and sums over states.

Diffusion

Molecules, either in the gas phase or the liquid phase, are always undergoing many collisions. Let us assume that the average distance traveled by a molecule between two successive collisions (the mean free path) is l. Whenever a molecule has a collision, its direction changes, depending upon the direction of the molecule hitting it. Starting from time zero, the total number, N, of collisions a given molecule has is proportional to the time t. There is a close relation between the diffusional process and the random walk. The fact that the individual "steps" in the diffusion process are not all of the same length turns out to be unimportant as long as the measurements are made over distances that are large compared with the mean free path. For a given spatial distribution of molecules (concentration gradient) at time zero, the spatial distribution of the molecules at time t can be obtained by considering each molecule as a random walker. The average value of the square of the displacement for the diffusional process based on Eq. (11.42) is expected to be proportional to N and, therefore, to the elapsed time. The proportionality constant is a measure of how fast the molecule diffuses and is directly related to the diffusion coefficient D, as was seen in Chapter 6.

Using the random-walk model, Einstein was able to derive the diffusion equation from the molecular point of view.

Average Dimension of a Linear Polymer

A flexible polymer can assume many conformations which differ little in energy. Consider a polypeptide,

for example. Along the backbone of the polymer, rotation around the amide bonds $N \cdots C$ has a high-energy barrier because of the partial double-bond character of these bonds (see Chapter 9, Problem 18). Rotation around the other bonds is relatively free, however, resulting in many polymer conformations of similar stability. Some of these conformations may be highly extended, while others may be much more compact. Thus, in discussing the dimensions of such a polymer, it is necessary to specify *average* dimensions.

A quantity frequently used to express the average dimension of a flexible polymer is the root-mean-square end-to-end distance, $\langle h^2 \rangle^{1/2}$. h is the straight-line distance between the ends of the molecule; the root-mean-square end-to-end distance is the square root of the average of its square. If a polymer molecule is highly extended, its end-to-end distance is much greater than for the same molecule in a compactly coiled form. A simple and useful model for the evaluation of the average end-to-end distance is the random-coil or Gaussian model. In this model the linear polymer is considered to be made of N segments each of length l, linked by $(N-1)$ universal joints (unrestricted bending) so that the angle between any pair of adjacent segments can take any value with equal probability.

The reader may recognize immediately that the random-coil model is exactly the same as the random-walk problem we have discussed. From Eq. (11.42), the mean-square end-to-end distance is $\langle h^2 \rangle = \langle d^2 \rangle = Nl^2$. For a random-coil polymer, $\langle h^2 \rangle$ is proportional to N and therefore proportional to the molecular weight of the polymer. We can also write

$$\langle h^2 \rangle = Nl^2 = Nl \cdot l = Ll \tag{11.44}$$

where $L \equiv Nl$ is the *contour length* of the molecule. The contour length of a polymer molecule is the length measured along the links of the polymer.

Real polymer molecules are, of course, made of monomers linked by chemical bonds rather than segments linked by universal joints. The relation between bond lengths and lengths of statistical segments is not simple. In general the length l of a statistical segment must be determined experimentally.

Let us consider polyethylene,

For simplicity, we consider only the carbon backbone chain

with a bond angle θ between two adjacent bonds and bond length b for each C—C bond. If θ could assume any value, one could imagine a universal joint at each carbon atom, and the mean-square end-to-end distance $\langle h^2 \rangle = Nl^2 = Nb^2$, where N is the number of carbon atoms per polymer chain. We know, however, that the bond angle θ cannot assume any value. Rather, the most stable conformation for

bonds has an angle θ close to the tetrahedral angle of 109°. The restriction in θ makes the molecule a little less flexible. If the random-coil model is used for polyethylene, we expect that $l > b$; the effective distance, l, between carbon atoms will be greater than the bond length, b. In fact, it can be shown that $l = \sqrt{2}b$ for polyethylene.

For a polymer such as a double-stranded DNA, the effective segment length, l, of an equivalent random coil cannot be related directly to the bond lengths. Hydrodynamic measurements give $l \approx 10^3$ Å for DNA. Thus for a flexible molecule such as polyethylene, the segment length is of the order of the bond length $[l = \sqrt{2}b \cong 2.2$ Å]; for a stiff molecule such as a double-stranded DNA, the effective length is several orders of magnitude larger than a bond length.

A quantity closely related to mean-square end-to-end distance is the mean-square radius, which is defined by

$$\langle R^2 \rangle = \frac{\Sigma_i\, m_i r_i^2}{\Sigma_i\, m_i} \tag{11.45}$$

for any collection of mass elements, where m_i is the mass of the ith element and r_i is the distance of this element from the center of mass of the collection.

It can be shown that for an open-ended random coil,

$$\langle R^2 \rangle = \frac{Nl^2}{6} \tag{11.46a}$$

and for a circular random coil (formed by linking the two ends of the open-ended chain),

$$\langle R^2 \rangle = \frac{Nl^2}{12} \tag{11.46b}$$

It was first shown by Debye that $\langle R^2 \rangle$ can be measured experimentally by light scattering.

Note that the random-coil model predicts that $\langle h^2 \rangle^{1/2}$ or $\langle R^2 \rangle^{1/2}$ is proportional to $\sqrt{N}$, or the square root of the molecular weight. This relation can be tested experimentally. If it is found to be true, the polymer is said to behave as a random coil, or a Gaussian chain.

For real polymers, in general, there are interactions between adjacent segments, and they have preferred orientations instead of equally probable orientations, as required by the random-coil model. At a given temperature the random-coil model is expected to be applicable only in solvents ("θ solvents") in which the interactions between polymer segments are balanced by solvent-polymer interactions. Otherwise, the interactions between the polymer segments will cause deviations from the random-coil model.

HELIX-COIL TRANSITIONS

Helix-Coil Transition in a Polypeptide

In a polypeptide chain the partial double-bond character of the $N \cdots C$ and $C \cdots O$ bonds for each amide residue in the chain requires the group of six atoms

to lie in a plane. It was first suggested by Pauling and his coworkers in 1951 that the chain of planar groups can be rotated around the α-carbons (C atoms between the CO group of one amino acid and the NH group of the next) into a helix such that a hydrogen bond is formed between each CO group and the fourth following NH group. The locations of the hydrogen and oxygen atoms involved in hydrogen bonding are illustrated in Fig. 11.6, and a projection of the resulting helix, called an α-*helix,* is shown in Fig. 3.6. The polypeptide backbone of the α-helix follows closely along the contour described by a helical spring made from a single strand of wire. The existence of the α-helix structure in many proteins has now been confirmed, mainly by x-ray diffraction studies of proteins. For example, in the enzyme adenyl kinase from porcine muscle, over one-half of the amide residues are in the α or distorted α-helix conformation.

For a number of synthetic polypeptides, the entire chain of the molecule can assume the α-helix structure, depending on the solvent, temperature, pH, and so on. In the non-hydrogen-bonded form, the polypeptide chain is flexible and can assume many conformations, as discussed previously. Such a chain is frequently said to be in the "coiled" form. The transition from the coiled form to the helical form (or vice versa) is referred to as a *coil-helix transition.* For polypeptides of high molecular weight, if one of the parameters (such as temperature) is changed, the transition from the coiled form to the α-helix form (or vice versa) generally occurs within a narrow range.

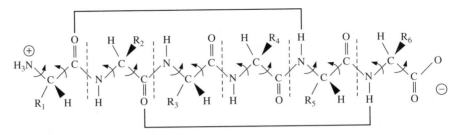

Fig. 11.6 A polypeptide chain showing the hydrogen bonding between the C=O and the NH groups in an α-helix. Residues are numbered starting from the amino terminal; the residues are separated by dotted lines. Arrows show the two bonds (on either side of each α-carbon) around which rotation can occur in the coil form.

We can understand the basic features of the helix-coil transition using a formulation that is similar to our discussion of the binding of small molecules by a polymer. The conformation of the chain can be specified by the states of the carbonyl oxygen atoms. We use the symbol h (for helix) to represent a carbonyl oxygen if it is hydrogen-bonded and the symbol c (for coil) if it is not. Whenever an oxygen is in a bonded state, it is to be understood that it is hydrogen-bonded to the NH group in the fourth residue ahead in the chain (see Fig. 11.6). We number starting at the amino end of the polypeptide; thus the CO in the residue 1 can bond to the NH in residue 5; the CO in residue i can bond to the NH in residue $i + 4$, and so forth. The last four residues in a chain by definition cannot be designated as h; their carbonyl groups cannot bond to an appropriate NH. It is further assumed that the sense of the helix is unique (either left-handed or right-handed, but not both). In this way, the conformation of a chain of N residues is represented by N letters, with the last four letters always c. For example, the conformation shown in Fig. 11.6 is designated hhccccc.

Now consider the transition from a state such as cccccccc . . . to a state cchcccccc . . . , which represents the transition of the third residue from the nonbonded state to the α-helix conformation. In the nonbonded state there are two covalent bonds in each residue around which rotation may occur. They are the bonds on each side of the α-carbon labeled with arrows in Fig. 11.6. In a helical form, however, the chain assumes a much more rigid conformation. Because the oxygen atom of the third residue is hydrogen-bonded to the NH group of the seventh residue, the transition of the third residue from the coil state to the helix state means that the fourth, fifth, and sixth residues must be ordered into the helical conformation (even though their oxygen atoms are not hydrogen bonded and they are designated c). In other words, the formation of the first helical element involves the ordering of three residues; the orientation around six bonds is fixed. The transition of the state cchcccccc . . . to the state cchhccccc . . . , however, requires the ordering of only one more residue, the seventh, because the fourth, fifth, and sixth are already ordered. The orientation around only two more bonds is fixed. Intuitively, then, we expect that the formation of the first helical element is more difficult than the formation of the next adjacent helical element.

Based on this discussion, we assign statistical weights according to the following rules:

1. For each c, the statistical weight is taken as unity.
2. For each h after an h, the statistical weight is s.
3. For each h after a c, the statistical weight is σs.

Note that s is the equilibrium constant for the reaction

$$cchccccc \ldots \longrightarrow cchhcccc \ldots$$

for the addition of a helical element to the end of a helix; σs is the equilibrium constant for the reaction

$$ccccccccc \ldots \longrightarrow cchccccc \ldots$$

for the *initiation* of a helical element. We expect σ to be less than 1, because initiation of an α-helix is more difficult than propagation of the helix.

The reader should recognize the similarity between the present case and the previously discussed binding problem with nearest-neighbor interactions. The similarity can be made more apparent if the parameters used in the previous discussion on binding, τ and S, are transformed into σ and s by the relation $\tau = 1/\sigma$ and $S = \sigma s$.

There is one difference between the two cases, however. For binding, we consider only nearest-neighbor interactions. For an α-helix formation, the very nature of the α-helix structure dictates more than nearest-neighbor influence: For example, if the third residue goes into a helical state, the fourth, fifth, and sixth residues are brought into the helix conformation. The consequence of this is that the conformations of three adjacent residues are closely related, and conformations with two h's separated by no more than two c's are expected to be rare. To see the last point more clearly, let us consider a reaction

$$\ldots hhhhccccc \ldots \longrightarrow \ldots hcchccccc \ldots$$
$$\quad 123456789 \qquad\qquad 123456789$$

as an example. For $\ldots$ hhhhccccc $\ldots$, the second to seventh residues are held in helical geometry by the hydrogen bonds from residues 1 to 5, 2 to 6, 3 to 7, and 4 to 8. In hcchccccc, the second to seventh residues are still held in helical geometry by hydrogen bonds from residues 1 to 5 and 4 to 8.

Thus for this reaction, although two hydrogen bonds are broken, there is no gain in freedom for any of the residues. Therefore, the reaction is not favored. This leads to the fourth rule for the assignment of statistical weights:

4. For two h's separated by no more than two c's, the statistical weight is zero.

To assign a statistical weight of zero is of course equivalent to saying that such conformations are not permitted. This is excessively strict but the rule is a good approximation for our purpose.

With these four rules, we can assign the statistical weight for any conformation. The statistical weight for the conformation hhhccccchccchhcccc is $\sigma^3 s^6$, for example.

Once the rules for the statistical weights have been formulated, the helix-coil transition problem can be handled by mathematical techniques similar to the ones we have discussed for the binding of small molecules to sites with nearest-neighbor interactions. The function Z can be formulated and, in a way similar to Eq. (11.19), $(\partial \ln Z/\partial \ln s)_\sigma$ gives the average number of monomer units in the helical conformation per polymer molecule. We will not give the mathematical details here. Certain features of the coil-helix transition will be discussed in the following examples.

Example 11.3 Results of the classical treatment of Zimm et al. (1959) for the coil-helix transition of poly-γ-benzyl-L-glutamate are shown in Fig. 11.7. The curves shown were obtained by the statistical analysis that we outlined. The value of σ used to calculate the curves is 2×10^{-4}. As we have discussed, the parameter s represents the equilibrium constant for the change of a coil element at the end of a helical segment into a helical segment:

$$\ldots \text{hhhhhc} \ldots \overset{s}{\rightleftharpoons} \ldots \text{hhhhhh} \ldots$$

If the polypeptide molecule is very long, the free energy difference between a monomer unit in the coil form and in the helix form is zero at the temperature corresponding to the midpoint of the transitions (T_m, the melting temperature) shown in Fig. 11.7. Thus s is equal to 1 at T_m. At any other temperature, s can be calculated from

$$\ln \frac{s_2}{s_1} = -\frac{\Delta H^0}{R}\left(\frac{1}{T_2} - \frac{1}{T_1}\right)$$

where ΔH^0 is the enthalpy change for the reaction shown above.

When N is large, the transition is very sharp. The polypeptide changes from 80% in the coil form to 80% in the helix form in a temperature range of ~7°. In this temperature

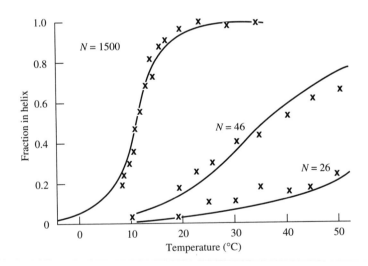

Fig. 11.7 Temperature dependence of the fraction of monomer units in the helix form for poly-γ-benzyl-L-glutamate (in a 7:3 mixture of dichloracetic acid and 1,2-dichloroethane as the solvent) with polymer lengths of 26, 46, and 1500 monomer units. In this solvent the helix is favored at high temperatures. [Data from B. H. Zimm and J. K. Bragg, *J. Chem. Phys. 31*, 526 (1959); B. H. Zimm, P. Doty, and K. Iso, *Proc. Natl. Acad. Sci. USA 45*, 1601 (1959).]

range the change in s is rather small, from 0.97 to 1.03. The cooperative transition is the result of a small σ. If we consider the formation of a helix from a coil, a small σ means that to start a helical region in the middle of a coil region is difficult, but once a helical region is initiated, it is much easier to extend it. This can be seen from the statistical weights of the species for the following transition:

$$\ldots ccccccc \ldots \rightleftarrows \ldots ccccchcc \ldots \rightleftarrows \ldots ccccchhc \ldots$$

| Statistical weight: | 1 | σs | σs^2 |

For the first reaction, the equilibrium constant is $\sigma s / 1$ or σs. For the second reaction, the equilibrium constant is $\sigma s^2 / \sigma s$ or s, which is much greater than σs.

Conversely, if we consider the formation of a coil from a helix, to start a coil region in the middle of a helix is difficult, as follows:

$$\ldots ccchhhhhh \ldots \rightleftarrows \ldots ccchhccchh \ldots \rightleftarrows \ldots ccchhcccch \ldots$$

| Statistical weight: | σs^7 | $\sigma^2 s^4$ | $\sigma^2 s^3$ |

The equilibrium constant for the first reaction is $\sigma^2 s^4 / \sigma s^7$ or σ / s^3. The equilibrium constant for the second reaction is $\sigma^2 s^3 / \sigma^2 s^4$ or $1/s$. Since in the region of interest, $s \approx 1$ and $\sigma \ll 1$, to initiate a coil region in the middle of a helix is more difficult. A consequence of this is that, for short polypeptides, coil regions are expected to be at the ends of the molecules. For long polypeptides there are so many sites in the middle of the molecules that, in the transition zone, coil regions are expected to be present in the middle of the molecules as well.

The sharpness of the transitions shown in Fig. 11.7 differs for different N, although all curves can be fitted by the same parameters. For the shorter polypeptides, the transition is broader. This is similar to the binding of small molecules to interacting sites as discussed previously.

Example 11.4   The unfolding of ribonuclease A by heating. As shown in Fig. 11.8, heating ribonuclease A (in 0.04 M glycine buffer, pH 3.15) causes the unfolding of the protein. The unfolding was monitored by the change in light absorption. The transition from the folded form to the unfolded form occurs in a narrow temperature range. The unfolding of a protein is more complex than the helix-coil transition of a simple polypeptide, however. The folded protein has structural features other than the α helix. In addition, factors such as disulfide bridges, ionic bonds, and interactions between the nonbonded groups can contribute to the stability of the folded form. However, because most of these factors contribute significantly only if the protein is in a correctly folded form, the unfolded to folded, or folded to unfolded, transition is highly cooperative. In fact, the all-or-none model has been used in such studies until very recently, when kinetic studies indicated that there are more than two states involved.

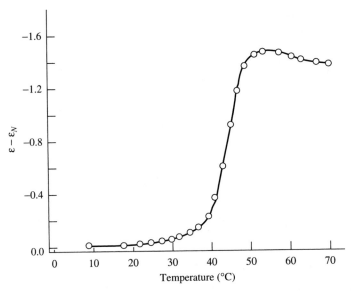

Fig. 11.8 Temperature dependence of $\varepsilon - \varepsilon_N$ of ribonuclease A. ε is the molar absorptivity of the protein at 287-nm wavelength and temperature T; ε_N is the same quantity when the protein is in its native (folded) form. [Data from J. Brandts and L. Hunt, *J. Am. Chem. Soc. 89*, 4826 (1967). Copyright by American Chemical Society.

Helix-Coil Transition in a Double-Stranded Nucleic Acid

In the preceding discussion on the helix-coil transition in an α-helix, we have seen that the difficulty in initiating the formation of a helical segment ($\sigma \ll 1$) is the essential reason for the cooperative transition. In a double-stranded nucleic acid, a somewhat similar situation exists. Double-stranded DNA is held together by the interactions between complementary bases in two single strands of DNA as illustrated in Fig. 3.7. Let us represent a base pair (in the hydrogen-bonded form) by the symbol 1, and a pair of bases in the nonbonded form by the symbol 0. Let s be the equilibrium constant for the formation of an additional bonded pair at the end of a helical segment:

$$11110000 \ldots \overset{s}{\rightleftharpoons} 11111000 \ldots$$

If we compare the reaction above with the reaction

$$11110000 \ldots \overset{\sigma s}{\rightleftharpoons} 11110100 \ldots$$

we note that there is an important difference. In the first reaction, the newly formed base pair is directly "stacked" on the base pair at the end of the original helix. For polynucleotides, either DNA or RNA, the stacking of base pairs is thermodynamically favored. In the second reaction, there is little stacking interaction between the newly formed pair and the last pair in the preceding helical segment. If the equilibrium constant is σs for the second reaction, we expect that $\sigma \ll 1$ if stacking interaction is thermodynamically favored. Thus, the helix-coil transition in a double-stranded nucleic acid is similar to that in a polypeptide.

There are several new features in the nucleic acid case, however. First, let us consider the two reactions

$$11110000 \ldots \rightleftharpoons 11110100 \ldots$$

and

$$11110000 \ldots \underbrace{0000}_{j \text{ zeros}} \ldots \rightleftharpoons 11110000 \ldots \underbrace{0010}_{j \text{ zeros}} \ldots$$

If j is very large, it means that the pair of bases which are to form the additional base pair are on the average far apart in the nonbonded form. From Eq. (11.46), the root-mean-square average distance between this pair of nonbonded bases is proportional to $\sqrt{j}$. It is therefore expected that the larger j is, the more difficult it is to bring the two together to form a base pair. Thus, unlike the case for a polypeptide, σ for a nucleic acid is not a constant but a function of j. The equilibrium constant for the first reaction is $\sigma_1 s$ and that for the second is $\sigma_j s$. This complicates the situation quite a bit. To compare this case with the binding of small molecules by a polymer, the dependence of σ on j is equivalent to the situation where we have to consider not only the nearest-neighbor interactions ($j = 0$) but also the next-nearest-neighbor interaction ($j = 1$), the next . . . , and so on.

Second, the formation of the very first base pair between two complementary chains means that the two chains, which are free to move about in solution independently of each other, must be brought together. After the formation of the first base pair one of the two chains can still be considered as free to be anywhere, but the second chain is constrained by the pairing to move with the first chain. Thus for the reaction

$$000 \ldots 000 \longrightarrow 0000100 \ldots 000$$
$$\text{all zeros} \qquad\qquad \text{all zeros except one}$$

the equilibrium constant is taken as κS, with κ expected to be much less than 1. The parameter κ is also expected to be dependent on total concentration. The lower the concentration, the harder it is to bring two chains together, and the smaller is κ.

Third, there are two major types of base pairs in a nucleic acid. For a DNA these are $A \cdot T$ pairs and $G \cdot C$ pairs; for an RNA they are $A \cdot U$ pairs and $G \cdot C$ pairs. Because the stabilities of the two kinds of base pairs are different, in general, we use the parameters s_A and s_G for these types rather than a single s. Strictly speaking, the stacking interaction between two adjacent pairs is dependent on what kinds of pairs they are, and therefore s_A and s_G are further dependent on at least the nearest-neighbor base sequence.

To include all these considerations in a theoretical analysis of the helix-coil transition is beyond the scope of this text. Nevertheless, from our analyses of the problem a number of features of the helix-coil transition in a nucleic acid can be understood:

1. When N is large, the transition from the completely helical form to the completely coiled form in a given solvent occurs within a narrow range of temperature. This cooperative process is frequently referred to as the *melting* of a nucleic acid. The temperature corresponding to the midpoint of the transition is designated as T_m, the *melting temperature*. The cooperativity of the transition is primarily a result of the stacking interactions between the adjacent base pairs, which make the initiation of a helical segment as well as the disruption of a base pair inside a helical segment difficult.

2. The enthalpy change for the reaction

$$11110000 \ldots \underset{}{\overset{s}{\rightleftarrows}} 11111000 \ldots$$

is negative. In most of the solvents $s_G > s_A$; thus T_m for a nucleic acid rich in G + C is higher than that for a nucleic acid rich in A + T (or A + U). If the base composition of a nucleic acid is intramolecularly heterogeneous, that is, some segments of each molecule are richer in A + T (or A + U) than the other portions, the melting profile is broadened, with the melting of the regions richer in A + T (or A + U) preceding the melting of the (G + C)-rich regions. Because of the cooperative nature of the transition, segments heterogeneous in base composition must be sufficiently long to show independent melting profiles.

3. For a large molecule, the parameter κ contributes negligibly to $\ln Z$. Because κ is the only concentration-dependent term in Z, T_m for a large molecule is expected to be independent of the concentration. For short helixes this is not true. T_m is lower at lower concentration. Similarly, at the same total concentration of nucleotides, T_m for a high-molecular-weight nucleic acid is higher than that of a low-molecular-weight nucleic acid of the same base composition. This dependence is evident only in the molecular-weight range where κ contributes significantly to $\ln Z$.

4. As discussed for the helix-coil transition in a polypeptide, if N is small, the coiled regions should be at the ends of the molecules. This is frequently referred to as "melting from the ends."

Example 11.5 The transition temperatures for the helix-coil transitions of DNAs correlate well with the base composition of the DNA. Plots of a large number of such measurements for two sets of experimental conditions are shown in Fig. 11.9. Analogously, for a particular DNA molecule the range of temperatures (transition width) over which melting occurs is related to the uniformity of base composition along major segments of the chain. Synthetic DNAs with regular repeating sequences will have sharp melting transitions.

Example 11.6 The DNA of coliphage 186 has a molecular weight of 20×10^6. The base composition of the DNA is not homogeneous along the molecule, and segments rich in AT pairs can be denatured before the denaturation of segments rich in G + C pairs. The consequences of such inhomogeneous denaturation are seen in Fig. 11.10. One can see short regions where the double-stranded DNA has opened into two single strands.

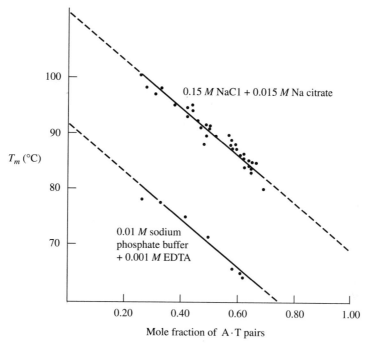

Fig. 11.9 Melting temperatures of double-stranded DNAs are plotted against the mole fraction of A·T pairs in the DNAs. In a given ionic medium, T_m decreases with increasing A·T content, indicating that $s_A < s_G$. Note also that for a given base composition, T_m is lower in a medium with a lower Na^+ concentration. This is because the repulsion between the negatively charged phosphate groups on the two complementary strands is decreased by the positively charged counterions (Na^+ in this case). Increasing the Na^+ concentration increases the stability of the double helix and, therefore, increases its T_m. [Data from J. Marmur and P. Doty, *J. Mol. Biol. 5,* 109 (1962).]

Example 11.7 The effect on T_m of shearing T2 DNA to reduce its length is seen in Fig. 11.12. Shorter fragments have lower T_m values and broader melting transitions.

Example 11.8 The difference in T_m for short helices of identical base composition but different sequence shows that s is dependent not only on whether a base pair is A·U or G·C but also on the *sequence* of the base pairs (Fig. 11.11). The nearest-neighbor approximation to the sequence dependence of formation of DNA double strands was discussed in Chapter 3. For large double-stranded DNAs and RNAs, the parameter s_A is therefore the *average s* of an A·T pair. Similarly, s_G is the *average s* of a G·C pair.

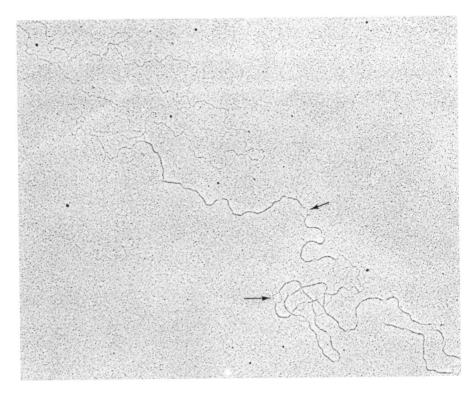

Fig. 11.10 Electron micrograph showing DNA that was partially denatured by exposure to a high pH (11.20) for 10 min, then processed for viewing. Denatured regions (such as marked by arrows) are shown as thinner and kinkier lines in the micrograph. (Courtesy of R. Inman.)

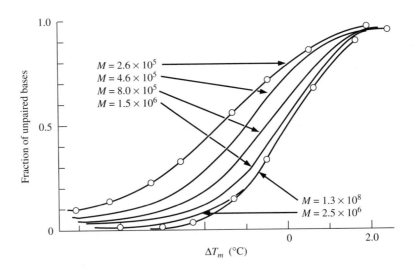

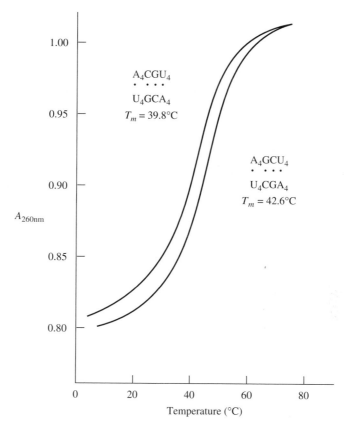

Fig. 11.12 Helix-coil transition curves for two short double-stranded RNA helices. A_{260nm} is the absorbance at a wavelength of 260 nm, which was measured to monitor the helix-coil transition. The two helices have identical base composition and the same concentration in media of identical salt concentrations. [Data from Professor Olke Uhlenbeck, U. of Colorado, Boulder.]

STATISTICAL THERMODYNAMICS

In previous sections we have illustrated the usefulness of statistical concepts in treating problems of biological interest. We stated at the beginning of this chapter that statistical methods also provide a bridge linking the thermodynamic properties of a macroscopic system and the molecular parameters of its microscopic constituents. In the sections below we will discuss certain aspects of statistical thermodynamics, which treats an equilibrium system by statistical methods.

Fig. 11.11 Fraction of unpaired bases plotted against ΔT_m, the difference between the melting temperature of a DNA of molecular weight M and that of a very large DNA. For $M > 2.5 \times 10^6$, ΔT_m is very small. For DNAs with molecular weights less than 2.5×10^6, ΔT_m is clearly dependent on the size of the DNA. [Data from D. M. Crothers, N. R. Kallenbach, and B. H. Zimm, *J. Mol. Biol. 11*, 802 (1965).]

To illustrate the basic concepts in simple terms, our *quantitative* discussions will be limited to a system of noninteracting particles (that is, an ideal gas). The purpose of these discussions is to provide a molecular basis for the thermodynamic laws. We will show how the macroscopic thermodynamic properties: energy, E, reversible heat, q_{rev}, reversible work, w_{rev}, and entropy, S, are related to the energy levels of the molecules. In the section on the statistical mechanical concept of entropy, some qualitative examples are provided for more complex systems.

Statistical Mechanical Internal Energy

In Chapter 2 we discussed the internal energy for a macroscopic system and the relation between the energy, E, the heat absorbed by the system, q, and the work done on the system, w. A *macroscopic* system, such as 1 L of air, 1 drop of water, or 1 carat of diamond, contains many *microscopic* particles. One liter of air at room temperature and atmospheric pressure contains approximately 5×10^{21} molecules of O_2, approximately four times as many molecules of N_2, and many other molecules. A tiny drop of water 1 μm (10^{-4} cm) in diameter contains about 2×10^{10} water molecules, and 1 carat (200 mg) of diamond contains about 10^{22} carbon atoms. In Chapter 9 we discussed the *energy levels* for a single molecule. For very simple systems (such as ideal gases) the energy levels can be calculated from quantum mechanics. For more complex systems, the energy levels can be obtained in principle, although the actual computation can be very difficult and in many cases is not yet possible.

For an ideal gas the energy levels of a molecule can be assigned with good approximation to translational, rotational, vibrational, and electronic motion. The spacing between energy levels is smallest for translational energy levels and largest for electronic levels. The values of the energy levels can be calculated from quantum mechanics. We derived in Chapter 9 that for a particle of mass m in a box of dimensions a, b, and c, its energy levels are

$$E(n_x, n_y, n_z) = \frac{h^2}{8m}\left(\frac{n_x^2}{a^2} + \frac{n_y^2}{b^2} + \frac{n_z^2}{c^2}\right) \tag{9.24}$$

where n_x, n_y, and n_z are integers. This equation gives the translational energy levels for a molecule contained in the box. For a monatomic ideal gas this equation characterizes the only significant energy levels (except at high temperatures) that contribute to its thermodynamic properties. At very high temperatures the electronic energy levels also become significant. For diatomic and polyatomic molecules in an ideal gas, rotational energy levels and vibrational energy levels are also important for thermodynamic properties.

At any particular time, each monatomic gas molecule will have an energy E_i given by Eq. (9.24). We can describe the energy distribution of the monatomic gas by specifying the number of molecules $N_1, N_2, \ldots, N_i, \ldots$ in each possible energy level $E_1, E_2, \ldots, E_i, \ldots$ Because the molecules of an ideal gas do not interact with one another, the total energy E of the system is simply the sum of the energies of the molecules:

$$E = N_1E_1 + N_2E_2 + N_3E_3 + \cdots + N_iE_i + \cdots$$
$$= \sum_i N_iE_i \tag{11.47}$$

The set of numbers $N_1, N_2, \ldots, N_i \ldots$ are called *occupation numbers*. Equation (11.47) tells us that the macroscopic energy of an ideal gas can be calculated from the energy levels of each molecule and the number of molecules in each energy level.

The energy of a system with interacting molecules (such as a real gas, a liquid, or solid) can still be written in the same form as Eq. (11.47). However, now the energies, E_i, are the energy levels of the entire system, and the N_i are replaced by the probabilities of finding the system in each energy level. Although it is impossible to do so, in principle the energy levels can be obtained by solving Schrödinger's equation for the whole system. The energy levels (E_i) obtained will depend on the contents of the system and the volume of the system, but not on the temperature. The temperature controls the N_i. The N_i characterize the distribution of the molecules among the energy levels for an ideal gas. For a system with interacting molecules, the N_i characterize the distribution of the whole system among its energy levels. To summarize our conclusions: Changing the temperature of a system changes the distribution of the system among its energy levels; it does not change the energy levels themselves. Changing the volume of a system changes its energy levels.

Work

Consider a gas in a container with a movable wall (a piston). Let the *external* force in the direction of movement of the piston be F_x. If we move the piston by a distance dx in the direction of the external force, the work done on the gas by the surroundings is

$$dw = F_x \, dx \tag{11.48}$$

The total force against the piston exerted by all molecules is

$$\sum_i N_i F_{ix} \tag{11.49}$$

where F_{ix} is the force in this direction exerted by a molecule in the ith energy level.

If the change dx is carried out *reversibly,* the external force is balanced by the forces exerted by the molecules:

$$F_x = \sum_i N_i F_{ix} \tag{11.50}$$

The work done by a reversible process

$$dw_{\text{rev}} = F_x \, dx = \sum_i N_i F_{ix} \, dx \tag{11.51}$$

For any molecule of the ideal gas, the force F_{ix} is related to the change in its energy, E_i, due to the change of the dimension of the container by dl. This is expressed by the equation

$$F_{ix} = -\frac{dE_i}{dl} \tag{11.52}$$

Substituting Eq. (11.52) into Eq. (11.51), we obtain

$$dw_{\text{rev}} = \sum_i N_i \cdot \left(-\frac{dE_i}{dl}\right) dx$$

but because $dx = -dl$ (a positive dx decreases the size of the container)

$$dw_{\text{rev}} = \sum_i N_i \, dE_i \tag{11.53}$$

Equation (11.53) states that the work done on the system for a reversible process is related to the change in the energy levels due to the change in the dimension of the container. This is true for any system. For an ideal gas changing the dimensions of the container affects only the translational energy.

As an example, the translational energy for a molecule in a one-dimensional box of length a is

$$E_i = \frac{h^2 n_x^2}{8ma^2} \tag{9.21}$$

The change in energy with change in dimension is

$$\frac{dE_i}{da} = \frac{-h^2 n_x^2}{4ma^3}$$

The energy levels are raised with decreasing dimension of the box. The force exerted by each molecule in energy level i is

$$F_{ix} = \frac{-dE_i}{da} = \frac{h^2 n_x^2}{4ma^3} \tag{11.54}$$

We can obtain the same expression for the force by a different method. As derived first in Chapter 6, for a molecule with average velocity u_x, the momentum change per collision with a wall is $2mu_x$. The number of collisions between the particle and one of the walls is $u_x/2a$ per unit time. From Newton's law the total momentum change with the wall per unit time is equal to the force exerted by the particle on the wall.

$$F_x = 2mu_x \cdot \frac{u_x}{2a} = \frac{mu_x^2}{a} \tag{11.55}$$

But the translational kinetic energy of the molecule is

$$\tfrac{1}{2} mu_x^2 = E_i = \frac{h^2 n_x^2}{8ma^2}$$

and thus

$$\frac{mu_x^2}{a} = \frac{h^2 n_x^2}{4ma^3}$$

Substituting this expression into Eq. (11.55) gives Eq. (11.54).

Heat

Starting with Eq. (11.47), we can write the differential expression for the change of E with change of E_i and N_i.

$$dE = \sum_i N_i \, dE_i + \sum_i E_i \, dN_i \tag{11.56}$$

Combining Eqs. (11.53) and (11.56), we obtain

$$dE = dw_{\text{rev}} + \sum_i E_i \, dN_i \tag{11.57}$$

but from the first law of thermodynamics, Eq. (2.19),

$$dE - dw_{\text{rev}} = dq_{\text{rev}}$$

Therefore,

$$dq_{\text{rev}} = \sum_i E_i \, dN_i \tag{11.58}$$

Equation (11.58) states that the heat absorbed by the system undergoing a reversible change is related to changes in the number of molecules in the various energy levels. This immediately suggests that dq_{rev} is the part of the total energy that is related to the *distribution* of the molecules among the various energy levels. As heat is absorbed the number of molecules in the higher energy levels increases relative to the number in the lower. At zero Kelvin, all the molecules are in the lowest energy level. As the temperature is raised, the population of the higher energy levels increases as heat is absorbed.

Most Probable (Boltzmann) Distribution

We have shown how the thermodynamic energy, the reversible heat, and the reversible work are related to the energy levels, E_i, and the number of molecules, N_i, in each energy level. In principle we can calculate the energy levels from quantum mechanics, but to obtain a thermodynamic property we must also be able to calculate the N_i.

Equation (11.47) was written for a system of ideal gas molecules, where it is clear that because the molecules do not interact, the energy of the system is the sum of the energies of the individual molecules. The same kind of equation applies to a system containing interacting molecules, but we have to be more careful how we define E_i. It is useful to think of a large number of identical systems in thermal equilibrium; all have the same T. This is called an *ensemble* of systems. The systems of the ensemble all have the same composition, the same volume, and the same temperature. Although all systems have the same average energy (equal to the total energy of the ensemble divided by the total number of systems), each system may have a different energy level, E_i. Now E_i is the energy of a system of interacting molecules. The energy of the ensemble is given by Eq. (11.47).

$$E = \sum_i N_i E_i \qquad (11.47)$$

but E_i is the energy of each system in the ensemble and N_i is the number of systems in energy level E_i. The average energy $\langle E \rangle$ (which is the thermodynamic energy) is

$$\langle E \rangle = \frac{\sum\limits_i N_i E_i}{N}$$
$$N = \sum_i N_i \qquad (11.59)$$

There are many combinations of N_i which will give the same $\langle E \rangle$, but we will find that the most probable distribution of N_i is so much more probable than all the others that we need to consider only the most probable distribution. For example, in an ensemble of ten systems, it is possible that nine systems have zero energy (the lowest energy level) and one system has 10 times the average energy. This corresponds to $N_i = 1$ for $E_i = 10\langle E \rangle$ and $N_i = 9$ for $E_i = 0$. This is very improbable. A much more probable distribution of energy is that a wide range of E_i values is represented among the systems of the ensemble. As the number of systems increases in the ensemble, all improbable distributions become negligible. We chose the ensemble as a thought experiment to allow us to consider the energies of interacting molecules in a system. Therefore, in our thought experiment we can have the number of systems become infinite. This means that the most probable distribution of energies among the systems is the only one we need to consider.

Let us consider a simple ensemble where it is easy to calculate all the possible distributions of N_i. We need to find all the distributions which give an arbitrary value of energy, ε_0. To simplify the example we choose a system in which the energy spacing is constant and equal to ε_0. Figure 11.13 shows the possible distributions of systems in an ensemble among energy levels. The first point to notice is that some distributions are more probable than others. For two systems in the ensemble the distribution with one system in the ground level ($E = 0$) and one system in the level $E = 2\varepsilon_0$ is twice as probable as the distribution with both systems in the level $E = \varepsilon_0$. There are ways of assigning two systems to two different levels; there is only one way of assigning two systems to one level. Similarly, for three systems in the ensemble the distribution with one system in each of the lowest three energy levels is the most probable; there are six ways of forming it. For any distribution the number of ways of assigning systems to energy levels is

$$t = \frac{N!}{N_1! \, N_2! \, N_3! \, \ldots} = \frac{N!}{\prod\limits_i N_i!} \tag{11.60}$$

where Π means the product over all i. This is the number of ways that N distinguishable objects can be arranged into boxes with N_1 in one box, N_2 in another, and so on. As seen in Fig. 11.13 the possible values of N_i depend on the value of the average energy chosen, and clearly the sum of N_i must equal N.

Exercise Find all the possible distributions of N_i for four systems in an ensemble ($N = 4$) for the energy-level pattern in Fig. 11.13 and $\langle E \rangle = \varepsilon_0$. There are five distributions. Use Eq. (11.60) to calculate the most probable distribution. There are two equally most probable distributions.

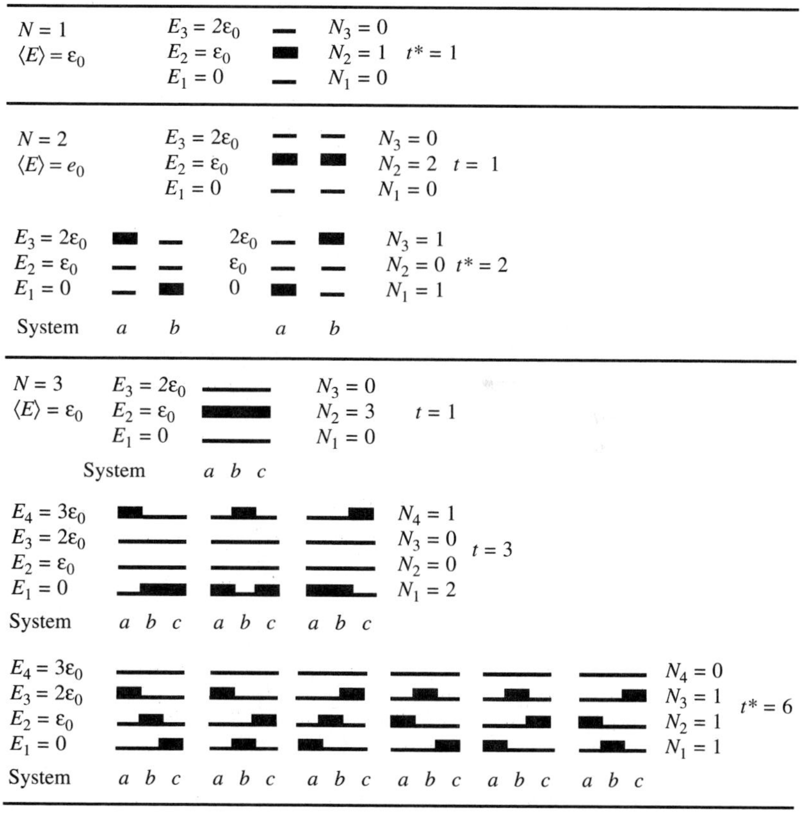

Fig. 11.13 Possible distributions of systems in an ensemble among energy levels. For simplicity the energy spacing for each system is made constant and equal to ε_0. All possible distributions are shown for three different numbers ($N = 1, 2, 3$) of systems in the ensemble. The average energy of all the systems is chosen to be ε_0. For each distribution the number of ways (t) of forming the distribution is shown. The most probable distribution is labeled t^*. The dark boxes depict which systems (a, b, c, etc.) are in which energy level (0, ε_0, $2\varepsilon_0$, $3\varepsilon_0$, etc.) Note that for each distribution $\sum_i N_i = N$, the total number of systems in the ensemble, and $\langle E \rangle = \varepsilon_0$.

The most probable distribution is the one with the largest number of ways of forming it. It is designated t^* in Fig. 11.13. As N gets large it becomes tedious to find all possible distributions and then to calculate all the t's to find the maximum. It is straightforward to find the maximum value of t in Eq. (11.60) by setting its derivative with respect to all N_i's equal to zero. We actually find the maximum in ln t. This tells us that the maximum value of t occurs for the distributions in which all $N_i = 1$; thus $t = $ N!. However, this is not the distribution we need, because it does not fit the necessary constraints that the average energy $\langle E \rangle$, and the total number of systems, N, are kept constant [Eq. (11.59)]. To find the maximum in t when these constraints are added requires more complicated mathematics. The result is that the most probable distribution is

$$\frac{N_i}{N} = \frac{e^{-\beta E_i}}{\sum\limits_i e^{-\beta E_i}}$$

with β a constant that depends on the value of the average energy of the system, $\langle E \rangle$. Boltzmann derived this equation and showed that β was proportional to the reciprocal of the absolute temperature. The Boltzmann distribution (or most probable distribution) is written

$$\frac{N_i}{N} = \frac{e^{-E_i/kT}}{\sum\limits_i e^{-E_i/kT}} \qquad (11.61)$$

where k is the Boltzmann constant. This equation is among the most important in statistical thermodynamics; it tells us the probability (N_i/N) of finding a system with a particular value of energy, E_i. It was obtained before quantum mechanics, but it applies to quantum mechanical systems with minor modification. We can use Eq. (11.61) as it stands, if we let the sum over i refer to *quantum states* instead of quantum energy levels. An equivalent way of writing it is by explicitly including the *degeneracy* of each energy level. *The degeneracy, g_i, is the number of quantum states having the same energy.* The Boltzmann distribution is now

$$\frac{N_i}{N} = \frac{g_i e^{-E_i/kT}}{\sum\limits_i g_i e^{-E_i/kT}} \qquad (11.62)$$

where the sum is over energy levels. This equation is also written as

$$\frac{N_i}{N} = \frac{g_i e^{-E_i/kT}}{Z} \qquad (11.62)$$

where Z is the *partition function*.

$$Z = \sum\limits_i g_i e^{-E_i/kT} \qquad (11.63)$$

When we apply Eqs. (11.61) and (11.62) to ideal gas molecules the E_i are the energies of individual gas molecules, and N_i is the number of molecules with each energy, E_i. Equations (11.61) and (11.62) also apply to solids and solutions in which there is strong interaction among molecules. Now E_i is the energy of the whole system, and (N_i/N) is the probability that the system will have that energy.

Equations (11.61) and (11.62) can be used in a very wide range of applications. From the energy-level distribution (obtained quantum mechanically) and the temperature we can calculate the average energy (the thermodynamic energy) from Eq. (11.59):

$$\langle E \rangle = \frac{\sum\limits_i N_i E_i}{N} = \frac{\sum\limits_i g_i E_i e^{-E_i/kT}}{\sum\limits_i g_i e^{-E_i/kT}} \tag{11.64}$$

The ratio of the number of molecules in two energy levels is

$$\frac{N_j}{N_i} = \frac{g_j e^{-E_j/kT}}{g_i e^{-E_i/kT}} = \frac{g_j e^{-(E_j-E_i)/kT}}{g_i} \tag{11.65}$$

We see that at zero absolute temperature only the ground level is populated; all molecules (or all systems) are in the lowest energy level. As the temperature is raised, higher energy levels are populated, and as the temperature approaches infinity, the distribution becomes independent of the energies. Equation (11.65) can be used to measure temperature. The ratio of the number of particles in two energy levels can be measured from the intensities and frequencies of emitted light. The temperatures of stars have been determined with this method.

The significance of the magnitude of kT is apparent from Eqs. (11.61) through (11.65). If the energy spacing $(E_j - E_i)$ *is small compared to kT*, many energy levels will be populated. This applies to translational energy and rotational energy levels at room temperature. If the energy-level spacing is large compared to kT, only the lowest energy level is significantly populated. This applies to electronic energy levels. Vibrational energy levels constitute an intermediate case, where several of the lowest energy levels are significantly populated at room temperature. At room temperature $(T = 300 \text{ K})$ kT is 4.14×10^{-21} J. It is more useful to remember the value per mole; $RT \cong 2.5$ kJ mol$^{-1} \cong$ 0.6 kcal mol^{-1}.

Quantum Mechanical Distributions

The discussion of the Boltzmann distribution in the preceding section used the formula [Eq. (11.60)] for distributing distinguishable objects among energy levels. However, for quantum systems (all real systems) many objects are indistinguishable and there are limitations on how many objects can be placed in each energy level.

For molecules, atoms, or subatomic particles, nature has set certain rules governing their distributions into the various quantum states. Particles with half-integral spin, such as electrons, protons, and ^{3}He nuclei, are subject to the Pauli exclusion principle; that is, no two particles can be in the same quantum state. Fermi-Dirac statistics therefore apply to these particles. Particles with integral spin, such as photons, deuterons, and ^{4}He nuclei, are not subject to the Pauli exclusion principle. Therefore, Bose-Einstein statistics represent their distributions. Boltzmann statistics was developed before the quantum mechanical rules were realized. No natural particles obey Boltzmann statistics strictly. However, except at temperatures close to absolute zero and a few other special cases, all of the distributions become indistinguishable. This means that the most probable Boltzmann distribution presented earlier is nearly always valid.

Statistical Mechanical Entropy

We have used the qualitative concept that entropy is a measure of the disorder of the system (Chapter 3). It is then reasonable that entropy is related to t, the number of ways of distributing the molecules in a system among their energy levels (or the number of ways of distributing the systems in an ensemble among their energy levels). Boltzmann showed that the entropy is related to the most probable distribution, t^*.

$$S = k \ln t^* \tag{11.66}$$

Equation (11.66) is a very important result; it provides a molecular interpretation of entropy. Let us see how this definition is consistent with the laws of thermodynamics.

As we have discussed, at ordinary temperatures t^* is very large. When the absolute temperature T approaches zero, however, the energy approaches a minimum, and all particles will be in the lowest energy levels available to them. Thus, t^* is either unity or a very small number compared with the values of t^* at ordinary temperatures. In other words, as T approaches zero, $\ln t^*$ is either rigorously zero or vanishingly small compared with $\ln t^*$ at ordinary temperatures. From the third law of thermodynamics, S is also zero as T approaches zero for a perfectly ordered crystalline substance. From the point of view of statistical mechanics, the third law of thermodynamics is a natural consequence of the occupation of the lowest available energy levels by the particles as T approaches zero.

One statement of the second law of thermodynamics is that for an isolated system, the equilibrium state is the one for which the entropy is a maximum. From the statistical mechanical point of view, the equilibrium state of an isolated system is one that represents the most probable distribution and has the maximum randomness.

Examples of Entropy and Probability

The relation

$$S = k \ln t^* \tag{11.66}$$

is not only important in providing a microscopic interpretation of entropy, but it also contains useful qualitative insights for many problems. In this section we will consider a number of quantitative examples.

Example 11.9 Consider two chambers, each of volume V, connected through a stopcock as shown in Fig. 11.14. Initially, the left chamber contains n moles of a gas at pressure P_0 and temperature T, the stopcock is closed, and the right chamber has been evacuated. The stopcock is opened, and the gas expands adiabatically into a vacuum. You may readily show from the discussions in Chapters 2 and 3 that for this process

$$q = 0 \quad \text{and} \quad w = 0$$

and therefore $\Delta E = 0$, $\Delta T = 0$, and $P_{final} = P_0/2$. To calculate ΔS from

$$\Delta S = \frac{q_{rev}}{T}$$

we must carry out a reversible process which arrives at the same final conditions: temperature T and pressure $P_0/2$. This can be done by an isothermal, reversible expansion:

$$\Delta E = 0$$

$$q_{rev} = w_{rev} = nRT \ln \frac{2V}{V} = nRT \ln 2$$

and

$$\Delta S = nR \ln 2$$

We will now see that Eq. (11.66) gives the same result. For the process under consideration,

$$\Delta S = S_2 - S_1 = k \ln t_2^* - k \ln t_1^*$$

$$= k \ln \frac{t_2^*}{t_1^*}$$

where the subscripts 1 and 2 refer to the initial and final states, respectively.

The ratio of the number of ways that the molecules can be arranged, t_2^*/t_1^*, is equivalent to the ratio of the probabilities. Because there are more ways to distribute the gas molecules in a volume $2V$ than in a volume V, the final state is more probable than the initial state. The ratio can be evaluated as follows. Imagine that we place the molecules one by one randomly into either of the two sides. The probability that the first molecule will be in the left side is $\frac{1}{2}$. The probability that both the first and the second molecules will be in the left side is $(\frac{1}{2})(\frac{1}{2}) = (\frac{1}{2})^2$. The probability that all nN_0 molecules are in the left side is $(\frac{1}{2})^{nN_0}$. This corresponds to t_1^*. The probability that all nN_0 molecules are in either side is obviously equal to 1; this corresponds to t_2^*. Thus $t_2^*/t_1^* = 2^{nN_0}$. Therefore,

$$\Delta S = k \ln 2^{nN_0} = nN_0 k \ln 2 = nR \ln 2$$

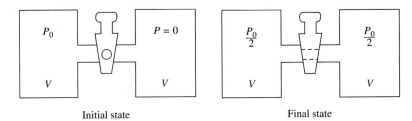

Fig. 11.14   Gas, initially all contained in a volume V at pressure P_0, expands adiabatically to volume $2V$ and pressure $P_0/2$ spontaneously, with a consequent increase in entropy.

The problem can also be looked at in a different way. Imagine that the volume V is divided into y boxes of equal volume. Initially, the nN_0 molecules are to be distributed among the y boxes. Because the molecules of an ideal gas occupy an insignificant volume themselves, we assume that there is no limit to how many molecules can occupy each box. The number of ways of placing one molecule in y boxes is y. The number of ways t_1^* of placing nN_0 molecules in y boxes is $t_1^* = y^{nN_0}$. For the final state there are $2y$ boxes; therefore, $t_2^* = (2y)^{nN_0}$. It follows then that

$$\Delta S = k \ln \frac{t_2^*}{t_1^*} = k \ln 2^{nN_0} = nR \ln 2$$

Example 11.10   Suppose that we have n_D moles of an ideal gas D and n_E moles of an ideal gas E, at the same temperature and pressure and separated initially by a partition, as shown in Fig. 11.15. If the barrier is withdrawn, mixing of the two gases occurs spontaneously. Let the volumes occupied initially by D and E be V_D and V_E, respectively. Because the gases are at the same temperature and pressure,

$$\frac{V_D}{V_E} = \frac{n_D}{n_E}$$

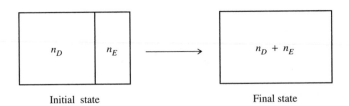

Fig. 11.15   Two different gases, initially separated by a barrier, will mix spontaneously, and the entropy of the system will increase when the barrier is removed.

To calculate ΔS for this process, we can divide V_D into y_D and V_E into y_E boxes of equal size. The number of ways of placing $n_D N_0$ molecules of D in y_D boxes is $y_D^{n_D N_0}$, and the number of ways of placing $n_E N_0$ molecules of E in y_E boxes is $y_E^{n_E N_0}$. Initially, with the barrier present, the total number of ways t_1^* is, therefore,

$$t_1^* = y_D^{n_D N_0} \cdot y_E^{n_E N_0}$$

If the barrier is removed, there are $y_D + y_E$ boxes for the gases, and

$$t_2^* = (y_D + y_E)^{n_D N_0} \cdot (y_D + y_E)^{n_E N_0}$$

Thus

$$\frac{t_1^*}{t_2^*} = \left(\frac{y_D}{y_D + y_E} \right)^{n_D N_0} \cdot \left(\frac{y_E}{y_D + y_E} \right)^{n_E N_0}$$

but

$$\frac{y_D}{y_E} = \frac{V_D}{V_E} = \frac{n_D}{n_E}$$

so

$$\frac{y_D}{y_D + y_E} = \frac{n_D}{n_D + n_E} = X_D$$

and

$$\frac{y_E}{y_D + y_E} = \frac{n_E}{n_D + n_E} = X_E$$

where X_D and X_E are the mole fractions of D and E, respectively, in the mixture. Thus

$$\frac{t_2^*}{t_1^*} = X_D^{-n_D N_0} \cdot X_E^{-n_E N_0}$$

and

$$\Delta S = k \ln \frac{t_2^*}{t_1^*}$$

$$= -(n_D k N_0 \ln X_D + n_E k N_0 \ln X_E)$$

$$= -(n_D R \ln X_D + n_E R \ln X_E)$$

This equation can be generalized to give the ideal entropy of mixing for any number of components

$$\Delta S = -R \sum_i n_i \ln X_i \qquad (11.67)$$

This is the same result presented in Chapter 3.

Example 11.11 Certain DNA molecules, such as those from phage λ, contain single-stranded ends of complementary base sequences which enable the molecules to circularize, as illustrated diagrammatically in Fig. 11.16.

Let the length of the double-stranded portion of the DNA be L. Because the ends are very short, L is essentially the length of the whole molecule. For a flexible molecule of linear DNA, we have seen that the mean-square end-to-end distance is

$$\langle h^2 \rangle = Ll \tag{11.44}$$

where l is the effective segment length.

Let us choose one end of the linear DNA molecule as the origin. The other end of the DNA will be found on the average in a sphere of volume $V = (4\pi/3)\langle h^2 \rangle^{3/2}$, whose center is the origin.

In the circular form, the two ends are constrained to be in a much smaller volume, v_i. Therefore, the circularization of each DNA molecule is associated with an unfavorable (negative) entropy term:

$$\Delta S_c = k \ln \frac{v_i}{V}$$

$$= k \ln v_i - k \ln \frac{4\pi}{3} - \frac{3}{2} k \ln \langle h^2 \rangle$$

$$= \left(k \ln v_i - k \ln \frac{4\pi}{3} - \frac{3k}{2} \ln l \right) - \frac{3k}{2} \ln L$$

For a group of DNA molecules with the same cohesive ends but of different lengths, the sum of the terms in the parentheses is a constant, and ΔS_c is more negative for larger L. Thus the longer such a DNA molecule is, the less favorable is ring formation. This has been observed experimentally, as shown in Fig. 11.17. The curve

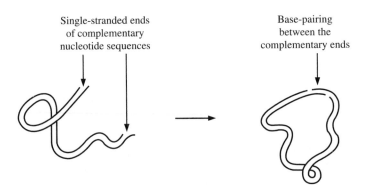

Single-stranded ends
of complementary
nucleotide sequences

Base-pairing
between the
complementary ends

Fig. 11.16 Cyclization of a DNA molecule with "cohesive" ends. Cyclization involves a loss of entropy proportional to the logarithm of the length of the DNA.

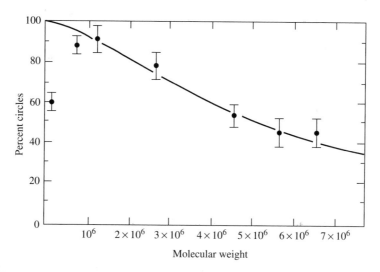

Fig. 11.17 Percentage of DNA molecules in the circular form as a function of the molecular weight of DNA. The DNA samples were obtained by treating an *E. coli* F-factor DNA with a restriction enzyme Eco RI which introduces two staggered breaks at sites with base sequence

-G ↓ AATTC-
-CTTAA ↑ G-

yielding DNA molecules with cohesive ends of sequences AATT-. [Data from J. E. Mertz and R. W. Davis, *Proc. Natl. Acad. Sci USA 69*, 3370 (1972).]

shown is the function predicted by the ΔS_c term. Note that at the low end of the molecular-weight scale, the discrepancy between the experimental results and the theoretical prediction is large. For such short DNA molecules, the equation $\langle h^2 \rangle = Ll$ (derived from the random-walk formula) is no longer applicable, because the stiffness of the double-stranded DNA makes it difficult for short segments to circularize.

Partition Function: Applications

In our treatment leading to Eq. (11.62), we have chosen an isolated system. The partition function

$$Z = \sum_i g_i e^{-E_i/kT} \tag{11.63}$$

sums over all the energy states that a molecule can occupy and is therefore called the *molecular partition function*.

For simple molecules the various energy levels can be obtained by quantum mechanical calculations or by spectroscopic measurements, and the partition functions can be obtained. If the partition function of a system is known, all of the thermodynamic properties of the system can be calculated. For a system of N noninteracting particles, the pressure P, internal energy E, and entropy S are related to the partition function Z by the relations

$$P = NkT \left(\frac{\partial \ln Z}{\partial V} \right)_T \qquad (11.68)$$

$$E = NkT^2 \left(\frac{\partial \ln Z}{\partial T} \right)_V \qquad (11.69)$$

$$S = kN \ln \frac{Z}{N} + \frac{E}{T} + kN \qquad (11.70)$$

Example 11.12   We will derive Eq. (11.69). The average energy of one molecule is given by Eqs. (11.63) and (11.64):

$$\langle E \rangle = \frac{\sum\limits_i g_i E_i e^{-E_i/kT}}{Z} \qquad (11.64)$$

$$Z = \sum\limits_i g_i e^{-E_i/kT} \qquad (11.63)$$

The derivative of the partition function Z with respect to T keeping the energy levels constant is specified by keeping V constant.

$$\left(\frac{\partial Z}{\partial T} \right)_V = \frac{1}{kT^2} \sum\limits_i g_i E_i e^{-E_i/kT}$$

We see that $\langle E \rangle$ can be written in terms of this derivative:

$$\langle E \rangle = \frac{kT^2}{Z} \left(\frac{\partial Z}{\partial T} \right)_V$$

But $dZ/Z = d \ln Z$, so

$$\langle E \rangle = kT^2 \left(\frac{\partial \ln Z}{\partial T} \right)_V$$

This is the average energy of one molecule; for N molecules we multiply by N to obtain Eq. (11.69). For derivations of Eqs. (11.68) and (11.70) a standard physical chemistry text should be consulted.

Another important property of the partition function can be seen better if we elect to sum over all quantum states rather than energy levels. That is, the g_i degenerate states are summed up individually:

$$Z = \sum\limits_{\substack{\text{all} \\ \text{quantum} \\ \text{states}}} e^{-E/kT} \qquad (11.71)$$

To a high degree of approximation the energy of a molecule in a particular state is the simple sum of various types of energy, such as translational energy, E_{tr}; rotational energy, E_{rot}; vibrational energy, E_{vib}; electronic energy, E_{el}; and so on. If

$$E = E_{tr} + E_{rot} + E_{vib} + E_{el} + \cdots$$

it follows immediately that

$$Z = \left(\sum e^{-E_{tr}/kT} \right) \left(\sum e^{-E_{rot}/kT} \right) \left(\sum e^{-E_{vib}/kT} \right) \left(\sum e^{-E_{el}/kT} \right)$$

$$= Z_{tr} Z_{rot} Z_{vib} Z_{el} \cdots$$

(11.72)

In other words, if the energy can be expressed as a sum of terms, the partition function can be partitioned into corresponding terms, the product of which gives the total partition function.

Systems other than an isolated system can also be treated by statistical mechanics. Because systems of chemical or biological interest are seldom isolated systems, it is more useful to obtain the partition functions for closed systems or open systems. It is beyond the scope of this book to treat such problems, however.

For a system consisting of complex molecules, it is not yet possible to obtain the partition function rigorously. Consider the example of a polypeptide molecule. Each amide residue in the chain is made of atoms, and each atom has its translational, etc., energy levels. Since the atoms interact, the energy levels of one are strongly affected by many others. It is obviously an impossible task to obtain the complete molecular partition function. On the other hand, if we are interested only in the helix-coil transition, there is no need to know the fine detail of the energy levels. All we really need are the *relative* contributions to the partition function of a residue in the helix and the coil states of the molecule. These relative contributions are what we have termed "statistical weights" in a number of discussions in the first part of this chapter. In fact, several of the functions that we denoted by Z are examples of partition functions. Therefore, in the first part of this chapter we have already given a number of examples involving the applications of statistical mechanics to complex systems of biological interest.

SUMMARY

Binding of Small Molecules by a Polymer

Polymer molecule with N identical and independent sites for the binding of A, a small molecule:

$$\frac{\nu}{[A]} = K(N - \nu)$$

(11.12)

or

$$\frac{f}{1 - f} = K[A]$$

(11.13)

v = number of bound A molecules per polymer molecule
f = fraction of sites occupied
N = number of sites per polymer molecule
K = intrinsic binding constant

Polymer with N identical sites in a linear array with nearest-neighbor interactions:

$$v = \left(\frac{\partial \ln Z}{\partial \ln S}\right)_\tau \qquad (11.19)$$

v = number of bound A molecules per polymer molecule
Z = sum of statistical weights of species
$S = K[A]$
τ = cooperativity parameter

Cooperative binding:

If $\tau > 1$, the binding of an A molecule to one site makes it easier to bind another A to an adjacent site. f versus A plot is sigmoidal in shape. If τ is very large, the binding approaches the all-or-none limit, and the predominant species have $f = 0$ and $f = 1$. In this limit,

$$\frac{d \log [f/(1 - f)]}{d \log [A]} = N \qquad (11.22)$$

Anticooperative binding:

If $\tau < 1$, the binding of an A molecule to one site makes it easier to bind another A to an adjacent site. In the limit $\tau = 0$, binding to one site *excludes* binding to adjacent sites.

Random Walk and Related Topics

$$\langle h^2 \rangle = Nl^2$$
$$= Ll \qquad (11.44)$$

$\langle h^2 \rangle$ = mean-square end-to-end distance
N = number of "segments" of a polymer molecule
l = length of a segment; l is a measure of the stiffness of the polymer molecule
L = contour length of a polymer

$$\langle R^2 \rangle = \frac{Nl^2}{6} \qquad \text{(open-ended random coil)} \qquad (11.46a)$$

$\langle R^2 \rangle$ = mean-square radius

$$\langle R^2 \rangle = \frac{Nl^2}{12} \qquad \text{(circular random coil)} \qquad (11.46b)$$

Helix-Coil Transitions

Simple polypeptides:
The α-helix is characterized by hydrogen bonds between CO groups and NH groups in the fourth residue ahead in a polypeptide chain. The *initiation* of a helical element in a coiled region is more difficult than the addition of a helical element to the end of a helix. For polypeptides of high molecular weight, helix-coil transitions can be very sharp. Statistical methods have been used successfully to treat the helix-coil transition.

Proteins:
Native proteins are in a highly ordered or folded structure. Unfolding of a protein involves the disruption of hydrogen bonds, disulfide bridges, ionic bonds, and interactions between the nonbonded groups. The transition from a folded to an unfolded form (and vice versa) is usually highly cooperative and approaches the all-or-none limit.

Double-stranded nucleic acids:
Owing to favorable interactions ("stacking interactions") between neighboring base pairs in a double helix, the formation of an additional base pair at the end of a helical region is favored compared with the formation of a base pair in the middle of a coiled region. This leads to a cooperative transition between the helical and coiled forms. The basic features of the transition, such as the sharpness of the transition, the temperature dependence, the effect of the molecular weight of the polynucleotide, and the effect of base composition, can be understood from statistical mechanical considerations of the problem.

Statistical Thermodynamics

The most probable distribution:
The properties of a system at equilibrium is represented, to a high degree of approximation, by the properties of the most probable distribution,

$$\frac{N_i}{N} = g_i e^{-E_i/kT}/Z \qquad (11.62)$$

$$Z = \sum_i g_i e^{-E_i/kT} = \text{partition function} \qquad (11.63)$$

$$\frac{N_i}{N_j} = \frac{g_i}{g_j} e^{-(E_i - E_j)/kT} \qquad (11.65)$$

$$N_i = \text{number of molecules in an energy level } E_i$$

N_i = number of molecules in an energy level E_i
N_j = number of molecules in an energy level E_j
N = total number of molecules
g_i, g_j = degeneracy, the number of states with energy E_i, E_j
k = Boltzmann constant
$\quad = R/N_0$ (R = gas constant and N_0 = Avogadro's number)
Z = molecular partition function, which sums up all the energy states of the molecule; for simple molecules Z can be calculated by quantum mechanics

Entropy:

$$S = k \ln t^* \tag{11.66}$$

t^* = maximum number of ways of distributing the molecules; the larger the t^*, the more probable the state is, and the larger is the entropy; Eq. (11.66) provides an interpretation of the second and third laws of thermodynamics

Molecular partition function and thermodynamic functions:
All thermodynamic functions, such as P, E, and S, can be obtained from the partition function Z. If the energy of a molecule can be expressed as a sum of terms (translational energy, rotational energy, vibrational energy, electronic energy, etc.), the partition function can be factored (partitioned) into corresponding terms, the product of which gives the total partition function.

MATHEMATICS NEEDED FOR CHAPTER 11

We use permutations and combinations in calculating distributions of systems among energy levels [Eq. (11.60)] and in random walks [Eq. (11.24b)]. The number of *permutations* of n objects is

$$P_n = n! = n \cdot (n - 1) \cdot (n - 2) \ldots \cdot 3 \cdot 2 \cdot 1$$

A simple example is the $3! = 6$ permutations of the three letters, a, b, c.
A combination of objects is a group of objects without respect to order. The three letters (a, b, c) form one combination. The number of *combinations* of n objects taken r at a time is

$$_nC_r = \frac{n!}{r!(n - r)!}$$

A simple example is the number of ways of arranging three pennies into groups of two heads and one tail:

$$_3C_2 = \frac{3!}{2! \, 1!} = 3$$

Stirling's approximation for factorials is useful for large N

$$\ln N! = N \ln N - N$$

If a group of particles can be arranged in t_1 ways and independently they can also be arranged in t_2 ways, the total number of arrangements is $t_1 \cdot t_2$. Similarly, if the probability of an event occurring is p_1 and if the probability of an independent event occurring is p_2, then the joint probability of the two independent events occurring is $p_1 \cdot p_2$. For n independent arrangements of n independent probabilities, the expressions are

$$t = \prod_{i=1}^{n} t_i = t_1 \cdot t_2 \cdot t_3 \ldots t_n$$

$$p = \prod_{i=1}^{n} p_i = p_1 \cdot p_2 \cdot p_3 \ldots p_n$$

REFERENCES

Textbooks on Statistical Mechanics

CHANDLER, D., 1987. *Introduction to Modern Statistical Mechanics,* Oxford University Press, New York.

DAVIDSON, N., 1962. *Statistical Mechanics,* McGraw-Hill, New York.

HILL, T. L., 1960. *An Introduction to Statistical Thermodynamics,* Addison-Wesley, Reading, Massachusetts.

MCQUARRIE, D. A., 1976. *Statistical Mechanics,* Harper & Row, New York.

NASH, L. K., 1970. *Introduction to Statistical Thermodynamics,* Addison-Wesley, Reading, Massachusetts.

For useful books that treat some of the material of this chapter in more detail, see

CANTOR, C. R., and P. R. SCHIMMEL, 1980. *Biophysical Chemistry,* Part III: *The Behavior of Biological Macromolecules,* W. H. Freeman, San Francisco.

The relation between entropy and probability is discussed entertainingly in

ATKINS, P. W., 1984. *The Second Law,* Scientific American Books, New York.

BENT, H. A., 1965. *The Second Law,* Oxford University Press, New York.

SUGGESTED READINGS

ACKERS, G. K., M. L. DOYLE, D. MYERS, and M. A. DAUGHERTY, 1992. Molecular Code for Cooperativity in Hemoglobin, *Science 255,* 54–63.

CREIGHTON, T. E., 1983. An Empirical Approach to Protein Conformation Stability and Flexibility, *Biopolymers 22,* 49–58.

CROTHERS, D. M., J. KRAK, J. D. KAHN, and S. D. LEVENE, 1992. DNA Bending, Flexibility, and Helical Repeat by Cyclization Kinetics, *Methods Enzymol. 212B,* 3–29.

LEVENE, S. D., and D. M. CROTHERS, 1986. Ring Closure Probabilities for DNA Fragments by Monte Carlo Simulation, *J. Mol. Biol. 189,* 61–72; Topological Distributions and the Torsional Rigidity of DNA: A Monte Carlo Study of DNA Circles, *J. Mol. Biol. 189,* 73–83.

McGhee, J. D., and P. H. von Hippel, 1974. Theoretical Aspects of DNA-Protein Interactions: Co-operative and Non-Cooperative Binding of Large Ligands to a One-Dimensional Homogenous Lattice, *J. Mol. Biol. 86*, 469–480.

Petsko, G. A., and D. Ringe, 1983. Fluctuations in Protein Structure from X-Ray Diffraction, *Annu. Rev. Biophys. Bioeng. 13*, 331–371.

Wang, J. C., 1982. DNA Topoisomerases, *Sci. Am. 247* (July), 94–109.

PROBLEMS

1. For a dicarboxylic acid HO_2C—R—CO_2H, where the two —CO_2H groups are far apart, using statistical methods show that the ratio of the acid dissociation constants K_1/K_2 is expected to be ~4. (The acid dissociation constants K_1 and K_2 are the equilibrium constants for the reactions

$$HO_2C—R—CO_2H \rightleftharpoons {}^-O_2C—R—CO_2H + H^+$$

and

$${}^-O_2C—R—CO_2H \rightleftharpoons {}^-O_2C—R—CO_2^- + H^+$$

respectively.)

2. A certain macromolecule P has four identical sites in a linear array for the binding of a small molecule A. Prepare a table listing all species with half of the sites occupied (species with two occupied sites and two unoccupied sites) and their statistical weights for the following cases:
   (a) The sites are independent.
   (b) Nearest-neighbor interactions are present.
   (c) Binding to one site excludes the binding to a site immediately adjacent to it.
   (d) Give also, for part (b), an expression for the concentration ratio $[PA_4]/[PA_2]$.

3. A DNA has short single-stranded ends which can join either intramolecularly to form a ring or intermolecularly to form aggregates. Discuss briefly and concisely under what condition you expect ring formation will predominate and under what condition you expect intermolecular aggregation will predominate.

4. An oligopeptide has seven amide linkages (or eight amino acid residues including the terminal carboxyl group). With the rules for the statistical weights that we have discussed for the formation of an α-helix, obtain the function Z, which is the sum of the statistical weights of all species. Also obtain an expression for ν, the average number of helical residues per molecule, as a function of σ and s.

5. The H^+ titration curve of 1,2,3,4-tetracarboxyl cyclobutane should be interesting.

We will try to estimate it by using a simple, nearest-neighbor interaction model with the binding constant for a single COO^-, $K = 10^5$, and $\tau = 10^{-2}$.

(a) There are 16 possible species involving 0, 1, 2, 3, and 4 H^+'s bound. Give the statistical weight for each in terms of τ and $S = K[H^+]$.

(b) Write an expression for Z, the sum over states, and use Z to obtain an expression for v, the number of H^+ bound per molecule.

(c) Calculate $f = v/4$ for pH 4, pH 5, and pH 6 and compare it with the titration curve (f versus pH) of a single COO^-.

6. Consider the coil-helix transition for a polypeptide containing 50 amides. At a certain temperature the equilibrium constant for adding an amide to a helical region is $s = 1$; the helix initiation parameter is $\sigma = 10^{-4}$. Statistical weights are relative to the species which is 100% coil.

(a) Write an expression in terms of σ and s for the statistical weight of the species which has all the possible amides in the α-helix conformation.

(b) Write an expression in terms of σ and s for the statistical weight of the species which has 20 amides in the α-helix conformation in three separate helical regions.

(c) Which of all the possible species is in the highest concentration? That is, which of all the possible species has the highest statistical weight?

(d) If s were 10, which of all possible species would have the highest concentration?

(e) Roughly sketch a plot of f = fraction of amides in the α-helix conformation versus temperature. The ΔH^0 for the reaction helix to coil is negative. Label which temperature region corresponds to the helix and which to the coil.

(f) Write an equation for f in terms of σ, s, and Z.

7. An RNA oligonucleotide has the sequence $A_6C_7U_6$. It can form a hairpin loop held together by a maximum of 6 A·U base pairs.

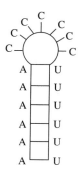

Assume that the helical region can melt only from either end. Use the notation s = equilibrium constant for adding a base pair to a helical region and $\sigma_j s$ = equilibrium constant for initiating the first base pair forming a loop. The subscript $j = 7, 9, 11, 13, 15, 17$ characterizes the number of unpaired bases in the loop.

(a) Calculate the statistical weight of each species which can be present.

(b) Calculate the mole fraction of each species which can be present.

(c) Assume that the molar absorptivity of each species depends only upon the number of base pairs formed: ε_0 = absorptivity per mole of mononucleotide for species with no base pairs; ε_1 = absorptivity of all species with 1 base pair; ε_2 = etc. Write an expression for the absorbance in a 1-cm cell of a solution of $A_6C_7U_6$ in terms of s, σ_j, ε_i, and c = concentration of $A_6C_7U_6$ in moles of nucleotides per liter.

8. The statistical effective segment length, l, of a DNA molecule is 100 nm. Calculate the mean-square end-to-end distance $\langle h^2 \rangle$ and the contour length L for a bacterial DNA with 10^7 base pairs. The distance between base pairs is 0.34 nm.

9. Calculate the degeneracy for each of the first five levels for a particle in a three-dimensional cubic box.

10. (a) How many three-letter words can be made from 26 letters? A word is any sequence of three letters from AAA to ZZZ.
    (b) How many different basketball teams of five players can be chosen from a group of 100 people?
    (c) How many different proteins containing 100 amino acids can be made from the 20 commonly occurring amino acids?

11. Consider two systems: one consists of an electron in a cubic box of 1 nm on a side; the other consists of a He atom in a cubic box of 1 cm on a side. For each system, calculate the ratio of the probabilities (N_2/N_1) of finding the particle in the first two energy levels at (a) 10 K; (b) 1000 K; (c) 10,000 K.

12. What is the change of entropy for the following reactions?
    (a) One hundred pennies are changed from all heads to all tails.
    (b) One hundred pennies are changed from all heads to fifty heads plus fifty tails.
    (c) One mol of heads is mixed with 1 mol of tails to give 2 mol of half heads and half tails.

13. A system of a particle in a magnetic field has a simple energy-level diagram that looks like this:

$$\text{———} \quad \varepsilon_2 = 0, \qquad g_2 = 1$$

$$\text{———} \quad \varepsilon_1 = -D, \qquad g_1 = 2$$

   (a) Write an expression for the partition function Z for the system. Your answer should be given in terms of D, T, and universal constants.
   (b) Write an expression for the average energy of the system. Your answer should be given in terms of D, T, and universal constants.
   (c) Calculate the average energy in the high-temperature limit.
   (d) Calculate the average energy in the low-temperature limit.
   (e) Write an expression for the ratio of the probabilities of finding the particle in the two states. Your answer should be given in terms of D, T, and universal constants.
   (f) Calculate the entropy of the particle in the high-temperature limit.

12

Macromolecular Structure and X-Ray Diffraction

CONCEPTS

X rays are electromagnetic radiation of short wavelength and therefore high energy. Visible light has a wavelength range from 4,000 Å to 8,000 Å; x rays used for diffraction studies have wavelengths in the 1-Å range. Like all electromagnetic radiation, x rays are absorbed, scattered, and diffracted by matter. The scattering and diffraction of x rays is caused by interaction with electrons. Electrons in molecules in gases, liquids, or disordered solids scatter x rays. However, electrons in ordered arrays of atoms in crystals scatter x rays only in particular directions; in other directions the scattering is negligible. *Diffraction* is the scattering of x rays in a few specific directions by crystals. The positions and intensities of the scattered beams produce a diffraction pattern. Diffraction occurs only when the wavelength of the radiation is of the same size as the periodicity in the crystal. Bond lengths in molecules, and x-ray wavelengths, are both in the 1-Å range, so diffraction can occur.

An electromagnetic wave of any wavelength is characterized by an amplitude and a phase. A lens can convert the amplitudes and phases of scattered or transmitted electromagnetic waves from an object into an image of the object. For visible light excellent lenses exist (including the lenses in our eyes) to produce images. However, at present no lenses exist which can produce images in the 1 Å x-ray wavelength region; we must use some other method. If we could measure the amplitudes and phases of the diffracted x-ray radiation from a crystal, we could calculate an image of the molecules. Detectors of x rays and other electromagnetic radiation measure only the intensity, which is the square of the amplitude. The phase information is lost in the measurement. To obtain information about molecular structure from a measured diffraction pattern, it is necessary to obtain the lost phase information by some means. This is the famous *phase problem* in x-ray diffraction. The type and size of the unit cell in a crystal can be determined from the measured diffraction pattern. The *unit cell* is the simplest repeating volume element which produces the crystal. To learn about the contents of the unit cell—the molecular structure—phase information is needed. One way to obtain this phase information is to introduce into the crystal a few heavy atoms which do not change the unit cell. This is the *isomorphous replacement method* for solving macromolecular structures.

There is a wavelength associated with any particle. Therefore, diffraction of electrons and neutrons can also be used to determine molecular structure. Electrons scatter from both electrons and nuclei; neutrons scatter only from nuclei. Magnetic lenses exist which allow focusing of electrons, thus providing images with a resolution in the 1-Å range.

APPLICATIONS

X-ray diffraction can provide molecular structures of molecules with a resolution of 0.001 Å for small molecules and 0.1 Å for macromolecules. The most difficult part of the project is to obtain crystals that have enough order to diffract the x rays. The structures of macromolecules determined by x-ray diffraction have led to understanding of the mechanism of their biological function. The binding of oxygen by hemoglobin, and the perturbation of binding caused by a change in an amino acid which occurs in some genetic diseases, was an early example. The structure of a flu virus revealed binding sites for a drug which could inhibit replication of the virus. Rational drug design involves determining the structure of a target enzyme, or enzyme-substrate complex, then designing a molecule that will bind to the active site and inhibit the enzyme. Target enzymes are those that exist in disease-causing bacteria or viruses, for example. Some cancers are caused by a single mutation in the base sequence of a DNA which leads to a single amino acid mutation in a protein. A normal protein is thus turned into a malignant protein. One example is a membrane protein that controls cell division. The structures of the normal and malignant proteins—products of the *ras* oncogene—have been determined. The one amino acid change which distinguishes between normal and malignant is involved in GTP hydrolysis.

The structure of the first nucleic acid determined—transfer RNA—was vital to understanding how protein synthesis occurs. It placed the anticodon, which reads the message of the messenger RNA, about 60 Å away from the amino acid coded for by the messenger RNA. There are 64 codons (4 bases read three at a time = 4^3 = 64); all but three code for amino acids. The structure of the complex between a transfer RNA and the enzyme which specifically attaches the correct amino acid to it helps explain how this important recognition step is accomplished.

VISIBLE IMAGES

If macroscopic objects can be seen by their scattering of light, it seems that we should also be able to see molecules, and the detailed atomic structures within the molecules, by light scattering. There are several different aspects about seeing an object, large or small: contrast, sensitivity, and resolution. *Contrast* is dependent on the amount of light scattered by the object and the amount of light scattered by other things around it. We have difficulty seeing a white fox in the snow, because the object and the background scatter about the same amount of light. *Sensitivity* is dependent on the absolute amount of light scattered by the object. A dark-adapted human eye can detect a pulse of visible light containing some 50 photons, and devices of even higher sensitivity have been constructed. *Resolution* is a measure of the spatial separation of two sources. For small enough separation the two no longer appear separated. The average human eye, for example, has an angular resolution of about 1 minute. Suppose that we write the number 11 on a wall with the 1's about 1 mm apart. At a distance of about 3.5 m from the wall it will be difficult for us to tell whether the number is an 11 or a fuzzy 1. At this distance the two 1's subtend an angle of 1×10^{-3} m/ 3.5 m $\cong 3 \times 10^{-4}$ rad $\cong$ 1 min at the eye.

Because of diffraction, even with the help of optically perfect lenses, the minimal resolvable resolution cannot be much less than the wavelength of light. Because the wavelength of light is several thousand angstroms and atomic separation in a molecule is of the order of several angstroms, to see molecular details, other types of radiation of much lower wavelengths must be employed.

X RAYS

In 1895 W. C. Roentgen accidentally discovered that when a beam of fast-moving electrons struck a solid surface, a new radiation was emitted. The new rays, which Roentgen called *x rays,* were found to cause certain minerals to fluoresce, to expose covered photographic plates, and to be transmitted through matter. One of the first experiments Roentgen did was to show that different kinds of matter were transparent to x rays to different degrees. The bones and flesh of the experimenter's hand could clearly be distinguished when the hand was placed in front of a screen made of a material that fluoresced when x-rayed. Three months after Roentgen's discovery, x rays were put to use in a surgical ward. Eighteen years after the discovery, the first crystal structure (that of NaCl) was solved by x-ray diffraction. Today x-ray diffraction is one of the most powerful techniques in studying biological structures.

Emission of X Rays

When a target is bombarded by electrons accelerated by an electric field, x rays are emitted. When electrons strike the target atoms, some are deflected by the field of the atoms with a loss of energy. This energy loss is accompanied by the emission of photons with energies corresponding to the energy differences of the incoming and the scattered electrons. Because the maximum energy of the photons emitted cannot exceed the energy of the incoming electrons, there is a sharp short-wavelength limit of the spectrum at the accelerating voltage. At longer wavelength there are spikes in the emission spectrum due to discrete energy levels of the orbital electrons of the target atoms. If the energy of an incoming electron is sufficiently high, it can eject an electron from an inner orbital. A photon of a characteristic wavelength is emitted when an electron in an outer orbital falls into the inner orbital to fill the vacancy left by the ejected electron. Thus x rays are photons (electromagnetic radiation) with wavelengths in the range from tenths of angstroms to a few angstroms. X-ray spectroscopists often characterize the x rays by their energy in kiloelectron volts (keV) instead of their wavelength. Applying the formula $\Delta E = hc/\lambda$, we find that

$$\text{energy (keV)} = \frac{12.39}{\lambda(\text{Å})}$$

Image Formation

Because the wavelengths of x rays are of the same order of magnitude as interatomic distances in molecules, we should be able to construct an x-ray camera to provide a detailed picture of a molecule. Let us analyze first how an optical image is formed. When an object is placed in a beam of light, light is scattered by the object in all directions. Because of the wave nature of light, the scattered light in any particular direction is characterized by an amplitude term and a phase term. For example, in Eq. (10.29), E_0 represents the amplitude and $\sin 2\pi(z - ct)/\lambda$ is the phase. The pattern of the scattered waves is called the *diffraction pattern* of the object. In the presence of a lens, part of the diffraction pattern is intercepted by the lens and is refocused to give an image in the image plane. Unfortunately, at present, x-ray lenses which are made of concentric rings of thin films (called Fresnel zone plates), have a resolution of only about 500 Å (Parsons, 1980). Another possibility is to make a holographic image of a molecule. Here the scattered x rays from the object are combined with a reference beam of x rays on a photographic film. A magnified image is obtained by projecting a beam of visible light through the developed photographic film (Solem and Baldwin, 1982). Very intense sources of x rays are needed, and other technological problems need to be solved. Nevertheless, it should be clear that all the information that can be deduced from the image of an object must be present in its diffraction pattern, because a lens or hologram cannot provide any information not present in the diffraction pattern. The determination of molecular structures by x-ray diffraction therefore depends on two aspects: obtaining the diffraction patterns and determining the structures.

Scattering of X Rays

X-ray photons are scattered by electrons in matter. The scattering can be elastic, in which the wavelengths of the incident and the scattered radiation are the same, or inelastic, in which the wavelengths differ. If inner orbital electrons are ejected by the incident radiation and secondary radiation is emitted when electrons in outer orbitals fall back to fill the vacancies, the emitted secondary radiation is referred to as fluorescence "scattering." In the discussions below we will consider only elastic scattering. Elastic scattering gives rise to diffraction patterns; inelastic scattered light does not.

In Fig. 12.1a, light is incident on a single electron at a point and is scattered. For light polarized perpendicular to the plane of scattering (the plane of the figure), the light is scattered with equal intensity in all directions in the plane. The scattered intensity (I_{sca}) decreases with the inverse square of the distance from the scattering point; it is proportional to the incident intensity, I_0. For one electron at a point

$$I_{sca} = \frac{CI_0}{r^2} \tag{12.1}$$

The distance from the scattering point to the detector is r; C is a proportionality factor which depends on the charge and mass of the electron.

In Fig. 12.1b light is scattered from two electrons. Each electron scatters the light independently in all directions, but the scattered intensity is not the sum of the intensities from each electron because of interference between the scattered waves. We must first sum

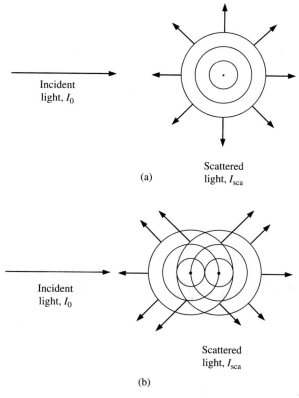

Incident
light, I_0

Scattered
light, I_{sca}

(a)

Incident
light, I_0

Scattered
light, I_{sca}

(b)

Fig. 12.1 (a) The scattering from a single point for light polarized perpendicular to the scattering plane is independent of angle in the scattering plane. The scattering intensity decreases inversely proportional to the square of the distance from the scattering point. (b) For scattering from two points the angle dependence of the scattering amplitude depends on the interference between the two scattered waves.

the amplitudes of the waves scattered by each electron; the square of this sum is the scattered intensity. The scattered intensity will still have an r^{-2} dependence, but it will have an angle dependence which depends on the distance between the electrons and on their orientation relative to the direction of incident light. Thus, from the scattering pattern we learn about the structure of the scattering object.

We can calculate the scattered intensity from two point electrons at any angle in the plane by simple geometric construction. We need to find the amplitude of the scattered wave at the detector, which is at a large distance, r, from the two electrons. The amplitude is the sum of the amplitudes contributed by the two electrons. The amplitude of the wave scattered from each electron will be different, because of the difference in distance traveled by each wave. From Eq. (10.29) we know that a light beam, or x-ray beam, of wavelength λ traveling along directions z and polarized along x can be written as

$$E = E_0 \sin \frac{2\pi}{\lambda} (z - ct) \qquad (10.29)$$

The intensity at the detector depends on the path difference for the two scattered beams. We can thus simplify our notation by setting $E_0 = 1$ and $t = 0$. The amplitude of the wave scattered from point 1 is chosen to be proportional to $\sin 2\pi z/\lambda$; then the amplitude scattered from point 2 is proportional to $\sin (2\pi/\lambda)(z + \Delta)$, where Δ is the path difference for the two rays between source and detector. Figure 12.2 shows how this path difference depends on the distance between the two points, R, the orientation of the two points relative to the incident beam, and the scattering angle, Θ. The phase difference between the two beams is $2\pi\Delta/\lambda$.

The two scattered rays add at the detector to give a wave of amplitude, A, and phase difference, $2\pi\delta/\lambda$.

$$A \sin \frac{2\pi}{\lambda} (z + \delta) = \sin \frac{2\pi z}{\lambda} + \sin \frac{2\pi}{\lambda} (z + \Delta) \qquad (12.2)$$

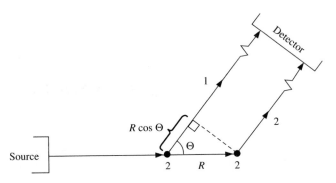

Path difference $\Delta = R \cos \Theta - R$

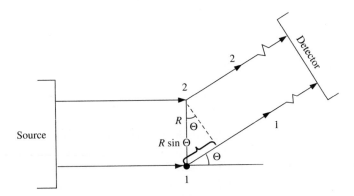

Path difference $\Delta = R \sin \Theta$

Fig. 12.2 Path difference for two beams scattered from two points separated by distance R. The angle between the incident beam and the scattered beam is Θ. At the top the two scatterers are on a line with the incident beam; at the bottom the scatterers are perpendicular to the incident beam direction.

Equation (12.2) states that a sum of sine waves is equal to a new sine wave whose amplitude, A, and phase, δ, depend on the amplitudes and phases of its components. However, the intensity, which is what the detector measures, is equal to A^2; therefore, to calculate a scattering pattern, we need only to find the amplitude of the combined beams, A. We use a trigonometric identity to rewrite Eq. (12.2).

$$\sin (\alpha + \beta) = \cos \beta \, \sin \alpha + \sin \beta \, \cos \alpha$$

$$A \cos \frac{2\pi\delta}{\lambda} \sin \frac{2\pi z}{\lambda} + A \sin \frac{2\pi\delta}{\lambda} \cos \frac{2\pi z}{\lambda}$$

$$= \left(1 + \cos \frac{2\pi\Delta}{\lambda}\right) \sin \frac{2\pi z}{\lambda} + \sin \frac{2\pi\Delta}{\lambda} \cos \frac{2\pi z}{\lambda} \tag{12.3}$$

Equation (12.2) and thus Eq. (12.3) must be true for all values of z, because z is a variable which can take on any value. This means that the coefficients of $\sin (2\pi z/\lambda)$ and $\cos (2\pi z/\lambda)$, respectively, in Eq. (12.3) are equal.

$$A \cos \frac{2\pi\delta}{\lambda} = 1 + \cos \frac{2\pi\Delta}{\lambda}$$

and

$$A \sin \frac{2\pi\delta}{\lambda} = \sin \frac{2\pi\Delta}{\lambda}$$

By squaring the two equations and adding them, we can obtain A^2.

$$A^2 \cos^2 \frac{2\pi\delta}{\lambda} + A^2 \sin^2 \frac{2\pi\delta}{\lambda} = \left(1 + \cos \frac{2\pi\Delta}{\lambda}\right)^2 + \sin^2 \frac{2\pi\Delta}{\lambda}$$

But

$$\sin^2 \alpha + \cos^2 \alpha = 1$$

Therefore,

$$A^2 = 2 \left(1 + \cos \frac{2\pi\Delta}{\lambda}\right) \tag{12.4}$$

We have found that the sum of two waves, each of which has amplitude equal to 1, gives a wave with an intensity (the square of the amplitude) depending on the phase difference between the waves. Therefore, the intensity scattered by two electrons is not twice Eq. (12.1); it is

$$I_{\text{sca}} = 2 \frac{CI_0}{r^2} \left(1 + \cos \frac{2\pi\Delta}{\lambda}\right) \tag{12.5}$$

where $2\pi\Delta/\lambda$ is the phase difference between the two points. Note that the phase difference depends on the path difference, Δ, divided by the wavelength. The path difference can be calculated geometrically as illustrated in Fig. 12.2.

Let us consider the general behavior of Eq. (12.5) because it illustrates how interference between scattered waves changes the intensity. If $\Delta = 0, \lambda, 2\lambda, \ldots$, then $\cos 2\pi\Delta/\lambda = 1$ and the intensity of scattering from two electrons is four times the scattering from one electron. The scattering amplitude is proportional to the number of electrons; the maximum scattering intensity is proportional to the square of the number of electrons. This is called total constructive interference, or total reinforcement. If $\Delta = \lambda/2, 3\lambda/2, 5\lambda/2, \ldots$, $\cos 2\pi\Delta/\lambda = -1$ and the intensity of scattering from two electrons is zero. This is total destructive interference. It is clear that it is the ratio of path difference to wavelength which determines the intensity. Figure 12.3 illustrates this summing of waves.

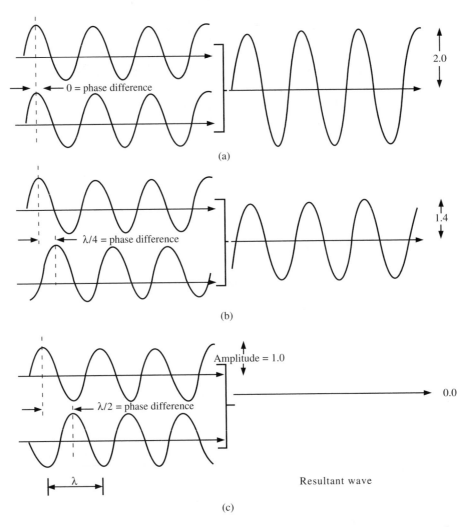

Fig. 12.3 Summing up of two waves: (a) waves completely in phase; (b) waves out of phase by one-quarter wavelength (partial reinforcement occurs); (c) waves out of phase by one-half wavelength (total destructive interference occurs). (From J. P. Glusker and K. N. Trueblood, *Crystal Structure Analysis: A Primer,* Oxford University Press, New York, 1972, p. 19, Fig. 5.)

Figure 12.4 shows the scattering as a function of angle for radiation incident along the line between the two points, as shown in Fig. 12.2a. For this geometry the path difference is $\Delta = R \cos \Theta - R$; R is the distance between the points and Θ is the angle between the incident beam and the scattered beam. Equation (12.5) was used to calculate the scattering pattern (angle dependence of the scattering) for these different ratios of R/λ. The patterns are symmetrical on either side of zero degrees, so only half the pattern is given. For $R = 0$, or very small relative to the wavelength, the phase difference is negligible, and the scattering is maximum and independent of angle. For $R = \lambda$, the scattering is maximum at $0°$, $90°$, and $180°$; it is zero at $60°$ and $120°$. For $R = \lambda/10$, a small angle dependence is seen. It should be evident that analysis of the scattering patterns of x rays with wavelengths near 1 Å (0.1 nm) can provide accurate measurements of interatomic distances.

Electrons are not points, but electrons in a molecule can be characterized by an electron density at each point in the molecule. The electron density is the number of electrons (or fraction of an electron) in a small volume element around the point. The amplitude of scattered light is proportional to the electron density at each point. To determine molecular structure we want to know the positions of the nuclei in the molecule. Each nucleus is surrounded by a cloud of electrons, so it is useful to know how this cloud will contribute to the scattering from a molecule.

The scattering from an atom is characterized by the atomic scattering factor f_0, which is the ratio of the amplitude scattered by the atom to that by a point electron. This is illustrated in Fig. 12.5 for a carbon atom, the relative amplitude of scattering, f_0, is plotted versus $(\sin \theta)/\lambda$; θ is half the scattering angle, Θ. A spherical electron density distribution always gives this type of angle dependence. At zero scattering angle f_0 is equal to the number

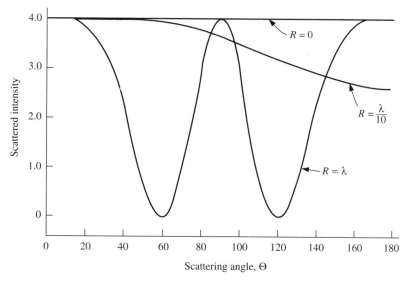

Fig. 12.4 Scattered intensity as a function of scattering angle Θ for two point charges collinear with the incident beam (Fig. 12.2, top). Equation (12.5) is used to do the calculation. For $R = 0$ the scattered intensity is independent of angle and is four times that of a single point. As R increases relative to λ, more structure appears in the scattering pattern. The patterns are symmetrical around zero degrees, so only half the pattern is shown. Note that at zero degrees there is always complete constructive interference.

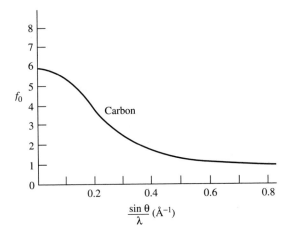

Fig. 12.5 Dependence of atomic scattering f_0 on $\sin \theta / \lambda$. The angle between the incident and scattered rays is $\Theta = 2\theta$. The curve shown is for carbon. Note that when θ approaches zero, the value of f_0 approaches the number of electrons per atom.

of electrons in the atom, because there is always total constructive interference in the forward direction. It is clear why hydrogen atoms will be difficult to locate accurately; their maximum scattering intensity is $\frac{1}{36}$ the scattering intensity of carbon atoms. Similarly, a heavy atom such as mercury (atomic number = 80) will have $(80/6)^2 = 178$ times the scattering intensity of carbon.

In a crystal, the atomic scattering factor is reduced further because of thermally induced vibrations of the atom. This temperature effect can be corrected by multiplying f_0 by a temperature-dependent factor. The symbol f is used for the corrected scattering factor.

Diffraction of X Rays by a Crystal

In the preceding section we showed how the scattered intensity depends on the distance between two points. We can use exactly the same reasoning to calculate the scattered intensity from a crystal. A crystal is a periodic array of identical scattering elements which differ only in their position within the crystal.

An x-ray diffraction experiment is done by placing a crystal, with dimensions of a mm or less, in the x-ray beam. The scattered intensity is measured as a function of angle from the incident beam. The positions of scattered intensity depend on the orientation of the crystal lattice relative to the direction of the incident x-ray beam. By rotating the crystal and measuring the angles and intensities of the scattered radiation, a diffraction pattern is obtained.

Let us consider first the diffraction of x rays by a linear array of identical *point* scatterers which are equally spaced, with a distance a between each adjacent pair. As illustrated in Fig. 12.6, we let α_0 be the angle the parallel incident rays make with the array. The incident rays are scattered in all directions. In a given direction α, the difference in path of rays scattered by two adjacent scatterers is

$$\text{path difference} = a(\cos \alpha - \cos \alpha_0)$$

as illustrated in Fig. 12.6. If this path difference is an integral multiple of the wavelength, then *all* the rays scattered by the points will reinforce. This condition is expressed by

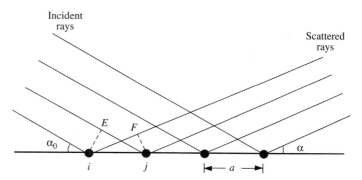

Fig. 12.6 Scattering of incident rays by a row of point scatterers of spacing a. The path difference of rays scattered by any two adjacent scatterers i and j is $(\overline{iF} - \overline{jE})$, which is equal to $a \cos \alpha - a \cos \alpha_0$ or $a(\cos \alpha - \cos \alpha_0)$.

$$a(\cos \alpha - \cos \alpha_0) = h\lambda \qquad (12.6)$$

where h is an integer. This equation tells us at what angles, α, diffraction occurs for a given spacing, a, between scattering points in the crystal and a given angle of incidence, α_0, of the x-ray beam. For a fixed value of a and a fixed incident angle α_0, there are several values of α at which constructive interference occurs, corresponding to $h = 0, 1, 2, \ldots$. For all other values of α, rays scattered from any two scatterers will be out of phase. The resultant wave from a large number of scattered rays with random phases has zero amplitude. Because the number of scattering points N is very large, we need consider only the angles of total constructive interference. At these angles the scattered intensity is proportional to N^2, at other angles the scattered intensity is negligible. In summary, then, when parallel monochromatic x rays strike a row of equidistant point scatterers at incident angle α_0, the scattered radiation will concentrate in a number of cones, each cone with a single α given by Eq. (12.6). A special case with $\alpha_0 = 90°$ is illustrated in Fig. 12.7. We might interject at this point that inelastically scattered rays, because of changes in λ, will be more or less randomly distributed spatially. They contribute to the background intensity but are not important for our discussions.

We can now generalize our treatment to the case of a three-dimensional crystal lattice. Figure 12.8 illustrates a simple lattice formed by three sets of equidistant planes. The points of intersection of these planes form the space lattice. The characteristic interplanar distances are a, b, and c, as indicated. We can also consider this simple lattice as generated by three sets of fundamental translations along three nonplanar axes a, b, and c.

There are only seven fundamental crystal lattices possible. For the most general case, the interaxial angles are unequal, and so are the values of a, b, and c. The lattice is called *triclinic*. If the interaxial angles are all 90° and $a \neq b \neq c$, the lattice is called *orthorhombic;* if the interaxial angles are 90° and $a = b = c$, the lattice is *cubic*. The other four lattices are *monoclinic* (two angles = 90°, $a \neq b \neq c$), *rhombohedral* (all angles equal, but not 90°, $a = b = c$), *tetragonal* (all angles = 90°, $a = b \neq c$), and *hexagonal* (one angle = 120°, two angles = 90°, $a = b \neq c$).

If we have an identical scatterer located at each lattice point in a triclinic lattice, it is easy to generalize Eq. (12.6), that total reinforcement occurs if the following *three* equations are satisfied:

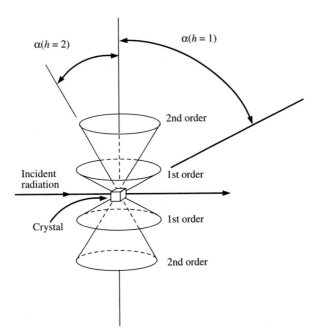

α(h = 2)   α(h = 1)

2nd order

Incident
radiation

1st order

Crystal

1st order

2nd order

Fig. 12.7 Equidistant point scatterers arranged in the vertical direction at the apices of the cones. With an incident angle $\alpha_0 = 90°$, Eq. (12.6) reduces to $a \cos \alpha = h\lambda$, or $\alpha = \cos^{-1}(h\lambda/a)$. Cones with $h = 1$ (first order) and $h = 2$ (second order) are shown. Note that $\cos \alpha = \cos(-\alpha)$; thus in this particular case, for each order there are two cones 180° apart. The zero-order diffraction ($h = 0$) is not shown.

$$a(\cos \alpha - \cos \alpha_0) = h\lambda \qquad (12.6)$$

$$b(\cos \beta - \cos \beta_0) = k\lambda \qquad (12.7)$$

$$c(\cos \gamma - \cos \gamma_0) = l\lambda \qquad (12.8)$$

These equations were first derived by M. von Laue and are referred to as von Laue's equations.

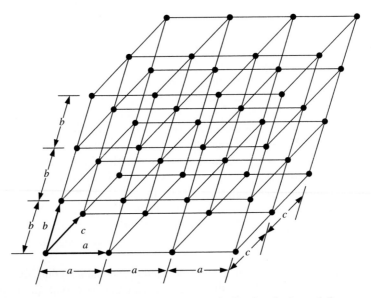

Fig. 12.8 Lattice of points. The lattice axes are depicted at the lower left corner.

The angles α_0, β_0, and γ_0 and α, β, and γ are the angles that the incident and diffracted beams make with the three axes, respectively.

The diffraction pattern from a crystal will thus consist of spots of scattered intensity whose positions depend on the crystal lattice. The diffraction pattern from a crystalline transfer RNA is shown in Fig. 12.9.

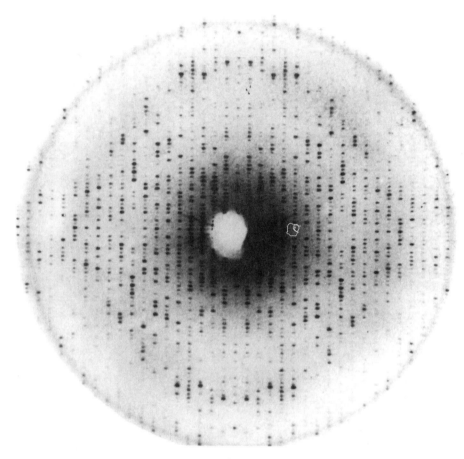

Fig. 12.9 Diffraction pattern for a crystal of yeast phenylalanine transfer RNA. The x-ray beam from a Cu K_α source ($\lambda = 1.54$ Å) is incident perpendicular to the film. A hole is cut in the film to reduce the overexposure caused by the very intense incident beam compared to the diffracted beams. The crystal has been rotated to produce what is technically labeled as a 15° precession photograph. The size of the unit cell is $a = 33$ Å, $b = 56$ Å, $c = 161$ Å. (The photograph was kindly supplied by Prof. Sung-Hou Kim, University of California, Berkeley.)

Measuring the Diffraction Pattern

Diffraction occurs only if all three von Laue's equations [Eqs. (12.6–12.8)] are satisfied. There is a further constraint on the diffraction direction because of geometrical constraints on the angles α, β, γ caused by the crystal lattice. For example, for an orthorhombic lattice

$$\cos^2 \alpha + \cos^2 \beta + \cos^2 \gamma = 1$$

Only at very special angles of beam incidence and scattering angles is constructive interference found; four equations must be satisfied.

This discussion shows that to measure the diffraction pattern, a special arrangement must be made. We have several choices:

1. Instead of using a monochromatic incident beam, we an use "white" x rays, which contain a broad distribution of wavelengths. For certain values of λ, the four equations may be satisfied. This option is of historical interest, because when von Laue and his coworkers did their first experiment on x-ray diffraction by a crystal, little was known about x rays, and the x-ray source they used gave a broad spectrum of radiation. Now monochromatic radiation is almost always used in x-ray diffraction studies. We will explain in later sections how to obtain monochromatic x rays.

2. If monochromatic radiation is used, characteristic diffraction rays can be observed if the orientation of the crystal relative to the incident beam is systematically changed. This can be achieved by rotating or oscillating a crystal about one axis of the crystal at a time, or by changing the position of the detector (a photographic film or an ionization or solid-state detector). Figure 12.10 illustrates one experimental arrangement. Sometimes the rotation or the oscillation of the crystal is coupled to the movement of detector so as to obtain a diffraction pattern which is easier to interpret. A particular mode of coupling can be achieved mechanically, or a computer can be programmed to control the movements.

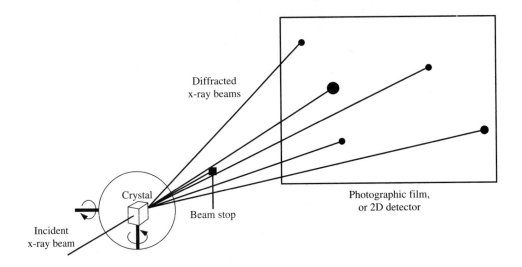

Fig. 12.10 Experimental arrangement for measuring the x-ray scattering from a crystal. A beam of x rays is incident on a crystal which is mounted so that it can be rotated around perpendicular axes. The intensities of the diffracted x-ray beams are measured with a photographic film, an equivalent two-dimensional detector, or a movable photomultiplier detector. A beam stop is used to attenuate the incident x-ray beam.

3. If monochromatic radiation is incident upon a powder of little crystals randomly oriented, a fraction of the crystals will have the correct orientation with respect to the incident beam to give characteristic diffractions. Such a picture is known as a powder pattern; it consists of a set of concentric rings.

Bragg Reflection of X Rays

So far we have mentioned only that the parameters *h*, *k*, and *l* are integers. The physical meaning of *h*, *k*, and *l* became clear primarily through the work of W. L. Bragg. He showed that the diffraction of an x-ray beam acts as if the x rays are reflected from planes in the crystal. Many groups of parallel planes can be drawn through the lattice points—the positions of the atoms—in a crystal, as shown in Fig. 12.11. Each group of parallel planes is characterized by three integers *h*, *k*, *l*; they are the same integers seen in von Laue's equations [Eqs. (12.6–12.8)]. These integers are known as the *Miller indices* of the planes. They are important because the position of each diffracted beam depends on the group of parallel planes which scatter constructively at this angle. Each spot in Fig. 12.9 can be labeled by three Miller indices; this is the first step in determining a structure from a diffraction pattern.

Let us see how the Miller indices are obtained. For simplicity, a two-dimensional lattice, and several groups of parallel lines going through the lattice points, are illustrated in Fig. 12.11. Each group of parallel lines can be indexed in the following way: pick any line such as the heavy line in the figure. The intercepts that this line makes with the *a* axis and *b* axis are 3*a* and 6*b*, respectively. The reciprocals of the coefficients are $(\frac{1}{3}, \frac{1}{6})$. Now multiply these reciprocals by the smallest integer that will give integral indices. In this example, the smallest integer is 6, and the indices obtained are (2, 1). These indices are called the *Miller indices.* The steps are summarized below:

	Axis	
Steps	*a*	*b*
1. Intercepts	3	6
2. Reciprocals	$\frac{1}{3}$	$\frac{1}{6}$
3. Miller indices	2	1

Exercise   Pick a few arbitrary lattice lines parallel to the one we discussed above. Convince yourself that these parallel lines all have the same Miller indices.

The distance d_{hk} between any two adjacent parallel lines is determined by their Miller indices. For example, if the *a* axis and *b* axis are perpendicular to each other, it is easy to show that

$$d_{hk} = \frac{1}{\sqrt{\dfrac{h^2}{a^2} + \dfrac{k^2}{b^2}}}$$

If the interaxial angle is different from 90°, the expression for d_{hk} is more cumbersome unless vector-algebra notation is used.

The situation is not much different for a three-dimensional lattice. Any group of parallel planes can be represented by three Miller indices (h, k, l). The interplanar spacing d_{hkl} can be calculated from the indices (for an orthorhombic lattice $d_{hkl} = 1/[(h/a)^2 + (k/b)^2 + (l/c)^2]^{1/2}$.

For a family of parallel crystal planes (h, k, l) with spacing d_{hkl}, a monochromatic incident beam of wavelength λ will *appear* to be reflected by the planes if the incident angle θ is

$$\theta = \sin^{-1} \frac{n\lambda}{2d_{hkl}} \qquad (12.9)$$

where n is an integer $0, 1, 2, \ldots$. To state it in different words, if the incident rays make an angle θ with a family of crystal planes (h, k, l) such that

$$2d_{hkl} \sin \theta = n\lambda \qquad (12.10)$$

reinforcement of the scattered rays occurs in a direction that also makes an angle θ with the planes. Equation (12.9) or (12.10) is called the *Bragg law;* it is illustrated in Fig. 12.12. The Bragg law and von Laue's equations can be shown to be equivalent.

We can write the Bragg law in the form

$$2 \cdot \frac{d_{hkl}}{n} \sin \theta = \lambda \qquad (12.11)$$

The quantity (d_{hkl}/n) can be viewed as the spacing between a family of planes with Miller indices (nh, nk, nl). The higher-order Bragg diffractions can be viewed as coming from these planes.

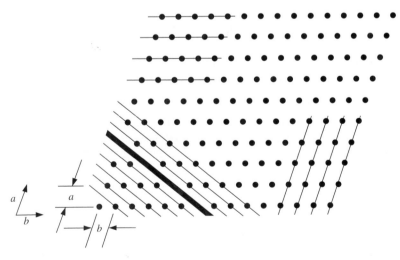

Fig. 12.11 Two-dimensional lattice with some possible lattice lines illustrated.

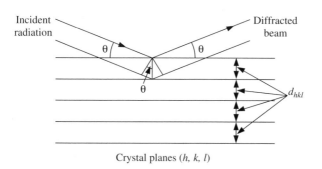

Fig. 12.12 Diffraction of radiation from a crystal. The parallel lines represent planes of atoms with Miller indices h, k, l. The Bragg condition for diffraction is for the incident beam and the diffracted beam to make an angle $\theta = \sin^{-1}(n\lambda/2d_{hkl})$ with the planes. Note that the scattering angle (Θ) between incident and scattered beams (defined before in Fig. 12.2) is 2θ. Simple geometric construction shows that the extra path length traveled by the ray scattered by the top layer relative to the next layer down is $2d_{hkl}\sin\theta$.

Incident radiation

Diffracted beam

Crystal planes (h, k, l)

Bragg condition: $n\lambda = 2d_{hkl}\sin\theta$

Intensity of Diffraction

So far we have considered the diffraction of x rays by a lattice of identical *point* scatterers. von Laue's equations, or the equivalent Bragg equation, provide the relation between the diffraction pattern and the lattice parameters. For a point scatterer the intensity of the scattered ray is independent of the scattering angle. If we have a real atom instead of a point scatterer, the intensity of the scattered ray is a function of $\sin\theta/\lambda$, as we have discussed already (see Fig. 12.5). Because constructive interference occurs only at $\sin\theta/\lambda = n/2d_{hkl}$ [Eq. (12.11)], the atomic scattering factor for a group of planes of indices (h, k, l) is independent of wavelength. The intensities of rays scattered by different groups of planes differ, of course, because of differences in $\sin\theta/\lambda$ values.

The situation is more complex when the lattice contains several kinds of atoms. The lattice still determines the diffraction pattern, but the intensity at each diffraction spot depends on the orientation of the atoms within the lattice. Analysis of this intensity distribution is what leads to the molecular structure of molecules in a crystal.

Suppose that we have a crystal of diatomic molecules AB as illustrated in Fig. 12.13. Consider the rows of A atoms first. The reinforcement of scattered rays occurs at angles

$$\theta = \sin^{-1}\frac{n\lambda}{2d}$$

where d is the spacing between the rows. For the rows of B atoms, the reinforcement of scattered rays occurs at the same angles

$$\theta = \sin^{-1}\frac{n\lambda}{2d}$$

because the spacing between the rows of B atoms must be the same as the spacing between the rows of A atoms in the crystal. The spacing between a row of A atoms and a row of B atoms is different from d, however. This means that rays scattered by rows of A and rows of B atoms are not in phase at the angle θ. This results in partial interference, and therefore the total intensity at the angle θ is less than the sum of the intensities due to independent

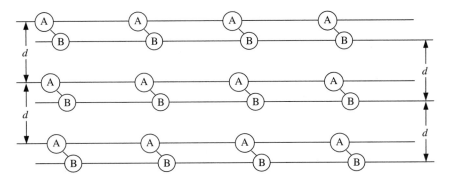

Fig. 12.13 Three rows of diatomic molecules AB in a crystal. The distance d between adjacent rows of A atoms is necessarily the same as the distance d between rows of B atoms. The distance between adjacent rows of A and B atoms is not the same as d.

scattering by A atoms and by B atoms. At a given Bragg angle, the intensity of rays diffracted by the lattice of A atoms is dependent on f_A, the atomic structure factor for A; the intensity of rays diffracted by the lattice of B atoms is dependent on f_B, the atomic structure factor for B. In addition to its dependence on f_A and f_B, the total intensity of rays diffracted by the two interpenetrating lattices is also dependent on the relative *phases* of rays diffracted by the two lattices. This is exactly analogous to the scattering from two electrons; the scattered intensity depends on the phase difference between the two scattered waves [Eq. (12.5)].

To generalize from the scattering of two different atoms to the scattering from a molecule it is convenient to use the concept of a unit cell.

Unit Cell

Any crystal can be considered as formed by placing a basic structural unit on every point of a lattice. (The crystal so formed is, of course, a perfect one. We shall not be concerned here with crystal imperfections.) An example is shown in Fig. 12.14.

We can consider the lattice as made of *unit cells*. The translations of the unit cell along the lattice axes generate the lattice. Figure 12.15 illustrates a two-dimensional lattice and several different unit cells that might be chosen. For our two-dimensional lattice, a unit cell is specified by two axes **a** and **b.** Let the dimensions of a unit cell be a and b in these directions; then the position of any atom in the unit cell is given by unitless coordinates (X, Y), where X and Y are given in fractions of the unit-cell dimensions. For example, an atom at the center of a unit cell has coordinates $(\frac{1}{2}, \frac{1}{2})$. For a three-dimensional lattice, a unit cell is specified by three axes **a, b,** and **c** and the position of an atom in the unit cell will be given by (X, Y, Z). If we determine the unit-cell dimensions and the coordinates of its atomic contents, we have all the information of the crystal structure.

The first step is to determine the shape of the unit cell (triclinic, cubic, etc.), and the unit cell dimensions. The number of molecules per unit cell and the symmetry of their arrangement is found next. Figure 12.14 shows the crystal structure of benzo[a]pyrene 4,5-oxide, a metabolite of the carcinogen benzo[a]pyrene (Glusker et al., 1976). The molecule

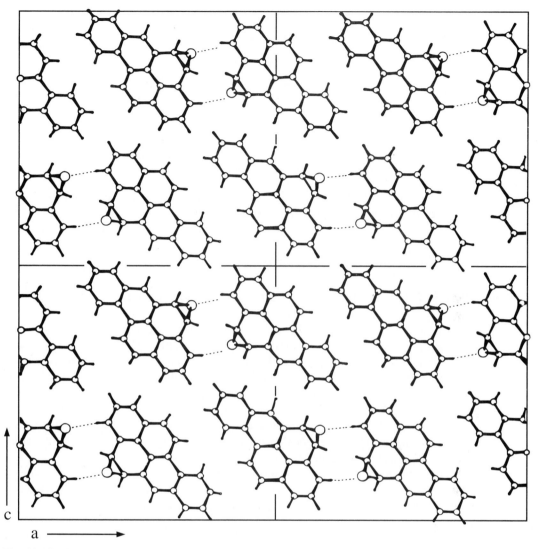

Fig. 12.14 Crystal structure of benzo[a]pyrene-4,5-oxide. The molecules form planar layers. Four unit cells are shown; there are four molecules per unit cell. See Glusker, et al., *Cancer Research 36,* 3951–3957 (1976). [Figure courtesy of Dr. Jenny P. Glusker.]

crystallizes in a monoclinic lattice (two angles = 90°, a ≠ b ≠ c) with dimensions $a = 17.341$ Å, $b = 4.095$ Å and $c = 17.847$ Å. The **b** axis forms an angle of 90.93° with respect to the perpendicular **a** and **c** axes shown in the figure. There are four molecules per unit cell, but only one is independent. The coordinates of the other three are related by the specific crystal symmetry (called the space group) found for these molecules within the monoclinic lattice. The information about size, shape, and symmetry of the unit cell is obtained from assigning Miller indices to the diffraction pattern. That is, the positions of the diffraction spots and their symmetries—which values of h, k, l have equal intensities or zero intensity—provide the unit-cell properties. To obtain the molecular structure is more difficult.

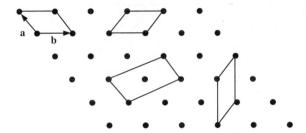

Fig. 12.15 Two-dimensional lattice of points and some of the unit cells that can be chosen. Each unit cell is defined by two axes, **a** and **b**, of length a and b, respectively, which are depicted for the unit cell at the upper left corner. For a three-dimensional lattice, a unit cell is defined by three axes, **a**, **b**, and **c**. Translation of a unit cell along its axes generates the lattice.

DETERMINATION OF MOLECULAR STRUCTURE

Calculation of Diffracted Intensities from Atomic Coordinates

Given the atomic coordinates of the molecules in the unit cell we can calculate the phase differences for the scattered waves from each atom at any angle. The phase differences control the total intensities at each angle. We did this calculation explicitly for two electrons. For two points the phase difference, $2\pi\Delta/\lambda$, depends on the path difference, Δ, which depends on the two coordinates and scattering angle as shown in Fig. 12.2. We now want to generalize this calculation to many atomic scatterers. We need to be able to add the scattered x-ray waves from all the atoms in a unit cell. Each wave is characterized by an amplitude and a phase. The amplitude depends on the atomic number of the scattering atom; the phase depends on its coordinates, and the scattering angle.

The summation of scattered waves can be done by several mathematical methods. The simplest uses a complex-number notation. Each scattered wave has two parameters: amplitude and phase. It can be represented by a complex number

$$a_j \, e^{i\alpha_j} \tag{12.12}$$

a_j = amplitude of the wave from scatterer j
α_j = phase of the wave from scatterer j
$i = \sqrt{-1}$

To calculate the intensity of the scattering, we calculate the amplitude of the sum of the scattered waves. The sum of the scattered waves is

$$\sum_{j=1}^{n} a_j \, e^{i\alpha_j} \tag{12.13}$$

where the sum is over the scatterers in the unit cell. From the mathematical identity

$$e^{ix} \equiv \cos x + i \sin x \tag{12.14}$$

we have for sum of the n waves,

$$\sum_{j=1}^{n} a_j \cos \alpha_j + i \sum_{j=1}^{n} a_j \sin \alpha_j$$

The magnitude of any complex number $A + iB$ is $(A^2 + B^2)^{1/2}$. The magnitude, or amplitude of the resultant wave is

$$\left[\left(\sum_{j=1}^{n} a_j \cos \alpha_j \right)^2 + \left(\sum_{j=1}^{n} a_j \sin \alpha_j \right)^2 \right]^{1/2}$$

The intensity of the scattered beam is

$$I = \left(\sum_{j=1}^{n} a_j \cos \alpha_j \right)^2 + \left(\sum_{j=1}^{n} a_j \sin \alpha_j \right)^2 \tag{12.15}$$

We see that the intensity of a scattered beam depends on both the amplitudes, a_j, and phases, α_j, of all the scattered waves.

The phases depend on the scattering angle, so if we knew the coordinates of the scattering atoms, we could calculate the intensities at all angles. However, significant intensities occur only where there is constructive interference of the waves. This means that for a crystal we calculate the intensities only at the Bragg angles specified by the Miller indices h, k, l.

Scattering from a Unit Cell: The Structure Factor

The waves scattered by a unit cell containing n atoms is the sum of the waves scattered by the atoms.

$$F(h, k, l) = \sum_{j=1}^{n} f_j e^{i\alpha_j} \tag{12.16}$$

$f_j =$ atomic scattering of an atom
$\alpha_j =$ the phase factor for scattering from the atom

$F(h, k, l)$ characterizes a scattered wave at the angle specified by Bragg's equation [Eq. (12.10)] and Miller indices h, k, l; it is called the *structure factor* of the planes h, k, l. From Eq. (12.14) it can be written as

$$F(h, k, l) = \sum_{j=1}^{n} f_j \cos \alpha_j + i \sum_{j=1}^{n} f_j \sin \alpha_j \tag{12.17}$$

Thus, we can write the structure factor in the form

$$F(h, k, l) = A(h, k, l) + i B(h, k, l) \tag{12.18}$$

with

$$A(h, k, l) = \sum_{j=1}^{n} f_j \cos \alpha_j \tag{12.19}$$

$$B (h, k, l) = \sum_{j=1}^{n} f_j \sin \alpha_j \qquad (12.20)$$

The magnitude of $F (h, k, l)$ is

$$| F (h, k, l) | = [A^2 (h, k, l) + B^2 (h, k, l)]^{1/2} \qquad (12.21)$$

The structure factor has an amplitude and a phase, just like any wave. It can also be represented in the form

$$F (h, k, l) = | F (h, k, l) | \; e^{i \alpha (h, k, l)} \qquad (12.22)$$

$| F (h, k, l) | =$ the amplitude of the scattered wave
$\alpha (h, k, l) =$ the phase of the scattered wave

We should remember that only the amplitude can be measured. Detectors measure the intensity of electromagnetic radiation; the intensity is the square of the amplitude. However, the phase of $F (h, k, l)$ —which we repeat cannot be measured—can be calculated from Eqs. (12.18–12.20) and the trigonometric identity $\tan \alpha = \sin \alpha / \cos \alpha$.

$$\alpha (h, k, l) = \tan^{-1} \frac{B (h, k, l)}{A (h, k, l)} \qquad (12.23)$$

The phase of α_j of the rays scattered by an atom j at X_j, Y_j, Z_j, in the direction of a Bragg diffraction angle, is $2\pi(hX_j + kY_j + lZ_j)$. The atomic coordinates (X_j, Y_j, Z_j) are expressed in fractions of the unit-cell dimensions, as we have defined them previously. Substituting into Eqs. (12.19) and (12.20) gives the components of the structure factor as

$$A (h, k, l) = \sum_{j=1}^{n} f_j \cos [2\pi(hX_j + kY_j + lZ_j)] \qquad (12.24a)$$

$$B (h, k, l) = \sum_{j=1}^{n} f_j \sin [2\pi(hX_j + kY_j + lZ_j)] \qquad (12.24b)$$

These equations tell us how to calculate the complete diffraction pattern—the intensity and phase of every scattered beam—from the coordinates of the atoms in the unit cell. For every group of parallel planes of Miller indices (h, k, l) there is scattered radiation at angle θ given by the Bragg equation ($n\lambda = 2 \, d_{hkl} \sin \theta$). From the X, Y, Z coordinates of the atoms in the unit cell, and using Eqs. (12.21) and (12.24), we can calculate the diffraction pattern.

Of course what we really want to be able to do is to go from the measured diffraction pattern to the coordinates of the atoms. We will learn about this in the next section. However, once coordinates are obtained, the way we check them and improve them is to calculate the diffraction patterns they predict. The coordinates are improved by minimizing the difference between the measured intensities and the calculated intensities from Eqs. (12.21) and (12.24).

Calculation of Atomic Coordinates from Diffracted Intensities

We have described how to calculate a diffraction pattern from a crystal structure; now it is time to learn how to calculate a structure from a diffraction pattern. The problem would be trivial if we could measure the phases of the scattered x-ray beams. Let us see why this is so.

We can rewrite the structure factor [Eq. (12.16)] as

$$F (h, k, l) = \sum_{j=1}^{n} f_j \, e^{+2\pi i \, (hX_j + kY_j + lZ_j)} \tag{12.25}$$

where the phase angles α_j have been replaced by their explicit values in terms of the coordinates of the scattering centers at X_j, Y_j, Z_j. The sum is over all the scattering centers in the unit cell. There is a standard mathematical method called *Fourier transformation* that solves Eq. (12.25) for the positions of the scattering centers. We used a Fourier transform to calculate an NMR spectrum as a function of frequency from a time-dependent free induction decay (FID) in Chapter 10 [see Eq. (10.36)]. Here we use a Fourier transform to calculate the electron density, a function of X, Y, Z from $F (h, k, l)$, a function of h, k, l.

$$\rho (X, Y, Z) = \frac{1}{V} \sum_h \sum_k \sum_l F (h, k, l) \, e^{-2\pi i \, (hX + kY + lZ)} \tag{12.26a}$$

Here $\rho (X, Y, Z)$ is the electron density (electrons per unit volume) at position X, Y, Z in the unit cell. X, Y, Z are in fractions of distances along the unit cell, and V is the volume of the unit cell. Equation (12.26a) tells us exactly how to calculate the electron density at every point in the unit cell. We choose a value of X, Y, Z in the unit cell. Then for every diffracted beam (which we have already assigned to a set of three Miller indices h, k, l) we calculate

$$F (h, k, l) \, e^{-2\pi i \, (hX + kY + lZ)}$$

Summing over all diffracted beams (that means summing over all Miller indices h, k, l), we calculate the electron density, $\rho (X, Y, Z)$, at X, Y, Z. Repeating this for many points in the unit cell, we obtain the distribution of the electrons. From a high resolution electron density map [see Fig. 12.16c], the position of each atom, and thus the molecular structure, is clear. For a lower resolution map [Fig. 12.16b], there will be more uncertainty in the placement of the atoms, but we can still obtain a structure. It is important to note that the entire diffraction pattern—all values of h, k, l—can contribute to the calculated electron density at X, Y, Z.

For every Fourier transform there is an inverse, which means that we can use the inverse of Eq. (12.26a) to calculate the diffraction pattern.

$$F (h, k, l) = \int dX \int dY \int dZ \, \rho (X, Y, Z) \, e^{+2\pi i \, (hX + kY + lZ)} \tag{12.26b}$$

Integrating over the electron density of the unit cell we can calculate each of the structure factors. Equations (12.26a) and (12.26b) allow us to calculate a structure—an electron density—from a known diffraction pattern, and to calculate a diffraction pattern from a known structure.

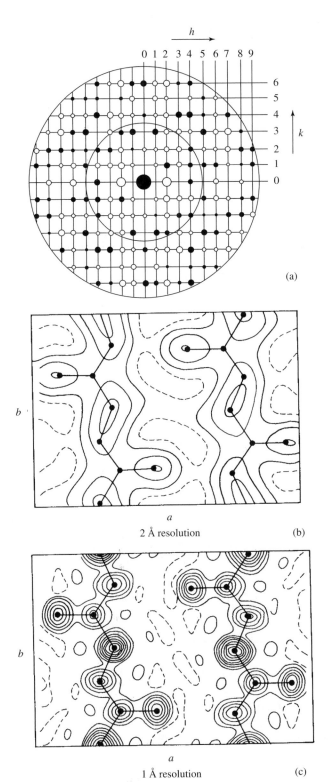

Fig. 12.16 An illustration of how x-ray diffraction data are used to generate a structure. In part (a) structure factor data are shown for a two-dimensional projection of a crystal of the polypeptide poly-L-alanine in antiparallel β-sheets. The dark spots represent $F(h, k)$ negative. The size of each spot is proportional to the magnitude of $F(h, k)$. The inner circle includes the data that would be sampled to obtain 2-Å resolution. The corresponding electron density map is shown in (b). The outer circle includes the data necessary to obtain 1-Å resolution, as shown in (c). The known position of the nuclei are shown in (b) and (c) superimposed on the electron density maps. Figure is adapted from Fig. 13-27 of C. R. Cantor and P. R. Schimmel in *Biophysical Chemistry,* Part II, 1980.

The Phase Problem

There is a fatal problem with using Eq. (12.26a); we cannot measure the structure factor $F(h, k, l)$. As shown in Eq. (12.22), the structure factor has an amplitude $|F(h, k, l)|$ and a phase $\alpha(h, k, l)$. We can measure only the amplitude, or its square—the intensity of each diffracted beam. The square of the amplitude loses the vital phase information, $\alpha(h, k, l)$.

Direct Methods

There are various methods for obtaining the missing phase factors. The method routinely used for small molecules (less than 200 to 300 non-hydrogen atoms) is called the *direct method* (Hauptman, 1989). Remember that the intensity of scattering from a hydrogen atom is only $(1/6)^2$ of that of a carbon atom, so the scattering from hydrogen atoms can be ignored in the first approximation. In the direct method you guess at the phases $\alpha(h, k, l)$ that you combine with the measured amplitudes $|F(h, k, l)|$ to give the structure factors $F(h, k, l)$, which allows calculation of the electron density. Of course the guesses of the phases are not random; only certain phases give reasonable results. For example, if a choice of phases gives a calculated negative electron density from Eq. (12.26a), we know it is a poor guess. Herbert Hauptman and Jerome Karle received the Nobel Prize in Chemistry in 1985 (see description by Hendrickson, 1986) for devising algorithms to find the phases that, when combined with the measured intensities, will give an electron density that corresponds to a collection of atoms—a molecule. The mathematical problem is equivalent to solving the simultaneous *non-linear* equations

$$|F(h, k, l)| = \left| \sum_{j=1}^{n} f_j\, e^{+2\pi i\, (hX_j + kY_j + lZ_j)} \right| \tag{12.27}$$

for the unknown positions of the atoms X_j, Y_j, Z_j. In Eq. (12.27) the measurable absolute values of the structure factors are related to the absolute values of the sums. All the parameters in Eq. (12.27) are known except for the three coordinates for each atom.

For 100 atoms ($n = 100$) there are 300 unknowns. The amplitudes $|F(h, k, l)|$ of the thousands of peaks in the diffraction pattern are measured. This means that there may be a few thousand equations specifying the 300 unknowns. The problem is overdetermined; there is enough information in the intensities alone to obtain the atomic coordinates. However, the solution of the simultaneous equations [Eq. (12.27)] is not straightforward, and it becomes increasingly difficult as the size of the molecule increases. Proteins and nucleic acids are at present too large to solve by direct methods.

Isomorphous Replacement

For macromolecules the most powerful method for determining phases is *isomorphous replacement* (isomorphous means same shape). A few heavy atoms are attached to the macromolecule in the crystal and their positions in the unit cell are determined. The scattering contribution to the intensity from an atom depends on the square of the atomic number, so a few heavy atoms can markedly change the diffraction pattern. Once the heavy atoms are positioned, and their contributions to the phases of the structure factors are calculated, the structure factors from the other atoms can be obtained. Now the electron density of the macromolecule can be calculated.

Before describing the concept of the isomorphous replacement method, we will introduce a useful function called the *Patterson function*. Patterson defined a function $P(X, Y, Z)$ which can be calculated from the experimental data.

$$P(X, Y, Z) = \frac{1}{V} \sum_h \sum_k \sum_l F^2(h, k, l)\, e^{-2\pi i\,(hX + kY + lZ)} \qquad (12.28)$$

The Patterson function, because it contains the square of the structure factor, gives information about vectors (specified by components X, Y, Z) between two atoms in the unit cell. This property is useful to us in the method of isomorphous replacement.

Consider a crystal of a protein and a crystal of a heavy atom derivative of the protein. The two crystals are isomorphous if they have essentially the same unit-cell dimensions and atomic arrangements: the main difference is the substitution of a heavy atom in the derivative with little distortion of the protein structure. The heavy atom can be weakly bound to a side chain, or a heavy metal ion can substitute for a lighter metal ion [Hg(II) for Zn(II)]. From a pair of isomorphous crystals, their *difference* Patterson map usually permits the location of the heavy atom in the unit cell, and therefore the phase of scattering contributions from the heavy atom can be calculated. Suppose that $F_{protein}(h, k, l)$, $F_{derivative}$ (h, k, l), and $f_{atom}(h, k, l)$ are the structure factors for the protein, the derivative and the heavy atom, respectively, for certain Miller indices (h, k, l). The magnitudes of $F_{protein}$ and $F_{derivative}$ are both known from the intensities, but their phase angles are not known. Both the magnitude and phase angle, α_{atom}, are known for the heavy atom structure factor, f_{atom}, because we know its position. Each of the three structure factors can be represented as vectors. For two vectors we know only their magnitudes; for the other—the heavy atom—we know its magnitude and direction.

If the heavy atom scatters much more than the light atom it replaces, then to a good approximation $F_{derivative}$ should be the vectorial sum of $F_{protein}$ and f_{atom},

$$\mathbf{F}_{derivative} = \mathbf{F}_{protein} + \mathbf{f}_{atom} \qquad (12.29)$$

We can therefore obtain the directions of $F_{derivative}$ and $F_{protein}$ by the geometric method illustrated in Fig. (12.17). The vector **AB** is drawn with length equal to the magnitude of f_{atom} and angle α equal to the phase angle of f_{atom}. A circle with radius equal to the magnitude of $F_{protein}$ and centered at A is first drawn. A vector drawn from any point on this circle to A gives a possible direction (and thus possible phase angle) for $F_{protein}$. We then draw a second circle with radius equal to the magnitude of $F_{derivative}$ and centered at B. A vector drawn from any point on this circle to B gives a possible direction (and thus possible phase angle) for $F_{derivative}$. The points of intersection of the two circles (at C and C' in Fig. 12.17) gives two solutions for the phase problem. The structure factor for the protein (magnitude and phase) is either the vector from C to A, or the one from C' to A. The ambiguity can be resolved from another isomorphous heavy atom derivative. Isomorphous replacement methods have played crucial roles in solving crystal structures of proteins. They allow us to deduce phase angles and thus structure factors from measured intensities. A Fourier transform of the structure factors gives the electron density [Eq. (12.26a)].

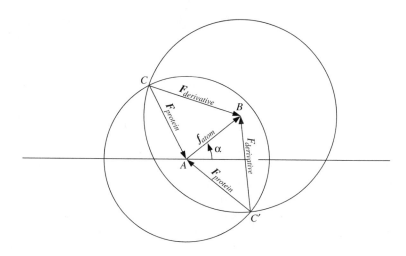

Fig. 12.17 Phase angle determination by isomorphous replacement. Once the position of a heavy atom is determined, the phase angle and amplitude of its structure factor can be calculated. This is the vector f_{atom} going from A to B in the figure; its phase angle is α. The structure factor for the heavy atom derivative of the protein, $\boldsymbol{F}_{derivative}$, is the vector sum of the structure factors of the protein and heavy atom, $\boldsymbol{F}_{protein}$, and $\boldsymbol{f}_{atom.}$ The magnitudes of $\boldsymbol{F}_{derivative}$ and $\boldsymbol{F}_{protein}$ are obtained from intensity measurements. The figure shows two possible vectors for $\boldsymbol{F}_{derivative}$ (from C to A and from C' to A) and two possible vectors for $\boldsymbol{F}_{protein}$ (from C to B and from C' to B). Another heavy atom derivative would allow us to choose between the two.

Determination of a Crystal Structure

As the x-ray diffraction method is the most powerful method available to determine structures of large biochemical molecules, such as proteins and nucleic acids, it is important to understand the procedure. The first step is to prepare suitable crystals. Crystals must be large enough to give sufficient scattered intensities, but not so large that the incident x-ray beam is greatly attenuated by absorption. Successful crystal growing depends mainly on trial and error. Many solvent and temperature conditions are tried until suitable crystals result. The next step, which is usually the easiest, is to obtain a diffraction pattern such as that shown in Fig. 12.9. From the positions of the spots, the crystal lattice and the dimensions of the unit cell can be determined.

Now the real work starts. Accurate intensity data for many different values of Miller indices h, k, l must be measured. Once the crystal class and unit-cell size are known, it is possible to calculate the crystal orientation and detector position necessary to produce a spot for any value of h, k, l. Because of the reciprocal relation in the Bragg condition [Eq.

(12.9)], the spots closest to the incident beam (θ small) correspond to large values of distances (d_{hkl} large). To calculate the electron density with a resolution of 6 Å, for example, all spots corresponding to this distance and larger distances must be measured. For higher resolution more spots must be measured. Figure 12.16 illustrates this for a two-dimensional projection of the electron density of two strands of poly-L-alanine antiparallel β-sheet. At 2-Å resolution the strands are barely visible; at 1-Å resolution the atoms are clearly seen. To obtain 2-Å resolution on a typical protein, about 10,000 values of F^2 (h, k, l) must be measured.

The electron density map is obtained by successive approximations. Heavy-atom derivatives provide the location of a few atoms in the unit cell. These are used to estimate phases and thus provide approximate values of F (h, k, l). These in turn provide approximate electron density maps, which are used to obtain locations of more atoms. Least-squares refinement methods are used to obtain the final coordinates. The sum of the squares of the difference between the observed and calculated | F (h, k, l) | values is minimized.

Figure 12.18 shows a structure for the transfer RNA molecule from yeast which adds a phenylalanine molecule to a polypeptide which is being synthesized. The location of the phenylalanine in the polypeptide sequence is determined by the base sequence of the corresponding messenger RNA. The three anticodon bases of the transfer RNA form base pairs with the three codon bases of the messenger RNA. The three-dimensional structure of the transfer RNA is a first step in understanding the mechanism of this complex reaction.

One example of a protein structure determined by x-ray crystallography will be presented. It is a transcription factor which binds to a specific sequence of DNA to control the synthesis of an RNA molecule—the transcription of the DNA sequence into an RNA sequence. The protein is the on-off switch for synthesis of the RNA. By understanding how transcription factors function, we may be able to control the synthesis of the messenger RNA, and thus the proteins they produce. The first step is to learn how the protein binds to the DNA. Nikola Pavletich and Carl Pabo (see their publication in the Suggested Readings) chose to study transcription factor IIIA from a frog (*Xenopus*). They used just the DNA-binding domain of the protein, 90 amino acids, and an 11-base-pair fragment of the DNA. Crystals of the polypeptide-oligonucleotide complex were prepared and diffraction data were obtained. For the native crystal 34,488 Bragg reflections were measured, but because of crystal symmetry only 9,458 were unique. Heavy-atom derivatives were made by replacing thymine in the DNA with iodouracil; this replaces a methyl group with an iodine atom. Two different heavy-atom derivatives were made and cocrystalized with the peptide. This provides values for the phases of the measured diffraction intensities and allows calculation of the structure using Eq. (12.26a). The resulting structure is shown in Fig. 12.19; it was obtained from a 2.1-Å resolution electron density map.

The structure shows a very common DNA-binding motif called a zinc finger (Rhodes and Klug, 1993). There are three zinc fingers binding to one turn of the DNA helix. Each zinc finger contains a zinc atom chelated to two cysteines in a β-sheet and two histidines in an α-helix. The three fingers contact a specific sequence of base pairs in the DNA by forming hydrogen bonds between five arginines and one histidine in the protein with six guanines in the nucleic acid. The ninety amino acids and three zinc atoms in the binding site serve to hold these six amino acids in the right place for recognizing the correct DNA sequence.

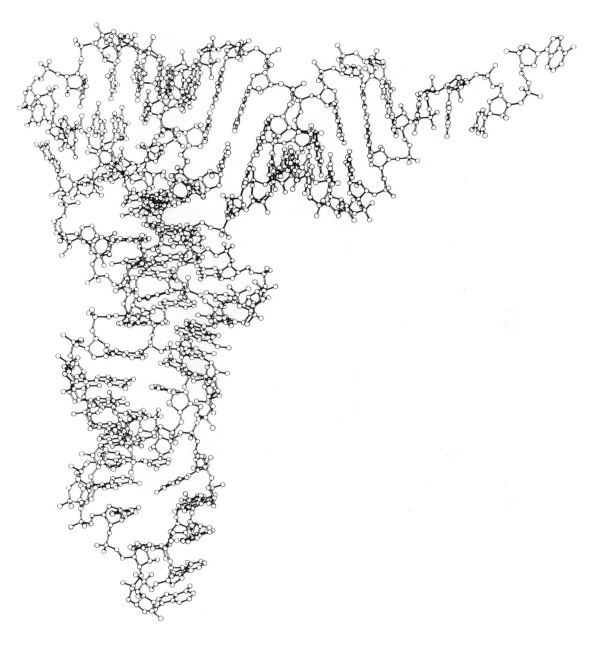

Fig. 12.18 A three-dimensional structure of phenylalanine transfer RNA from yeast. The structure was deduced from x-ray diffraction data such as shown in Fig. 12.9. The anticodon is made up of the three bases at the lower right of the molecule. The specific amino acid (phenylalanine) corresponding to the three bases of the complementary codon can be attached to the terminal nucleotide of the tRNA at the upper right of the molecule. [Holbrook et al., *J. Mol. Biol. 123*, 631–660 (1978).] The figure was kindly supplied by Prof. Sung-Hou Kim, University of California, Berkeley.

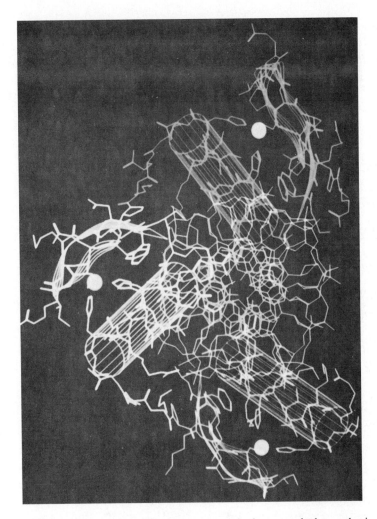

Fig. 12.19 Part of a transcription factor—a protein that controls the synthesis of RNA on a DNA template—bound to a double-stranded DNA. Three zinc fingers binding to one turn of the DNA helix are shown. Each zinc finger contains a zinc ion chelated between an α-helix (shown as a cylinder) and a β-sheet. The three fingers contact a specific sequence of base pairs in the DNA by forming hydrogen bonds. We are looking down the helix axis of the DNA in this figure. The structure was determined by x-ray crystallography by Pavletich and Pabo (*Science 252,* 809 (1991).

Scattering of X Rays by Noncrystalline Materials

Because of the long-range three-dimensional order in a single crystal, detailed atomic arrangement can be deduced from x-ray diffraction studies of the crystal. X-ray scattering by noncrystalline materials also yields structural information, although in general, much less detail is obtainable.

For long, threadlike macromolecules such as DNA, it is often possible to obtain fibers of a bundle of molecules with their long axes parallel or nearly parallel to the fiber axes. The orientation of the molecules in a fiber about the fiber axis is random. When such a fiber is placed in an x-ray beam, any long-range regular feature of the fiber will stand out in the diffraction pattern; the disordered features of the fiber will contribute only to the general background scattering. For example, for a DNA fiber with the molecules in the B-form conformation, since the helix axes of the DNA double helices are all parallel to the fiber axis, the strongest long-range regularity is the helical repeat of $p = 34$ Å per helical turn and $h = 3.4$ Å per base pair. These features are reflected in the diffraction pattern shown in Fig. 12.20.

For solutions of macromolecules, several types of information can be obtained by x-ray scattering measurements. First, as mentioned earlier, at zero scattering angle, constructive interference always occurs for rays scattered by different parts of a scatterer, and the intensity is dependent on the total number of electrons. Therefore, when extrapolated to zero scattering angle, and after subtracting scattering due to the solvent, the scattered intensity of a solution of macromolecules is related to the number of electrons per macromolecule. Since we can easily determine the elemental composition of the macromolecule, we can calculate the number of electrons per unit mass, and therefore obtain from the corrected zero-angle scattering intensity the molecular weight of the macromolecule. Second, although the macromolecules are randomly oriented in a solution so that little information on the atomic arrangement of the molecule can be deduced from the diffraction pattern, the angle dependence of the scattering intensity does provide us with information on the size of the macromolecule. It can be shown that at very low angles (of the order of milliradians),

$$\ln i(\Theta) = \ln i(0) - \frac{16\pi^4 R^2}{3\lambda^2} \sin^2 \frac{\Theta}{2} \qquad (12.30)$$

where Θ is the angle between the incident and the scattered beams, $i(\Theta)$ is the scattering intensity at Θ, λ is the wavelength of the x rays, and R^2 is the mean-square radius of the macromolecule. Thus if the experimentally measured $\ln i(\Theta)$ is plotted as a function of $\sin^2 (\Theta/2)$, in the low-angle range a straight line results. The intercept gives $\ln i(0)$, from which the molecular weight of the macromolecule can be calculated as we have discussed, and the slope gives $-16\pi^4 R^2/3\lambda^2$, from which we can calculate the mean-square radius. The angle dependence of scattering intensity at larger angles is also dependent on the general shape of the macromolecule.

Absorption of X Rays

Because of coherent, incoherent, and fluorescence scattering, the intensity of a beam of x rays is reduced after passing through any material. Similar to Beer's law behavior which we discussed in Chapter 10, the absorbance for monochromatic x rays of a given wavelength is proportional to the thickness of the sample and to the number of electrons per unit volume. For a given material, the absorbance is dependent on the energy (wavelength) of the x rays. Figure 12.21 illustrates this dependence. There are several abrupt rises in the

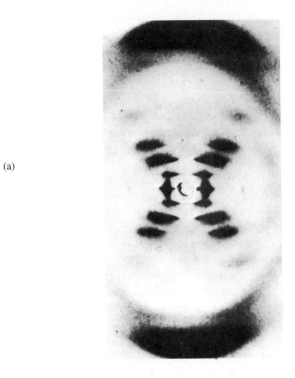

(a)

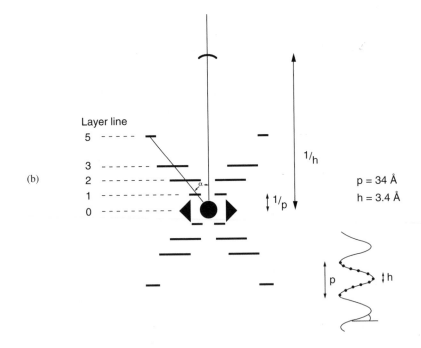

(b)

Layer line

5

3
2
1
0

α

$1/h$

$1/p$

$p = 34\ \text{Å}$
$h = 3.4\ \text{Å}$

p

h

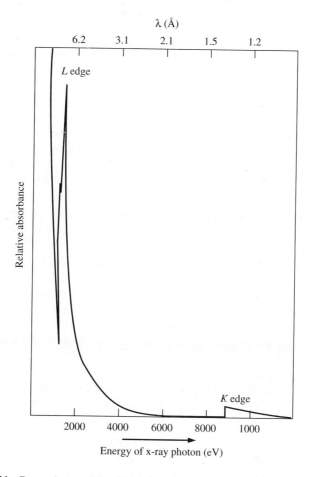

λ (Å)

| 6.2 | 3.1 | 2.1 | 1.5 | 1.2 |

L edge

Relative absorbance

K edge

| 2000 | 4000 | 6000 | 8000 | 1000 |

Energy of x-ray photon (eV)

Fig. 12.21   Dependence of the absorption of x rays by Cu on the energy (wavelength) of the x-ray photon. There are three barely discernible maxima in the *L* edge.

curve going from left to right: three in a region a little over 1000 eV and another near 9000 eV. These absorption edges are due to the ejection of orbital electrons from the *L* (2*s, 2p* electrons) and *K* (1*s* electrons) shells, respectively, and are labeled in the figure as *L* and *K* edges. The energy at which an edge absorption occurs is determined by the energy of the orbital electron, which is different for each element in the periodic table.

Fig. 12.20   (a) X-ray diffraction pattern of a NaDNA fiber in the B-helical conformation. Note the characteristic cross and the very strong meridian spots. (b) Line drawings of the pattern above, illustrating the relation between the helix dimensions (lower right) and the layer lines and meridian spots. The DNA helix has repeating distances of $h = 3.4$ Å, the distance between successive base pairs, and pitch, $p = 34$ Å, the distance between successive turns. The angle α characterizes the ratio of the pitch to the radius; $\tan \alpha \cong \text{pitch}/2\pi$ radius. (From J. P. Glusker and K. N. Trueblood, *Crystal Structure Analysis: A Primer,* Oxford University Press, New York, 1972, p. 137, Fig. 39b.)

Extended Fine Structure of Edge Absorption

In Fig. 12.21, we depicted an edge absorption (the K edge, for example) as a sharp rise followed by a smooth gradual drop. Careful experimental measurements showed, however, that the drop is not smooth. For a given material there are characteristic wiggles on the short-wavelength side of the absorption edge, as illustrated in Fig. 12.22. Qualitatively speaking, such wiggles appear because the probability of absorption of an x-ray photon by an orbital electron depends on the initial and final energy states of the electron. The initial energy state of the electron (in the K shell, for example) is, of course, quantized (Chapter 9). If we have an isolated atom, the final energy state of the ejected electron is not quantized, because it can possess different kinetic energies. Indeed, for such an isolated atom no wiggles would appear in the absorption curve. If the atom that absorbs the x-ray photon is not isolated but is surrounded by neighboring atoms, the ejected electron is back-scattered by the neighboring atoms. This situation is reminiscent of the particle-in-a-box problem (Chapter 9). We recall that the energy of a particle in a box is quantized, with the energy levels dependent on the dimensions of the box. Similarly, in the presence of the neighboring atoms, the energy of the electron ejected by an x-ray photon is also quantized, with energy levels dependent on the distances between the atom at which absorption occurs and its neighboring atoms. This quantization of the final energy state of the ejected electron results in the characteristic wiggles in the fine structure of the edge absorption curve. By analyzing the edge absorption fine structure, it is possible to deduce the neighboring atomic arrangement around the absorbing atom.

Because the K absorption edge occurs at different wavelengths for different elements, it is possible to select a particular wavelength so that the atomic arrangement around a particular atom can be deduced. For example, if an enzyme contains a heavy metal atom,

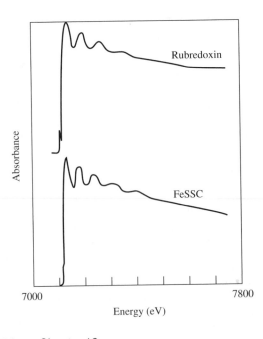

Fig. 12.22 Extended x-ray absorption fine structure (EXAFS) of the iron-sulfur protein *Peptococcus aerogenes* rubredoxin and a model compound tris (pyrrolidine carbodithioate −S, S′) iron (III), FeSSC. The x-ray source was synchrotron radiation from the Stanford Synchrotron Radiation Laboratory. The protein sample was a freeze-dried powder. Analysis of the data leads to a determination of the iron-sulfur distances in the protein. Figure from R. G. Shulman, P. Eisenberger, W. E. Blumberg, and N. A. Stombaugh, *Proc. Natl. Acad. Sci. USA* **72**, 4003 (1975).

the wavelength of the incident x ray can be chosen to correspond to the absorption edge of the heavy metal. Analysis of the fine structure of the absorption profile can then yield information on the atomic environment of the heavy metal. It should be noted that structural analysis by edge absorption fine structure does not require crystalline material. It is the close neighbors of the absorbing atom that back-scatter the ejected electron; therefore, long-range order is not necessary for the phenomenon.

X Rays from Synchrotron Radiation

An intense source of x rays is obtainable from the storage rings of electrons associated with an electron accelerator such as a synchrotron. In these storage rings the circulating electrons produce electromagnetic radiation over a broad range, including the x-ray region.

The intensity of x rays from a synchrotron storage ring can be tens of thousands times higher than that from a conventional source. This high intensity permits the use of crystal monochromators which permit the selection of intense monochromatic radiation of the desired wavelength. In cases where high intensity is desirable, the synchrotron radiation is advantageous. For example, if diffraction patterns can be obtained in a very short time because of the high beam intensity, time-dependent structural changes can be studied. Another important application is to obtain the fine structure of edge absorption discussed in the preceding section. Here one must be able to tune the wavelength of the x-ray source to cover the regions of interest for a particular element, such as a heavy metal in an enzyme. Especially in biological systems, where the concentrations of such absorbing atoms are very low, the high intensity of x rays from a synchrotron source is advantageous. The application of synchrotron radiation to structural studies of biological systems is increasingly important.

X-ray lenses are being developed to utilize the higher resolving power obtainable in principle from the shorter wavelengths. One strong impetus is the desire to make smaller integrated circuits. At present visible and ultraviolet light is used to project an image of the circuit onto photosensitive plastic, which is then developed to produce the circuit. In principle, 100-Å x rays can give much higher resolution images, and therefore smaller circuits, than 2000-Å ultraviolet light. These x-ray lenses can also provide better resolution images of cells, cell nuclei, and chromosomes (Parsons, 1980). All of the spectroscopic techniques based on light, which we described in Chapter 10, can be extended into the x-ray region with synchrotron sources (Bienenstock and Winick, 1983).

ELECTRON DIFFRACTION

According to de Broglie's equation ($\lambda = h/p$), a beam of electrons should exhibit wavelike properties. If an electron initially at rest at a point where the electrostatic potential is zero is accelerated by an electric field to a point where the potential is V, its kinetic energy at that point is

$$\frac{p^2}{2m_e} = eV \qquad (12.31)$$

where p is the momentum, m_e the rest mass of an electron, and e the charge of an electron. The corresponding wavelength is

$$\lambda = \frac{h}{p} = \frac{h}{\sqrt{2m_e eV}} \qquad (12.32)$$

Actually, if the electrons are moving at a velocity not negligible compared with the velocity of light c, a relativistic correction is needed:

$$\lambda = \frac{h}{\sqrt{2m_e eV_{corrected}}} \qquad (12.33)$$

where

$$V_{corrected} = V \cdot \left(1 + \frac{eV}{2m_e c^2}\right) \qquad (12.34)$$

It is easy to show that if the potential is of the order of 10^4 volts, the wavelength is of the same order of magnitude as x rays. Therefore, electron diffraction can also be used for structural determinations of crystals. Electrons are scattered much more strongly by atoms than are x-ray photons. This is a blessing in some ways and a nuisance in others. Because of the strong scattering, a much smaller crystal can be used, and higher diffracted intensity means that less time is needed in data acquisition. Especially for low-energy electrons (V of the order of a few hundred volts or less), scattering is due to the first few layers of the atoms at the surface, and therefore much information can be obtained about the *surface* structures. But also because of the strong scattering, the path for the beam of electrons must be evacuated; otherwise, the beam will be scattered by the gas molecules. This creates problems for many biological samples, which frequently contain water and cannot be kept in a vacuum. Also, effects due to interaction between the diffracted rays with the incident rays, which are not important for x-ray diffraction, are much more pronounced in electron diffraction. They make the interpretation of the electron diffraction patterns more difficult.

NEUTRON DIFFRACTION

In Chapter 6 we gave the translational kinetic energy of a gas as

$$U_{tr} = \frac{3}{2} RT$$

This equation is not limited to molecular or atomic particles; it applies to neutrons as well. If high-velocity neutrons from an atomic reactor are slowed down by collisions with molecules (D_2O is usually used) at around room temperature, it can be readily shown that their energy corresponds to a wavelength of the order of 1 Å. Such thermal neutrons can be used in diffraction studies of crystals and in low-angle-scattering studies of solutions, similarly to the use of x rays.

There are several important differences. X rays are scattered primarily by electrons; neutrons are scattered by nuclei. Because the size of the nucleus is much smaller compared with the wavelength of the thermal neutrons, the scattering factor f_0 of an atom is not much affected by the value of $(\sin \theta)/\lambda$. (Thermal motions of atoms in a crystal, however, do cause the scattering intensity to decrease at larger scattering angles, and it is necessary to correct f_0 for this.)

The scattering factor f_0 of an atom for neutrons is not related to the atomic number or mass in any simple way. The scattering of neutrons by hydrogen is quite high, unlike the scattering of x rays, for which hydrogen is the weakest scatterer. Therefore, the positions of hydrogens can be more readily deduced from the neutron scattering data of a crystal. The scattering of neutrons by deuterium is quite different from that by hydrogen. Thus deuterium can be used to replace hydrogen to give isomorphous crystals. For small-angle-scattering studies, neutron diffraction has one important advantage. It can be calculated that the relative scattering factors for H_2O and D_2O (deuterium oxide) are -0.0056 and $+0.064$, respectively. The negative sign means that neutron scattering from H_2O (and from ^1H) is $180°$ out of phase with the scattering from D_2O and ^2H. Most of the biological substances, such as nucleic acids, proteins, and lipids, have scattering factors between these values. Therefore, by using mixtures of H_2O and D_2O as the solvent, the background scattering of the solvent can be made to be the same as one of the components. Only components that scatter differently will stand out and contribute to the scattering contrast. Thus if we are studying a protein-nucleic acid complex, we can in essence look at either the protein or the nucleic acid component in the complex by selecting the appropriate H_2O—D_2O mixtures. Similarly, the contrast between membrane and cytoplasm can be varied from positive to negative by increasing the D_2O content of the solvent.

A disadvantage of neutron scattering studies is that there are very few high-flux neutron sources available. If the flux is low, a larger crystal is needed to give a diffraction pattern in a reasonable time. Therefore, neutron-scattering studies are under way at only a few places.

ELECTRON MICROSCOPY

In our discussion of x-ray diffraction, we commented on the fact that the lack of high-resolution x-ray lenses makes it necessary to deduce the structure mathematically from the diffraction pattern. A beam of electrons, however, can be focused by electrostatic or magnetic lenses. Therefore, a beam of electrons, after interacting with a specimen, can be focused to give an image of the specimen on a fluorescent screen or a photographic plate. Because of the short wavelength attainable for a beam of electrons, the resolution of an electron microscope can be made several hundredfold higher than that of a light microscope. As a consequence, electron microscopy has developed into one of the most powerful techniques for studying macromolecules, assemblies of macromolecules, cellular organizations, and so on. One might expect, because the wavelength of electrons can easily be made much less than atomic dimensions, it should be possible to see detailed atomic arrangement within a molecule by electron microscopy. This is difficult, if not impossible, to achieve. The reasons will become clearer after brief discussions on resolution, contrast, and specimen damage by electron bombardment.

Resolution

For an ideal lens, a luminous point source should be imaged into a point in the image plane. In practice, lenses for focusing rays of electrons cannot be constructed to achieve such ideality. A point source will give instead a circle of radius Δr_i at the image plane. If M is the magnification factor, the point source will have an apparent radius $\Delta r = \Delta r_i / M$. This lens aberration is called *spherical aberration;* it depends on the lens aperture. If α (in radians) is the angle made by the point source at the lens aperture (Fig. 12.23), then

$$\Delta r = C_s \alpha^3 \tag{12.35}$$

where C_s is a proportionality constant. The value of C_s for a good electron microscope is about 2 mm. If we wish to make $\Delta r \approx 1$ Å, we must use a small aperture of several milliradians. There are other lens aberrations, but they are less important for an electron microscope.

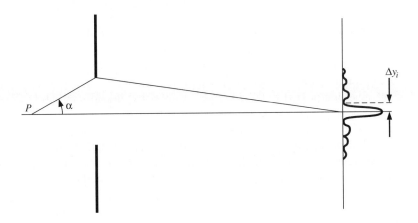

Fig. 12.23 Diffraction by a lens aperture. An axial point source P will give a diffraction pattern at the image plane as depicted. The central bright spot of radius Δy_i at the plane is called the *Airy disk,* after G. Airy.

It appears that we can make spherical aberration as small as possible by making α very small. We must keep in mind, however, that when an aperture is very small, diffraction effects become important. This is illustrated in Fig. 12.23. A point source will give the diffraction pattern illustrated. At the image plane there is a central spot surrounded by concentric circular maxima of decreasing brightness. The distance from the lens axis to the position of the first minimum, Δy_i, can be taken as the radius of the central spot. Thus, because of diffraction, a point will appear to have an apparent radius $\Delta y = \Delta y_i / M$, with M being the magnification as before. For small apertures it can be shown that

$$\Delta y \approx \frac{0.6\lambda}{\alpha} \tag{12.36}$$

The smaller the aperture, or the larger the wavelength, the greater will be the diffraction effect on resolution.

As a compromise, α can be chosen so that the magnitudes of diffraction error and spherical aberration are about the same. Equating Eqs. (12.35) and (12.36), we obtain

$$\alpha = \left(\frac{0.6\lambda}{C_s}\right)^{1/4} \qquad (12.37)$$

and

$$\Delta r = \Delta y \approx \frac{0.6\lambda}{(0.6\lambda/C_s)^{1/4}} \approx 0.7 C_s^{1/4}\lambda^{3/4} \qquad (12.38)$$

At the present time, the point resolution (the smallest separation between two points that can be resolved) of a good electron microscope is about 3 Å. For crystalline materials with periodic spacings, spacing close to 1 Å can be resolved.

Contrast

At the beginning of this chapter, we mentioned that contrast is an important factor in seeing an object. To see a macromolecular species by electron microscopy, molecules are usually mounted on a supporting film, such as a carbon film (typically about 100-Å thick, although much thinner films can be used). Contrast between a macromolecule and the background is due to a number of phenomena. One important factor is that some of the incident electrons are scattered by the specimen and will not reach the image plane, either because of the beam aperture or because of lens aberration. A strong scatterer will therefore appear as a dark spot in the image. Diffraction of the incident rays by a crystalline material or at the edges of dense particles, and interference between incident and scattered rays, also contribute to contrast. Detailed discussions can be found in books on electron microscopy. Contrast is frequently much more of a problem than resolution in electron microscopy.

Several methods have been developed to enhance the contrast. The macromolecules can be "shadowed" by impinging evaporated metal obliquely onto them. More metal will be deposited on the sides of the macromolecules than on the supporting film. The macromolecules can also be preferentially "stained" with an electron-dense stain (positive staining), or the supporting film can be stained preferentially with an electron-dense stain (negative staining). Some examples are shown in Fig. 12.24. Contrast is also improved by using a dark-field arrangement: either the incident electron beam is tilted, or the aperture is moved off center, so that only the scattered electrons will reach the image.

Radiation Damage

We have mentioned the problem of keeping a biological specimen in an evacuated chamber. Procedures used in sample preparation for viewing in the microscope can also cause

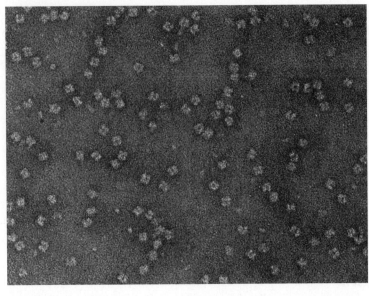

(a)

Fig. 12.24 (a) Electron micrograph of the enzyme δ-aminolevulinic acid dehydratase from bovine liver. The enzyme molecules were negatively contrasted with sodium phosphotungstate. The magnification factor is 250,000. The particles have the appearance of four discrete lobes arrayed at the corners of a square. Sedimentation and gel electrophoresis studies of the enzyme indicate that it is an octamer. A model consistent with all these observations is one in which the subunits of the octameric enzyme are arranged at the corners of a cube. [For details, see W. H. Wu, D. Shemin, K. E. Richards, and R. C. Williams, *Proc. Natl. Acad. Sci. USA 71*, 1767 (1974). Photomicrograph courtesy of the Virus Laboratory, University of California at Berkeley.] (b) An electron micrograph of a part of a "heteroduplex" DNA. When two strands of complementary base sequences associate by base pairing, a duplex or double-stranded molecule is formed (see the section on DNA helix-coil transition in Chapter 11). If the two strands are not completely complementary, the noncomplementary regions will not pair, and will show as single-stranded loops. Such a molecule is called a heteroduplex. For the particular heteroduplex shown, there is a single-stranded loop. One side of the loop contains two tyrosine transfer RNA genes of the bacterium *E. coli*. To determine the positions of these transfer RNA genes, molecules of an iron-containing protein, ferritin, are linked chemically to the ends of the transfer RNA molecules. The ferritin-carrying transfer RNA and the heteroduplex DNA are annealed to allow the binding of the transfer RNA. The ferritin molecules are essentially opaque to the electron beam and therefore show as dark dots on the micrograph. To see the rest of the molecule, positive staining with uranyl acetate and shadowing with platinum were employed. [For details, see M. Wu and N. Davidson, *J. Mol. Biol. 78,* 1 (1973). Micrograph courtesy of N. Davidson.]

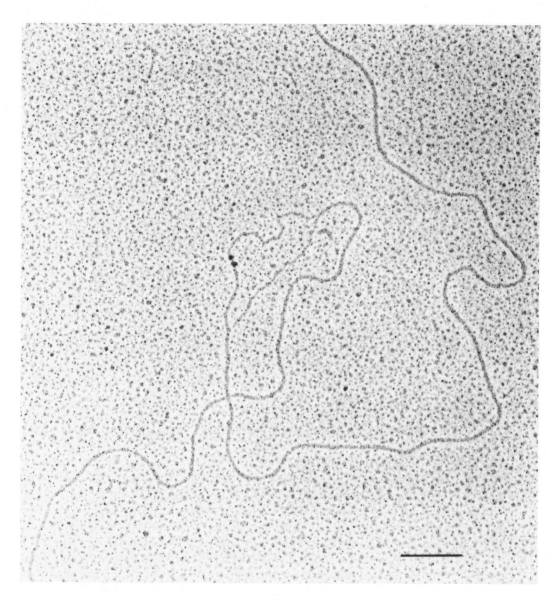

(b)

deformation of the species. Another problem of electron microscopy is radiation damage. Energy absorbed by a macromolecule from the incident beam causes heating of the molecule, breakage of chemical bonds, and rearrangement of atoms. If the specimen is stained, movement of the stain may also occur. Because the final image of the molecule on a photographic film is the composite picture of the molecule during the period when it is subject to electron bombardment, the image might be quite different from the original structure. Radiation damage is a serious problem when high-resolution results are attempted on individual molecules.

Transmission and Scanning Electron Microscopes

In a transmission electron microscope, electrons from a source (usually a hot tungsten filament) are first accelerated by an electric field to give the required wavelength. After passing through the specimen, the electrons are detected by a fluorescent screen for viewing or by a photographic plate or videotape for recording. Lenses are used to guide the paths of the electrons. In a scanning electron microscope, electrons from a source, after acceleration, are first focused by a lens to give a very small spot. The specimen is then scanned by moving the focused spot across it point by point and line by line. This scanning is achieved by the use of scanning coils, in much the same way as in a television camera. A suitable detector analyzes either the transmitted electrons, or back-scattered and secondary electrons ejected from the sample by the incident electrons. The signal detected can be displayed on an oscilloscope (television tube), point by point and line by line, in synchronization with the movement of the focused spot.

It was pointed out by A. Crewe that the electron optics for the two types of microscopes are actually rather similar, as illustrated in Fig. 12.25. Transmission electron microscopes have had a longer history of development, and commercial instruments capable of a point resolution of several angstroms are available. Before 1970, the resolution of scanning electron microscopes was about 100 Å, limited primarily by the size of the focused spot. The scanning microscope has several advantages, however, in certain applications. When secondary electrons are detected, the contrast can be made very high for the surface features of a specimen by coating it with a thin layer of gold. Surface topography can be vividly revealed, as illustrated in Fig. 12.26. One can also use a detector to measure the energy of the x rays emitted from each point of an uncoated specimen when irradiated by the electrons. Because the energy of the x rays emitted is characteristic of the elements irradiated, a scanning microscope in this mode can be used as a microanalyzer for the elemental composition of the specimen.

A scanning electron microscope with a much improved resolution (about 5 Å) was designed by Crewe around 1970. Instead of using a heated filament as the source, electrons emitted from a cold tungsten tip under the influence of a high electric field (field emission) were used. This greatly reduces the size of the focused spot. Contrast is achieved by analyzing both the elastically and inelastically scattered electrons. Excellent micrographs of biological materials have been obtained with this type of high-resolution electron microscope.

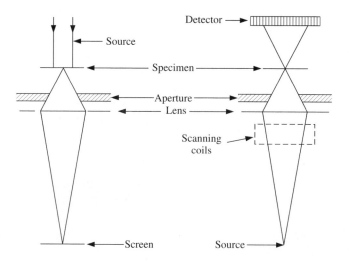

Fig. 12.25 Schematic diagram showing the similarity between a conventional microscope (on the left) and a scanning microscope (on the right). To make the comparison more obvious, the scanning microscope diagram has been drawn so that the electrons move from the bottom of the diagram to the top. [From A. V. Crewe and J. Wall, *J. Mol. Biol. 48*, 377 (1970).]

Scanning Tunneling and Atomic Force Microscopy

A microscope that uses electrons to produce an image is the scanning tunneling microscope. If two conducting surfaces are separated by an insulator a few Å thick, a current tunnels across the intervening insulator when a voltage is applied. The electron current depends exponentially on the distance between the two conductors; it is thus a very sensitive measure of the distance between them. If an atomically sharp conducting tip is scanned over a conducting surface in air or vacuum, individual surface atoms can be seen. Atomic holes in the surface decrease the tunneling current, and atomic bumps increase the current. Experimentally, the tunneling current is kept constant (at 0.1 nanoampere, for example) and the variation of a few Å in the height of the tip is measured. Molecules deposited on the surface can also be imaged. The main difficulty in studying biological molecules by scanning tunneling microscopy (STM) is in the sample preparation. For small molecules it is possible to deposit only a few desired molecules in a high vacuum on the 100 nm by 100 nm square that will be scanned. For macromolecules the sample is usually in a buffer solution which is applied to the conducting gold or graphite support. The solvent is evaporated to leave the molecules of interest. It is not always easy to distinguish a few macromolecules from impurities, or from imperfections in the surface of the support.

Atomic force microscopy (AFM) uses the repulsive force between all atoms (the London-van der Waals interaction) to image molecules. A tip on a cantilever arm is scanned over the surface of the sample and the angstrom-size displacement is measured. It is surprising, but true, that the spring force constants for atomic repulsion—of order

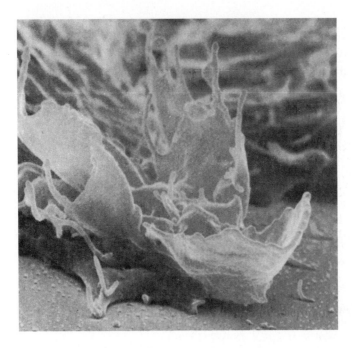

Fig. 12.26 Electron photomicrograph of a cultured hamster kidney cell adhered to a glass surface. Cells grown on glass microscope slide coverslips were "fixed" in glutaraldehyde, dried, uniformly coated with gold, and examined in a scanning electron microscope. The magnification is about 14,000 times. Photograph courtesy of J. P. Revel. [More details about this micrograph can be found in J. P. Revel, P. Hoch, and D. Ho, *Exp. Cell Res. 84,* 207 (1974).]

10 Newtons per meter—are larger than that of a cantilever arm made from household aluminum foil 4 mm long and 1 mm wide (as described by Rugar and Hansma, 1990). Thus, the slight up and down motion of the tip as it is scanned over the sample provides a surface image of the sample. The advantage for biological samples is that the atomic force scanning can be done in aqueous solution. No drying, staining, metal deposition, or other fixing need be done (Radmacher et al., 1992). The polymerization of fibrinogen to form fibrin, the main component of blood clots, has been followed by AFM (Drake et al., 1989). The atomic force microscope cannot only visualize macromolecules, the tip can be used to cut a DNA molecule at a specific place (Hansma et al., 1992), such as 25% of the distance from one end. Figure 12.27 shows an atomic force microscope image of a double-stranded DNA plasmid (psK 31) on a mica support under propanol. The apparent width of the DNA is 3 nm; B-form DNA has a width of about 2.4 nm.

Image Enhancement and Reconstruction

If a specimen is composed of a lattice of subunits, the periodicity of the structure should be present in its image recorded on an electron micrograph. We have already discussed direct studies of periodic structures by diffraction methods. Powerful mathematical methods have been developed to analyze diffraction patterns of periodic structures. Diffraction techniques can also be applied to study the *image* recorded on a micrograph. Because the image has been magnified many times from the original molecular assembly, its diffraction pattern must be studied with radiation of a much longer wavelength. The wavelength of visible light is convenient for studying the diffraction pattern of the image. Certain structural

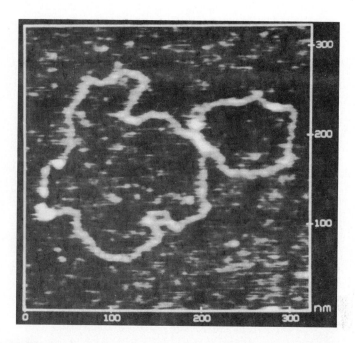

Fig. 12.27 An atomic force microscope picture of a double-stranded DNA plasmid (psK 31). The DNA molecules were deposited on freshly cleaved mica pre-treated with magnesium acetate. The DNA on mica was stored over a desiccant then imaged under propanol. The sharp tip on a narrow silicon nitride cantilever arm is scanned over the 400 nm × 400 nm area; about one minute is required to obtain an image. The apparent width of the DNA is about 3 nm; B-form DNA has a width of about 2.4 nm. The picture was kindly provided by Professor Carlos Bustamante, U. of Oregon; further description of the results is given in Hansma et al., *Science 256*, 1180 (1992).

features not readily discernible from the original electron micrograph image can be better characterized from the optical diffraction of the image. Undesirable features of the image, such as noise (which has no periodicity), can be filtered out. The filtered diffraction pattern can be reconstructed if necessary to give an improved image by the use of an appropriate lens. Diffraction, filtration, and reconstruction of the image can also be done mathematically. The intensity of the image is sampled at a grid of points covering the image and digitized. The other operations are then carried out mathematically by the use of a computer (see Fig. 12.28).

Because the image recorded on an electron micrograph is a two-dimensional superposition of different layers in a three-dimensional structure, it should be possible to deduce the three-dimensional structure from several images obtained with different viewing angles. Such three-dimensional image reconstruction is facilitated if the specimen is composed of symmetrically arranged subunits. The image of a structure composed of helically arranged subunits, for example, contains many different views of the subunit, and therefore

(a)

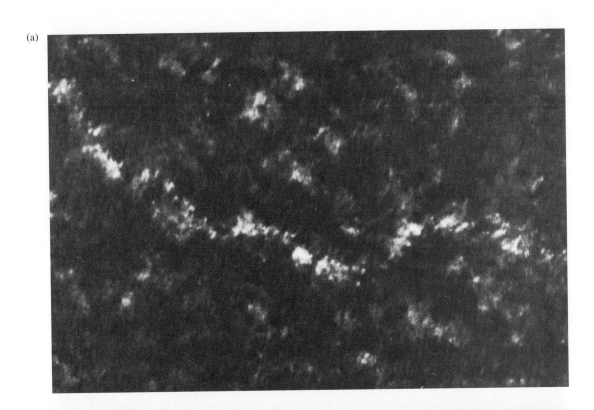

(b)

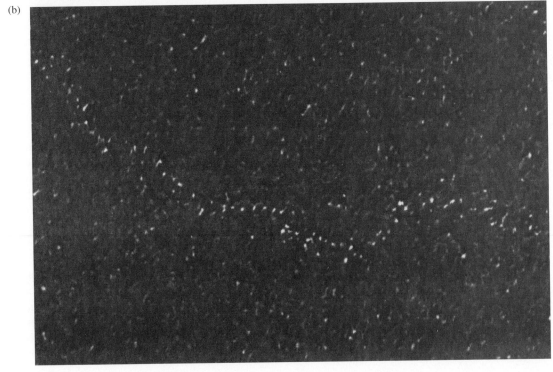

our image is sufficient for deducing the three-dimensional structure of a subunit. Three-dimensional image reconstruction has been done in recent years for a number of ordered structures, and resolutions in the range 4 to 20 Å have been achieved.

SUMMARY

X-ray Diffraction

X rays are photons (electromagnetic radiation) with wavelengths in the range from a few tenths of 1 Å to several angstroms. X rays are usually produced by bombarding a metal target with electrons. Synchrotrons produce radiation covering the entire range from hard x rays (1 Å) to soft x rays (100 Å) to the ultraviolet (1000 Å) and beyond.

Scattering from one point:

$$I_{sca} = \frac{CI_0}{r^2} \tag{12.1}$$

I_0 = intensity of incident plane-polarized light
I_{sca} = intensity of scattered light in plane perpendicular to plane of polarization
r = distance from scattering point to detector
C = proportionality constant

Scattering from two points:

$$I_{sca} = \frac{2CI_0}{r^2} \left(1 + \cos \frac{2\pi\Delta}{\lambda} \right) \tag{12.5}$$

Δ = path difference between the rays scattered from the two points; the path difference is the difference in distance traveled from source to each scattering point to detector
λ = wavelength of radiation
$2\pi\Delta/\lambda$ = phase difference between the rays scattered from the two points

von Laue's equations:

For a crystal made of identical scatterers at the lattice points, the conditions for diffraction are

---

Fig. 12.28 Electron micrographs of a segment of a polyuridylic acid molecule stained by an osmium compound. A high-resolution scanning transmission electron microscope was used in obtaining these micrographs. (a) Micrograph prior to noise filtering. (b) Micrograph after noise filtering (see the section on image enhancement and reconstruction). Individual osmium atoms attached to the uridine bases can be clearly seen. [Micrographs courtesy of M. Beer. For detailed discussions, see M. Cole, W. Wiggins, and M. Beer, *J. Mol. Biol. 117*, 387 (1978).]

$$a(\cos \alpha - \cos \alpha_0) = h\lambda \qquad (12.6)$$

$$b(\cos \beta - \cos \beta_0) = k\lambda \qquad (12.7)$$

$$c(\cos \gamma - \cos \gamma_0) = l\lambda \qquad (12.8)$$

λ = wavelength of x rays
a, b, c = lattice spacings along the axes **a, b, c,** respectively
$\alpha_0, \beta_0, \gamma_0$ = angles the incident light makes with the three axes **a, b,** and **c**
α, β, γ = angles the diffracted beam makes with the three axes **a, b,** and **c**
h, k, l = integers

Miller indices:

Any plane of Miller indices (h, k, l) is parallel to one that makes intercepts a/h, b/k, and c/l with the lattice axes **a, b,** and **c.** For an orthorhombic lattice, the interplanar spacing for planes of Miller indices (h, k, l) is

$$d_{hkl} = \left[\left(\frac{h}{a} \right)^2 + \left(\frac{k}{b} \right)^2 + \left(\frac{l}{c} \right)^2 \right]^{-1/2}$$

Bragg equation:

For a family of crystal planes of Miller indices (h, k, l) and an interplanar spacing d_{hkl}, the incident x-ray beam of wavelength λ appears to be reflected by the planes when the incident angle θ with respect to the planes satisfies the equation

$$\theta = \sin^{-1} \frac{n\lambda}{2d_{hkl}} \qquad n = 0, 1, 2, \ldots \qquad (12.9)$$

$$n\lambda = 2d_{hkl} \sin \theta \qquad (12.10)$$

The Bragg law is equivalent to von Laue's equations. We can also write

$$\theta = \sin^{-1} \frac{\lambda}{2(d_{hkl}/n)} \qquad (12.11)$$

d_{hkl}/n can be considered to be the interplanar spacing of planes of Miller indices (nh, nk, nl).

Unit cell:

The unit cell is the basic unit of a crystal structure. Translation of the unit cell along the three lattice axes generates the crystal.

Structure factor:

At a Bragg angle $\theta = \sin^{-1}(\lambda/2d_{hkl})$ corresponding to diffraction by planes of indices (h, k, l), the amplitude of the rays scattered by a unit cell can be represented by a complex number

$$F(h, k, l) = A(h, k, l) + i B(h, k, l) \tag{12.18}$$

$$A(h, k, l) = \sum_{j=1}^{n} f_j \cos [2\pi(hX_j + kY_j + lZ_j)] \tag{12.24a}$$

$$B(h, k, l) = \sum_{j=1}^{n} f_j \sin [2\pi(hX_j + kY_j + lZ_j)] \tag{12.24b}$$

f_j = atomic scattering factor of atom j

X_j, Y_j, Z_j = unitless coordinates of atom j, fractions of the unit-cell dimensions

n = total number of atoms in the unit cell

$F(h, k, l)$ is the structure factor. The intensity of the diffracted rays by planes of indices (h, k, l)) is proportional to $A^2 + B^2$. The phase of F is

$$\alpha(h, k, l) = \tan^{-1} \frac{B}{A} \tag{12.23}$$

Phase problem:

A critical step in obtaining the structure of the crystal from its diffraction pattern is to determine the phases of the structure factors. An important method for phase determination of macromolecules is the isomorphous replacement method.

Equations (12.18), (12.24a), and (12.24b) can be written in an equivalent form

$$F(h, k, l) = \sum_{j=1}^{n} f_j \, e^{+2\pi i \,(hX_j + kY_j + lZ_j)} \qquad i = \sqrt{-1} \tag{12.25}$$

The Fourier transform of this equation is the electron density of the unit cell.

$$\rho(X, Y, Z) = \frac{1}{V} \sum_{h} \sum_{k} \sum_{l} F(h, k, l) \, e^{-2\pi i \,(hX + kY + lZ)} \tag{12.26a}$$

$\rho(X, Y, Z)$ = electron density (electrons volume^{-1}) at coordinate X, Y, Z of the unit cell

V = volume of unit cell

$F(h, k, l)$ is needed to calculate the electron density; $F^2(h, k, l)$ or $|F(h, k, l)|$ is not sufficient.

Scattering by noncrystalline materials:

Some molecular parameters, such as molecular weight, mean-square radius, and neighboring atomic arrangement around an atom from which an electron has been ejected, can be deduced for noncrystalline materials. Structures of fibers can also be studied by x-ray diffraction.

Neutron Diffraction

The de Broglie wavelength of thermal neutrons is of the order of interatomic distances in crystals. Neutrons are scattered by nuclei; x-ray photons are scattered primarily by electrons.

The neutron scattering factor of an atom is not related to the atomic number or mass in any simple way. The scattering of neutrons by hydrogen is quite high, making neutron diffraction a valuable tool for determining the positions of the hydrogens in a crystal.

The isotopes hydrogen and deuterium scatter neutrons quite differently. This fact is useful in structural studies of macromolecules by neutron diffraction.

Electron Microsopy

The de Broglie wavelength of electrons accelerated by a voltage V is

$$\lambda = \frac{h}{\sqrt{2m_e e V_{\text{corrected}}}} \tag{12.33}$$

$$V_{\text{corrected}} = V \cdot \left(1 + \frac{eV}{2m_e c^2}\right) \tag{12.34}$$

c = speed of light in vacuum
m_e = rest mass of an electron
e = charge of an electron
h = Planck's constant

Electrostatic or magnetic lenses can be used to control the paths of a beam of electrons. Electron microscopes with a point resolution of a few angstroms have been constructed. In addition to resolution, contrast and specimen damage by the electron beam are also important considerations in electron microscopy.

MATHEMATICS NEEDED FOR CHAPTER 12

Trigonometric identities:

$$\sin^2 \theta + \cos^2 \theta = 1$$

$$\sin(\alpha + \beta) = \sin \alpha \cos \beta + \sin \beta \cos \alpha$$

$$\sin n\pi = 0 \quad \text{for } n = \text{integer}$$

$$\sin\left(\frac{\pi}{2} + n\pi\right) = 1 \quad \text{for } n = \text{even integer}$$

$$\sin\left(\frac{\pi}{2} + n\pi\right) = -1 \quad \text{for } n = \text{odd integer}$$

$$\cos n\pi = 1 \qquad \text{for } n = \text{even integer}$$

$$\cos n\pi = -1 \qquad \text{for } n = \text{odd integer}$$

$$\cos \frac{n\pi}{2} = 0 \qquad \text{for } n = \text{odd integer}$$

If $y = \sin \theta$, $\theta = \sin^{-1} y$

Complex numbers:

A complex number an be written as a sum of a real part and an imaginary part.

$a + ib$

$a = $ real part

$b = $ imaginary part

$i = \sqrt{-1}$

The magnitude of a complex number, $a + ib$, is equal to $\sqrt{a^2 + b^2}$.

A complex number can also be written in exponential form.

$Fe^{i\theta}$

$F = $ magnitude

$\theta = $ phase angle

$e^{i\theta} = \cos \theta + i \sin \theta$

By definition of the magnitude of a complex number, we understand why F is called the magnitude.

$$Fe^{i\theta} = F \cos \theta + iF \sin \theta$$

$$\text{magnitude} = \sqrt{F^2 \cos^2 \theta + F^2 \sin^2 \theta} = \sqrt{F^2} = F$$

REFERENCES

STOUT, G. H., and L. H. JENSEN, 1989. *X-Ray Structure Determination,* John Wiley, New York.
LADD, M. F. C., and R. A. PALMER, 1985. *Structure Determination by X-Ray Crystallography,* 2nd ed., Plenum Press, New York.

SUGGESTED READINGS

BUGG, C. E., W. C. CARSON, and J. A. MONTGOMERY, 1993. Drugs by Design, *Sci. Am.,* December, 92–98.
CHAPMAN, M. S., I. MINOR, M. G. ROSSMANN, G. D. DIANA, et al., 1991. Human Rhinovirus-14 Complexed with Antiviral Compound r-61837, *J. Mol. Biol. 217,* 455–463.
DICKERSON, R. E., and I. GEIS, 1983. *Hemoglobin: Structure, Function, Evolution, and Pathology,* Benjamin/Cummings, Menlo Park, California.

DRAKE, B., C. B. PRATER, A. L. WEISENHORN, S. A. C. GOULD, T. R. ALBRECHT, C. F. QUATE, D. S. CANNEL, H. G. HANSMA, and P. K. HANSMA, 1989. Imaging Crystals, Polymers, and Processes in Water with the Atomic Force Microscope, *Science 243,* 1586–1589.

ENGEL, A., 1991. Biological Applications of Probe Microscopes, *Annu. Rev. Biophys. Biophys. Chem. 20,* 79–108.

GLUSKER, J. P., D. E. ZACHARIAS, H. L. CARRELL, P. P. FU, and R. G. HARVEY, 1976. Molecular Structure of Benzo[a]pyrene 4,5-Oxide, *Cancer Research 36,* 3951–3957.

HAUPTMAN, H. A., 1989. The Phase Problem of X-Ray Crystallography, *Physics Today,* November, 24–29.

HANSMA, H. G., J. VESENKA, C. SIEGERIST, G. KELDERMAN, H. MORRETT, R. L. SINSHEIMER, V. ELLINGS, C. BUSTAMANTE, and P. K. HANSMA, 1992. Reproducible Imaging and Dissection of Plasmid DNA Under Liquid with the Atomic Force Microscope, *Science 256,* 1180–1184.

HENDRICKSON, W., 1986. The 1985 Nobel Prize in Chemistry, *Science 231,* 362–364. A description of the direct method for determining phases in x-ray diffraction.

HOLBROOK, S. R., J. L. SUSSMAN, R. W. WARRANT, and S.-H. KIM, 1978. Crystal Structure of Yeast Phenylalanine Transfer RNA. II. Structural Features and Functional Implications. *J. Mol. Biol. 123,* 631–660.

PAVLETICH, N. P., and C. O. PABO, 1991. Zinc Finger-DNA Recognition: Crystal Structure of a Zif268-DNA Complex at 2.1 Å, *Science 252,* 809–817.

PAVLETICH, N. P., and C. O. PABO, 1993. Crystal Structure of a Five-Finger GLI-DNA Complex: New Perspectives on Zinc Fingers, *Science 261,* 1701–1707.

PERUTZ, M., 1992. *Protein Structure: New Approaches to Disease and Therapy,* W. H. Freeman, New York. The first chapter is entitled "Diffraction Without Tears: A Pictorial Introduction to X-Ray Analysis of Crystal Structures."

RADMACHER, M., R. W. TILLMAN, M. FRITZ, and H. E. GAUB, 1992. From Molecules to Cells: Imaging Soft Samples with the Atomic Force Microscope, *Science 257,* 1900–1905.

RHODES, D., and A. KLUG, 1993. Zinc Fingers, *Sci. Am.* (February), 56–65.

RUGAR, D., and P. HANSMA, 1990. Atomic Force Microscopy, *Physics Today,* October, 23–29.

TONG, L., A. M. DEVOS, M. V. MILBURN, and S.-H. KIM, 1991. Crystal Structures at 2.2 Å Resolution of the Catalytic Domains of Normal Ras Protein and an Oncogenic Mutant Complexed with GDP, *J. Mol. Biol. 217,* 503–516.

ROSSMANN, M. G., and J. E. JOHNSON, 1989. Icosahedral RNA Virus Structure, *Annu. Rev. Biochem. 58,* 533–573.

PROBLEMS

1. The energy of an electron upon acceleration by a potential of V volts is eV, where e is the charge. In units of electron volts, the energy is just V electron volts.
   (a) Calculate the energy in joules of an electron accelerated by a potential of 40 kV.
   (b) When the electron strikes a target atom, the maximum energy of the photon emitted corresponds to the conversion of all the energy of the incoming electron to that of the photon. Calculate the wavelength of such photons.
   (c) Show that in general,

$$\lambda_s = \frac{12,390}{V} \text{ Å}$$

where λ_s is the shortest wavelength emitted when the electron strikes the target.

2. (a) For Cu metal, the K absorption edge is at $\lambda = 1.380$ Å (Fig. 12.21). For Ag metal, the absorption edge is at $\lambda = 0.4858$ Å. Calculate the energy required for the ejection of an electron from the K shell of Cu and that from the K shell of Ag.

(b) The wavelengths of four emission lines from Ag are 1.54433 Å (α_2), 1.54051 Å (α_1), 1.39217 Å (β) (which is actually a doublet at high resolution; the value given is the average of the doublet), and 1.38102 Å (γ). These lines result from the emission of x-ray photons when the electrons in the L-shell energy levels fall into the K shell vacated by the ejected electrons. Calculate the L-shell energy levels for Ag corresponding to these emission lines.

3. (a) For the arrangement shown in Fig. 12.7 calculate α for the first- and second-order diffraction cones if the wavelength of the incident x rays is 1.54 Å and the spacing between the equidistant scattering centers is 4 Å.

(b) A cylindrically shaped film, with its axis coincident with the row of scatterers, is used to record the diffracted rays. It is easy to see that the diffraction cones will intersect the film to give layer lines. Calculate the distance between the layer lines formed by the first and the second diffraction cones. The radius of the cylindrical film is 3 cm and all other parameters are the same as in part (a).

4. In Fig. 12.2 the path difference for the scattered rays is calculated for two points parallel and perpendicular to the incident beam. Derive an equation for the general case in which the incident beam makes an angle ϕ with the line joining the two points. Obtain $\Delta = f(R, \Theta, \phi)$.

5. Assume that you can measure scattered intensities to 1%. What is the smallest separation between two points which you could distinguish from a single point using 1.54-Å radiation? That is, what value of R gives a scattering pattern that differs by at least 1% from that of a point scatterer?

6. For $\lambda = 1.54$-Å x rays, obtain the first-order ($n = 1$) Bragg angles for interplanar spacings of 5 Å, 10 Å, and 1000 Å.

7. From Eq. (12.11) it can be seen that for a fixed λ, the interplanar spacing (d_{hkl}/n) is inversely proportional to $\sin \theta$. In other words scattering corresponding to the smallest spacing occurs at the maximum value of $\sin \theta$ (when $\theta = 90°$). If $\lambda = 1.54$-Å x rays are used, what is the theoretical limit of resolution? (That is, what is the minimum spacing observable?)

8. (a) The lattice of NaCl is cubic with a unit-cell dimension of 5.64 Å. Each unit cell contains 4 Na^+ and 4 Cl^- ions. The density of NaCl is 2.163 g cm^{-3}. Calculate Avogadro's number using this information.

(b) The dimensions of the unit cell of crystalline lysozyme are $a = 79.1$ Å, $b = 79.1$ Å, and $c = 37.9$ Å. The interaxial angles are all 90°. Each unit cell contains eight lysozyme molecules. The density of crystalline lysozyme is 1.242 g cm^{-3}, and chemical analysis showed that 64.4% (by weight) is lysozyme, the rest being water and salt. From these data, calculate the molecular weight of lysozyme.

9. (a) Show that for an orthorhombic lattice, the interplanar spacing d_{hkl} between planes of Miller indices (h, k, l) is

$$d_{hkl} = \frac{1}{\sqrt{\dfrac{h^2}{a^2} + \dfrac{k^2}{b^2} + \dfrac{l^2}{c^2}}}$$

where a, b, and c are the lattice spacings.

(b) Show that for a cubic lattice, $d_{100} : d_{110} : d_{111} = 1 : 0.707 : 0.578$.

10. (a) A cubic unit cell contains one atom at the origin. Calculate the relative intensities of the diffractions by the (1,0,0), (2,0,0), and (1,1,0) planes.
    (b) Do the same calculations for a unit cell containing two identical atoms, one at the origin and one at $X = Y = Z = \frac{1}{2}$.

11. Consider a two-dimensional rectangular lattice with unit cell dimensions 50 Å × 30 Å along the a and b axes, respectively.
    (a) Sketch an accurate picture of this lattice giving at least nine lattice sites.
    (b) On the drawing in part (a) sketch and label three representative planes having the Miller indices $(h, k) = (1, 0)$. Repeat this operation for the (0, 1) and (1, 1) planes.
    (c) Make another drawing of the lattice. Sketch and label (2, 0), (1, 2), and (2, 1) planes.
    (d) What are the distances between the planes in the six cases above?
    (e) If we use 1.54-Å radiation, what will be the angle of the (1, 0), (0, 1), (1, 1), and (2, 2) reflections in first order?
    (f) Suppose that the unit cell contains one copy of a protein that has been labeled with two lead atoms at (x, y) positions given by (12 Å, 7 Å) and (37 Å, 18 Å). The atomic scattering factor for Pb is 82. Determine the intensity of the (1, 0) and (2, 0) reflections and the phase of the respective structure factors. Discuss the physical basis behind the difference between these two intensities with reference to the drawings in parts (b) and (c).

12. Calculate the de Broglie wavelengths of electrons accelerated by a voltage of:
    (a) 10 kV.
    (b) 100 kV.

13. Calculate the de Broglie wavelength of thermal neutrons at 300 K.

14. To convince yourself that you understand x-ray diffraction:
    (a) Describe a diffraction pattern. How is it experimentally measured?
    (b) How is the electron density determined from a diffraction pattern? Describe what this has to do with phases, amplitudes, and intensities (if anything).
    (c) How are coordinates for the atoms of a molecule obtained?
    (d) What, if anything, does the structure of a molecule determined in a crystal have to do with biology?

15. Consider the crystal structure of flatein, a two-dimensional protein. The unit cell is a 2-D rectangle with cell dimensions 80 Å × 90 Å along a and b, respectively.
    (a) Sketch the unit cell and draw and label three planes having Miller indices: (i) 0,1 (ii) 2,0 (iii) 1,0 (iv) 1,1.
    (b) What are the distances between the planes above?
    (c) What is the first order Bragg diffraction angle using 1.75 Å radiation for a distance between planes of 7.00 Å?
    (d) You use heavy-atom isomorphous substitution to solve the phase problem. There is one protein per unit cell with one heavy atom per protein. The atom is Au which has an atomic scattering factor of 74. Determine the intensity of the diffraction spot for the 2,0 reflection, and the phase of the structure factor. The location of the Au atom within the unit cell is $x = 20$ Å, along a and $y = 30$ Å along b.

16. (a) An x-ray beam of wavelength 1.54 Å shines on a crystal. Two beams scattered from the crystal are exactly in phase at the detector. What are the possible path differences in Å for the two scattered beams? One answer is zero. Give three other possible answers.
    (b) The two beams in (a) are exactly out of phase at the detector. Give three possible values in Å for their possible path differences.

(c) A Bragg reflection occurs in first order at 5.00 degrees. What is the distance in Å between the scattering planes causing this reflection?

(d) A cubic crystal has a unit cell which is 5 Å on a side. Many different scattering planes can be drawn through the crystal; they are labeled by Miller indices h, k, l. What is the largest distance (d_{hkl}) between two scattering planes that can occur in this crystal?

(e) What is the smallest value of d_{hkl} that can be detected with 1.54 Å x rays?

(f) Consider a one-dimensional crystal of carbon atoms; the unit cell has a spacing of 3 Å. There are two atoms per unit cell at coordinates $X = 0$ and $X = \frac{2}{3}$. The scattering intensities will depend on one Miller index, h. What are the possible values of h? Write an expression for the structure factor, F_h, and calculate the intensity, I_h, for $h = 1$. The atomic scattering factor for C is $f_c = 6$.

17. The ovals below represent four proteins which form a complex in solution which was crystallized and studied by x-ray diffraction.

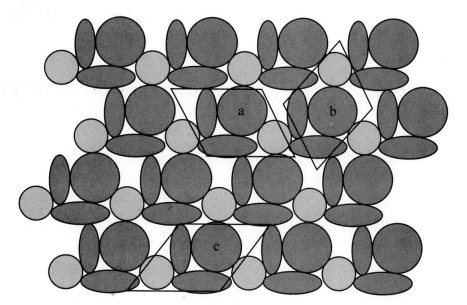

(a) Three possible unit cells are shown and labeled a, b, c. Define a unit cell and state which, if any, of the parallelograms can represent correct unit cells. Assume that the third axis is chosen correctly for all the unit cells.

(b) From the crystal structure can you tell whether the group of proteins in the unit cell corresponds to the complex present in solution?

(c) Sketch the diffraction pattern that you would expect from a two-dimensional crystal of these proteins. The symmetry of the diffraction pattern is the part to emphasize.

(d) If the lightly shaded spheres were crystallized separately, how would the spacing of the diffraction pattern change?

(e) If the wavelength of the x rays were decreased by a factor of two, how would the spacing of the diffraction pattern change?

(f) If the lightly shaded spheres were removed, but the packing of the other three proteins remained the same so that the unit cell size and symmetry did not change, how would the spacing of the diffraction pattern change?

Appendix

Table A.1 Useful physical constants

		SI units	cgs-esu units
Gas constant	R	8.3145 J K^{-1} mol^{-1}	8.3145 $\times$ 10^7 erg deg^{-1} mol^{-1}
			1.987 cal deg^{-1} mol^{-1}
			0.08205 L atm deg^{-1} mol^{-1}
Avogadro's number	N_0	6.0221 $\times$ 10^{23} mol^{-1}	6.0221 $\times$ 10^{23} molecules mol^{-1}
Boltzmann constant	k	1.3807 $\times$ 10^{-23} J K^{-1}	1.3807 $\times$ 10^{-16} erg deg^{-1}
Faraday constant	F	9.6485 $\times$ 10^4 C mol^{-1}	9.6485 $\times$ 10^4 C mol^{-1}
Speed of light	c	2.9979 $\times$ 10^8 m s^{-1}	2.9979 $\times$ 10^{10} cm s^{-1}
Planck constant	h	6.6261 $\times$ 10^{-34} J s	6.6261 $\times$ 10^{-27} erg s
Elementary charge	e	1.6022 $\times$ 10^{-19} C	4.8030 $\times$ 10^{-10} esu
Electron mass	m_e	9.1094 $\times$ 10^{-31} kg	9.1094 $\times$ 10^{-28} g
Proton mass	m_p	1.6726 $\times$ 10^{-27} kg	1.6726 $\times$ 10^{-24} g
Standard gravity	g	9.8066 m s^{-2}	980.66 cm s^{-2}
Permittivity of vacuum	ε_0	8.8542 $\times$ 10^{-12} C^2 N^{-1} m^{-2}	

C	= coulomb	L	= liter
g	= gram	m	= meter
J	= joule	N	= newton
K	= Kelvin	s	= second
kg	= kilogram		

Table A.2   Definition of prefixes

tera ≡ T	means multiply by 10^{12}
giga ≡ G	means multiply by 10^{9}
mega ≡ M	means multiply by 10^{6}
kilo ≡ k	means multiply by 10^{3}
centi ≡ c	means multiply by 10^{-2}
milli ≡ m	means multiply by 10^{-3}
micro ≡ μ	means multiply by 10^{-6}
nano ≡ n	means multiply by 10^{-9}
pico ≡ p	means multiply by 10^{-12}
femto ≡ f	means multiply by 10^{-15}
atto ≡ a	means multiply by 10^{-18}

Table A.3   Energy conversion factors

To convert from:	To:	Multiply by:
calories	ergs	4.184×10^{7}
calories	joules	4.184
calories	kilowatt-hours	1.162×10^{-6}
electron volts	kcal mol^{-1}	23.058
electron volts	ergs	1.602×10^{-12}
electron volts	joules	1.602×10^{-19}
electron volts	kilojoules mol^{-1}	96.474
ergs	calories	2.390×10^{-8}
ergs	joules	10^{-7}
ergs molecule^{-1}	joules mol^{-1}	6.022×10^{16}
ergs molecule^{-1}	kcal mol^{-1}	1.439×10^{13}
joules	calories	0.2390
joules	ergs	10^{7}
joules	kilowatt-hours	2.778×10^{-7}
joules mol^{-1}	ergs molecule^{-1}	1.661×10^{-17}
joules mol^{-1}	kcal mol^{-1}	2.390×10^{-4}
kilowatt-hours	calories	8.606×10^{5}
kilowatt-hours	joules	3.600×10^{6}
kcal mol^{-1}	electron volts	0.04337
kcal mol^{-1}	ergs molecule^{-1}	6.949×10^{-14}
kcal mol^{-1}	joules mol^{-1}	4.184×10^{3}
kilojoules mol^{-1}	electron volts	0.01036
kilojoules mol^{-1}	ergs molecule^{-1}	1.661×10^{-14}
kilojoules mol^{-1}	kcal mol^{-1}	0.2390

Table A.4   Miscellaneous conversions and abbreviations

Length

1 ångstrom $= 10^{-10}$ m $= 10^{-8}$ cm $= 0.1$ nm
1 inch $= 2.54$ cm
1 foot $= 30.48$ cm
1 mile $= 5280$ ft $= 1609$ m

Mass

1 pound $= 453.6$ g
1 kg $= 2.205$ pounds

Energy

1 erg $= 1$ g cm^2 s^{-2}
1 joule $= 1$ kg m^2 s^{-2}

Force

1 newton $= 1$ kg m s^{-2} $= 10^5$ dyn
1 dyne $= 1$ g cm s^{-2} $= 10^{-5}$ N

Pressure

1 atmosphere $= 760$ mmHg (Torr.) $= 14.70$ lb in^{-2}
     $= 1.013 \times 10^6$ dyn cm^{-2}
     $= 1.013 \times 10^5$ newton m^{-2}
     $= 1.033 \times 10^4$ kg-force m^{-2}
     $= 1.013 \times 10^5$ pascals
1 newton m^{-2} $= 1$ pascal

Power

1 watt $= 1$ J s^{-1} $= 1$ VA

Volume

1 gallon $= 3.785$ L

Viscosity

1 poise (P) $= 1$ g cm^{-1} s^{-1} $= 10^{-1}$ Pa s
1 pascal sec (Pa s) $= 1$ kg m^{-1} s^{-1} $= 10$ P
1 mPa s $= 1$ cP

Å	= ångstrom	P	= poise
A	= ampere	rad	= radian
cal	= calorie	V	= volt
hertz (Hz)	= cycles per second	W	= watt
hr	= hour	y	= year

Table A.5 Inorganic compounds*

	$\Delta H_f^0 \equiv \bar{H}^0$ (kJ mol^{-1})	$\bar{S}^0$ (J K^{-1} mol^{-1})	$\Delta G_f^0 \equiv \bar{G}^0$ (kJ mol^{-1})
Ag(s)	0	42.55	0
Ag$^+$(aq)†	105.579	72.68	77.107
AgCl(s)	−127.068	96.2	−109.789
C(g)	716.682	158.096	671.257
C(s, graphite)	0	5.740	0
C(s, diamond)	1.895	2.377	2.900
Ca(s)	0	41.42	0
CaCO$_3$(s, calcite)	−1206.92	92.9	−1128.79
Cl$_2$(g)	0	223.066	0
Cl$^-$(aq)	−167.159	56.5	−131.228
CO(g)	−110.525	197.674	−137.168
CO$_2$(g)	−393.509	213.74	−394.359
CO$_2$(aq)	−413.80	117.6	−385.98
HCO$_3^-$(aq)	−691.99	91.2	−586.77
CO$_3^{2-}$(aq)	−677.14	−56.9	−527.81
Fe(s)	0	27.28	0
Fe$_2$O$_3$(s)	−824.2	87.40	−742.2
H$_2$(g)	0	130.684	0
H$_2$O(g)	−241.818	188.825	−228.572
H$_2$O(l)	−285.830	69.91	−237.129
H$^+$(aq)	0	0	0
OH$^-$(aq)	−229.994	−10.75	−157.244
H$_2$O$_2$(aq)	−191.17	143.9	−134.03
H$_2$S(g)	−20.63	205.79	−33.56
N$_2$(g)	0	191.61	0
NH$_3$(g)	−46.11	192.45	−16.45
NH$_3$(aq)	−80.29	111.3	−26.50
NH$_4^+$(aq)	−132.51	113.4	−79.31
NO(g)	90.25	210.761	86.55
NO$_2$(g)	33.18	240.06	51.31
NO$_3^-$(aq)	−205.0	146.4	−108.74
Na$^+$(aq)	−240.12	59.0	−261.905
NaCl(s)	−411.153	72.13	−384.138
NaCl(aq)	−407.27	115.5	−393.133
NaOH(s)	−425.609	64.455	−379.494
O$_2$(g)	0	205.138	0
O$_3$(g)	142.7	238.93	163.2
S(rhombic)	0	31.80	0
SO$_2$(g)	−296.830	248.22	−300.194
SO$_3$(g)	−395.72	256.76	−371.06

* Standard thermodynamic values at 25°C (298 K) and 1 atm pressure. Values for ions refer to an aqueous solution at unit activity on the molarity scale. Standard enthalpy of formation, ΔH_f^0, third-law entropies, S^0, and standard Gibbs free energy of formation, ΔG_f^0, are given.

† The standard state for all ions and for species labeled (aq) is that of a solute on the molarity scale.

SOURCE: Data from *The NBS Tables of Thermodynamic Properties*, D. D. Wagman et al., eds., *J. Phys. Chem. Ref. Data, 11*, Suppl. 2 (1982).

Table A.6 Hydrocarbons*

	$\Delta H_f^0 \equiv \bar{H}^0$ (kJ mol^{-1})	$\bar{S}^0$ (J K^{-1} mol^{-1})	$\Delta G_f^0 \equiv \bar{G}^0$ (kJ mol^{-1})
Acetylene, $C_2H_2(g)$	226.73	200.94	209.20
Benzene, $C_6H_6(g)$	82.93	269.20	129.66
Benzene, $C_6H_6(l)$	49.04	173.26	124.35
n-Butane, $C_4H_{10}(g)$	−126.15	310.12	−17.15
Cyclohexane, $C_6H_{12}(g)$	−123.14	298.24	31.76
Ethane, $C_2H_6(g)$	−84.68	229.60	−32.82
Ethylene, $C_2H_4(g)$	52.26	219.56	68.15
n-Heptane, $C_7H_{16}(g)$	−187.78	427.90	7.99
n-Hexane, $C_6H_{14}(g)$	−167.19	388.40	−0.25
Isobutane, $C_4H_{10}(g)$	−134.52	294.64	−20.88
Methane, $CH_4(g)$	−74.81	186.264	−50.72
Napthalene, $C_{10}H_8(g)$	150.96	335.64	223.59
n-Octane, $C_8H_{18}(g)$	−208.45	466.73	16.40
n-Pentane, $C_5H_{12}(g)$	−146.44	348.95	−8.37
Propane, $C_3H_8(g)$	−103.85	269.91	−23.47
Propylene, $C_3H_6(g)$	20.42	266.94	62.72

* Standard thermodynamic values at 25°C (298 K) and 1 atm pressure. Standard enthalpy of formation, ΔH_f^0, third-law entropies, S^0, and standard Gibbs free energy of formation, ΔG_f^0, are given.

SOURCE: Data from D. R. Stull, E. F. Westrum, Jr., and G. C. Sinke, *The Chemical Thermodynamics of Organic Compounds,* John Wiley, New York (1969).

Table A.7 Organic compounds*

	$\Delta H_f^0 \equiv \bar{H}^0$ (kJ mol^{-1})	$\bar{S}^0$ (J K^{-1} mol^{-1})	$\Delta G_f^0 \equiv \bar{G}^0$ (kJ mol^{-1})	$\Delta G_f^0 \equiv \bar{G}^0$ (1 M activity, aq) (kJ mol^{-1})
Acetaldehyde CH$_3$CHO(g)	−166.36	264.22	−133.30	−139.24
Acetate$^-$(aq)	—	—	—	−372.334
Acetic acid CH$_3$CO$_2$H(l)	−484.1	159.83	−389.36	−396.60
Acetone CH$_3$COCH$_3$(l)	−248.1	200.4	−155.39	−161.00
Adenine C$_5$H$_5$N$_5$(s)	95.98	151.00	299.49	—
L-Alanine CH$_3$CHNH$_2$COOH(s)	−562.7	129.20	−370.24	−371.71
L-Alanylglycine C$_5$H$_{10}$N$_2$O$_3$(s)	−826.42	195.05	−532.62	—
L-Aspartate^{+--}(aq)	—	—	—	−698.69
Aspartic acid C$_4$H$_7$NO$_4$(s)	−973.37	170.12	−730.23	−719.98
Butyric acid C$_3$H$_7$COOH(s)	−533.9	226.4	−377.69	—
Citrate^{3-}(aq) C$_6$H$_5$O$_7$	—	—	—	−1168.34
Creatine C$_4$H$_9$N$_3$O$_2$(s)	−537.18	189.5	−264.93	—
L-Cysteine HSCH$_2$CHNH$_2$COOH(s)	−533.9	169.9	−343.97	−340.33
L-Cystine C$_6$H$_{12}$N$_2$O$_4$S$_2$(s)	−1051.9	280.58	−693.33	−674.29
Ethanol C$_2$H$_5$OH(l)	−276.98	160.67	−174.14	−180.92
Formaldehyde CH$_2$O(g)	−115.90	218.78	−109.91	−130.5
Formamide HCONH$_2$(g)	−186.2	248.45	−141.04	—
Formic acid HCOOH(l)	−424.76	128.95	−361.46	—
Fumarate$^-$(aq)	—	—	—	−604.21
Fumaric acid trans-(=CHCOOH)$_2$(s)	−811.07	166.1	−653.67	−646.05
α-D-Galactose C$_6$H$_{12}$O$_6$(s)	−1285.37	205.4	−919.43	−924.58
α-D-Glucose C$_6$H$_{12}$O$_6$(s)	−1274.4	212.1	−910.52	−917.47
L-Glutamate^{+--}(aq)	—	—	—	−694.00
L-Glutamic acid C$_5$H$_9$NO$_4$(s)	−1009.68	118.20	−731.28	−722.70
Glycerol HOCH$_2$CHOHCH$_2$OH(l)	−668.6	204.47	−477.06	−488.52
Glycine H$_2$CNH$_2$COOH(s)	−537.2	103.51	−377.69	−379.9
Glycylglycine C$_4$H$_8$N$_2$O$_3$(s)	−745.25	189.95	−490.57	—

Table A.7 Organic compounds* *(cont.)*

	$\Delta H_f^0 \equiv \bar{H}^0$ (kJ mol^{-1})	$\bar{S}^0$ (J K^{-1} mol^{-1})	$\Delta G_f^0 \equiv \bar{G}^0$ (kJ mol^{-1})	$\Delta G_f^0 \equiv \bar{G}^0$ (1 M activity, aq) (kJ mol^{-1})
Guanine $C_5H_5N_5O(s)$	−183.93	160.2	47.40	—
L-Isoleucine $C_6H_{13}NO_2(s)$	−638.1	207.99	−347.15	—
Lactate$^-$(aq)	—	—	—	−517.812
L-Lactic acid $CH_3CHOHCOOH(s)$	−694.08	142.26	−522.92	—
β-Lactose $C_{12}H_{22}O_{11}(s)$	−2236.72	386.2	−1566.99	−1569.92
L-Leucine $C_6H_{13}NO_2(s)$	−646.8	211.79	−357.06	−353.09
Maleic acid cis-(=CHCOOH)$_2$(s)	−790.61	159.4	−631.20	—
Methanol $CH_3OH(l)$	−238.57	126.8	−166.23	−175.23
L-Methionine $C_5H_{11}NO_2S(s)$	−758.6	231.08	−505.76	—
Oxalic acid (—COOH)$_2$(s)	−829.94	120.08	−701.15	—
Oxaloacetate^{--}(aq) $C_4H_2O_5$	—	—	—	−797.18
L-Phenylalanine $C_9H_{11}NO_2(s)$	−466.9	213.63	−211.59	—
Pyruvate$^-$(aq)	—	—	—	−474.33
Pyruvic acid $CH_3COCOOH(l)$	−584.5	179.5	−463.38	—
L-Serine $HOCH_2CHNH_2COOH(s)$	−726.3	149.16	−509.19	—
Succinate^{2-}(aq)	—	—	—	−690.23
Succinic acid (—CH$_2$COOH)$_2$(s)	−940.90	175.7	−747.43	−746.22
Sucrose $C_{12}H_{22}O_{11}(g)$	−2222.1	360.2	−1544.65	−1551.76
L-Tryptophan $C_{11}H_{12}N_2O(s)$	−415.0	251.04	−119.41	—
L-Tyrosine $C_9H_{11}NO_3(s)$	−671.5	214.01	−385.68	−370.83
Urea $NH_2CONH_2(s)$	−333.17	104.60	−197.15	−203.84
L-Valine $C_5H_{11}NO_2(s)$	−617.98	178.86	−358.99	—

* Standard thermodynamic values at 25°C (298 K) and 1 atm pressure. Values for ions refer to an aqueous solution at unit activity on the molarity scale. Standard enthalpy of formation, ΔH_f^0, third-law entropies, S^0, and standard Gibbs free energy of formation, ΔG_f^0, are given.

SOURCES: Data from D. R. Stull, E. F. Westrum, Jr., and G. C. Sinke, *The Chemical Thermodynamics of Organic Compounds,* John Wiley, New York (1969) and from J. T. Edsall and J. Wyman, Biophysical Chemistry, Vol. 1, Academic Press, New York (1958).

Table A.8 Atomic weights of the elements*

Element	Symbol	Atomic number	Atomic weight	Element	Symbol	Atomic number	Atomic weight
Actinium	Ac	89	(227)	Lutetium	Lu	71	174.97
Aluminum	Al	13	26.98	Magnesium	Mg	12	24.312
Americium	Am	95	(243)	Manganese	Mn	25	54.94
Antimony	Sb	51	121.75	Mendelevium	Md	101	(256)
Argon	Ar	18	39.948	Mercury	Hg	80	200.59
Arsenic	As	33	74.92	Molybdenum	Mo	42	95.94
Astatine	At	85	(210)	Neodymium	Nd	60	144.24
Barium	Ba	56	137.34	Neon	Ne	10	20.183
Berkelium	Bk	97	(249)	Neptunium	Np	93	(237)
Beryllium	Be	4	9.012	Nickel	Ni	28	58.71
Bismuth	Bi	83	208.98	Niobium	Nb	41	92.91
Boron	B	5	10.81	Nitrogen	N	7	14.007
Bromine	Br	35	79.909	Nobelium	No	102	(253)
Cadmium	Cd	48	112.40	Osmium	Os	76	190.2
Calcium	Ca	20	40.08	Oxygen	O	8	15.9994
Californium	Cf	98	(251)	Palladium	Pd	46	106.4
Carbon	C	6	12.011	Phosphorus	P	15	30.974
Cerium	Ce	58	140.12	Platinum	Pt	78	195.09
Cesium	Cs	55	132.91	Plutonium	Pu	94	(242)
Chlorine	Cl	17	35.453	Polonium	Po	84	(210)
Chromium	Cr	24	52.00	Potassium	K	19	39.102
Cobalt	Co	27	58.93	Praseodymium	Pr	59	140.91
Copper	Cu	29	63.54	Promethium	Pm	61	(147)
Curium	Cm	96	(247)	Protactinium	Pa	91	(231)
Dysprosium	Dy	66	162.50	Radium	Ra	88	(226)
Einsteinium	Es	99	(254)	Radon	Rn	86	(222)
Erbium	Er	68	167.26	Rhenium	Re	75	186.23
Europium	Eu	63	151.96	Rhodium	Rh	45	102.91
Fermium	Fm	100	(253)	Rubidium	Rb	37	85.47
Fluorine	F	9	19.00	Ruthenium	Ru	44	101.1
Francium	Fr	87	(223)	Samarium	Sm	62	150.35
Gadolinium	Gd	64	157.25	Scandium	Sc	21	44.96
Gallium	Ga	31	69.72	Selenium	Se	34	78.96
Germanium	Ge	32	72.59	Silicon	Si	14	28.09
Gold	Au	79	196.97	Silver	Ag	47	107.870
Hafnium	Hf	72	178.49	Sodium	Na	11	22.9898
Helium	He	2	4.003	Strontium	Sr	38	87.62
Holmium	Ho	67	164.93	Sulfur	S	16	32.064
Hydrogen	H	1	1.0080	Tantalum	Ta	73	180.95
Indium	In	49	114.82	Technetium	Tc	43	(99)
Iodine	I	53	126.90	Tellurium	Te	52	127.60
Iridium	Ir	77	192.2	Terbium	Tb	65	158.92
Iron	Fe	26	55.85	Thallium	Tl	81	204.37
Krypton	Kr	36	83.80	Thorium	Th	90	232.04
Lanthanum	La	57	138.91	Thulium	Tm	69	168.93
Lawrencium	Lw	103	(257)	Tin	Sn	50	118.69
Lead	Pb	82	207.19	Titanium	Ti	22	47.90
Lithium	Li	3	6.939	Tungsten	W	74	183.85

Table A.8 Atomic weights of the elements* *(cont.)*

Element	Symbol	Atomic number	Atomic weight	Element	Symbol	Atomic number	Atomic weight
Uranium	U	92	238.03	Yttrium	Y	39	88.91
Vanadium	V	23	50.94	Zinc	Zn	30	65.37
Xenon	Xe	54	131.30	Zirconium	Zr	40	91.22
Ytterbium	Yb	70	173.04				

* Based on mass of ^{12}C at 12.000 . . . . The ratio of these weights to those on the older chemical scale (in which oxygen of natural isotopic composition was assigned a mass of 16.000 . . .) is 1.000050. (Values in parentheses represent the most stable known isotopes.)

Table A.9 Biochemical compounds

Amino acids found in proteins:

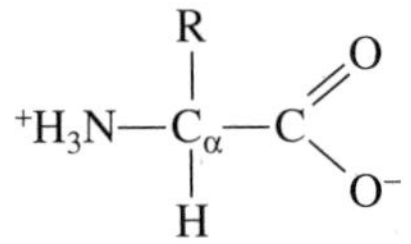

R groups:

Glycine	H—
Alanine	CH_3—
Valine	(isopropyl group)
Leucine	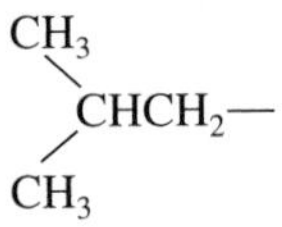
Isoleucine	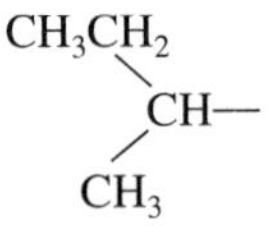
Phenylalanine	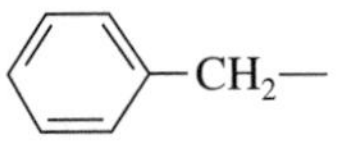

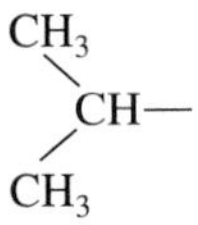

Tyrosine

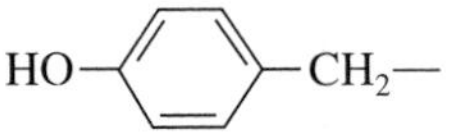

Tryptophan

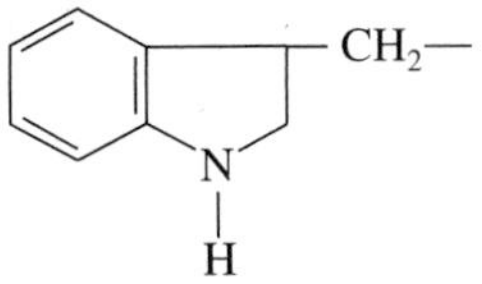

Threonine

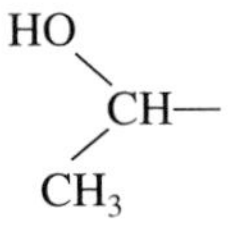

Methionine $CH_3SCH_2CH_2$—

Cysteine $HSCH_2$—

Proline (amino acid)

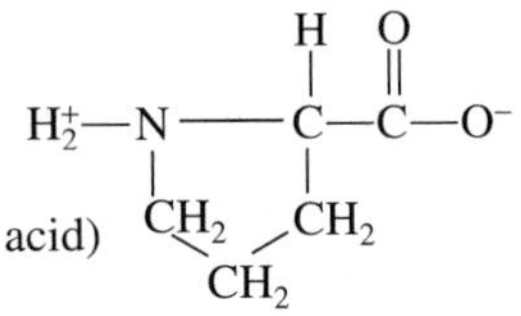

Table A.9 Biochemical compounds *(cont.)*

R groups *(cont.)*		Lysine	$H_2NCH_2CH_2CH_2CH_2-$

R groups (cont.)

Aspartic acid

$$\underset{\bar{O}}{\overset{O}{\overset{\|}{C}}}CH_2-$$

Glutamic acid

$$\underset{\bar{O}}{\overset{O}{\overset{\|}{C}}}CH_2CH_2-$$

Asparagine

$$\underset{NH_2}{\overset{O}{\overset{\|}{C}}}CH_2-$$

Glutamine

$$\underset{NH_2}{\overset{O}{\overset{\|}{C}}}CH_2CH_2-$$

Lysine $\quad H_2NCH_2CH_2CH_2CH_2-$

Histidine

$$\overset{N}{\underset{H}{\bigsqcup_{N}}}-CH_2-$$

Arginine

$$\underset{NH}{\overset{}{H_2N\overset{\|}{C}NHCH_2CH_2CH_2-}}$$

Serine $\quad HOCH_2-$

A polypeptide chain:

$$^+H_3N\overset{R_1}{\underset{|}{C}}H\overset{O}{\overset{\|}{C}}NH-\left[\overset{R_n}{\underset{|}{C}}H\overset{O}{\overset{\|}{C}}NH\right]_{\!\!n}-\overset{R_2}{\underset{|}{C}}HCO_2^-$$

The amino terminal residue is R_1; the carboxyl terminal is R_2.

Components of DNA: Nucleotides

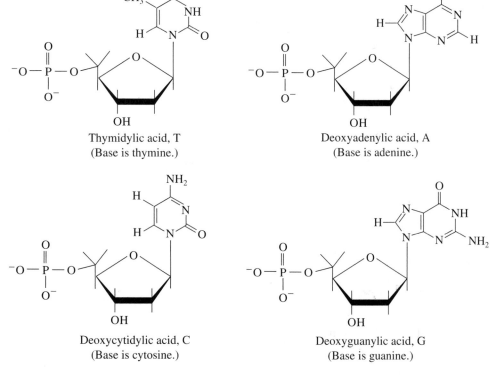

Thymidylic acid, T
(Base is thymine.)

Deoxyadenylic acid, A
(Base is adenine.)

Deoxycytidylic acid, C
(Base is cytosine.)

Deoxyguanylic acid, G
(Base is guanine.)

Components of RNA: Nucleotides *(cont.)*

Uridylic acid, U
(Base is uracil.)

Cytidylic, adenylic and guanylic acid are similar to uridylic acid,
but with cytosine, adenine, and guanine replacing uracil.

A polynucleotide chain, a single strand of DNA. (A single strand of RNA would have a
hydroxyl group at each 2′ position.)

Base 1 is at the 5′ terminal end of the chain; base 2 is at the 3′ terminal end.

Table A.9 Biochemical compounds *(cont.)*

Miscellaneous:

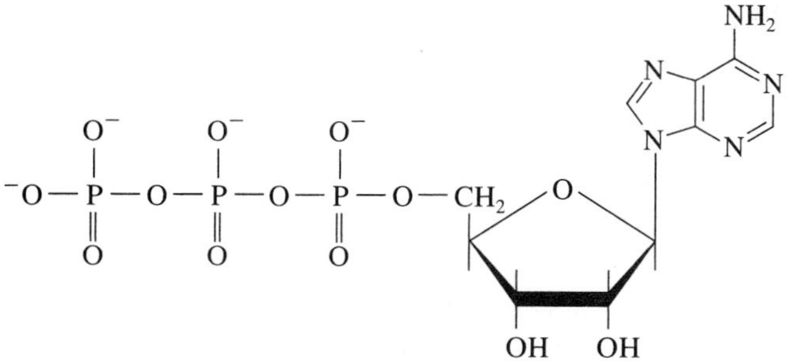

Adenosine triphosphate (ATP)

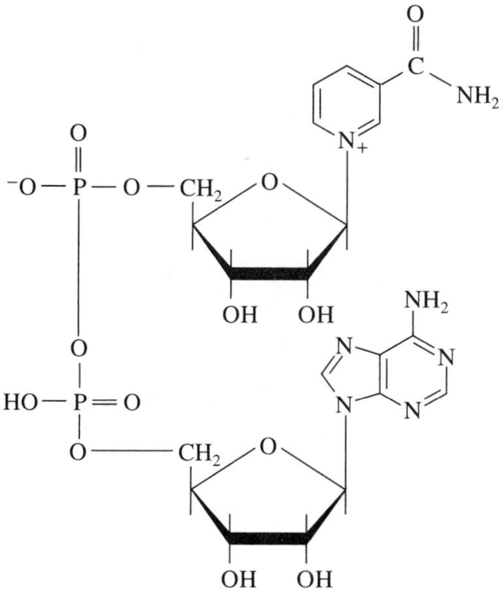

Nicotinamide adenine dinucleotide (NAD⁺)

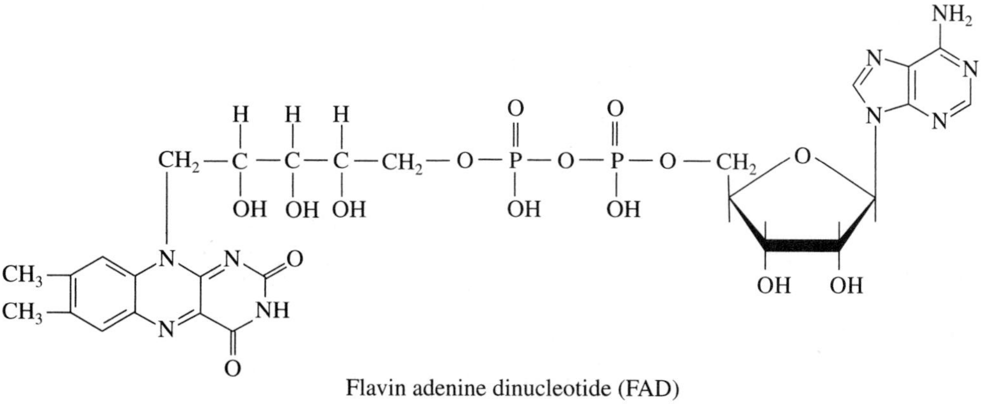

Flavin adenine dinucleotide (FAD)

Miscellaneous *(cont.)*

Heme, Fe II-protoporphyrin IX
(Hematin has Fe III.)

Chlorophyll *a* (R = phytyl, $C_{20}H_{39}-$)

Coenzyme Q (Ubiquinone)

Answers to Selected Problems

Chapter 2

2. (a) 6.8×10^6 L; (c) 1.3 yr. **3.** (d) -203 J **6.** (a) 41.84 kJ; (b) -33.34 kJ **7.** (a) -2.49 kJ
10. (a) $q = 0$, $w < 0$, $\Delta E < 0$, $\Delta H < 0$ **11.** (a) 40.66 kJ **12.** (a) 761 J **15.** (b) $q = 0$, $w > 0$,
$\Delta E > 0$, $\Delta H > 0$ **17.** 1 degree per hour. About 3 hours. **20.** 40°C **22.** (a) $q = -66.6$ kJ mol^{-1};
(b) 2.4% **26.** (a) -66.6 kJ mol^{-1}; (b) -1052.5 kJ mol^{-1}; (c) -2802 kJ mol^{-1} **29.** (a) -141.8 kJ g^{-1};
(b) -48.3 kJ g^{-1}; (c) on a weight basis, $H_2(g)$ produces 3 times as much heat on burning as does
n-octane. **33.** (a) -217.28 kJ mol^{-1}; (b) 140.5 kJ mol^{-1}; (c) -189.3 kJ mol^{-1}; (d) 25.23°C

Chapter 3

1. (b) $\Delta E = -1.247$ kJ; $\Delta S = -0.217$ J K^{-1} **2.** (a) $q_1 = 133$ kJ; $q_3 = -33$ kJ; (b) $q_1 = -133$ kJ;
$q_3 = 33$ kJ; (c) $q_1' = 125$ kJ; $q_3' = -25$ kJ **5.** (a) 100°C; (b) 8.3×10^{-4} mol; (c) 4.83×10^{-7} kJ K^{-1}
8. (a) $\Delta H_{298}^0 = -141,780$ kJ; $\Delta S_{298}^0 = -80.94$ kJ K^{-1}; $\Delta G_{298}^0 = -117,650$ kJ;
(b) $\Delta H_{298}^0 = -48,254$ kJ; $\Delta S_{298}^0 = -6.087$ kJ K^{-1}; $\Delta G_{298}^0 = -46.440$ kJ **10.** (a) irreversibly;
(c) enthalpy change **14.** (a) -6.007 kJ mol^{-1}; (b) -6.007 kJ mol^{-1}; (c) -22.2 J K^{-1} mol^{-1}; (d) zero;
(e) -6.007 kJ mol^{-1}; heat is released; (f) -0.17 J mol^{-1}; work is done by the system
15. (a) decrease; (b) remain unchanged; (g) remain unchanged; (h) more negative than
17. (a) If more solvent is bound by the coil than the helix, the entropy and enthalpy can decrease.
Increasing the temperature will not favor the helix-coil transition. (b) Yes, $\Delta G^0 = -0.26$ kJ (mol amide)$^{-1}$
(c) $T_m = 60$°C (d) No **24.** (a) temperature, internal energy; (c) increases; (e) greater than;
(g) q, decreases in magnitude **27.** 2668 atm

Chapter 4

1. (a) $K = 77$, $\Delta G_{298}^{0'} = -10.8$ kJ mol^{-1}; (b) $\Delta G_{298}^{0'} = 20.2$ kJ mol^{-1}, $K = 2.9 \times 10^{-4}$ **3.** (c) 1.2×10^{-5}
5. (a) 0.22 **8.** (a) -3.4 kJ mol^{-1}; (b) 0; (c) -3.4 kJ mol^{-1}; (d) -6.8 kJ mol^{-1}; (e) 52.9 kJ mol^{-1};
(f) 189 J K^{-1} mol^{-1} **10.** 5×10^{-49} atm. Do not worry.
13. (b) $[H^+] = 1.0 \times 10^{-7}$ M, $[OH^-] = 1.0 \times 10^{-7}$ M, $[Na^+] = 0.138$ M, $[H_2PO_4^-] = 0.062$ M,
$[HPO_4^{2-}] = 0.038$ M, $[PO_4^{3-}] = 1.7 \times 10^{-7}$ M, $[H_3PO_4] = 8.7 \times 10^{-7}$ M **16.** (a) -46.25 kJ mol^{-1};
(b) -1.72 kJ mol^{-1}; (c) $[GTP] = 1.9 \times 10^{-8}$ M, $[GDP] = 0.055$ M, $[P_i] = 0.065$ M **19.** (a) $+0.363$ V
21. (c) -145.8 kJ mol^{-1} **22.** (a) 5×10^{-8}; (b) -0.439 V **25.** (a) -1.38 V; (b) 133 kJ mol^{-1};
(c) -13.5 kJ mol^{-1} **27.** (a) -215 kJ mol^{-1} **31.** (a) For the sequence 5'GGGCCC3', $\Delta G^0 =$
-44.1 kJ mol^{-1}, $\Delta H^0 = -230.4$ kJ mol^{-1}, $\Delta S^0 = -625.2$ J K^{-1} mol^{-1}. For the sequence 5'-GGTTCC-3' +
5'-GGAACC-3', $\Delta G^0 = -25.1$ kJ mol^{-1}, $\Delta H^0 = -180.7$ kJ mol^{-1}, $\Delta S^0 = -521.8$ J K^{-1} mol^{-1}.
(b) For the sequence 5'-GGGCCC-3' $T_m = 55$°C; for the sequence 5'-GGTTCC-3' + 5'-GGAACC-3'
$T_m = 26$°C.

Chapter 5

1. (a) 7.99 kg m^{-2}; (b) 1.57×10^5 L; (c) 5.28 kg rain, 12,920 kJ, therefore temperature rises;
(d) 15.9 atm   **3.** (a) 9.3 kJ mol^{-1} **5.** (a) 45.9 kJ mol^{-1}; (b) 22.07 kJ mol^{-1}; not enough; (c) 2.1
7. $n = 5$, $K = 1.1 \times 10^5$ M^{-1} **13.** (a) $-76°C$; (b) 97.2 J K^{-1} mol^{-1}. Hydrogen bonding and association in liquid decreases its entropy. (c) 128.1 kJ mol^{-1}. Pure liquid is standard state.
(d) NH_4Cl dissociates completely in NH_3. Boiling point corresponds to a 2 molal solution.
16. (a) true; (c) false; (e) true   **18.** (a) 17.52 Torr; (b) 1015 Torr; (c) 0.999; (d) 1548 Torr
21. (a) -2.79 kJ; (b) -22.2 kJ; (c) -7.0 kJ; (d) Yes, the free energy of transfer of the side chain of
valine to the interior is negative; (e) $K(gly)/K(val) = 16.9$ **25.** 0.84 atm   **26.** (a) -0.062 V;
(b) 13.8 kJ mol^{-1} **29.** (a) 0.30 M sucrose; (b) 4.9 atm

Chapter 6

1. (a) 1.84×10^5 cm s^{-1}; (b) 3.4 kJ mol^{-1}; (c) 2.7×10^{19} molecule cm^{-3}; (d) 1.34×10^{-5} cm;
(e) 1.26×10^{10} s^{-1}; (f) 1.70×10^{29} cm^{-3} s^{-1}; (g) 1.5 cm s^{-1} **2.** (a) 0 K; (b) ∞ K; (c) 320 K
3. (b) 6.7×10^{-8} g s^{-1} **6.** (a) 5.05×10^7 g mol^{-1}; (b) 2.58×10^7 g mol^{-1}

7. $\dfrac{s_2}{s_1} = \left(\dfrac{M_2}{M_1}\right)^{2/3}$; $\dfrac{D_2}{D_1} = \left(\dfrac{M_1}{M_2}\right)^{1/3}$; $\dfrac{[\eta_2]}{[\eta_1]} = 1$

10. (a) 271×10^3 g mol^{-1}; (b) 132×10^3 g mol^{-1}; (c) 17.4×10^3 g mol^{-1} **13.** (a) 7.3×10^{-9} s
17. (a) 68,400   **18.** ring/linear = 0.328   **23.** (a) 62.2 Å; (b) 60.1 Å;
(c) Either shape is not spherical, or hydration changes on ligation.   **25.** 2.02 cm^3 g^{-1}

Chapter 7

1. (a) I_2 is zero order; ketone is first order; H^+ is first order; (b) 0.034 M^{-1} s^{-1}
5. (e) $A + A \rightarrow A_2$ (slow), $A_2 + B \rightarrow C + D + A$ (fast)   **7.** (a) $d(D)/dt = k_1 k_3 (A)(C)/[k_2 + k_3(C)]$
10. (a) 1.0%; (b) 5.3%   **13.** (a) A 29-year-old wine usually commands a premium price.

16. (a) $\dfrac{d[U]}{dt} = k_0 - k_1[U]$; (b) $\dfrac{d[U]}{dt} = k_0 e^{k_1 t}$ from derivative of expression for [U] given in (b);
substitute expression for [U] given in (b) into answer for (a) to show derivatives are equal; (c) Max.
occurs at $t \rightarrow \infty$; $[U]_{max} = k_0/k_1 = 2.60$ μM; (d) 1260 s; (e) 3817 s; (e) 0.385 nM s^{-1}.

20. 52.9 kJ   **22.** (a) $\ln [(A)/(A)_0] = -k_1(B)t$, first order in (A) and in (B); (b) $(A)_0 - (A) = k_2(B)t$,
zero order in (A) and first order in (B).   **24.** (a) second order; (b) $A + A \rightarrow P$; (c) 5 M^{-1} min^{-1};
(d) 67 min   **28.** 26.0°C   **32.** (b) $k_1 = 8.5 \times 10^8$ M^{-1} s^{-1}; $k_{-1} = 2.0 \times 10^{-6}$ s^{-1}
34. (a) (I) $d(O_2)/dt = 3k(O_3)^2$, (II) $d(O_2)/dt = 2(k_2 k_1/k_{-1})(O_3)^2/(O_2)$; (c) No. For mechanism (I),
$\Delta H^{\neq}$ must be > 0; for mchanism (II), $\Delta H^{\neq}$ must be > 107 kJ.   **40.** (a) 4.8 hr   **42.** (c) 4 mW   **45.**
1.4×10^{17}

Chapter 8

1. $K_M = 0.44$ M, V $= 0.35$ μmol CO_2 min^{-1} **2.** (c) $K'_{eq} = 3.75$ at pH 7.1, $K_{eq} = 3.0 \times 10^{-7}$
5. (a) 10^{-3} M s^{-1}; (b) 24 kJ mol^{-1}; (c) 10^4 M; (d) -14 kJ mol^{-1}

8. $\dfrac{d(P)}{dt} = \dfrac{k_2(E)_0}{1 + \dfrac{K_M}{(S)}\left(1 + \dfrac{(I)}{K_I}\right)}$, where $K_I = \dfrac{k_{-3}}{k_3}$ **11.** (a) 44 mM; (b) noncompetitive

15. (b) $\dfrac{1}{V_{max}}$ and $\dfrac{1}{K_M} = 0$; (c) $k_2 \gg (k_1 + k_{-1})$; (d) first order; (e) 1.25×10^9 M^{-1} s^{-1}; (b) 2;

(h) $k_1 = 1.8 \times 10^9$ M^{-1} s^{-1} **18.** (b) 0.025 M, 6 hr; (c) 0.050 M, 6 hr, 12 hr; (d) 12 hr;
(e) 10 mL hr^{-1} **20.** (b) $K_M(C_2H_5OH) = 0.44$ mM, $K_M(NAD^+) = 18$ μM; (c) competitive;
(d) noncompetitive   **23.** 2.61 mM **26.** (a) 38.4 min; (b) 0.9 s; (c) 76.8 min, 1.8 s

Chapter 9

2. (a) 52.9 kJ mol^{-1}; (b) 23.3 kJ mol^{-1}; (c) $\phi = 0°, 60°, 120°$ **4.** (a) $E = 6.02 \times 10^{-20}$ J, $\lambda = 1100$ nm;
(c) $E = 12.04 \times 10^{-20}$ J, $\lambda = 1100$ nm

6. (b) $\dfrac{5h^2}{4\pi^2 ma^2}$ **7.** (d) 660 nm   **10.** (c) $E_0 = 4.12 \times 10^{-20}$ J, $E_1 = 12.36 \times 10^{-20}$ J, etc.

12. (a) 1597 N m^{-1}; (b) 1870 cm^{-1}; (c) 1390 cm^{-1}; (d) 1148 N m^{-1};
(e) $v(c) = 1343$ cm^{-1}, $v(d) = 1586$ cm^{-1}
15. (b) Bond energies: NO$^-$ < NO < NO$^+$
Bond lengths: NO$^+$ < NO < NO$^-$
21. (a) Ground state: $\mu_{\parallel} = 1.95$ D, $\mu_{\perp} = 1.70$ D
Excited state: $\mu_{\parallel} = 0.28$ D, $\mu_{\perp} = 0.41$ D

Chapter 10

1. (a) 2.91×10^{-19} J photon^{-1} = 1.82 eV = 176.2 kJ einstein^{-1}; (b) 2.76; (c) 30–35%
4. (a) 0.01 M, (b) 0.02 M; (c) 9.8×10^{-4} cm
6. c(Chl a, μg ml^{-1}) = 12.80 A_{663} − 2.585 A_{645}
c(Chl b, μg ml^{-1}) = 22.88 A_{645} − 4.67 A_{663}
c(Chl total, μg ml^{-1}) = 8.13 A_{663} + 20.30 A_{645}
8. (a) 4.0; (b) 0.400   **10.** (a) 6.7 ns; (b) 1.05×10^8 s^{-1} **15.** (a) −0.50 L mol^{-1} cm^{-1};
(b) (DNA) = 0.109×10^{-4} M, (RNA) = 0.52×10^{-4} M; (c) decrease; (d) increase
18. (a) $[\alpha]$ = 2320 deg dm^{-1} cm^3 g^{-1}; $[\theta]$ = 9894 deg M^{-1} cm^{-1};
(b) A = 0.80, $A_L − A_R = 3 \times 10^{-4}$, α = 0.0075 deg;
(c) $\alpha = 1.31 \times 10^{-4}$ radians cm^{-1}, $n_L − n_R = 1.16 \times 10^{-9}$, $\psi = 1.73 \times 10^{-4}$ radians cm^{-1}
23. (a) CH$_3$CHO; (b) CH$_2$

$$\begin{matrix} CH_2 \\ | \\ CH_2 \end{matrix} \Big\rangle C{=}CH_2$$

30. (c) α-helix
33. (a) Peak ratio $\cong 3/1$; $\Delta G^0 = -2.72$ kJ mol^{-1}

Chapter 11

2. (d) $\dfrac{\tau^3 s^2}{3(1 + \tau)}$ **5.** (b) $Z = 1 + 4S + 4\tau S^2 + 2S^2 + 4\tau^2 S^3 + \tau^4 S^4$; (c) $f = 0.091$ at pH 4

7. (c) $A = \left(\dfrac{c}{Z}\right)[\varepsilon_0 + \varepsilon_1 s(\sigma_7 + \sigma_9 + \sigma_{11} + \sigma_{13} + \sigma_{15} + \sigma_{17}) + \varepsilon_2 s^2(\sigma_7 + \sigma_9 + \sigma_{11} + \sigma_{13} + \sigma_{15})$
$+ \varepsilon_3 s^3(\sigma_7 + \sigma_9 + \sigma_{11} + \sigma_{13}) + \varepsilon_4 s^4(\sigma_7 + \sigma_9 + \sigma_{11}) + \varepsilon_5 s^5(\sigma_7 + \sigma_9) + \varepsilon_6 s^6 \sigma_7]$
Z = term in brackets with at ε_i set equal to 1.   **10.** (a) $26^3 = 17,576$
11. (a) Electron: $N_2/N_1 = 0$, He: 3.0

Chapter 12

1. (a) 6.4×10^{-15} J; (b) 0.31 Å   **3.** (a) 67.4° and 39.6°, respectively; (b) 2.38 cm   **7.** 0.77 Å
8. (a) 6.025×10^{23}; (b) 14,300   **10.** (a) 1:1:1   **13.** 1.45 Å

16. (a) $\lambda, 2\lambda, 3\lambda$; (b) $\dfrac{\lambda}{2}, \dfrac{3\lambda}{2}, \dfrac{5\lambda}{2}$; (c) 8.83 Å; (d) 5.0 Å; (e) $\dfrac{\lambda}{2}$

Index